LEVEL 5 VICTORIAN ESSENTI… LEARNING S…

MATHS Quest 7

For Victoria

SECOND EDITION

Sonja **Stambulic**

Chris **Evangelou**

Lyn **Elms**

Contributing authors

Roger Blackman • Steve Craven • Caroline Denney • Louisa Doherty
Elena Iampolsky • Linda Johnston • Carolyn Mews • Mario Panaccio • David Phillips
Colin Shnier • Andrew Skourdoumbis • Helen Warner • Jenny Watson

Support material

Brett Barber • Stephen Broderick • Robert Cahn • Beth Caruso
Andrea Dineen • John Dowsey • Dennis Fitzgerald • Stephen Heames
Robert Johnston • David Tynan • Jill Vincent • Don Wagstaff

jacaranda plus

Second edition published 2006 by
John Wiley & Sons Australia, Ltd
42 McDougall Street, Milton, Qld 4064

First edition published 2000

Typeset in 10.5/12.5pt Times

National Library of Australia
Cataloguing-in-Publication data

Maths Quest 7 for Victoria 2nd ed. VELS level 5

Includes index.
For secondary school students.
ISBN 978 0 7314 0330 1 (student edition)
ISBN 978 0 7314 0334 9 (teacher edition)

1. Mathematics — Textbooks. I. Stambulic, Sonja. (Series: Maths Quest series).

510

Illustrated by the Wiley Art Studio

Cover photograph and internal design images: © Digital Vision

Printed in Singapore by
Craft Print International Ltd

10 9 8 7 6

Contents

CHAPTER 4

▶ Decimals 135

CHAPTER 5

▶ Patterns and algebra 179

CHAPTER 6

▶ Equations 221

CHAPTER 7

▶ Length and perimeter 257

CHAPTER 8

Area, volume and capacity 291

CHAPTER 9

Time, mass and temperature 341

CHAPTER 10

Angles 387

CHAPTER 11

CHAPTER 12

CHAPTER 13

CHAPTER 14

Chance 613

Strategies for problem solving 653

Introduction

The second edition of ***Maths Quest 7 for Victoria*** is one of the exciting award-winning *Maths Quest* resources specifically designed for the Victorian Essential Learning Standards (VELS).

This resource contains:

- a student textbook with accompanying eBookPLUS
- a teacher edition with accompanying eGuidePLUS
- a student homework book
- a teacher support book.

Student textbook

The key features added in the second edition include:

- an introductory set of questions at the start of each chapter to establish the students' current level of understanding. Each question is supported with a *SkillSHEET* to explain the concept involved and provide extra practice if needed.
- a new chapter on problem-solving strategies to assist students' skills in working mathematically
- *eTHINKING* multi-level interactive problem-solving games to encourage deeper understanding of concepts and to challenge students' mathematical thinking and reasoning (see page x for details)
- updated and revised interactive technology files to foster Information and Communications Technology (ICT) skills and aid conceptual learning
- a glossary to enhance students' mathematical literacy.

The second edition textbook continues to offer the following award-winning features.

The use of *full colour* throughout produces clearer graphs and headings, provides bright, stimulating photographs and makes navigation through the text easier.

Clear, concise *theory sections* contain *worked examples*, *graphics calculator tips* (Texas Instruments in the text and Casio online) and *highlighted important text* and *remember boxes*.

Worked examples in a Think/Write format provide clear explanations of key steps and suggestions for presentation of solutions.

Exercises contain many carefully graded skills and application problems, including multiple choice questions. Cross-references to relevant worked examples appear beside the first 'matching' question throughout the exercises.

Many exercises finish with *Maths Quest challenge* questions designed to extend students. *Code puzzles* reinforce mathematical skills and *History of Mathematics* sections highlight mathematical concepts in context.

Two sets of *10 Quick Questions* in each chapter provide a cumulative review of concepts covered.

Investigations focus on links to the interdisciplinary strands, providing further opportunities for discovery learning.

Each chapter concludes with a *summary* and *chapter review* exercise helping consolidate students' learning of new concepts.

Technology is fully integrated (in line with VELS recommendations) and features graphics calculators, computer algebra systems, spreadsheets, dynamic geometry software and several graphing packages. *Maths Quest* continues to promote these technologies as learning tools, providing demonstration versions of the programs (with the exception of Microsoft Excel) as well as hundreds of supporting files on the *bonus* accompanying online resources.

Programs included

Mathcad Client: a computer algebra system and graphing program

Graphmatica: an excellent graphing utility

GrafEq: graphs any relation, including complicated inequalities

Poly: for visualising 3D polyhedra and their nets

Tess: for producing tessellations and other symmetric planar illustrations

TI Connect and Casio software FA-123: calculator screen capture and transfer programs

Cabri Geometry II PLUS: dynamic geometry program

Adobe® Acrobat® Reader v7

Teacher edition with eGuidePLUS

The teacher edition textbook contains everything in the student textbook and more. To support teachers assisting students in class, *answers* appear in red next to most questions in the exercises. Each exercise is annotated with relevant standards. The readily accessible and comprehensive *VELS planning document*, *Work program* and curriculum grid list available resources and curriculum coverage.

The accompanying online resources for teachers contain everything in the student eBookPLUS and more. Four tests per chapter, fully worked solutions to *WorkSHEETs*, interactive *Test yourself* tests in editable Word format, the work program and other curriculum advice in editable Word 2000 format are provided.

Student Homework Book

The *Homework Book* is a valuable new addition to the *Maths Quest* series and complements the student textbook with a variety of activities including *WorkSHEETs*, rich tasks and fun code puzzles.

Teacher Support Book

This fully photocopiable resource contains all the *WorkSHEETs* in the *Homework Book*, plus teacher support material that includes assessment sheets, a VELS planning grid and extra activities for students to further develop their skills in working mathematically. All answers to the *WorkSHEETs* and additional teacher support materials are also provided.

eTHINKING Software developed by cooperation between Greygum Software Pty Ltd and John Wiley and Sons Australia, Ltd.

e-THINKING

Ten interactive problem-solving games are vibrantly presented. Each game is content-based and contains 7 to 10 multi-level tasks, increasing in complexity and difficulty.

Due to the multi-level nature of the *eTHINKING* problem-solving games, students can attempt these at various stages throughout the chapter. Titles, placement and descriptions of these games follow.

Game	Description
Number puzzles *Chapter 1 Number skills and the history of number*	Eight puzzles ranging from straightforward calculations involving the four basic operations (×, ÷, +, −) to magic squares
Adding fractions *Chapter 3 Fractions*	Nine puzzles involving the addition of fractions, recognition of fractions and equivalent fractions
Function machines *Chapter 5 Patterns and algebra*	Nine puzzles involving the recognition of input, output numbers, observing patterns and formulating rules
Flow charts *Chapter 6 Equations*	Nine puzzles involving various concepts related to flow charts including: following a sequence of operations performed on an input number, interpreting written instructions and solving equations of varying degrees of difficulty
Area and perimeter *Chapter 8 Area, volume and capacity*	Nine puzzles involving the concepts of perimeter and area, and the relationship between the two
Directions and bearings *Chapter 10 Angles*	Seven puzzles involving the concept of true bearings, where participants are challenged to apply their knowledge of bearings and test their reflexes in navigational tasks
Transformations *Chapter 11 Plane and solid shapes*	Nine puzzles involving single transformations; such as translations, reflections and rotations, as well as a combination of these
Stepping stones *Chapter 12 Maps, networks and coordinates*	Seven puzzles involving the concept of Eulerian paths and circuits
Mean, mode, median *Chapter 13 Data*	Eight puzzles to consolidate and extend the participant's knowledge of the mean, mode and median
Probability *Chapter 14 Chance*	Eight puzzles to consolidate the concept of chance, experiments, simulations, theoretical probability and long-run proportion; as well as revise 'chance' words and their equivalent probabilities

jacaranda plus

Next generation teaching and learning

About eBookPLUS

This book features eBookPLUS: an electronic version of the entire textbook and supporting multimedia resources. It is available for you online at the JacarandaPLUS website (www.jacplus.com.au).

These additional resources include:

- Word documents designed for easy customisation and editing
- interactive activities and a wealth of ICT resources
- weblinks to other useful resources and information on the internet.

Using the JacarandaPLUS website

To access your eBookPLUS resources, simply log on to www.jacplus.com.au. There are three easy steps for using the JacarandaPLUS system.

Step 1. Create a user account

The first time you use the JacarandaPLUS system, you will need to create a user account. Go to the JacarandaPLUS home page (www.jacplus.com.au) and follow the instructions on screen.

Step 2. Enter your registration code

Once you have created a new account and logged in, you will be prompted to enter your unique registration code for this book, which is printed on the inside front cover of your textbook.

LOGIN

Once you have created your account, you can use the same email address and password in the future to register any JacarandaPLUS books.

Step 3. View or download eBookPLUS resources

Your eBook and supporting resources are provided in a chapter-by-chapter format. Simply select the desired chapter from the drop-down list and navigate through the tabs to locate the appropriate resource.

Key to the icons in the book

In this student text, the resource icons, such as the one shown, indicate that digital support materials for this section of the text are available online at JacarandaPLUS.

Troubleshooting

- Go to the JacarandaPLUS help page at www.jacplus.com.au
- Contact John Wiley & Sons Australia, Ltd.
  Email: support@jacplus.com.au
  Phone: 1800 777 474

Minimum requirements

- Internet Explorer 7, Mozilla Firefox 1.5 or Safari 1.3
- Adobe Flash Player 9
- Javascript must be enabled (most browsers are enabled by default).

Acknowledgements

The authors and publisher would like to thank the following copyright holders, organisations and individuals for their assistance and for permission to reproduce copyright material in this book.

Illustrative material

• Attorney-General's Department, Weather Maps from Bureau of Metorology pages 521, 527, 556 Copyright Commonwealth of Australia reproduced by permission • AAP Image pages 17/Pierre Philippe Marcou, 294 (lower right), 508/AP Photo/Chitose Suzuki • Courtesy of ACT Land Information Centre, ACT Planning & Land Authority page 528 • ANTPhoto.com.au/David Carter page 596 • APL, Australian Picture Library pages 109 (lower)/Sharpshooters, 289/John Carnemolla, 341/Mark E. Gibson, 451, 456 (bottom left)/Fukuhara/WL and (top right)/John Carnemolla, 475 (right)/LucidioStudio Inc • © Banana Stock page 291 • © Brand X Pictures pages 260 (upper right), 271 (upper), 297 (upper right), 323 (second bottom), 620 • Cabrilog/Cabri Geometry II screen used with permission of Cabrilog, 7 av. Felix Viallet, 38000 Grenoble, FRANCE, http://www.cabri.com • Coo-ee Picture Library pages 322 (right), 456 (bottom right), 539, 628 • © Corbis Corporation pages 10 (upper), 19, 45, 68 (lower), 94 (upper left), 122, 240, 259 (upper right), 260 (upper left), 269 (right), 284 (upper), 293 (lower), 332 (middle), 358, 385, 410, 456 (second bottom right), 465 (lower), 507, 612, 633 • © Digital Stock/Corbis Corporation pages 47, 85 (top), 135, 144, 252, 270, 300, 311, 346 (upper left), 367, 394 (top right), 401 (upper left and right), 411 (lower left), 443 (upper right), 456 (second bottom left), 473 (bottom and upper right), 573, 588, 652 • © Digital Vision pages 59 (lower), 219, 372 (upper right), 651, 664 • © EyeWire Images pages 150, 411 (lower right), 460 • © Flat Earth page 317 (lower left) • Getty Images pages 23/Altrendo, 115/Stone/Christel Rosenfeld, 268 (left)/Stone/Tim Flach • HJ Heinz Company Aust Ltd page 197 'HEINZ' is a registered trademark and used with the permission of H.J. Heinz Company Australia Limited • Image Addict © www.image-addict.com.au pages 259 (upper left), 405 (upper), 489 (right), 593 (lower), 625, 645 • © Image Disk Photography pages 8 (upper left), 179, 215, 260 (lower left and lower right), 274 (lower), 275, 346 (lower left), 402 (two) • IT StockFree © IT Stock page 401 (lower left) • John Wiley & Sons Australia © John Wiley & Sons Australia pages 8/Jo Patterson (upper right, lower left and right), 94/Jennifer Wright (lower left and lower right), 257/Jo Patterson, 266/Jo Patterson (eight), 267/Jo Patterson (two), 288/Jo Patterson (five), 294/Jennifer Wright (middle), 323/Vikki Steele (second top), 387/Carolyn Mews, 394/Carolyn Mews (bottom), 405/Jennifer Wright (lower), 425/Carolyn Mews, 431/Carolyn Mews, 452/Carolyn Mews (lower), 484/Carolyn Mews (upper right), 492/John Nolan, 534/Jo Patterson, 561/Maggy Saldais, 565/Jo Patterson, 587/Jennifer Wright, 590/Jo Patterson, 591/Narelle Kremmer, 637/Jennifer Wright • MAPgraphics Pty Ltd, Brisbane pages 269 (upper), 516, 555, 556 (lower) • Mary Evans Picture Library page 262 • © Melbourne Aquarium pages 501, 549 • Melbourne 2006 Commonwealth Games Corporation pages 522 (two), 557 • Copyright Melway Publishing Pty Ltd Reproduced with permission pages 424, 501, 544 • Microsoft Corporation screenshots reproduced throughout with permission from Microsoft Corporation • Newspix pages 1/James Elsby, 32/Mark Evans, 37 (upper)/ James Elsby, 205 (lower)/Ray Titus, 221, 263/Todd Martyn-Jones, 278 (right)/ Phil Hillyard, 293 (upper)/AFP/Stan Honda, 563 (lower)/Adam Head, 594/ Cameron L'estrange, 609 (upper)/Kelly Barnes, 613/Frank Violi, 619/Kelly Barnes, 621/Michael Dodge, 649/Frank Violi • © Peter Storer page 40 • © PhotoAlto/Patrick Sheándell O'Carroll pages 297 (upper left), 666 • © PhotoDisc, Inc pages 4, 9, 14 (two), 16 (three), 22, 27, 35, 36 (two), 37 (lower), 38, 43 (two), 59 (upper left and right), 60, 67, 69, 81 (two), 85 (bottom and middle), 87, 89 (two), 90 (two), 92 (two), 94 (upper right), 95, 99, (100 (two), 101, 103, 104, 109 (upper), 110, 111, 116, 117, 120, 127 (two), 130, 132, 134, 143, 155, 163, 167, 174, 178, 198, 207 (three), 216, 220, 224, 256, 259 (lower left and lower right), 261 (three), 264 (two), 268 (upper), 271 (lower), 274 (upper), 278 (left), 284 (lower), 290 (two), 294 (upper), 297 (lower right), 305, 310 (three), 317 (lower right and upper left), 322 (left), 323 (top and bottom), 329 (two), 330, 332 (bottom), 333, 336, 339, 340 (top and bottom), 343 (two), 344, 346 (lower and upper right), 351, 355, 359, 361 (two), 369 (two), 372 (upper and lower left, lower right), 375, 377 (three), 381 (two), 386, 391, 394 (second bottom left and right, second top left and right, top left), 401 (lower right), 411 (upper left and right), 413, 417, 418, 420, 426, 438, 441, 443 (upper left, lower left and right), 452 (upper), 455, 456 (second top left and right), 463 (four), 465 (upper), 471, 472, 473 (bottom, middle and upper left, middle right), 475 (left), 478, 480, 484 (upper and lower left, lower right), 488 (four), 489 (left and centre), 515, 524, 525, 540 (two), 559, 563 (upper), 564, 566, 568, 589, 592, 593 (upper), 607, 609 (lower), 611 (two), 615, 618, 623, 626, 631, 642 (two), 647, 654, 657, 658 • photolibrary.com pages 362/Science Photo Library/Pascal Goetgheluck, 610/Index Stock Imagery/Vohra Aneal • © Stockbyte pages 68 (upper), 182, 205 (upper), 332 (top), 340 (middle), 253 • © Stockdisc page 317 (upper right) • Texas Instruments screens used throughout with permission • Tourism Victoria pages 433/Tim Webster, 579/Gavin Hansford • Universal Publishers Pty Ltd page 514 Map reproduced with permission of Gregorys • © Viewfinder Australia Photo Lib pages 268 (lower right), 269 (bottom), 661, 669 • © Weis page 46 • © Earth Vistas/WorldSat International Inc. overlay designed by MAPgraphics page 510.

Text

Attorney-General's Department page 364/ Tide Predictions for Western Port (Stony Point), November 2005 Table, http://www.bom.gov.au/cgi-bin/oceanography/tides/tide_prediction.cgi. Copyright Commonwealth of Australia reproduced by permission • *Herald & Weekly Times* page 594 (From The *Herald Sun* 10 March 2000. Reprinted with permission of the *Herald & Weekly Times)* • V/Line Passenger Pty Ltd, page 363 © V/Line Passenger Pty Ltd.

Software

The authors and publisher would like to thank the following software providers for their assistance and for permission to use their materials. However, the use of such material does not imply that the providers endorse this product in any way.

Third party software — registered full version ordering information

Full versions of third party software may be obtained by contacting the companies listed below.

Texas Instruments TI Connect™ and TI-GRAPHLINK software

TI Connect™ and TI-GRAPHLINK software reproduced with permission of the publisher Texas Instruments Incorporated.

TI Connect software available from Texas Instruments

Web: http://education.ti.com/us/product/software.html

Note: The TI Connectivity cable can be purchased from educational booksellers or calculator suppliers.

Casio FA-123

Software used with permission of Casio Computer Co. Ltd. © 2001

Distributed by Shriro Australia Pty Ltd

23–27 Chaplin Drive

Lane Cove

NSW 2066

Web: http://www.casioed.net.au and find the calculator product range

Mathcad® Client

Included with permission of MathSoft Engineering & Education, Inc.

Web: http://www.mathsoft.com

Distributed in Australia by Hearne Scientific Software Pty Ltd

Level 6, 552 Lonsdale Street, Melbourne 3000

e-mail: info@hearne.com.au

Web: http://www.hearne.com.au/

Phone: (03) 9602 5088

Mathcad is a registered trademark of Math Soft Engineering & Education, Inc.

Graphmatica

Reproduced with permission of kSoft, Inc.

345 Montecillo Dr., Walnut Creek, CA 94595-2654.

e-mail: ksoft@graphmatica.com

Web: http://www.graphmatica.com

Software included is for evaluation purposes only. The user is expected to register share-ware if use exceeds 30 days. Order forms are available at www.graphmatica.com/register. txt

Cabri Geometry™ II PLUS

Reproduced with permission of Cabrilog.

Cabrilog 6, Robert Schuman Place

38000 Grenoble FRANCE

Web: http://www.cabri.com

1. Due to copyright restrictions, the demo version of Cabri Geometry™ II Plus must not be used in the classroom for presentation on a regular basis.
2. For site licences contact Cabrilog — Grenoble-France at «sales@cabri.com» or www.cabri.com

GrafEq, Poly and Tess

Evaluation copies of GrafEq™, Poly™ and Tess™ have been included with permission from Pedogoguery Software, Inc.

e-mail: peda@peda.com

Web: http://www.peda.com

Microsoft® Excel

Microsoft Excel is a registered trademark of the Microsoft Corporation in the United States and/or other countries.

Every effort has been made to trace the ownership of copyright material. Information that will enable the publisher to trace the copyright holders or to rectify any error or omission in subsequent reprints will be welcome. In such cases, please contact the Permission Section of John Wiley & Sons Australia, who will arrange for the payment of the usual fee.

eThinking Software

Developed by cooperation between Greygum Software Pty Ltd and John Wiley and Sons Australia, Ltd.

345 Kelvin Grove Rd
Kelvin Grove
Queensland 4059
Phone: (07) 3352 4940
Fax: (07) 3352 4950
Email: support@greygum.com.au

eTHINKING 1

Number skills and the history of number

1

Look at the photograph of people at a concert in a mall. How many people are there? How quickly were you able to work this out? You have used your understanding of numbers and counting.

In this chapter, you will reinforce your skills in adding, subtracting, multiplying and dividing whole numbers to find faster ways of solving problems like this as well as more difficult ones.

Are you ready?

Try the questions below. If you have difficulty with any of them, extra help can be obtained by completing the matching SkillSHEET. Either click on the SkillSHEET icon next to the question on the *Maths Quest 7* CD-ROM or ask your teacher for a copy.

SkillSHEET 1.1

Place value

1 What is the place value of the digit shown in red?

a 362 b 59 472 c 7308 d 238 946

SkillSHEET 1.2

Ascending and descending order

2 a Place the following numbers in ascending order.

919, 99, 991, 199, 19, 91

b Place the following numbers in descending order.

12, 102, 21, 120, 201, 112, 121

SkillSHEET 1.3

Adding and subtracting whole numbers less than 20

3 Write the answer to each of the following.

a $7 + 8$ b $12 + 15$ c $9 + 17$

d $9 - 5$ e $18 - 11$ f $12 - 3$

SkillSHEET 1.4

Times tables

4 Find the answer to each of these as quickly as possible.

a 4×8 b 6×5 c 9×7

d 12×3 e 8×2 f 11×6

SkillSHEET 1.5

Multiplying whole numbers

5 Work out the answer to each of the following.

a 45×7 b 23×14 c 157×36

SkillSHEET 1.6

Dividing whole numbers

6 Work out the answer to each of the following.

a $56 \div 4$ b $6\overline{)979}$ c $\frac{651}{7}$

SkillSHEET 1.7

Order of operations

7 Find the value of each of the following.

a $3 + 2 \times 8$ b $6 \times 5 - 4$ c $8 \div 2 + 3 \times 6$

SkillSHEET 1.8

Rounding to the first (leading) digit

8 Round each of the following numbers to the first digit.

a 463 (*Hint:* Is 463 closer to 400 or 500?)

b 2401

c 68

Number systems from the past

All ancient civilisations developed methods to count and measure number. Some began by using pebbles, knots tied in a rope, or notches cut in a stick to count or record numbers. As the need arose to use larger numbers, many civilisations developed their own number systems. Some systems were more efficient than others and some are still in use today.

Our number system is based on the number 10 and is known as the Hindu–Arabic system. It is believed that it was used by the Hindus and brought to Spain by the Moors in the eighth or ninth century AD. The symbols used today, called *digits*, are 0, 1, 2, 3, 4, 5, 6, 7, 8 and 9.

The advantage of the Hindu–Arabic system over other number systems is that *place value*, or the position of the **digit**, is important. Each symbol represents a different value depending on where it is placed in the number. This makes the four number operations — adding, subtracting, multiplying and dividing — much easier than they are in other systems.

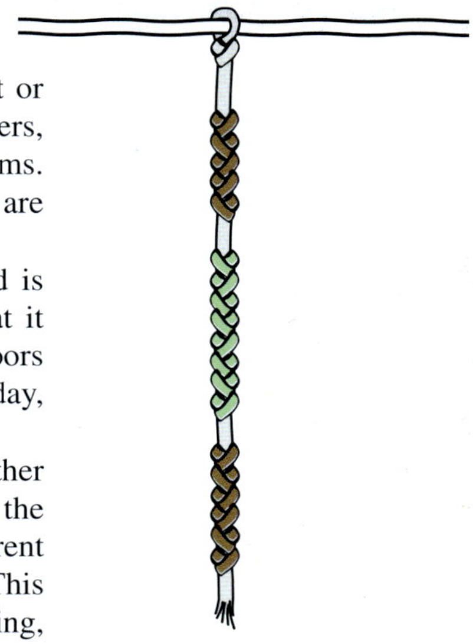

The Incas, an ancient South American civilisation, recorded numerical information by tying knots in strings called quipu. The diagram above illustrates how the number 586 would be recorded on a quipu. Take note of the spacing between the units, tens, and hundreds positions.

The table below illustrates the evolution of numbers from the past to the present. The earliest numbers had no zero.

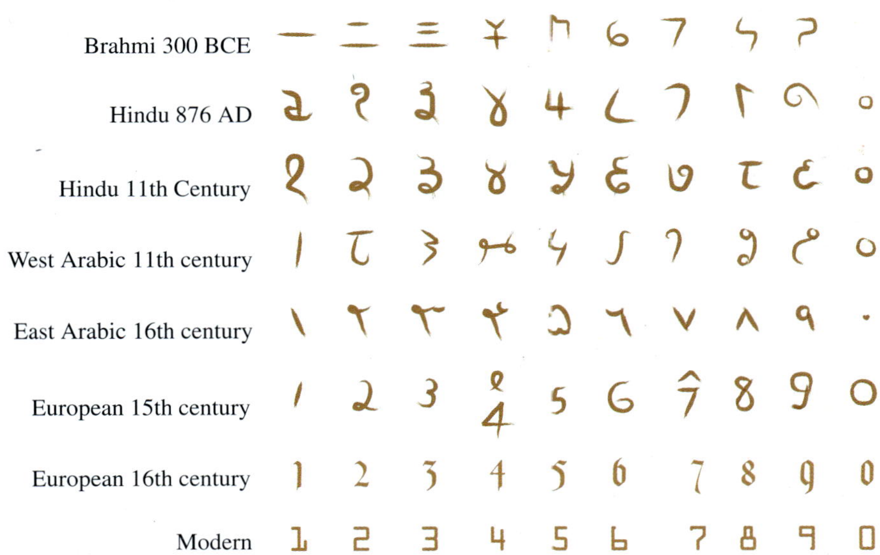

Our number system is called the Hindu–Arabic system. It uses 10 digits: 0, 1, 2, 3, 4, 5, 6, 7, 8 and 9. The order in which the digits are placed is important.

THINKING

The Roman numeral system

The ancient number system with which you are probably most familiar is the Roman system. You may have seen Roman numerals on clock faces, in prefaces of books or in the credits of a movie to show the year in which it was produced. Can you think of other places you have seen them?

The Roman symbols are based on fives. The symbols used are shown in the table below.

Number	1	5	10	50	100	500	1000
Symbol	I	V	X	L	C	D	M

Numbers are written by combining symbols.

The order in which the Roman numerals are written is important.

The Roman numerals for 1 to 10 are: I II III IV V VI VII VIII IX X.

1 What happens when I is just before V in the number?
2 What happens when I is just after V in the number?
3 What happens when I is just before X in the number?
4 What would you expect to happen if I is straight after X in the number?
5 Write the Roman numerals for 11 to 20 using I, V and X.
6 What numbers are represented by the following Roman numerals:
   a XXV? b XXXIV?
7 Write the Roman numerals for the following numbers: a 26 b 39.
8 What numbers are represented by the following Roman numerals?
   a LV b XL c CC d DC
   e CM f XLII g LXXIV h MDCCCXXIII
9 Write the Roman numerals for the following numbers.
   a 356 b 1650 c 94 d 179
   e 2243 f 931 g 428 h 1085
10 What time is indicated on the clock face shown at right?
11 In the credits at the end of a film the date of the production is often shown in Roman numerals. For the following films, state what year each film was produced.
   a *Grease* MCMLXXVII
   b *Titanic* MCMXCVII
   c *ET* MCMLXXXII
   d *Snow White and the Seven Dwarfs* MCMXXXVII

12 If the production of a film is completed this year, write the date that would appear in the credits in Roman numerals.
13 Investigate how addition and subtraction was performed using Roman numerals.

COMMUNICATION

Number systems of the past

Below is a brief introduction to some number systems used in the past.

The Chinese number system

One of the oldest number systems, the Chinese number system, was represented by the horizontal and vertical arrangement of sticks.

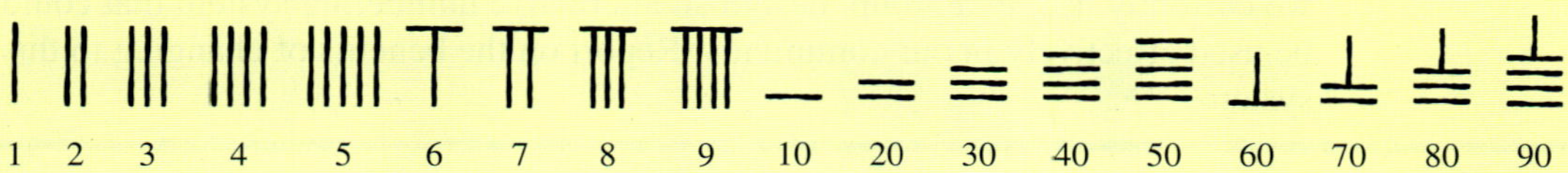

The Babylonian number system

The Babylonian number system was based on 10 and 60, and was represented by wedge-shaped symbols called cuneiform. The unit symbol was represented by a vertical wedge and the tens symbol was represented by a corner wedge.

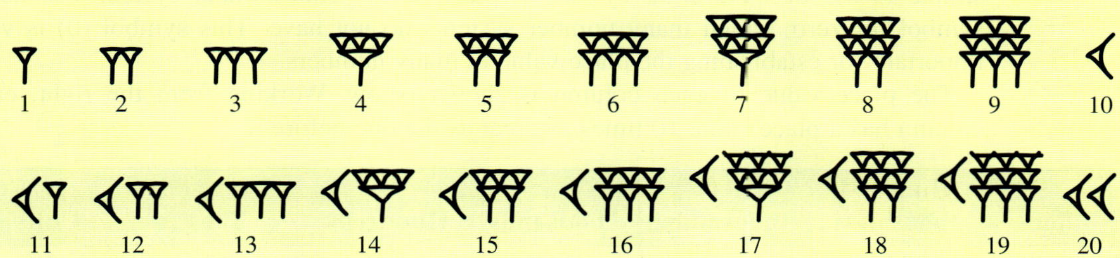

The Mayan number system

The Mayan number system was based on 20 and used a series of dots and bars to represent numerals. The value of zero was denoted by a special symbol.

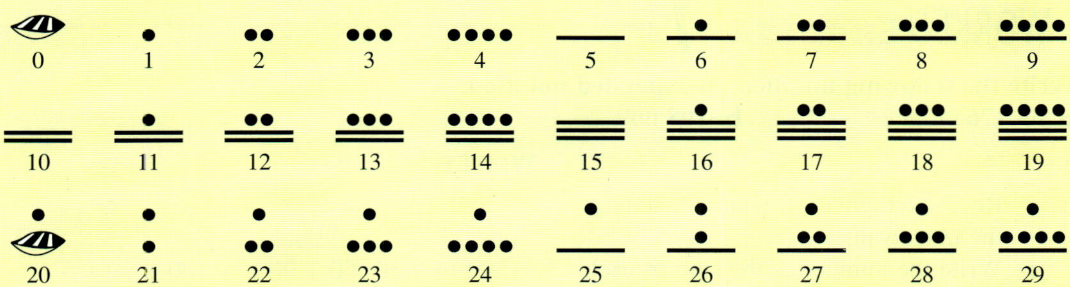

1 Australian Aborigines' systems

Australian Aboriginal people had many different systems among different tribes to represent numbers. Choose one system and show:

a the symbols that were used

b how the system benefited their society.

2 Other systems

Elaborate on one of the ancient civilisations mentioned above or alternatively choose another ancient civilisation that we have not discussed. Using an encyclopedia or the Internet, find out the following.

a What symbols were used to represent what numbers?
b Was there a symbol used to represent zero?
c What base did the number system use?
d Did this number system have any way of representing place value?
e Demonstrate how addition and subtraction would be performed under this system.
f Discuss the advantages for this civilisation in using this system.
g Discuss why this system might have fallen out of use.

3 You have looked at past numbering systems and the Hindu–Arabic system that we currently use. In a group, design an alternative numbering system that could be used effectively in our community. Report on the benefits of changing to this system.

Place value

As stated on page 3, our Hindu–Arabic number system has the advantage of being simple to use because of its system of place value. This number system also has a symbol for zero, which many number systems do not have. This symbol (0) is very important for establishing the place value in many numbers.

The place value of each column is shown below. Working from the right, each column has a place value 10 times as great as the one before it.

Millions	Hundred thousands	Ten thousands	Thousands	Hundreds	Tens	Units
1 000 000	100 000	10 000	1000	100	10	1

Numbers can be written in expanded notation by breaking them up into their place values. This is demonstrated in the worked example below.

WORKED Example 1

Write the following numbers in expanded notation.
a 59 176 **b 108 009**

THINK	WRITE
a 1 Read the number to yourself, stating the place values.	**a**
2 Write the number as the sum of each place value.	59 176 = 50 000 + 9000 + 100 + 70 + 6
b 1 Read the number to yourself, stating the place values.	**b**
2 Write the number as the sum of each place value.	108 009 = 100 000 + 8000 + 9

Numbers are ordered according to their place values. For whole numbers, the number with the most digits is the greatest in value because it will have the highest place value. If two numbers have the same number of digits, then the digits with the highest place value are compared. If they are equal, the next higher place values are compared, and so on.

WORKED Example 2

Write the following numbers in descending order.

858 58 85 8588 5888 855

THINK	WRITE
1 Look for the numbers with the most digits.	
2 There are two numbers with 4 digits. The number with the higher digit in the thousands column is larger; the other is placed second.	
3 Compare the two numbers with 3 digits. Both have the same hundreds and tens values so compare the units values.	
4 Compare the two 2-digit numbers.	
5 Write the answer.	8588, 5888, 858, 855, 85, 58

remember

1. Numbers are organised by place value.
2. The first place value is the units. Each place value to the left of this column has a place value 10 times as great as the value in the previous column.
3. Numbers can be placed in order by comparing their place values. For whole numbers, the more digits, the greater the number. If two numbers have the same number of digits, they are ordered by comparing the digits with the highest place value.

EXERCISE 1A Place value

WORKED Example 1

1 Write the following numbers in expanded notation.

a 925 b 1062 c 28 469 d 43
e 502 039 f 800 002 g 1 080 100 h 22 222

2 Write the following numbers in words.

a 765 b 9105 c 90 450 d 100 236

3 Write the numeral for each of the following.

a Four hundred and ninety-five.
b Two thousand, six hundred and seventy.
c Twenty-four thousand.
d One hundred and nine thousand, six hundred and five.

4 Give the value of the digits (in brackets) in the distances to the following destinations in these road signs. Use words to state these values.

a 

i Broken Hill (2)
ii Yancannia (4)
iii White Cliffs (8)

b 

i Geelong (2)
ii Ballarat (1)
iii Bendigo (3)

c 

i Sunbury (1)
ii Romsey (4)
iii Lancefield (9)

d 

i East Warburton (4)
ii Upper Yarra Reservoir (2)
iii Woods Point (1)

SkillSHEET 1.1 Place value

5 **multiple choice**

Which of the following numbers is the largest?

A 4884 B 4488 C 4848 D 4844 E 4888

6 **multiple choice**

Which of the following numbers is the smallest?

A 4884 B 4488 C 4848 D 4844 E 4888

SkillSHEET 1.2 Ascending and descending order

WORKED Example 2

7 In each of the following, write the numbers in **descending** order.

a 8569, 742, 48 987, 28, 647
b 47 890, 58 625, 72 167, 12 947, 32 320
c 6477, 7647, 7476, 4776, 6747
d 8088, 8800, 8080, 8808, 8008, 8880

8 In each of the following, write the numbers in **ascending** order.

a 58, 9, 743, 68 247, 1 258 647
b 78 645, 58 610, 60 000, 34 108, 84 364
c 9201, 2910, 1902, 9021, 2019, 1290
d 211, 221, 212, 1112, 222, 111

9 Did you know that we can use the abbreviation K to represent 1000? For example, \$50 000 can be written as \$50 K.

a What amounts do each of the following represent?
i \$6 K ii \$340 K iii \$58 K

b Write the following using K as an abbreviation.
i \$430 000 ii \$7000 iii \$800 000

c Find a job or real estate advertisement that uses this notation.

History of mathematics

THE ABACUS: c. 500 BC TO NOW!

The abacus is a primitive computer that when used properly can perform the four main operations of addition, subtraction, multiplication and division as fast as a pocket calculator. It has been around for about 2500 years and is still used in some countries today.

The original abacus was a board with sand used to record the numbers. The name *abacus* comes from the Greek word 'abax' which means calculating board or from the Phoenician word 'abak' which means sand. History records that Archimedes was killed by a soldier while working with figures drawn in the sand — it is thought he may have been looking at an abacus.

At the next stage of development, an abacus had grooves for the stones that became the number markers used for calculations. Eventually these were replaced by rods or wires similar to the present style of abacus. The abacuses used by the Greeks and Romans had a position for the zero value but the concept of zero as a written place holder was not introduced in writing until about AD 1200. This was about 2000 years after it had been seen on an abacus.

The abacus was used in most parts of the world. The European abacus that we are familiar with has 5 counters below the 'crossbar' each representing one unit, and 2 counters above the crossbar each representing 5 units. (The column on the right is the 'ones' column, the next column to the left is the 'tens' column, and so on.) In Japan it is called the Sorabon and has 1 counter above the bar and 4 below it. The Aztecs called their device the Nepohualtzitzin. It had 3 counters above and 4 below, and was made of strings of maize kernels attached to a wooden frame. It dated back to about AD 1000. The Chinese have been using their Suan Pan, which translates as 'calculating plate', since about 500 BC.

The Japanese device was based on the Chinese one and then improved. This is still used in many areas today and can perform at least as fast as a calculator. A contest was held in 1946 between the champion user (Thomas Wood) of an American calculating device, and Kiyoshi Matsuzaki who was a champion with the abacus. The competition involved a series of tests with complex examples of the 4 operations. The abacus won in 4 out of 5 tests. Mr. Matsuzaki had spent most of his life working with the abacus every day for his calculations.

The world's smallest abacus

When most people think of an abacus they think either of the toy ones that are often used as an ornament or the larger wooden ones that are used in some shops, but there is an even smaller one.

In 1996 the IBM Research Division built an abacus with the counters being made from individual molecules so that the counters were approximately one millionth of a millimetre (1 nanometre) in size. The counters were moved by a single atom using a scanning tunnelling microscope. This abacus has no commercial value and was built as a method of controlling very small molecules.

However, similar principles are being used to develop nanotechnology which may have numerous benefits to us.

Questions

1. Where does the word *abacus* come from?
2. What is an abacus called in Japan?
3. What was the Aztecs' abacus called?
4. Who won the contest between the abacus and the calculating machine in 1946?

Research

1. Make your own abacus and use it to do addition and subtraction.
2. Use the Internet to find out more about the abacus and how it is used for various mathematical operations.

COMMUNICATION

Numbers as identifiers

ISBN 0-918894-28-X

00028

0 29129 00495

A place value is associated with most of the numbers we encounter. However, there are sets of numbers where a sequence of digits represents a decimal number which is just an identifier. Telephone numbers and bar codes are examples of this.

1 In a telephone number such as (03) 9999 1234, the 03 indicates that this is a Victorian number. What digits identify telephone numbers in other states of Australia? Investigate the significance of the other digits.

2 Investigate the significance of the sequence of digits in the mobile phone number +610411123456.

3 A bar of Cadbury's hazelnut truffle chocolate has the bar code 9300617310969 on its packaging. What do these digits represent?

4 A cheque has groups of digits such as 123456 123-456 9999-99999 written at the bottom. These three sets of numbers identify the writer of the cheque and the bank of issue. Investigate how this is so.

5 Research another case in which a sequence of digits acts as an identifier. Explain the significance of the digits.

9 300657 006020 >

Adding and subtracting whole numbers

SkillSHEET 1.3 Adding and subtracting whole numbers less than 20

You should be familiar with the facts for addition and subtraction of two numbers when the numbers are less than 20. If you are not, click on the SkillSHEET icon to practise these facts.

As we stated on page 3 and in the previous section, our Hindu–Arabic number system uses ten digits but their value depends on their place in the number. We shall now look at the methods for adding and subtracting larger numbers, which involve writing them in columns so that the numbers line up according to place value.

Addition of whole numbers

To add larger numbers, write them in columns according to place value and then add them.

WORKED Example 3

Arrange these numbers in columns, then add them.

1462 + 78 + 316

THINK

1. Set out the sum in columns.
2. Add the digits in the units column in your head (2 + 8 + 6 = 16). Write the 6 in the units column of your answer and carry the 1 to the tens column as shown in red.
3. Now add the digits in the tens column (1 + 6 + 7 + 1 = 15). Write the 5 in the tens column of your answer and carry the 1 to the hundreds column as shown in orange.
4. Add the digits in the hundreds column (1 + 4 + 3 = 8). Write 8 in the hundreds column of your answer as shown in green. There is nothing to carry.
5. There is only a 1 in the thousands column. Write 1 in the thousands column of your answer.

WRITE

$$\begin{array}{r} 1\,4\,6\,2 \\ 7\,8 \\ +\ 3_1 1_1 6 \\ \hline 1\,8\,5\,6 \\ \hline \end{array}$$

Graphics Calculator tip! Adding, using a calculator

A calculator can be used to check your answers. On a graphics calculator, remember to press ENTER to obtain the answer after the last digit has been entered.

Adding, using a calculator

When you are adding numbers, note that the order in which they are added is not important.
As an example,

$$29 + 46 = 46 + 29$$

By reordering an addition we are often able to add pairs of numbers that will allow us to calculate the answer mentally. Consider the following example.

$$\begin{aligned} 14 + 78 + 36 + 22 &= (14 + 36) + (78 + 22) \\ &= 50 + 100 \\ &= 150 \end{aligned}$$

WORKED Example 4

Mentally perform the addition 27 + 19 + 141 + 73 by finding suitable pairs of numbers.

THINK

1. Write the question.
2. Look for pairs of numbers that can be added to make a multiple of 10. Reorder the sum, pairing these numbers.
3. Add the number pairs.
4. Complete the addition.

WRITE

27 + 19 + 141 + 73
= (27 + 73) + (141 + 19)
= 100 + 160
= 260

Subtraction of whole numbers

For subtracting numbers, the digits are lined up vertically according to place value, as we saw for addition. The most commonly used method of subtraction is called the **decomposition method**. This is because the larger number is decomposed or 'taken apart'.

The 10 which is added to the top number is taken from the previous column of the same number.

So 32 – 14 is written as

$$\begin{array}{r} 32 \\ -\ 14 \\ \hline \\ \hline \end{array} \qquad \text{or} \qquad \begin{array}{r} 30 + 2 \\ -\ (10 + 4) \\ \hline \\ \hline \end{array}$$

and becomes

$$\begin{array}{r} {}^{2}\not{3}{}^{1}2 \\ -\ 1\ 4 \\ \hline \\ \hline \end{array} \qquad\qquad \begin{array}{r} 20 + 12 \\ -\ (10 + \ \ 4) \\ \hline \\ \hline \end{array}$$

Now 4 can be taken from 12 and 10 from 20 to give 18.

$$\begin{array}{r} {}^{2}\not{3}{}^{1}2 \\ -\ \ 1\ 4 \\ \hline 1\ 8 \\ \hline \end{array}$$

WORKED Example 5

Evaluate: a 6892 – 467 b 3000 – 467.

THINK

a
1. Since 7 cannot be subtracted from 2, take one ten from the tens column of the larger number and add it to the units column of the same number. So the 2 becomes 12, and the 9 tens become 8 tens.
2. Subtract the 7 units from the 12 units (12 – 7 = 5).
3. Subtract 6 tens from the 8 remaining tens (8 – 6 = 2).
4. Subtract 4 hundreds from the 8 hundreds (8 – 4 = 4).
5. Subtract 0 thousands from the 6 thousands (6 – 0 = 6).

WRITE

a

$$\begin{array}{r} 6\ 8\ {}^{8}\not{9}\ {}^{1}2 \\ -\ 4\ 6\ 7 \\ \hline 6\ 4\ 2\ 5 \\ \hline \end{array}$$

THINK	WRITE
b 1 Since 7 cannot be taken from 0, 0 needs to become 10.	**b** ${}^{2}\not{3}\,{}^{9}\not{0}\,{}^{9}\not{0}\,{}^{1}\not{0}$
2 We cannot take 10 from the tens column, as it is also 0. The first column that we can take anything from is the thousands, so 3000 is decomposed to 2 thousands, 9 hundreds, 9 tens and 10 units.	$-\ \ \ 4\ 6\ 7$
3 Subtract the units (10 – 7 = 3).	$2\ 5\ 3\ 3$
4 Subtract the tens (9 – 6 = 3).	
5 Subtract the hundreds (9 – 4 = 5).	
6 Subtract the thousands (2 – 0 = 2).	

We can check answers by performing the reverse operation. In worked example 5 part **a**, perform the operation 6425 + 467. The answer is 6892, which shows our original subtraction to be correct.

remember

1. When we are adding and subtracting, it is important to line up the numbers vertically so that the digits of the same place value are in the same column.
2. When we are adding numbers, the order in which they are added is not important. To simplify an addition, we can find suitable pairs of numbers that will add to a multiple of 10, 100 and so on.
3. Large numbers are subtracted by decomposing the larger number into parts.

EXERCISE 1B Adding and subtracting whole numbers

Mathcad: Adding numbers

EXCEL Spreadsheet: Adding numbers

1 Answer these questions, doing the working in your head.

a 7 + 8 = **b** 18 + 6 = **c** 20 + 17 =
d 80 + 41 = **e** 195 + 15 = **f** 227 + 13 =
g 420 + 52 = **h** 1000 + 730 = **i** 7300 + 158 =
j 17 000 + 1220 = **k** 125 000 + 50 000 = **l** 2 + 8 + 1 + 9 =
m 6 + 8 + 9 + 3 + 2 + 4 + 1 + 7 = **n** 12 + 5 + 3 + 7 + 15 + 8 =

2 Add these numbers, setting them out in columns as shown. Check your answers using a calculator.

a $\begin{array}{r} 34 \\ +\ 65 \\ \hline \\ \hline \end{array}$ **b** $\begin{array}{r} 482 \\ +\ 517 \\ \hline \\ \hline \end{array}$ **c** $\begin{array}{r} 1418 \\ +\ 2765 \\ \hline \\ \hline \end{array}$

d
```
   68 069
      317
        8
 +  4 254
 ________
```

e
```
       696
 3 421 811
 +  63 044
 _________
```

f
```
    399
   1489
   2798
 + 8943
 ______
```

WORKED Example 3

3 Arrange these numbers in columns, then add them.

a 137 + 841 **b** 723 + 432
c 149 + 562 + 55 **d** 47 + 198 + 12
e 376 + 948 + 11 **f** 8312 + 742 + 2693
g 8 + 12 972 + 59 + 1423 **h** 465 + 287 390 + 45 012 + 72 + 2
i 1 700 245 + 378 + 930 **j** 978 036 + 67 825 + 7272 + 811

Check your answers with a calculator.

4 Julia was performing in a ballet and needed to buy a tutu, pointe shoes and white tights. How much did she spend in total on her costume?

5 The Melbourne telephone directory has 1544 pages in the A–K book and 1488 pages in the L–Z book. How many pages does it have in total?

6 Nathan has given his parents this surprise dinner for their wedding anniversary. What will be the cost of their meal at right?

WORKED Example 4

7 Mentally perform each of the following additions by pairing numbers together.

a 56 + 87 + 24 + 13 **b** 69 + 45 + 55 + 31
c 74 + 189 + 6 + 11 **d** 254 + 187 + 6 + 13
e 98 + 247 + 305 + 3 + 95 + 42 **f** 180 + 364 + 59 + 141 + 47 + 20 + 16

8 Hussein's family drove from Melbourne to Perth. In the first 2 days they drove from Melbourne to Adelaide, a distance of 738 kilometres. After a couple of days' sightseeing in Adelaide, Hussein's family took a day to drive 321 kilometres to Port Augusta, and another to drive the 468 kilometres to Ceduna. They drove 476 kilometres to Norseman the following day, then took 3 more days to travel the remaining 1489 kilometres to Perth.

a How many days did it take for Hussein's family to drive from Melbourne to Perth?
b How far is Norseman from Melbourne?
c How many kilometres is Perth from Melbourne?
d How far is Adelaide from Perth?

9 Of all the world's rivers, the Amazon in South America and the Nile in Africa are the two longest. The Amazon is 6437 kilometres in length and the Nile is 233 kilometres longer than the Amazon. How long is the Nile River?

10 An **arithmagon** is a triangular figure in which the two numbers at the end of each line add to the number along the line. An example is shown at right. Work in pairs to try and solve each of these arithmagons.

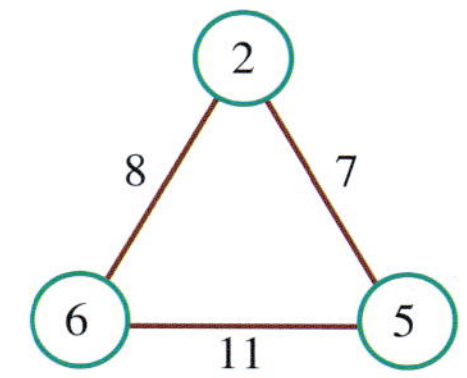

a

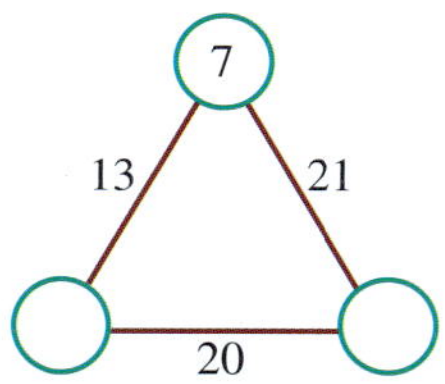

b

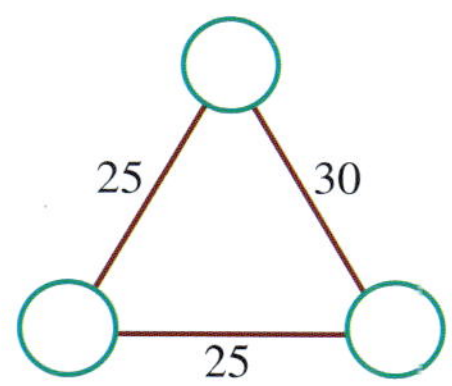

c

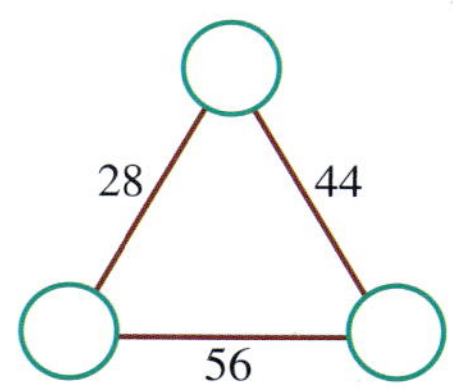

11 Answer these questions without using a calculator.

Mathcad
Subtracting numbers

a 11 – 5	**b** 20 – 12	**c** 53 – 30
d 100 – 95	**e** 87 – 27	**f** 150 – 25
g 820 – 6	**h** 1100 – 200	**i** 1700 – 1000
j 22 000 – 11 500	**k** 100 – 20 – 10	**l** 75 – 25 – 15
m 1000 – 50 – 300 – 150	**n** 80 – 8 – 4 – 5	**o** 24 – 3 – 16
p 54 – 28	**q** 78 – 39	

EXCEL Spreadsheet
Subtracting numbers

12 Answer these questions which involve adding and subtracting whole numbers.

a 10 + 8 – 5 + 2 – 11	**b** 40 + 15 – 35	**c** 16 – 13 + 23
d 120 – 40 – 25	**e** 53 – 23 + 10	**f** 15 + 45 + 25 – 85
g 100 – 70 + 43	**h** 1000 – 400 + 250 + 150 + 150	

WORKED Example 5

13 Find:

a 98 – 54	**b** 167 – 132	**c** 47 836 – 12 713
d 149 – 63	**e** 642 803 – 58 204	**f** 3642 – 1811
g 664 – 397	**h** 12 900 – 8487	**i** 69 000 – 3561

Check your answers using a calculator.

14 Find:

a 406 564 – 365 892	**b** 2683 – 49	**c** 70 400 – 1003
d 64 973 – 8797	**e** 27 321 – 25 768	**f** 518 362 – 836
g 812 741 – 462 923	**h** 23 718 482 – 4 629 738	

Check your answers using a calculator.

15 Hayden received a box of 36 chocolates. He ate 3 on Monday, 11 on Tuesday and gave 7 away on Wednesday. How many did he have left?

16 A crowd of 90 414 attended the 2005 AFL match between Carlton and Collingwood. If 57 492 people supported Carlton and the rest supported Collingwood, how many supporters did Collingwood have?

17 A school bus left Laurel High School with 31 students aboard. Thirteen of these passengers alighted at Hardy Railway Station. The bus collected 24 more students at Hardy High School and a further 11 students disembarked at Laurel swimming pool. How many students were still on the bus?

18 The most commonly spoken language in the world is Mandarin, spoken by approximately 575 000 000 people (in north and east central China). Approximately 360 000 000 people speak English and 140 000 000 Spanish.

a How many more people speak Mandarin than English?

b How many more people speak English than Spanish?

19 The photographs show three of the highest waterfalls in the world.

How much higher are the:

a Victoria Falls than the Iguazu Falls?

b Iguazu Falls than the Niagara Falls?

c Victoria Falls than the Niagara Falls?

d Explain how you obtained your answers.

20 Lucy and Ty were driving from Melbourne to Sydney for a holiday. The distance via the Hume Highway is 867 kilometres, but they chose the more scenic Princes Highway even though the distance is 1039 kilometres. They drove to Lakes Entrance the first day (339 kilometres), a further 347 kilometres to Narooma on the second day and arrived in Sydney on the third day.

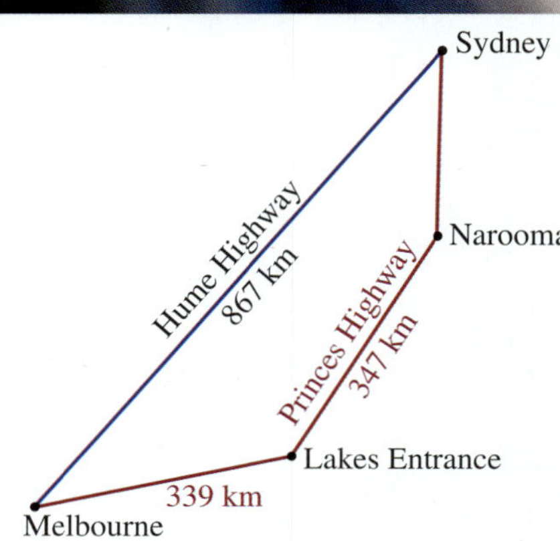

a How much further is Melbourne to Sydney via the Princes Highway than via the Hume Highway?

b How far did Lucy and Ty travel on the third day?

21 The following table shows how many medals Australia has won at each Olympic Games from the 1956 Melbourne Olympics to the 2004 Athens Olympics.
Copy and complete the table by filling in the missing numbers.

	Year	Location of Olympic Games	Medals won by Australia			
			Gold	Silver	Bronze	Total
a	2004	Athens, Greece		16	16	49
b	2000	Sydney, Australia	16		17	58
c	1996	Atlanta, U.S.A.	9	9		41
d	1992	Barcelona, Spain	7	9	11	
e	1988	Seoul, South Korea	3		5	14
f	1984	Los Angeles, U.S.A.		8	12	24
g	1980	Moscow, U.S.S.R.	2	2		9
h	1976	Montreal, Canada		1	4	5
i	1972	Munich, Germany	8	7	2	
j	1968	Mexico City, Mexico	5	7	5	
k	1964	Tokyo, Japan		2	10	18
l	1960	Rome, Italy	8		6	22
m	1956	Melbourne, Australia	13	8	14	

GAME time: Number skills and the history of number — 001

WorkSHEET 1.1

MATHS QUEST CHALLENGE

1 Can you fill in the blanks? The * can represent any digit.

a
```
  6*8 *2*
– 488 417
---------
 *49 9*4
```

b
```
  3*9*
– *6*5
------
  1*07
```

2 Without using a calculator, and in less than 10 seconds, find the answer to 6 849 317 – 999 999.

3 A beetle has fallen into a hole that is 15 metres deep. It is able to climb a distance of 3 metres during the day but at night the beetle is tired and must rest. However, during the night it slides back 1 metre. How many days will it take the beetle to reach the top of the hole to freedom?

4 Each letter below stands for a digit from 0 to 9. Find the value of each letter so that the addition statement is true.

BIG + JOKE = HAHA

Multiplying whole numbers

SkillSHEET 1.4 Times tables

The key to being able to multiply numbers is to know your times tables up to 12. If you need practice with these, click on the SkillSHEET icon.

There are two multiplication algorithms with which you should be familiar — short multiplication and long multiplication.

Short multiplication

Short multiplication can be used when multiplying a large number by a single digit number.

WORKED Example 6

Calculate 1456 × 5.

THINK

1. Multiply the units ($5 \times 6 = 30$). Write the 0 and carry the 3 to the tens column.
2. Multiply the tens digit of the question by 5 and add the carried number ($5 \times 5 + 3 = 28$). Write the 8 in the tens column and carry the 2 to the hundreds column.
3. Multiply the hundreds digit by 5 and add the carried number ($5 \times 4 + 2 = 22$). Write the last 2 in the hundreds column and carry the other 2 to the thousands column.
4. Multiply the thousands digit by 5 and add the carried number ($5 \times 1 + 2 = 7$). Write 7 in the thousands column of the answer.

WRITE

$$\begin{array}{r} {}^{2}1\;{}^{2}4\;{}^{3}5\;6 \\ \times \qquad 5 \\ \hline 7\;2\;8\;0 \\ \hline \end{array}$$

Long multiplication

We use long multiplication to multiply larger numbers. The process is the same as in short multiplication, but repeated for each digit. Remember to add the extra zero when multiplying by each new digit (1 zero when multiplying by the 'tens' digit, 2 zeros for the 'hundreds' digit and so on).

WORKED Example 7

Calculate 1456 × 132 using long multiplication.

THINK

1. Multiply the first number by 2 using short multiplication ($1456 \times 2 = 2912$).
   Write the answer directly below the question as shown.
   Move to the next row.

WRITE

THINK

2. Place a zero in the units column when multiplying 1456 by the tens digit; that is, when multiplying 1456 by 3. This is because we are really working out $1456 \times 30 = 43\,680$. Write the answer directly below the previous answer as shown. Move to the next row.
3. Place zeros in the units and tens columns when multiplying 1456 by the hundreds digit; that is, when multiplying 1456 by 1. This is because we are really working out $1456 \times 100 = 145\,600$. Write the answer directly below the previous answer as shown.
4. Add the respective columns of each of the rows.

WRITE

$$\begin{array}{r} 1456 \\ \times\ 132 \\ \hline 2\,912 \\ 43\,680 \\ 145\,600 \\ \hline 192\,192 \\ \hline \end{array}$$

Mental strategies for multiplication

In many cases it is not practical for us to use pen and paper or even a calculator to perform a multiplication. We need some mental strategies to make a multiplication easier.

As with addition, the order in which we multiply numbers is not important. For example,

$$3 \times 8 \times 6 = 6 \times 3 \times 8.$$

We can simplify a multiplication if we can find a pair of numbers to make a multiplication that will simplify the next step.

Suppose that we are asked to multiply $2 \times 17 \times 5$. Rather than do 2×17 then multiply the result by 5, we can rearrange the question:

$$\begin{aligned} &2 \times 17 \times 5 \\ &= 17 \times (2 \times 5) \\ &= 17 \times 10 \\ &= 170 \end{aligned}$$

By finding the multiplication pair that made 10, we could easily complete 17×10.

WORKED Example 8

Use mental strategies to calculate $4 \times 23 \times 25$.

THINK

1. Write the question.
2. Rearrange it looking for a number pair that makes a simpler multiplication.
3. Mentally calculate 4×25.
4. Mentally calculate the final answer.

WRITE

$$\begin{aligned} &4 \times 23 \times 25 \\ &= 23 \times (4 \times 25) \\ &= 23 \times 100 \\ &= 2300 \end{aligned}$$

Another strategy that can be used is to break a number into two separate multiplications. Consider the case where we are multiplying by 20. We can first multiply by 2 and then by 10.

WORKED Example 9

Use a mental strategy to calculate 34 × 200.

THINK	WRITE
1 Write the question.	34×200
2 Break 200 into 2 × 100.	$= 34 \times 2 \times 100$
3 Calculate 34 × 2.	$= 68 \times 100$
4 Calculate 68 × 100.	$= 6800$

If both numbers are **multiples** of 10, 100 and so on, we can multiply by ignoring the zeros, multiply the remaining numbers then add the total number of zeros to the answer. For example,

$$900 \times 6000 = 5\,400\,000$$

Consider now the multiplication 9 × 58. We can regard this multiplication as:

$$10 \times 58 - 1 \times 58$$

Using this we can mentally calculate the answer by multiplying 58 by 10 then subtracting 58 from the answer.

WORKED Example 10

Use mental strategies to calculate 77 × 9.

THINK	WRITE
1 Write the question.	77×9
2 Use the strategy of 'multiply by 10'.	$= 77 \times 10 - 77$
3 Calculate 77 × 10 and subtract 77.	$= 770 - 77$
	$= 693$

remember

1. The basis for all multiplication work is to know your multiplication tables up to 12.
2. You should know how to perform both short and long multiplication.
3. Some multiplications can be done mentally using various strategies.
   (a) Look for a multiplication pair that will make a multiple of 10, 100, and so on.
   (b) To multiply numbers that are multiples of 10, ignore the zeros, perform the multiplications, and then add the total number of zeros to your answer.
   (c) To multiply by a number such as 9, multiply by 10 then subtract the number.

EXERCISE 1C Multiplying whole numbers

1.5 SkillSHEET Multiplying whole numbers

Mathcad Multiplying numbers

EXCEL Spreadsheet Multiplying numbers

EXCEL Spreadsheet Tangle tables

GC program—TI Tables

GC program—Casio Tables

1 Write the answer to each of the following without using a calculator.
a 4×3 **b** 9×5 **c** 2×11 **d** 8×7 **e** 12×8
f 10×11 **g** 6×9 **h** 12×11 **i** 9×8

2 Multiply the following without using a calculator.
a 13×2 **b** 15×3 **c** 25×2 **d** 3×13
e 25×4 **f** 45×2 **g** 16×2 **h** 35×2
i 14×3 **j** 21×3 **k** 54×2 **l** 25×3
m $3 \times 4 \times 6$ **n** $2 \times 5 \times 9$ **o** $3 \times 3 \times 3$ **p** $5 \times 6 \times 3$
q $5 \times 4 \times 5$ **r** $8 \times 5 \times 2$

WORKED Example 6

3 Calculate these using short multiplication.
a 16×8 **b** 29×4 **c** 137×9 **d** 857×3
e 4920×5 **f** $15\,984 \times 7$ **g** 7888×8 **h** 472×4
i 2015×8 **j** $10\,597 \times 6$ **k** $34\,005 \times 11$ **l** $41\,060 \times 12$
Check your answers using a calculator.

WORKED Example 7

4 Calculate these using long multiplication.
a 52×44 **b** 97×31 **c** 59×28 **d** 16×57
e 173×41 **f** 850×76 **g** 407×53 **h** 47×2074
i $80\,055 \times 27$ **j** $19 \times 256\,340$ **k** $57\,835 \times 476$ **l** 8027×215
Check your answers using a calculator.

WORKED Example 8

5 Use mental strategies to calculate each of the following.
a $2 \times 8 \times 5$ **b** $2 \times 4 \times 5 \times 6$ **c** $4 \times 19 \times 25$
d $50 \times 45 \times 2$ **e** $2 \times 9 \times 50$ **f** $4 \times 67 \times 250$

WORKED Example 9

6 Use mental strategies to calculate each of the following.
a 45×20 **b** 61×30 **c** 62×50
d 84×200 **e** 500×19 **f** 86×2000

7 Find each of the following.
a 200×40 **b** 30×700 **c** 600×800
d 90×80 **e** 120×400 **f** 1100×5000
g $900\,000 \times 7000$ **h** $120\,000 \times 1200$ **i** 800×7000
j 9000×6000 **k** 4000×110 **l** $12\,000 \times 1100$

WORKED Example 10

8 Use mental strategies to calculate each of the following.
a 34×9 **b** 83×9 **c** 628×9
d 75×99 **e** 24×19 **f** 26×8

9 **a** Calculate 56×100. **b** Calculate 56×10.
c Use your answers to parts **a** and **b** to calculate the answer to 56×90.

10 Use the method demonstrated in question **9** to calculate each of the following.
a 48×90 **b** 74×90 **c** 125×90
d 32×900 **e** 45×80 **f** 72×800

11 **a** Calculate 25×6. **b** Multiply your answer to part **a** by 2.
c Now calculate 25×12.
d Use the answers to parts **a**, **b** and **c** to describe a method for mentally multiplying by 12.

12 Use the method that you discovered in question 11 to mentally calculate the value of each of the following.

a 15×12 b 70×12 c 18×12 d 105×12
e 25×14 f 40×16 g 11×18 h 34×20

13 a What is the value of 9×10? b What is the value of 9×3?
c Calculate the value of 9×13.
d Use the answers to parts a, b and c above to describe a method for mentally multiplying by 13.

14 Use the method that you discovered in question 13 to calculate the value of each of the following.

a 25×13 b 30×13 c 24×13 d 102×13

15 John wants to make a telephone call to his friend Rachel who lives in San Francisco. The call will cost him \$3 per minute. If John speaks to Rachel for 24 minutes:
a what will the call cost?
b what would John pay if he made this call every month for 2 years?

16 Chris is buying some generators. The generators cost \$12 000 each and she needs 11 of them. How much will they cost her?

17 Julie was saving money to buy a digital camera. She was able to save \$75 each month.
a How much did she save after 9 months?
b How much did she save over 16 months?
c If Julie continued to save at the same rate, how much will she save over a period of 3 years?

18 A car can travel 14 kilometres using 1 litre of fuel. How far could it travel with 35 litres?

19 As Todd was soaking in the bath, he was contemplating how much water was in the bath. If Todd used 85 litres of water each time he bathed and had a bath every week:
a how much bath water would Todd use in 1 year?
b how much would he use over a period of 5 years?

20 In 1995, a team of British soldiers at Hameln, Germany, constructed the fastest bridge ever built. The bridge spanned an 8-metre gap and it took the soldiers 8 minutes and 44 seconds to build it. How many seconds did it take them to build it?

21 You are helping your Dad build a fence around your new swimming pool. He estimates that each metre of fence will take 2 hours and cost \$65 to build.
a How long will it take you and your Dad to build a 17-metre fence?
b How much will it cost to build a 17-metre fence?
c How much would it cost for a 29-metre fence?

22 Narissa does a paper round each morning before school. She travels 2 kilometres each morning on her bicycle, delivers 80 papers and is paid \$35. She does her round each weekday.
a How far does she travel in 1 week?
b How much does she get paid in 1 week?
c How far does she travel in 12 weeks?
d How much would she be paid over 52 weeks?
e How many papers would she deliver in 1 week?
f How many papers would she deliver in 52 weeks?

10 QUICK QUESTIONS 1

1 Write 3200 in Roman numerals.

2 Write DXV in Hindu–Arabic numbers.

3 Evaluate 3297 + 6287.

4 Evaluate 23 + 84 + 7 + 16 by finding suitable pairs of numbers.

5 Find 3182 – 958.

6 Use short multiplication to evaluate 428×7.

7 Use long multiplication to evaluate 184×39.

8 Use a mental strategy to evaluate 79×9.

9 Use a mental strategy to evaluate 92×20.

10 Use a mental strategy to evaluate $2 \times 56 \times 5$.

MATHS QUEST CHALLENGE

1 **In AFL football, a goal scores 6 points and a behind scores 1 point. Find a score which is the same as the product of the number of goals and the number of behinds. For example, 2 goals 12 behinds = $2 \times 6 + 12 = 24$ points. Also $2 \times 12 = 24$. Find two other similar results.**

2 **a** **Consider numbers with 2 identical digits multiplied by 99. Work out each of the following.**
$11 \times 99 =$
$22 \times 99 =$
$33 \times 99 =$
Can you see a pattern? Without using long multiplication or a calculator, write down the answers to 44×99, 55×99, 66×99, 77×99, 88×99 and 99×99.

b **Try it again but this time multiply numbers with 3 identical digits by 99. Use only long multiplication or a calculator with the first 3 calculations. Look for a pattern and then write down the answers to the remaining multiplications.**

c **What about numbers with 4 or 5 identical digits which are multiplied by 99? Try these as well.**

Dividing whole numbers

Short division

We can use short division when dividing by numbers up to 12.

WORKED Example 11

Calculate 89 656 ÷ 8.

THINK

1. Divide 8 into the first digit and carry the remainder to the next digit; 8 goes into 8 once. Write 1 above the 8 as shown. There is no remainder.
2. Divide 8 into the second digit and carry the remainder to the next digit; 8 goes into 9 once with 1 left over. Write 1 above the 9 and carry 1 to the hundreds column.
3. Divide 8 into the third digit and carry the remainder to the next digit; 8 goes into 16 twice with no remainder. Write 2 above the 6 as shown.
4. Divide 8 into the fourth digit and carry the remainder to the next digit; 8 doesn't go into 5. Write the 0 above the 5. Carry 5 to the next digit.
5. Divide 8 into 56; 8 goes into 56 seven times.
6. Write the answer.

WRITE

$$\begin{array}{r} 1\ 1\ 2\ 0\ 7 \\ 8\overline{)8\ 9\ {}^{1}6\ 5\ {}^{5}6} \end{array}$$

89 656 ÷ 8 = 11 207

The answer to a division can be checked using multiplication. In the above worked example, 11 207 × 8 = 89 656.

Long division

Long division is used when the divisor is larger than 12. It involves the same process as short division, but all working is shown. The divisor is the number that you are dividing by.

For larger numbers the process is repeated until the problem is completed.

WORKED Example 12

Use long division to calculate 356 ÷ 15.

THINK

1. Divide 15 into the first digit. If it doesn't go, write 0 above the first digit.
2. Divide 15 into the first two digits. 15 goes into 35 twice. Write the 2 above the second digit.
3. Multiply (15 × 2 = 30). Write 30 below the first two digits.
4. Subtract 30 from 35. The answer is the 5 remaining from the division in step 2.

WRITE

$$\begin{array}{r} 02 \\ 15\overline{)356} \end{array}$$

$$\begin{array}{r} 02 \\ 15\overline{)356} \\ -30 \\ \hline 5 \end{array}$$

THINK	WRITE
5 Bring down the third digit; that is, bring down the 6. The process is repeated.	$\begin{array}{r} 02 \\ 15\overline{)356} \\ -30 \\ \hline 56 \end{array}$
6 Divide 15 into the last number which is 56. 15 goes into 56 three times. Multiply ($15 \times 3 = 45$) and write the 3 above the third digit and 45 below 56. 7 Subtract 45 from 56 as shown.	$\begin{array}{r} 023 \\ 15\overline{)356} \\ -30 \\ \hline 56 \\ -45 \\ \hline 11 \end{array}$
8 Write the answer.	$356 \div 15 = 23$ remainder 11

The answer to worked example 12 can also be written as a fraction. Rather than writing 23 remainder 11, we may write $23\frac{11}{15}$.

The result of a division can also be written as a multiplication. Again in worked example 12, we said $356 \div 15 = 23$ remainder 11. We can therefore say:

$356 = 23 \times 15 + 11$. Check this result for yourself.

Dividing numbers that are multiples of ten

WORKED Example 13

Calculate 48 000 ÷ 600.

THINK	WRITE
1 Write the question.	$48\,000 \div 600$
2 Write the question as a fraction.	$= \frac{48\,0\not{0}\not{0}}{6\not{0}\not{0}}$
3 Cancel as many zeros as possible, crossing off the same number in both numerator and denominator.	$= \frac{480}{6}$
4 Perform the division.	$\begin{array}{r} 080 \\ 6\overline{)480} \end{array}$
5 Write your answer.	$48\,000 \div 600 = 80$

remember

1. Use short division when dividing by numbers up to 12 (or higher, if you know the tables for it, for example 13, 15, 20).
2. Use long division when you are dividing by a number larger than 12. Repeat the same process — divide, multiply, subtract, bring down.
3. When dividing numbers that are multiples of 10, write the question as a fraction, cancel as many zeros as possible and then divide.

EXERCISE 1D Dividing whole numbers

SkillSHEET 1.6 Dividing whole numbers

EXCEL Spreadsheet Dividing numbers

Mathcad Dividing numbers

1 Evaluate these divisions without using a calculator. There should be no remainder.

a $24 \div 6$ b $24 \div 8$ c $36 \div 9$ d $72 \div 8$
e $49 \div 7$ f $96 \div 12$ g $108 \div 9$ h $56 \div 7$
i $16 \div 4$ j $28 \div 7$ k $40 \div 2$ l $26 \div 2$
m $45 \div 15$ n $32 \div 16$ o $27 \div 3 \div 3$ p $96 \div 8 \div 6$
q $48 \div 12 \div 2$ r $72 \div 2 \div 9$ s $56 \div 7 \div 4$ t $100 \div 2 \div 10$
u $90 \div 3 \div 2$

Check your answers using multiplication.

2 Perform these calculations which involve a combination of multiplication and division. Always work from left to right.

a $4 \times 5 \div 2$ b $9 \times 8 \div 12$ c $80 \div 10 \times 7$
d $45 \div 9 \times 7$ e $144 \div 12 \times 7$ f $120 \div 10 \times 5$
g $4 \times 9 \div 12$ h $121 \div 11 \times 4$ i $81 \div 9 \times 6$

WORKED Example 11

EXCEL Spreadsheet Four operations (DIY)

3 Calculate each of the following using short division.

a $3\overline{)1455}$ b $4\overline{)27\,768}$ c $7\overline{)43\,456}$ d $9\overline{)515\,871}$
e $11\overline{)30\,371}$ f $8\overline{)640\,360}$ g $3\overline{)255\,194}$ h $6\overline{)516\,285}$
i $12\overline{)103\,717}$ j $7\overline{)6\,328\,530}$ k $5\overline{)465\,777}$ l $8\overline{)480\,594}$

Check your answers using a calculator.

WORKED Example 12

4 Calculate each of these using long division. Write any remainders as fractions.

a $16\overline{)4144}$ b $21\overline{)20\,328}$ c $25\overline{)2\,075\,375}$
d $32\overline{)214\,496}$ e $43\overline{)26\,703}$ f $13\overline{)27\,989}$
g $18\overline{)11\,557}$ h $24\overline{)725\,916}$ i $14\overline{)75\,383}$

Check your answers using a calculator.

WORKED Example 13

5 Divide these numbers, which are multiples of ten.

a $4200 \div 6$ b $700 \div 70$ c $210 \div 30$
d $720\,000 \div 800$ e $8100 \div 900$ f $4\,000\,000 \div 8000$
g $600\,000 \div 120$ h $560 \div 80$ i $880\,000 \div 1100$

6 Earlier, we saw that $356 \div 15 = 23$ remainder 11. We then said $356 = 23 \times 15 + 11$. Perform each of the following divisions, then write the answer in this form.

a $27 \div 5$ b $68 \div 5$ c $82 \div 10$ d $156 \div 6$
e $784 \div 11$ f $230 \div 15$ g $458 \div 20$ h $3625 \div 37$

7 a What is the value of $580 \div 10$? b Halve your answer to part a.
c Now find the value of $580 \div 20$.
d Use the answers to parts a, b and c to describe a method for mentally dividing by 20.

8 Use the method that you discovered in question 7 to evaluate:

a $280 \div 20$ b $1840 \div 20$ c $790 \div 20$ d $960 \div 30$

9 Spiro travels 140 kilometres per week travelling to and from work. If Spiro works 5 days per week:

a how far does he travel each day? b what distance is his work from home?

10 Kelly works part time at the local pet shop. Last year she earned $2496.

a How much did she earn each month? **b** How much did she earn each week?

11 David makes kites from a special lightweight fabric. An Australian company is able to supply this fabric, but only in rolls of 50 metres. It is worth buying this roll only if he can make more than 18 kites from one roll. He needs to decide whether he should order from this company.

Each kite requires 250 cm of fabric from a roll.

a How many centimetres of fabric are in a roll if there are 100 centimetres in 1 metre?

b How many kites could he make with the fabric from one roll?

c Will he order fabric from this company?

12 At the milk processing plant, the engineer asked Farid how many cows he had to milk each day. Farid said he milked 192 cows because he obtained 1674 litres of milk each day and each cow produced 9 litres. Does Farid really milk 192 cows each day? If not, calculate how many cows he does milk.

13 When Juan caters for a celebration such as a wedding he fills out a form for the client to confirm the arrangements. Juan has been called to answer the telephone so it has been left to you to fill in the missing details. Copy and complete this planning form.

Celebration type	Wedding
Number of guests	152
Number of people per table	8
Number of tables required	
Number of courses for each guest	4
Total number of courses to be served	
Number of courses each waiter can serve	80
Number of waiters required	
Charge per guest	$55
Total charge for catering	

14 Janet is a land developer and has bought 10 450 square metres of land. She intends to subdivide the land into 11 separate blocks.

a How many square metres will each block be?

b If she sells each block for $72 250, how much will she receive for the subdivided land?

15 Shea has booked a beach house for a week over the summer period for a group of 12 friends. The house costs $1344 for the week. If all 12 people stayed for 7 nights, how much will the house cost each person per night?

GAME time
Number skills and the history of number — 002

16 Mario is a farmer who has to shear 4750 sheep.

a If each sheep produces 5 kilograms of wool, how much wool will Mario have to sell?

b If Mario packs 250 kilograms of wool into each bale, how many bales will he have?

c If he sells the wool for $4 per kilogram, how much money will Mario receive for the wool?

WorkSHEET 1.2

Order of operations

In mathematics, conventions are followed so that we all have a common understanding of mathematical operations.

Tran and Liz discovered that they had different answers to the same question. The question was $6 + 6 \div 3$. Tran thought the answer was 8, but Liz thought it was 4. Who do you think is right?

There is a set order in which mathematicians calculate problems. The order is:
1. **brackets**
2. **multiplication and division (from left to right)**
3. **addition and subtraction (from left to right).**

WORKED Example 14

Calculate $6 + 12 \div 4$.

THINK	WRITE
1 Write the question.	$6 + 12 \div 4$
2 Perform the division before the addition.	$= 6 + 3$
3 Calculate the answer.	$= 9$

WORKED Example 15

Calculate:

a $12 \div 2 + 4 \times (4 + 6)$ **b** $80 \div \{[(11 - 2) \times 2] + 2\}$

THINK	WRITE
a 1 Write the question.	**a** $12 \div 2 + 4 \times (4 + 6)$
2 Remove the brackets by working out the addition inside.	$= 12 \div 2 + 4 \times 10$
3 Perform the division and multiplication next, working from left to right.	$= 6 + 40$
4 Complete the addition last and calculate the answer.	$= 46$
b 1 Write the question.	**b** $80 \div \{[(11 - 2) \times 2] + 2\}$
2 Remove the innermost brackets by working out the difference of 11 and 2.	$= 80 \div \{[9 \times 2] + 2\}$
3 Remove the next pair of brackets by working out the multiplication inside them.	$= 80 \div \{18 + 2\}$
4 Remove the final pair of brackets by working out the addition inside them.	$= 80 \div 20$
5 Perform the division last and calculate the answer.	$= 4$

WORKED Example 16

Insert one set of brackets in the appropriate place to make the following statement true.
$3 \times 10 - 8 \div 2 + 4 = 7$

THINK	WRITE
1 Write the left-hand side of the equation.	$3 \times 10 - 8 \div 2 + 4$
2 Place one set of brackets around the first two values.	$= (3 \times 10) - 8 \div 2 + 4$
3 Perform the multiplication inside the bracket.	$= 30 - 8 \div 2 + 4$
4 Perform the division.	$= 30 - 4 + 4$
5 Perform the subtraction and addition working from left to right. *Note:* Since this is not the answer, the above process must be repeated.	$= 26 + 4$ $= 30$ Since this is not equal to 7 we must place the brackets in a different position.
6 Place one set of brackets around the second and third values.	$3 \times 10 - 8 \div 2 + 4$ $= 3 \times (10 - 8) \div 2 + 4$
7 Perform the subtraction inside the bracket.	$= 3 \times 2 \div 2 + 4$
8 Perform the multiplication and division working from left to right.	$= 6 \div 2 + 4$ $= 3 + 4$
9 Perform the addition last and calculate the answer. This is correct, so the brackets are in the correct place.	$= 7$

Graphics Calculator tip! Order of operations

A graphics calculator will automatically calculate the answer using the correct order of operations. You need to enter the numbers and operations as they are written from left to right and then press ENTER to obtain the answer. Also include brackets if required. Notice that the multiplication sign × is shown as ∗ and the division sign ÷ is shown as / on the screen. For the calculation in worked example 15a, the screen at right would be obtained.

Order of operations

remember

1. The operations inside brackets are always calculated first.
2. If there is more than one set of brackets, calculate the operations inside the innermost brackets first.
3. Multiplication and division operations are calculated in the order that they appear.
4. Addition and subtraction operations are calculated in the order that they appear.

EXERCISE 1E Order of operations

SkillSHEET 1.7 Order of operations

Mathcad Order of operations

EXCEL Spreadsheet Four operations (DIY)

1 Tran and Liz discovered that they had different answers to the same question, which was to calculate $6 + 6 \div 3$. Tran thought the answer was 8. Liz thought the answer was 4. Who was correct, Tran or Liz?

WORKED Example 14, 15

2 Calculate each of these, following the order of operations rules.

a $3 + 4 \div 2$
b $8 + 1 \times 12$
c $24 \div (12 - 4)$
d $15 \times (17 - 15)$
e $11 + 6 \times 8$
f $30 - 45 \div 9$
g $56 \div (7 + 1)$
h $12 \times (20 - 12)$
i $3 \times 4 + 23 - 10 - 5 \times 2$
j $42 \div 7 \times 8 - 8 \times 3$
k $10 + 40 \div 5 + 14$
l $81 \div 9 + 108 \div 12$
m $16 + 12 \div 2 \times 10$
n $(18 - 15) \div 3 \times 27$
o $4 + (6 + 3 \times 9) - 11$
p $52 \div 13 + 75 \div 25$
q $(12 - 3) \times 8 \div 6$
r $88 \div (24 - 13) \times 12$
s $(4 + 5) \times (20 - 14) \div 2$
t $(7 + 5) - (10 + 2)$
u $\{[(16 + 4) \div 4] - 2\} \times 6$
v $60 \div \{[(12 - 3) \times 2] + 2\}$

WORKED Example 16

3 Insert one set of brackets in the appropriate place to make these statements true.

a $12 - 8 \div 4 = 1$
b $4 + 8 \times 5 - 4 \times 5 = 40$
c $3 + 4 \times 9 - 3 = 27$
d $3 \times 10 - 2 \div 4 + 4 = 10$
e $12 \times 4 + 2 - 12 = 60$
f $17 - 8 \times 2 + 6 \times 11 - 5 = 37$
g $10 \div 5 + 5 \times 9 \times 9 = 81$
h $18 - 3 \times 3 \div 5 = 9$

4 **multiple choice**

$20 - 6 \times 3 + 28 \div 7$ is equal to:

A 46 B 10 C 6 D 4 E 2

5 **multiple choice**

The two signs marked with * in the equation $7 * 2 * 4 - 3 = 12$ are:

A $-, +$ B $\times, +$ C $-, \div$ D $+, \times$ E $\times, \div$

6 Insert brackets *if necessary* to make each statement true.

a $6 + 2 \times 4 - 3 \times 2 = 10$
b $6 + 2 \times 4 - 3 \times 2 = 26$
c $6 + 2 \times 4 - 3 \times 2 = 16$
d $6 + 2 \times 4 - 3 \times 2 = 8$

MATHS QUEST CHALLENGE

1 What is the smallest number of pebbles greater than 10 for which grouping them in heaps of 7 leaves 1 extra and grouping them in heaps of 5 leaves 3 extra?

2 Choose a digit from 2 to 9. Write it six times. For example, if 4 is chosen the number is 444 444. Divide the six-digit number by 33. Next divide the result by 37 and finally divide this last result by 91. What is the final result?

Try this again with another six-digit number formed as before. (Divide by 33, then 37, then 91.) What is your final result in this case? Try to explain how this works.

What's special about the speed 370 km/h?

The letter beside each question and its answer gives the puzzle solution code.

A 7 + 6 ÷ 2 =	**W** 12 + 8 + 9 =	**A** 30 x 20 − 520 =	**N** 8 − 18 ÷ 6 + 10 =
C 15 − 9 + 6 =	**A** 47 − 12 − 4 =	**D** 79 ÷ 1 + 8 =	**S** 69 − 13 + 8 =
D 8 x 7 + 5 =	**D** 6 x 12 + 18 =	**E** 12 x 2 x 3 =	**T** 5 + 20 ÷ 2 x 3 =
E 2 + 9 x 4 =	**E** 75 ÷ 5 + 7 =	**H** 19 − 8 ÷ 4 =	**E** 8 + 21 ÷ 7 − 9 =
G 3 x 4 x 5 =	**G** 5 x 6 x 7 =	**I** 71 − 52 + 8 =	**H** 12 x 11 − 33 =
H 15 + 21 − 6 =	**H** 120 − 40 − 10 =	**N** 8 x 9 x 0 =	**I** 200 ÷ 5 + 15 =
I 8 ÷ 2 + 2 =	**I** 45 ÷ 5 + 7 =	**R** 80 ÷ 4 ÷ 5 =	**S** 1 + 16 x 7 x 0 =
M 100 ÷ 5 x 2 =	**M** 8 + 37 − 3 =	**S** 42 ÷ 6 + 6 =	**T** 13 + 7 x 7 =
N 7 x 9 + 12 =	**N** 128 − 48 x 2 =	**T** 90 ÷ 18 + 3 =	**E** 8 + 17 + 9 − 1 =
O 63 − 18 ÷ 2 =	**O** 5 + 38 + 16 =	**W** 54 x 2 − 12 =	**H** 11 x 3 + 6 =
P 90 − 20 + 15 =	**P** 72 − 13 + 6 =	**A** 3 + 7 x 9 =	**I** 48 ÷ 8 + 5 =
R 10 + 20 x 2 =	**R** 82 ÷ 2 + 2 =	**D** 15 + 16 x 2 =	**S** 34 x 2 + 6 =
S 63 ÷ 9 x 7 =	**S** 8 x 12 ÷ 4 =	**E** 18 ÷ 9 + 5 =	**T** 15 x 5 + 14 =
T 20 x 5 − 14 =	**T** 13 x 4 − 4 =	**H** 5 + 8 − 7 + 14 =	**E** 6 x 8 ÷ 4 − 9 =
U 400 ÷ 20 ÷ 4 =	**W** 250 ÷ 5 + 18 =	**I** 73 x 1 ÷ 1 =	**E** 14 ÷ 2 ÷ 7 + 8 =

16 8 6 24 35 17 22 39 27 60 20 33 49 62

29 73 0 61 64 85 7 2 47 4 9 12 54 43 87 3 90

80 86 42 48 • 68 10 1 30 11 15 210 89 59 75,

32 72 96 70 66 40 65 74 99 55 50 38, 5 13 31

Estimation

A football commentator makes the following announcement: 'As today's game begins, there are 58 271 people sitting in the stands waiting to watch their heroes play'. How do you think listeners react to such an exact number? If you asked them, could they repeat the figure?

It is more usual to hear that there are 58 000 or 60 000 spectators as it is often not necessary to know the exact number of people or things. An estimate is enough, so the nearest rounded number is used.

An **estimation** is not the same as a guess because it is based on information. For example, we may know how many people are able to fit into the football ground and the approximate percentage of seats filled. We can use this information to produce our estimate.

Estimation is also useful when we are working with calculators. By mentally estimating an approximate answer, we increase our chances of noticing if we have pressed a wrong button on the calculator.

To estimate the answer to a mathematical problem, round the numbers to the first digit and find an approximate answer. This can be done in your head and used to check your calculations. If the exact answer is not required, then this estimate can be calculated with very little effort.

Rounding

If the second digit is 0, 1, 2, 3 or 4, the first digit stays the same.
If the second digit is 5, 6, 7, 8 or 9, the first digit is rounded up.
If rounding to the first digit:
6512 would be rounded to 7000 as it is actually closer to 7000
6397 would be rounded to 6000 as it is actually closer to 6000
6500 would be rounded to 7000. It is exactly halfway between 6000 and 7000. So to avoid confusion, if it is halfway (if the second digit is 5) the number is rounded up.

Estimations can be made when multiplying, dividing, adding or subtracting. They can also be used when there is more than one operation in the same question.

WORKED Example 17

Estimate 48 921 × 823 by rounding to the first digit.

THINK	WRITE
1 Write the question.	$48\,921 \times 823$
2 Round each part of the question to the first digit.	$\approx 50\,000 \times 800$
3 Multiply.	$= 40\,000\,000$

The actual answer is 40 261 983, which is higher than the estimation. However, the estimate is accurate enough when an exact answer is not needed.

remember

1. An estimation can be used when the exact answer is not required.
2. An estimation can be used to check a calculation.
3. A useful estimation can be made by rounding each number to the first digit and then performing the appropriate calculation.
4. If the second digit is 0, 1, 2, 3 or 4, the first digit stays the same. If the second digit is 5, 6, 7, 8 or 9, the first digit is increased by 1 or rounded up.

EXERCISE 1F Estimation

Mathcad: Estimation

EXCEL Spreadsheet: Four operations (DIY)

1.8 SkillSHEET: Rounding to the first (leading) digit

WORKED Example 17

1 Estimate 67 451 × 432 by rounding to the first digit.

2 Copy and complete the following table by rounding the numbers to the first digit. The first row has been completed as an example. In the column headed 'Prediction', guess whether the actual answer will be higher or lower than your estimate. Then use a calculator to work out the actual answer and record it in the final column titled 'Calculation' to determine whether it was higher or lower than your estimate.

		Estimate	Estimated answer	Is the actual answer higher or lower than the estimate?	
				Prediction	Calculation
Example	4129 ÷ 246	4000 ÷ 200	20	Lower	16.784 553 so lower
a	487 + 962				
b	33 041 + 82 629				
c	184 029 + 723 419				
d	1127 + 6302				
e	29 + 83				
f	55 954 + 48 312				
g	93 261 − 37 381				
h	321 − 194				
i	468 011 − 171 962				
j	942 637 − 389 517				

(continued)

		Estimate	Estimated answer	Is the actual answer higher or lower than the estimate?	
				Prediction	Calculation
k	64 064 − 19 382				
l	89 830 − 38 942				
m	36 × 198				
n	8631 × 9				
o	87 × 432				
p	623 × 12 671				
q	29 486 × 39				
r	222 × 60				
s	31 690 ÷ 963				
t	63 003 ÷ 2590				
u	867 910 ÷ 3300				
v	8426 ÷ 3671				
w	69 241 ÷ 1297				
x	37 009 ÷ 180				

3 **multiple choice**

a The best estimate of 4372 + 2587 is:
A 1000 B 5527 C 6000 D 7000 E 7459

b The best estimate of 672 × 54 is:
A 30 000 B 35 000 C 36 000 D 40 000 E 42 000

c The best estimate of 67 843 ÷ 365 is:
A 150 B 175 C 200 D 230 E 250

4 Estimate the answers to each of these.

a 5961 + 1768
b 432 − 192
c 48 022 ÷ 538
d 9701 × 37
e 98 631 + 608 897
f 6501 + 3790
g 11 890 − 3642
h 83 481 ÷ 1751
i 112 000 × 83
j 66 501 ÷ 738
k 392 × 113 486
l 12 476 ÷ 24

5 Su-Lin was using her calculator to answer some mathematical questions, but found she obtained a different answer each time she performed the same calculation. Using your estimation skills, predict which of Su-Lin's answers is most likely to be correct.

a 217 × 489
  i 706 ii 106 113 iii 13 203 iv 19 313

b 89 344 ÷ 256
  i 39 ii 1595 iii 89 088 iv 349

c 78 × 6703
  i 522 834 ii 52 260 iii 6781 iv 56 732 501

d 53 669 ÷ 451
  i 10 ii 1076 iii 53 218 iv 119

6 Julian is selling tickets for his school's theatre production of South Pacific. So far he has sold 439 tickets for Thursday night's performance, 529 for Friday's and 587 for Saturday's. The costs of the tickets are \$9.80 for adults and \$4.90 for students.

a Round the figures to the first digit to estimate the number of tickets Julian has sold so far.

b If approximately half the tickets sold were adult tickets and the other half were student tickets, estimate how much money has been received so far by rounding the cost of the tickets to the first digit.

7 During the show's intermission, Jia is planning to run a stall selling hamburgers to raise money for the school. She has priced the items she needs and made a list in order to estimate her expenses.

a By rounding the item price to the first digit, use the table below to estimate how much each item will cost Jia for the quantity she requires.

Item	Item price	Quantity required	Estimated cost
Bread rolls	\$2.90/dozen	25 packets of 12	
Hamburgers	\$2.40/dozen	25 packets of 12	
Tomato sauce	\$1.80/litre	2 litres	
Margarine	\$2.20/tub	2 tubs	
Onions	\$1.85/kilogram	2 kilograms	
Tomatoes	\$3.50/kilogram	2 kilograms	
Lettuce	\$1.10 each	5 lettuces	

b Estimate what Jia's total shopping bill will be.

c If Jia sells 300 hamburgers over the 3 nights for \$2 each, how much money will she receive for the hamburgers?

d Approximately how much money will Jia raise through selling hamburgers over the 3 nights?

WorkSHEET 1.3

COMMUNICATION

Estimating

Estimating skills can be used to work out large totals that would be impractical to count separately. An estimate is not a guess, it is based on information.

Look at the photograph below. Can you estimate how many sweets are shown?

1 The following steps will guide you in solving this problem.

a Lightly draw a grid in pencil over the photograph. We need to divide the photograph into equal-sized sections. (You may like to draw lines which make sections that are squares of side length 2 centimetres.)

b How many equal-sized sections do you have?
c Select one section and count the number of bubble gum sweets in this section.
d What calculation needs to be performed to work out the number of bubble gum sweets in all the sections?
e Perform the calculation and write out your answer to this problem in a sentence.

2 Repeat this estimating process for the following photograph. Estimate the number of people in this crowd. Compare this method with that which you used to calculate the number of people in the crowd in the photograph on page 1.

3 Estimate the number of graduating students shown in the photograph below. If the hall holds 12 times this number, estimate the total capacity of the hall. Show all your working and write a sentence explaining how you solved this problem.

10 QUICK QUESTIONS 2

1 Calculate 62×45.

2 Calculate 46×99.

3 Calculate $345 \div 5$.

4 Calculate $83\,466 \div 18$.

5 The members of a youth group are having a pizza night. If they order 12 pizzas that are each cut into 8 pieces and there are 15 people attending the night, how many pieces would each person get? Would there be any slices of pizza left over?

6 Calculate $32 - 32 \div 4$.

7 Calculate $(12 + 60) \div 9 + 3$.

8 Insert grouping symbols to make the following statement true: $7 + 2 \times 11 - 4 = 21$.

9 Estimate 494×61.

10 Estimate $71\,569 \div 911$.

Subsets (special groups) of numbers

In this section, we shall look at several groups of numbers called subsets that are of special interest. These include the *figurative* numbers (based on geometric figures like squares and triangles), the Fibonacci numbers, Pascal's triangle, and the group of numbers called **palindromes**.

Square numbers

Square numbers are numbers that can be arranged in a square, as shown below.

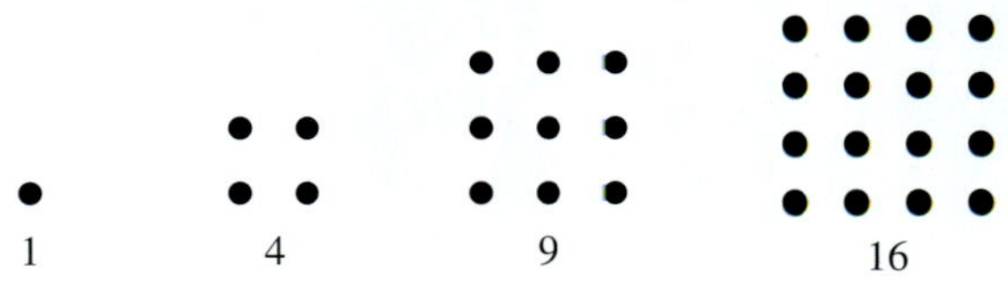

By looking at the pattern formed we can see that:
1. the first square number, 1, equals 1×1
2. the second square number, 4, equals 2×2
3. the third square number, 9, equals 3×3
4. the fourth square number, 16, equals 4×4.

We can see that if this pattern is continued we can find any square number by multiplying the position of the square number by itself. This is known as *squaring a number* and is written in shorthand by an index (or power) of 2. For example, 3×3 can be written as 3^2.

Any number that is multiplied by itself produces a square number and can be written using an index (or power) of 2.

WORKED Example 18

Find the sixth square number.

THINK	WRITE
1 Write the number that shows the square number you are looking for. Multiply this number by itself.	6×6
2 Find the answer.	$= 36$
3 Write your answer in a sentence.	The sixth square number is 36.

Triangular numbers

Triangular numbers are numbers that can be arranged in a triangle as shown below.

1 3 6

By looking at the pattern we can see that:

1. the first triangular number is 1
2. the second triangular number, 3, is $1 + 2$
3. the third triangular number, 6, is $1 + 2 + 3$.

Each triangular number can be found by adding all numbers up to the position of that triangular number.

WORKED Example 19

Find the tenth triangular number.

THINK	WRITE
1 Add all the numbers from 1 up to the triangular number you are looking for.	$1 + 2 + 3 + 4 + 5 + 6 + 7 + 8 + 9 + 10 = 55$
2 Answer the question in a sentence.	The tenth triangular number is 55.

DESIGN More number patterns in shapes

There are several number patterns that fit geometric shapes.

1. Find out what the 'hexagonal numbers' are. Draw a diagram to represent the number pattern.
2. Find out what the 'cubic numbers' are. Can you describe how to find a given cubic number?

Fibonacci numbers

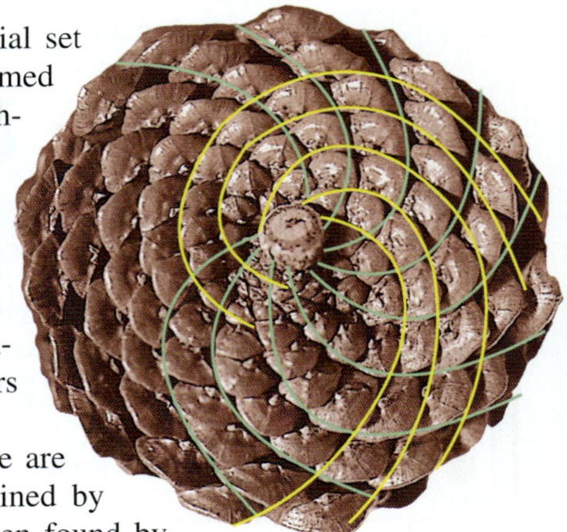

Another number sequence that produces a special set of numbers is the Fibonacci sequence. It was named after Leonardo Fibonacci (c. 1170–1250), a mathematician from Pisa in Italy. Examples of the Fibonacci sequence can be found among plants and animals in the environment. Try counting the number of spirals on the picture of the pine cone (or on a pineapple). You can often see two series of spirals, cutting across each other. Compare the numbers that you find with the numbers in the series described below.

The first two terms of the Fibonacci sequence are both 1. The next term in the sequence is obtained by adding these. Each of the following terms is then found by adding the two previous terms.

1		1		2		3		5		8		13		21
1	+	1	=	2										
		1	+	2	=	3								
				2	+	3	=	5						
						3	+	5	=	8				
								5	+	8	=	13		
										8	+	13	=	21

Since the third term is 2 we can say that $F_3 = 2$. Thirteen is the seventh Fibonacci number and so we can say that $F_7 = 13$.

Pascal's triangle

A number pattern that occurs in many branches of mathematics is Pascal's triangle. It was named after Blaise Pascal (1623–1662), a French philosopher, mathematician and physicist. Pascal's triangle begins with a single 1 in what is called Row 0 followed by two ones in what is called Row 1. Each row from there on begins and ends with a 1 with numbers in between found by adding the two numbers directly above it.

Pascal's triangle is begun for you below.

Row 0					1				
Row 1				1		1			
Row 2			1		2		1		
Row 3		1		3		3		1	
Row 4	1		4		6		4		1

Palindromes

Palindromes are words, sentences or numbers that read the same backwards as they do forwards. For example, the word DAD is a palindrome and the number 14541 is a palindrome (or palindromic number).

Numbers that are not palindromes can produce palindromes by a process called 'reverse and add'; that is, we reverse the digits and then add this new number to the original number. For example, starting with 17, which is not a palindrome, we get:

$$17 + 71 = 88$$

The number 88 is a palindrome.

You will have a chance to look more closely at palindromes in the next exercise.

remember

1. Square numbers are numbers that can be arranged to form a square. Each square number is found by multiplying the position of the square number by itself.
2. Triangular numbers are numbers that can be arranged to form a triangle. Each triangular number is found by adding the numbers from 1 up to the position of the triangular number.
3. Fibonacci numbers are numbers in the Fibonacci sequence. The first two numbers are 1 and each number after that is found by adding the two previous numbers.
4. Pascal's triangle is a number pattern that is found by beginning and ending each row with a 1. The middle numbers are then found by adding the two numbers directly above them.
5. Palindromes are words, sentences or numbers that read the same backwards as they do forwards.

EXERCISE 1G Subsets (special groups) of numbers

WORKED Example 18

1 Find the ninth square number.

2 Find each of the following square numbers.
a fifth **b** seventh **c** fourteenth
d twentieth **e** fiftieth **f** one hundredth

3 From the list below select the numbers that are square numbers.
12 16 20 25 30 40 64 81 100 200 300 400 640 1000

4 Use your knowledge of square numbers to find the value of:
a 8^2 **b** 11^2 **c** 15^2 **d** 30^2

5 The fourth square number is 16. Find the position of the following square numbers.
a 9 **b** 49 **c** 144 **d** 2500

6 As 16 is the fourth square number, we can say that $\sqrt{16} = 4$. This is expressed as 'The square root of 16 equals 4'. Find:
a $\sqrt{25}$ **b** $\sqrt{81}$ **c** $\sqrt{169}$ **d** $\sqrt{225}$

WORKED Example 19

7 Find the eighth triangular number.

8 Find the first ten triangular numbers.

9 You now should know how to generate the Fibonacci sequence. Copy and complete the Fibonacci table shown below.

EXCEL Spreadsheet: Fibonacci numbers

$F_1 = 1$	$F_2 = 1$	$F_3 = 2$	$F_4 = 3$	$F_5 = 5$
$F_6 = 8$	$F_7 = 13$	$F_8 = 21$	$F_9 =$	$F_{10} =$
$F_{11} =$	$F_{12} =$	$F_{13} =$	$F_{14} =$	$F_{15} =$
$F_{16} =$	$F_{17} =$	$F_{18} =$	$F_{19} =$	$F_{20} =$

10 Below are sets of three consecutive Fibonacci numbers. Perform the stated calculations and see if you can find a pattern in the answers. What is the pattern?

a 5, 8, 13
Calculate: **i** 5×13 **ii** 8^2

b 3, 5, 8
Calculate: **i** 3×8 **ii** 5^2

c 34, 55, 89
Calculate: **i** 34×89 **ii** 55^2

11 In each of the questions below there are four consecutive Fibonacci terms. Evaluate the expression next to the four numbers.

a 2, 3, 5, 8 $2 \times 8 - 3 \times 5$

b 13, 21, 34, 55 $13 \times 55 - 21 \times 34$

c 89, 144, 233, 377 $89 \times 377 - 144 \times 233$

12 To add up Fibonacci numbers we can use a special rule. See if you can work out what it is by completing the following.

a $F_1 + F_2 =$ $F_4 - 1 =$

b $F_1 + F_2 + F_3 =$ $F_5 - 1 =$

c $F_1 + F_2 + F_3 + F_4 =$ $F_6 - 1 =$

13 Complete Pascal's triangle down to row 10.

14 a Find the total of each of the rows of Pascal's triangle as far as row 10.

b What do you notice about the pattern of the answers?

c Use this pattern to find the sum of row 11 of Pascal's triangle without actually adding it up.

15 a Rearrange the first 10 rows of Pascal's triangle from question **13** so that they resemble the following pattern.

1
1 1
1 2 1
1 3 3 1
1 4 6 4 1

b Draw in diagonal rows as shown in the diagram below.

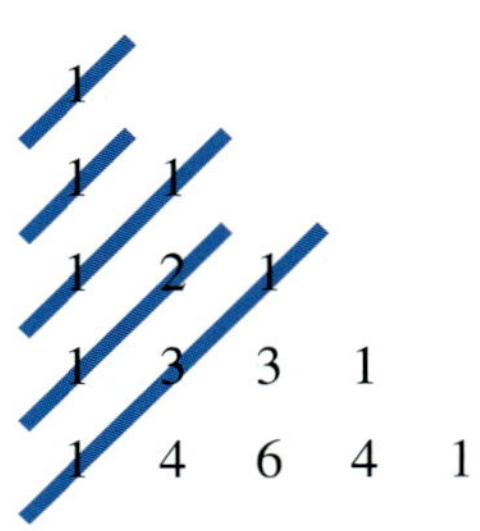

c Find the sum of each diagonal row. *Note:* The diagonal row must begin with 1.

d Study the sum of each diagonal row carefully and comment on the sequence of numbers obtained.

16 Count the petals on each of the flowers shown. What do you notice about these numbers?

a

b

17 As we have seen, palindromes are words, sentences or numbers which read the same backwards as they do forwards (for example, DAD and 14541).

GC program—TI: Palindromes

- **a** List two other words that are palindromes.
- **b** List five numbers that are palindromes.
- **c** How many palindromes are there between 100 and 250? List them.

18 Numbers that are not palindromes can produce palindromes if we reverse the digits and then add this new number to the original number.

- **a** Produce palindromes starting with the following numbers.
  i 34 **ii** 27 **iii** 521
- **b** Apply a 'reverse and add' step to 84. Does this produce a palindrome? Try another 'reverse and add' step. Have you now produced a palindrome? How many steps did it take to produce a palindrome?
- **c** Produce palindromes starting with the following numbers. In each case, write down how many steps it took to achieve this.
  i 75 **ii** 153 **iii** 97 **iv** 381 **v** 984 **vi** 7598
- **d** Choose five different starting numbers. Produce palindromes from these numbers. Check your answers by clicking on the Excel icon shown at right when using the *Maths Quest 7* CD-ROM. The file 'Palindromes' will produce a palindrome from any starting number and show you how many steps are needed.

EXCEL Spreadsheet: Palindromes

MATHS QUEST CHALLENGE

1. **Write the largest 4 digit number that has 3 and 8 as two of its digits.**
2. **Write the smallest 5 digit number which has one 0, one 7 and no digit is repeated. Be careful where you place the zero.**
3. **What number am I? I am a whole number between 10 and 99. The sum of my digits is 8. My units digit is three times my tens digit.**
4. **Use each of the digits 1, 2, 3, 4, 5, 6, 7, 8 and 9 once only to create an addition problem with the total ninety-nine thousand, nine hundred and ninety-nine.**
5. **Use any one of the numbers from 1 to 10 any number of times and any mathematical symbols to make an expression equal to 7. For example, $(5 + 5) \div 5 + 5 = 7$. See how many different symbols you can use.**

summary

Copy the sentences below. Fill in the gaps by choosing the correct word or expression from the word list that follows.

1 The Roman system of numbers is based on ___________. In Roman numerals a smaller symbol to the left of a larger one means to ___________ its value, while when it is written on the right we ___________ its value.

2 Our number system is based on the number ___________ and is known as the ___________ system.

3 When adding or subtracting, line up the numbers vertically so that digits of the same ___________ are in the same column.

4 A strategy that can be used to simplify an addition is to look for ___________ that add to a multiple of 10.

5 Short multiplication and division are used when we are multiplying or dividing by a number up to ___________. Otherwise, long multiplication or division is used.

6 To multiply numbers which are multiples of ten, disregard the zeros, perform the multiplication, then add the total number of ___________ to the answer.

7 Rules for the order of operations: ___________; Multiplication and division (from left to right); Addition and ___________ (from left to right).

8 One method of ___________ answers to mathematical questions is to round the numbers to the first digit then calculate an approximate answer.

9 ___________ numbers are found by multiplying the position in the sequence by itself.

10 ___________ numbers are found by adding up all the numbers from 1 to the position in the sequence.

11 The Fibonacci sequence of numbers begins with two 1s and then each number is found by adding the ___________ previous numbers.

12 A triangular series of numbers where each row begins and ends with a 1 and the middle numbers are the sum of the two numbers above is called ___________ triangle.

WORD LIST

estimating	Hindu–Arabic	place value	two
10	zeros	number pairs	fives
brackets	subtract	triangular	add
subtraction	Pascal's	square	12

CHAPTER review

1 State the place value of the digit shown in red in each of the following. 1A

a 74 037 b 541 910 c 1 904 000 d 290

2 Write each of the following numbers using expanded notation. 1A

a 392 b 4109 c 42 001 d 120 000

3 List the numbers 394, 349, 943, 934, 3994, 3499 in ascending order. 1A

4 List the numbers 1011, 101, 110, 1100, 1101 in descending order. 1A

5 Add these numbers. 1B

a 43 + 84
b 139 + 3048
c 3488 + 91 + 4062
d 3 486 208 + 38 645 + 692 803

6 Uluru is a sacred Aboriginal site. The map below shows some roads between Uluru and Alice Springs. The distances (in kilometres) along particular sections of road are indicated. 1B

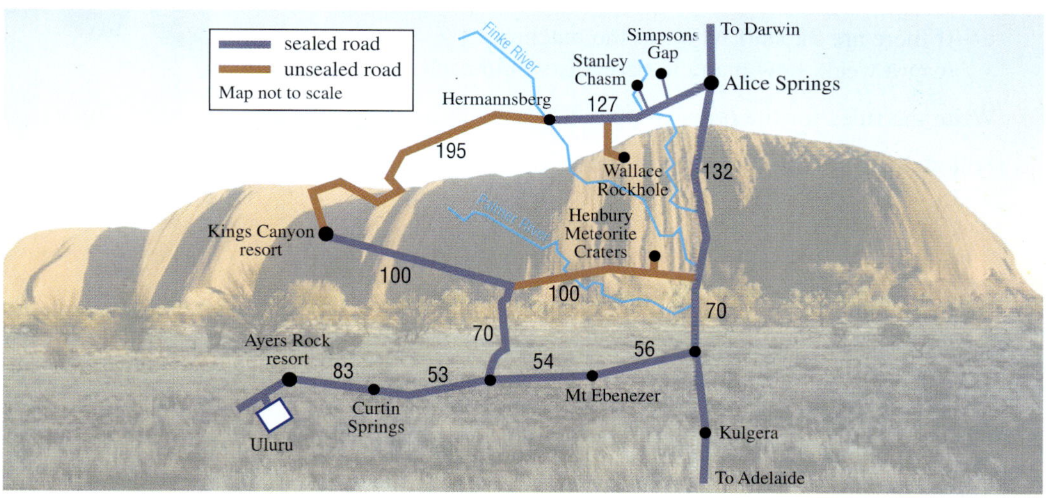

a How far is Kings Canyon resort from Ayers Rock resort near Uluru?
b What is the shortest distance by road if you are travelling from Kings Canyon resort to Alice Springs?
c If you are in a hire car, you must travel only on sealed roads. Calculate the distance you need to travel if driving from Kings Canyon resort to Alice Springs.

7 Calculate each of the following. 1B

a 20 − 12 + 8 − 14
b 35 + 15 + 5 − 20
c 300 − 170 + 20
d 18 + 10 − 3 − 11

8 Complete these subtractions. 1B

a 688 − 273
b 400 − 183
c 68 348 − 8026
d 46 234 − 8476
e 286 005 − 193 048
f 1 370 000 − 1 274 455

1C **9** Use mental strategies to multiply each of the following.

a $2 \times 15 \times 5$ **b** $4 \times 84 \times 25$ **c** 62×20
d 56×300 **e** 67×9 **f** 31×19

1D **10** Calculate each of these using short division.

a $4172 \div 7$ **b** $101\,040 \div 12$ **c** $15\,063 \div 3$

1D **11** Calculate each of these.

a $6 \times 4 \div 3$ **b** $4 \times 9 \div 12$ **c** $49 \div 7 \times 12$
d $81 \div 9 \times 5$ **e** $6 \times 3 \div 9 \div 2$ **f** $12 \div 2 \times 11 \div 3$

1D **12** Calculate these using long division.

a $8910 \div 22$ **b** $14\,756 \div 31$ **c** $34\,255 \div 17$

1D **13** Divide these multiples of 10.

a $84\,000 \div 120$ **b** $4900 \div 700$ **c** $12\,300 \div 30$

1C, D **14** In summer, an ice-cream factory operates 16 hours a day and makes 28 ice-creams each hour.

a How many ice-creams are produced each day?

b If the factory operates 7 days a week, how many ice-creams are produced in one week?

c If there are 32 staff who run the machines over a week, how many ice-creams would each person produce?

1E **15** Write the rules for the order of operations.

1E **16** Follow the rules for the order of operations to calculate each of the following.

a $35 \div (12 - 5)$ **b** $11 \times 3 + 5$ **c** $8 \times 3 \div 4$
d $5 \times 12 - 11 \times 5$ **e** $(6 + 4) \times 7$ **f** $6 + 4 \times 7$
g $3 \times (4 + 5) \times 2$ **h** $5 + [21 - (5 \times 3)] \times 4$

1F **17** By rounding each number to its first digit, estimate the answer to each of the calculations.

a $6802 + 7486$ **b** $8914 - 3571$ **c** $5304 \div 143$
d 5706×68 **e** $49\,581 + 73\,258$ **f** $17\,564 - 10\,689$
g $9480 \div 2559$ **h** 289×671

1G **18** Write down the first 10 square numbers.

1G **19** Find:

a 11^2 **b** 16^2 **c** $\sqrt{16}$ **d** $\sqrt{81}$

1G **20** Find the seventh triangular number.

1G **21** Write down the first ten Fibonacci numbers.

1G **22** Give an example of a palindromic number.

1G **23** Use the ‘reverse and add’ process to produce a palindromic number starting with 437.

Number theory

2

Have you ever wondered why the runners in a 100-m race are lined up next to each other while those in a 400-m race are staggered around the track? What is the distance between each of the starting blocks? Is there a pattern?

This chapter will provide you with skills to answer these questions more easily.

Are you ready?

Try the questions below. If you have difficulty with any of them, extra help can be obtained by completing the matching SkillSHEET. Either click on the SkillSHEET icon next to the question on the *Maths Quest 7* CD-ROM or ask your teacher for a copy.

SkillSHEET 2.1

Finding the square of a number

1 Find the value of each of the following.

a 4×4 **b** 12×12 **c** 7×7

SkillSHEET 2.2

Finding the square root of a number

2 Find the value of the missing number represented by the symbol Δ in each of the following.

a If $6 \times \Delta = 36$ then $\Delta = ?$ **b** If $\Delta \times 9 = 81$ then $\Delta = ?$

c If $\Delta \times \Delta = 25$ then $\Delta = ?$ **d** If $\Delta \times \Delta = 64$ then $\Delta = ?$

SkillSHEET 2.3

Finding the cube of a number

3 Find the value of each of the following.

a $5 \times 5 \times 5$ **b** $11 \times 11 \times 11$ **c** $3 \times 3 \times 3$

SkillSHEET 2.4

Finding the cube root of a number

4 Find the value of the missing number represented by the symbol Δ in each of the following.

a If $6 \times \Delta \times \Delta = 216$ then $\Delta = ?$ **b** If $\Delta \times 9 \times 9 = 729$ then $\Delta = ?$

c If $\Delta \times \Delta \times \Delta = 8$ then $\Delta = ?$ **d** If $\Delta \times \Delta \times \Delta = 64$ then $\Delta = ?$

SkillSHEET 2.5

Order of operations with squares and square roots

5 **Evaluate** each of the following.

a $10 \times 10 + 7$ **b** $9 \times 9 + 2 \times 2$ **c** $35 - 3 \times 3$

SkillSHEET 2.6

Multiples

6 **a** Write the first five numbers for a sequence where you are counting by fours.

b Write the first five numbers for a sequence where you are counting by sixes.

SkillSHEET 2.7

Factor pairs

7 Find the missing factor in each of the following factor pairs of 16:

a ____ and 16 **b** 2 and ____ **c** 4 and ____.

SkillSHEET 2.9

Consecutive numbers

8 **a** Write three consecutive numbers, beginning with 5.

b Write four consecutive even numbers, beginning with 14.

c Write three consecutive odd numbers, beginning with 37.

Squares, square roots, cubes and cube roots

Square numbers

In chapter 1 we learned about **square numbers**. Square numbers (or 'squares') are numbers that can be arranged to form a square. We saw that the first four square numbers are 1, 4, 9 and 16.

We also saw that square numbers can be written using an index (or power) of 2. For example, the fifth square number can be written as 5^2 and is equal to 25. The index or power indicates how many of the same number, 5 in this case, are multiplied together.

Any number that is multiplied by itself produces a square number and can be written using an index (or power) of 2.

WORKED Example 1

Find the square numbers between 90 and 150.

THINK	WRITE
1 Use your knowledge of tables to find the first square number after 90.	$10^2 = 10 \times 10 = 100$
2 Find the square numbers which come after that one but before 150.	$11^2 = 11 \times 11 = 121$ $12^2 = 12 \times 12 = 144$ $13^2 = 13 \times 13 = 169$ (too big)
3 Write the answer in a sentence.	The square numbers between 90 and 150 are 100, 121 and 144.

WORKED Example 2

Write the two whole numbers between which 5.7^2 will lie.

THINK	WRITE
1 Write the whole numbers either side of 5.7.	5.7 is between 5 and 6.
2 Consider the square of each whole number.	5.7^2 is between 5^2 and 6^2.
3 Simplify 5^2 and 6^2 then write the answer in a sentence.	So 5.7^2 is between 25 and 36.
4 Verify your answer with a calculator.	

Graphics Calculator tip!
Calculating the square of a number

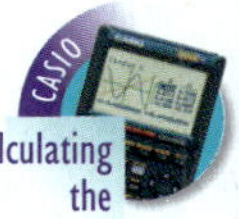
Calculating the square of a number

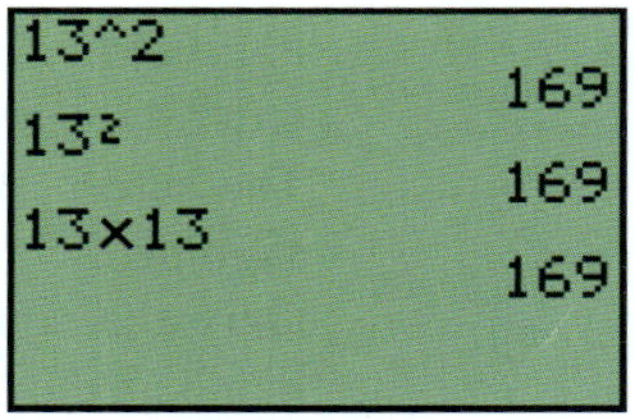

There are 3 ways to obtain the square of a number using a graphics calculator.

1. Type in the base, press (^), then type in the index which is 2. Press (ENTER) to complete the calculation.
2. Type in the base, press the key marked (x^2), then press (ENTER).
3. Type in the number, press the (×) sign then type in the same number again. Remember to press (ENTER). All 3 methods are shown in the screen opposite.

Square roots

Finding the **square root** of a number is the opposite of squaring a number.
$4^2 = 16$ means that 4 multiplied by itself is equal to 16.

The opposite of this is to say that $\sqrt{16} = 4$; so $\sqrt{16} = 4$ means that we are finding a number that multiplies by itself to equal 16.

To find the square roots of larger numbers, it helps to break the number up as a product of two smaller square roots with which we are more familiar. For example,

$$\begin{aligned}\sqrt{900} &= \sqrt{9} \times \sqrt{100}\\ &= 3 \times 10\\ &= 30\end{aligned}$$

WORKED Example 3

Find: **a** $\sqrt{49}$ **b** $\sqrt{3600}$.

THINK	WRITE
a Find a number which when multiplied by itself gives 49 ($7 \times 7 = 49$).	**a** $\sqrt{49} = 7$
b 1 Write 3600 as the product of two smaller numbers for which we can find the square root.	**b** $\sqrt{3600} = \sqrt{36} \times \sqrt{100}$
2 Take the square root of each of these numbers.	$= 6 \times 10$
3 Find the product.	$= 60$

Graphics Calculator tip! Calculating the square root of a number

Calculating the square root of a number

To calculate the square root of a number, first press 2nd [$\sqrt{\ }$] then the number. Remember to press ENTER to complete the calculation. The $\sqrt{\ }$ operation is listed in yellow above the key marked x^2. This is why we need to first press the key marked 2nd. The screen shows the calculations needed for worked example 3.

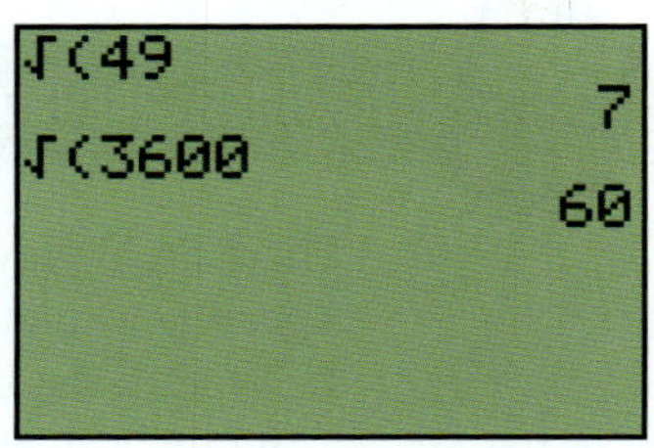

Only the square numbers have a square root that can be found exactly. For other numbers the calculator will give us a decimal approximation. In practice it is important that we are able to estimate the square root of a number that is not a square number. We do this by looking at the square number either side of the original number.

WORKED Example 4

Between which two numbers will $\sqrt{74}$ lie?

THINK	WRITE
1 Write the square numbers either side of 74.	74 is between 64 and 81.
2 Consider the square root of each number.	$\sqrt{74}$ is between $\sqrt{64}$ and $\sqrt{81}$.
3 Simplify $\sqrt{64}$ and $\sqrt{81}$.	So $\sqrt{74}$ is between 8 and 9.
4 Verify your answer with a calculator.	

Cubes and cube roots

We have seen that a square number can be found by multiplying a number by itself; that is, the number is written twice and multiplied. For example, $4^2 = 4 \times 4$. The index (or power) used for squaring a number is 2.

A cubic number is generated when a number is written 3 times and multiplied. For example, $4^3 = 4 \times 4 \times 4 = 64$. The index (power) used for cubing a number is 3.

WORKED Example 5

Find the value of 5^3.

THINK	WRITE
1 Write 5^3 as the product of three lots of 5.	$5^3 = 5 \times 5 \times 5$
2 Evaluate.	$= 125$

Graphics Calculator tip! Calculating the cube of a number

To calculate a number in index form, first type in the base, then (^), then the index and press (ENTER). The screen at right shows the calculation for 5^3 (from worked example 5).

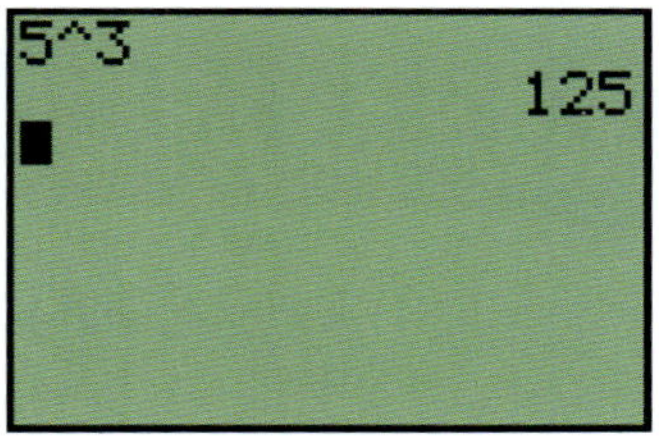

Calculating the cube of a number

The opposite of cubing a number is finding the cube root of a number. The cube root of a given number is found by looking for a number that can be written three times and multiplied to produce the given number. The cube root symbol is similar to the square root symbol but with a small 3 written in front as shown below.

$$\sqrt[3]{}$$

From worked example 5 we can see that $\sqrt[3]{125} = 5$.

WORKED Example 6

Find $\sqrt[3]{27}$.

THINK	WRITE
Look for a number that when written three times and multiplied gives 27.	$27 = 3 \times 3 \times 3$ $\sqrt[3]{27} = 3$

Graphics Calculator tip! Calculating the cube root of a number

Calculating the cube root of a number

To find the **cube root** of a number, press MATH, select **4:** $\sqrt[3]{}$ **(**, enter the number under the root sign and press ENTER.

The screen to the right shows the calculations for the following cube roots: $\sqrt[3]{27}$, $\sqrt[3]{2197}$, $\sqrt[3]{12.56}$.

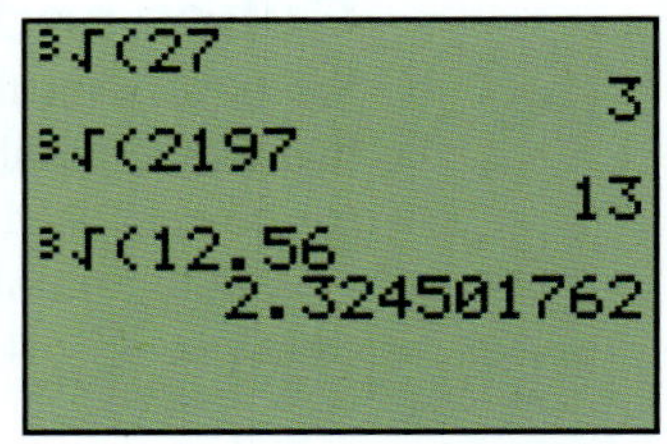

remember

1. Square numbers are written using an index (or power) of 2 and are found by multiplying the number by itself.
2. The opposite of squaring a number is finding the square root of a number.
3. To find the square root of a number we can use our knowledge of square numbers or use a calculator.
4. When a number is written three times and multiplied, this is called cubing a number and is written using an index (or power) of 3.
5. The opposite of cubing a number is finding the cube root of a number. This is calculated using our knowledge of cubic numbers or using a calculator.

EXERCISE 2A Squares, square roots, cubes and cube roots

SkillSHEET 2.1 Finding the square of a number

1 Evaluate and verify your answers with a calculator.
a 8^2 **b** 11^2 **c** 15^2 **d** 25^2

2 Use a mental strategy to evaluate 99^2.

Mathcad Squares

WORKED Example 1

3 a Find the square numbers between 50 and 100.
b Find the square numbers between 160 and 200.

4 a Find the even square numbers between 10 and 70.
b Find the odd square numbers between 50 and 120.

WORKED Example 2

5 Write two whole numbers between which each of the following will lie.

a 6.4^2 b 7.8^2 c 9.2^2 d 12.5^2

2.2 SkillSHEET Finding the square root of a number

WORKED Example 3a

6 Find:

a $\sqrt{25}$ b $\sqrt{81}$ c $\sqrt{144}$ d $\sqrt{400}$

WORKED Example 3b

7 Find and verify your answers with a calculator.

a $\sqrt{4900}$ b $\sqrt{14\,400}$ c $\sqrt{360\,000}$ d $\sqrt{160\,000}$

Mathcad Square roots

WORKED Example 4

8 Between which two numbers will $\sqrt{60}$ lie?

9 Write the two whole numbers between which each of the following will lie.

a $\sqrt{14}$ b $\sqrt{90}$ c $\sqrt{200}$ d $\sqrt{2}$

10 **multiple choice**

For which of the following square roots can we calculate an exact answer?

A $\sqrt{10}$ B $\sqrt{25}$ C $\sqrt{50}$ D $\sqrt{75}$ E $\sqrt{82}$

11 **multiple choice**

For which of the following square roots can we NOT calculate the exact value?

A $\sqrt{160}$ B $\sqrt{400}$ C $\sqrt{900}$ D $\sqrt{2500}$ E $\sqrt{3600}$

WORKED Example 5

12 Find the value of 4^3.

13 Find the value of:

a 2^3 b 3^3 c 6^3 d 10^3

2.3 SkillSHEET Finding the cube of a number

14 Write the first 10 cubic numbers.

WORKED Example 6

15 Find $\sqrt[3]{8}$.

16 Find each of the following. Verify your answers with a calculator.

a $\sqrt[3]{64}$ b $\sqrt[3]{216}$ c $\sqrt[3]{343}$ d $\sqrt[3]{729}$

(*Hint:* Use your answer to question **14**.)

2.4 SkillSHEET Finding the cube root of a number

17 Evaluate the following. Verify your answers with a calculator.

a $2^2 + \sqrt{25}$ b $9^2 - \sqrt{36}$ c $5^2 \times 2^2 \times \sqrt{49}$

d $3^2 + 2^2 \times \sqrt{16}$ e $3^2 - 2^2 \div \sqrt{4} + \sqrt{49}$ f $\sqrt{9} \times 4^2 - \sqrt{144} \div 2^2$

2.5 SkillSHEET Order of operations with squares and square roots

MATHS QUEST CHALLENGE

1 **Megan has 3 game scores that happen to be square numbers. The first 2 scores have the same three digits. The total of the 3 scores is 590. What are the 3 scores?**

2 **The difference of the squares of two consecutive odd numbers is 32. What are the two odd numbers?**

Index notation

In the previous section we looked at square numbers and cubic numbers. We saw that a square number occurred when a number was multiplied by itself and could be denoted by an **index** (or **power**) of 2. That is to say that 4×4 could be written as 4^2.

Similarly, a cubic number occurs when a number is written three times and multiplied. That is to say that $4 \times 4 \times 4$ could be written as 4^3.

A multiplication written in this form is said to be in **index notation**. In the case of 4^3, 4 is said to be the *base* and 3 is the *index* (or power).

An index is a shorthand way of writing a repeated multiplication. To write a repeated multiplication in index notation the base is the number that is being repeatedly multiplied and the index (or power) is the number of times this number is written.

WORKED Example 7

Write the following using index notation.

a $5 \times 5 \times 5 \times 5 \times 5 \times 5 \times 5$ **b** $3 \times 3 \times 3 \times 3 \times 7 \times 7$

THINK	WRITE
a 1 Write the multiplication.	**a** $5 \times 5 \times 5 \times 5 \times 5 \times 5 \times 5$
2 Write the number being multiplied as the base and the number of times it is written as the index.	$= 5^7$
b 1 Write the multiplication.	**b** $3 \times 3 \times 3 \times 3 \times 7 \times 7$
2 Write the number being multiplied as the base and the number of times it is written as the index.	$= 3^4 \times 7^2$

You can use your calculator to calculate the value of an expression written in index notation by using the ^ button or the x^y function. Enter the base first, followed by the index.

WORKED Example 8

Use your calculator to evaluate 4^6.

THINK	WRITE
On your calculator press (4) (^) (6) or (4) (x^y) (6).	$4^6 = 4096$

WORKED Example 9

Simplify each of the following by first writing as a repeated multiplication and then using index notation.

a $3^4 \times 3^6$ **b** $(4^3)^3$ **c** $(3 \times 5)^2$

THINK	WRITE
a 1 Write the question.	**a** $3^4 \times 3^6$
2 Write the question using repeated multiplication (that is, in expanded form).	$= (3 \times 3 \times 3 \times 3) \times (3 \times 3 \times 3 \times 3 \times 3 \times 3)$ $= 3 \times 3 \times 3 \times 3 \times 3 \times 3 \times 3 \times 3 \times 3 \times 3$
3 Write the repeated multiplication using index notation. *Note:* $3^4 \times 3^6 = 3^{10}$.	$= 3^{10}$
b 1 Write the question.	**b** $(4^3)^3$
2 Write the question using repeated multiplication (that is, in expanded form).	$= 4^3 \times 4^3 \times 4^3$ $= (4 \times 4 \times 4) \times (4 \times 4 \times 4) \times (4 \times 4 \times 4)$ $= 4 \times 4 \times 4 \times 4 \times 4 \times 4 \times 4 \times 4 \times 4$
3 Write the repeated multiplication using index notation. *Note:* $(4^3)^3 = 4^9$.	$= 4^9$
c 1 Write the question.	**c** $(3 \times 5)^2$
2 Write the question using repeated multiplication (that is, in expanded form).	$= (3 \times 5) \times (3 \times 5)$ $= 3 \times 5 \times 3 \times 5$ $= 3 \times 3 \times 5 \times 5$
3 Write the repeated multiplication using index notation. *Note:* $(3 \times 5)^2 = 3^2 \times 5^2$.	$= 3^2 \times 5^2$

remember

1. Index notation is a shorthand way of writing a repeated multiplication.
2. The number being multiplied is the base of the expression; the number of times it has been written is the index (or power).
3. Your calculator can be used to evaluate an expression in index notation by using the ^ or x^y function.

EXERCISE 2B Index notation

Mathcad: Index notation

WORKED Example 7a

1 Write $7 \times 7 \times 7 \times 7$ using index notation.

2 Write each of the following using index notation.
a $2 \times 2 \times 2 \times 2$
b $8 \times 8 \times 8 \times 8 \times 8 \times 8$
c $10 \times 10 \times 10 \times 10 \times 10$
d $3 \times 3 \times 3 \times 3 \times 3 \times 3 \times 3 \times 3 \times 3$
e $6 \times 6 \times 6 \times 6 \times 6 \times 6 \times 6$
f $13 \times 13 \times 13$
g $12 \times 12 \times 12 \times 12 \times 12 \times 12$
h $9 \times 9 \times 9 \times 9 \times 9 \times 9 \times 9 \times 9$

WORKED Example 7b

3 Write $4 \times 4 \times 4 \times 4 \times 4 \times 6 \times 6 \times 6$ using index notation.

4 Write the following using index notation.
a $2 \times 2 \times 3$
b $3 \times 3 \times 3 \times 3 \times 2 \times 2$
c $5 \times 5 \times 2 \times 2 \times 2 \times 2$
d $7 \times 2 \times 2 \times 2$
e $5 \times 11 \times 11 \times 3 \times 3 \times 3$
f $13 \times 5 \times 5 \times 5 \times 7 \times 7$
g $2 \times 2 \times 2 \times 3 \times 3 \times 5$
h $3 \times 3 \times 2 \times 2 \times 5 \times 5 \times 5$

5 Write 6^5 using a repeated multiplication (that is, in expanded form).

6 Write each of the following using repeated multiplication (that is, in expanded form).
a 11^3
b 4^9
c 5^6

WORKED Example 8

7 Use your calculator to evaluate 3^5.

8 Use a calculator to evaluate each of the following.
a 4^5
b 7^4
c 9^3
d 2^{10}

9 **multiple choice**

Which of the following expressions has the greatest value?
A 2^8
B 8^2
C 3^4
D 4^3
E 9^2

10 **multiple choice**

The value of 4^4 is:
A 8
B 16
C 64
D 256
E 484

11 Find the value of $2^3 \times 3^2$.

12 Evaluate each of the following.
a $3^4 \times 4^3$
b $3^5 + 9^3$
c $8^3 \div 2^5$
d $6^4 - 9^3$
e $5^3 + 2^5 \times 9^2$
f $2^7 - 4^5 \div 2^6$

WORKED Example 9

13 Simplify each of the following by first writing as a repeated multiplication and then using index notation.
a $5^8 \times 5^3$
b $7^2 \times 7^6$
c $4^8 \times 4^5$
d $(2^7)^3$
e $(11^3)^8$
f $(9^5)^3$
g $(3 \times 13)^6$
h $(2 \times 5)^4$
i $(5 \times 7)^5$

Multiples

A **multiple** of a number is the answer obtained when that number is multiplied by another whole number.

All numbers in the 5 times table are multiples of 5; so 5, 10, 15, 20, 25, ... are all multiples of 5.

The number 5 has been multiplied by one other number to find each of the numbers in the 5 times table, so they are all multiples of 5.

WORKED Example 10

List the first five multiples of 7.

THINK	WRITE
1 First multiple is the number × 1; that is, 7 × 1.	
2 Second multiple is the number × 2; that is, 7 × 2.	
3 Third multiple is the number × 3; that is 7 × 3.	
4 Fourth multiple is the number × 4; that is, 7 × 4.	
5 Fifth multiple is the number × 5; that is, 7 × 5.	
6 Answer the question.	The first five multiples of 7 are 7, 14, 21, 28, 35.

WORKED Example 11

Write the numbers in the list that are multiples of 8.
18, 8, 80, 100, 24, 60, 9, 40

THINK	WRITE
1 The biggest number in the list is 100. List multiples of 8 using the 8 times table just past 100; that is, 8 × 1 = 8, 8 × 2 = 16, 8 × 3 = 24, 8 × 4 = 32, 8 × 5 = 40, and so on.	8, 16, 24, 32, 40, 48, 56, 64, 72, 80, 88, 96, 104
2 Write any multiples that appear in the list.	Numbers in the list that are multiples of 8 are 8, 24, 40, 80.

An alternative method would be to divide each number by 8 and see if it goes evenly (without a remainder).

remember

1. A *multiple* of a number is the answer obtained when that number is multiplied by another whole number. For example, all numbers in the 5 times table are multiples of 5; that is, 5, 10, 15, 20, 25...
2. If two or more numbers have the same multiple, it is called a common multiple.

EXERCISE 2C Multiples

SkillSHEET 2.6 Multiples

WORKED Example 10

1 List the first five multiples of the following numbers.

a 3 b 6 c 100 d 11
e 15 f 4 g 21 h 25
i 14 j 12 k 50 l 30
m 33 n 120 o 45 p 72

Mathcad Multiples

WORKED Example 11

2 Write the numbers in the following list that are multiples of 10.
10, 15, 20, 100, 38, 62, 70

3 Write the numbers in the following list that are multiples of 7.
17, 21, 7, 70, 47, 27, 35

4 Write the numbers in the following list that are multiples of 16.
16, 8, 24, 64, 160, 42, 4, 32, 1, 2, 80

5 Write the numbers in the following list that are multiples of 35.
7, 70, 95, 35, 140, 5, 165, 105, 700

6 The numbers 16, 40 and 64 are all multiples of 8. Find three more multiples of 8 that are less than 100.

7 List the multiples of 9 that are less than 100.

8 List the multiples of 6 between 100 and 160.

9 **multiple choice**

a The first three multiples of 9 are:
A 1, 3, 9 B 3, 6, 9 C 9, 18, 27 D 9, 18, 81 E 18, 27, 36

b The first three multiples of 15 are:
A 15, 30, 45 B 30, 45, 60 C 1, 15, 30 D 45 E 3, 5, 15

10 a List the first ten multiples of 4.
b List the first ten multiples of 6.
c In your lists, circle the multiples that 4 and 6 have in common (that is, circle the numbers that appear in both lists).
d What is the *lowest multiple* that 4 and 6 have in *common*? This is the Lowest Common Multiple of 4 and 6, known as the LCM.

11 a List the first six multiples of 3.
b List the first six multiples of 9.
c Circle the multiples that 3 and 9 have in common.
d What is the lowest common multiple of 3 and 9?

12 Find the LCM of each of the following pairs of numbers.
a 3 and 6 b 6 and 15 c 7 and 10 d 12 and 16 e 4 and 15

13 Answer true (T) or false (F) to each of the following statements.
a 20 is a multiple of 10 and 2 only.
b 15 and 36 are both multiples of 3.
c 60 is a multiple of 2, 3, 6, 10 and 12.
d 100 is a multiple of 2, 4, 5, 10, 12 and 25.

14 Kate goes to the gym every second evening, while Ian goes every third evening.

a How many days will it be before both attend the gym again on the same evening?

b Explain how this answer relates to the multiples of 2 and 3.

15 Vinod and Elena are riding around a mountain bike trail. Each person completes one lap in the time shown on the stopwatches.

a If they both begin cycling from the starting point at the same time, how long will it be before they pass this starting point again at exactly the same time?

b Relate your answer to the multiples of 5 and 7.

16 A warehouse owner employs Bob and Charlotte as security guards. Each security guard checks the building at midnight. Bob then checks the building every 4 hours, and Charlotte every 6 hours.

a How long will it be until both Bob and Charlotte are next at the warehouse at the same time?

b Relate your answer to the multiples of 4 and 6.

17 Two smugglers, Bill Bogus and Sally Seadog have set up signal lights that flash continuously across the ocean. Bill's light flashes every 5 seconds and Sally's light flashes every 4 seconds. If they both start together, how long will it take for both lights to flash again at the same time?

18 Alex and Nadia were having races running down a flight of stairs. Nadia took the stairs two at a time while Alex took the stairs three at a time. In each case, they reached the bottom with no steps left over.

- a How many steps are there in the flight of stairs? List three possible answers.
- b What is the smallest number of steps there could be?
- c If Alex can also take the stairs five at a time with no steps left over, what is the smallest number of steps in the flight of stairs?

19 Twenty students in Year 7 were each given a different number from 1 to 20 and then asked to sit in numerical order in a circle. Three older girls — Milly, Molly and Mandy — came to distribute jelly beans to the class. Milly gave red jelly beans to every second student, Molly gave green jelly beans to every third student and Mandy gave yellow jelly beans to every fourth student.

- a Which student had jelly beans of all 3 colours?
- b How many students received exactly 2 jelly beans?
- c How many students did not receive any jelly beans?

GAME time Number theory — 001

WorkSHEET 2.1

20 Place the first 6 multiples of 3 into the triangle at right, so the numbers in each side of the triangle add up to 27. Use each number once only.

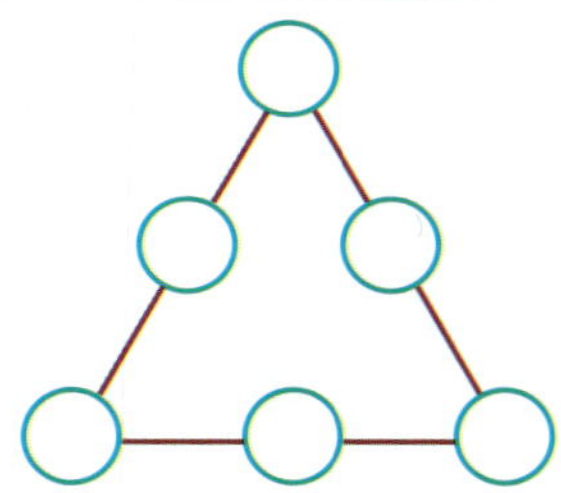

Maths Quest Challenge

1 **I am a 2-digit number that can be divided by 3 with no remainder. The sum of my digits is a multiple of 4 and 6. My first digit is double my second digit. What number am I?**

2 **Find a 2-digit number such that if you subtract 3 from it, the result is a multiple of 3; if you subtract 4 from it, the result is a multiple of 4 and if you subtract 5 from it, the result is a multiple of 5.**

3 **In a class election with 3 candidates, the winner beat the other 2 candidates by 3 and 6 votes respectively. If 27 votes were cast, how many votes did the winner receive?**

COMMUNICATION

A staggering thought

At the beginning of this chapter on page 47 you were asked to consider why the runners in a 100-m race are lined up next to each other while those in a 400-m race are staggered around the track. Look at the diagram of a standard 400-m athletics track and take note of both the starting position and finish line for a 100-m race.

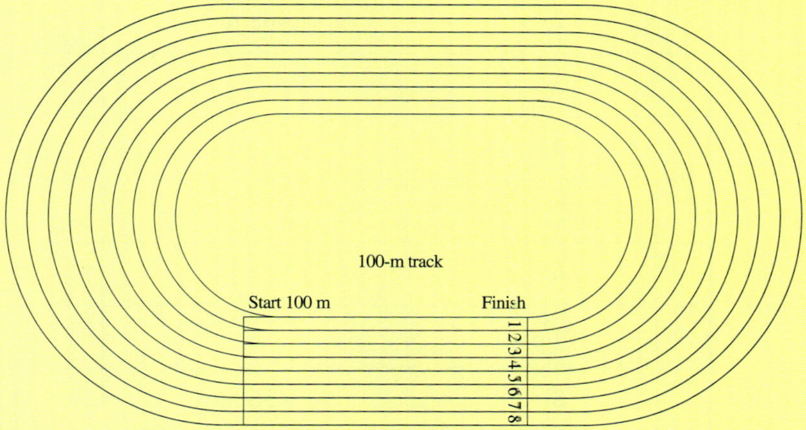

1 What two shapes is the track made up of?

2 Explain why the starting line for a 100-m race has been positioned in the extended section.

3 Explain why the finish line for a 100-m race has been positioned where it is.

Track events such as the 200-m, 400-m and 800-m races have staggered starting positions. The diagram illustrates the position of the starting blocks S_1–S_8 for each lane in the 400-m race.

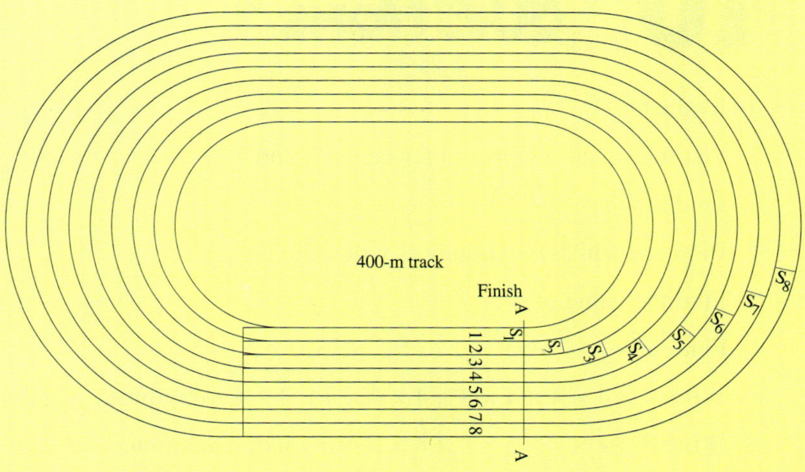

4 Why is there a need to stagger the starting blocks in the 200-m, 400-m and 800-m track events?

5 A runner completing one lap of the inside lane (lane 1) runs a distance of 400 m, while a runner completing one lap in the next lane (lane 2) runs a distance of 408 m. If this pattern continues, determine how far runners in lanes 3–8 run. Enter these results in the appropriate column in the table.

Lane number	Distance travelled (m)	Difference
1	400 m	
2	408 m	
3		
4		
5		
6		
7		
8	456 m	

6 Calculate the difference between the distances travelled by the runners in each of the lanes compared to the distance travelled by the runner in lane 1. Enter these results in the appropriate column in the table.

7 What do you notice about the values obtained in question 6?

10 QUICK QUESTIONS 1

1 Evaluate 13^2.

2 List the square numbers between 100 and 200.

3 Find $\sqrt{36}$.

4 Between which two numbers will $\sqrt{20}$ lie?

5 Find the value of 6^3.

6 Find $\sqrt[3]{64}$.

7 Write $4 \times 4 \times 4 \times 4 \times 4 \times 4 \times 4 \times 4 \times 4 \times 4$ using index notation.

8 Write $7 \times 7 \times 7 \times 8 \times 8 \times 8 \times 8$ using index notation.

9 Write the first 5 multiples of 15.

10 Find the smallest number that is a multiple of both 4 and 6.

In 1893, which country was the first to let women vote?

24	44	27	48	10	60	40	14	65	28

Which people were not allowed to vote until 1924?

18	36	6	8	12	15	100	9	34	30	45	20	22	42	21

Find the code to answer these questions by matching the letter beside each pair of numbers below with their lowest common multiple (LCM).

A 2, 7	D 5, 9	I 6, 4	N 8, 6	C 3, 15
D 7, 4	A 11, 2	N 3, 9	I 10, 4	L 8, 10
E 2, 5	N 10, 6	A 5, 12	M 4, 18	I 17, 2
N 6, 14	E 2, 3	R 4, 8	A 9, 2	S 7, 3
W 3, 27	Z 16, 3	E 4, 11	N 13, 5	A 4, 25

Factors

A **factor** is a whole number that divides exactly into another whole number, with no remainder.

If one number is divisible by another number, the second number divides exactly into the first number.

For example, 8 is divisible by 4 because $8 \div 4 = 2$. The number 4 is a factor of 8 because 4 divides into 8 twice with no remainder, or $8 \div 4 = 2$. The number 2 is also a factor of 8 because $8 \div 2 = 4$.

The number 3 is a factor of 15 because 3 goes into 15, or $15 \div 3 = 5$.

WORKED Example 12

Find all the factors of 14.

THINK	WRITE
1 1 is a factor of every number and the number itself is a factor; that is, $1 \times 14 = 14$.	1, 14
2 14 is an even number so 14 is divisible by 2; therefore 2 is a factor. Divide the number by 2 to find the other factor ($14 \div 2 = 7$).	2, 7
3 Write a sentence placing the factors in order from smallest to largest.	The factors of 14 are 1, 2, 7 and 14.

Factor pairs

It is often easiest to write factors in pairs called **factor pairs**. These are pairs of numbers which multiply to equal a certain number.

WORKED Example 13

List the factor pairs of 30.

THINK	WRITE
1 1 and the number itself are factors; that is, $1 \times 30 = 30$.	1, 30
2 30 is an even number so 2 and 15 are factors; that is, $2 \times 15 = 30$.	2, 15
3 Divide the next smallest number into 30. Therefore, 3 and 10 are factors; that is, $3 \times 10 = 30$.	3, 10
4 30 ends in 0 so 5 divides evenly into 30, that is, $5 \times 6 = 30$.	5, 6
5 List the factor pairs.	The factor pairs of 30 are 1, 30; 2, 15; 3, 10 and 5, 6.

Common factors

A *common factor* is a number which is a factor of two or more given numbers.

If the numbers 6 and 10 are given, then 2 is a common factor because 2 is a factor of 6 and 2 is a factor of 10.

The **Highest Common Factor** or **HCF** is the largest of the common factors.

WORKED Example 14

a Find the common factors of 8 and 24 by:
i listing the factors of 8 **ii listing the factors of 24**
iii listing the factors common to both 8 and 24.
b State the highest common factor of 8 and 24.

THINK	WRITE
a i 1 Find the pairs of factors of 8.	a i 1, 8 2, 4
2 Write them in order.	Factors of 8 are 1, 2, 4, 8.
ii 1 Find the pairs of factors of 24.	ii 1, 24 2, 12 3, 8 4, 6
2 Write them in order.	Factors of 24 are 1, 2, 3, 4, 6, 8, 12, 24.
iii Write the common factors.	iii Common factors are 1, 2, 4, 8.
b Find the highest common factor.	b HCF is 8.

Factors can make it easier to multiply numbers mentally. The following examples show the thought processes required.

WORKED Example 15

Find 12 × 15 using factors.

THINK	WRITE
1 Write the question.	12×15
2 Write one of the numbers as a factor pair.	$= 12 \times 5 \times 3$
3 Multiply the first two numbers.	$= 60 \times 3$
4 Find the answer.	$= 180$

WORKED Example 16

Use factors to evaluate 32 × 25.

THINK	WRITE
1 Write the question.	32×25
2 Rewrite the question so that one pair of factors is easy to multiply. The number 4 is a factor of 32 and $4 \times 25 = 100$.	$= 8 \times 4 \times 25$
3 Find the answer.	$= 8 \times 100$ $= 800$

With practice this could be done in your head. Some useful **products** are:
$2 \times 50 = 100$, $8 \times 125 = 1000$, $4 \times 250 = 1000$.

remember

1. A *factor* is a whole number that divides exactly into another whole number, with no remainder.
2. The factors of a number can be found using *factor pairs*. One of the pairs may be a single number which multiplies by itself. The number 5 is a factor of 25 because $5 \times 5 = 25$. The number itself and 1 are always factors of a number.
3. If 2 or more numbers have the same factor, it is called a *common factor.*
4. The *highest common factor* or HCF of 2 or more numbers is the largest factor which divides into all of the given numbers.

EXERCISE 2D Factors

SkillSHEET 2.7 Factor pairs

Mathcad Factors

Mathcad Highest common factor

GC program—TI Highest common factor

GC program—Casio Highest common factor

WORKED Example 12

1 Find all the factors of each of the following numbers.

a 12 b 8 c 40 d 35
e 28 f 60 g 100 h 72
i 39 j 85 k 76 l 69
m 99 n 250 o 51 p 105

WORKED Example 13

2 List the factor pairs of:

a 20 b 18 c 36 d 132

3 If 3 is a factor of 12, state the smallest number greater than 12 which has 3 as one of its factors.

4 **multiple choice**

a A factor pair of 24 is:
A 2, 4 B 4, 6 C 6, 2 D 2, 8 E 3, 9

b A factor pair of 42 is:
A 6, 7 B 20, 2 C 21, 1 D 16, 3 E 6, 8

5 Which of the numbers 3, 4, 5 and 11 are factors of 2004?

WORKED Example 14

6 a Find the common factors of 15 and 35 by:
i listing the factors of 15
ii listing the factors of 35
iii listing the factors common to both 15 and 35.
b State the highest common factor of 15 and 35.

7 a List the factors of 21. b List the factors of 56.
c Find the highest common factor of 21 and 56.

8 a List the factors of 27. b List the factors of 15.
c Find the highest common factor of 27 and 15.

9 a List the factors of 7. b List the factors of 28.
c Find the highest common factor of 7 and 28.

10 a List the factors of 48. b List the factors of 30.
c Find the HCF of 48 and 30.

11 Find the highest common factor of 9 and 36.

12 Find the highest common factor of 42 and 77.

13 Find the highest common factor of 36 and 64.

14 Find the highest common factor of 45, 72 and 108.

WORKED Example 15

15 Find the following using factors.

a 12×25 **b** 12×35 **c** 12×55
d 11×16 **e** 11×14 **f** 11×15
g 20×15 **h** 20×18 **i** 30×21

WORKED Example 16

16 Use factors to evaluate the following.

a 36×25 **b** 44×25 **c** 24×25
d 72×25 **e** 124×25 **f** 132×25
g 56×50 **h** 48×125 **i** 52×250

17 Connie Pythagoras is trying to organise her Year 4 class into rows for their class photograph. If Ms Pythagoras wishes to organise the 20 students into rows containing equal numbers of students, what possible arrangements can she have? *Note:* Ms Pythagoras will not be in the photograph.

18 **multiple choice**

Tilly Tyler has 24 green bathroom tiles left over. If she wants to use them all on the wall behind the kitchen sink (without breaking any) which of the following arrangements would be suitable?

I 4 rows of 8 tiles II 2 rows of 12 tiles III 4 rows of 6 tiles
IV 6 rows of 5 tiles V 3 rows of 8 tiles

A I and II **B** I, II and III **C** II, IV and V
D II, III and V **E** V

19 Wah needs to cut tubing into the largest pieces of equal length that he can, without having any offcuts left over. He has three sections of tubing; one 6 metres long, another 9 metres long and the third 15 metres long.

a How long should each piece of tubing be?
b How many pieces of tubing will Wah end up with?

GAME time Number theory — 002

20 Mario, Luigi, Zoe and Daniella are playing on a playstation. Mario takes 2 minutes to play one level of the game, Luigi takes 3 minutes, Zoe takes 4 minutes and Daniella takes 5 minutes. They have 12 minutes to play.

a If they play continuously, which of the players would be in the middle of a game as time ran out?

b After how many minutes did this player begin the last game?

MATHS QUEST CHALLENGE

1 **What number am I? I am a multiple of 5 with factors of 6, 4 and 3. The sum of my digits is 6.**

2 **My age is a multiple of 3 and a factor of 60. The sum of my digits is 3. How old am I? (There are two possible answers.)**

3 **Find the highest common factor of 462, 504 and 630.**

THINKING

How many tiles?

A room measures 550 centimetres by 325 centimetres. What would be the side length of the largest square tile that can be used to tile the floor without any cutting?

1 Try this question now. (*Hint:* Find the factors of 550 and 325.)

2 How many tiles would fit on the floor along the wall 550 centimetres long?

3 How many tiles would fit on the floor along the wall 325 centimetres long?

4 How many floor tiles would be needed for this room?

History of mathematics

ERATOSTHENES OF CYRENE (c. 270 BC to 194 BC)

During his life ...
The Roman Empire introduces a minted coin, the denarius.
Hannibal crosses the Alps.
The Great Wall of China is built.
Terracotta Soldiers are buried in a tomb at Xian (China).

Eratosthenes was a Greek astronomer and mathematician. He was born in Cyrene (now Shahhat, Libya). Some sources suggest he was born in 270 BC and died in 194 BC.

Eratosthenes was first educated by his father, Aglaos, and was then educated in Athens. He eventually became the main librarian at the Alexandrian museum.

Eratosthenes was the first person to work out an accurate measurement of the circumference of the Earth (when many people believed the Earth was flat!). He did this by comparing the angles of the shadows of sticks at two distant locations, Alexandria and Athens, at noon on the same day.

Eratosthenes was one of the most famous mathematicians of his day and was asked by the ruler of Egypt, Ptolemy III, to tutor his son Ptolemy Philadelphus. He also took on the job of being the chief librarian at the University of Alexandria where other mathematicians, such as Euclid, had held this position. As chief librarian he was able to borrow and copy books from the library.

Because he worked on many types of research, some people thought that he was not a specialist in any area. However, the work that he did was always of the highest standard.

Eratosthenes made discoveries in many areas that still benefit us now. He tried to improve the accuracy of calendars and realised there was a need for leap years to keep calendars in time with the seasons. He published the first map of the world based on longitude and latitude. He compiled a star catalogue containing 675 stars.

Eratosthenes discovered a way of finding prime numbers, called the Sieve of Eratosthenes (see page 70). A recently discovered prime is $2^{1\,398\,269} - 1$, which has 420 921 digits, but new primes are now being discovered regularly with the help of computers.

As Eratosthenes grew old he became blind. He was so upset by this disability that he took his own life.

Questions

1. What was Eratosthenes' main job in life?
2. What physical property was he the first person to measure accurately?
3. What was different about Eratosthenes' map of the world?
4. Which ruler did he work for?
5. What is the largest known prime at the time of this publication?
6. Why did Eratosthenes become depressed and end his life?

Research

1. Look on the Internet for information about primes and find the largest prime number known at present. Compare it with the one reported above. How much bigger is it now?
2. Why do we need leap years (they occur once every 4 years normally) and what is the difference between the years 1900 and 2000 in terms of leap years?

Prime numbers

A **prime number** is a counting number that has exactly 2 factors, itself and 1. (Counting numbers are 1, 2, 3, 4, …)
The number 3 is a prime number. Its only factors are 3 and 1.
The number 7 is a prime number. Its only factors are 7 and 1.

A **composite number** is one which has more than two factors.
The number 10 is a composite number; its factors are 1, 2, 5, 10.
The number 16 is also a composite number; its factors are 1, 2, 4, 8, 16.

The number 1 is a special number. It is neither a prime number nor a composite number because it has only one factor, 1.

COMMUNICATION The sieve of Eratosthenes

An easy way to find prime numbers is to use the 'Sieve of Eratosthenes'. Eratosthenes discovered a simple method of sifting out all of the composite numbers so that only prime numbers are left.

EXCEL Spreadsheet Sieve of Eratosthenes

You can follow the steps below to find all prime numbers between 1 and 100. Alternatively you can use the Excel file provided on the *Maths Quest* CD-ROM to simulate this process.

1 Copy the numbers from 1 to 100 in a grid as shown below. Use 1-centimetre square grid paper.

~~1~~	2	3	4	5	6	7	8	9	10
11	12	13	14	15	16	17	18	19	20
21	22	23	24	25	26	27	28	29	30
31	32	33	34	35	36	37	38	39	40
41	42	43	44	45	46	47	48	49	50
51	52	53	54	55	56	57	58	59	60
61	62	63	64	65	66	67	68	69	70
71	72	73	74	75	76	77	78	79	80
81	82	83	84	85	86	87	88	89	90
91	92	93	94	95	96	97	98	99	100

2 Cross out 1 as shown. It is not a prime number.

3 Circle the first prime number, 2. Then cross out all of the multiples of 2.

4 Circle the next prime number, 3. Now cross out all of the multiples of 3 that have not already been crossed out.

5 The number 4 is already crossed out. Circle the next prime number, 5. Cross out all of the multiples of 5 that are not already crossed out.

6 The next number that is not crossed out is 7. Circle 7 and cross out all of the multiples of 7 that are not already crossed out.

Continue until you have circled all of the prime numbers and crossed out all of the composite numbers.

Once we have an understanding of prime numbers we are able to answer questions that relate to prime numbers.

WORKED Example 17

List all the prime numbers between 50 and 100.

THINK	WRITE
Use the list of primes generated by the sieve of Eratosthenes to list all the numbers required.	53, 59, 61, 67, 71, 73, 79, 83, 89, 97

WORKED Example 18

State whether the following numbers are prime or composite.

a 45 **b** 37 **c** 86

THINK	WRITE
a Factors of 45 are 1, 3, 5, 9, 15 and 45.	**a** 45 is composite
b The only factors of 37 are 1 and 37.	**b** 37 is prime
c All even numbers except 2 are composite.	**c** 86 is composite

remember

1. A *prime number* is a whole number which has exactly two factors, itself and 1.
2. A *composite number* is one which has more than two factors.
3. The number 1 is neither a prime number nor a composite number.

EXERCISE 2E Prime numbers

Mathcad: Prime numbers

1 a List the factors for each number between 1 and 20.
  b Use the factor lists to write down all the prime numbers up to 20.

WORKED Example 17

2 Find four prime numbers that are between 20 and 40.

3 Can you find four prime numbers that are even? Explain.

WORKED Example 18

4 State whether each of the following numbers is prime or composite.

a 9	b 13	c 27	d 55
e 41	f 64	g 49	h 93
i 51	j 79	k 243	l 101

SkillSHEET 2.8 Odd and even numbers

5 Answer true (T) or false (F) for each of the following.
  a All **odd numbers** are prime numbers.
  b No **even numbers** are prime numbers.
  c 1, 2, 3 and 5 are the first four prime numbers.
  d A prime number has two factors only.
  e 2 is the only even prime number
  f The sum of two prime numbers is always even.
  g The product of two prime numbers is always odd.
  h There are no consecutive prime numbers.

6 **multiple choice**

  a The number of primes less than 10 is:
    A 4 B 3 C 5 D 2 E 1
  b The first three prime numbers are:
    A 1, 3, 5 B 2, 3, 4 C 2, 3, 5 D 3, 5, 7 E 2, 5, 7
  c The number 15 can be written as the sum of two prime numbers. These are:
    A 3 + 12 B 1 + 14 C 13 + 2 D 7 + 8 E 9 + 6
  d Factors of 12 that are prime numbers are:
    A 1, 2, 3, 4 B 2, 3, 6 C 2, 3
    D 2, 4, 6, 12 E 1, 2, 3, 4, 6, 12

7 Twin primes are pairs of primes that are separated from each other by one even number. For example, 3 and 5 are twin primes. Find two more pairs of twin primes.

SkillSHEET 2.9 Consecutive numbers

8 a Which of the numbers 2, 3, 4, 5, 6 and 7 cannot be the difference between two consecutive prime numbers? Explain.
  b For each of the numbers that can be a difference between two consecutive primes, give an example of a pair of primes less than 100 with such a difference.

WorkSHEET 2.2

9 The following numbers are not primes. Each of them is the product of two primes. Find the two primes in each case.
  a 365 b 187

MATHS QUEST CHALLENGE

1 What is the largest three-digit prime number in which each digit is a prime number?

2 Find a prime number greater that 10 where the sum of the digits equals 11.

3 My age is a prime number. I am older than 50. The sum of the digits in my age is also a prime number. If you add a multiple of 13 to my age the result is 100. How old am I?

4 I am a three-digit number. I am divisible by 6. My middle digit is a prime number. The sum of my digits is 9. I am between 400 and 500. My digits are in descending order. What number am I?

5 Angus is the youngest in his family and today he and his Dad share a birthday. Both their ages are prime numbers. Angus's age has the same two digits as his Dad's but in reverse order. In 10 years' time, Dad will be three times as old as Angus. How old will each person be when this happens?

6 What is the largest five-digit number you can write if each digit must be different and no digit may be prime?

7 Use the following clues to find the values of A, B, C and D.
- A is a factor of B.
- C is the second multiple of B and the fourth multiple of A.
- D is a prime number between 10 and 20.
- All numbers are different, and add up to 41.

Prime factors and factor trees

A **factor tree** shows the *prime factors* of a composite number.

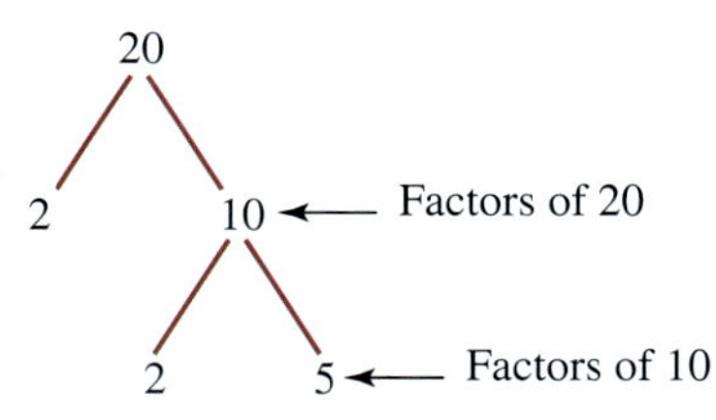

Each branch shows a factor of all numbers above it. The last numbers are all prime numbers; therefore, they are prime factors of the original number. From the factor tree shown, 2, 2 and 5 are prime factors of 20.

Note that the factor 2 need only be written once, so we can write that the prime factors of 20 are 2 and 5.

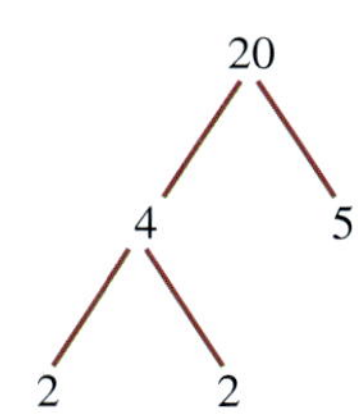

Prime factors are 2 and 5

If we had chosen different factors of 20 to start with, would we end up with different prime factors?

Choosing 4 and 5 instead of 10 and 2 did not change the prime factors of 20. If all the prime factors are multiplied together, the answer will be the original number.

$2 \times 2 \times 5 = 20$

WORKED Example 19

a Find the prime factors of 50 by drawing a factor tree.
b Write 50 as a product of its prime factors.

THINK

a 1 Find a factor pair of the given number and begin the factor tree ($50 = 5 \times 10$).

WRITE

a

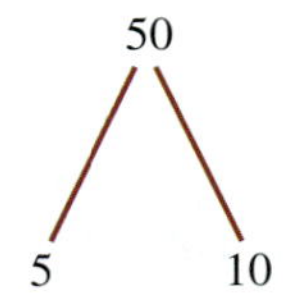

2 If a branch is prime, no other factors can be found (5 is prime). If a branch is composite, find factors of that number; 10 is composite so $10 = 5 \times 2$.

3 Continue until all branches end in a prime number, then stop.

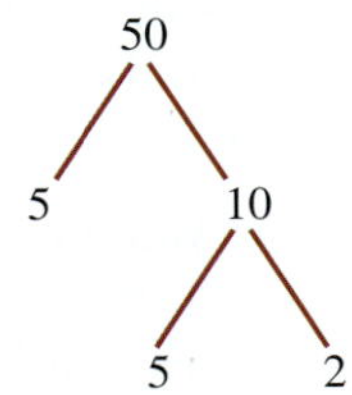

4 Write the prime factors.

The prime factors of 50 are 2 and 5.

b Write 50 as a product of prime factors found in part **a**.

b $50 = 5 \times 5 \times 2$

WORKED Example 20

Find the prime factors of 72.

THINK

1 Draw a factor tree. When all factors are prime numbers you have found the prime factors.

WRITE

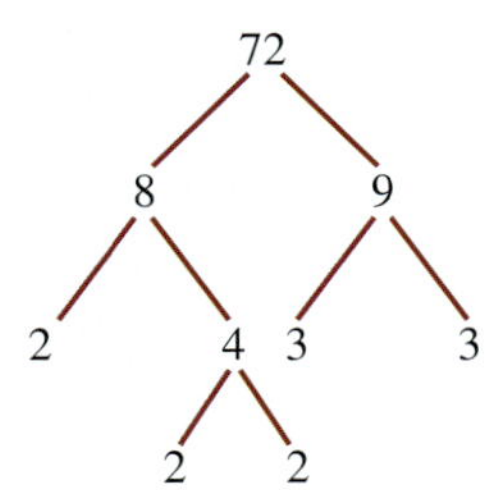

2 Write the prime factors.

The prime factors of 72 are 2 and 3.

We can simplify an expression involving prime factors using the work covered earlier in the chapter on index notation.

Suppose that we write 200 as a product of its prime factors. The result will be $200 = 2 \times 2 \times 2 \times 5 \times 5$. Using index notation we can write this as $200 = 2^3 \times 5^2$.

WORKED Example 21

Write 360 as a product of prime factors using index notation.

THINK / **WRITE**

1. Find a factor pair and begin the factor tree. If the number on the branch is a prime number, stop. If not, continue until a prime number is reached.

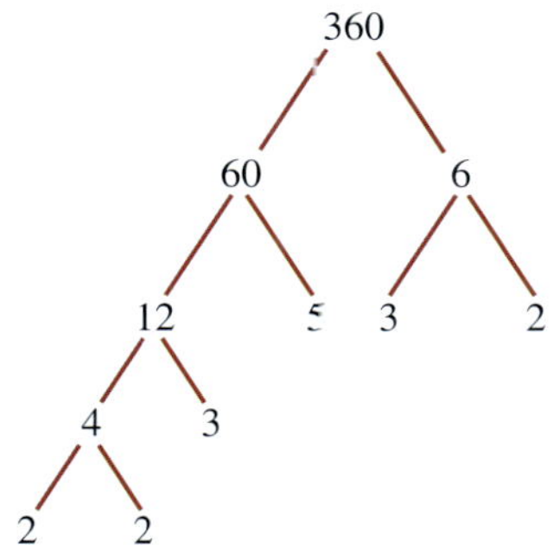

2. Write the number as a product of prime factors.
   $360 = 2 \times 2 \times 2 \times 3 \times 3 \times 5$
3. Write your answer using index notation.
   $360 = 2^3 \times 3^2 \times 5$

remember

1. A *factor tree* shows the *prime factors* of a composite number.
2. The last numbers in the factor tree are all prime numbers, therefore they are prime factors of the original number.
3. Every composite number can be written as a product of prime factors; for example, $20 = 2 \times 2 \times 5$.
4. The product of prime factors can be written in shorter form using index notation.

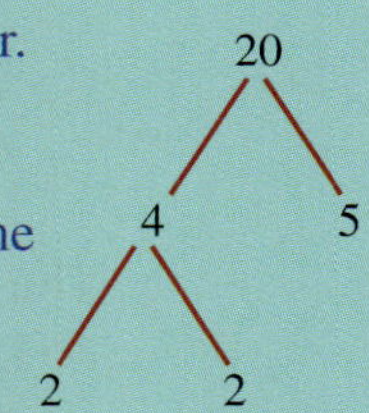

EXERCISE 2F Prime factors and factor trees

Mathcad: Prime factors
EXCEL Spreadsheet: Prime factors
GC program—TI: Prime factors
GC program—Casio: Prime factors

WORKED Example 19

1 i Find the prime factors of each of the following numbers by drawing a factor tree.
ii Write each one as a product of its prime factors.

a 15	b 30	c 24
d 100	e 49	f 72
g 18	h 56	i 45
j 84	k 98	l 112

2 i Find the prime factors of the following numbers by drawing a factor tree.
ii Express the number as a product of its prime factors.

a 40	b 35	c 32
d 121	e 110	f 150
g 3000	h 64	i 96
j 196	k 90	l 75

3 multiple choice

a A factor tree for 21 is:

A
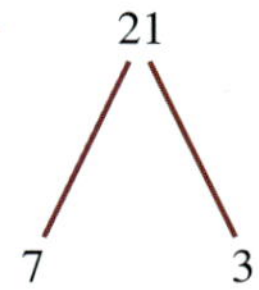

B
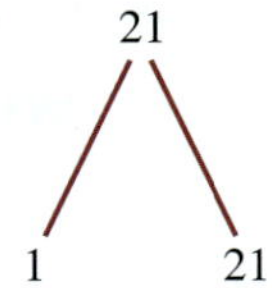

C
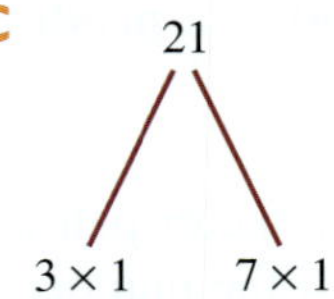

D
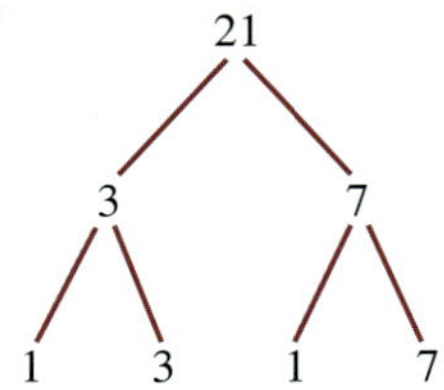

E
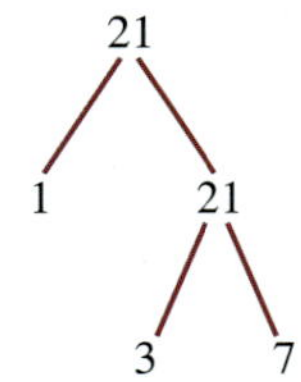

b A factor tree for 36 is:

A
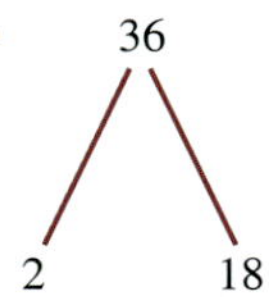

B
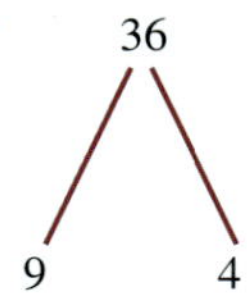

C
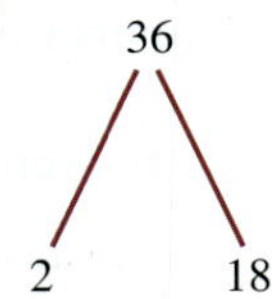

D
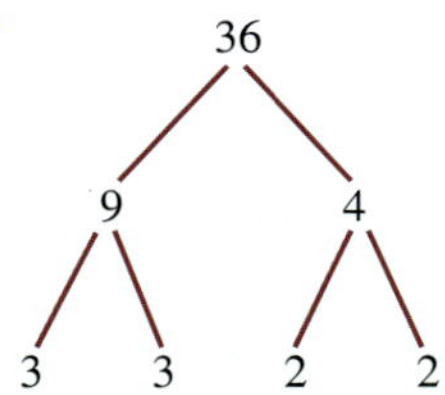

E
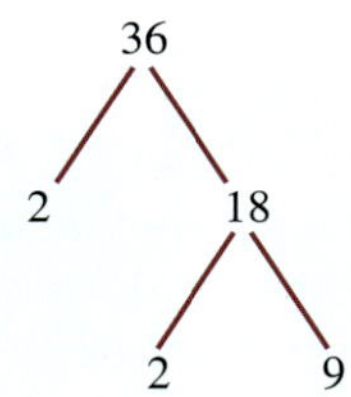

c The prime factors of 16 are:
A 1, 2 **B** 1, 2, 4 **C** 2 **D** 1, 2, 4, 8, 16 **E** 2, 4, 8

d The prime factors of 28 are:
A 1, 28 **B** 2, 7 **C** 1, 2, 14 **D** 1, 2, 7 **E** 2, 7, 14

WORKED Example 20

4 Find the prime factors of each of the following numbers.
a 48 **b** 200 **c** 42 **d** 81
e 18 **f** 39 **g** 27 **h** 300
i 60 **j** 120 **k** 50 **l** 80

5 State whether each of the following is true (T) or false (F).
a The number 3 is the only prime factor of 9.
b No two numbers can have the same prime factors.
c The numbers 2, 3, 5 and 7 are the prime factors of 210.
d The numbers 1, 2 and 5 are the prime factors of 40.
e The prime factors of 220 are 2, 5 and 11.
f All numbers have exactly two prime factors.

WORKED Example 21

6 Write the following as a product of prime factors using index notation.
a 60 **b** 50 **c** 75 **d** 220
e 192 **f** 72 **g** 124 **h** 200

10 QUICK QUESTIONS 2

1 Find all the factors of 45.

2 List all the factors of 80.

3 List the factor pairs of 100.

4 List the prime numbers less than 40.

5 State whether 21 is a prime number or a composite number.

6 List the first prime number greater than 100.

7 a Write 40 as the product of its prime factors.
  b Write 70 as the product of its prime factors.

8 Write 96 as the product of its prime factors.

9 Write 500 as the product of its prime factors.

10 Complete the following factor trees to show that the same product of prime factors will be obtained regardless of the starting factors.

a
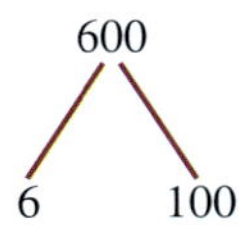

b
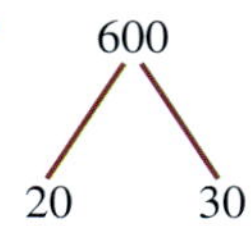

c
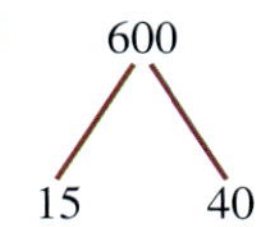

MATHS QUEST CHALLENGE

1 A whole number is 'perfect' if it equals the sum of all its proper factors. The proper factors of a number are all factors smaller than the number. So 6 is perfect since its proper factors are 1, 2 and 3 and $6 = 1 + 2 + 3$.

a Find all the proper factors of 28 and show that 28 is a perfect number.

b Do the same for 496.

2 Find the seven proper factors of the number 999. Is 999 a perfect number?

3 Use a calculator to find the 13 proper factors of 8128. Is 8128 a perfect number?

4 A whole number is 'abundant' if the sum of its proper factors is greater than the number itself.

a List the first 10 abundant numbers.

b How many of the abundant numbers obtained are odd?

Divisibility

It is often difficult to find all the factors of a number, or to work out if a number is prime or composite. To make this task easier we can use a range of tests that tell us if a number is divisible by 2, 3, 4, 5, 6, 8, 9 and 10.

Number	Divisibility test
2	The last digit is 0, 2, 4, 6, or 8.
3	The sum of the digits in the number is divisible by 3.
4	The last two digits are divisible by 4.
5	The last digit is 5 or 0.
6	The number is divisible by both 2 and 3.
8	The last three digits are divisible by 8.
9	The sum of the digits in the number is divisible by 9.
10	The last digit is 0.

There is a divisibility test to test if a number is divisible by 7. However, the test is quite cumbersome and it is simpler to just divide by 7 and see if it goes evenly (without a remainder).

WORKED Example 22

State whether 72 564 is divisible by:

a 2 **b** 4 **c** 8.

THINK	WRITE
a The last digit is a 4 and so the number is divisible by 2.	**a** 72 564 is divisible by 2.
b The last two digits are 64, which is divisible by 4. Therefore 72 564 is divisible by 4.	**b** 72 564 is divisible by 4.
c The last three digits are 564, which is not divisible by 8. Therefore 72 564 is not divisible by 8.	**c** 72 564 is not divisible by 8.

WORKED Example 23

Determine if 648 816 is divisible by:

a 3 **b** 6 **c** 9.

THINK

a 1. Find the sum of the digits.
2. Check whether this sum is divisible by 3. It is, so the number 648 816 is divisible by 3.

WRITE

a $6 + 4 + 8 + 8 + 1 + 6 = 33$
As 33 is divisible by 3, 648 816 is divisible by 3.

THINK

b 1. Find the sum of the digits.
2. Check whether the sum is divisible by 6. It is not, so the number 648 816 is not divisible by 6.

WRITE

b $6 + 4 + 8 + 8 + 1 + 6 = 33$
33 is not divisible by 6 therefore 648 816 is not divisible by 6.

THINK

c 1. Find the sum of the digits.
2. Check whether the sum is divisible by 9. It is not, so the number 648 816 is not divisible by 9.

WRITE

c $6 + 4 + 8 + 8 + 1 + 6 = 33$
As 33 is not divisible by 9, 648 816 is not divisible by 9.

A knowledge of the divisibility tests will help us to find the prime factors of larger numbers where it is otherwise difficult to find an initial factor pair.

WORKED Example 24

Find the prime factors of 2160.

THINK

1. The number 2160 is divisible by 10 because 2160 ends in 0. Use 10 as a factor in the factor tree.
2. The number 216 is divisible by 9 because $2 + 1 + 6 = 9$ and 9 is divisible by 9. Use 9 as a factor in the factor tree.
3. The number 24 is divisible by 3 because $2 + 4 = 6$ and 6 is divisible by 3. Use 3 as a factor.
4. The number 8 is even so use 2 as a factor and continue until all prime factors are found.

WRITE

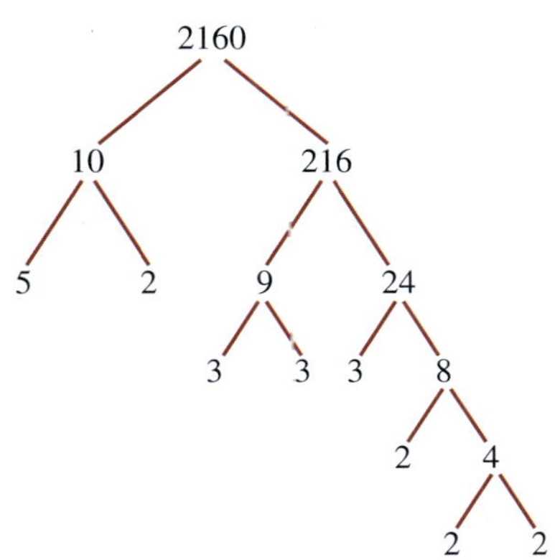

5. Write the number as a product of prime factors using index notation.

WRITE

$2160 = 2^4 \times 3^3 \times 5$

remember

1. There are divisibility tests that will allow us to determine if a number is divisible by 2, 3, 4, 5, 6, 8, 9 and 10. These tests are shown in the table on page 78.
2. The divisibility tests can be used to help to find an initial factor pair when writing a number as the product of prime factors.

EXERCISE 2G Divisibility

For questions 1 to 8 write 'yes' if divisible and 'no' if not divisible.

WORKED Example 22a

1 State whether each of the following is divisible by 2.
a 38 b 17 c 26 d 3093
e 288 f 100 g 370 h 4131
i 2020 j 719 k 4533 l 64 512

WORKED Example 22b

2 State whether each of the following numbers is divisible by 4.
a 44 b 28 c 328 d 212
e 917 f 3041 g 6084 h 68
i 149 j 70 232 k 69 478 l 324 636

WORKED Example 22c

3 State whether each of the following is divisible by 8.
a 64 b 45 c 4064 d 4008 e 71 045
f 1020 g 112 h 55 112 i 42 391 016 j 3102

WORKED Example 23a

4 State whether each of the following numbers is divisible by 3.
a 27 b 51 c 38 d 126
e 4017 f 943 g 362 h 84
i 69 j 9462 k 523 l 763 836

WORKED Example 23b

5 State whether each of the following is divisible by 6.
a 42 b 63 c 54 d 124 e 96
f 312 g 4311 h 3106 i 9 246 000 j 72 144

WORKED Example 23c

6 State whether each of the following is divisible by 9.
a 81 b 108 c 362 d 261
e 918 f 1602 g 34 901 h 60 201
i 7096 j 8 446 520 k 356 892 l 459 258

7 State whether each of the following is divisible by 5.
a 5 b 30 c 47 d 105 e 5032
f 1175 g 943 h 77 i 39 260 j 645

8 State whether each of the following is divisible by 10.
a 30 b 45 c 600 d 750
e 891 f 8910 g 365 h 4120
i 77 602 j 714 980 k 1245 l 3 205 400

9 **multiple choice**

A 250-gram block of Dairy Milk Chocolate has 60 pieces. It can be shared evenly between the following number of people:

A 2, 5, 7 **B** 3, 5, 9 **C** 2, 5, 11 **D** 3, 5, 12 **E** 3, 10, 14

10 If there are 220 students in Year 7, could they form groups of:

a 2 with no students left over?
b 3 with no students left over?
c 4 with no students left over?

11 A tennis team consists of 4 players. If a tennis club has 326 members who want to play in a team, will they be able to make teams of exactly 4 players, or will there be some players left over?

12 Could 142 students be broken up into volleyball teams, with no one left out, if each team has 6 players exactly?

WORKED Example 24

13 Write the following numbers as the product of prime factors using index notation. Use divisibility tests to help.

a 1575 b 616
c 1080 d 2646

14 Use a divisibility test to show that the following numbers are composite.

a 7667 b 9219

15 Use divisibility tests to decide which of the following numbers are primes.

a 119 b 209
c 103 d 91

WorkSHEET 2.3

THINKING

Finding square roots and cube roots without a calculator

Writing a number as the product of prime factors can help us to find the square root or cube root of a number without a calculator.

1 Write 3600 as a product of its prime factors in index form.

2 Use a calculator to find $\sqrt{3600}$.

3 Write your answer to part 2 as a product of prime factors in index form.

4 Compare your answers to parts 1 and 3. What do you notice about the indices in each answer?

5 Repeat steps 1 to 4 using 28 224.

6 Use this method to find each of the following without a calculator.
a $\sqrt{196}$ b $\sqrt{576}$ c $\sqrt{11\,664}$ d $\sqrt{3\,504\,384}$

7 Write 27 000 as a product of prime factors.

8 Use a calculator to find $\sqrt[3]{27\,000}$.

9 Write your answer to part 8 as a product of prime factors.

10 Compare your answers to parts 7 and 9. What do you notice about the indices in each answer?

11 Repeat steps 7 to 10 using 74 088.

12 Use this method to find the following without a calculator.
a $\sqrt[3]{1000}$ b $\sqrt[3]{512}$ c $\sqrt[3]{729}$ d $\sqrt[3]{42\,875}$

MATHS QUEST CHALLENGE

1 My age is a 2-digit number. It is divisible by 2, 3, 6 and 9. I am older than 20 but younger than 50 years old. How old am I?

2 My age is greater than 10 and less than 20. It is a composite number that is divisible by 3 and 9. How old am I?

3 My age is greater than 80 and less than 100. It is a composite number that is divisible by 2, 3, 4, 6, 8. How old am I?

summary

Copy the sentences below. Fill in the gaps by choosing the correct word or expression from the word list that follows.

1 Square numbers are written using an index of ___________ and ___________ numbers are written using an index of 3.

2 The opposite of finding the square of a number is finding the ___________, and the opposite of finding the cube of a number is finding the ___________ .

3 ___________ is used to write a repeated multiplication of the same number.

4 A ___________ is obtained by multiplying a number by another number.

5 A ___________ is a whole number that divides exactly into another whole number, with no remainder.

6 A ___________ has two factors only: itself and 1.

7 A ___________ has more than 2 factors.

8 The number ___________ is neither a prime number nor a composite number.

9 Composite numbers can be written as the product of ___________ factors. This is best done by drawing a ___________.

10 Tests of divisibility

A number is divisible by:

2 if it ends in an ___________ number.

3 if the sum of its digits is divisible by ___________.

4 if the last ___________ digits are divisible by 4.

5 if it ends in 0 or ___________.

6 if it is ___________ by 2 and 3.

8 if the last three digits are divisible by ___________.

9 if the ___________ of the digits is divisible by 9.

10 if it ends in ___________.

WORD LIST

one	3	multiple
prime number	sum	factor
square root	index notation	0
composite number	two	prime
2	8	divisible
even	5	factor tree
cube root	cubic	

CHAPTER review

2A **1** Evaluate:
a 6^2 b 14^2 c 19^2 d 80^2

2A **2** Write the two whole numbers between which each of the following will lie.
a 3.8^2 b 5.1^2 c 10.6^2 d 15.2^2

2A **3** Find:
a $\sqrt{49}$ b $\sqrt{256}$ c $\sqrt{900}$ d $\sqrt{1369}$

2A **4** Find the two whole numbers between which $\sqrt{71}$ will lie.

2A **5** Explain why we cannot find the exact value of $\sqrt{10}$.

2A **6** Evaluate:
a 4^3 b 7^3 c 15^3 d 30^3

2A **7** Find each of the following.
a $\sqrt[3]{27}$ b $\sqrt[3]{125}$ c $\sqrt[3]{1000}$ d $\sqrt[3]{8000}$

2A **8** Write the following using index notation.
a $3 \times 3 \times 3 \times 3$ b $5 \times 5 \times 5 \times 5 \times 5 \times 5$
c $7 \times 7 \times 7 \times 7 \times 7 \times 7 \times 7$

2A **9** Evaluate the following. Verify your answers with a calculator.
a $7^2 \times 3^2 - 5^2$ b $\sqrt{144} \times \sqrt{36} \div \sqrt{81}$ c $4^3 \times 3^2 \div \sqrt{144} + 5^2$

2B **10** Write the following using index notation.
a $2 \times 3 \times 3 \times 3$ b $5 \times 5 \times 6 \times 6 \times 6 \times 6$
c $2 \times 5 \times 5 \times 5 \times 9 \times 9 \times 9$

2B **11** Use your calculator to evaluate:
a 3^5 b 7^3 c 8^4 d 11^5

2B **12** Simplify each of the following by first writing as a repeated multiplication and then using index notation.
a $7^3 \times 7^9$ b $3^7 \times 3^6$ c $(13^2)^8$
d $(9^4)^5$ e $(4 \times 6)^8$ f $(12 \times 15)^3$

2C **13** List the first 5 multiples of each number.
a 11 b 100 c 5
d 20 e 13 f 35

2C **14** Find the lowest common multiple (LCM) of the following pairs of numbers.
a 3 and 12 b 6 and 15 c 4 and 7 d 5 and 8

15 In a race, one trail bike rider completes each lap in 40 seconds while another completes it in 60 seconds. How long after the start of the race will the two bikes pass the starting point together?

16 Joe, Claire and Daniela were having races up and down a flight of stairs. Joe took the stairs three at a time, Daniela took the stairs two at a time while Claire took the stairs four at a time. In each case, they reached the bottom with no steps left over.

a How many steps are there in the flight of stairs? List three possible answers.

b What is the smallest number of steps there could be?

c If Joe can also take the stairs five at a time with no steps left over, what is the smallest number of steps in the flight of stairs?

17 Find all the factors of each of the following numbers.

a 16 **b** 27
c 50 **d** 42
e 36 **f** 72

18 List the factor pairs of the following numbers.

a 24 **b** 40 **c** 48 **d** 21 **e** 99 **f** 100

19 Dhiba wants to cut equal lengths of streamers to decorate her room for a 'pink' party. She wants them to be as long as possible. If she has a roll containing 15 metres and another containing 35 metres, what should be the length of each streamer to have no leftover sections?

2E 20 List all of the prime numbers less than 30.

2E 21 How many single digit prime numbers are there?

2E 22 Find the prime number which comes next after 50.

2F 23 Find the prime factors of:

a 99
b 63
c 125
d 124

2F 24 Express 280 as a product of prime factors.

2F 25 Write the following as prime factors using expanded form.

a 44
b 132
c 150
d 360

2G 26 Find the two smallest numbers that are divisible by both 2 and by 3.

2G 27 State whether the following are divisible by both 3 and by 4.

a 120
b 155
c 76
d 252

2G 28 a State the test of divisibility for:

i 2
ii 5
iii 10.

b What do these three tests have in common?

2G 29 Which of the following numbers are divisible by 6?

a 65
b 121
c 90
d 294

2G 30 Which of the following numbers are divisible by 9?

a 162
b 488
c 459
d 49 725

2G 31 State true (T) or false (F) for each of the following.

a 146 is divisible by 2.
b 3100 is divisible by 5 and 10.
c 435 is divisible by 2 and 5.
d 144 is divisible by 3.
e 7650 is divisible by 8.
f 3124 is divisible by 4.
g 24 264 is divisible by 3 and 4.
h 6045 is divisible by 5 and 8.
i 234 is divisible by 6.
j 345 098 is divisible by 11.
k 240 is divisible by 2, 3, 4, 5, 6 and 8.

Fractions

3

A vegetarian pizza is cut into 6 equal pieces while a sausage pizza of the same size is cut into 8 equal pieces. If Martin eats 2 pieces of the vegetarian pizza and Olivia eats 3 pieces of the sausage pizza, who has eaten more? How can you prove this?

In this chapter, you will learn to compare, order and calculate with fractions.

Are you ready?

Try the questions below. If you have difficulty with any of them, extra help can be obtained by completing the matching SkillSHEET. Either click on the SkillSHEET icon next to the question on the *Maths Quest 7* CD-ROM or ask your teacher for a copy.

SkillSHEET 3.1

Types of fractions

1 State whether each of the following is a proper fraction, an improper fraction or a mixed number.

a $\frac{5}{6}$ **b** $\frac{9}{4}$ **c** $3\frac{1}{2}$

d $\frac{3}{2}$ **e** $1\frac{3}{4}$ **f** $\frac{11}{20}$

SkillSHEET 3.2

Equivalent fractions I

2 a Form an equivalent fraction to $\frac{2}{3}$ by multiplying the numerator and denominator by 5.

b Form an equivalent fraction to $\frac{24}{28}$ by dividing the numerator and denominator by 4.

SkillSHEET 3.3

Equivalent fractions II

3 Copy and complete the following equivalent fractions by filling in the gaps.

$\frac{1}{3} = \frac{2}{\_} = \frac{\_}{9} = \frac{\_\_}{12}$

SkillSHEET 3.4

Lowest common multiple

4 a List the first eight multiples of 3.
b List the first eight multiples of 4.
c Which multiples are common (the same) in both lists?
d What is the lowest common multiple of 3 and 4?

SkillSHEET 3.5

Ordering fractions

5 Write these fractions in ascending order (from smallest to largest).

a $1, \frac{4}{5}, \frac{1}{5}, \frac{3}{5}, \frac{7}{5}$ **b** $\frac{11}{12}, 0, \frac{7}{12}, \frac{13}{12}, \frac{5}{12}$

SkillSHEET 3.6

Highest common factor

6 a List the factors of 18.
b List the factors of 12.
c Which factors are common (the same) in both lists?
d What is the highest common factor of 18 and 12?

SkillSHEET 3.7

Converting an improper fraction to a mixed number

7 Use a diagram to express $\frac{5}{2}$ as a mixed number.

SkillSHEET 3.8

Converting a mixed number to an improper fraction

8 Use a diagram to express $1\frac{1}{3}$ as an improper fraction.

SkillSHEET 3.9

Adding and subtracting fractions with the same denominator

9 Find:

a $\frac{5}{6} + \frac{3}{6}$ **b** $\frac{7}{12} - \frac{5}{12}$

c $1 - \frac{1}{3}$ **d** $4 - \frac{1}{4}$

Understanding fractions

Antonio makes pizzas.

Besides selling whole pizzas, he often sells part of a pizza to his customers. For example, this pizza has been cut into three equal pieces. If a customer wants only two of these slices, then Antonio is selling $\frac{2}{3}$ of the pizza to the customer.

Fractions are used to describe parts of a whole. The **numerator**, or top number of the fraction, shows how many parts are required, while the **denominator** or bottom number shows the number of parts the whole has been divided into. So $\frac{2}{3}$ means that the whole thing has been divided into three equal parts and that we are concerned with two parts. The horizontal bar separating the numerator from the denominator is called the **vinculum**.

A **proper fraction** is a fraction with the numerator smaller than the denominator, like $\frac{2}{3}$.

An **improper fraction** has a numerator that is larger than the denominator, like $\frac{5}{4}$.

A **mixed number** is made up of a number and a fraction, like $1\frac{3}{5}$.

WORKED Example 1

What fraction of the rectangle shown has been shaded?

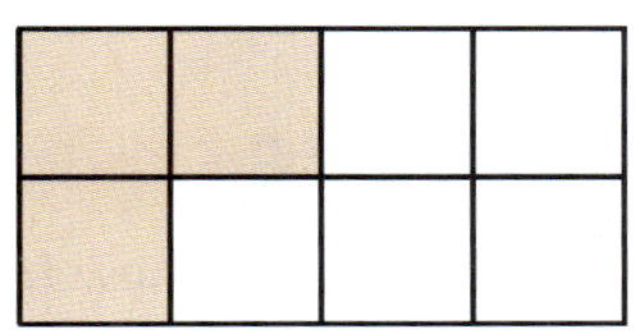

THINK	WRITE
1 Count how many equal parts the rectangle has been divided into.	Total number of parts = 8
2 State how many parts are shaded.	3 parts are shaded.
3 Write the answer in a sentence.	$\frac{3}{8}$ of the rectangle has been shaded.

Finding equivalent fractions

Antonio cuts a pizza into two equal pieces as shown. Each piece is exactly half of the pizza.

A lady with a small child comes into the shop and wants to buy some pizza. She decides to cut one of the halves into two smaller, equal pieces as shown.

Now the child eats two quarters of a pizza and the mother eats half a pizza. If you look carefully you will see that each person has the same amount of pizza. We can say that $\frac{1}{2}$ is equivalent to $\frac{2}{4}$ or $\frac{1}{2} = \frac{2}{4}$.

Equivalent fractions can be found by multiplying or dividing both the numerator and the denominator by the same number as long as the number is not zero. Therefore,

$$\frac{1}{2} = \frac{1 \times 2}{2 \times 2} = \frac{2}{4} \quad \text{and} \quad \frac{1}{2} = \frac{1 \times 3}{2 \times 3} = \frac{3}{6}.$$

$$\text{So} \quad \frac{1}{2} = \frac{2}{4} = \frac{3}{6}.$$

WORKED Example 2

For the equivalent fractions $\frac{4}{5} = \frac{8}{10}$ find the number which has been used to multiply both numerator and denominator.

THINK	WRITE
1 What number is 4 multiplied by to equal 8? (4 × _ = 8) What number is 5 multiplied by to equal 10? (5 × _ = 10)	$\frac{4}{5} = \frac{4 \times 2}{5 \times 2} = \frac{8}{10}$
2 Write the answer.	The number used is 2.

WORKED Example 3

From the list, find those fractions which are equivalent to $\frac{1}{2}$: $\frac{3}{9}, \frac{3}{6}, \frac{9}{18}, \frac{10}{15}, \frac{7}{14}, \frac{17}{34}$.

THINK	WRITE
1 Multiply the numerator and denominator of $\frac{1}{2}$ by the numerator of the first fraction (3) to check whether the new fraction is in the list.	$\frac{1 \times 3}{2 \times 3} = \frac{3}{6}$

THINK

WRITE

2. Multiply the numerator and denominator of $\frac{1}{2}$ by the next different numerator (9) and check whether the new fraction is in the list.

$$\frac{1 \times 9}{2 \times 9} = \frac{9}{18}$$

3. Continue until all fractions have been considered.

$$\frac{1 \times 10}{2 \times 10} = \frac{10}{20}$$
$$\frac{1 \times 7}{2 \times 7} = \frac{7}{14}$$
$$\frac{1 \times 17}{2 \times 17} = \frac{17}{34}$$

4. Write the equivalent fractions.

From the given list the equivalent fractions of $\frac{1}{2}$ are $\frac{3}{6}, \frac{9}{18}, \frac{7}{14}, \frac{17}{34}$.

WORKED Example 4

Write the sequence of the first three equivalent fractions for $\frac{2}{3}$.

THINK

WRITE

1. Find the first three equivalent fractions in the sequence by multiplying both the numerator and denominator by 2, 3 and 4.

$$\frac{2}{3} = \frac{2 \times 2}{3 \times 2} = \frac{2 \times 3}{3 \times 3} = \frac{2 \times 4}{3 \times 4}$$

2. Write the sequence of equivalent fractions.

$$\frac{2}{3} = \frac{4}{6} = \frac{6}{9} = \frac{8}{12}$$

Note: There are other fractions which are equivalent to $\frac{2}{3}$. These can be found by multiplying the numerator and denominator by other numbers.

Equivalent fractions can also be shown using diagrams. Equivalent fractions for $\frac{2}{3}$ are shown below.

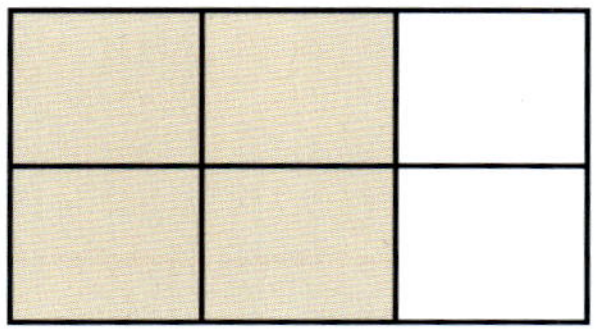

$\frac{2}{3} = \frac{4}{6}$

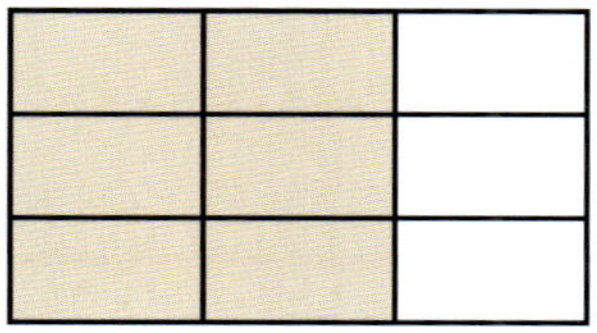

$\frac{2}{3} = \frac{6}{9}$

$\frac{2}{3} = \frac{8}{12}$

Notice that the same amount of the rectangle has been shaded in each case.

Comparing fractions

If Antonio's pizza is cut into quarters it would be easy to see that 1 piece is less than 2 pieces, or that $\frac{1}{4}$ of the pizza is less than $\frac{2}{4}$.

If Antonio has another pizza of the same size but cuts it into thirds, it can still be easily seen that $\frac{1}{4}$ is less than $\frac{1}{3}$.

This is because a pizza divided into four equal pieces will give smaller pieces than one divided into three equal pieces.

If the numerators are the same, the smaller fraction is the one with the larger denominator. For example, $\frac{1}{6}$ is greater than $\frac{1}{7}$ and $\frac{3}{10}$ is less than $\frac{3}{4}$. If the numerators are different, another method is required to compare fractions.

Mathcad: Lowest common multiple

GC program—TI: Lowest common multiple

GC program—Casio: Lowest common multiple

Fractions which do not have the same denominator can be compared by using equivalent fractions. We need to rewrite the fractions so that they have the same denominator.

To do this we look at the multiples of both denominators and choose the lowest multiple that is common to both.

For example, to compare the fractions $\frac{3}{4}$ and $\frac{2}{3}$, we need to find a number that can be used as the new denominator for each equivalent fraction. We start by listing the multiples of 4 and 3.

Multiples of 4 are 4, 8, 12, 16, 20, 24, 28, …

Multiples of 3 are 3, 6, 9, 12, 15, 18, 21, 24, …

From these lists we can see that the lowest number that is the same in each is 12. (We can also see that 24 is the same, but it is not the lowest.) So the **lowest common multiple (LCM)** of 4 and 3 is 12. As 12 is divisible by both 4 and 3, each fraction can be written as an equivalent fraction using 12 as the denominator.

WORKED Example 5

Which is the bigger fraction, $\frac{2}{3}$ or $\frac{3}{5}$?

THINK	WRITE
1 Find the lowest common multiple of the denominators. First list the multiples of 3 and 5. Identify the lowest number that is common to both lists.	Multiples of 3 are 3, 6, 9, 12, 15, 18, … Multiples of 5 are 5, 10, 15, 20, … The lowest common multiple is 15.
2 Write each fraction as an equivalent fraction using the lowest common multiple (15) as the denominator.	$\frac{2}{3} = \frac{2 \times 5}{3 \times 5} = \frac{10}{15}$ and $\frac{3}{5} = \frac{3 \times 3}{5 \times 3} = \frac{9}{15}$
3 Decide which is bigger by comparing the numerators of the equivalent fractions ($\frac{10}{15}$ and $\frac{9}{15}$).	
4 Answer the question.	$\frac{2}{3}$ is bigger than $\frac{3}{5}$.

The answer to worked example 5 can be written as $\frac{2}{3} > \frac{3}{5}$.

The symbol '>' means *is bigger than* or *is greater than*. The symbol '<' means *is less than*.

For example, $\frac{1}{2} < \frac{3}{4}$ means that $\frac{1}{2}$ is less than $\frac{3}{4}$. An easy way to remember this is that < looks a little bit like the capital letter L (for less than).

WORKED Example 6

Insert the appropriate symbol, < or >, between each pair of fractions to make a true statement.

a $\frac{6}{7}$ $\frac{7}{8}$ **b** $\frac{7}{12}$ $\frac{5}{9}$

THINK	WRITE
a 1 Find the lowest common multiple of the denominators.	**a** Multiples of 7 are: 7, 14, 21, 28, 35, 42, 49, 56, 63, … Multiples of 8 are: 8, 16, 24, 32, 40, 48, 56, 64, … The lowest common multiple is 56.
2 Write each fraction as an equivalent fraction using the lowest common multiple (56) as the denominator.	$\frac{6}{7} = \frac{6 \times 8}{7 \times 8} = \frac{48}{56}$ and $\frac{7}{8} = \frac{7 \times 7}{8 \times 7} = \frac{49}{56}$
3 Decide which is bigger by comparing the numerators of the equivalent fractions.	$\frac{6}{7}$ is less than $\frac{7}{8}$
4 Answer the question.	$\frac{6}{7} < \frac{7}{8}$
b 1 Find the lowest common multiple of the denominators.	**b** Multiples of 12 are: 12, 24, 36, 48, … Multiples of 9 are: 9, 18, 27, 36, 45, … The lowest common multiple is 36.
2 Write each fraction as an equivalent fraction using the lowest common multiple (36) as the denominator.	$\frac{7}{12} = \frac{7 \times 3}{12 \times 3} = \frac{21}{36}$ and $\frac{5}{9} = \frac{5 \times 4}{9 \times 4} = \frac{20}{36}$
3 Decide which is bigger by comparing the numerators of the equivalent fractions.	$\frac{7}{12}$ is greater than $\frac{5}{9}$
4 Answer the question.	$\frac{7}{12} > \frac{5}{9}$

remember

1. *Equivalent fractions* can be found by multiplying or dividing both the numerator and the denominator by the same non-zero number.
2. The *numerator* is the top part of the fraction.
3. The *denominator* is the bottom part of the fraction.
4. Different fractions can be compared or ordered by writing each one with the same denominator. This is called finding the lowest common multiple (LCM).
5. The symbol '>' means *is greater than*, the symbol '<' means *is less than*.

EXERCISE 3A Understanding fractions

SkillSHEET 3.1 Types of fractions

WORKED Example 1

1 What fraction of each of the following rectangles has been shaded?

a 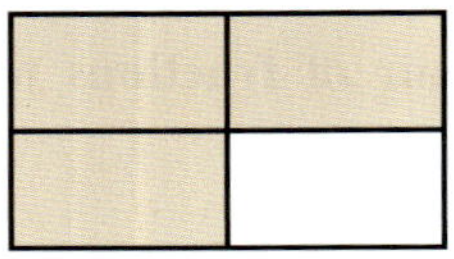

b 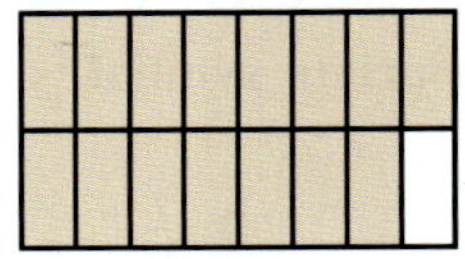

c 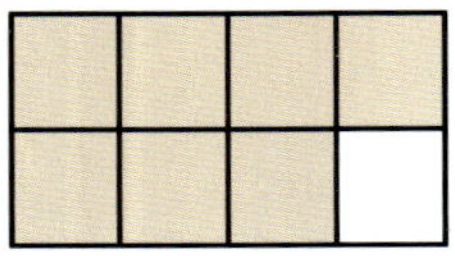

d 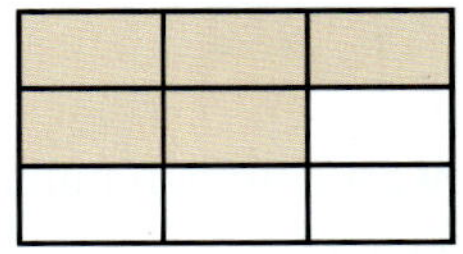

2 What fraction of each of these flags is coloured red?

a 

b 

c Do you recognise these flags? Can you name the country to which each flag belongs?

3 a What fraction of the total number of pieces of fruit is made up of:

i bananas? **ii** oranges?

b What fraction of chocolate biscuits in this packet has been eaten?

SkillSHEET 3.2 Equivalent fractions I

SkillSHEET 3.3 Equivalent fractions II

EXCEL Spreadsheet Equivalent fractions (DIY)

WORKED Example 2

4 For the equivalent fractions below find the number which has been used to multiply both numerator and denominator.

a $\frac{3}{8} = \frac{9}{24}$ **b** $\frac{1}{4} = \frac{3}{12}$ **c** $\frac{3}{5} = \frac{12}{20}$

d $\frac{2}{3} = \frac{10}{15}$ **e** $\frac{5}{6} = \frac{30}{36}$ **f** $\frac{9}{10} = \frac{81}{90}$

g $\frac{7}{8} = \frac{77}{88}$ **h** $\frac{7}{8} = \frac{84}{96}$ **i** $\frac{2}{5} = \frac{14}{35}$

WORKED Example 3

5 From the list, find those fractions which are equivalent to the fractions marked in red.

a $\frac{2}{3}$ $\frac{7}{9}$ $\frac{20}{30}$ $\frac{5}{8}$ $\frac{12}{16}$ $\frac{14}{21}$ $\frac{40}{60}$

b $\frac{4}{5}$ $\frac{12}{15}$ $\frac{15}{20}$ $\frac{36}{45}$ $\frac{16}{20}$ $\frac{28}{35}$ $\frac{80}{100}$

c $\frac{7}{8}$ $\frac{17}{18}$ $\frac{40}{45}$ $\frac{56}{64}$ $\frac{14}{18}$ $\frac{18}{19}$ $\frac{21}{24}$

d $\frac{7}{10}$ $\frac{18}{25}$ $\frac{35}{50}$ $\frac{14}{21}$ $\frac{21}{30}$ $\frac{14}{20}$ $\frac{140}{200}$

6 Fill in the gaps.

a $\frac{3}{5} = \frac{9}{\_\_} = \frac{\_\_}{20} = \frac{\_\_}{35}$

b $\frac{3}{4} = \frac{\_\_}{40} = \frac{18}{\_\_} = \frac{\_\_}{36}$

c $\frac{1}{4} = \frac{\_\_}{28} = \frac{6}{\_\_} = \frac{3}{\_\_}$

d $\frac{5}{6} = \frac{15}{\_\_} = \frac{\_\_}{42} = \frac{40}{\_\_}$

WORKED Example 4

7 Write a sequence of the first three equivalent fractions for each of the following.

a $\frac{5}{6}$ **b** $\frac{5}{8}$ **c** $\frac{3}{10}$ **d** $\frac{2}{3}$ **e** $\frac{1}{8}$ **f** $\frac{7}{8}$

WORKED Example 5

8 Which is the bigger fraction?

a $\frac{2}{5}$ or $\frac{3}{5}$ **b** $\frac{5}{8}$ or $\frac{7}{8}$ **c** $\frac{1}{5}$ or $\frac{1}{6}$ **d** $\frac{1}{8}$ or $\frac{1}{10}$ **e** $\frac{1}{2}$ or $\frac{3}{10}$

f $\frac{3}{4}$ or $\frac{3}{5}$ **g** $\frac{2}{5}$ or $\frac{5}{8}$ **h** $\frac{3}{5}$ or $\frac{5}{8}$ **i** $\frac{4}{5}$ or $\frac{5}{8}$

3.4 SkillSHEET Lowest common multiple

WORKED Example 6

9 Insert the appropriate symbol, < or >, between each pair of fractions to make a true statement.

a $\frac{5}{8}$ $\frac{3}{8}$ **b** $\frac{5}{6}$ $\frac{7}{6}$ **c** $\frac{3}{5}$ $\frac{1}{4}$ **d** $\frac{3}{4}$ $\frac{3}{5}$ **e** $\frac{1}{5}$ $\frac{1}{4}$ **f** $\frac{3}{10}$ $\frac{2}{5}$

GC program — TI Comparing fractions

10 multiple choice

a Which fraction is smaller than $\frac{5}{8}$?

A $\frac{7}{8}$ **B** $\frac{17}{24}$ **C** $\frac{11}{12}$ **D** $\frac{3}{5}$ **E** $\frac{13}{16}$

b Which fraction is equivalent to $\frac{3}{4}$?

A $\frac{3}{8}$ **B** $\frac{6}{8}$ **C** $\frac{9}{16}$ **D** $\frac{12}{24}$ **E** $\frac{15}{24}$

GC program — Casio Comparing fractions

11 Write the following fractions with the same denominator and then write them in ascending order. (Ascending means from smallest to largest.)

a $\frac{3}{10}, \frac{1}{2}, \frac{1}{5}$ **b** $\frac{3}{8}, \frac{1}{2}, \frac{1}{3}$ **c** $\frac{2}{3}, \frac{4}{5}, \frac{6}{15}$ **d** $\frac{3}{4}, \frac{2}{3}, \frac{7}{15}$

3.5 SkillSHEET Ordering fractions

12 Four hungry teenage girls ordered pizzas, but none could eat the whole pizza.

Sally Seafood ate $\frac{6}{8}$ of her seafood pizza.

Hanna Hawaiian ate $\frac{1}{2}$ of her Hawaiian pizza.

Sonya Special ate $\frac{3}{4}$ of her special pizza.

Vanessa Vegetarian ate $\frac{5}{6}$ of her vegetarian pizza.

a Draw the four pizzas and shade the amount that each girl ate.

b Which girl ate the most pizza?

c Who ate more pizza, Sally Seafood or Sonya Special?

GAME time Fractions — 001

Flatback turtle facts

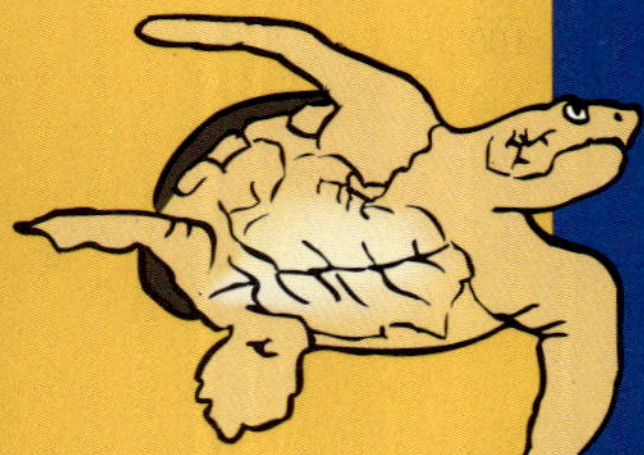

Connect the dots next to equivalent fractions. Each straight line drawn will pass through a letter and number, giving the puzzle answer code.

1	2	3	4	5	3	3	6	7	2	8	9	7	5	2	7	10	3	11	1			
10	8	12	7	1	1	13	14	11	15	8	5	16	17	7	13	3	18	18	14	18	5	12
3	11	13	8	11	19	12	15	7	8	16	17	7	13	3	18	18	5	17	7			
3	20	5	7	2	15	8	2	2	14	7	2	2	7	7	18	8	11	19				
5	8	21	7	13	7	6	7	11	4	7	7	21	13	5	3	17	8	5	16	17		

Simplifying fractions

Usually the best way to write fractions is in the simplest form. In other words, we reduce the fraction to its lowest equivalent form. To do this, it is necessary to divide both the numerator and the denominator by the same number.

The best number to divide by is the highest common factor of both the numerator and the denominator. The highest common factor (HCF) of two or more numbers is the largest factor that is the same for the given numbers. For example, the highest common factor of 8 and 12 is 4. This is because:

- 8 has the factors 1, 2, 4 and 8
- 12 has the factors 1, 2, 3, 4, 6 and 12.

The factors that are common to both 8 and 12 are 1, 2 and 4. The highest of these is 4. This means that 4 is the highest number that divides evenly into 8 and 12 without a remainder.

Dividing both the numerator and denominator by the same number is called *cancelling*.

Mathcad: Highest common factor

GC program–TI: Highest common factor

GC program–Casio: Highest common factor

WORKED Example 7

Write $\frac{3}{6}$ in simplest form.

THINK	WRITE
1 Write the fraction and find the highest common factor (HCF) or the largest number that is a factor of both the numerator and the denominator (3).	$\frac{3}{6}$
2 Cancel by dividing the numerator and denominator by this factor (3 ÷ 3 = 1, 6 ÷ 3 = 2).	$= \frac{\cancel{3}^{1}}{\cancel{6}_{2}}$
3 Write the answer in simplest form.	$= \frac{1}{2}$

If the largest factor is not used the first time, the process can be repeated to find the simplest fraction. To simplify a mixed number, leave the whole number and simplify the fraction.

WORKED Example 8

Simplify $7\frac{40}{64}$.

THINK	WRITE
1 Write the mixed number and think of the HCF or largest number that divides into both numerator and denominator; that is, the number 8.	$7\frac{40}{64}$
2 Divide both numerator and denominator by that factor (40 ÷ 8 = 5, 64 ÷ 8 = 8).	$= 7\frac{\cancel{40}^{5}}{\cancel{64}_{8}}$
3 Write your answer as a mixed number in simplest form.	$= 7\frac{5}{8}$

Graphics Calculator tip! Simplifying fractions

Simplifying fractions

To reduce a fraction into its simplest form using a TI graphics calculator, type in the fraction using the ÷ key. Press MATH then select **1:Frac** and finally ENTER to obtain the answer. For the fraction $\frac{40}{64}$ in worked example 8, you would press the following keys: 40 ÷ 64 MATH then select **1:Frac** and press ENTER.

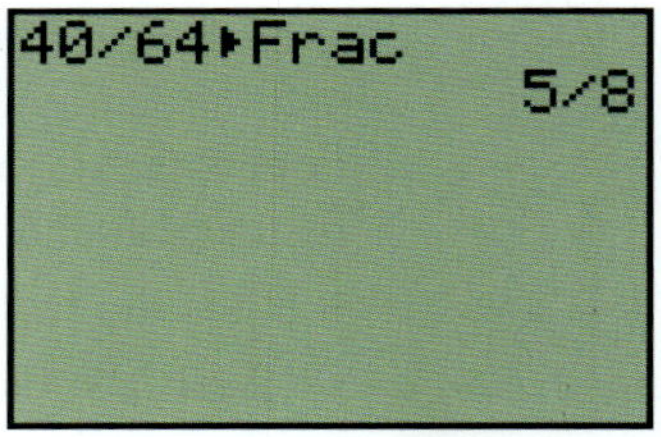

WORKED Example 9

Dipper's team scored 24 goals. Dipper scored 18 of the team's goals. What fraction of the team's goals did Dipper score? Simplify the answer.

THINK	WRITE
1 Write a fraction with Dipper's goals in the numerator and the team's scores in the denominator.	$\frac{18}{24}$
2 Simplify the fraction by dividing top and bottom by the HCF (6).	$= \frac{3}{4}$
3 Write the answer in a sentence.	Dipper scored $\frac{3}{4}$ of the team's goals.

remember

1. To simplify a fraction, divide both the numerator and denominator by the highest common factor or HCF.
2. To simplify a mixed number, leave the whole number and simplify the fraction.

EXERCISE 3B Simplifying fractions

SkillSHEET 3.6 Highest common factor

WORKED Example 7

1 Write the following fractions in simplest form.

a $\frac{5}{10}$ b $\frac{8}{12}$ c $\frac{21}{24}$ d $\frac{48}{60}$ e $\frac{28}{35}$

f $\frac{18}{24}$ g $\frac{81}{90}$ h $\frac{49}{56}$ i $\frac{100}{120}$ j $\frac{48}{50}$

k $\frac{63}{72}$ l $\frac{49}{70}$ m $\frac{33}{36}$ n $\frac{22}{50}$ o $\frac{21}{56}$

WORKED Example 8

2 Simplify.

a $4\frac{21}{35}$ b $7\frac{45}{54}$ c $10\frac{10}{20}$ d $5\frac{7}{42}$ e $3\frac{56}{64}$

f $1\frac{44}{48}$ g $6\frac{21}{28}$ h $5\frac{16}{48}$ i $3\frac{11}{55}$ j $2\frac{16}{64}$

EXCEL Spreadsheet: Simplifying a fraction

Mathcad: Simplifying a fraction

3 **multiple choice**

The fraction $\frac{26}{39}$ is the same as:

A $\frac{2}{3}$ B $\frac{2}{6}$ C $\frac{3}{8}$ D $\frac{9}{12}$ E $\frac{2}{13}$

WORKED Example 9

4 Kylie's netball team scored 28 goals. Kylie scored 21 of her team's goals. What fraction of the team's goals did Kylie score? Simplify the answer.

5 David's basketball team scored 36 goals. David scored 20 of the team's goals.
a What fraction of the team's goals did David score? Simplify the answer.
b What fraction of the team's goals did the rest of the team score. Simplify the answer.

6 Year 7 students at Springfield High School ran a car wash to raise money for the local hospital. They raised a total of $1000 and drew up a table showing how much money each class raised.

7A $200 7B $150 7C $320
7D $80 7E $250

Express as a simple fraction how much of the total each class raised.

7 Below are the results of a book-reading competition.

Mark	5 books
David	10 books
Samantha	8 books
Jules	6 books
Ahmed	4 books
Darren	7 books

a How many books were read in total?
b For each contestant, record the number of books read as a fraction of the total number of books read. Where possible, reduce the fractions to their simplest forms.

MATHS QUEST CHALLENGE

1 Can you arrange the digits 1, 2, 3, 4, 5, 6, 7, 8 and 9 to form a fraction which is equivalent to $\frac{1}{2}$? For example, one way is $\frac{7329}{14\,658}$. Can you find two others?
Hint: **Try starting with a numerator beginning with 72 for one of them and 67 for the other.**

2 Can you find a way to arrange the nine digits so that they form a fraction equivalent to $\frac{1}{3}$?

Improper fractions and mixed numbers

An improper fraction has a numerator larger than the denominator; for example, $\frac{3}{2}$.

A mixed number is made up of a whole number and a fraction; for example, $1\frac{1}{2}$, $2\frac{3}{4}$.

Changing improper fractions to mixed numbers

An improper fraction can be changed to a mixed number by dividing the denominator into the numerator and writing the remainder over the same denominator.

WORKED Example 10

Draw a diagram to show $\frac{5}{4}$ as pieces of pizza, then write the improper fraction as a mixed number.

THINK	WRITE
1 Draw a whole pizza and divide it into the number of parts shown by the denominator.	 This is 4 quarters or $\frac{4}{4}$.
2 Find the number of parts left over and draw them.	 Extra $\frac{1}{4}$ needed.
3 Write the improper fraction as a mixed number.	$\frac{5}{4} = 1\frac{1}{4}$

If the improper fraction had a larger numerator it might be necessary to draw more pizzas and the whole number part would then be greater than 1.

WORKED Example 11

Express $\frac{11}{5}$ as a mixed number.

THINK	WRITE
1 Write the improper fraction.	$\frac{11}{5}$
2 Determine how many times the denominator can be divided into the numerator and what the remainder is. The whole number is part of the answer and the remainder becomes the numerator of the fractional part with the same denominator as the original improper fraction.	$= 11 \div 5$ $= 2$ remainder 1
3 Write the answer.	$= 2\frac{1}{5}$

Changing mixed numbers to improper fractions

A mixed number can be changed to an improper fraction by first multiplying the whole number with the denominator then adding the numerator. The denominator stays the same.

WORKED Example 12

Draw a diagram to show $2\frac{1}{3}$ as pieces of pizza and then write $2\frac{1}{3}$ as an improper fraction.

THINK

1. Draw two whole pizzas and $\frac{1}{3}$ of a pizza.
2. Divide the whole pizzas into thirds.
3. Count the number of thirds and write the mixed number as an improper fraction.

WRITE

$2\frac{1}{3} = \frac{7}{3}$

WORKED Example 13

Express $2\frac{3}{4}$ as an improper fraction.

THINK

1. Write the mixed number.
2. Multiply the whole number with the denominator then add the numerator. The result becomes the numerator and the denominator stays the same.
3. Evaluate the top line of the fraction.
4. Write the answer.

WRITE

$2\frac{3}{4}$

$= \frac{2 \times 4 + 3}{4}$

$= \frac{8 + 3}{4}$

$= \frac{11}{4}$

Graphics Calculator tip! Changing a mixed number to an improper fraction

To change a mixed number to an improper fraction, enter the mixed number as the sum of the whole number and the fraction. For example, to change $2\frac{3}{4}$ to an improper fraction, press (2) (+) (3) (÷) (4) (MATH), select **1:Frac** and press (ENTER).

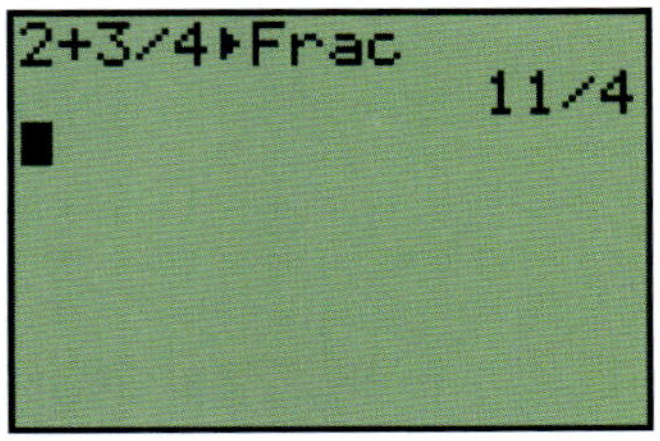

Changing a mixed number to an improper fraction

To obtain a fraction answer to any calculation when using a TI graphics calculator, always remember to press (MATH), select **1:Frac** and press (ENTER) after all numbers and operations have been typed in.

remember

1. An *improper fraction* has a numerator that is larger than the denominator.
2. A *mixed number* is made up of a whole number and a fraction.
3. An improper fraction can be changed to a mixed number by dividing the denominator into the numerator and writing the remainder over the same denominator.
4. A mixed number can be changed to an improper fraction by first multiplying the whole number with the denominator then adding the numerator. The denominator stays the same.

EXERCISE 3C Improper fractions and mixed numbers

SkillSHEET 3.7 Converting an improper fraction to a mixed number

WORKED Example 10

1 Draw a diagram to show the following improper fractions as pieces of pizza, then write each improper fraction as a mixed number.

a $\frac{5}{2}$ b $\frac{4}{3}$ c $\frac{11}{6}$ d $\frac{13}{8}$ e $\frac{12}{4}$ f $\frac{7}{6}$

2 Copy these statements, filling the gaps with the appropriate expression from the list below.

mixed number proper fraction improper fraction

a An ________________ has a numerator that is larger than the denominator.
b A ________________ consists of a whole number and a fraction.
c A ________________ has a denominator that is larger than the numerator.

3 a Sarah ate $\frac{3}{2}$ or 3 halves of an apple. Draw the amount of apple that Sarah ate and express $\frac{3}{2}$ as a mixed number.

b Dean ate $\frac{9}{4}$ or 9 quarters of pizza. Draw the amount of pizza that Dean ate and express $\frac{9}{4}$ as a mixed number.

EXCEL Spreadsheet Converting improper fractions to mixed numbers

WORKED Example 11

Mathcad Converting improper fractions to mixed numbers

4 Express these improper fractions as mixed numbers.

a $\frac{7}{5}$ b $\frac{11}{4}$ c $\frac{21}{2}$ d $\frac{39}{10}$
e $\frac{51}{12}$ f $\frac{92}{11}$ g $\frac{29}{13}$ h $\frac{23}{8}$
i $\frac{100}{3}$ j $\frac{25}{2}$ k $\frac{20}{3}$ l $\frac{10}{3}$

EXCEL Spreadsheet Converting improper fractions to mixed numbers (DIY)

5 Change these improper fractions to mixed numbers.

a $\frac{8}{5}$ b $\frac{27}{8}$ c $\frac{11}{2}$ d $\frac{97}{9}$ e $\frac{58}{7}$
f $\frac{117}{10}$ g $\frac{67}{8}$ h $\frac{74}{9}$ i $\frac{31}{12}$ j $\frac{89}{6}$

6 **multiple choice**

a Which of the following is the same as $\frac{61}{8}$?

A $5\frac{5}{8}$ B $6\frac{5}{8}$ C $7\frac{5}{8}$ D $8\frac{5}{8}$ E $9\frac{5}{8}$

b Which of the following is the same as $\frac{73}{10}$?

A $3\frac{7}{10}$ **B** $7\frac{3}{10}$ **C** $7\frac{7}{10}$ **D** $3\frac{3}{10}$ **E** $7\frac{4}{10}$

c Which of the following is the same as $\frac{48}{5}$?

A $8\frac{4}{5}$ **B** $5\frac{5}{9}$ **C** $9\frac{5}{9}$ **D** $9\frac{3}{5}$ **E** $9\frac{4}{9}$

7 Kim and Carly arrived home from school quite hungry and cut up some fruit. Kim ate 9 quarters of an apple and Carly ate 11 quarters. How many apples did they each eat?

8 Daniel was selling pieces of quiche at the school fete. Each quiche was divided into 8 pieces. If Daniel sold 51 pieces, how many quiches did Daniel sell?

9 It was Cecilia's responsibility to supply oranges for her soccer team at half-time. If 19 players ate $\frac{1}{4}$ orange each, how many oranges were eaten?

WORKED Example 12

10 Draw a diagram to show the following mixed numbers as pieces of pizza and then write each one as an improper fraction.

a $1\frac{1}{8}$ **b** $3\frac{3}{4}$ **c** $2\frac{1}{6}$

d $1\frac{1}{6}$ **e** $5\frac{1}{4}$ **f** $1\frac{3}{5}$

3.8 SkillSHEET Converting a mixed number to an improper fraction

WORKED Example 13

11 Express the following mixed numbers as improper fractions.

a $1\frac{1}{7}$ **b** $1\frac{4}{5}$ **c** $2\frac{2}{3}$ **d** $3\frac{1}{3}$ **e** $6\frac{1}{4}$

f $4\frac{1}{9}$ **g** $11\frac{1}{2}$ **h** $3\frac{7}{8}$ **i** $7\frac{4}{5}$ **j** $9\frac{2}{5}$

EXCEL Spreadsheet Converting mixed numbers to improper fractions

12 Change these mixed numbers to improper fractions.

a $3\frac{1}{2}$ **b** $4\frac{1}{5}$ **c** $5\frac{4}{5}$ **d** $6\frac{5}{7}$ **e** $2\frac{9}{10}$

f $3\frac{7}{12}$ **g** $5\frac{3}{5}$ **h** $9\frac{4}{7}$ **i** $1\frac{10}{11}$ **j** $8\frac{5}{6}$

Mathcad Converting mixed numbers to improper fractions

13 multiple choice

EXCEL Spreadsheet Converting mixed numbers to improper fractions (DIY)

a Which of the following is the same as $10\frac{2}{7}$?

A $\frac{27}{10}$ **B** $\frac{72}{10}$ **C** $\frac{10}{27}$ **D** $\frac{72}{7}$ **E** $\frac{72}{2}$

b Which of the following is the same as $8\frac{5}{6}$?

A $\frac{85}{6}$ **B** $\frac{58}{6}$ **C** $\frac{56}{8}$ **D** $\frac{53}{6}$ **E** $\frac{68}{5}$

c Which of the following is the same as $7\frac{9}{11}$?

A $\frac{86}{11}$ **B** $\frac{99}{7}$ **C** $\frac{79}{11}$ **D** $\frac{97}{11}$ **E** $\frac{77}{9}$

14 **a** Amy made apple pies for her Grandma's 80th birthday party. She divided each pie into 6 pieces, and after the party she noted that $4\frac{1}{6}$ pies had been eaten. How many pieces had been eaten?

b Amy's cousin Ria provided cordial for the same party, and calculated that she could make 20 drinks from each bottle. At the end of the party $3\frac{17}{20}$ bottles had been used. How many drinks had been consumed?

15 Insert the appropriate < or > sign between each pair of fractions to make a true statement.

WorkSHEET 3.1

a $\frac{8}{5}$ $\quad 1\frac{2}{5}$	**b** $\frac{19}{2}$ $\quad 8\frac{1}{2}$	**c** $\frac{33}{32}$ $\quad 1\frac{3}{32}$
d $\frac{15}{2}$ $\quad 5\frac{1}{2}$	**e** $3\frac{3}{4}$ $\quad \frac{19}{4}$	**f** $7\frac{1}{7}$ $\quad \frac{65}{7}$
g $2\frac{1}{3}$ $\quad \frac{8}{3}$	**h** $1\frac{16}{17}$ $\quad \frac{35}{17}$	**i** $4\frac{3}{8}$ $\quad \frac{36}{8}$

10 QUICK QUESTIONS 1

1 Write down the equivalent fractions from this list:
$\frac{1}{4}, \frac{2}{6}, \frac{3}{12}, \frac{6}{25}, \frac{8}{32}, \frac{10}{41}, \frac{16}{64}, \frac{19}{76}, \frac{31}{124}, \frac{123}{490}$.

2 Write a sequence of the first three equivalent fractions for $\frac{2}{5}$.

3 Write $\frac{12}{20}$ in simplest form.

4 On a recent quiz, Simone answered 6 out of the 50 questions incorrectly. What fraction of the test did Simone answer correctly? Express your answer in simplest form.

5 Write $\frac{26}{3}$ as a mixed number.

6 Change $1\frac{1}{4}$ to an improper fraction.

7 Write $\frac{12}{9}$ as a mixed number in simplest form.

8 Change $3\frac{6}{7}$ to an improper fraction.

9 Rewrite these fractions in ascending (from smallest to largest) order:
$\frac{2}{5}, \frac{14}{5}, 1\frac{3}{5}, 3\frac{3}{5}, \frac{15}{5}, 4\frac{2}{5}, \frac{28}{5}, 6\frac{1}{5}$.

10 A catering company provided 12 quiches which were each cut into 8 pieces for a luncheon. At the end of the function there were 12 pieces of quiche left over. How many whole quiches were eaten?

Adding and subtracting fractions

Fractions can be added and subtracted if they have the same denominator. Remember to simplify your answer if possible.

WORKED Example 14

Find $\frac{5}{9} + \frac{1}{9}$.

THINK	WRITE
1 Write the question.	$\frac{5}{9} + \frac{1}{9}$
2 Add the numerators since the denominators are the same.	$= \frac{6}{9}$
	$= \frac{\cancel{6}^{2}}{\cancel{9}^{3}}$
3 Simplify the answer if possible by cancelling the highest common factor (6 ÷ 3 = 2, 9 ÷ 3 = 3).	$= \frac{2}{3}$

If fractions do not have the same denominator, use equivalent fractions to make the denominators the same.

WORKED Example 15

Find $\frac{5}{6} - \frac{1}{12}$, expressing the answer in simplest form.

THINK	WRITE
1 Find multiples of each denominator.	Multiples of 6 are 6, 12 . . . Multiples of 12 are 12 . . .
2 Find the lowest common multiple (LCM) of 6 and 12 (12).	LCM is 12.
3 Rewrite the fractions if necessary so that they both have the LCM as the new denominator.	$\frac{5}{6} - \frac{1}{12}$ $= \frac{5 \times 2}{6 \times 2} - \frac{1}{12}$ $= \frac{10}{12} - \frac{1}{12}$
4 Subtract the numerators since the denominators are the same.	$= \frac{9}{12}$
5 Simplify your answer if possible by cancelling the highest common factor (9 ÷ 3 = 3, 12 ÷ 3 = 4).	$= \frac{\cancel{9}^{3}}{\cancel{12}^{4}}$ $= \frac{3}{4}$

With practice many of these steps can be done in your head and the solution written as follows.

$$\frac{5}{6} - \frac{1}{12} = \frac{10}{12} - \frac{1}{12}$$
$$= \frac{9}{12}$$
$$= \frac{3}{4}$$

Adding and subtracting mixed numbers

There are a number of ways of adding and subtracting mixed numbers. In this chapter we will use the method shown in the worked examples. All mixed numbers can be added and subtracted by first changing them to improper fractions. Remember to simplify the answer if possible.

WORKED Example 16

Find $2\frac{3}{4} + 1\frac{1}{2}$.

THINK	WRITE
1 Write the question.	$2\frac{3}{4} + 1\frac{1}{2}$
2 Change each mixed number to an improper fractions.	$= \frac{11}{4} + \frac{3}{2}$
	$= \frac{11}{4} + \frac{3 \times 2}{2 \times 2}$
3 Write both fractions with the same denominator using equivalent fractions.	$= \frac{11}{4} + \frac{6}{4}$
4 Add the fractions.	$= \frac{17}{4}$
5 Write the answer as a mixed number if appropriate.	$= 4\frac{1}{4}$

WORKED Example 17

Find $3\frac{1}{5} - 1\frac{3}{4}$.

THINK	WRITE
1 Write the question.	$3\frac{1}{5} - 1\frac{3}{4}$
2 Change each mixed number to an improper fraction.	$= \frac{16}{5} - \frac{7}{4}$
3 Write both fractions with the same denominator using equivalent fractions.	$= \frac{16 \times 4}{5 \times 4} - \frac{7 \times 5}{4 \times 5}$
	$= \frac{64}{20} - \frac{35}{20}$
4 Subtract the second fraction from the first.	$= \frac{39}{20}$
5 Write the answer as a mixed number if appropriate.	$= 1\frac{19}{20}$

Graphics Calculator tip! Adding and subtracting fractions

As with any calculation involving fractions, if you wish to have an answer expressed as a fraction then each calculation needs to end by pressing MATH, selecting **1:Frac** and pressing ENTER. The calculation for worked example 15 would be entered as 5 ÷ 6 – 1 ÷ 12 then press MATH, select **1:Frac** and press ENTER.

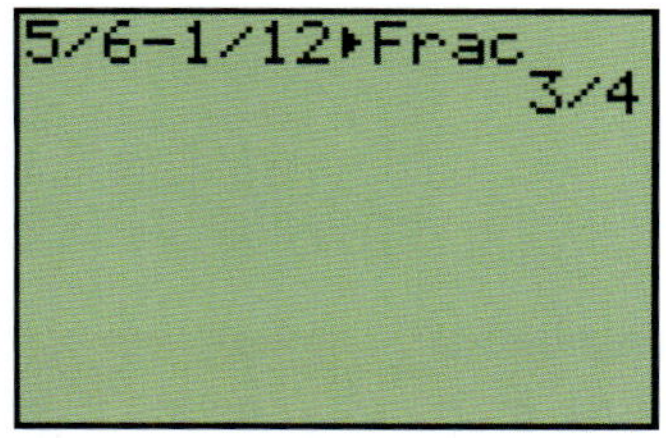

When entering mixed numbers, it is a good idea to use brackets. This allows the correct order of operations to occur, particularly when subtracting.

The calculations for worked examples 16 and 17 can be viewed in the screen shown. Note that the answers are given as improper fractions.

Adding and subtracting fractions

Estimation can be used to find approximate answers to addition and subtraction of mixed numbers. This allows you to easily check whether your answer to a calculation looks reasonable.

WORKED Example 18

Find $3\frac{1}{3} + 2\frac{1}{2}$ and check the answer by estimation.

THINK	WRITE
1 Write the question.	$3\frac{1}{3} + 2\frac{1}{2}$
2 Change each mixed number to an improper fraction.	$= \frac{10}{3} + \frac{5}{2}$
3 Write both fractions with the same denominator using equivalent fractions.	$= \frac{10 \times 2}{3 \times 2} + \frac{5 \times 3}{2 \times 3}$ $= \frac{20}{6} + \frac{15}{6}$
4 Add the fractions.	$= \frac{35}{6}$
5 Write the answer as a mixed number if appropriate.	$= 5\frac{5}{6}$
6 Check the answer by adding the whole numbers to find an approximation.	$3\frac{1}{3} + 2\frac{1}{2} > 3 + 2$ $3\frac{1}{3} + 2\frac{1}{2} > 5$ The approximation shows that the answer we obtain should be greater than 5.

remember

1. Fractions can be added and subtracted if they have the same denominator.
2. If fractions do not have the same denominator, use equivalent fractions to make the denominators the same.
3. Mixed numbers can be added or subtracted by first changing them to improper fractions.

EXERCISE 3D Adding and subtracting fractions

SkillSHEET 3.9 Adding and subtracting fractions with the same denominator

WORKED Example 14

1 Find:

a $\frac{1}{5} + \frac{3}{5}$ **b** $\frac{9}{11} - \frac{3}{11}$ **c** $\frac{3}{6} + \frac{2}{6}$

d $\frac{20}{50} + \frac{11}{50}$ **e** $\frac{15}{25} - \frac{6}{25}$ **f** $\frac{12}{12} - \frac{1}{12}$

g $\frac{5}{8} - \frac{1}{8} + \frac{3}{8}$ **h** $\frac{4}{7} + \frac{5}{7} - \frac{3}{7}$ **i** $\frac{33}{100} + \frac{7}{100} - \frac{11}{100}$

EXCEL Spreadsheet Adding and subtracting fractions

2 Find the following, expressing the answers in simplest form.

a $\frac{5}{12} + \frac{1}{12}$ **b** $\frac{7}{16} - \frac{3}{16}$ **c** $\frac{9}{15} - \frac{4}{15}$

d $\frac{37}{100} + \frac{13}{100}$ **e** $\frac{9}{4} - \frac{7}{4}$ **f** $\frac{4}{7} + \frac{5}{7} - \frac{2}{7}$

g $\frac{13}{16} + \frac{9}{16} - \frac{10}{16}$ **h** $\frac{11}{28} + \frac{10}{28}$ **i** $\frac{21}{81} + \frac{21}{81} + \frac{30}{81}$

Mathcad Adding fractions

WORKED Example 15

3 Find the following, expressing the answers in simplest form.

a $\frac{1}{2} + \frac{1}{4}$ **b** $\frac{1}{2} - \frac{1}{4}$ **c** $\frac{1}{3} + \frac{2}{21}$

d $\frac{1}{8} + \frac{20}{32}$ **e** $\frac{9}{18} + \frac{1}{6}$ **f** $\frac{17}{40} + \frac{1}{2}$

g $\frac{7}{9} - \frac{8}{27}$ **h** $\frac{31}{35} - \frac{3}{7}$ **i** $\frac{8}{10} - \frac{28}{70}$

Mathcad Subtracting fractions

4 Find the lowest common multiple of each of the following pairs of numbers.

a 2 and 3 **b** 3 and 5 **c** 5 and 10

d 4 and 6 **e** 5 and 6 **f** 9 and 6

g 4 and 8 **h** 6 and 8 **i** 7 and 5

SkillSHEET 3.10 Lowest common multiple

5 Using the lowest common multiples that were found in question **4** as denominators, add or subtract these fractions.

a $\frac{1}{2} - \frac{1}{3}$ **b** $\frac{2}{3} - \frac{2}{5}$ **c** $\frac{3}{5} + \frac{3}{10}$

d $\frac{3}{4} - \frac{1}{6}$ **e** $\frac{4}{5} - \frac{1}{6}$ **f** $\frac{1}{9} + \frac{5}{6}$

g $\frac{1}{4} + \frac{3}{8}$ **h** $\frac{5}{6} - \frac{3}{8}$ **i** $\frac{5}{7} - \frac{2}{5}$

6 Answer the following questions by first finding the lowest common denominator. Simplify your answer if necessary.

a $\frac{2}{3} + \frac{1}{4}$ **b** $\frac{5}{7} - \frac{3}{8}$ **c** $\frac{4}{5} - \frac{3}{4}$ **d** $\frac{7}{10} + \frac{2}{15}$

e $\frac{5}{8} - \frac{5}{12}$ **f** $\frac{7}{9} - \frac{1}{2}$ **g** $\frac{5}{12} + \frac{2}{9}$ **h** $\frac{7}{11} + \frac{1}{4}$

i $\frac{4}{5} - \frac{2}{11}$ **j** $\frac{5}{6} + \frac{7}{10}$ **k** $\frac{3}{10} - \frac{1}{20}$ **l** $\frac{7}{8} - \frac{3}{5}$

m $\frac{2}{3} + \frac{4}{5}$ **n** $\frac{5}{7} + \frac{3}{4}$ **o** $\frac{5}{9} - \frac{2}{7}$ **p** $\frac{2}{3} - \frac{1}{13}$

q $\frac{11}{12} - \frac{7}{10}$ **r** $\frac{4}{7} + \frac{5}{6}$ **s** $\frac{4}{5} - \frac{2}{3}$ **t** $\frac{6}{11} - \frac{1}{3}$

7 **multiple choice**

a The lowest common denominator of $\frac{1}{3}$, $\frac{1}{4}$ and $\frac{1}{6}$ is:

A 16 **B** 24 **C** 18 **D** 12 **E** 72

b The lowest common denominator of $\frac{1}{2}$, $\frac{1}{4}$ and $\frac{1}{5}$ is:

A 8 **B** 10 **C** 20 **D** 30 **E** 40

8 Answer the following:

a $\frac{1}{2} + \frac{1}{3} - \frac{1}{4}$ **b** $\frac{3}{10} + \frac{4}{15} - \frac{1}{6}$ **c** $\frac{7}{8} - \frac{3}{4} + \frac{4}{6}$ **d** $\frac{11}{12} - \frac{8}{15} - \frac{3}{20}$

e $\frac{1}{16} + \frac{1}{8} + \frac{1}{4}$ **f** $\frac{5}{6} - \frac{7}{18} - \frac{1}{9}$ **g** $\frac{1}{2} + \frac{2}{3} + \frac{3}{4}$ **h** $\frac{3}{4} + \frac{1}{3} + \frac{5}{6}$

Mathcad — Adding and subtracting fractions

9 Angela was a little hungry the day before her birthday, and spied the cake to be eaten the following day. She decided that if she ate only $\frac{1}{10}$ of it, her Mum might not notice. However, she was still hungry and took another $\frac{1}{6}$.

a How much of the cake did Angela eat before her birthday?

b How much of the cake was left for her birthday?

10 Bread needed to be cooked at 200°C for $\frac{3}{4}$ hour then a further $\frac{1}{2}$ hour at 150°C. What is the total cooking time in hours?

11 Robert, Jason and Luke play football in the same team. Last Saturday, Robert kicked $\frac{1}{4}$ of the team's goals, Jason kicked $\frac{1}{3}$ and Luke $\frac{5}{12}$. What fraction of the team's goals were kicked by:

a Jason and Luke? **b** Robert and Jason? **c** the three boys together?

12 Jacinta and her friend decided to climb the tree in her Nan's front yard. It took her 10 minutes to climb halfway up, 10 more minutes to climb another sixth of the way and yet another 10 minutes to reach a further tenth of the way. Here she sat admiring the view for 5 minutes before beginning her descent.

a What fraction of the height of the tree had Jacinta climbed after 20 minutes?

b What fraction of the tree did Jacinta climb in total?

WORKED Example 16

13 Find:

a $1\frac{1}{5} + 3\frac{1}{2}$ **b** $3\frac{2}{3} - 2\frac{1}{2}$ **c** $5\frac{2}{3} - 2\frac{1}{4}$

d $9\frac{3}{8} + 4$ **e** $2\frac{3}{5} - 1\frac{1}{2}$ **f** $3\frac{1}{2} + 2\frac{3}{4}$

WORKED Example 17

14 Find:

a $7 - 5\frac{2}{3}$ **b** $2\frac{1}{3} - \frac{5}{6}$ **c** $3\frac{1}{2} - 1\frac{7}{8}$

d $4\frac{3}{5} - 2\frac{9}{10}$ **e** $5\frac{1}{4} - 1\frac{7}{12}$ **f** $5\frac{3}{4} - 3\frac{5}{6}$

15 Find:

a $\frac{1}{5} + 2\frac{2}{7}$ **b** $\frac{5}{8} + 3\frac{3}{4} - \frac{1}{2}$ **c** $2\frac{2}{9} + 1\frac{1}{3} + \frac{1}{6}$ **d** $4\frac{2}{3} + 1\frac{3}{8} - 3$

WORKED Example 18

16 Find the following and check the answers by estimation.

a $5\frac{1}{3} + 2\frac{1}{4}$ **b** $1\frac{1}{6} + 2\frac{1}{2}$ **c** $3\frac{1}{3} + 1\frac{1}{6}$

d $6\frac{1}{2} - 2\frac{1}{4}$ **e** $4\frac{3}{5} - 1\frac{3}{10}$ **f** $7\frac{5}{6} - 3\frac{1}{3}$

17 One-third of a litre of cordial is mixed with $1\frac{1}{2}$ litres of water. How many litres of drink have been made?

18 At a school dance-a-thon, the number of hours each student danced was recorded.

Matthew $1\frac{3}{4}$ hours, Leigh $2\frac{2}{3}$ hours, Daniel $1\frac{5}{6}$ hours

How many hours did the boys dance in total?

19 The lengths of bolts and nails are often measured in inches. For the bolt shown, find the length of the threaded section if the total length of the bolt is $3\frac{1}{4}$ inches.

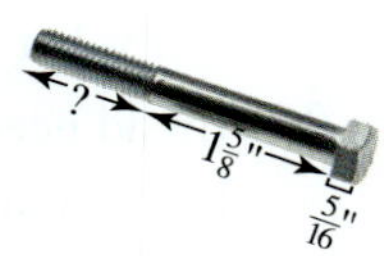

20 To make the casing for a simple collapsible telescope you need two cylinders, one of which fits inside the other. One cylinder has measurements as shown below left.

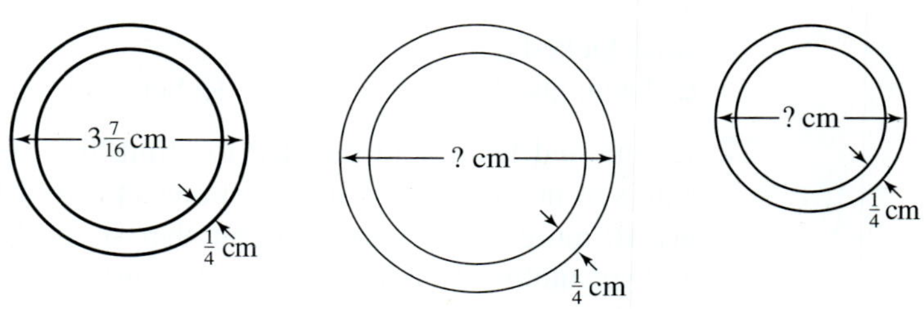

a What is the width of a cylinder (see middle diagram above) which would slide over the one shown above left?

b What is the width of a cylinder (see diagram above right) which would slide into the one shown above left?

COMMUNICATION

Egyptian fractions

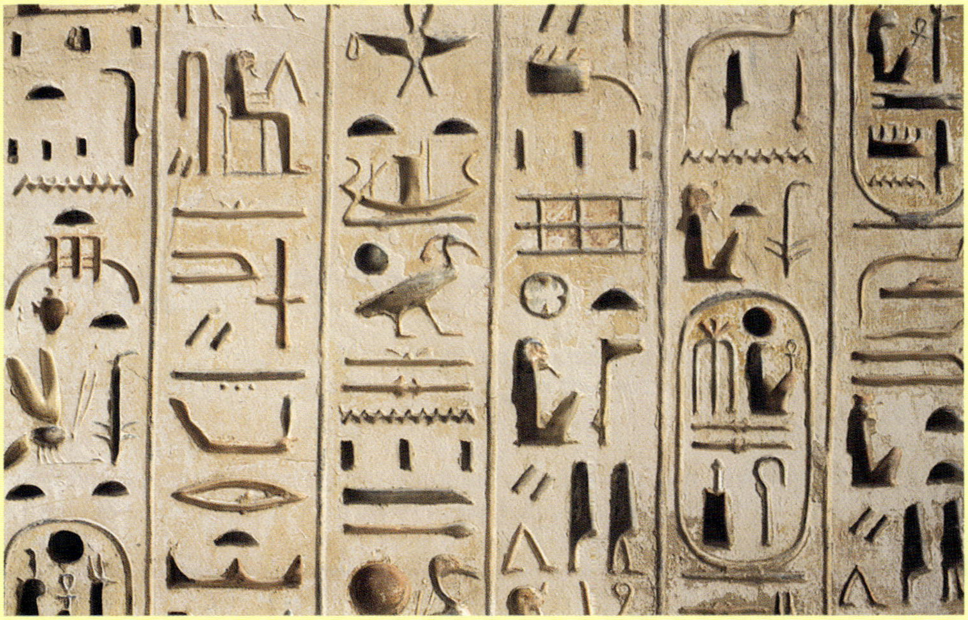

The ancient Egyptians (from 3100 BC to 332 BC) used different symbols or hieroglyphs to show numbers.

1 10 100 1000 10 000

Using these symbols, the number 23 462 would be shown as:

23 462

1 Work out each number using the description given above.

a  **b**  **c**  **d** 

As well as expressing words and whole numbers with hieroglyphics, the ancient Egyptians devised a way of expressing fractions. They placed a 'mouth' (pronounced 'er') over their symbols for whole numbers. For example:

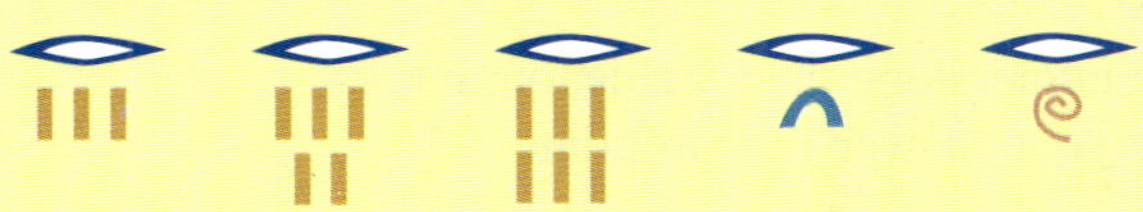

$\frac{1}{3}$ $\frac{1}{5}$ $\frac{1}{6}$ $\frac{1}{10}$ $\frac{1}{100}$

Apart from some special fractions, all numerators were of the value 1, so they would express a fraction like $\frac{3}{5}$ as shown at right.

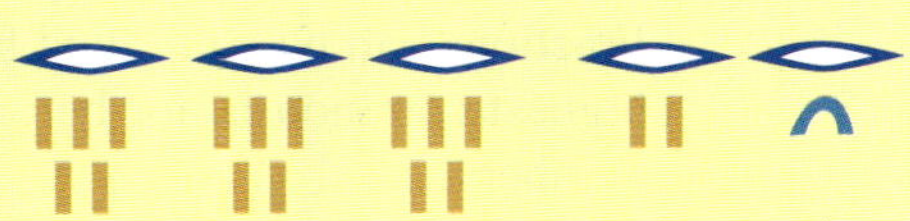

$\frac{1}{5} + \frac{1}{5} + \frac{1}{5}$ or $\frac{1}{2} + \frac{1}{10}$

2 What do these symbols represent?

a 

b 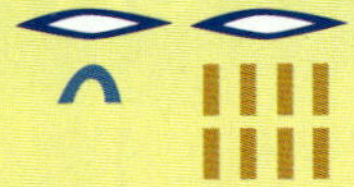

b 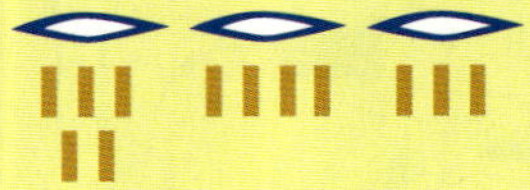

d 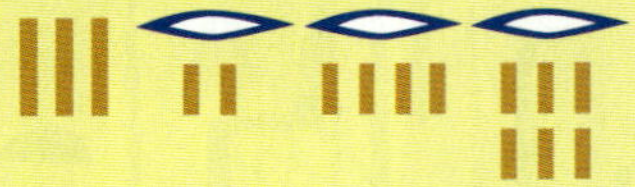

3 Perform the following calculations. Give your answer in Egyptian hieroglyphics and in Hindu–Arabic numbers.

a Add these two numbers. 

b Subtract the second number from the first.

4 Write four of your own questions using Egyptian hieroglyphics. Swap these with a partner. Calculate the answers to your partner's set of questions, showing clearly how you were able to work them out.

Multiplying fractions

To multiply fractions, multiply the numerators and then multiply the denominators.

It is important to note that when multiplying fractions the denominators do not need to be the same.

WORKED Example 19

Find $\frac{1}{5} \times \frac{3}{4}$.

THINK	WRITE
1 Write the question.	$\frac{1}{5} \times \frac{3}{4}$
2 Write as one fraction.	$= \frac{1 \times 3}{5 \times 4}$
3 Multiply the numerators, then the denominators and simplify your answer if appropriate.	$= \frac{3}{20}$

Fractions can be simplified by dividing both the numerator and denominator with their highest common factor before being multiplied. This is easier than working with large numbers.

Remember that dividing both numerator and denominator by the same number is often called *cancelling*.

WORKED Example 20

Find $\frac{2}{3} \times \frac{9}{10}$.

THINK	WRITE
1 Write the question.	$\frac{2}{3} \times \frac{9}{10}$
2 Write the two fractions as one keeping the multiplication signs.	$= \frac{{}^{1}\cancel{2} \times \cancel{9}^{3}}{{}^{1}\cancel{3} \times \cancel{10}^{5}}$
3 Divide the terms in the numerator and denominator by their highest common factor. *Note:* This may be done vertically or diagonally.	
4 Multiply the numerators, then the denominators and simplify your answer if appropriate.	$= \frac{3}{5}$

WORKED Example 21

Find $\frac{1}{3} \times 36$.

THINK	WRITE
1 Write the question.	$\frac{1}{3} \times 36$
2 Express the whole number as a fraction with a denominator of 1 and cancel if possible.	$= \frac{1}{{}^{1}\cancel{3}} \times \frac{\cancel{36}^{12}}{1}$
3 Multiply the numerators and then the denominators.	$= \frac{12}{1}$
4 Simplify your answer.	$= 12$

Mixed numbers must be changed to improper fractions before multiplying.

WORKED Example 22

Find $1\frac{1}{2} \times 3\frac{3}{5}$.

THINK	WRITE
1 Write the question.	$1\frac{1}{2} \times 3\frac{3}{5}$
2 Change mixed numbers to improper fractions and cancel if possible.	$= \frac{3}{{}^{1}\cancel{2}} \times \frac{\cancel{18}^{9}}{5}$
3 Multiply the numerators and then the denominators.	$= \frac{27}{5}$
4 Simplify your answer.	$= 5\frac{2}{5}$

Using the word 'of'

Sometimes the word *of* can mean the same as *multiply*. Antonio might take $\frac{1}{3}$ of an hour (60 minutes) to cook his pizzas. A third of an hour, or $\frac{1}{3}$ of 60 minutes, is the same as $\frac{1}{3} \times \frac{60}{1} = 20$ minutes. So it takes Antonio 20 minutes to cook a pizza.

A customer wants to buy $\frac{1}{4}$ of the 8 remaining pizzas for the family's evening meal. A quarter of 8 is the same as $\frac{1}{4} \times \frac{8}{1} = 2$. So the customer buys 2 pizzas.

WORKED Example 23

Find $\frac{2}{3}$ of $\frac{1}{4}$.

THINK	WRITE
1 Write the question.	$\frac{2}{3}$ of $\frac{1}{4}$
2 Change 'of' to '×' and cancel if appropriate.	$= \frac{\cancel{2}^1}{3} \times \frac{1}{\cancel{4}^2}$
3 Perform the multiplication.	$= \frac{1}{6}$

WORKED Example 24

Mum put a 2-litre carton of flavoured milk in the refrigerator. When the children came home from school, Joanna drank one quarter of the milk, Kieren drank one half of it and Daisy drank one sixth of it. What fraction of the milk, in litres, did each person drink?

THINK	WRITE
1 Write the fraction of the milk that Joanna drank and simplify the answer.	Joanna: $\frac{1}{4}$ of $2 = \frac{1}{4} \times 2$
2 Perform the multiplication.	$= \frac{1}{{}_2\cancel{4}} \times \frac{\cancel{2}^1}{1}$ $= \frac{1}{2}$
3 Write the fraction of the milk that Kieren drank and simplify the answer.	Kieren: $\frac{1}{2}$ of $2 = \frac{1}{2} \times 2$
4 Perform the multiplication.	$= \frac{1}{{}_1\cancel{2}} \times \frac{\cancel{2}^1}{1}$ $= 1$
5 Write the fraction of the milk that Daisy drank and simplify the answer.	Daisy: $\frac{1}{6}$ of $2 = \frac{1}{6} \times 2$
6 Perform the multiplication.	$= \frac{1}{{}_3\cancel{6}} \times \frac{\cancel{2}^1}{1}$ $= \frac{1}{3}$
7 Write a sentence giving the answer in litres.	Joanna drank a half a litre, Kieren drank 1 litre and Daisy drank a third of a litre of milk.

remember

1. To multiply fractions, multiply the numerators and then multiply the denominators.
2. When multiplying fractions the denominators do not need to be the same.
3. Fractions can be simplified (or cancelled) before being multiplied.
4. A whole number can be written as a fraction with a denominator of 1.
5. Change mixed numbers to improper fractions before multiplying.
6. In a problem involving fractions, 'of' can be replaced with 'multiply', then the fractions can be multiplied in the usual way.

EXERCISE 3E Multiplying fractions

Mathcad: Multiplying fractions

EXCEL Spreadsheet: Multiplying fractions

WORKED Example 19

1 Find:

a $\frac{1}{2} \times \frac{1}{4}$ b $\frac{1}{3} \times \frac{2}{3}$ c $\frac{2}{5} \times \frac{3}{5}$ d $\frac{5}{9} \times \frac{4}{3}$

e $\frac{11}{13} \times \frac{1}{2}$ f $\frac{6}{7} \times \frac{9}{11}$ g $\frac{11}{12} \times \frac{11}{12}$ h $\frac{8}{9} \times \frac{5}{9}$

i $\frac{5}{8} \times \frac{1}{3}$ j $\frac{2}{3} \times \frac{2}{3}$ k $\frac{5}{6} \times \frac{11}{12}$ l $\frac{5}{11} \times \frac{5}{12}$

m $\frac{7}{8} \times \frac{7}{8}$ n $\frac{2}{3} \times \frac{4}{13}$ o $\frac{3}{5} \times \frac{3}{10}$ p $\frac{4}{7} \times \frac{2}{3}$

WORKED Example 20

2 Find:

a $\frac{4}{8} \times \frac{3}{9}$ b $\frac{2}{5} \times \frac{1}{2}$ c $\frac{3}{7} \times \frac{5}{6}$

d $\frac{3}{8} \times \frac{4}{15}$ e $\frac{10}{11} \times \frac{22}{25}$ f $\frac{24}{27} \times \frac{9}{8}$

g $\frac{18}{32} \times \frac{64}{72}$ h $\frac{48}{56} \times \frac{24}{60}$ i $\frac{15}{27} \times \frac{36}{45}$

j $\frac{8}{49} \times \frac{14}{16}$ k $\frac{4}{9} \times \frac{6}{8}$ l $\frac{6}{12} \times \frac{8}{10}$

m $\frac{42}{54} \times \frac{12}{49}$ n $\frac{5}{25} \times \frac{12}{48}$ o $\frac{18}{24} \times \frac{12}{36}$

3 Mark's uncle gave him a family-size block of chocolate. He divided it into thirds to share with his two sisters. His friends Tom and Nick then arrived to visit, and Mark shared his portion of the chocolate with them, so that they each had the same amounts.

a Draw a block of chocolate with 9×6 pieces, and shade Mark's third of the block. Then draw lines to represent Tom's share.

b What fraction of the block of chocolate did Nick receive?

WORKED Example 21

4 Find:

a $\frac{2}{3} \times 3$ b $\frac{5}{14} \times 7$ c $\frac{15}{22} \times 11$ d $\frac{5}{16} \times 4$

WORKED Example 22

5 Find:

a $1\frac{1}{4} \times \frac{2}{3}$ **b** $\frac{3}{4} \times 1\frac{5}{8}$ **c** $1\frac{1}{2} \times 1\frac{1}{2}$ **d** $2\frac{5}{8} \times 1\frac{3}{4}$ **e** $2\frac{1}{5} \times 3\frac{1}{2}$

f $10\frac{9}{10} \times \frac{2}{3}$ **g** $2\frac{1}{4} \times 2\frac{1}{2}$ **h** $\frac{4}{5} \times 1\frac{6}{7}$ **i** $\frac{9}{10} \times 3\frac{2}{3}$ **j** $4\frac{1}{3} \times \frac{3}{5}$

k $1\frac{1}{2} \times 12$ **l** $2\frac{3}{4} \times 6$ **m** $5 \times 3\frac{1}{2}$ **n** $8 \times 3\frac{1}{4}$ **o** $4\frac{3}{5} \times 2\frac{1}{10}$

p $3\frac{5}{6} \times 9\frac{5}{7}$ **q** $5\frac{1}{5} \times 6\frac{2}{3}$ **r** $3\frac{3}{4} \times 2\frac{2}{3}$ **s** $\frac{7}{8} \times 3\frac{1}{3}$ **t** $6\frac{2}{5} \times 4\frac{1}{7}$

6 Tim was looking for something to do one Saturday afternoon and his dad suggested he cook something for afternoon tea.

Tim found the recipe for peanut butter muffins.

Ingredients:

$\frac{1}{4}$ cup sugar, $\frac{1}{4}$ cup margarine,

$\frac{1}{2}$ cup peanut butter, 2 eggs,

$1\frac{1}{2}$ cups milk, $2\frac{1}{2}$ cups self-raising flour,

$\frac{1}{4}$ teaspoon baking soda

Method:

Blend the sugar, margarine and peanut butter. Beat in the eggs and milk. Add the self-raising flour and baking soda. Place the mixture in greased muffin pans and sprinkle with cinnamon sugar. Bake for 15–20 minutes in a 200°C oven.

This recipe makes 10 muffins, but Tim wants enough mixture to make 15. He needs to multiply the quantities by $1\frac{1}{2}$. Write the quantities of each ingredient Tim needs to make 15 muffins.

Mathcad: Fraction of a number

GAME time: Fractions — 002

WORKED Example 23

7 Find:

a $\frac{1}{2}$ of $\frac{1}{4}$ **b** $\frac{3}{4}$ of $\frac{2}{3}$ **c** $\frac{5}{6}$ of $\frac{5}{6}$ **d** $\frac{9}{10}$ of $\frac{5}{7}$

e $\frac{8}{9}$ of $\frac{1}{4}$ **f** $\frac{5}{7}$ of $\frac{7}{25}$ **g** $\frac{4}{5}$ of 20 **h** $\frac{7}{8}$ of 32

i $\frac{9}{10}$ of 50 **j** $\frac{2}{5}$ of 45 **k** $\frac{3}{5}$ of 25 **l** $\frac{8}{9}$ of 81

m $\frac{3}{4}$ of 6 **n** $\frac{9}{10}$ of 15 **o** $\frac{1}{12}$ of 10 **p** $\frac{3}{7}$ of 12

q $\frac{1}{2}$ of $1\frac{1}{2}$ **r** $\frac{3}{4}$ of $8\frac{2}{3}$ **s** $\frac{2}{3}$ of $\frac{3}{2}$ **t** $\frac{5}{6}$ of $\frac{18}{25}$

8 A car's petrol tank holds 48 litres of fuel.

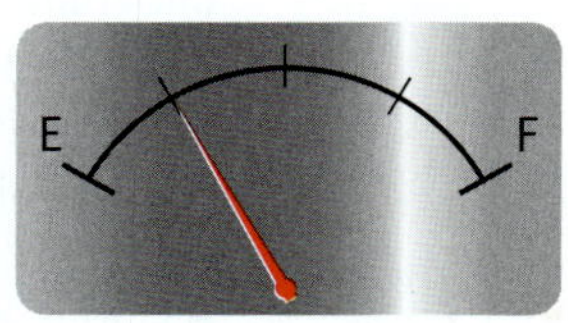

a If the tank was full at the start of a trip, what fraction of the tank has been used?

b How many litres of petrol have been used?

WORKED Example 24

9 Jamie's mum made 6 litres of cordial for his birthday party. Emily drank $\frac{1}{3}$ of the cordial, Tracy drank $\frac{1}{12}$ and Jonathan drank $\frac{1}{18}$. How much cordial, in litres, did each person drink?

10 Zoe and Sarah play basketball with the Sharp Shooters. The games are played for 40 minutes. Zoe played $\frac{4}{5}$ of last week's game and Sarah played $\frac{7}{8}$. How many minutes of the game did:

a Zoe play? **b** Sarah play?

11 Gustave's monthly take-home pay is \$2400. From this he spends a quarter on his home loan payments, one half on food and drink and one sixth on clothing. One half of the remainder goes into his savings account. How much money does Gustave put into his savings account each month?

12 To make a loaf of fruit bread, the fruit needs to be added after the second rise stage and the top of the bread needs to be glazed two-thirds of the way through the bake stage. The recipe shows you how long each stage of the total cycle takes.

Stage	**Length of time**
First knead	5 minutes
Second knead	20 minutes
First rise then punch down	40 minutes
Second rise then punch down	25 minutes
Third rise	50 minutes
Bake	40 minutes

a After what length of time should the fruit be added?
b Express this as a fraction of the total time in the bread maker.
c After what length of time should the top of the bread be glazed?

13 **a** Draw a quarter of a pie.
b Shade half of this piece of pie.
c What fraction of the whole pie is shaded?
d Complete the mathematical sentence $\frac{1}{4} \times \frac{1}{2} = \_\_$ to show how the answer was found.

14 **a** Find how many half pizzas in 1 pizza.
b Find how many half pizzas in 2 whole pizzas.
c Find how many half pizzas in 5 whole pizzas.

DESIGN

Musical fractions

In music, particular symbols tell us how long a note is to be held for. The time value of each note is shown below.

4 beats	2 beats	1 beat	$\frac{1}{2}$ beat	$\frac{1}{4}$ beat
(semibreve)	(minim)	(crotchet)	(quaver)	(semi-quaver)

More than 1 quaver is shown as 

More than 1 semi-quaver is shown as 

1 What is the total number of beats indicated by these notes?

a 
b 
c 
d 

When musicians decide they want a note held for a little longer, they put a dot after the note. The dot adds an extra half of the number of beats to the original time value. For example, a dotted minim 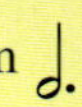 has a time value of

$2 + \frac{1}{2} \times 2 = 3$ beats.

2 What is the time value indicated by a dotted crotchet?

3 What is the time value indicated by a dotted quaver?

4 What is the total number of beats indicated by these notes?

a 
b 

5 Write one note which is equivalent in time value to each of the following.

a 
b 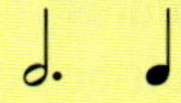
c 
d 
e 
f 
g 
h 

To make music more organised, time signatures were added to the beginning of a piece of music so that each section or bar was divided up into equal parts. For example, $\frac{4}{4}$ tells us that we need 4 crotchet beats (or notes that will make up that amount) in a bar. The 4 on the top shows us the number of beats per bar, while the 4 on the bottom shows us that the note values we use are the ones that belong to crotchet time. In $\frac{3}{4}$ time, we need 3 beats to make up a bar and in $\frac{2}{4}$ time we need 2 beats to make up a bar.

6 Add one note to complete each of the following bars.

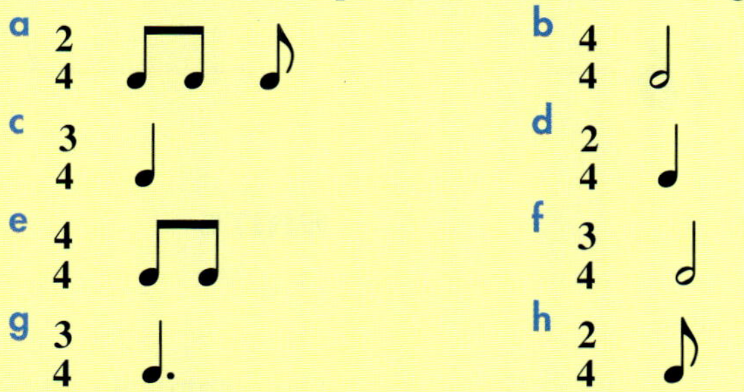

7 Add 2 notes to complete each of the following bars.

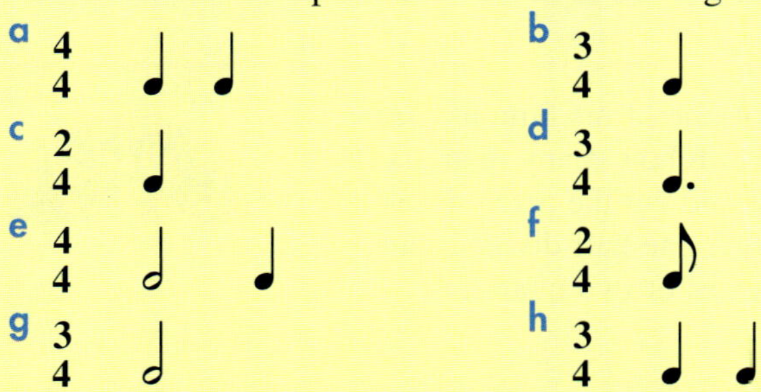

8 Put time signatures to each of the following rhythm patterns. (They are crotchet beats.)

a 

b 

9 Write your own rhythm patterns 4 bars long, using a different pattern in each bar for each of the following time signatures.

a $\frac{3}{4}$ b $\frac{4}{4}$

Dividing fractions

The **reciprocal** of a fraction is found by tipping the fraction upside down. So $\frac{3}{5}$ is the reciprocal of $\frac{5}{3}$.

A whole number can be written as a fraction by putting it over 1 so that $\frac{1}{4}$ is the reciprocal of $\frac{4}{1}$.

WORKED Example 25

Find the reciprocal of $\frac{2}{3}$.

THINK	WRITE
Turn the fraction upside down and write the answer in a sentence.	The reciprocal of $\frac{2}{3}$ is $\frac{3}{2}$.

To find the reciprocal of a mixed number, express it as an improper fraction first.

WORKED Example 26

Find the reciprocal of $1\frac{2}{3}$.

THINK	WRITE
1 Write $1\frac{2}{3}$ as an improper fraction.	$1\frac{2}{3} = \frac{5}{3}$
2 Turn it upside down and write the answer in a sentence.	The reciprocal of $1\frac{2}{3}$ is $\frac{3}{5}$.

Antonio had a half of one of his famous pizzas left. A small child asked him how many quarters there were in his half. Antonio cut the pizza as shown, so that it was clear that one half divided into quarters gave two pieces or two quarters.

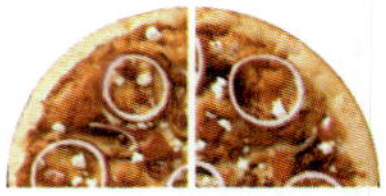

$$\frac{1}{2} \div \frac{1}{4} = 2$$

The answer could also have been found by changing the division sign to a multiplication sign and tipping the second fraction upside down.

$$\frac{1}{2} \div \frac{1}{4} = \frac{1}{2} \times \frac{4}{1}$$
$$= 2$$

Division by a number is the same as multiplication by its reciprocal.
To divide fractions:
1. **change the division sign to a multiplication sign**
2. **tip the second fraction (that is, take the reciprocal of the fraction that follows the division sign)**
3. **perform the multiplication.**

WORKED Example 27

Find $\frac{2}{3} \div \frac{4}{9}$.

THINK	WRITE
1 Write the question.	$\frac{2}{3} \div \frac{4}{9}$
2 Change the division sign to a multiplication sign and tip the second fraction.	$= \frac{2}{3} \times \frac{9}{4}$
3 Cancel if possible.	$= \frac{\not{2}^{1}}{\not{3}^{1}} \times \frac{\not{9}^{3}}{\not{4}^{2}}$
4 Perform the multiplication.	$= \frac{3}{2}$
5 Simplify your answer.	$= 1\frac{1}{2}$

To divide mixed numbers, first change them to improper fractions and then follow the steps in the previous worked example.

WORKED Example 28

Find $1\frac{1}{2} \div 3\frac{2}{5}$.

THINK	WRITE
1 Write the question.	$1\frac{1}{2} \div 3\frac{2}{5}$
2 Change the mixed numbers to improper fractions.	$= \frac{3}{2} \div \frac{17}{5}$
3 Change the ÷ to × and tip the second fraction.	$= \frac{3}{2} \times \frac{5}{17}$
4 Multiply the numerators, then the denominators and simplify your answer if necessary.	$= \frac{15}{34}$

Graphics Calculator tip! Dividing a fraction by a fraction

With the TI graphics calculator we use the (÷) key to enter a fraction as well as when we need to divide two numbers. Using brackets around each fraction helps to keep the operations clear, particularly with mixed numbers.

The calculation for $1\frac{1}{2} \div 3\frac{2}{5}$ is shown.

Casio: Dividing a fraction by a fraction

remember

1. The *reciprocal* of a fraction is found by tipping the fraction upside down.
2. The reciprocal of a mixed number is found by expressing it as an improper fraction, then tipping it upside down.
3. A whole number can be written as a fraction by putting it over 1.
4. Division by a number is the same as multiplication by its reciprocal.
5. To divide fractions:
   (a) change the division sign to a multiplication sign
   (b) tip the second fraction (that is, take the reciprocal of the fraction that follows the division sign)
   (c) perform the multiplication.
6. To divide mixed numbers, first change them to improper fractions then follow the steps in part 5 above.

EXERCISE 3F Dividing fractions

Mathcad: Reciprocals

WORKED Example 25

1 Find the reciprocals of each of the following.

a $\frac{3}{4}$ b $\frac{2}{7}$ c $\frac{5}{3}$ d $\frac{5}{12}$

e $\frac{6}{5}$ f $\frac{9}{2}$ g $\frac{5}{3}$ h $\frac{10}{3}$

i $\frac{1}{5}$ j $\frac{1}{12}$ k $\frac{3}{15}$ l $\frac{2}{22}$

m 5 n 20 o 13 p 1

WORKED Example 26

2 Find the reciprocals of these mixed numbers.

a $1\frac{2}{3}$ b $4\frac{3}{7}$ c $7\frac{1}{2}$ d $4\frac{2}{5}$ e $9\frac{2}{7}$

f $3\frac{6}{7}$ g $3\frac{3}{8}$ h $5\frac{1}{3}$ i $10\frac{2}{9}$ j $6\frac{9}{10}$

3 Multiply each of these numbers by its reciprocal.

a $\frac{5}{7}$ b $\frac{2}{9}$ c $\frac{3}{5}$ d $3\frac{2}{7}$ e $2\frac{1}{8}$ f $3\frac{3}{4}$

4 Copy and complete the following statement:

When any number is multiplied by its ____________, the answer is always ____________.

Mathcad: Dividing fractions

WORKED Example 27

5 Find:

a $\frac{1}{4} \div \frac{2}{3}$ b $\frac{8}{9} \div \frac{7}{6}$ c $\frac{9}{10} \div \frac{1}{2}$ d $\frac{1}{12} \div \frac{2}{3}$ e $\frac{4}{11} \div \frac{5}{6}$

f $\frac{7}{12} \div \frac{9}{11}$ g $\frac{10}{3} \div \frac{15}{21}$ h $\frac{2}{7} \div \frac{20}{21}$ i $\frac{8}{15} \div \frac{16}{15}$ j $\frac{4}{9} \div \frac{4}{9}$

k $\frac{4}{7} \div 10$ l $\frac{5}{8} \div 20$ m $\frac{4}{3} \div 12$ n $\frac{15}{11} \div 5$ o $\frac{7}{9} \div 30$

EXCEL Spreadsheet: Dividing fractions

WORKED Example 28

6 Find:

a $1\frac{1}{4} \div \frac{3}{4}$ b $9 \div 4\frac{1}{2}$ c $\frac{9}{4} \div 3\frac{1}{2}$ d $\frac{10}{3} \div 3\frac{1}{3}$ e $5 \div 2\frac{3}{4}$

f $7\frac{4}{9} \div \frac{1}{3}$ g $\frac{1}{8} \div 2\frac{1}{2}$ h $1\frac{1}{5} \div 2\frac{1}{3}$ i $5\frac{6}{7} \div 2\frac{3}{4}$ j $1\frac{1}{2} \div \frac{1}{2}$

k $10\frac{9}{10} \div 4$ l $4\frac{1}{2} \div 3\frac{6}{7}$ m $3\frac{2}{7} \div 2\frac{1}{7}$ n $4\frac{2}{3} \div \frac{7}{6}$ o $3\frac{5}{8} \div 1\frac{3}{4}$

7 Felicity is dividing $1\frac{1}{2}$ kilograms of mince steak into 8 hamburger portions. How many kilograms of mince will be in each hamburger portion?

8 Ned spends his school holidays helping his father with the shearing.

a It takes Ned $\frac{1}{4}$ hour to shear one sheep. How many sheep will Ned shear in $5\frac{1}{2}$ hours?

b Ned's father, Wesley, shears a sheep every $\frac{1}{12}$ of an hour. How many sheep would Wesley shear if he worked continuously for 8 hours?

WorkSHEET 3.2

9 Michelle can run one lap of a cross-country course in $\frac{1}{5}$ hour.

a How many laps could she complete in $1\frac{1}{3}$ hours? (Assume she can keep up the same pace.)

b How many minutes does it take Michelle to run one lap?

This part of AUSTRALIA was first settled in 1804.

Answer these fraction questions to discover the puzzle answer code.

M $\frac{5}{8} \div \frac{3}{16}$
=

A $\frac{5}{6} \div \frac{1}{3}$
=

N Divide $\frac{3}{4}$ by $\frac{5}{8}$

E $1\frac{2}{7} \div \frac{3}{7}$
=

N $1\frac{7}{10} \div \frac{2}{5}$
=

S How many $\frac{1}{4}$ s are there in $\frac{2}{3}$?

I $5 \div \frac{5}{8}$
=

V $3\frac{1}{2} \div 2$
=

D $5\frac{1}{2} \div 1\frac{4}{7} =$

L $1\frac{3}{4} \div \frac{7}{8}$
=

E $2\frac{1}{2} \div \frac{2}{3}$
=

D Divide $4\frac{1}{4}$ by $1\frac{1}{2}$

A $\frac{3}{7} \div 2$
=

N $5 \div 1\frac{2}{5}$
=

$1\frac{3}{4}$	$2\frac{1}{2}$	$1\frac{1}{5}$	$2\frac{5}{6}$	8	$3\frac{3}{4}$	$3\frac{1}{3}$
3	$4\frac{1}{4}$	$2\frac{2}{3}$	2	$\frac{3}{14}$	$3\frac{4}{7}$	$3\frac{1}{2}$

10 QUICK QUESTIONS 2

Calculate each of the following, expressing your answer in simplest form.

1 $\frac{12}{15} - \frac{4}{15}$

2 $\frac{5}{6} + \frac{3}{10}$

3 $3\frac{1}{4} + 2\frac{2}{5}$

4 $11\frac{5}{8} - 3\frac{1}{2}$

5 $\frac{12}{20} \times \frac{4}{18}$

6 $3\frac{5}{9} \times \frac{15}{20}$

7 $\frac{2}{7}$ of $\frac{21}{40}$

8 $\frac{5}{9} \div \frac{2}{3}$

9 $8 \div 1\frac{3}{4}$

10 $4\frac{2}{5} \div \frac{2}{11}$

Mixed operations with fractions

If more than one operation is involved in a question with fractions, the same set order is used as with whole numbers. Therefore, the order of operations is:

1. brackets
2. multiplication and division (from left to right)
3. addition and subtraction (from left to right).

WORKED Example 29

Find $\frac{6}{7} \times 1\frac{5}{6} \div \frac{1}{4}$.

THINK	WRITE
1 Write the question.	$\frac{6}{7} \times 1\frac{5}{6} \div \frac{1}{4}$
2 Change the mixed number to an improper fraction.	$= \frac{6}{7} \times \frac{11}{6} \div \frac{1}{4}$
3 Change the division sign to multiplication sign and tip the fraction that follows (multiply and tip). Cancel where appropriate.	$= \frac{\cancel{6}^{1}}{7} \times \frac{11}{\cancel{6}_{1}} \times \frac{4}{1}$
4 Perform the multiplication.	$= \frac{44}{7}$
5 Simplify your answer.	$= 6\frac{2}{7}$

WORKED Example 30

Find: $\frac{5}{2} \div (6\frac{1}{2} + \frac{5}{6})$.

THINK / **WRITE**

1. Write the equation.

$\frac{5}{2} \div (6\frac{1}{2} + \frac{5}{6})$

2. Change the mixed number to an improper fraction.

$= \frac{5}{2} \div \left(\frac{13}{2} + \frac{5}{6}\right)$

3. Perform the operation in the brackets first.
   (a) Write both fractions with the same denominator using equivalent fractions.
   (b) Add the fractions.

$= \frac{5}{2} \div \left(\frac{13 \times 3}{2 \times 3} + \frac{5}{6}\right)$

$= \frac{5}{2} \div \left(\frac{39}{6} + \frac{5}{6}\right)$

$= \frac{5}{2} \div \frac{44}{6}$

4. Change the division sign to a multiplication sign and tip the second fraction. Cancel where appropriate.

$= \frac{5}{\cancel{2}_1} \times \frac{\cancel{6}^3}{44}$

5. Perform the multiplication and simplify where necessary.

$= \frac{15}{44}$

6. Verify your answer using a calculator.

Graphics Calculator tip! Mixed operations with fractions

On a TI graphics calculator, the calculation in worked example 29 can be entered as $(6 \div 7) \times (1 + 5 \div 6) \div (1 \div 4)$, then pressing MATH, selecting 1:Frac and pressing ENTER.

CASIO Mixed operations with fractions

remember

The order of operations is the same for fractions as it is for whole numbers:

1. brackets
2. multiplication and division (left to right)
3. addition and subtraction (left to right).

EXERCISE 3G Mixed operations with fractions

Mathcad: Mixed operations

EXCEL Spreadsheet: Four operations with fractions

GC program—TI: Mixed operations

GC program—Casio: Mixed operations

1 Find the following. Verify each answer using a calculator.

a $\frac{5}{8} + \frac{5}{6}$ **b** $\frac{7}{8} \div \frac{21}{32}$ **c** $\frac{3}{4} \times \frac{9}{10}$

d $\frac{4}{5} - \frac{3}{8}$ **e** $1\frac{3}{5} + 9\frac{4}{7}$ **f** $6\frac{1}{3} \div 4\frac{1}{2}$

g $1\frac{4}{5} \times \frac{9}{7}$ **h** $\frac{5}{2} - 1\frac{7}{8}$ **i** $5\frac{1}{4} - 2\frac{1}{5}$

WORKED Example 29

2 Find the following. Verify each answer using a calculator.

a $\frac{3}{5} \times 2\frac{1}{3} \div \frac{1}{2}$ **b** $1\frac{3}{4} \times 5 \div 3\frac{1}{2}$ **c** $4\frac{9}{10} + 2\frac{1}{4} - \frac{3}{5}$

d $5\frac{1}{3} + 7\frac{3}{8} - 8\frac{1}{2}$ **e** $2\frac{5}{6} \times 2 - 3\frac{1}{2}$ **f** $\frac{1}{2} + \frac{2}{3} \times \frac{1}{5}$

g $\frac{4}{5} \times 10 + 2\frac{3}{4}$ **h** $\frac{1}{3} + \frac{2}{9} \times \frac{3}{4}$ **i** $\frac{11}{12} - \frac{1}{2} \div \frac{3}{4}$

WORKED Example 30

3 Calculate these, remembering to solve the brackets first. Verify each answer using a calculator.

a $\left(\frac{5}{4} + \frac{2}{3}\right) \times \frac{1}{2}$ **b** $\left(5 - 3\frac{6}{7}\right) \times 4\frac{1}{3}$ **c** $\left(1\frac{1}{2} + \frac{3}{4}\right) \times \left(5 - 3\frac{3}{4}\right)$

d $6 - \left(7\frac{5}{6} \div 2\right)$ **e** $\frac{3}{2} \div \left(4\frac{1}{2} + \frac{5}{6}\right)$ **f** $\frac{1}{11} \times \left(8\frac{2}{9} - 5\frac{1}{6}\right)$

g $\left(\frac{2}{3} \times 4\frac{1}{12}\right) - \left(7\frac{1}{2} \times \frac{1}{10}\right)$ **h** $4 \times \left(2 - 1\frac{3}{7}\right)$ **i** $6 \div \left(3\frac{2}{3} + 2\frac{7}{12}\right)$

j $\left(\frac{2}{3} - \frac{1}{6}\right) \div \left(3\frac{1}{8} - 2\frac{5}{6}\right)$

MATHS QUEST CHALLENGE

1 Without using a calculator find three-quarters of four-ninths of 78.

2 Without using a calculator find the answer to $\dfrac{\frac{1}{2} + \frac{1}{3} + \frac{1}{4}}{\frac{1}{4} + \frac{1}{9}}$.

3 The fraction $\frac{1}{4}$ is a third of what number?

4 According to a will, a \$120 000 estate is to be left to three children. Sam is to receive $\frac{2}{5}$ of the money, Georgia is to receive $\frac{1}{3}$ of the money and Olivia is to receive the remainder of the money. How much money is Olivia to receive?

5 When any positive number is multiplied by a number between zero and 1, what is the result?

6 Find the fraction that is equivalent to $\frac{3}{5}$ with the product of its numerator and denominator being equal to 375.

COMMUNICATION

You want more pizza?

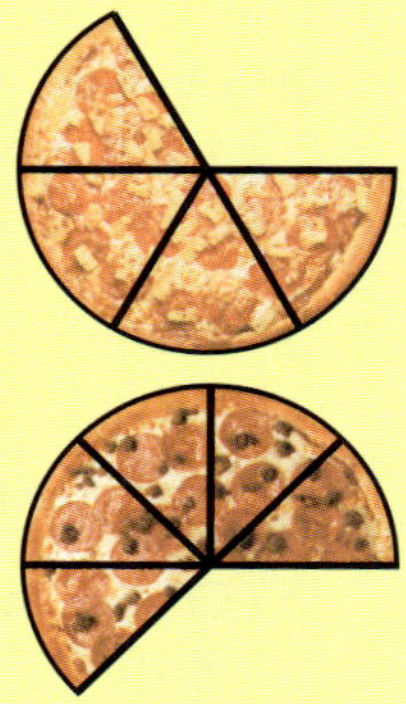

Do you remember the question posed at the start of this chapter? Martin eats 2 slices of a vegetarian pizza which has been cut into 6 equal pieces while Olivia eats 3 pieces of the sausage pizza which has been cut into 8 equal pieces. Who has eaten more?

1 Try this question again, showing your calculations.

2 How many pieces of each pizza are left?

3 What fraction of each pizza is left?

4 What is the total amount of pizza left?

Olivia then takes a slice of vegetarian pizza and Martin takes a slice of sausage pizza. However, Martin complains that his piece is smaller than Olivia's.

5 What is the difference in size between the larger and the smaller slice? (Write your answer as a fraction of a whole pizza.)

Olivia continues to eat the vegetarian pizza and Martin continues to eat the sausage pizza.

6 How many slices of pizza should Olivia and Martin each have so as to eat the same amount?

7 What fraction of each pizza is now left over?

Powers and square roots of fractions

Fractions and mixed numbers can be squared (index or power of 2) or cubed (index or power of 3) by using index notation just as with whole numbers.

For example, $\left(\frac{2}{3}\right)^2 = \frac{2}{3} \times \frac{2}{3} = \frac{2^2}{3^2} = \frac{4}{9}$.

WORKED Example 31

Find: **a** $\left(\frac{1}{4}\right)^2$ **b** $\left(\frac{2}{5}\right)^3$ **c** $(1\frac{1}{3})^2$.

THINK	WRITE
a 1 Write the question.	**a** $\left(\frac{1}{4}\right)^2$
2 Square the numerator and the denominator.	$= \frac{1^2}{4^2}$
	$= \frac{1 \times 1}{4 \times 4}$
3 Simplify.	$= \frac{1}{16}$

Continued over page

THINK	WRITE
b 1 Write the question.	**b** $\left(\frac{2}{5}\right)^3$
2 Cube the numerator and the denominator.	$= \frac{2^3}{5^3}$
	$= \frac{2 \times 2 \times 2}{5 \times 5 \times 5}$
3 Simplify.	$= \frac{8}{125}$
c 1 Write the question.	**c** $(1\frac{1}{3})^2$
2 Change the mixed number to an improper fraction.	$= \left(\frac{4}{3}\right)^2$
3 Square both the numerator and the denominator.	$= \frac{4^2}{3^2}$
	$= \frac{4 \times 4}{3 \times 3}$
	$= \frac{16}{9}$
4 Change back to a mixed number.	$= 1\frac{7}{9}$

It is also possible to find the square roots of fractions and mixed numbers.

WORKED Example 32

Find: **a** $\sqrt{\frac{9}{16}}$ **b** $\sqrt{6\frac{1}{4}}$.

THINK	WRITE
a 1 Write the question.	**a** $\sqrt{\frac{9}{16}}$
2 Write the numerator and denominator as separate square roots.	$= \frac{\sqrt{9}}{\sqrt{16}}$
3 Calculate the square root of both the numerator and the denominator.	$= \frac{3}{4}$
b 1 Write the question.	**b** $\sqrt{6\frac{1}{4}}$
2 Change the mixed number to an improper fraction.	$= \sqrt{\frac{25}{4}}$
3 Write the numerator and the denominator as separate square roots.	$= \frac{\sqrt{25}}{\sqrt{4}}$
4 Calculate the square root of both the numerator and the denominator.	$= \frac{5}{2}$
5 Change back to a mixed number.	$= 2\frac{1}{2}$

Graphics Calculator tip! Powers and square roots

As with whole numbers, we can use the TI graphics calculator to find different powers and square roots of fractions. To enter the index or power we use the ^ key or if we are squaring a number we can also use the x^2 key. Remember to press the appropriate keys to obtain your answer as a fraction.

The top screen at right shows the calculations for worked example 31.

To obtain the square root of a fraction, first enter the square root sign by pressing 2nd [√] (that is, press 2nd then x^2) then follow with the fraction. The screen obtained for worked example 32 is shown at right.

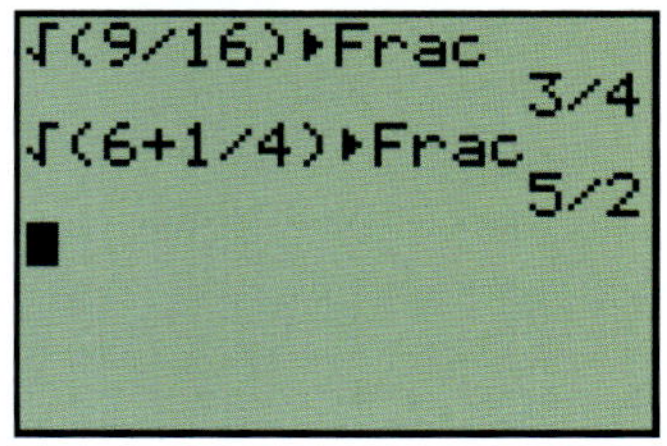

CASIO Powers and square roots

remember

1. To square a fraction, square both numerator and denominator (base is raised to an index or power of 2).
2. To cube a fraction, cube both numerator and denominator (base is raised to an index or power of 3).
3. To find the square root of a fraction, take the square root of both numerator and denominator.
4. To find the square root of a mixed number, change the mixed number to an improper fraction, then find the square root of both the numerator and the denominator.

EXERCISE 3H Powers and square roots of fractions

Mathcad Squaring and cubing fractions

WORKED Example 31a

1 Find:

a $\left(\frac{1}{2}\right)^2$ b $\left(\frac{1}{3}\right)^2$ c $\left(\frac{2}{5}\right)^2$

d $\left(\frac{5}{6}\right)^2$ e $\left(\frac{3}{4}\right)^2$ f $\left(\frac{7}{10}\right)^2$

g $\left(\frac{5}{12}\right)^2$ h $\left(\frac{3}{11}\right)^2$ i $\left(\frac{3}{8}\right)^2$

WORKED Example 31b

2 Find:

a $\left(\frac{1}{4}\right)^3$ b $\left(\frac{1}{2}\right)^3$ c $\left(\frac{2}{5}\right)^3$

d $\left(\frac{3}{10}\right)^3$ e $\left(\frac{2}{3}\right)^3$ f $\left(\frac{1}{6}\right)^3$

WORKED Example 31c

3 Find:

a $\left(1\frac{1}{3}\right)^2$ b $\left(2\frac{1}{4}\right)^2$ c $\left(1\frac{1}{2}\right)^2$

d $\left(1\frac{1}{10}\right)^2$ e $\left(2\frac{1}{5}\right)^2$ f $\left(3\frac{2}{3}\right)^2$

WORKED Example 32

Mathcad: Square root of a fraction

4 Find:

a $\sqrt{\frac{16}{25}}$ b $\sqrt{\frac{49}{100}}$ c $\sqrt{\frac{9}{64}}$

d $\sqrt{\frac{25}{36}}$ e $\sqrt{2\frac{1}{4}}$ f $\sqrt{1\frac{7}{9}}$

g $\sqrt{1\frac{9}{16}}$ h $\sqrt{2\frac{14}{25}}$ i $\sqrt{1\frac{15}{49}}$

WorkSHEET 3.3

5 Calculate:

a $\left(\frac{1}{2}\right)^2 + \left(\frac{3}{4}\right)^2$ b $\left(\frac{2}{3}\right)^2 - \left(\frac{1}{2}\right)^2$ c $\left(\frac{3}{5}\right)^2 \times \left(\frac{5}{9}\right)^2$

d $\left(\frac{4}{7}\right)^2 \div \left(\frac{3}{7}\right)^2$ e $\sqrt{\frac{4}{9}} + \left(\frac{2}{3}\right)^2 - \frac{1}{2}$ f $\sqrt{\frac{16}{25}} \times \left(\frac{1}{2}\right)^2 - \frac{1}{10}$

MATHS QUEST CHALLENGE

1 **Find the missing power for this calculation: $\left(\frac{3}{5}\right)^{\square} = \frac{2187}{78\,125}$.**

2 **A length of rope is cut in half and one half is used. Then one third of the other half is cut off and used. If the remaining piece of rope is 10 m long, how long was the original rope?**

3 **Solve this expression.**

$$\frac{\left(\frac{1}{2}\right)^3}{\left(1 + \frac{1}{1+3}\right)^2}$$

4 **Look for a pattern in this list of fractions and write down the next two in the sequence.**

$\frac{2}{3} \rightarrow \frac{9}{5} \rightarrow \frac{16}{14} \rightarrow \frac{23}{30} \rightarrow \frac{30}{53}$

5 **The sum of two fractions is $1\frac{1}{2}$. Their difference is $\frac{5}{12}$. What are the fractions?**

6 **If a clock takes 5 seconds to strike 6 o'clock, how many seconds will it take to strike 12 o'clock?**

7 **Jeff wrote $\frac{3}{4} + \frac{1}{5} = \frac{4}{9}$. Is he correct? Explain your answer.**

summary

Copy the sentences below. Fill in the gaps by choosing the correct word or expression from the word list that follows.

1 The ______________ is the top part of the fraction.

2 The ______________ is the bottom part of the fraction.

3 Different fractions can be compared or ordered by writing each one with the ______________ denominator.

4 ______________ order is from smallest to largest.

5 ______________ order is from largest to smallest.

6 ______________ are found by multiplying or dividing both the numerator and the denominator by the same number as long as the number is not equal to zero.

7 To ______________ fractions, find the largest number that is a factor of the numerator and the denominator and divide both by that number.

8 ______________ fractions can be changed to mixed numbers by dividing the numerator by the denominator. The ______________ becomes the numerator and the denominator remains the same.

9 Mixed ______________ can be changed to improper fractions by multiplying the whole number and the denominator then adding the numerator. The result becomes the numerator and the denominator remains the same.

10 Fractions must have the same denominator to be added or subtracted. This is done by finding a ______________ denominator. Once the denominators are the same, the numerators can be added or subtracted.

11 To multiply fractions, simply ______________ the numerators and multiply the denominators. It is often easier to ______________ or simplify numerators and denominators before multiplying. Remember to change mixed numbers to improper fractions first.

12 To find the ______________ of a fraction, swap the numerator and denominator.

13 To divide fractions, change the ÷ to × and ______________ the second fraction. Remember to change ______________ numbers to improper fractions first.

WORD LIST

denominator	common	descending
ascending	same	remainder
mixed	equivalent fractions	numbers
tip	cancel	reciprocal
multiply	numerator	
improper	simplify	

CHAPTER review

3A **1** What fraction of the eggs in this carton are brown?

3A **2** Write the sequence of the first three equivalent fractions for the following.

a $\frac{3}{8}$ b $\frac{4}{9}$ c $\frac{5}{11}$ d $\frac{6}{7}$

3A **3** Fill the gaps to make equivalent fractions.

a $\frac{1}{2} = \frac{\;}{4} = \frac{3}{\;}$ b $\frac{1}{3} = \frac{3}{\;} = \frac{\;}{15}$ c $\frac{3}{4} = \frac{9}{\;} = \frac{\;}{28}$

d $\frac{2}{5} = \frac{\;}{10} = \frac{16}{\;}$ e $\frac{1}{4} = \frac{5}{\;} = \frac{\;}{100}$ f $\frac{3}{10} = \frac{\;}{100} = \frac{300}{\;}$

3A **4** Ricky, Leo and Gabriel were the highest goal scorers for their local football team throughout the season. Ricky scored $\frac{2}{7}$ of the team's goals, Leo scored $\frac{1}{3}$ of the goals and Gabriel scored $\frac{1}{4}$ of the goals.

a Who scored most of the goals?
b Who scored more goals; Ricky or Gabriel?

3B **5** Simplify these fractions.

a $\frac{42}{49}$ b $\frac{81}{108}$ c $\frac{18}{20}$ d $\frac{21}{28}$

e $\frac{36}{72}$ f $\frac{50}{150}$ g $\frac{75}{125}$ h $\frac{2}{80}$

3B **6** Simplify the following mixed numbers.

a $6\frac{8}{10}$ b $3\frac{6}{9}$ c $4\frac{15}{20}$ d $1\frac{20}{100}$

3B **7** What fraction of the months of the year begin with the letter J? Express your answer in simplest form.

3B **8** Year 7 students at Camberwell Secondary College raised a total of \$1600 for the Royal Childrens Hospital Good Friday appeal. The table below shows how much money each class raised.

Class	7.1	7.2	7.3	7.4	7.5	7.6
Amount raised (\$)	\$200	\$400	\$250	\$250	\$280	\$220

Express as a simple fraction how much of the total each class raised.

9 Convert these improper fractions to mixed numbers. 3C

a $\frac{16}{3}$ b $\frac{21}{5}$ c $\frac{12}{7}$ d $\frac{11}{2}$

e $\frac{5}{4}$ f $\frac{80}{7}$ g $\frac{55}{9}$ h $\frac{72}{10}$

10 Express these mixed numbers as improper fractions. 3C

a $2\frac{3}{4}$ b $9\frac{7}{8}$ c $3\frac{5}{7}$ d $5\frac{5}{6}$

e $3\frac{2}{3}$ f $9\frac{9}{10}$ g $3\frac{11}{12}$ h $8\frac{1}{2}$

11 Marcella was in charge of baking and selling lemon slices at the school fete. Each tray she baked contained 15 lemon slices. If Marcella sold 68 lemon slices, how many empty trays were there? 3C

12 Find: 3D

a $\frac{5}{6}+\frac{1}{8}$ b $\frac{7}{9}-\frac{2}{5}$ c $\frac{12}{5}-\frac{4}{3}$

d $\frac{5}{4}+\frac{11}{2}$ e $1\frac{7}{8}+2\frac{5}{6}$ f $10\frac{3}{7}-8\frac{1}{2}$

g $6\frac{1}{3}-2\frac{5}{9}$ h $5\frac{1}{2}-2\frac{5}{6}$ i $3\frac{5}{8}+2\frac{7}{10}$

j $8\frac{2}{9}-4\frac{5}{8}$ k $12\frac{4}{7}-10\frac{9}{10}$ l $12\frac{6}{11}-11\frac{7}{9}$

13 Calculate the following and verify each answer using a calculator. 3D

a $1\frac{1}{5}+3\frac{1}{3}-\frac{13}{15}$ b $4\frac{7}{9}-1\frac{5}{6}+3\frac{1}{3}$

14 Tony caught 3 fish with masses: $2\frac{7}{10}$ kg, $2\frac{1}{2}$ kg and $2\frac{4}{5}$ kg. 3D

a Which fish is the heaviest?
b What is the combined mass of the fish?
c What is the difference in mass between the heaviest and the lightest fish?

15 Find: 3E

a $\frac{4}{7}\times\frac{5}{9}$ b $\frac{2}{5}\times\frac{3}{5}$ c $\frac{25}{36}\times\frac{18}{20}$ d $\frac{56}{81}\times\frac{18}{64}$

e $4\frac{2}{5}\times 3\frac{2}{11}$ f $10\frac{2}{7}\times 6\frac{2}{9}$ g $\frac{4}{5}$ of 55 h $\frac{2}{11}$ of 121

i $\frac{7}{9}$ of $\frac{18}{84}$ j $\frac{1}{12}$ of 200

16 Find: 3E

a $\frac{3}{5}\times\frac{5}{6}\times\frac{2}{7}$ b $\frac{3}{4}\times\frac{16}{21}\times\frac{7}{20}$ c $\frac{5}{11}\times\frac{3}{10}\times 1\frac{1}{10}$

d $\frac{4}{9}\times 2\frac{1}{4}\times\frac{3}{5}$ e $3\frac{5}{12}\times 1\frac{3}{5}\times 2\frac{1}{2}$ f $\frac{6}{17}\times 1\frac{1}{4}\times 3\frac{2}{5}$

17 Anthony's monthly take-home pay is $2800. From this he spends a quarter on his home loan payments, one-half on food and drink and one-seventh on clothing. One-half of the remainder goes into his savings account. How much money does Anthony put into his savings account each month? 3E

3F

18 Write the reciprocals of:

a $\frac{5}{2}$ **b** $\frac{2}{7}$ **c** $\frac{1}{9}$ **d** 12

e 124 **f** $\frac{1}{4}$ **g** $5\frac{3}{4}$ **h** $8\frac{1}{2}$

i $7\frac{2}{5}$ **j** $2\frac{4}{9}$ **k** $3\frac{5}{8}$ **l** $4\frac{1}{5}$

3F

19 Find:

a $\frac{4}{5} \div \frac{8}{15}$ **b** $\frac{9}{10} \div \frac{27}{40}$ **c** $\frac{7}{9} \div 4$

d $\frac{3}{7} \div 10$ **e** $1\frac{1}{2} \div 4\frac{4}{5}$ **f** $6\frac{2}{3} \div 4\frac{1}{6}$

g $4\frac{1}{5} \div 2\frac{7}{8}$ **h** $10\frac{1}{3} \div 12\frac{5}{7}$ **i** $6\frac{2}{7} \div 4\frac{1}{5}$

3G

20 George can run one lap of a cross-country course in one-sixth of an hour.

a How many laps could he complete in $1\frac{1}{2}$ hours? (Assume he can keep up the same pace.)

b How many minutes does it take George to run one lap?

3G

21 Calculate the following and verify each answer using a calculator.

a $\left(\frac{4}{9} + \frac{5}{6}\right) \div 3$ **b** $\frac{5}{14} \times \left(\frac{3}{10} + \frac{2}{5}\right)$

c $\left(\frac{2}{3} - \frac{1}{6}\right) \times \left(\frac{3}{4} + \frac{5}{8}\right)$ **d** $\frac{3}{4} \div \frac{9}{10} + \frac{5}{7} \times \frac{14}{15}$

e $2 + 4\frac{1}{2} \div 2\frac{1}{4}$ **f** $7 - 2\frac{1}{4} \div 1\frac{1}{2}$

3H

22 Calculate:

a $\left(\frac{4}{5}\right)^2$ **b** $\left(\frac{9}{10}\right)^2$

c $\left(\frac{3}{4}\right)^2$ **d** $\left(\frac{7}{8}\right)^2$

e $\left(\frac{1}{3}\right)^3$ **f** $\left(\frac{5}{6}\right)^3$

g $\left(1\frac{1}{2}\right)^2$ **h** $\left(2\frac{1}{4}\right)^2$ **i** $\left(1\frac{3}{5}\right)^2$

3H

23 Calculate:

a $\sqrt{\frac{81}{100}}$ **b** $\sqrt{\frac{25}{49}}$ **c** $\sqrt{2\frac{7}{9}}$ **d** $\sqrt{2\frac{9}{36}}$

test yourself CHAPTER 3

Decimals

4

At a school athletics event, five boys threw the javelin. Aaron threw 22.09 metres, Tyler threw 21.98 metres, David threw 22.1 metres, Anton threw 21.19 metres and Jai threw 21.80 metres. Can you arrange these distances in descending order? Who threw the javelin the furthest?

In this chapter you will learn how to use decimals so that you can compare the distances that each javelin was thrown.

Are you ready?

Try the questions below. If you have difficulty with any of them, extra help can be obtained by completing the matching SkillSHEET. Either click on the SkillSHEET icon next to the question on the *Maths Quest 7* CD-ROM or ask your teacher for a copy.

SkillSHEET 4.1

Writing decimals

1 Write each of the following as a decimal.
a 3 tenths   b 5 hundredths   c 7 thousandths

SkillSHEET 4.2

Comparing decimals

2 Find the largest number in each of the following.
a 1.25, 1.35, 1.32   b 0.438, 0.483, 0.484

SkillSHEET 4.4

Rounding to the nearest whole number

3 Round each of the following to the nearest whole number.
a 4.7   b 18.29   c 56.51

SkillSHEET 4.5

Adding decimals (same number of decimal places)

4 Rewrite the following in columns, then add.
a $4.3 + 2.4$ b $0.52 + 0.85$ c $7.43 + 6.59$

SkillSHEET 4.6

Adding decimals (different number of decimal places)

5 Rewrite the following in columns, then add.
a $2.76 + 1.2$ b $5.2 + 0.376$ c $15.37 + 11.4$

SkillSHEET 4.7

Subtracting decimals (same number of decimal places)

6 Rewrite the following in columns, then subtract.
a $9.6 - 4.3$ b $0.87 - 0.55$ c $12.31 - 7.14$

SkillSHEET 4.8

Subtracting decimals (different number of decimal places)

7 Rewrite the following in columns, then subtract.
a $8.25 - 2.7$ b $1.3 - 0.85$ c $26.796 - 19.3$

SkillSHEET 4.9

Multiplying decimals by a single-digit number

8 Calculate the following.
a 2.5×2 b 3.68×3 c 12.71×5

SkillSHEET 4.10

Multiplying decimals by 10, 100, 1000

9 Calculate the following.
a 1.456×10 b 0.351×100 c 16.039×1000

SkillSHEET 4.12

Dividing decimals by a single-digit number

10 Calculate the following.
a $6.9 \div 3$ b $8.32 \div 2$ c $15.735 \div 5$

SkillSHEET 4.13

Dividing decimals by 10, 100, 1000

11 Calculate the following.
a $14.35 \div 10$ b $27.1 \div 100$ c $38.664 \div 1000$

Place value

Whole numbers

The position of a digit within a number indicates the value of the digit. The further the digit is to the left in a number, the larger the place value.

Hundred thousands	Ten thousands	Thousands	Hundreds	Tens	Units
100 000	10 000	1000	100	10	1

Each place to the left of another has a value which is 10 times larger. Each place to the right has a value which is $\frac{1}{10}$ of the previous position.

Look at the value of the 3 in each of the following.

Number	Value of 3 in number
132	3 tens or 30
3217	3 thousands or 3000
4103	3 units (ones) or 3

Decimal parts

Whole numbers have units as their smallest place value. It is possible to keep going to smaller values than units. To show values smaller than units, a decimal point, or dot, is placed after the units. The value of the positions to the left and right of the decimal point are shown in the table below.

Thousands	Hundreds	Tens	Units	.	Tenths	Hundredths	Thousandths	Ten thousandths
1000	100	10	1	.	$\frac{1}{10}$	$\frac{1}{100}$	$\frac{1}{1000}$	$\frac{1}{10\,000}$

Look at the value of the 3 in these numbers:

14.32 the 3 is 3 tenths or $\frac{3}{10}$

106.013 the 3 is 3 thousandths or $\frac{3}{1000}$

0.000 03 the 3 is 3 hundred thousandths or $\frac{3}{100\,000}$.

WORKED Example 1

Find the value of the 7 in each of the following.

a 10.74 b 0.173 c 321.037

THINK	WRITE
a The value of the first place to the right of the decimal point is tenths, so the digit is tenths.	**a** $\frac{7}{10}$
b The second place after the decimal point is hundredths, so the digit is hundredths.	**b** $\frac{7}{100}$
c The digit is in the thousandths place, so the digit is thousandths.	**c** $\frac{7}{1000}$

WORKED Example 2

For the number 76.204 write the value of each digit in words and numbers.

THINK	WRITE
1 7 is in the tens position.	seventy, 70
2 6 is in the units position.	six, 6
3 2 is in the first position after the decimal point, so it is tenths.	two tenths, $\frac{2}{10}$
4 0 is in the hundredths position.	zero hundredths, $\frac{0}{100}$
5 4 is in the thousandths position.	four thousandths, $\frac{4}{1000}$

The number 76.204 can be written in expanded notation as:

$$(7 \times 10) + (6 \times 1) + (2 \times \tfrac{1}{10}) + (4 \times \tfrac{1}{1000}).$$

WORKED Example 3

Write 3.4501 in expanded notation.

THINK	WRITE
1 Write the decimal.	3.4501
2 Find the place value of each digit. 3: 3 units = 3, 4: 4 tenths = $\frac{4}{10}$, 5: 5 hundredths = $\frac{5}{100}$, 0: 0 thousandths, 1: 1 ten thousandth = $\frac{1}{10\,000}$.	= 3 units + 4 tenths + 5 hundredths + 0 thousandths + 1 ten thousandth $= 3 + \frac{4}{10} + \frac{5}{100} + \frac{0}{1000} + \frac{1}{10\,000}$
3 Write the number in expanded notation.	$= (3 \times 1) + (4 \times \frac{1}{10}) + (5 \times \frac{1}{100}) + (1 \times \frac{1}{10\,000})$

WORKED Example 4

Add one tenth to each of the following numbers.

a 3.4 b 0.179 c 13

THINK	WRITE
a Tenths is the first place after the decimal point so add 1 to the first digit after the decimal point.	**a** 3.4 + one tenth = 3.4 + 0.1 = 3.5
b Add 1 to the first digit after the decimal point.	**b** 0.179 + one tenth = 0.279 + 0.1 = 0.279
c There is no tenths place shown (zero tenths), because 13 = 13.0; so add a decimal point and write 1 after the decimal point.	**c** 13 + one tenth = 13.0 + 0.1 = 13.1

Decimal places

The number of decimal places is the number of digits after the decimal point.

10.3014 has 4 decimal places
0.002 has 3 decimal places
51 has no decimal places
51.0 has one decimal place

Zeros at the end of a decimal do not change the size of the number.

remember

1. The position of a digit within a number indicates the value of the digit. Each place is 10 times larger than the one immediately to the right. The positions are shown in the table below.

Thousands	Hundreds	Tens	Units	.	Tenths	Hundredths	Thousandths	Ten thousandths

2. The number of decimal places is the number of digits after the decimal point.
3. Zeros at the end of a decimal do not change the size of the number.

EXERCISE 4A Place value

4.1 SkillSHEET Writing decimals

WORKED Example 1

1 Find the value of the 2 in each of the following.
a 5.2 **b** 19.12 **c** 0.02 **d** 100.29
e 0.982 047 **f** 491.7521 **g** 6.1342 **h** 90.0002
i 27.003 **j** 12.14 **k** 1.8902 **l** 897.014 12

2 Find the value of the 9 in each of the following.
a 0.9 **b** 14.98 **c** 6.1937 **d** 18.89
e 12.090 **f** 0.009 14 **g** 3.4629 **h** 1.276 89
i 39.214 **j** 9 **k** 900.76 **l** 90.612

WORKED Example 2

3 For the following numbers write the value of each digit in *words* and *numbers*.
a 4.1 **b** 1.85 **c** 0.271 **d** 9.020
e 16.001 **f** 3.402 07

4 For each of the following numbers write the value of each digit in *numbers*.
a 0.4 **b** 2.7 **c** 6.80 **d** 5.23
e 0.763 **f** 2.108 **g** 19.910 **h** 0.1101
i 7.2964 **j** 0.330 24 **k** 300.03 **l** 12.276 403

WORKED Example 3

5 Write the following numbers in expanded notation.
a 2.47 **b** 3.69 **c** 1.25 **d** 56.01
e 39.01 **f** 16.07 **g** 7.123 **h** 5.987
i 13.482 **j** 0.3062 **k** 0.1305 **l** 0.5002

WORKED Example 4

6 Add one tenth to each of the following numbers.
a 5.7 **b** 3.6 **c** 0.24 **d** 0.15
e 10.0 **f** 103.568 **g** 28.02 **h** 0.248
i 65.772 **j** 1.729 021 **k** 0.899 99 **l** 1.9

7 Add four hundredths to each of the following numbers.

a 2.11 b 7.24 c 512.35 d 0.107
e 41.10 f 100.002 g 13.251 43 h 0.335 12
i 0.9 j 11.4 k 31.69 l 0.08

8 Add two thousandths to the following numbers.

a 9.413 b 11.305 c 84.2114 d 17.7751
e 0.6102 f 2.000 g 0.004 3678 h 49.27
i 3.21 j 0.2 k 0.229 l 4.168

9 State how many decimal places are in each of the following numbers.

a 3.15 b 10.2 c 100.019 d 4.60
e 8.999 f 14.3021 g 1869.0567 h 7.0
i 9.246 013 90 j 112 k 30 l 1

10 **multiple choice**

a Four tenths, 1 hundredth and 3 thousandths equals:
A 4.13 B 3.14 C 0.413 D 0.314 E 314

b Five hundredths, 2 thousandths and 7 ten thousandths equals:
A 527 B 52.7 C 5.27 D 0.0527 E 0.527

11 For each of the following numbers:

i state the place value of the zero.
ii Would the value of the number change if the zero wasn't there? (Write yes or no.)

a 6.02 b 10.49 c 7.360 d 13.10 e 4.0
f 133.027 g 0.65 h 17.809 i 20 j 108.62

12 Write true (T) or false (F) for each of the following.

a 76.34 has 4 decimal places.
b 109.651 plus two tenths equals 109.851.
c $\frac{6}{10} + \frac{3}{100} + \frac{4}{10\,000}$ is the same as 0.6304.
d 4.03 has the same value as 4.3.
e 29.60 has the same value as 29.6.
f 1.2804 could be written as $1 + \frac{3}{10} + \frac{8}{100} + \frac{4}{1000}$.
g 1090.264 51 has 5 decimal places.
h If one hundredth is added to 0.627, the answer is 0.727.

13 Copy and complete the table by putting only one digit in each box.

		Tens	Units		Tenths	Hundredths	Thousandths
Example	37.684	3	7	.	6	8	4
a	0.205			.			
b	1.06			.			
c	74.108			.			
d	0.108			.			
e	50.080			.			

Comparing decimals

To compare the size of numbers that include decimal digits ('decimal numbers' or just 'decimals'), it is necessary to compare the size of the digits with the same place value. First compare the whole number parts or the digits to the left of the decimal point. If they are the same for each number, move to the first digit after the decimal point, then the second, the third and so on until the digits are different. The larger digit will belong to the larger number.

To work out which is larger, 1.214 or 1.223, look at the digits carefully. In this example the whole number part is 1 for both numbers. There are 2 tenths in each number. The number 1.214 has 1 hundredth, 1.223 has 2 hundredths and so 1.223 is the larger number.

WORKED Example 5

Find the largest number in each of the following.
a 0.126, 0.216, 0.122 b 2.384, 2.388, 2.138 c 0.506, 0.605, 0.612

THINK	WRITE
a 1 As the units digit is 0 in each number, compare the tenths. The number 0.216 has 2 tenths, the others have 1 tenth so 0.216 is the largest number.	**a** 0.216 is larger than 0.126 and 0.122.
2 Answer the question.	The largest number is 0.216.
b 1 As the units digits are the same, compare the tenths and eliminate the smallest number.	**b** 2.384 and 2.388 are both larger than 2.138.
2 The hundredths are the same so compare the thousandths and decide which number is bigger.	2.388 is larger than 2.384 and 2.138.
3 Answer the question.	The largest number is 2.388.
c 1 As the unit digit is 0 compare the tenths and eliminate the smallest number.	**c** 0.605 and 0.612 are larger than 0.506.
2 Compare the hundredths and find the biggest number.	0.612 is larger than 0.605 and 0.506.
3 Answer the question.	The largest number is 0.612.

When comparing two numbers, it is easier to use symbols instead of words, as shown in chapter 3 on fractions. So '0.825 is less than 0.897' can be written as $0.825 < 0.897$.

Similarly, '0.486 is greater than 0.156' can be written as $0.486 > 0.156$.

WORKED Example 6

Insert the appropriate $<$ or $>$ sign between the following pairs of numbers to make true statements.
a 0.312 0.318 b 0.0246 0.0168

THINK	WRITE
a Compare the numbers. Both numbers have the same number of tenths and the same number of hundredths, so compare thousandths and insert the correct sign.	**a** $0.312 < 0.318$
b Compare the numbers. Both numbers have no tenths, so compare hundredths and insert the correct sign.	**b** $0.0246 > 0.0168$

remember

1. To compare the size of decimals, it is necessary to compare the size of the digits with the same place value.
2. The symbol ‘$<$’ means ‘is less than’.
3. The symbol ‘$>$’ means ‘is greater than’.

EXERCISE 4B Comparing decimals

WORKED Example 5

SkillSHEET 4.2 Comparing decimals

1 Find the largest number in each of the following.

a 0.24, 0.32, 0.12
b 0.76, 0.68, 0.91
c 0.57, 0.51, 0.59
d 0.92, 0.99, 0.93
e 0.192, 0.191, 0.901
f 0.660, 0.666, 0.662
g 0.1440, 0.1436, 0.1433
h 0.0392, 0.039 90, 0.0039
i 2.506, 2.305, 2.559
j 10.0023, 10.0231, 10.0233
k 0.110 43, 0.110 49, 0.110 40
l 0.102 36, 0.100 23, 0.101 09

WORKED Example 6

2 Insert the appropriate $<$ or $>$ sign between each of the following pairs of numbers to make true statements.

a	3.2	2.9	b	8.6	8.9
c	1.27	1.28	d	0.64	0.67
e	1.37	1.41	f	0.29	0.39
g	13.103	13.112	h	0.427	0.424
i	0.580	0.508	j	0.0101	0.0120
k	0.048 01	0.4801	l	1.3830	1.3824

3 Write the following in order from smallest to largest (ascending order).

a 0.21, 0.39, 0.17, 0.45, 0.33
b 0.16, 0.19, 0.27, 0.12, 0.22
c 0.314, 0.413, 0.420, 0.391, 0.502
d 0.613, 0.624, 0.690, 0.710, 0.677
e 0.821, 0.803, 0.811, 0.807, 0.902
f 0.1164, 0.1245, 0.1033, 0.1002, 0.1196
g 0.9864, 0.9812, 0.9943, 0.9087, 0.9189
h 0.4004, 0.4139, 0.4826, 0.4100, 0.4076
i 4.6249, 4.5097, 4.802, 4.6031, 4.0292
j 13.0294, 13.0291, 13.0229, 13.0299, 13.0929
k 0.004 65, 0.005 02, 0.003, 0.0056, 0.009
l 0.507, 0.61, 0.595, 0.5079, 0.617

4 Write the following in order from largest to smallest (descending order).

a 0.36, 0.31, 0.39, 0.48, 0.19
b 0.27, 0.38, 0.16, 0.02, 0.35
c 0.91, 0.97, 0.90, 0.95, 0.99
d 0.02, 0.29, 0.07, 0.13, 0.09
e 1.264, 1.279, 1.273, 1.291, 1.288
f 0.442, 0.437, 0.491, 0.406, 0.433
g 0.372, 0.318, 0.390, 0.309, 0.317
h 0.502, 0.556, 0.573, 0.602, 0.591

i 0.8207, 0.8889, 0.8823, 0.8217, 0.8448
j 0.7657, 0.6024, 0.0307, 0.1079, 0.7695
k 1.349 54, 1.486 59, 1.702 96, 1.843 21, 1.486 13
l 12.289 50, 12.208 64, 12.392, 12.002 36, 12.9092

5 **multiple choice**

The largest number in the following list; 0.4261, 0.4265, 0.4273, 0.4199, 0.3999 is:
A 0.4261 B 0.4199 C 0.4265 D 0.3999 E 0.4273

6 **multiple choice**

The smallest number in the following list; 0.4261, 0.4265, 0.4273, 0.4199, 0.3999 is:
A 0.4261 B 0.4199 C 0.4265 D 0.3999 E 0.4273

7 **multiple choice**

The following list; 0.4261, 0.4265, 0.4273, 0.4199, 0.3999 when arranged from smallest to largest is:
A 0.4273, 0.4265, 0.4261, 0.4199, 0.3999
B 0.4273, 0.4261, 0.4265, 0.4199, 0.3999
C 0.3999, 0.4199, 0.4265, 0.4261, 0.4273
D 0.3999, 0.4199, 0.4261, 0.4273, 0.4265
E 0.3999, 0.4199, 0.4261, 0.4265, 0.4273

8 Year 7 girls competing in their school swimming sports recorded the following times in the 50-metre freestyle, backstroke and breaststroke events.

	Time (seconds) recorded by contestants						
Event	**Carolyn**	**Jessica**	**Mara**	**Jenika**	**Robyn**	**Shelley**	**Kyah**
Freestyle	37.23	39.04	40.90	38.91	37.45	37.02	37.89
Backstroke	40.23	43.87	44.08	42.55	41.98	40.29	41.05
Breaststroke	41.63	42.7	41.10	41.21	42.66	41.33	41.49

a Who won the freestyle event? How much did she win it by?
b Who won the backstroke event? How much did she win it by?
c Who won the breaststroke event? How much did she win it by?
d List the first 3 placings of the freestyle event.
e List the first 3 placings of the backstroke event.
f List the first 3 placings of the breaststroke event.
g Did any students obtain a placing in all three events?

COMMUNICATION

The javelin event

At a school athletics event, five boys threw the javelin. Aaron threw 22.09 metres, Tyler threw 21.98 metres, David threw 22.1 metres, Anton threw 21.19 metres and Jai threw 21.80 metres.

1 Arrange these distances in descending order.

2 Who threw the javelin the furthest?

The student who threw the furthest went on to compete in the Combined Schools Athletics Competition. The results sheet from the boys' javelin event is shown below (each student had three throws).

	Distance (metres)		
Student	**Throw 1**	**Throw 2**	**Throw 3**
David	22.25	22.34	22.17
Brenton	22.46	22.78	22.65
Lloyd	22.09	21.98	22.21
Eli	22.12	22.32	22.32
Daniel	22.67	22.56	22.65
Ari	21.89	21.78	22.05
Aaron	22.23	22.30	22.43

3 Write the longest throw for each boy.

4 Who won the competition?

5 Who came second?

6 Who came third?

7 How much further did the winner throw the javelin than the boy who came second?

8 How much further did the winner throw the javelin than the boy who came third?

9 Compare the difference between the longest and shortest distance of the three throws made by each of the top three players.

10 Which of the top three players had the most consistent results?

Converting decimals to fractions

In addition to expressing decimals in expanded notation, decimals can also be written as single fractions. This can be done by following the steps given below.

Step 1 Write every number after the decimal point as the numerator of the fraction.

Step 2 The denominator is the place value of the last digit.

Once decimals have been written as a fraction or mixed number, they can be simplified or cancelled. To simplify a fraction, divide both the numerator and the denominator by the highest common factor.

WORKED Example 7

Write the following decimals as fractions, then simplify where appropriate.

a 0.2 **b** 0.86 **c** 0.6021

THINK	WRITE
a 1 Write the decimal.	**a** 0.2
2 The numerator is 2 and the last decimal place is tenths so the denominator is 10.	$= \frac{2}{10}$
3 Divide the numerator and the denominator by the highest common factor.	$= \frac{1}{5}$
4 Simplify the fraction.	
b 1 Write the decimal.	**b** 0.86
2 The numerator is 86. The last decimal place is hundredths so the denominator is 100.	$= \frac{86}{100}$
3 Repeat steps 3 and 4 of part **a**.	$= \frac{43}{50}$
c 1 Write the decimal.	**c** 0.6021
2 The numerator is 6021. The last place is tens of thousandths so the denominator is 10 000.	$= \frac{6021}{10\,000}$

WORKED Example 8

Write each of the following as a mixed number in its simplest form.

a 3.041 **b** 7.264

THINK	WRITE
a 1 Write the decimal.	**a** 3.041
2 Write the whole number part and change the decimal part to a fraction. The numerator is 41. The last decimal place is thousandths so the denominator is 1000.	$= 3\frac{41}{1000}$

Continued over page

THINK	WRITE
b 1 Write the decimal.	**b** 7.264
2 Write the whole number part and change the decimal part to a fraction. The numerator is 264 and the denominator is 1000.	$= 7\frac{264}{1000}$
3 Divide the numerator and the denominator by the highest common factor.	$= 7\frac{264 \div 8}{1000 \div 8}$
4 Simplify the fraction.	$= 7\frac{33}{125}$

Note: Sometimes the highest common factor may not be obvious when cancelling, so extra steps will be required to obtain the answer in simplest form; for example, in worked example 8, $7\frac{264}{1000} = 7\frac{132}{500} = 7\frac{66}{250} = 7\frac{33}{125}$.

remember

To write decimals as single fractions:
1. write the whole number part first
2. write all decimal places as the numerator of the fraction
3. write the denominator as the place value of the last digit
4. simplify the fraction if required.

EXERCISE 4C Converting decimals to fractions

SkillSHEET 4.3 Simplifying fractions (denominators powers of 10)

EXCEL Spreadsheet Converting decimals to fractions

Mathcad Converting decimals to fractions

WORKED Example 7

1 Write the following decimals as fractions, then simplify where appropriate.

a 0.3	**b** 0.5	**c** 0.9	**d** 0.21
e 0.4	**f** 0.8	**g** 0.24	**h** 0.44
i 0.49	**j** 0.63	**k** 0.502	**l** 0.617
m 0.12	**n** 0.30	**o** 0.64	**p** 0.28
q 0.882	**r** 0.9456	**s** 0.9209	**t** 0.4621
u 0.75	**v** 0.120	**w** 0.286	**x** 0.468

WORKED Example 8

2 Write the following decimals as mixed numbers in simplest form.

a 1.3	**b** 1.6	**c** 2.7	**d** 9.4
e 1.2	**f** 2.8	**g** 4.2	**h** 8.5
i 2.13	**j** 6.48	**k** 5.27	**l** 19.182
m 12.42	**n** 3.15	**o** 6.25	**p** 9.140
q 12.843	**r** 16.682	**s** 2.4917	**t** 4.3386
u 37.205	**v** 18.645	**w** 24.345	**x** 100.0048

3 **multiple choice**

a 0.13 as a fraction is:

A $\frac{13}{10}$ B $\frac{13}{100}$ C $\frac{13}{1000}$ D $\frac{1.3}{100}$ E $1\frac{3}{10}$

b 0.207 as a fraction is:

A $\frac{207}{1000}$ B $\frac{207}{100}$ C $2\frac{7}{10}$ D $20\frac{7}{10}$ E 207

c 0.52 as a fraction in simplest form is:

A $\frac{52}{100}$ B $\frac{26}{50}$ C $\frac{13}{25}$ D $\frac{26}{100}$ E $\frac{13}{50}$

d 0.716 as a fraction in simplest form is:

A $\frac{716}{10\,000}$ B $\frac{368}{500}$ C $\frac{716}{1000}$ D $\frac{179}{250}$ E $\frac{358}{1000}$

e 5.325 as a fraction in simplest form is:

A $\frac{5325}{1000}$ B $\frac{325}{1000}$ C $5\frac{325}{1000}$ D $5\frac{65}{200}$ E $5\frac{13}{40}$

Rounding

Often when a calculator is used to solve a problem, many decimal places are given in the answer. When this happens it is necessary to round the number. For example, when calculating interest on an investment, an answer of $152.3479 would be meaningless because the first 2 decimal places indicate the number of cents. The answer would be rounded to 2 decimal places and the interest would be $152.35. An answer in centimetres could be rounded to 1 decimal place because a tenth of a centimetre is 1 millimetre and it would be difficult to measure more accurately than to the nearest millimetre.

Rounding 3.243 1096 to 2 decimal places would give 3.24.

Rounding 4.328 1334 to 2 decimal places gives 4.33.

Rounding decimals

When **rounding** decimals look at the first digit past the number of decimal places required.

- If this number is *less than* 5, write the number with the required number of decimal places.
- If this number is *5 or more*, add 1 to the last decimal place being kept.

For example if 2.6137 is rounded to 2 decimal places, we would think of the number with the required decimal places (2.61). Then we would look at the next digit, which in this case is 3. As this is less than 5, we leave the number as it is. The answer is approximately 2.61. This can be written as $2.6137 \approx 2.61$.

If we were rounding 9.8367 to 2 decimal places, we would think of the number with the required decimal places (9.83). Then we would look at the next digit, which in this case is 6. Whenever the next digit is a 5, 6, 7, 8 or 9, we increase the last digit that we are keeping by 1, so the 3 becomes 4. The answer is $9.8367 \approx 9.84$ when rounded to 2 decimal places.

WORKED Example 9

Round the following to 2 decimal places. a 3.641 883 b 18.965 402 0

THINK	WRITE
a 1 Write the number and underline the required decimal place.	**a** 3.641 883
2 Circle the next digit and round according to the rule. *Note:* Since the circled digit is less than 5, we leave the number as it is.	= 3.64(1) 883 ≈ 3.64
b 1 Write the number and underline the required decimal place.	**b** 18.965 402 0
2 Circle the next digit and round according to the rule. *Note:* Since the circled digit is greater than or equal to 5, add 1 to the last decimal place that is being kept.	= 18.96(5)402 0 ≈ 18.97

If you add 1 to the last decimal place and the number in this position is a 9, the result is 10. The 0 is put in the last required place and the 1 is added to the digit in the next place to the left. So 0.298 rounded to 2 decimal places is 0.30.

WORKED Example 10

Round the following to the number of decimal places shown in the brackets.
a 27.462 973 (4) b 0.009 94 (3)

THINK	WRITE
a 1 Write the number and underline the required decimal place.	**a** 27.462 973
2 Circle the next digit and round according to the rule. *Note:* Since the circled digit is greater than 5, add 1 to the last decimal place that is being kept. As 1 is being added to 9, write 0 in the last place and add 1 to the previous digit.	= 27.462 9(7)3 ≈ 27.4630
b Repeat steps 1 and 2 of part **a**.	**b** 0.009 94 = 0.009(9)4 ≈ 0.010

WORKED Example 11

Round 8.672 to the nearest unit.

THINK	WRITE
1 Write the decimal and think of the number with the whole number part only.	8.672
2 Look at the first digit after the decimal point and, if it is greater than or equal to 5, add 1 to the whole number.	≈ 9

It can be seen that 8.672 is closer to 9 than it is to 8, so this is a way of finding a suitable approximation.

WORKED Example 12

Melinda had \$51.67 in her bank account. She wanted to withdraw all her money so the bank rounded the amount to the nearest 5 cents. How much money did the teller give to Melinda?

THINK	WRITE
1 Write the actual amount she had in her account.	\$51.67
2 Determine whether the last digit is closer to 5 or closer to 10, then rewrite the approximate value. *Note:* Alternatively it can be seen that 67 cents is closer to 65 cents than 70 cents.	≈ \$51.65
3 Write a sentence.	Melinda will receive \$51.65 from the bank.

remember

1. To round a decimal.
   Look at the first digit past the number of decimal places required.
   (a) If this number is *less than 5*, write the number with the number of decimal places required.
   (b) If this number is *5 or more*, add 1 to the last decimal place being kept.
   If you add 1 to the last decimal place and the number in this position is a 9, the result is 10. The 0 is put in the last required place and the 1 is added to the digit in the next place to the left.
2. Rounding an answer gives a useful approximation of the original decimal. Use the symbol ≈ instead of = to show that the answer is an approximation.

EXERCISE 4D Rounding

WORKED Example 9

EXCEL Spreadsheet: Rounding

1 Round the following to 2 decimal places.

a 0.3241	**b** 0.863	**c** 1.246 10	**d** 13.049 92
e 7.128 63	**f** 100.813 82	**g** 71.260 39	**h** 0.0092
i 0.185 00	**j** 19.6979	**k** 0.3957	**l** 0.999

EXCEL Spreadsheet: Rounding (DIY)

2 Round the following to 1 decimal place.

a 0.410	**b** 0.87	**c** 9.27	**d** 25.25
e 300.06	**f** 12.82	**g** 99.91	**h** 8.88
i 17.610 27	**j** 0.8989	**k** 93.994	**l** 0.959 027

WORKED Example 10

3 Round the following to the number of decimal places shown in the brackets.

a 2.386 214 (2)	**b** 14.034 59 (1)	**c** 0.027 135 (2)
d 0.876 4903 (4)	**e** 64.295 18 (4)	**f** 0.382 04 (3)
g 96.28 049 (1)	**h** 3.0409 (2)	**i** 8.902 (2)
j 47.879 69 (3)	**k** 0.099 498 632 (2)	**l** 0.486 2590 (2)

4 **multiple choice**

a 13.179 rounded to 2 decimal places is equal to:
A 13.17 **B** 13.20 **C** 13.18 **D** 13.27 **E** 13.19

b 0.2465 rounded to 1 decimal place is equal to:
A 0.3 **B** 0.25 **C** 1.2 **D** 0.2 **E** 0.5

c 1.7688 rounded to 3 decimal places is equal to:
A 1.768 **B** 1.770 **C** 1.778 **D** 1.769 **E** 1.800

d 2.998 rounded to 1 decimal place is equal to:
A 3.0 **B** 2.9 **C** 2.8 **D** 3.1 **E** 3.9

SkillSHEET 4.4 Rounding to the nearest whole number

WORKED Example 11

5 Round the following to the nearest unit.
a 10.7 **b** 8.2 **c** 3.6 **d** 92.7
e 112.1 **f** 21.76 **g** 42.0379 **h** 2137.50
i 0.12 **j** 0.513 **k** 0.99 **l** 40.987

6 Round the following to the nearest ten.
a 13 **b** 76 **c** 47 **d** 138
e 262 **f** 175 **g** 306.2 **h** 1484
i 10 024 **j** 209 718.5 **k** 18.6 **l** 5.92

7 Round the following to the nearest hundred.
a 320 **b** 190 **c** 894 **d** 138
e 125 **f** 6751 **g** 875.2 **h** 9750.051
i 1724 **j** 1 462 836.5 **k** 71 **l** 47

8 Round the following to the nearest thousand.
a 3426 **b** 5890 **c** 12 300 **d** 18 640
e 28 000 **f** 9462 **g** 1098 **h** 496 830
i 12 780 **j** 862 **k** 129 980 **l** 49 899

WORKED Example 12

9 In the supermarket Christine's shopping bill came to \$27.68. As there are no 1- or 2-cent pieces, this amount must be rounded to the nearest 5 cents. How much will Christine pay for her shopping?

10 The maximum temperature was recorded as 24.7°C. In the news broadcast, the presenter quoted this to the nearest degree. What temperature was quoted?

WorkSHEET 4.1

11 Using a calculator, Greg worked out that the piece of timber required to finish making a support for a gate should be 3.567 82 metres. Realistically, the timber can be measured only to the nearest millimetre (nearest thousandth of a metre). What measurement should be used for the length of the timber? Explain why 3.567 82 m is unreasonable as a measurement for timber.

10 QUICK QUESTIONS 1

1 What is the value of 5 in 12.654?

2 For the number 1.9632, write the value of each digit in words.

3 Add two hundredths to 4.308.

4 Use < or > between 4.098 and 4.908 to make a true statement.

5 Write the following in order from smallest to largest.
7.03, 4.15, 24.5, 6.08, 0.0709, 0.8, 324.8, 1.61

6 Write 0.67 as a fraction.

7 Write 0.4 as a fraction in simplest form.

8 Write 7.082 as a mixed numeral.

9 Round 13.628 to 2 decimal places.

10 Round 13.628 to 1 decimal place.

Adding and subtracting decimals

Adding decimals

When adding numbers with decimals:

1. Line up the decimal points so that digits of the same place value are underneath each other. It may be necessary to add zeros to fill in the blank spaces.
2. Add numbers with decimals as though adding whole numbers, working from right to left.
3. Make sure that the decimal point in the answer is directly underneath the decimal points in the question.

WORKED Example 13

Calculate:

a $\begin{array}{r} 1.3 \\ +\,0.5 \\ \hline \\ \hline \end{array}$ **b** $\begin{array}{r} 12.84 \\ +\,2.33 \\ \hline \\ \hline \end{array}$ **c** $\begin{array}{r} 1.25 \\ 3.146 \\ +\,7.0 \\ \hline \\ \hline \end{array}$

THINK

a Copy the question exactly and add the digits as for whole numbers, working from right to left. Write the decimal point directly below the decimal points in the question.

WRITE

a $\begin{array}{r} 1.3 \\ +\,0.5 \\ \hline 1.8 \\ \hline \end{array}$

Continued over page

THINK

b Copy the question exactly and add the digits as for whole numbers, working from right to left. Write the decimal point directly below the decimal points in the question.

WRITE

b
$$\begin{array}{r} 12.84 \\ +\ 2.33 \\ \hline 15.17 \\ \hline \end{array}$$

c 1 Write the question, replacing the spaces with zeros.

c
$$\begin{array}{r} 1.250 \\ 3.146 \\ +\ 7.000 \\ \hline \end{array}$$

2 Add the digits as for whole numbers, working from right to left. Write the decimal point directly below the decimal points in the question.

$$\begin{array}{r} 11.396 \\ \hline \end{array}$$

It is useful to check answers by rounding them to the nearest first digit which is not zero and adding them in your head. Using this approach in worked example 13c gives 1 + 3 + 7 = 11 which is close to 11.396.

If the question is not written in columns, it is necessary to rewrite it with the decimal points one under the other.

WORKED Example 14

Rewrite in columns, then add 0.26 + 1.8 + 12.214.

THINK

1 Write the question in columns with the decimal points directly beneath each other with the zeros included.

2 Add the digits as for whole numbers. Write the decimal point directly below the decimal points in the question.

3 Check the answer by rounding and then adding 0.3 + 2 + 12 = 14.3 which is close to 14.274.

WRITE

$$\begin{array}{r} 0.260 \\ 1.800 \\ +\ 12.214 \\ \hline 14.274 \\ \hline \end{array}$$

You may like to check your answers by using one of the technology files on the *Maths Quest* 7 CD-ROM (see the icons on page 154).

Subtracting decimals

When subtracting numbers containing decimals:

1. Line up the decimal points so that the digits of the same place value are underneath each other. It may be necessary to add zeros to fill in the blank spaces.
2. Work from right to left, subtracting digits as though subtracting whole numbers.
3. Make sure that the decimal point in the answer is directly underneath the decimal points in the question.

WORKED Example 15

Calculate: $\begin{array}{r} 0.56 \\ -\,0.14 \\ \hline \end{array}$

THINK

1. Copy the question exactly and subtract the digits as for whole numbers, working from right to left. Write the decimal point directly below the decimal points in the question.
2. Check the answer by rounding; 0.6 – 0.1 = 0.5 which is close to 0.42. (This step can be done in your head.)

WRITE

$$\begin{array}{r} 0.56 \\ -\,0.14 \\ \hline 0.42 \\ \hline \end{array}$$

Make sure when subtracting decimals that you fill in the blank decimal places with zeros.

WORKED Example 16

Rewrite the following in columns, then subtract.

a **1.82 – 0.57** **b** **2.641 – 0.85**

THINK

a
1. Write in columns with the decimal points directly under each other. Subtract, and insert the decimal point directly below the other decimal points in the question.
2. Check the answer by rounding; 2 – 1 = 1 which is close to 1.25.

b
1. Write in columns with the decimal points directly under each other, adding zeros as appropriate. Subtract as for whole numbers and insert the decimal point directly below the other decimal points.
2. Check the answer by rounding; 3 – 1 = 2 which is close to 1.791.

WRITE

a
$$\begin{array}{r} 1.\overset{7}{\not 8}{}^{1}2 \\ -\,0.5\;7 \\ \hline 1.2\;5 \\ \hline \end{array}$$

b
$$\begin{array}{r} 2.\overset{5}{\not 6}{}^{1}41 \\ -\,0.8\;50 \\ \hline 1.7\;91 \\ \hline \end{array}$$

remember

1. To add numbers containing decimals, write the numbers in columns, making sure that the decimal points are directly underneath each other. If there are blank places, replace them with zeros.
2. Add numbers with decimal points as you would for whole numbers and insert the decimal point directly below the decimal points in the question.
3. To subtract numbers containing decimals, write the numbers in columns, making sure that the decimal points are directly underneath each other. If there are blank places, replace them with zeros.
4. Subtract as you would for whole numbers and insert the decimal point directly below the decimal points in the question.

EXERCISE 4E Adding and subtracting decimals

SkillSHEET 4.5 Adding decimals (same number of decimal places)

WORKED Example 13a, b

1 Calculate the following.

a $\begin{array}{r} 1.2 \\ +\ 2.3 \\ \hline \\ \hline \end{array}$ **b** $\begin{array}{r} 1.67 \\ +\ 1.02 \\ \hline \\ \hline \end{array}$ **c** $\begin{array}{r} 8.062 \\ +\ 5.177 \\ \hline \\ \hline \end{array}$ **d** $\begin{array}{r} 10.0364 \\ +\ 92.1494 \\ \hline \\ \hline \end{array}$

SkillSHEET 4.6 Adding decimals (different number of decimal places)

WORKED Example 13c

2 Calculate the following, after filling the blank spaces.

a $\begin{array}{r} 6.27 \\ +\ 0.5 \\ \hline \\ \hline \end{array}$ **b** $\begin{array}{r} 3.26 \\ +\ 18.6460 \\ \hline \\ \hline \end{array}$ **c** $\begin{array}{r} 4.2 \\ 62.013 \\ +\ 1946.12 \\ \hline \\ \hline \end{array}$ **d** $\begin{array}{r} 48.129\,06 \\ 9 \\ +\ 204.32 \\ \hline \\ \hline \end{array}$

EXCEL Spreadsheet Adding decimals

WORKED Example 14

3 Rewrite the following in columns, then add.

a 1.4 + 3.2 **b** 6.5 + 0.4 **c** 0.22 + 1.37
d 3.261 + 0.21 **e** 15.987 + 1.293 **f** 8.027 + 0.9415
g 10.8271 + 6.5 **h** 1.8 + 18.6329 **i** 26.29 + 1030.4963

Mathcad Adding decimals

4 Rewrite the following sums, then add.

a 0.24 + 3.16 + 8.29 **b** 14.23 + 1.06 + 86.29 + 3.64
c 40.271 + 0.36 + 1.4 **d** 5.27 + 1.381 + 12.3
e 100 + 4.3 + 0.298 + 1.36 **f** 82.3 + 100.6 + 0.9949 + 9
g 3.026 + 5.9938 + 8.7718 + 3.2 **h** 126 + 372.8 + 100.0264 + 2020.13

EXCEL Spreadsheet Adding decimals (DIY)

5 **multiple choice**

a 1.6 + 4.8 equals:
A 5.4 **B** 6.4 **C** 0.54 **D** 0.64 **E** 64

b 3.26 + 0.458 equals:
A 3.718 **B** 0.784 **C** 0.037 18 **D** 3.484 **E** 7.84

c 1.84 + 0.61 + 4.07 equals:
A 6.52 **B** 6.42 **C** 5.42 **D** 5.52 **E** 0.652

d 216 + 1.38 + 0.002 64 equals:
A 217.4064 **B** 0.618 **C** 217.644 **D** 217.382 64 **E** 21.7644

6 Josh deposited \$27.60 into his bank account. If his balance before the deposit was \$139.40, what is Josh's new bank balance?

7 Jessica bought the following items at the school canteen: 1 can of Coke for \$1.60, 1 sausage roll for \$1.20, 1 packet of chips for \$1.50 and 2 Redskins for \$0.40 (Redskins cost 20 cents each). How much did Jessica spend?

8 A triathlon consists of a 0.5-kilometre swim, a 15.35-kilometre ride and a 4.2-kilometre run. How far do the competitors have to travel altogether?

9 In one day Amy walked 3.6 kilometres to school, 0.8 kilometres from school to the shops, 1.2 kilometres from the shops to a friend's house and finally 2.5 kilometres from her friend's house to her home. How far did Amy walk?

10 For lunch Paula ordered 1 potato cake, 1 dim sim, the minimum of chips and a milkshake from the menu shown below. How much did Paula spend on her lunch?

MENU			
Flake	$3.50	Coffee	$2.20
Whiting	$3.50	Tea	$2.20
Dim sims	$0.60	Soft drinks	$1.80
Potato cakes	$0.50	Milkshakes	$3.00
Minimum chips	$2.50	Water	$1.80

WORKED Example 15

11 Calculate the following, filling the spaces with zeros as required.

a 6.87 − 6.27

b 12.231 − 8.026

c 0.6301 − 0.5495

d 3.0091 − 1.6723

e 31.02 − 26

f 98.26 − 9.07

g 146 − 58.91

h 3.2 − 0.467

4.7 SkillSHEET Subtracting decimals (same number of decimal places)

WORKED Example 16

12 Rewrite the following in columns, then subtract.

a 5.64 − 2.3
b 12.07 − 6.14
c 13.869 − 10.02
d 0.687 − 0.36
e 15.226 − 11.08
f 42.83 − 15
g 6.734 − 4.8
h 12.2 − 8.911
i 13.9009 − 12.65
j 100.562 − 86.0294
k 38 − 21.234
l 47 − 8.762
m 5 − 0.8864
n 0.2 − 0.0049
o 3.279 − 2.506 84

4.8 SkillSHEET Subtracting decimals (different number of decimal places)

Mathcad Subtracting decimals

EXCEL Spreadsheet Subtracting decimals

13 **multiple choice**

a 0.39 − 0.15 equals:

A 0.0024 B 0.024 C 0.24 D 2.4 E 24

b 1.4 − 0.147 would be rewritten as:

A 1.4 − 0.417

B − 1.400 − 0.147

C 1.40 − 1.47

D 1.004 − 0.147

E 1.040 − 0.147

c 0.3 − 0.024 equals:

A 0.06 B 0.276 C 0.7 D 0.76 E 76

d 150.278 − 0.99 equals:

A 150.728 B 149.288 C 1.492 88 D 159.388 E 1.593 88

14 Ryan works in a newsagency. A customer buys $9.65 worth of goods and gives Ryan a $20 note. How much change should Ryan give the customer?

15 A jockey has a mass of 52.3 kilograms. After exercising and training for 2 days and spending time in a sauna, the jockey has lost 1.82 kilograms. What is the jockey's mass now?

16 If 1.27 metres is cut from a piece of material that is 13 metres long, how much material is left?

17 Cathy Freeman won a particular 400 metres race in 51.35 seconds. In her next race, her time was 2.97 seconds faster than this. What was Cathy's time for this race?

18 Gary and Liz are replacing the skirting boards in their lounge room. They know the perimeter of the room is 34.28 metres. If there is a door 0.82 metres wide and a fireplace 2.18 metres wide that do not require skirting boards, how much wood will they need to buy for their lounge room?

19 The following table shows the times recorded for each swimmer in the under-13, 50-metre freestyle relay for 6 teams.

	Times for each swimmer (seconds)			
Team	**Swimmer 1**	**Swimmer 2**	**Swimmer 3**	**Swimmer 4**
1	36.7	41.3	39.2	35.8
2	38.1	46.5	38.8	35.9
3	34.6	39.2	39.9	35.2
4	41.6	40.8	43.7	40.5
5	37.9	40.2	38.6	39.2
6	38.3	39.1	40.8	37.6

GAME time Decimals — 001

a Find the total time for each team. Put your results in a table.
b Which team won the relay?
c What was the difference in time between the first and second placed teams?

THINKING

Palindromic decimals

Palindromic numbers read the same forwards and backwards. For example 121, 23 532 and 6776 are all palindromic numbers.

Palindromic decimal numbers read the same from left to right and right to left. For example, 75.57 is a palindromic decimal number whereas 6.96 and 722.7 are not. If a decimal number is not palindromic we can follow a procedure to obtain a palindromic decimal number.

Start with the decimal number 6.8, then add its reverse (8.6) to obtain a new decimal number.

$$\begin{array}{r} 6.8 \\ +\ 8.6 \\ \hline 15.4 \\ \hline \end{array}$$

The new decimal number is not palindromic, so continue the process until a palindromic decimal number is obtained.

Next add 15.4 and its reverse.

$$\begin{array}{r} 15.4 \\ +\ 4.51 \\ \hline 19.91 \\ \hline \end{array}$$

The number 19.91 is a palindromic decimal number. This process took 2 steps.

1 How many steps are required until a palindromic decimal number results after starting with 5.9?

2 How many steps are required until a palindromic decimal number results after starting with 9.7?

3 How many steps are required until a palindromic decimal number results after starting with 8.9?

4 Which decimal numbers of the form 1.1, 2.2, 3.3, 4.4, 5.5, 6.6, 7.7, 8.8, 9.9 require the most number of steps to form another palindromic decimal number?

Multiplying decimals by a whole number

When multiplying a decimal by a whole number:

Step 1 set the question out as for multiplying whole numbers

Step 2 ignore the decimal points and multiply the numbers

Step 3 count the total number of decimal places in all parts of the question. The answer has the same number of decimal places as the question.

WORKED Example 17

Calculate: a 12.6 × 7 b 4.361 × 32.

THINK	WRITE
a 1 Rewrite and multiply digits as for whole numbers, ignoring the decimal point. Count the number of decimal places altogether (1) and put in the decimal point.	**a** $\begin{array}{r} 12.6 \\ \times\ 7 \\ \hline 88.2 \\ \hline \end{array}$
2 Check the answer by rounding; 10 × 7 = 70 which is close to 88.2.	
b 1 Rewrite and multiply digits as for whole numbers, ignoring the decimal point. Count the number of decimal places altogether (3) and put in the decimal point.	**b** $\begin{array}{r} 4.361 \\ \times\ 32 \\ \hline 8\,722 \\ 130\,830 \\ \hline 139.552 \\ \hline \end{array}$
2 Check the answer by rounding; 4 × 30 = 120 which is close to 139.552.	

WORKED Example 18

Calculate: a 5.1 × 600 b 0.0364 × 24 000.

THINK	WRITE
a 1 Multiplying by 600 is the same as first multiplying by 6 then multiplying by 100. Calculate 5.1 × 6.	**a** $\begin{array}{r} 5.1 \\ \times\ 6 \\ \hline 30.6 \\ \hline \end{array}$
2 Multiply the result by 100. Move the position of the decimal point 2 places to the right.	30.6 × 100 = 3060
3 Write the final answer.	5.6 × 600 = 3060
b 1 Multiplying by 24 000 is the same as first multiplying by 24 then multiplying by 1000. Calculate 0.0364 × 24.	**b** $\begin{array}{r} 0.0364 \\ \times\ 24 \\ \hline 1456 \\ 7280 \\ \hline 0.8736 \\ \hline \end{array}$
2 Multiply the result by 1000. Move the position of the decimal point 3 places to the right.	0.8736 × 1000 = 873.6
3 Write the final answer.	0.0364 × 24 000 = 873.6

remember

1. When multiplying a decimal by a whole number, set the question out as for multiplying whole numbers. Ignore the decimal points and multiply the numbers then count the total number of decimal places in all parts of the question. The answer has the same number of decimal places as the question.
2. When multiplying by a multiple of 10, move the position of the decimal point the same number of places to the right as there are zeros. When multiplying by a multiple of 10, the answer is always larger.

EXERCISE 4F Multiplying decimals by a whole number

SkillSHEET 4.9 Multiplying decimals by a single-digit number

WORKED Example 17a

1 Calculate the following.

a 3.5 × 4	b 15.7 × 8	c 16.3 × 9
d 10.2 × 6	e 22.34 × 5	f 47.63 × 9
g 27.18 × 7	h 64.87 × 8	i 3.724 × 7
j 1.064 × 6	k 0.264 81 × 3	l 14.192 683 × 8

WORKED Example 17b

2 Calculate the following. (You may use a calculator for parts g to o.)

a 14.9 × 21	b 21.8 × 15	c 0.62 × 17
d 2.38 × 23	e 4.93 × 31	f 8.62 × 27
g 4.621 × 58	h 3.921 × 82	i 0.6438 × 47

j 86.324×39 k 94.9286×12 l 87.652×18
m 347.016×21 n 824.095×84 o 9632.1048×96

Mathcad: Multiplying decimals by whole numbers

3 Calculate the following.
a 14.23×10 b 8.652×10 c 3.149×10
d 0.99×10 e 27.4815×10 f 0.0023×10

4 Copy and complete the following sentence.
Multiplying a decimal by 10 is the same as moving the position of the decimal point ______________ place to the ______________.

SkillSHEET 4.10: Multiplying decimals by 10, 100, 1000

5 Calculate.
a 8.278×100 b 4.962×100 c 0.2784×100
d 13.026×100 e 5.2943×100 f 296.845×100

6 Copy and complete the following sentence.
Multiplying a decimal by 100 is the same as moving the position of the decimal point ______________ places to the ______________.

7 Calculate the following.
a 0.2498×1000 b $0.602\,49 \times 1000$ c 1.4861×1000 d $83.020\,91 \times 1000$

8 Copy and complete the following sentence.
Multiplying a decimal by 1000 is the same as moving the position of the decimal point ______________ places to the ______________.

9 State the number of places that the position of the decimal point would be moved to the right when multiplying by:
a 10 000 b 100 000 c 1 000 000.

10 Calculate the following by moving the position of the decimal point.
a 13.27×10 b 0.264×10 c 8.049×10
d 0.486×10 e 7.457×10 f 127.729×10
g 82.3×10 h 197×10 i 3.0×10

11 Calculate the following by moving the position of the decimal point.
a 6.48×10 b 13.896×100
c 589.0643×100 d 0.2708×1000
e $217.148\,96 \times 1000$ f $8.426\,19 \times 100\,000$
g $0.820\,496\,783 \times 100\,000$ h $32.689\,043\,267 \times 100\,000$
i $0.984\,326\,641 \times 1\,000\,000$ j $59.027\,683\,017 \times 1\,000\,000$
k $0.000\,278\,498\,32 \times 1\,000\,000$ l 0.46×1000
m $529 \times 10\,000$ n $39.486 \times 1\,000\,000$

12 Change the following amounts of money to cents. (*Hint:* There are 100 cents in one dollar.)
a \$35 b \$127 c \$11 d \$25.35 e \$58.20 f \$110.15

WORKED Example 18

13 Calculate the following.
a 3.64×300 b 7.58×600 c 26.9×500 d 42.6×900
e 0.127×8000 f $0.543 \times 11\,000$ g $4.6 \times 32\,000$ h $8.1 \times 54\,000$

14 One thousand Year 7 students contributed 75 cents each to the bushfire appeal. How many dollars did they contribute altogether?

15 Benjamin and Robyn were providing ice-cream for 600 children. How much ice-cream would be needed if each child was expected to eat 0.18 litres?

MATHS QUEST CHALLENGE

1 **If a bottle and its cork cost $1.10 and the bottle costs $1.00 more than the cork, how much does the cork cost? (*Hint:* The answer is not 10 cents.)**

2 **What decimal gives the same result when multiplied by 5 as it does when 5 is added to it?**

3 **Identify the pattern in this sequence of numbers:**
0.1, 0.2, 0.3, 0.6, 1.1, 2.0.
What are the next three numbers in the sequence?

Multiplying, squaring and obtaining square roots of decimals

To multiply numbers containing decimals, ignore the decimal point and follow the same steps as if multiplying by whole numbers. Count the number of decimal places in all parts of the question to find the number of decimal places in the answer.

WORKED Example 19

Calculate the following.

a $\begin{array}{r} 3.26 \\ \times\ 0.4 \\ \hline \\ \hline \end{array}$

b $\begin{array}{r} 0.4629 \\ \times\ 2.6 \\ \hline \\ \hline \end{array}$

THINK

a 1 Multiply, ignoring the decimal places.

2 Count the number of digits after the point in both the decimals being multiplied and insert the decimal point in the answer. There are 2 decimal places in 3.26 and 1 in 0.4 so there will be 3 decimal places in the answer.

3 Check the answer by rounding; $3 \times 0.4 = 1.2$ which is close to 1.304.

b 1 Multiply, ignoring the decimal places.

WRITE

a $\begin{array}{r} 326 \\ \times 4 \\ \hline 1304 \\ \hline \end{array}$

$3.26 \times 0.4 = 1.304$

b $\begin{array}{r} 4629 \\ \times 26 \\ \hline 27\,774 \\ 92\,580 \\ \hline 120\,354 \\ \hline \end{array}$

THINK	WRITE
2 Count the number of digits after the point in both the decimals being multiplied and insert the decimal point in the answer. There are 4 decimal places in 0.4629 and 1 in 2.6 so there will be 5 decimal places in the answer.	$0.4629 \times 2.6 = 1.203\,54$
3 Check the answer by rounding; $0.5 \times 3 = 1.5$ which is close to 1.203 54.	

Squaring decimals

To square a decimal number, multiply the number by itself, ignoring the decimal places. Count the number of digits after the decimal point in both of the decimal numbers being multiplied and insert the decimal point in the answer. The number of decimal places in the answer will be the same as the total number in both parts of the question.

WORKED Example 20

Calculate the following.

a 0.5^2 **b** 1.2^2

THINK	WRITE
a **1** Multiply the number by itself, ignoring the decimal places.	**a** $5 \times 5 = 25$
2 Count the number of digits after the point in both the decimals being multiplied and insert the decimal point in the answer. There will be 2 decimal places in the answer.	$0.5 \times 0.5 = 0.25$
3 Write the answer.	$0.5^2 = 0.25$
b **1** Multiply the number by itself, ignoring the decimal places.	**b** $12 \times 12 = 144$
2 Count the number of digits after the point in both the decimals being multiplied and insert the decimal point in the answer. There will be 2 decimal places in the answer.	$1.2 \times 1.2 = 1.44$
3 Write the answer.	$1.2^2 = 1.44$

Square roots of decimals

The opposite of squaring a number is taking the square root. When taking the square root of a decimal, ignore the decimal point and solve as for whole numbers. The number of decimal places in the answer will be half the number of decimal places in the original number. The next example looks at this process. Only decimal numbers with exact square roots will be considered.

WORKED Example 21

Calculate the following.

a $\sqrt{0.81}$ **b** $\sqrt{0.0121}$

THINK	WRITE
a 1 Take the number under the square root symbol, ignoring the decimal places. Find the square root of this number.	**a** $\sqrt{81} = 9$
2 Count the number of decimal places in the original number (2 in this case). The number which needs to be squared to form this decimal will have half the number of decimal places. So, the final answer will have 1 decimal place. Write the answer.	$\sqrt{0.81} = 0.9$
3 Check the result using a calculator or by squaring the answer. ($0.9^2 = 0.81$)	
b 1 Take the number under the square root symbol, ignoring the decimal places. Find the square root of this number.	**b** $\sqrt{121} = 11$
2 Count the number of decimal places in the original number (4 in this case). The number, which needs to be squared to form this decimal, will have half this number of decimal places. So, the final answer will have 2 decimal places. Write the answer.	$\sqrt{0.0121} = 0.11$
3 Check the result using a calculator or by squaring the answer. ($0.11^2 = 0.0121$)	

remember

1. To multiply numbers containing decimals, ignore the decimal points and multiply them in the same way as for whole numbers. Count the number of decimal places in all parts of the question. Place a decimal point in the answer so that the number of decimal places is the same as the total number in the question.
2. Check calculations by rounding.
3. To square a decimal number, multiply the number by itself, ignoring the decimal places. Count the number of digits after the point in both the decimal numbers being multiplied and insert the decimal point in the answer. The number of decimal places in the answer must be the same as the total number in both parts of the question.
4. The opposite of squaring a number is taking the square root.
5. When taking the square root of a decimal, ignore the decimal point and solve as for whole numbers. The number of decimal places in the answer will be half the number of decimal places in the original number.

EXERCISE 4G Multiplying, squaring and obtaining square roots of decimals

WORKED Example 19a

4.11 SkillSHEET Multiplying decimal numbers

1 Calculate the following.

a 1.4×0.6 **b** 4.2×0.7 **c** 0.8×0.4
d 9.7×0.8 **e** 0.35×0.4 **f** 0.64×0.3
g 0.77×0.5 **h** 0.49×0.9 **i** 1.63×0.2
j 5.38×0.8 **k** 0.347×0.6 **l** 0.498×0.7
m 4.832×0.6 **n** 12.2641×0.4 **o** $20.032\,79 \times 0.5$

Mathcad Multiplying decimals

2 Calculate the following.

a 0.002×0.05 **b** 0.003×0.004 **c** 0.7×0.09
d 0.037×0.006 **e** $0.000\,061 \times 0.04$ **f** 0.004×0.09
g 0.56×0.7 **h** $0.030\,31 \times 0.02$ **i** 0.0006×0.007

WORKED Example 19b

EXCEL Spreadsheet Multiplying decimals

3 Calculate the following. (You may use a calculator for parts **g** to **o**.)

a 0.25×1.2 **b** 0.37×2.3 **c** 0.47×5.4
d 0.79×8.3 **e** 4.68×3.6 **f** 8.04×7.5
g 11.64×4.8 **h** 15.08×1.9 **i** 35.17×0.35
j 67.73×0.28 **k** 90.65×0.88 **l** 46.96×0.76
m 13.56×0.31 **n** 3.694×0.46 **o** 12.41×1.2

4 **multiple choice**

a When calculating 8.32×0.64, the number of decimal places in the answer is:
A 0 **B** 1 **C** 2 **D** 3 **E** 4

b 0.2×0.2 equals:
A 0.004 **B** 0.04 **C** 0.4 **D** 4 **E** 40

c 1.4×0.8 equals:
A 1.12 **B** 8.2 **C** 82 **D** 11.2 **E** 112

d 0.0312×0.51 equals:
A 0.001 5912 **B** 0.015 912 **C** 0.156 312 **D** 0.159 12 **E** 1.5912

5 Michael bought 0.65 kilograms of cubed steak at this butcher's shop. This meat costs $8.50 a kilogram. How much did Michael pay for the steak?
Note: Round to the nearest 5c.

6 Judy bought 34.5 litres of petrol at 92.9 cents per litre. How much did she pay for her petrol:

a in cents? **b** in dollars?

7 James is using the recipe for chocolate chip muffins to make 1.5 times the given amount. If the recipe lists0.25 litres of milk in the ingredients, how much milk should James use for his muffins?

8 Find the area of the tennis court shown if area = length × width.

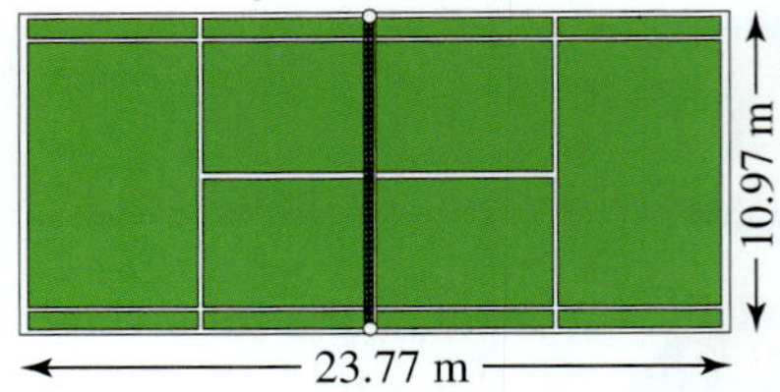

WORKED Example 20, 21

9 Calculate each of the following.

a 0.02^2 b 1.3^2 c 2.05^2

d $\sqrt{0.09}$ e $\sqrt{0.16}$ f $\sqrt{0.0049}$

MATHS QUEST CHALLENGE

1 A grasshopper can jump about 59 times its length. How far can a 3.25 cm long grasshopper jump?

2 What number can you multiply by itself to get an answer of 0.36?

3 A ball is dropped from a height of 1 metre. If it reaches 0.7 times the height of the previous bounce at each bounce, work out how many times the ball bounces until the height is less than 1 centimetre.

Dividing decimals by whole numbers

To divide a decimal by a whole number:

1. set the question out as for division of whole numbers
2. divide as for whole numbers
3. the decimal point in the answer must be exactly in line with the decimal point in the question.

WORKED Example 22

Calculate: a 1.48 ÷ 2 b 124.968 ÷ 12.

THINK

a 1 Set up the division. Write the decimal point in the answer directly above the decimal point in the question and divide as for short division.

2 Check by rounding; 1 ÷ 2 = 0.5 which is close to 0.74.

b 1 Set up the division. Write the decimal point in the answer directly above the decimal point in the question and divide as for short division.

2 Check by rounding; 120 ÷ 12 = 10 which is close to 10.414.

WRITE

a
```
   0 . 7 4
2)1 . ¹4 8
```

b
```
     1 0 . 4 1 4
12)1 2 4 . ⁴9 ¹6 ⁴8
```

Sometimes, when you are dividing numbers, you will find that there is a remainder.

For example 15.3 ÷ 4: $4\overline{)15.3}$ = 3.8 remainder 1

Instead of leaving a remainder, you can sometimes add zeros to the end of the decimal and keep dividing until there is no remainder.

$4\overline{)15.300}$ = 3.825

WORKED Example 23

Calculate 21.76 ÷ 5. Add zeros and keep dividing until there is no remainder.

THINK

1. Set up the division. Write the decimal point in the answer directly above the decimal point in the question and divide as for short division adding zeros as required.
2. Check the answer by rounding; 20 ÷ 5 = 4 which is close to 4.352.

WRITE

$5\overline{)21.{}^{1}7{}^{2}6{}^{1}0}$ = 4.352

WORKED Example 24

Calculate: a 4.8 ÷ 40 b 19.2 ÷ 6000.

THINK

a
1. Dividing by 40 is the same as first dividing by 4 then dividing by 10.
2. To divide by 10, move the position of the decimal point 1 place to the left.
3. Write your final answer.

b
1. Dividing by 6000 is the same as dividing by 6 then dividing by 1000.
2. To divide by 1000, move the position of the decimal point 3 places to the left.
3. Write your final answer.

WRITE

a $4\overline{)4.8}$ = 1.2

1.2 ÷ 10 = 0.12

4.8 ÷ 40 = 0.12

b $6\overline{)19.2}$ = 3.2

3.2 ÷ 1000 = 0.0032

19.2 ÷ 6000 = 0.0032

remember

1. To divide a decimal by a whole number, set the question out as for division of whole numbers and divide as for whole numbers.
2. The decimal point in the answer must be exactly in line with the decimal point in the question.
3. When there is a remainder, add zeros to the end of the decimal and keep dividing until there is no remainder.
4. When dividing a decimal by a multiple of 10, move the position of the decimal point to the left the same number of places as there are zeros. When dividing by a multiple of 10, the answer is always smaller.

EXERCISE 4H Dividing decimals by whole numbers

SkillSHEET 4.12 Dividing decimals by a single-digit number

WORKED Example 22

1 Calculate:

a 3.6 ÷ 6	**b** 21.7 ÷ 7	**c** 17.4 ÷ 6	**d** 4.86 ÷ 9
e 8.05 ÷ 5	**f** 14.13 ÷ 3	**g** 9.68 ÷ 4	**h** 1.576 ÷ 2
i 17.847 ÷ 9	**j** 8.029 ÷ 7	**k** 32.5608 ÷ 8	**l** 41.8645 ÷ 5
m 20.5782 ÷ 3	**n** 126.4704 ÷ 4	**o** 37.56 ÷ 12	**p** 46.80 ÷ 15
q 24.541 ÷ 11	**r** 17.108 ÷ 14	**s** 77.052 ÷ 12	**t** 121.3421 ÷ 11

Mathcad Dividing decimals by whole numbers

WORKED Example 23

2 Calculate the following. In each case, add zeros and keep dividing until there is no remainder.

a 3.7 ÷ 2 **b** 9.5 ÷ 2 **c** 7.3 ÷ 5 **d** 9.8 ÷ 4 **e** 7.5 ÷ 6 **f** 55.6 ÷ 8

3 Calculate the following.

a 67.2 ÷ 10	**b** 35.9 ÷ 10	**c** 81.4 ÷ 10
d 102.5 ÷ 10	**e** 99.9 ÷ 10	**f** 0.546 ÷ 10
g 0.989 ÷ 10	**h** 102.04 ÷ 10	**i** 35.492 ÷ 10

SkillSHEET 4.13 Dividing decimals by 10, 100 and 1000

4 Copy and complete the following sentence.
Dividing a decimal by 10 is the same as moving the position of the decimal point ____________ place to the ____________ .

5 Calculate the following.

a 156.9 ÷ 100	**b** 24.9 ÷ 100	**c** 10.6 ÷ 100
d 607.7 ÷ 100	**e** 2340.1 ÷ 100	**f** 34.8 ÷ 100
g 24.80 ÷ 100	**h** 0.312 ÷ 100	**i** 0.3581 ÷ 100

6 Copy and complete the following sentence.
Dividing a decimal by 100 is the same as moving the position of the decimal point ____________ places to the ____________.

7 Change the following to dollars ($) by dividing by 100.

a 365 cents	**b** 170 cents	**c** 5685 cents
d 75 cents	**e** 90 cents	**f** 6350 cents

8 Calculate the following.

a 13.98 ÷ 1000	**b** 29.876 ÷ 1000	**c** 504.87 ÷ 1000
d 767.9062 ÷ 1000	**e** 140.09 ÷ 1000	**f** 6001.65 ÷ 1000

9 State the number of decimal places the position of the decimal point would be moved to the left when dividing by:

a 1000 **b** 10 000 **c** 100 000 **d** 1 000 000.

10 Calculate the following by changing the position of the decimal point.

a 14.07 ÷ 10	**b** 968.13 ÷ 100	**c** 985.06 ÷ 100
d 620.8 ÷ 1000	**e** 3592.87 ÷ 1000	**f** 2349.78 ÷ 100 000
g 5332.0667 ÷ 100 000	**h** 9.0769 ÷ 100 000	**i** 103 454.97 ÷ 1 000 000
j 802 405.6 ÷ 1 000 000	**k** 152.70 ÷ 1 000 000	**l** 0.583 ÷ 1000
m 0.7205 ÷ 10 000	**n** 0.0032 ÷ 1 000 000	**o** 0.0487 ÷ 1 000 000

WORKED Example 24

11 Calculate the following.

a 15.9 ÷ 60	**b** 23.7 ÷ 30	**c** 164.5 ÷ 700
d 238 ÷ 400	**e** 8.79 ÷ 6000	**f** 5.22 ÷ 3000

12 **multiple choice**

a 5.2 ÷ 4 equals:
A 0.013 B 0.13 C 1.3 D 1.5 E 13

b 14.025 ÷ 3 equals:
A 0.4675 B 4.675 C 4.6 D 40.675 E 46.75

c 0.27 ÷ 8 equals:
A 0.033 75 B 0.361 25 C 0.3612 D 0.3125 E 3.612

d 28.149 ÷ 100 equals:
A 0.028 149 B 0.281 49 C 2.8149 D 281.49 E 2814.9

13 Stephanie spent $6.95 on these chocolates from The Chocolate Box. What was the cost of each chocolate? Give your answer to the nearest 5 cents.

14 If you have $22.50 for bus fares to school for the week, how much would you spend on each of the 5 days?

15 Emily wants to make 10 cushions from 6.75 metres of material that she found on a table of remnants at Costlight Fabrics. How much material would she have for each cushion?

WorkSHEET 4.2

10 QUICK QUESTIONS 2

Calculate each of the following.

1 3.86 + 2.15

2 7.4 + 1.1 + 12.78

3 3.55 − 2.61

4 4075.2 − 12.81

5 8.921 × 15

6 56.332 × 0.024

7 1.25^2

8 6.784 ÷ 5

9 $\sqrt{0.0036}$

10 In preparation for a major cross-country event, the school team was required to complete the following training runs: 6.2, 8.7, 3.45, 5.5 and 12.8 kilometres. How far did the team run in total?

Converting fractions to decimals and recurring decimals

To change any fraction into a decimal, divide the denominator into the numerator. For example, to change $\frac{1}{4}$ into a decimal, divide 4 into 1:

$$4\overline{)1.00}\quad 0.25$$

Add a decimal point and as many zeros as required.

WORKED Example 25

Change the following fractions into finite decimals. **a** $\frac{2}{5}$ **b** $\frac{1}{8}$

THINK	WRITE
a 1 Set out the question as for division of whole numbers, adding a decimal point and the required number of zeros. *Note:* $\frac{2}{5} = 2 \div 5$.	**a** $5\overline{)2.0}$ = 0.4
2 Divide, writing the answer with the decimal point exactly in line with the decimal point in the question.	$\frac{2}{5} = 0.4$
b 1 Set out the question as for division of whole numbers, adding a decimal point and the required number of zeros. *Note:* $\frac{1}{8} = 1 \div 8$.	**b** $8\overline{)1.000}$ = 0.125
2 Divide, writing the answer with the decimal point exactly in line with the decimal point in the question.	$\frac{1}{8} = 0.125$

When a decimal number ends (terminates) after any number of decimal places the number is said to be a *finite decimal*. The decimal number 0.125 is an example of a finite decimal.

Sometimes when dividing the denominator into the numerator, the decimal places in the answer keep repeating and the amount left over each time also keeps repeating. When this happens, the answer is called an *infinite recurring decimal*.

WORKED Example 26

Convert $\frac{1}{11}$ to a decimal. Continue dividing until a pattern emerges, then round the answer to 2 decimal places.

THINK	WRITE
1 Set out the question as for division of whole numbers, adding a decimal point and enough zeros to see a pattern emerging.	$11\overline{)1\,.\,{}^{1}0\,{}^{10}0\,{}^{1}0\,{}^{10}0\,{}^{1}0\,{}^{10}0\,{}^{1}0\,{}^{10}0\,{}^{1}0}$ = 0 . 0 9 0 9 0 9 0 9...
2 Divide, writing the answer with the decimal point exactly in line with the decimal point in the question. (The amount left over each time is 10 then 1 then 10 then 1 again. The decimal answer is also repeating.)	
3 Write the approximate answer rounded to 2 decimal places.	$\frac{1}{11} \approx 0.09$

Note that the answer in worked example 26 shows the whole pattern. If a recurring decimal is rounded, it usually includes one complete cycle of the pattern.

Recurring decimals can be written in one of the following shorter ways for an exact answer:

- 4.6666 … could be written as $4.\dot{6}$ (with a dot above the repeating part of the decimal)
- 3.512 512 … could be written as $3.\dot{5}1\dot{2}$ (with a dot above the first and last digits of the repeating part)
- 6.121 212 … could be written as $6.\overline{12}$ (with a line above the repeating part of the decimal).

By knowing the decimal equivalent of any fraction, it is easy to determine the decimal equivalent of any multiple of that fraction. The following worked example illustrates this.

WORKED Example 27

Use the results of worked examples 25 and 26 to find decimal equivalents for:

a $\frac{3}{8}$ **b** $4\frac{5}{8}$ **c** $\frac{6}{11}$

THINK	WRITE
a 1 Write the decimal equivalent for the fraction with 1 as the numerator.	**a** $\frac{1}{8} = 0.125$
2 Multiply both sides of this equation by the appropriate multiple (3 in this case).	$\frac{1}{8} \times 3 = 0.125 \times 3$
3 Write the answer.	$\frac{3}{8} = 0.375$
b 1 Consider only the fraction part of the mixed number. Write the decimal equivalent of this fraction with 1 as the numerator.	**b** $\frac{1}{8} = 0.125$
2 Multiply both sides of this equation by the appropriate multiple (5 in this case).	$\frac{1}{8} \times 5 = 0.125 \times 5$
3 Simplify both sides.	$\frac{5}{8} = 0.625$
4 Combine with the whole number and write the answer.	$4\frac{5}{8} = 4.625$
c 1 Write the decimal equivalent for the fraction with 1 as the numerator. In this case it is an infinite recurring decimal therefore 0.090909… can be written as $0.\overline{09}$.	**c** $\frac{1}{11} = 0.090\,909\ldots$
2 Multiply both sides of this equation by the appropriate multiple (6 in this case).	$\frac{1}{11} \times 6 = 0.090\,909\ldots \times 6$ $= 0.545\,454\ldots$
3 Simplify and write the answer.	$\frac{6}{11} = 0.\overline{54}$

remember

1. To change any fraction into a decimal, divide the denominator into the numerator.
2. A *finite decimal* terminates after a particular number of decimal places.
3. An *infinite recurring decimal* is obtained when the decimal places in the answer keep repeating and the amount left over each time also keeps repeating. An approximate answer can be found by rounding so that the answer includes one complete cycle of the pattern.
4. An *infinite recurring decimal* can be written exactly by placing a dot above the first and last digits of the repeating part or by drawing a line above the repeating part.

EXERCISE 4I Converting fractions to decimals and recurring decimals

GC program—TI: Converting fractions to decimals

GC program—Casio: Converting fractions to decimals

EXCEL Spreadsheet: Converting fractions to decimals

Mathcad: Converting fractions to decimals

WORKED Example 25

1 Change the following fractions to finite decimals.

a $\frac{3}{4}$ **b** $\frac{1}{2}$ **c** $\frac{4}{5}$ **d** $\frac{1}{20}$ **e** $\frac{3}{15}$

f $\frac{3}{12}$ **g** $\frac{3}{8}$ **h** $\frac{1}{50}$ **i** $\frac{8}{25}$

2 Write the following infinite recurring decimals using one of the short forms.

a 2.555 . . . **b** 0.666 . . . **c** 12.888 88 . . .

d 49.111 11 . . . **e** 0.262 626 . . . **f** 0.414 141 . . .

g 0.913 913 . . . **h** 8.641 864 18 . . . **i** 0.040 121 21 . . .

j 133.946 2462 . . . **k** 1.833 333 . . . **l** 0.127 7777 . . .

WORKED Example 26

3 Convert each of the following to a decimal. Continue dividing until a pattern emerges, then round the answer to the number of decimal places indicated in the brackets.

a $\frac{1}{6}$ (2) **b** $\frac{1}{3}$ (1) **c** $\frac{1}{9}$ (1)

d $\frac{2}{15}$ (2) **e** $\frac{2}{11}$ (2) **f** $\frac{4}{9}$ (1)

g $\frac{5}{12}$ (3) **h** $\frac{1}{7}$ (6) **i** $\frac{7}{15}$ (2)

4 **multiple choice**

a $\frac{3}{5}$ as a decimal is:

A 0.3 **B** 0.6 **C** 0.2 **D** 0.9 **E** 0.5

b 1.8888 . . . written as an exact answer is:

A $1.\dot{8}$ **B** 1.888 **C** 1.88 **D** 1.9 **E** 1.889

c 12.412 412 . . . written as an exact answer is:

A 12.412 **B** $12.41\dot{2}$ **C** 12.412 412 **D** $12.\dot{4}1\dot{2}$ **E** $12.4\overline{12}$

d $\frac{3}{7}$ as a decimal in exact form is:

A $0.\dot{4}2857\dot{1}$ **B** $0.\overline{428}$ **C** 0.4 **D** 0.428 5174 **E** $2.\dot{3}$

5 Find decimal equivalents for the following fractions. Give your answer in exact form as either a finite decimal or an infinite recurring decimal.

a $\frac{1}{2}$ b $\frac{1}{3}$ c $\frac{1}{4}$ d $\frac{1}{5}$ e $\frac{1}{9}$

WORKED Example 27

6 Using the answers from question 5, find decimal equivalents for the following fractions.

a $\frac{7}{2}$ b $4\frac{2}{3}$ c $\frac{3}{4}$ d $3\frac{4}{5}$ e $\frac{7}{9}$

THINKING

Fractions to decimals

1 Complete the following table. This table will be useful in helping you answer later questions.

2^1	2^2	2^3	2^4	2^5	2^6	2^7	2^8	2^9	2^{10}
2	4	8							

2 Convert each of the following to decimals.

a $\frac{1}{2}$ b $\frac{1}{4}$ c $\frac{1}{8}$ d $\frac{1}{16}$ e $\frac{1}{32}$

3 For each decimal obtained in question 2, write down the number of decimal places it has.

4 Can you see a pattern? Write down what you think the pattern is.

5 Predict how many decimal places the following fractions have when expressed as decimals.

$\frac{1}{128}, \frac{1}{64}, \frac{1}{512}$

6 Check your answers to question 5 by converting each fraction to a decimal.

7 How many decimal places will the fraction $\frac{1}{16\,384}$ have when expressed as a decimal?

8 How many decimal places will the fraction $\frac{1}{262\,144}$ have when expressed as a decimal?

9 Investigate what happens to fractions like $\frac{1}{5}, \frac{1}{25}, \frac{1}{125}, \ldots$ when expressed as decimals. Write a short report on what you find.

10 Super challenge

a Investigate what happens to fractions like $\frac{1}{10}, \frac{1}{20}, \frac{1}{50}, \frac{1}{100}, \frac{1}{200}, \frac{1}{400}, \frac{1}{5000}, \frac{1}{50\,000} \ldots$ when expressed as decimals. (*Hint:* Write each denominator as a product of prime factors.) Explain the pattern you observe. Check whether this pattern works with another fraction of this type.

b Without calculating the decimal, predict how many decimal places the fraction $\frac{1}{2^7 \times 5^9}$ will have.

c Predict the number of decimal places that the following products will have when expressed as decimals. Note that all denominators have factors of 2 and 5.

i $\frac{1}{25} \times \frac{1}{64}$ ii $\frac{1}{16} \times \frac{1}{625}$ iii $\frac{1}{128} \times \frac{1}{125}$ iv $\frac{1}{1024} \times \frac{1}{625}$

Dividing decimals

To divide one decimal by another, it is necessary to change the second decimal (the divisor) to a whole number. This can be done by rewriting the question as a fraction then multiplying both numerator and denominator by whichever power of 10 makes the denominator a whole number. So:

$$14.62 \div 0.4 \text{ would be rewritten as } \frac{14.62}{0.4}.$$

The fraction can be multiplied by $\frac{10}{10}$ without changing the value of the fraction (remember that $\frac{10}{10} = 1$) . This is the same as multiplying the numerator and denominator by 10.

$$\frac{14.62}{0.4} = \frac{14.62}{0.4} \times \frac{10}{10} = \frac{146.2}{4}$$

The calculation now becomes $146.2 \div 4$.

Alternatively. the decimal point could be moved until the second decimal (the divisor) is a whole number; for example $14.62 \div 0.4 = 146.2 \div 4$.

WORKED Example 28

Calculate: **a** $\mathbf{26.724 \div 0.4}$ **b** $\mathbf{3.0276 \div 0.12}$.

THINK / **WRITE**

a

1. Rewrite the question as a fraction.
   $26.724 \div 0.4 = \dfrac{26.724}{0.4}$
2. Multiply both the numerator and the denominator by the appropriate multiple of 10.
   $= \dfrac{26.724}{0.4} \times \dfrac{10}{10}$
   $= \dfrac{267.24}{4}$
3. Divide the decimal by the whole number.
   $$\begin{array}{r} 6\;6\;.\;81 \\ 4\;\overline{)\;26^{2}7\;.^{3}24} \end{array}$$
4. Write the answer.
   $26.724 \div 0.4 = 66.81$

b

1. Rewrite the question as a fraction.
   $3.0276 \div 0.12 = \dfrac{3.0276}{0.12}$
2. Multiply both the numerator and the denominator by the appropriate multiple of 10.
   $= \dfrac{3.0276}{0.12} \times \dfrac{100}{100}$
   $= \dfrac{302.76}{12}$
3. Divide the decimal by the whole number.
   $$\begin{array}{r} 2\;5\;.\;2\;3 \\ 12\;\overline{)\;30^{6}2\;.^{2}7^{3}6} \end{array}$$
4. Write the answer.
   $3.0276 \div 0.12 = 25.23$

Using the alternative method, worked example 28a and 28b could be initially set out as 26.724 ÷ 0.4 = 267.24 ÷ 4 and 3.0276 ÷ 0.12 = 302.76 ÷ 12 respectively.

WORKED Example 29

How many litres of petrol could be purchased for $50.88 if 1 litre costs $1.06?

THINK	WRITE
1 Write the problem.	50.88 ÷ 1.06
2 Rewrite the problem as a fraction.	$= \frac{50.88}{1.06}$
3 Multiply the numerator and denominator by the appropriate multiple of 10, in this case 100. Alternatively, the decimal point could be moved twice to the right in both numbers so that the divisor is a whole number (that is, 50.88 ÷ 1.06 = 5088 ÷ 106).	$= \frac{50.88}{1.06} \times \frac{100}{100}$ $= \frac{5088}{106}$
4 Divide the decimal by the whole number.	$= 106 \overline{)\, 508^{84}8}$ = 48 50.88 ÷ 1.06 = 48
5 Answer the question.	Forty-eight litres of petrol could be purchased for $50.88.

remember

To divide a decimal by a decimal, change the second decimal (the divisor) to a whole number. This can be done by rewriting the question as a fraction, then multiplying both numerator and denominator by whichever power of 10 makes the denominator a whole number.

EXERCISE 4J Dividing decimals

4.14 SkillSHEET Dividing decimals

EXCEL Spreadsheet Dividing decimals

WORKED Example 28a

1 Calculate each of the following.

a 2.5 ÷ 0.5	**b** 4.2 ÷ 0.6	**c** 6.4 ÷ 0.8
d 8.1 ÷ 0.9	**e** 2.8 ÷ 0.7	**f** 6.66 ÷ 0.6
g 0.248 ÷ 0.8	**h** 3.57 ÷ 0.7	**i** 1.32 ÷ 0.6
j 39.6 ÷ 0.6	**k** 57.68 ÷ 0.8	**l** 4.924 ÷ 0.4
m 0.2556 ÷ 0.3	**n** 0.067 354 ÷ 0.2	**o** 0.5468 ÷ 0.4

Mathcad Dividing decimals

WORKED Example 28b

2 Calculate:

a 172.0488 ÷ 0.11
b 0.510 48 ÷ 0.12
c 6.4032 ÷ 0.32
d 2.5473 ÷ 0.21
e 21.470 10 ÷ 0.15
f 142.888 ÷ 0.08
g 0.028 692 ÷ 0.06
h 473.159 61 ÷ 0.03
i 32.619 ÷ 0.02

3 **multiple choice**

a To calculate 9.84 ÷ 0.8, rewrite it as:
A 9.84 ÷ 8 B 0.984 ÷ 0.8 C 98.4 ÷ 0.8 D 98.4 ÷ 8 E 984 ÷ 4

b To calculate 151.368 ÷ 1.32, rewrite it as:
A 151.368 ÷ 132 B 151.368 ÷ 13.2 C 1513.68 ÷ 132
D 15 136.8 ÷ 132 E 151 368 ÷ 132

c 0.294 ÷ 0.7 equals:
A 0.042 B 0.42 C 4.2 D 42 E 420

d 21.195 ÷ 0.15 equals:
A 0.1413 B 1.413 C 14.13 D 141.3 E 1413

WORKED Example 29

4 How many 1.25-litre bottles of water could be poured into a 25-litre drink dispenser?

5 The area of Tanya's lounge room floor is 85.8 square metres. How many people could she fit in the lounge room if each person takes up 1.2 square metres?

GAME time Decimals — 002

WorkSHEET 4.3

6 How many compact discs can be stacked on a shelf that is 28.6 centimetres high if each compact disc case is 1.1 centimetres high?

7 How many Big Burgers could be bought for $562.80 if each Big Burger costs $2.80?

Unusual journeys to the North Pole

Calculate the decimal divisions below to find the puzzle solution code.

A = 6.8 ÷ 0.5

B = 7.68 ÷ 0.8

C = 3.7 ÷ 0.2

E = 9 ÷ 0.6

F = 6.45 ÷ 0.3

G = 6.66 ÷ 0.9

H = 7.84 ÷ 0.7

I = 16.17 ÷ 1.1

J = 9.96 ÷ 1.2

K = 0.648 ÷ 0.04

L = 1.308 ÷ 0.12

1987:

19.8 19.1 7.8 19.1 20.9 18.5 6.6 18.5 10.9 15

9.6 6.6 21.5 3.75 16.2 13.6 5.1 11.2 14.7

16.2 13.6 8 13.6 19.8 14.7 19.1 21.5 8.3 13.6 14.3 13.6 12.7

M = 118.8 ÷ 6

N = 25.4 ÷ 2

O = 152.8 ÷ 8

P = 71.5 ÷ 5

R = 146.3 ÷ 7

S = 45.9 ÷ 9

T = 312.0 ÷ 40

U = 112.5 ÷ 30

W = 482.4 ÷ 60

Y = 0.462 ÷ 0.07

Z = 1.76 ÷ 0.22

1989:

8.04 13.6 10.9 16.2 14.7 12.7 7.4 9.6 6.6 20.9 19.1 9.6 15 20.9 7.8

5.1 8.04 13.6 12.7 19.1 21.5 9.6 20.9 14.7 7.8 13.6 14.7 12.7

summary

Copy the sentences below. Fill in the gaps by choosing the correct word from the word list that follows.

1 The number of ____________ places is the number of digits after the decimal point.

2 To compare the size of decimals, it is necessary to compare the size of the digits with the same ____________ value.

3 The symbol < means 'is ____________ than'.

4 The symbol > means 'is ____________ than'.

5 When rounding a decimal, consider the ____________ digit after the number of decimal places required.

6 To multiply decimals by multiples of 10, the decimal point should be moved to the ____________, one place for each ____________.

7 To show a decimal is recurring, a ____________ or a ____________ is placed above the repeating part of the decimal.

8 To simplify a fraction, divide the ____________ and ____________ by the same number.

9 When multiplying decimals, the number of decimal places in the ____________ is the same as the number of decimal places in the ____________.

10 When adding or subtracting decimals, the decimal point in the answer should be ____________ with the decimal point in the question.

11 To change any fraction into a decimal, ____________ the numerator by the denominator.

12 To divide decimals by multiples of ____________, the position of the decimal point should be moved to the ____________, one place for each zero.

WORD LIST

greater	dot	right	decimal
question	divide	numerator	line
lined up	left	ten	zero
denominator	answer	less	first
			place

CHAPTER review

1 Give the value of the 7 in each of the following. **4A**
a 1.719 b 3.0726 c 4.7218 d 0.2078
e 23.1487 f 0.002 57 g 17.592 h 50.007

2 Write the following numbers in expanded notation. **4A**
a 2.64 b 0.369 c 18.406 d 96.3428

3 Add 2 tenths to the following. **4A**
a 6.2 b 0.743 c 12.06 d 3.91

4 Add 3 thousandths to the following. **4A**
a 0.456 b 12.803 c 1.6 d 2.79

5 Put < or > between the following. **4B**
a 8.72 ________ 8.27 b 0.35 ________ 0.37
c 1.06 ________ 1.27 d 10.214 ________ 10.219
e 0.021 ________ 0.018 f 13.0496 ________ 13.149
g 0.804 06 ________ 0.804 17 h 0.000 879 ________ 0.000 876

6 Write the following decimals in order from smallest to largest. **4B**
a 0.13, 0.86, 0.34, 0.71, 0.22 b 0.247, 0.274, 0.124, 0.258, 0.285
c 0.834, 0.826, 0.859, 0.888, 0.891 d 0.356, 0.358, 0.365, 0.385, 0.217

7 Write the following decimals as fractions in simplest form. **4C**
a 0.8 b 0.17 c 0.36 d 0.187
e 0.125 f 0.568 g 0.205 h 0.950

8 Write the following decimals as mixed numerals in simplest form. **4C**
a 1.5 b 4.60 c 3.48 d 5.25
e 2.75 f 2.625 g 1.56 h 8.32

9 Round the following to the number of decimal places shown in the brackets. **4D**
a 1.29 (1) b 2.047 (2) c 13.8649 (2)
d 0.0482 (3) e 1.925 96 (4) f 17.898 193 (2)

10 Round the following to the nearest unit. **4D**
a 13.6 b 29.02 c 86.99 d 100.09

11 Calculate the following. **4E**
a 1.8 + 7.3 b 4.21 + 5.88 c 6.75 + 0.243
d 12.047 + 3.6 e 194 + 18.62 + 3.1 f 34.1 + 7.629 + 0.008 45

12 Jim saved the following amounts of pocket money to take away on holidays: $12.50, $15.00, $9.30, $5.70, $10.80. How much money did Jim have to spend on holidays? **4E**

13 Calculate the following. **4E**
a 9.6 − 4.3 b 18.25 − 9.18 c 3.92 − 1.88
d 100 − 9.341 e 4.876 − 3.927 f 1.6 − 0.025

4E

14 Mandie poured 0.375 litres from a 1.5-litre bottle of juice. How much juice was left in the bottle?

4F

15 Calculate the following.

a 6.2×3 b 4.67×9
c 13.2036×5 d 0.7642×7

4F

16 Calculate the following.

a 0.23×11 b 16.28×41
c 182.94×28 d $0.028\,94 \times 32$

4F

17 Calculate.

a 0.26×10 b 1.345×10
c 0.0645×100 d 1.8294×100
e 146.6281×100 f $0.048\,0643 \times 1000$
g $0.839\,204 \times 1000$ h 0.368×1000

4G

18 Calculate.

a 3.2×0.41 b 1.72×0.3 c 0.87×0.9 d 0.03×0.006
e 0.58×1.5 f 2.83×0.96 g 11.468×1.3 h 1.248×0.82

4G

19 Tara bought 0.350 kilogram of shaved ham at $10.50 per kilogram. How much did Tara pay for the ham?

4G

20 Calculate each of the following.

a 2.5^2 b 0.03^2 c $\sqrt{0.64}$ d $\sqrt{0.0025}$

4H

21 Calculate each of the following.

a $2.4 \div 8$ b $1.64 \div 4$ c $12.48 \div 6$
d $147.24 \div 2$ e $1.76 \div 11$ f $264.88 \div 8$

4H

22 Calculate each of the following.

a $14.623 \div 10$ b $102.36 \div 10$ c $9612.347 \div 1000$
d $20.032 \div 100$ e $264\,983.0026 \div 1000$ f $3462.94 \div 100$

4I

23 Write as a finite decimal or as an infinite recurring decimal.

a $\frac{1}{20}$ b $\frac{5}{8}$ c $\frac{5}{16}$ d $\frac{2}{16}$

4I

24 Write these infinite recurring decimals using a short form.

a 4.555 … b 0.8282 … c 19.278 127 81 … d 83.016 262 62 …

4I

25 The decimal equivalent of $\frac{3}{13}$ is $0.\overline{230\,769}$.

a What is the decimal equivalent of $\frac{1}{13}$?

b Find the decimal equivalent of $\frac{5}{13}$.

4J

26 Calculate.

a $4.8 \div 0.6$ b $35.7 \div 0.7$ c $12.1 \div 1.1$
d $13.72 \div 0.4$ e $17.8946 \div 0.02$ f $372.045\,72 \div 0.06$
g $0.289\,56 \div 0.12$ h $3214.0170 \div 0.15$

5 Patterns and algebra

Surfside beach huts are constructed using wood attached to metal framing struts. The diagrams on the top of the photograph show the number of struts needed to support the back walls of one and two beach huts.

Surfside Constructions has a contract to build a row of 34 beach huts at Golden Beach. To save space and materials, the huts will be joined in a row and share common walls. How many struts will be required for the back walls of a row of three beach huts?

In this chapter we will be using rules and formulas involving pronumerals (letters to represent numbers) to help solve problems like this more easily.

Are you ready?

Try the questions below. If you have difficulty with any of them, extra help can be obtained by completing the matching SkillSHEET. Either click on the SkillSHEET icon next to the question on the *Maths Quest 7* CD-ROM or ask your teacher for a copy.

SkillSHEET 5.1

Number patterns

1 For each of the following sequences of numbers, describe the pattern in words and then write down the next three numbers in the pattern.

a 7, 9, 11, 13, …
b 28, 24, 20, 16, …
c 3, 6, 12, 24, …
d 100 000, 10 000, 1000, …

SkillSHEET 5.2

Worded operations

2 Use 8 as your starting number in each case and find the result when you follow each instruction.

a Multiply by 2
b Add 15
c Subtract 6
d Divide by 4

SkillSHEET 5.3

Describing a number pattern from a table

3 Describe each number pattern shown in the tables.

a

First number	1	2	3	4	5
Second number	7	8	9	10	11

b

First number	1	3	5	7	9
Second number	3	9	15	21	27

SkillSHEET 5.5

Alternative expressions used to describe the four operations

4 Rewrite each of the following as a mathematical sentence (that is, rewrite each using numbers and one of the four operations +, −, × or ÷).

a The sum of 3 and 5
b The product of 7 and 8
c The difference between 6 and 2
d 9 more than 5

SkillSHEET 5.6

Using tables to show number patterns

5 a Complete the table shown, using the diagrams below as a guide.

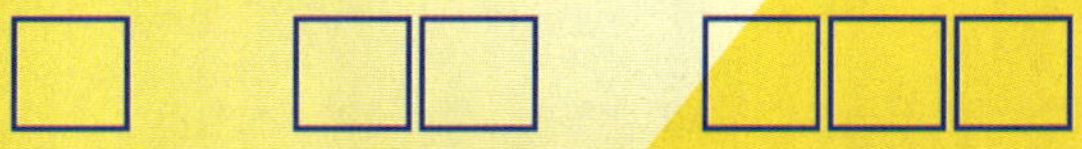

1 square 2 squares 3 squares

Number of squares	1	2	3	4	5	6
Number of sides	4	8				

b How many sides are there for 10 squares?

Using rules

We often use mathematics to describe relationships in the world about us. Mathematicians study these patterns found in nature to discover rules that describe how the relationships are produced. These rules can then be applied in other, more general situations.

Patterns and rules

Let's look at some patterns in numbers and shapes, and see if we can discover the rules that describe them.

Number patterns

If we look at the number pattern 1, 4, 7, 10, … we can see that by adding 3 to any of these numbers, we obtain the next number. Hence, the next three numbers would be 13, 16 and 19. This number pattern is called a *sequence*. Each number in the sequence is called a *term*. This sequence (like all sequences) has a **rule** that describes the pattern. The rule here is 'add 3'.

THINKING

Number patterns

5.1 SkillSHEET Number patterns

Copy the patterns below, describe the pattern in words and then write the next three numbers in the pattern.

1 2, 4, 6, 8, …

2 3, 8, 13, 18, …

3 27, 24, 21, 18, …

4 1, 3, 9, 27, …

5 128, 64, 32, 16, …

6 1, 4, 9, 16, …

Geometric patterns

We can also find patterns in geometric shapes. If we examine the three shapes below, we can see patterns by investigating the changes from one shape to the next — for example, look at the number of matchsticks in each set of triangles.

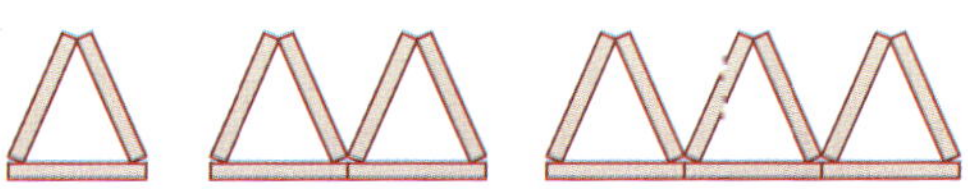

By using a table of values we can see a number pattern developing:

Number of triangles	1	2	3	4	5	6
Number of matchsticks	3	6	9	12	15	18

The pattern in the bottom row is 3, 6, 9, . . .; we can see that the rule here is 'add 3'.

We can also look for a relationship between the number of triangles and the number of matchsticks in each shape. If you examine the table, you will see that a relationship can be found. In words, the relationship is 'the number of matchsticks equals 3 times the number of triangles'.

THINKING

Geometric patterns

For each of the sets of shapes below:

a Construct a table, as shown on the previous page, to show the relationship between the number of shapes in each figure and the number of matchsticks used to construct it.

b Devise a rule in words that describes the pattern relating the number of shapes in each figure and the number of matchsticks used to construct it.

c Use your rule to work out the number of matchsticks required to construct a figure made up of 20 such shapes.

1 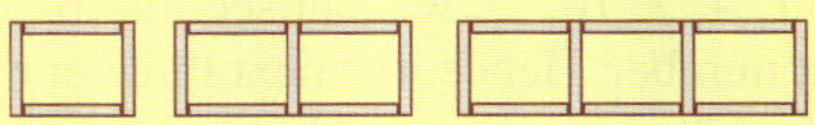

2 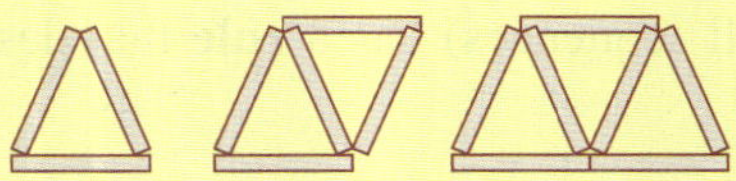

3 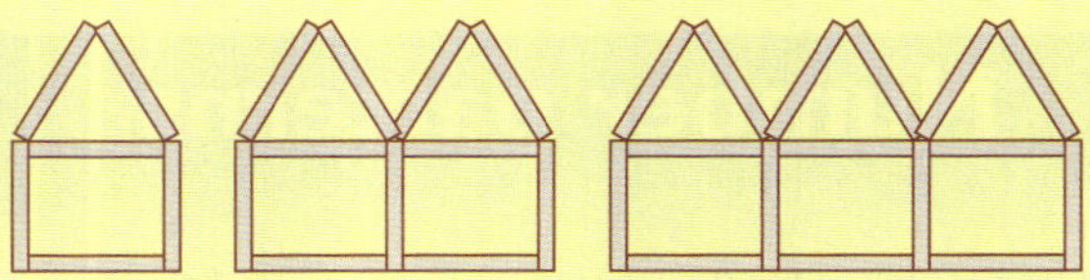

4 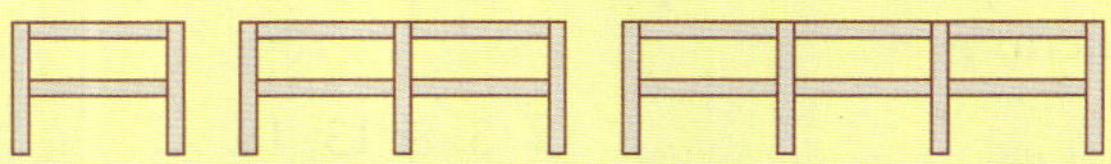

Geometry patterns have a rule that can describe the pattern. This rule can be used to make predictions.

Algebra

The branch of mathematics that deals with the patterns and rules discussed above is called *algebra*.

It allows us to investigate the rules which govern the behaviour of everyday things, such as the arrangement of petals in flowers, the speed that a ball will reach when it is dropped from a certain height and the movement of the earth around the sun.

An example of a simple mathematical rule is 'add 5'. Imagine an 'algebra machine' which takes numbers (input numbers) and processes them according to a particular rule to produce a set of output numbers. The input and output numbers resulting from this rule are shown below.

Input numbers	Rule	Output numbers
3, 5, 89, 222	Add 5	8, 10, 94, 227

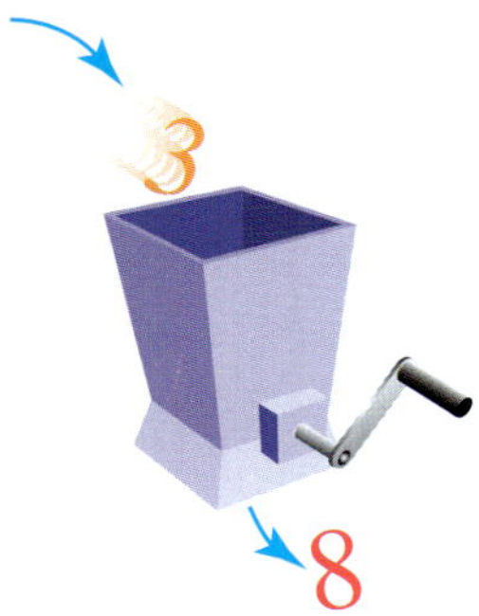

The algebra machine takes any input number and changes it according to the rule given. In this case, 5 is added to each input number. Any number can be used as an input number. The algebra machine might print out a table like this:

Input	3	5	89	222
Output	8	10	94	227

The machine can also be reprogrammed to work with different rules. The rule can be as simple or as complicated as you like. For example, you could change the rule to 'multiply each input number by 2' or 'subtract 26 from each input number, then multiply by 143'.

In algebra, a rule can be used to produce a set of output numbers from a set of input numbers.

WORKED Example 1

Complete the table at right using the given rules to work out the correct output numbers.
a Rule: Subtract 3 from each input number.
b Rule: Divide each input number by 2.

Input	4	10	38	144
Output				

THINK

a 1 Take the first input number (4) and apply the rule; that is, subtract 3.
2 Repeat for the other input numbers.
3 Enter these output values in the table.

b 1 Take the first input number (4) and apply the rule; that is, divide by 2.
2 Repeat for the other input numbers.
3 Enter these output values in the table.

WRITE

a $4 - 3 = 1$

$10 - 3 = 7$
$38 - 3 = 35$
$144 - 3 = 141$

Input	4	10	38	144
Output	1	7	35	141

b $4 \div 2 = 2$

$10 \div 2 = 5$
$38 \div 2 = 19$
$144 \div 2 = 72$

Input	4	10	38	144
Output	2	5	19	72

WORKED Example 2

Complete the table below by using the following rule to work out the correct output numbers in each case.
Rule: Multiply each input number by 8, then subtract 2.

Input	1	3	5	24
Output				

THINK

1. Take the first input number (1) and apply the rule.
   Note: First multiply the input number by 8 and then subtract 2 from the answer.
2. Repeat for the other input numbers.
3. Enter the output values in the table.

WRITE

1. $1 \times 8 = 8$ then $8 - 2 = 6$
2. $3 \times 8 = 24$ then $24 - 2 = 22$
   $5 \times 8 = 40$ then $40 - 2 = 38$
   $24 \times 8 = 192$ then $192 - 2 = 190$
3.

Input	1	3	5	24
Output	6	22	38	190

WORKED Example 3

Use the rule given below to work out the missing input and output numbers.
Rule: Add 4 to each input number.

Input	1	5	25			
Output				6	38	190

THINK

1. Consider the input numbers that are given and apply the rule; that is, add 4.
2. Consider the output numbers that are given and perform the opposite operation; that is, subtract 4.
3. Enter the new output and input values in the table.

WRITE

1. $1 + 4 = 5$
   $5 + 4 = 9$
   $25 + 4 = 29$
2. $6 - 4 = 2$
   $38 - 4 = 34$
   $190 - 4 = 186$
3.

Input	1	5	25	2	34	186
Output	5	9	29	6	38	190

remember

1. By observing number patterns, we can discover the rules by which a pattern is produced.
2. In algebra, a rule can be used to produce a set of output numbers from a set of input numbers.

EXERCISE 5A Using rules

WORKED Example 1

5.2 SkillSHEET Worded operations

1 Copy and complete the following tables. Use the rule given in each case to work out the correct output numbers.

a Subtract 4 from each input number.

Input	4	5	6	10	14
Output					

b Add 12 to each input number.

Input	1	2	6	13	107
Output					

c Multiply each input number by 2.

Input	1	10	51	60	144
Output					

d Divide each input number by 3.

Input	3	12	21	66	141
Output					

WORKED Example 2

2 Copy and complete the following tables. Use the rule given in each case to work out the correct output numbers.

a Multiply each input number by 2, then add 5.

Input	2	3	4	10	17
Output					

b Multiply each input number by 8, then subtract 4.

Input	2	4	5	20	100
Output					

c Add 3 to each input number, then multiply by 5.

Input	2	3	5	12	43
Output					

d Subtract 3 from each input number, then multiply by 11.

Input	3	4	5	8	25
Output					

e Multiply each input number by itself.

Input	0	1	5	8	10
Output					

f Multiply each input number by itself, then add 4.

Input	0	1	4	7	12
Output					

WORKED Example 3

3 Copy and complete the following tables. Use the rule given in each case to work out the missing input and output numbers.

a Add 2 to each input number.

Input	1	3	27			
Output				4	55	193

b Add 20 to each input number.

Input	3	25	56			
Output				20	94	1773

c Subtract 5 from each input number.

Input	7	15	96			
Output				4	12	104

d Multiply each input number by 4.

Input	1	6	321			
Output				8	28	412

e Multiply each input number by 2, then add 2.

Input	3	13	21			
Output				4	22	102

f Multiply each input number by 5, then subtract 7.

Input	2	5	16			
Output				8	18	93

g Divide each input number by 3.

Input	3	15	273			
Output				2	21	54

h Multiply each input number by itself.

Input	1	5	17			
Output				4	49	144

i Multiply each input number by 20, then subtract 6.

Input	2	7	15			
Output				54	174	214

j Subtract 3 from each input number, then divide by 4.

Input	7	19	31			
Output				5	12	15

MATHS QUEST CHALLENGE

1 Using the six consecutive numbers from 4 to 9, complete the magic square at right so that each row, column and diagonal totals 15.

	1	
3		
		2

2 Using 20 cubes, make four piles so that the first pile contains 4 more cubes than the second pile, the second pile contains 1 cube less than the third pile, and the fourth pile contains twice as many cubes as the second pile. How many cubes are in each pile?

History of mathematics

THE DEVELOPMENT OF ALGEBRA

Algebra has been used by mathematicians for thousands of years. The earliest recorded use of algebra is found in clay tablets from the Mesopotamian civilisation which are over 3500 years old. The Mesopotamians, who lived in the fertile valley between the Tigris and Euphrates rivers, in what is now Iraq, used algebra to solve many practical problems.

In the seventh century AD the Eastern Roman Empire, whose capital was Byzantium (now Istanbul), was conquered by the Arabs, who were followers of the new prophet Mohammed (AD 570–632). The Muslims had a strong trading tradition and they were interested in mathematics as a way of calculating shares in goods and transactions. In AD 830, Al–Khwarizmi, one of the greatest Arab mathematicians, published a book called *Al–jabr wa'l muqabala.* Al-jabr, meaning 'reunion of broken parts' or 'balance', is the word from which the modern term 'algebra' is derived.

One of Al–Khwarizmi's famous successors was the Persian poet and mathematician Omar Khayyam (shown below left), who was also interested in equations. Knowledge of algebra came to Europe from Latin translations of Al–Khwarizmi's work.

Centuries later, the English scientist and mathematician Isaac Newton (1642–1727) showed that the behaviour of most things in the known universe could be described by mathematical rules involving algebra. It is only in the last 400 years that we have developed symbols to use with algebraic problems. Before then, words and sentences were used.

Questions

1. What is the origin of the word 'algebra'?
2. What is the name of the Persian poet who followed after Al-Khwarizmi?
3. What is the modern name for the city of Byzantium?
4. Which country now occupies the area formerly known as Mesopotamia?

Writing and finding formulas

One of the features of algebra which distinguishes it from other branches of mathematics is the use of letters or **pronumerals**.

A pronumeral is a letter or symbol that is used in place of a number.

Pronumerals allow us to write a rule more simply. Look at the following table.

Input	1	2	3	4	20
Output	4	5	6	7	23

The rule for this table can be written as output = input + 3. If we use the letter a to represent the input numbers and the letter b to represent the output numbers, the table looks like the one on the right.

a	1	2	3	4	20
b	4	5	6	7	23

The rule can now be written more simply as $b = a + 3$.

When the rule is written using pronumerals, it is called a formula.

Many students find algebra confusing because of the way letters are chosen at random to represent numbers. Any letter from the alphabet can be used to represent the input and output numbers for the algebra machine. For example, we could use x for the input numbers and m for the output numbers, making the formula $m = x + 3$. The actual letter you choose is not important; what is important is that the rule is shown correctly. When you write a formula using pronumerals, be careful to put the input and output numbers in the right place. For example, $x = m + 3$ is not the same formula as $m = x + 3$.

The value of a pronumeral is not determined by its place in the alphabet. If we use b to represent a number, this does not necessarily mean that it is a smaller number than x or c.

When multiplying numbers and pronumerals, we do not show the multiplication sign. We place numbers in front of pronumerals, so $h = k \times 5$ is written as $h = 5k$.

WORKED Example 4

Rewrite the following formulas, leaving out the multiplication (×) sign.

a $m = q \times 4 + 3$ **b** $b = (m + 2) \times 5$ **c** $g = (2 \times w - 6) \times 3$

THINK	WRITE
a 1 Write the formula.	**a** $m = q \times 4 + 3$
2 Leave out the multiplication sign. (Remember to write the number being multiplied with the pronumeral in front of the pronumeral.)	$m = 4q + 3$
b 1 Write the formula.	**b** $b = (m + 2) \times 5$
2 Leave out the multiplication sign. (Remember to write the number being multiplied in front of the bracket.)	$b = 5(m + 2)$

THINK

c 1 Write the formula.
2 Leave out the multiplication signs. (Remember to write the number being multiplied in front of the respective pronumeral and bracket.)

WRITE

c $g = (2 \times w - 6) \times 3$
$g = 3(2w - 6)$

WORKED Example 5

Look at the table at right and complete the formula by inserting a number in the gap.
$g = f + \_\_$

f	3	6	8	13	20
g	5	8	10	15	22

THINK	WRITE
1 Look at the first pair of numbers, 3 and 5. Look at the formula and try to guess the number which must be added to 3 to get 5.	$5 = 3 + 2$
2 Look at the next pair of numbers to see if adding 2 works again.	$8 = 6 + 2$
3 Check the other number pairs to see if this works every time.	$10 = 8 + 2$ $15 = 13 + 2$ $22 = 20 + 2$
4 Write the formula.	$g = f + 2$

WORKED Example 6

Use the pronumerals given to write a formula for the table at right.

a	4	5	6	7	13
b	3	5	7	9	21

THINK	WRITE
1 Look for two consecutive numbers in the top row, for example, $a = 4$ and $a = 5$, and compare the difference between their corresponding b values.	$5 - 3 = 2$
2 Repeat step 1 for a different pair of consecutive numbers in the top row, for example, $a = 6$ and $a = 7$.	$9 - 7 = 2$

Continued over page →

THINK	WRITE	
3 If the value obtained in step 1 is the same as the one in step 2, use it along with the basic operations (+, −, ×, ÷) or a combination of operations to obtain the correct result. *Note:* The b-value increases by 2 for every increase of 1 in the a-value. This means the formula will contain $a \times 2$. Therefore possible rules may be 'Multiply a by 2 and add a number to obtain b' or 'Multiply a by 2 and subtract a number to obtain b'.	Attempt 1 $a = 4$ and $b = 3$ Attempt 2 $a = 4$ and $b = 3$	Multiply a by 2 and add a number $4 \times 2 = 8$, 8 + any number does not equal 3 Multiply a by 2 and subtract a number $4 \times 2 = 8$, $8 - 5 = 3$ ✓
4 Check that the combination of operations (rule) works for the other number pairs. *Note:* If the rule does not work for all pairs of values in the table further attempts must be used.	$a = 5$ and $b = 5$ $a = 6$ and $b = 7$ $a = 7$ and $b = 9$ $a = 13$ and $b = 21$	$5 \times 2 = 10$, $10 - 5 = 5$ $6 \times 2 = 12$, $12 - 5 = 7$ $7 \times 2 = 14$, $14 - 5 = 9$ $2 \times 13 = 26$, $26 - 5 = 21$
5 Write the formula carefully, being sure to put the correct pronumerals in the right places. Remember when the rule is written using pronumerals, it is called a formula.	$b = 2a - 5$	

remember

1. A *pronumeral* is a letter or symbol that is used in place of a number.
2. When a rule is written using pronumerals it is called a *formula*.
3. When multiplying numbers and pronumerals we don't need to show the multiplication sign, and we place numbers being multiplied in front of pronumerals.

EXERCISE 5B Writing and finding formulas

WORKED Example 4

1 Rewrite the following formulas, leaving out the multiplication (×) sign.

a $b = 4 \times h$
b $m = f \times 4$
c $r = a \times 5$
d $m = t \times 4$
e $x = (k + 4) \times 5$
f $k = 6 \times w - 2$
g $t = 4 \times (20 - g)$
h $b = 10 \times a - 5$
i $d = 6 \times f + 7$
j $h = (x + 5) \times 9$
k $y = (b \times 3 + 6) \times 8$
l $y = 8 \times p - 6$
m $g = 2 \times (3 \times r + 17)$
n $j = (h \times 5 - 4) \times 18$

WORKED Example 5

2 Look at the following tables and complete the formula for each table by inserting a number in the gap.

5.3 SkillSHEET Describing a number pattern from a table

a

f	0	1	3	6	8
g	9	10	12	15	17

$g = f +$ ____

b

a	3	4	7	8	11
b	0	1	4	5	8

$b = a -$ ____

c

k	1	3	5	6	12
t	3	7	11	13	25

$t = 2k +$ ____

d

x	1	3	5	8	11
y	2	16	30	51	72

$y = 7x -$ ____

e

a	1	4	6	9	12
g	6	18	26	38	50

$g = 4a +$ ____

f

m	1	2	3	4	11
t	20	29	38	47	110

$t = 9m +$ ____

g

p	0	4	5	6	13
w	2	14	17	20	41

$w =$ ____ $\times p + 2$

h

t	2	3	7	9	12
x	2	7	27	37	52

$x =$ ____ $\times t - 8$

i

m	1	2	4	7	15
p	2	5	11	20	44

$p = 3m -$ ____

j

s	1	5	6	7	12
b	3	43	53	63	113

$b = s \times$ ____ $- 7$

WORKED Example 6

3 Look at the following tables and use the pronumerals given to write a formula for each table.

GC program—TI Finding the formula

GC program—Casio Finding the formula

a

a	2	4	5	8	12
b	5	7	8	11	15

b

t	1	2	3	8	15
w	6	7	8	13	20

c

m	4	5	7	11	12
a	0	1	3	7	8

d

s	0	1	2	5	12
t	3	5	7	13	27

e

f	1	3	4	6	11
g	1	5	7	11	21

f

s	1	3	7	9	12
c	4	10	22	28	37

g

d	2	4	5	7	12
a	5	13	17	25	45

h

s	1	2	5	6	10
g	1	6	21	26	46

i

f	2	3	4	8	11
e	2	13	24	68	101

j

p	1	2	5	6	12
q	24	44	104	124	244

k

b	20	55	60	75	100
t	80	220	240	300	400

l

u	5	10	20	100	7000
p	14	29	59	299	20 999

For questions **4** to **12**, a formula has been used by an algebra machine to produce each table.

4 multiple choice

a	0	1	3	6	36
b	2	3	5	8	38

The formula used is:

A $b = a + 5$
B $b = 3a$
C $b = a + 2$
D $b = a - 2$
E $b = a - 5$

5 multiple choice

g	1	2	4	18	56
k	0	1	3	17	55

The formula used is:

A $k = g + 1$
B $k = 2g$
C $k = g - 1$
D $k = 2g - 2$
E $k = 2g + 2$

6 multiple choice

r	5	6	8	9	17
w	0	1	3	4	12

The formula used is:

A $w = r + 5$
B $w = 5r$
C $w = 2r - 5$
D $w = r - 5$
E $w = r + 6$

7 multiple choice

p	0	6	7	9	21
m	0	12	14	18	42

The formula used is:

A $m = p$
B $m = p + 6$
C $m = p + 2$
D $m = 12p$
E $m = 2p$

8 multiple choice

m	1	2	5	7	98
w	8	9	12	14	105

The formula used is:

A $w = 8m$
B $w = 2m + 6$
C $w = m + 7$
D $w = 2m + 5$
E $w = 2m$

9 multiple choice

p	2	3	4	11	78
t	8	11	14	35	236

The formula used is:

A $t = p + 6$
B $t = 2p + 3$
C $t = 4p$
D $t = 3p + 2$
E $t = 3p + 1$

10 multiple choice

g	2	3	4	11	78
k	10	15	20	55	390

The formula used is:

A $k = 5g$
B $k = 4g + 2$
C $k = g + 8$
D $k = 2g + 6$
E $k = 4g - 2$

11 multiple choice

b	1	2	9	14	65
a	9	11	25	35	137

The formula used is:

A $a = 8b + 1$
B $a = 3b + 5$
C $a = b + 8$
D $a = 2b + 7$
E $a = b - 8$

12 multiple choice

m	2	3	5	7	11
e	25	37	61	85	133

The formula used is:

A $e = m + 78$
B $e = 10m + 11$
C $e = 12m + 1$
D $e = 6m + 43$
E $e = 10m - 11$

WorkSHEET 5.1

10 QUICK QUESTIONS 1

1 Complete the following table using the rule given. Rule: Add 4 to each input number.

Input	2	5	14	20	23	36
Output						

2 Complete the following table using the rule given. Rule: Subtract 2 from each input number, then multiply by 3.

Input	6	8	11	16	24	32
Output						

3 Complete the following table using the rule given to work out the missing input and output numbers. Rule: Divide each input number by 5.

Input			30	40	65	
Output	3	5				14

4 Find the formula used by the algebra machine to complete this table.

Input	1	2	3	4	5	6
Output	7	8	9	10	11	12

5 Find the formula used by the algebra machine and then use this to complete the rest of the table.

Input	1	4	7	12	13	18	24	31
Output		16	28	48	52			

6 Find the formula used by the algebra machine to complete this table. (*Hint:* The rule has two steps.)

Input	2	8	10	16	23	35
Output	15	45	55	85	120	180

7 Rewrite $k = 8 \times w + 12$ without a multiplication sign.

8 Rewrite $g = 7 \times m - 4 \times (t + 11)$ without multiplication signs.

9 Complete the rule for the table by inserting a number in the space provided.

a	4	5	8	11	26	31
b	2	3	6	9	24	29

$b = a -$ ___

10 Complete the rule for the table by inserting a number in the space provided.

t	4	6	10	11	17	25
m	20	28	44	48	72	104

$m = 4t +$ ___

Substitution

If we know the rule or formula for a particular table, we can take on the role of the 'algebra machine' and find the numbers that are needed to fill any gaps in the table. Look at the table below.

g	2	3	5	7	11
e	8	13	23		

Were you able to work out that the formula (rule) for this table is $e = 5g - 2$? This formula can be used to find the missing numbers in the table.

For example, to find the first missing 'e' number, write the formula: $e = 5g - 2$. Now replace g with its value from the table; that is, 7. This gives:

$$\begin{aligned} e &= 5 \times 7 - 2 \text{ (remember, } 5g \text{ means } 5 \times g\text{)} \\ &= 35 - 2 \\ &= 33 \end{aligned}$$

The other gap in the table can be filled the same way.

$$\begin{aligned} e &= 5g - 2 \\ e &= 5 \times 11 - 2 \\ &= 55 - 2 \\ &= 53 \end{aligned}$$

If the value of g is 20, let us consider how we can find the value of e. First we would write the rule to find e and then replace g with the given value of 20.

$$\begin{aligned} e &= 5g - 2 \\ \text{If } g &= 20 \\ e &= 5 \times 20 - 2 \\ &= 100 - 2 \\ &= 98 \end{aligned}$$

When a pronumeral in a formula is replaced by a number, we say that the number is *substituted* into the formula.

In the example above, g was replaced by 20; that is, 20 was substituted for g. Numbers can be substituted for pronumerals whenever the formula is known. Look at the following worked example.

WORKED Example 7

If $x = 3t - 6$, substitute the given values of t into the formula to find the value of x in each case.

a $t = 5$ **b** $t = 12$

THINK	WRITE
a 1 Write the formula.	**a** $x = 3t - 6$
2 Substitute 5 for t.	If $t = 5$
3 Perform the multiplication.	$x = 3 \times 5 - 6$
4 Subtract 6 from the above result.	$x = 15 - 6$
5 Write the answer.	$x = 9$

THINK	WRITE
b 1 Write the formula.	**b** $x = 3t - 6$
2 Substitute 12 for t.	If $t = 12$
3 Perform the multiplication.	$x = 3 \times 12 - 6$
4 Subtract 6 from the above result.	$x = 36 - 6$
5 Write the answer.	$x = 30$

WORKED Example 8

Find the value of m by substituting the given value of x into the formula below.
$m = 3(2x + 3)$, $x = 4$

THINK	WRITE
1 Write the formula.	$m = 3(2x + 3)$
2 Substitute 4 for x.	If $x = 4$ $m = 3(2 \times 4 + 3)$
3 Work inside the brackets first: (a) Multiply 2 by 4 (remember that $2x$ means $2 \times x$). (b) Add 3 to the above result.	$m = 3(8 + 3)$ $m = 3(11)$
4 Multiply by the number outside the brackets.	$m = 3 \times 11$
5 Write the answer.	$= 33$

remember

When a pronumeral in a formula is replaced by a number, we say that the number is substituted into the formula.

EXERCISE 5C Substitution

EXCEL Spreadsheet Substitution

1 Use the given rule to complete the table in each case.

a $y = x + 5$

x	2	3	11	18	40
y					

b $m = n - 17$

n	17	20	22	43	700
m					

c $y = x + 1$

x	1	3	40	70	123
y					

d $p = 2h + 4$

h	1	5	11	22	62
p					

Mathcad Substitution

EXCEL Spreadsheet Substitution game

e $t = 3b - 7$

b	5	15	20	60	100
t					

f $s = 4t + 4$

t	1	3	7	8	40
s					

g $y = x - 1$

x	1	3	40	70	123
y					

h $s = 10g - 8$

g	1	7	8	19	154
s					

i $t = 12 - m$

m	3	4	7	11	12
t					

j $p = 6f$

f	1	8	11	12	111
p					

k $m = 12 - 2y$

y	0	1	3	4	6
m					

l $k = 12r - 30$

r	4	11	12	16	804
k					

WORKED Example 7

2 Substitute the given values into each formula to find the value of m in each case.

a $m = g - 2$
- **i** $g = 4$ **ii** $g = 5$ **iii** $g = 2$ **iv** $g = 102$

b $m = 2t - 3$
- **i** $t = 7$ **ii** $t = 2$ **iii** $t = 100$ **iv** $t = 8$

c $m = 12h + 7$
- **i** $h = 1$ **ii** $h = 0$ **iii** $h = 5$ **iv** $h = 20$

d $m = 25 - 4w$
- **i** $w = 1$ **ii** $w = 3$ **iii** $w = 6$ **iv** $w = 0$

WORKED Example 8

SkillSHEET 5.4 Substitution into rules

3 Find the value of m by substituting the given value of the pronumeral into the formula.

a $m = 2(g + 1)$
- **i** $g = 1$ **ii** $g = 0$ **iii** $g = 12$ **iv** $g = 75$

b $m = 5(x - 2)$
- **i** $x = 6$ **ii** $x = 10$ **iii** $x = 11$ **iv** $x = 2$

c $m = 4(12 - p)$
- **i** $p = 2$ **ii** $p = 3$ **iii** $p = 12$ **iv** $p = 11$

d $m = 5(2g - 3)$
- **i** $g = 2$ **ii** $g = 14$ **iii** $g = 5$ **iv** $g = 9$

e $m = 2(d + 2) - 3$
- **i** $d = 3$ **ii** $d = 0$ **iii** $d = 7$ **iv** $d = 31$

f $m = 3(f - 1) + 17$
- **i** $f = 1$ **ii** $f = 3$ **iii** $f = 6$ **iv** $f = 21$

g $m = 4s - s$
- **i** $s = 3$ **ii** $s = 1$ **iii** $s = 101$ **iv** $s = 72$

WorkSHEET 5.2

h $m = 3(y + 5) - 2$
- **i** $y = 1$ **ii** $y = 5$ **iii** $y = 0$ **iv** $y = 12$

i $m = 50 - 6v$
- **i** $v = 4$ **ii** $v = 7$ **iii** $v = 1$ **iv** $v = 8$

COMMUNICATION

Nutrition panels

Most tinned and packaged foods that we buy have labels which list the nutritional content of the food. Most of the energy (measured in calories or kilojoules) in food comes from fat, protein and carbohydrates.

It is estimated that there are 9 calories in each gram of fat and 4 calories in each gram of protein and each gram of carbohydrate. Hence, if a particular food contained 1.6 grams of fat, 2.1 grams of protein and 16.3 grams of carbohydrate, it would be equal to $1.6 \times 9 = 14.4$ calories of fat, $2.1 \times 4 = 8.4$ calories of protein and $16.3 \times 4 = 65.2$ calories of carbohydrate. The total number of calories would be 88 (that is, $14.4 + 8.4 + 65.2$).

1 From the label shown, write the number of grams of fat, protein and carbohydrate in 100 grams of baked beans.

2 Estimate how many calories there would be in 100 grams of baked beans using the information supplied above. *Note:* To convert kilojoules to calories, divide by 4.2.

3 Compare this to the value stated on the tin.

If we wanted to look at a wide variety of foods, it would be more convenient to have a rule or formula into which we could substitute numbers to easily obtain our answer.

4 Clearly explain how you obtained your answer in question 2.

5 Summarise your explanation from question 4 and, in words, write a rule that could be used to calculate the number of calories in food.

6 Choose a pronumeral for each of the following quantities:

a number of grams of fat
b number of grams of protein
c number of grams of carbohydrate
c number of calories.

7 Using the pronumerals you have chosen, write a formula to calculate the number of calories.

8 The table below contains data on the fat, protein and carbohydrate content of selected foods. Copy this table into your workbook and write in your chosen pronumerals in the first row within the brackets. Use your formula to work out the calorie content of each of these foods. (The last column will be completed later on.)

Food	Number of grams of fat ( )	Number of grams of protein ( )	Number of grams of carbohydrate ( )	Number of calories ( )	Number of kilojoules ( )
1 hot cross bun	3	3	19		
100 grams of chocolate cake	16	4	56		
100 grams of roast chicken	14	26	0		
70 grams of bacon	8	21	0		
2 grilled sausages	17	13	15		
1 piece of fish (flake), no batter	1	21	0		
10 grams of sultanas	0	0	6		
25 grams of dried apricots	0	0	17		
1 banana	0	1	20		
1 apple	0	0	17		
1 serve of carrots	0	0	5		
1 serve of potatoes	0	0	17		
250 millilitres of milk	10	8	12		
210 grams of tinned tomato noodle soup	0.8	2.6	14.9		

9 Which quantity of food in this list has the highest number of calories? Is this what you expected?

10 What types of food in this list generally have a lower number of calories?

A calorie is an energy measurement unit in the imperial system. Nowadays we mostly use the metric system unit of kilojoules to measure energy. It is estimated that the number of kilojoules is equivalent to the number of calories multiplied by 4.2.

11 Use a pronumeral to represent each of the quantities 'number of calories' and 'number of kilojoules' and write a formula connecting them.

12 Complete the final column of your table by using the formula to find the number of kilojoules for each listed food. (Round your answers to the nearest kilojoule.)

13 Obtain three labels from three types of packaged food. Use the formula for converting calories to kilojoules to obtain estimates of the energy for each food item in both calories and kilojoules. Compare your answers to the data supplied on each of the three nutrition labels.

10 QUICK QUESTIONS 2

1 Rewrite $c = 6 \times k + 3 \times t$ without multiplication signs.

2 Insert multiplication signs wherever possible in $r = 17y + 8h - 2m$.

3 Complete the rule for the table by inserting a number in the space provided.

y	3	8	15	22	46	62
w	5	15	29	43	91	123

$w = 2y -$ ___

4 Study the following table and use the pronumerals to write a formula.

f	2	6	8	15	19	23
c	6	18	24	45	57	69

5 Use the rule given to complete the table.
$h = n + 5$

n	1	5	9	16	18	23
h						

6 Use the rule given to complete the table.
$c = 4r + 12$

r	2	8	15	16	32	41
c						

7 Substitute $k = 4$ into the formula $g = 5(k + 2)$ to find the value of g.

8 Substitute $t = 6$ into the formula $m = 4t - 3$ to find the value of m.

9 Substitute $f = 11$ into the formula $d = 5(2f - 7)$ to find the value of d.

10 If $k = 3$, $w = 5$ and $y = 7$, find the value of $2y - w + 4k$.

MATHS QUEST CHALLENGE

1 **If a mother is 37 years old and her daughter is 9 years old, in how many years will the mother be three times as old as the daughter?**

2 **A class of 25 students has 7 more boys than girls. How many boys are there?**

3 **Two men and two boys wish to cross a river from one side to the other. Their small canoe will carry only one man or two boys. What is the minimum number of canoe trips needed to get everyone across? List the number of people transported during each canoe trip and the destination.**

Terms, expressions and equations

As mentioned before, a pronumeral is a letter that is used in place of a number. Sometimes a pronumeral has a number written in front, for example, $7x$ or $3y$. These numbers are known as *coefficients*. So the coefficient in the term $6xyz$ is 6. The value of a pronumeral can vary, so it is known as a *variable*. The value of a number is constantly the same, so it is known as a *constant*. The number 6 always has a value of 6, whereas the variable x may take the value 6, $\frac{1}{5}$, 2.3, ….

Expressions are parts of a rule made up of pronumerals and numbers. For example, in the rule $y = 3k - 8$, we call '$3k - 8$' an expression. Here are some examples of expressions: $3t + 4$, $7 - 2k$, $6(g - 9)$, $3r + 6t$, $4y + 2y$, $9p - 3y + 5$.

When we want to solve problems with quantities that may vary, we can start building an expression by representing the variables with a pronumeral. Let's look at an example.

Molly and Vinh had a pile of 20-cent coins and some moneyboxes. They filled one moneybox before starting the next. When they had finished, they had two full moneyboxes and five coins left over.

To work out how many coins Molly and Vinh had, we would need to find out how many coins fitted into a moneybox. To help us write down an expression for this problem, we use a pronumeral to represent the number of coins in each moneybox.

Let c represent the number of coins in a moneybox.

We have two moneyboxes plus 5 coins = 2 lots of c plus 5 coins

$= 2 \times c + 5$

$= 2c + 5$

Therefore, the number of coins Molly and Vinh had is $2c + 5$ coins.

WORKED Example 9

Write expressions to represent the total number of coins in each of the following situations where  **represents a full moneybox and**  **represents one coin. Use c to represent the number of coins in a moneybox.**

a Mary fills one moneybox and has two coins left over. 

b Valentino and Loris decide to combine their coins. Valentino has three moneyboxes with four coins left over and Loris has two moneyboxes with six coins left over. That is,

Valentino has 

Loris has 

THINK	WRITE
a 1 The total number of coins will be found by adding the number of coins in one moneybox plus two coins left over.	**a** $1 \times c + 2$
2 Remove the multiplication sign to write the expression.	$= c + 2$

THINK | **WRITE**

b ① Add the number of coins Valentino and Loris have. — **b** $(3c + 4) + (2c + 6)$

② Count the moneyboxes first, and then the coins. — $= (3c + 2c) + (4 + 6)$

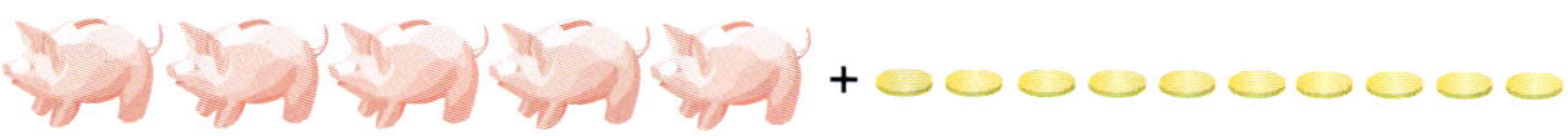

③ Simplify. — $= 5c + 10$

Equations always contain an equals sign, whereas expressions do not. Here are some examples of equations:

$$y = 4t - 6, \quad m + 3 = y - 8, \quad 2(g + 5) = 5, \quad y = 6t, \quad w = e - 87, \quad 9 = 7 + 4r.$$

The expressions and equations shown above are made up of **terms**. Terms may contain one or more pronumerals, such as $6t$ or $5axy$ or they may consist of a number only. Here are some examples of terms: $3t$, $2y$, 7, $5gh$, m, brt. Terms are separated by + or – signs.

This can be summarised as shown below.

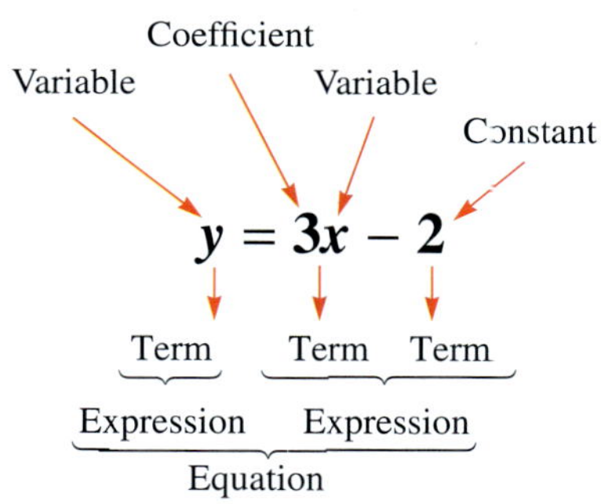

As seen earlier in worked example 9, information written in words can often be displayed more clearly as an expression. Look at further examples that follow.

WORKED Example 10

Write an expression for the sum of T and G.

THINK	**WRITE**
The word 'sum' means to add together.	The expression is $T + G$ or $G + T$.

When writing expressions or equations, think about which operations are being used and the order in which they occur.

WORKED Example 11

If Y represents any number, write expressions for:
a 5 times that number
b 2 less than that number
c 8 more than that number
d the number divided by 4 (or the quotient of Y and 4)
e the next consecutive number (that is, the counting number which comes after Y).

THINK	WRITE
In each case think about which operations are being used and the order in which they occur.	
a When multiplying we don't show the multiplication sign. Remember to put the number first.	a $5 \times Y$ or $5Y$
b An expression with 2 less means 'subtract 2'.	b $Y - 2$
c An expression with 8 more means 'add 8'.	c $Y + 8$
d A number divided by 4 means 'write as a quotient (fraction)'.	d $\frac{Y}{4}$ or $Y \div 4$
e The next consecutive number means 'add 1 to the number'.	e $Y + 1$

WORKED Example 12

Write expressions for the following rules.
a Take a number and add another number.
b Multiply 2 numbers.
c Add 2 numbers and multiply the answer by 6.
d Take a number and multiply it by 4, then subtract 8 from that answer.

THINK	WRITE
1 In each case choose any pronumeral to represent an unknown number.	
2 In each case think about which operations are being used and the order in which they occur.	
a Add two pronumerals.	a $B + G$ or $G + B$
b Multiply two pronumerals but don't write a multiplication sign (×).	b xy
c Add two pronumerals and place the expression in brackets. Multiply by 6. (Remember to write the numbers first.)	c $6(t + f)$
d First multiply a pronumeral by 4, writing the number first ($4b$). Then subtract 8.	d $4b - 8$

WORKED Example 13

Write equations for the following rules.

a If 15 is added to a number the result is 21.

b The quotient of a number and 7 is equal to 5.

THINK	WRITE
a 1 Choose a pronumeral to represent the number.	**a** Let x = the number
2 Add 15 to the pronumeral then place an equals sign followed by 21.	$x + 15 = 21$
b 1 Choose a pronumeral to represent the number.	**b** Let z = the number
2 Divide the pronumeral by 7, (that is, write it as a fraction) then place an equals sign followed by 5.	$\frac{z}{7} = 5$

remember

1. A pronumeral is a letter used in place of a number.
2. A pronumeral is a *variable*, whereas a number is a *constant*.
3. A number in front of a variable is known as a *coefficient*.
4. *Expressions* are parts of a rule made up of pronumerals and numbers.
5. *Equations* always contain an equals sign whereas expressions do not.
6. Expressions and equations are made up of *terms*.
7. When writing expressions, think about which operations are being used, and the order in which they occur.
8. If pronumerals are not given in a question, you can choose any letters you like.

EXERCISE 5D Terms, expressions and equations

WORKED Example 9a

1 Write an expression for the total number of coins in each of the following, using c to represent the number of coins in a moneybox.

a 

b 

c 

WORKED Example 9b

d 

e 

2 Christie and Jane both had two full moneyboxes and seventeen coins each. They combined their money and spent every cent on a day out in town.

a Write an expression to represent the number of coins they had in total.

b Christie opened the first moneybox and counted 52 twenty-cent pieces inside. Use your expression to calculate how many coins the girls had in total, assuming that all the moneyboxes held the same number of coins.

c What is the total amount that the girls spent on their day out?

3 Luke has three macadamia nut trees in his backyard. He saves takeaway containers to store the nuts in. He has two types of containers, rectangular 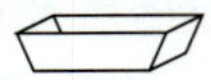 and round . Using m to represent the number of nuts in a rectangular container and n to represent the number of nuts in a round container, write expressions for the following.

a 

b 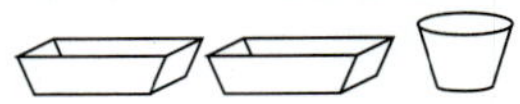

c 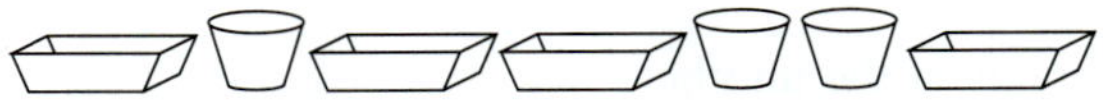

d  where  represents one nut.

e 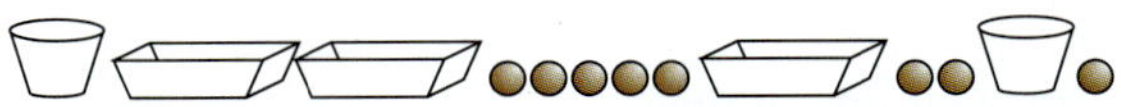

4 Luke found that a rectangular container holds 17 nuts and a round container holds 12 nuts. Calculate how many nuts Luke would have for each part in question **3** by substituting the appropriate values into each expression.

SkillSHEET 5.5 Alternative expressions used to describe the four operations

WORKED Example 10

5 Write an expression for each of the following.

a The sum of B and 2
b 3 less than T
c 6 added to D
d 5 taken away from K
e The sum of G, N and W
f D increased by H
g N increased by N
h H added to C
i G subtracted from 12
j D multiplied by 4
k 6 added to H
l The difference between Z and G
m B multiplied by F
n Y added to the product of 3 and M

6 Answer true or false for each of the statements below.

a $3x$ is a term.
b $3mn$ is a term.
c $g = 23 - t$ is an expression.
d $g = 5t - 6$ is an equation.
e $rt = r \times t$
f $5 + A = A + 5$
g $3d + 5$ is an expression.
h $7 - B = B - 7$
i $3x = 9$ is an expression.
j The expression $g + 2t$ has two terms.
k $2f + 4 = 4 + 2f$
l $5a - 7t + h$ is an equation.

7 **multiple choice**

In each of the following, the letter N has been used to represent a number. For each expression written in words, choose the answer that you think matches:

a five times the number
A $N + 5$ **B** $5N$ **C** $N - 5$ **D** $N \div 5$ **E** $5 - N$

b the sum of the number and 52
A $52 - N$ **B** $52N$ **C** $N + 52$ **D** $N - 52$ **E** $N \div 52$

c the next consecutive number
A O **B** $2N$ **C** $N - 1$ **D** $N + 1$ **E** $12N$

d the number multiplied by another number
A $N + T$ **B** $N \div S$ **C** $N + 4$ **D** NS **E** $N - 5$

e half the number
A $2N$ **B** $N \div 2$ **C** $N + \frac{1}{2}$ **D** $N - 2$ **E** $2 + N$

f four more than the number
A $N + 4$ **B** $4N$ **C** $N - 4$ **D** $4 - N$ **E** $N \div 4$

g the number plus the product of another number and 2
A $N + 2D$ **B** $D(N + 2)$ **C** $2N + 2B$ **D** $2N + D$ **E** $N + T$

WORKED Example 11

8 If A, B and C represent any 3 numbers, write an expression for each of the following.

a The sum of all 3 numbers
b The difference between A and C
c The product of A and B
d The product of all 3 numbers
e The quotient of B and C (that is, B divided by C)
f The sum of A and C, divided by B
g 3 more than A

WORKED Example 12

9 Write expressions for the following rules.

a The number of students left in the class if X students leave for the canteen out of a total group of T students
b The amount of money earned by selling B cakes where each cake is sold for $4.00
c The total number of sweets if there are G bags of sweets with 45 sweets in each bag
d The cost of one concert ticket if 5 tickets cost $$T$
e The number of students in a class that contains R boys and M girls
f The number of insects if there are Y legs altogether
g The cost of M movie tickets if each ticket costs $11.00
h The total amount of fruit sold if A apples and H bananas are sold
i The number of cards left in a pack of 52 cards if J cards are removed
j The total number of floor tiles if there are B boxes of tiles with 12 tiles in each box
k The average of B and G
l The total runs scored by 3 batsmen if they have scores of A, H and K respectively
m The total number of legs if there are R chairs

WORKED Example 13

10 Write equations for the following rules.

a If 8 is added to a number the result is 11.
b If 11 is subtracted from a number the result is 63.
c The product of a number and 8 is equal to 128.
d The quotient of a number and 7 is equal to 5.
e The total cost of A apples and H bananas is $6.49.
f When Q cards are removed from a pack of 52 cards only 19 remain.
g The total cost of M concert tickets at $99 each was $792.
h The total number of legs if there are R chairs and D tables is 164.

What was the 'Model T' in 1908?

Match up the everyday situations with the appropriate formula. The number and letter beside each give the puzzle answer code.

- **S** Fish and chips cost \$2.80 per serve.
- **F** At a hotel, each room can accommodate 4 people.
- **U** Each person at the grand final pays a \$12 entry fee.
- **N** Mrs Min buys 5 eggs for each member of her family.
- **M** Table tennis bats cost \$18 each and balls are worth \$6 per person.
- **A** To go to the movies costs \$7 per person.
- **Y** Mark takes 6 minutes to wash each window of the house.
- **C** 'Bo' eats 3 cans of dog food each week.
- **H** Sally was paid \$15 per hour for a week.
- **O** A can of paint will cover an area of 16 square metres.
- **P** Sausages cost \$5/kg and chops cost \$6/kg.
- **D** A \$50 prize is to be split evenly between the members of the winning family.
- **I** Petrol for the car is worth 82 cents per litre.
- **R** Yvonne has 2 sandwiches per day on her holidays.
- **E** Ben's pocket money consists of \$10 plus \$1 for each time he has washed his dad's car.
- **T** Rolls cost 28 cents each.

1. $P = 15H$
2. $P = 10 + C$
3. $E = 5N$
4. $S = 2D$
5. $T = 6W$
6. $P = 4R$
7. $C = \frac{A}{16}$
8. $A = \frac{50}{N}$
9. $C = 82L$
10. $C = 28R$
11. $A = 12N$
12. $C = 3W$
13. $C = 2.8N$
14. $C = 7N$
15. $C = 5S + 6C$
16. $C = 18B + 6D$

1	2	3	4	5	6	7	4	8

9	3	10	4	7	8	11	12	2	8	10	1	2

6	9	4	13	10	12	1	2	14	15	16	14	13	13

15	4	7	8	11	12	2	8	12	14	4

Like terms

Terms containing exactly the same pronumerals regardless of order are called **like terms**. For example $5e$ and $11e$, and $8ast$ and sat are like terms. But $3bad$ and $9bd$ are not like terms, since the pronumerals are not exactly the same.

Further examples of like and non-like terms are:

Like terms

$5e$ and $6e$
x and $13x$
$2bc$ and $4cb$
$3g^2$ and $45g^2$

If c represents the number of chicks, then the diagram above illustrates an example of like terms.

Non-like terms

$5x$ and $5y$
$12xy$ and $21x$
$15abc$ and $4bcd$
$16g$ and $45g^2$

If b represents the number of tennis balls and e represents the number of emus, then the diagram at right illustrates an example of non-like terms.

Simplifying expressions containing like terms

Expressions can often be written in a more simple form. For example, the expression $3x + 8x$ can be written more simply as $11x$. Notice that the expression was simplified (put into a more simple form) even though we did not know the value of the pronumeral x.

When simplifying expressions, we can collect (add or subtract) only like terms.

Here are some more examples of simplifying:

$6b + 4b = 10b$ $\quad$ $5xy + 8xy = 13xy$
$12y + 3y - 5y = 10y$ $\quad$ $4t - 3t = t$
$7t - 5t = 2t$

Remember that simplifying can be done only with like terms. (The pronumerals must be exactly the same.) For example:

$5g + 5c$ cannot be simplified $\quad$ $4t - 4$ cannot be simplified
$3xy + 4x$ cannot be simplified $\quad$ $14abc - 6bc$ cannot be simplified
$45b - 43$ cannot be simplified $\quad$ $4g - 4n$ cannot be simplified.

WORKED Example 14

Where possible, simplify the following expressions by adding or subtracting like terms.

a $4g + 6g$ **b** $11ab - ab$ **c** $6ad + 5da$ **d** $4t + 7t - 5$ **e** $8x + 3y$

THINK	WRITE
a The pronumerals are the same. The terms can be added.	**a** $4g + 6g = 10g$
b The pronumerals are the same. The terms can be subtracted. Note that *ab* is the same as 1*ab*.	**b** $11ab - ab = 10ab$
c Although the order of the pronumerals is different, the terms are like terms and can be added. *Note:* Pronumerals may be rearranged in alphabetical order.	**c** $6ad + 5da = 6ad + 5ad$ $= 11ad$
d The pronumerals are the same and the first two terms can be added. Do not add the other term.	**d** $4t + 7t - 5 = 11t - 5$
e The pronumerals are not the same. These are not like terms and therefore cannot be simplified.	**e** $8x + 3y$ cannot be simplified.

In the next worked example we will see that substituting known values for pronumerals in an expression allows us to evaluate the expression.

WORKED Example 15

Simplify the following expressions first, then find the value of the expression if $a = 4$.

a $5a + 2a$ **b** $7a - a + 5$

THINK	WRITE
a 1 Add the like terms. 2 Substitute 4 for *a* in the simplified expression. 3 Evaluate.	**a** $5a + 2a = 7a$ If $a = 4$ $7a = 7 \times a$ $= 7 \times 4$ $= 28$
b 1 Subtract the like terms. 2 Substitute 4 for *a* in the simplified expression. 3 Evaluate.	**b** $7a - a + 5 = 6a + 5$ If $a = 4$ $6a + 5 = 6 \times a + 5$ $= 6 \times 4 + 5$ $= 24 + 5$ $= 29$

remember

1. Terms containing the same pronumerals are called *like terms*.
2. The process of adding or subtracting like terms is called *simplifying*.
3. Substituting known values for pronumerals in an expression allows us to evaluate the expression.

EXERCISE 5E Like terms

1 Answer True or False to each of the following statements.

a $4t$ and $6t$ are like terms.
b $3x$ and x are like terms.
c $5g$ and $5t$ are like terms.
d $7t$ and $5i$ are like terms.
e $5a$ and $6a$ are like terms.
f $4a$ and $4ab$ are like terms.
g $9fg$ and $4gf$ are like terms.
h $6gh$ and $7gk$ are like terms.
i $6r$ and $5r$ are like terms.
j yz and $45zy$ are like terms.
k $3acd$ and $6cda$ are like terms.
l $5g$ and $5fg$ are like terms.
m $8gefh$ and $3efgh$ are like terms.
n $6ab$ and $3ba$ are like terms.
o $8xy$ and $5xy$ are like terms.
p $12ep$ and $4pe$ are like terms.
q $7eg$ and $7g$ are like terms.
r $7y$ and $18yz$ are like terms.

2 **multiple choice**

Patterns and algebra — 001

a Which one of the following is a like term for $7t$?
A $7g$ **B** $5tu$ **C** 2 **D** $6f$ **E** $5t$

b Which one of the following is a like term for $3a$?
A $5ab$ **B** $6a$ **C** $3w$ **D** $a + 2$ **E** $6a$

c Which one of the following is a like term for $5ab$?
A $6abe$ **B** $4b$ **C** $2ba$ **D** 5 **E** $5abe$

d Which one of the following is a like term for $7bgt$?
A $7t$ **B** $4gt$ **C** $2gb$ **D** btg **E** $b + g + t$

e Which one of the following is a like term for $2fgk$?
A $fg + k$ **B** $5fg$ **C** $2fgk$ **D** $12f$ **E** $12p$

f Which one of the following is a like term for $20m$?
A $11m$ **B** $20mn$ **C** $20f$ **D** $m + 2$ **E** 20

g Which one of the following is a like term for $9xyz$?
A $9x$ **B** $2yz$ **C** $8xz$ **D** xzy **E** $9z$

3 Answer True or False to each of the following statements.

a The equation $y = 4x + 3x$ can be simplified to $y = 7x$.
b The equation $k = 8y + 4y$ can be simplified to $k = 10y$.
c The equation $y = 4x + 3$ can be simplified to $y = 7x$.
d The equation $b = 3a - a$ can be simplified to $b = 2a$.
e The equation $k = 7a + 4d$ can be simplified to $k = 11ad$.
f The equation $y = 5x - 3x$ can be simplified to $y = 2x$.
g The equation $m = 7 + 2x$ can be simplified to $m = 9x$.
h The equation $p = 3x - 2x$ can be simplified to $p = x$.
i The equation $t = 3h + 12h + 7$ can be simplified to $t = 15h + 7$.
j The equation $y = 16g + 6g - 7g$ can be simplified to $y = 15g$.

WORKED Example 14

Mathcad Simplifying expressions

4 Where possible, simplify the following expressions by adding or subtracting like terms.

a $3a + 2a$ b $9y + 5y$ c $3c + 12c$
d $4u - 2u$ e $7e + 13e$ f $7t - 2t$
g $12ab + 2ab$ h $8fg - 2fg$ i $4e - e$
j $6t + t$ k $f + 4f$ l $6y - 6y$
m $4x + 14x$ n $3m + 16m - 7m$ o $6a + 4a - 2a$

5 Simplify the following expressions if possible.

a $24ab + ab - 7$ b $5y + y - 3y$ c $5t + 5s$
d $18i + 12i - 2$ e $4t + 8t - 3 + 2t$ f $7r + 2r + 5r - r$
g $2x + 2y$ h $7y + 6$ i $18f - 2f + 5$
j $4 + 3g - g$ k $6t - 5t$ l $18bg - 18bg$
m $13mno - 11mno$ n $11pq + 3qp$ o $6pr + 2 + 5rp$
p $9e - 9t - 1$ q $7t + 4t - 5$ r $32t - 31t$
s $11aw - 3aw$ t $7xy - 7x$ u $5t + 6t - 8$
v $5t - t + 3$ w $3g + 7g - 2$ x $5r + 17r + 4 + 2$

6 Simplify each of the following.

a $3t - 3t$ b $18r - 18r$ c $12ab - 12ab$
d $5x - 5x + 8$ e $6t + 7 - 6t$ f $9g - 9g + 2$
g $13xyz - 13xyz$ h $5x + 7 - 5x$ i $5y + 2y - y$

WORKED Example 15

GAME time Patterns and algebra — 002

7 Simplify the following expressions first, then find the value of the expression if $a = 7$:

a $3a + 2a$ b $7a + 2a$ c $6a - 2a$
d $9a + a$ e $13a + 2a - 5a$ f $3a + 7a$
g $17 + 5a + 3a$ h $6a - a + 2$ i $a + a$
j $7a - 6a$ k $7a - 7a$ l $12a + 5a - 16$

MATHS QUEST CHALLENGE

A *cryptarithm* is a puzzle where each digit of a number has been replaced by a pronumeral or letter. Solve the following cryptarithms by determining the digit that corresponds to each letter to make the calculation true.

1
```
   A B
+  B A
 C A C
```

2
```
   D
×  D
  E D
```

3 This cryptarithm has several solutions. Find at least 4 solutions.
```
   T W O
+  T W O
 F O U R
```

4
```
    S E N D
+   M O R E
  M O N E Y
```

5
```
     O N E
+    T W O
+  F O U R
 S E V E N
```

Problem solving using algebra

Finding a pattern or rule can be a very useful problem-solving technique. When solving problems involving algebra, it is important to identify what your chosen pronumeral represents. Once you have found a rule or formula, **substitution** allows you to use the same rule with many different input numbers. An example follows.

WORKED Example 16

A farmer constructs temporary sheep pens by making a square, as shown in the diagrams below. A wire fence is attached to the posts which are placed at regular intervals around the pen.

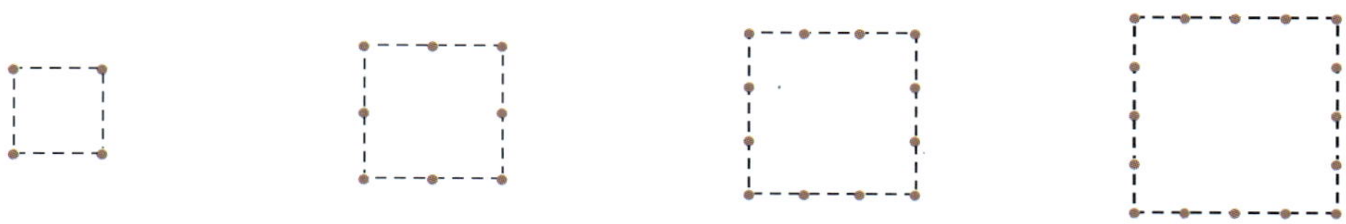

Number of posts on each side = 2
Total number of posts = 4

Number of posts on each side = 3
Total number of posts = 8

Number of posts on each side = 4
Total number of posts = 12

Number of posts on each side = 5
Total number of posts = 16

How many posts will be required for a square pen which has 45 posts on each side? (*Hint:* The answer is not 4×45 or 180 posts!)

THINK

1. Draw a table with a row for the number of posts on each side (p) and another row for the total number of posts (T). Using the diagrams above, write in the pairs of numbers.

WRITE

Let p represent the number of posts on each side and T represent the total number of posts.

p	2	3	4	5
T	4	8	12	16

2. Look at two consecutive numbers in the top row, for example, $p = 2$ and $p = 3$, and compare the difference between their corresponding T values.

$8 - 4 = 4$

3. Repeat step 2 for another two consecutive numbers in the top row, for example, $p = 4$ and $p = 5$.

$16 - 12 = 4$

4. If the value obtained in step 2 is the same as the one in step 3, use it along with the basic operations (+, –, ×, ÷) or a combination of operations to obtain the correct result.
Note: The T-value increases by 4 for every increase of 1 in the p-value. This means the rule will contain $T \times 4$.

Attempt 1
$p = 2$ and $T = 4$ $4 \times 2 = 8$, $8 +$ any number does not equal 4

Attempt 2
$p = 2$ and $T = 4$ $4 \times 2 = 8$, $8 - 4 = 4$ ✓

Continued over page

THINK	WRITE
5 Check that the combination of operations (rule) works for the other number pairs. *Note:* If the rule does not work for all pairs of values in the table, further attempts must be used.	$p = 3$ and $T = 8$ $\quad 4 \times 3 = 12, 12 - 4 = 8$ $p = 4$ and $T = 12$ $\quad 4 \times 4 = 16, 16 - 4 = 12$ $p = 5$ and $T = 16$ $\quad 4 \times 5 = 20, 20 - 4 = 16$
6 Write the rule carefully, being sure to put the correct pronumerals in the right places.	$T = 4p - 4$
7 Substitute $p = 45$ into the rule and answer the question.	If $p = 45$ Then $T = 4p - 4$ becomes $T = 4 \times 45 - 4$ $= 180 - 4$ $= 176$ A pen with 45 posts on each side requires a total of 176 posts.

Notice in the previous worked example that once we have found the formula which connects the two quantities, we can use it to find the number of posts that would be needed for a square pen of any size. For example, by substituting 20 for L, we could find the number of posts needed for a pen with 20 posts on each side. Try it.

remember

1. Identify what your chosen pronumeral represents.
2. Once you have found a rule or formula, substitution allows you to use the same rule with many different input numbers.

EXERCISE 5F Problem solving using algebra

SkillSHEET 5.6 Using tables to show number patterns

WORKED Example 16

1 The dining tables at the 'Trio' theatre restaurant are triangular in shape. Diners are seated at the tables in the arrangements shown below.

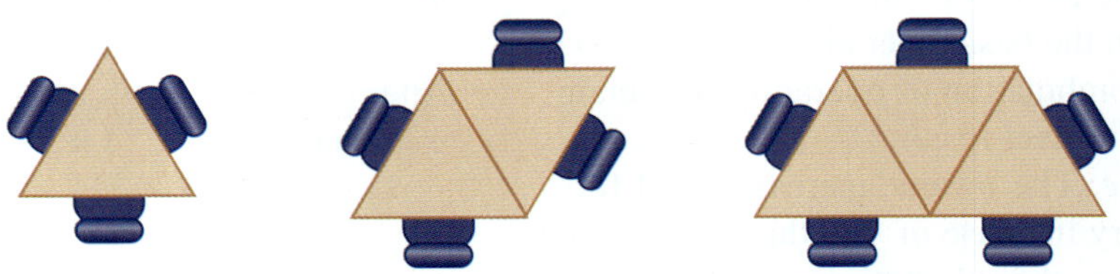

One table, 3 diners Two tables, 4 diners Three tables, 5 diners

The manager of the restaurant has received a booking from a large party. She sets out a row of 24 tables. How many diners can be seated at this row of tables?

Follow the steps below to solve this problem.

a Draw a table with a column for the number of tables (T) and another column for the number of diners (D). Using the diagrams above, write in the pairs of numbers.

b Draw a diagram to work out the number of diners who could sit at a row of 4 tables. Now draw another diagram to show the number of diners who could sit at a row of 5 tables. Write these pairs of values in your table.

c Work out the formula which connects the number of diners and the number of tables. Write the formula in the form $D = \ldots$

d Substitute $T = 24$ into the formula to find out how many guests could be seated at the row of 24 tables.

GC program—TI: Finding the formula

GC program—Casio: Finding the formula

EXCEL Spreadsheet: Substitution

Mathcad: Substitution

2 Jane is tiling the floor in her bathroom. She has decided to use blue and white tiles in the following pattern:

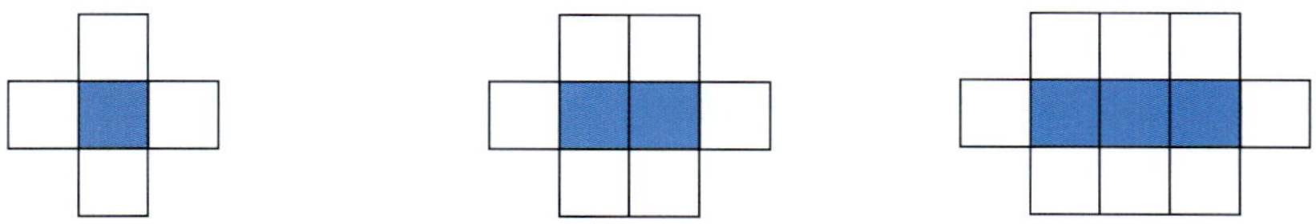

Number of blue tiles = 1, Number of white tiles = 4 | Number of blue tiles = 2, Number of white tiles = 6 | Number of blue tiles = 3, Number of white tiles = 8

Find out how many white tiles Jane would need to complete the pattern for a row of 10 blue tiles by following the steps below:

a Draw a table with two columns; one headed B showing the number of blue tiles and one headed W showing the number of white tiles required. Count the numbers of blue and white tiles from each of the tiling patterns above and put them into the table in the correct column.

b Draw a diagram of the tiling pattern for a row of 4 blue tiles and work out the number of white tiles needed to complete the pattern. Then draw another pattern for a row of 5 blue tiles. Add these two sets of values for B and W into the table.

c Look carefully at the table and work out the formula which connects W and B. Write the formula in the form $W = \ldots$

d Substitute $B = 10$ into the formula to find out how many white tiles would be needed for a row of 10 blue tiles, using Jane's tiling pattern.

To answer questions **3** to **5**, use steps similar to those you used in questions **1** and **2**.

3 Michael is constructing a timber fence at his stud farm. The sections of fence are shown below:

1 section, 4 pieces | 2 sections, 7 pieces | 3 sections, 10 pieces

Michael calculates that he will need 220 sections to fence off his first paddock.

How many pieces of timber will Michael need? Find the cost of the fence if each piece of timber costs \$3.85.

Follow the steps below.

a Draw a table with 2 columns, one headed S showing the number of sections and one headed P showing the number of pieces of timber required. Write in the numbers from the diagrams above.

b Draw a sketch of the fence with 4 sections, following the same pattern. Count the number of pieces of timber required. Draw a sketch for 5 sections and count the pieces required. Enter both pairs of numbers into your table in the correct columns.

c Use trial and error to find the formula connecting P and S. Write the formula in the form $P = \ldots$

d Substitute $S = 220$ into the formula to find out how many pieces of timber Michael needs.

e Find the cost of the fence.

4 A T-shirt shop uses the letter T as its logo. The company wishes to make up a large advertising sign from square lights. Several small T-shirt shop signs are shown below:

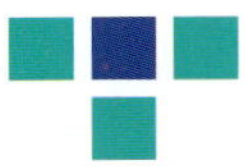

Arm length = 1
Number of lights = 4

Arm length = 2
Number of lights = 7

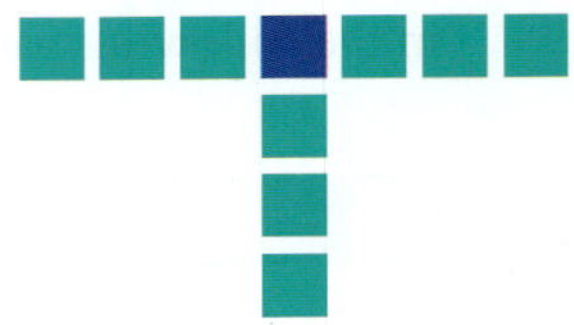

Arm length = 3
Number of lights = 10

Find out how many square light panels would be needed to make up a large sign with an arm length of 75. Follow the steps below.

a Draw a table with columns headed A (arm length), and L (number of lights needed).

b Write in the values of A and L from the diagrams above.

c Complete the diagrams showing the number of lights needed for a sign with an arm length of 4, and a sign with an arm length of 5. Add these values to your table.

d Look at the table to determine the formula which connects L and A. Write the formula in the form $L = \ldots$

e Substitute $A = 75$ into the formula to find the number of lights needed for the large display sign.

5 Johanna (a landscape gardener) is installing hexagonal flowerbeds in a park. The flowerbeds, which are surrounded by wooden sleepers, are arranged in rows as shown below:

1 bed, 6 sleepers required

2 beds, 11 sleepers required

3 beds, 16 sleepers required

How many wooden sleepers will Johanna need to install a row of 12 flowerbeds? Follow the steps below:

a Draw a table with columns headed B (the number of flowerbeds), and S (the number of sleepers needed). Write in the numbers from the diagrams above.

b Complete a diagram showing the number of sleepers needed for a row of 4 flowerbeds, and a row of 5 flowerbeds. Add these values to your table.

c Look at the table to determine the formula which connects S and B. Write the formula in the form $S = \ldots$

d Substitute $B = 12$ into the formula to find the number of sleepers needed for the row of flowerbeds.

WorkSHEET 5.3

THINKING

How many struts?

The diagrams below show the number of struts planned for the back walls of the Surfside beach huts, which are to be built in rows.

1 hut, 8 struts   2 huts, 15 struts   3 huts, 22 struts

Surfside Constructions has a contract to build a row of 34 beach huts at Golden Beach. Based on the plans above, how many struts will be required for the back walls of this row of beach huts? Follow the steps below to solve this problem.

1 Draw a table with columns headed H (the number of huts), and S (the number of struts needed). Write in the numbers from the diagrams above.

2 Following the same pattern, complete a diagram showing the number of struts needed for a row of 4 huts and a row of 5 huts. Add these values to your table.

3 Look at the table to determine the formula that connects S and H. Write the formula in the form $S = \ldots$

4 Substitute $H = 34$ into the formula to find the number of metal struts needed for the row of huts.

The engineers at Surfside Constructions have advised builders that only one of the metal strut cross bracings are required. Only 5 struts are required for the back wall of each hut. The diagram below is the amended plan for the huts' struts.

1 hut, 5 struts   2 huts, 9 struts

5 Draw a diagram showing the number of struts needed for a row of 3 huts.

6 Repeat steps 1 to 4 for a beach hut containing 5 struts as illustrated in the diagram above.

7 How many metal struts will now be required for the back wall of a row of 34 beach huts?

COMMUNICATION

How high will it grow?

A seed you have planted in a pot is starting to grow. When you first notice the seedling, you measure it to be 2.4 centimetres high. Each day after this you continue to measure its height.

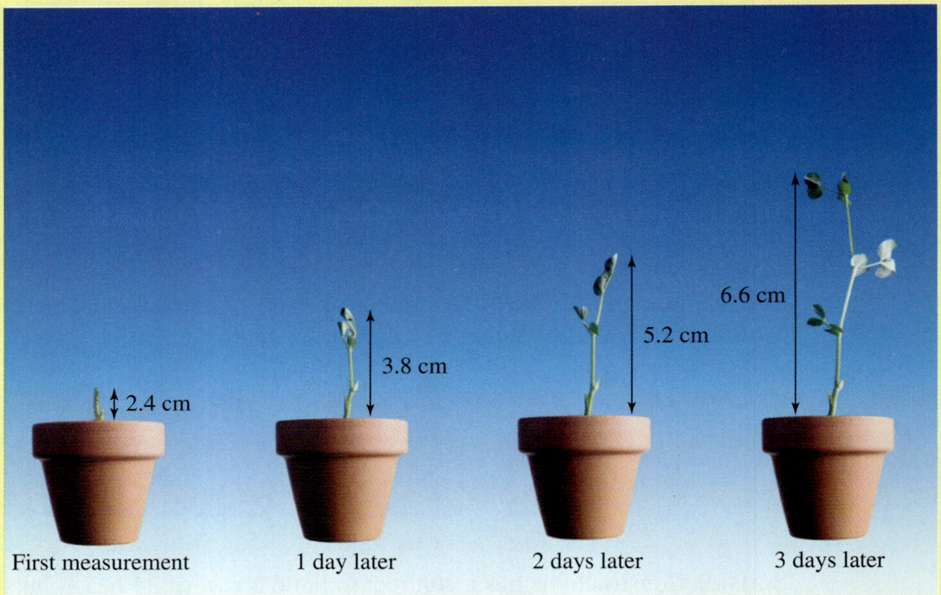

1 If the seedling continues to grow at the same rate, how tall will it be 4 days later?

2 How tall will it be 5 days later?

3 Write a formula connecting the height of the plant with the number of days after the plant height was first measured. Remember to identify what each pronumeral represents in your formula.

4 Check your formula by calculating the height of the plant after 7 days.

5 What is the height of the plant after 18 days?

6 What would be the height of the plant after 25 days?

7 Challenge: Use your formula to estimate when the plant would be approximately 32 centimetres high.

8 Discuss and give reasons why this formula may not be accurate to predict the plant height after a period of time.

summary

Copy the sentences below. Fill in the gaps by choosing the correct word or expression from the word list that follows.

1 In algebra, a rule can be used to produce a set of ____________ numbers from a set of input numbers.

2 A letter or symbol which is used in place of a number is called a ____________.

3 The statement $m = b + 2$, which shows the way in which two sets of numbers (m and b) are related is called a ____________ or ____________.

4 When ____________ numbers and pronumerals we can leave out the multiplication sign.

5 We place numbers in ____________ of pronumerals when multiplying.

6 Replacing a pronumeral with a number is called ____________.

7 The expression $8y - 4$ contains two ____________.

8 Equations are different from expressions because equations always include an ____________ sign.

9 Like terms contain the same ____________.

10 The process of adding together like terms is called ____________.

11 The ____________ $3x + 4y - 2x$ contains two ____________.

12 Substituting known values for pronumerals in an expression allows us to ____________ the expression.

WORD LIST

multiplying	expression	pronumeral	terms
equals	formula	simplifying	pronumerals
equation	like terms	evaluate	
front	output	substitution	

CHAPTER review

5A

1 Copy and complete the tables below. For each table, use the rule to work out the correct output numbers.

a Multiply each input number by 4.

Input	1	3	4	7	13
Output					

b Add 7 to each input number.

Input	1	4	6	7	15
Output					

c Multiply each input number by 2, then subtract 3.

Input	3	4	7	10	13
Output					

d Divide each input number by 4.

Input	0	4	8	16	20
Output					

5B

2 Copy and complete the tables below. Use the rule given in each case to work out the missing input and output numbers.

a Subtract 7 from each input number.

Input	15	16	33		
Output				3	7

b Multiply each input number by 3, then add 7.

Input	1	4	15		
Output				22	106

c Add 5 to each input number, then multiply by 2.

Input	1	3	6		
Output				18	96

d Multiply each input number by 2, then subtract 5.

Input	8	11	15		
Output				21	71

5B

3 **multiple choice**

Which formula has the algebra machine used in each of the following tables?

a

a	1	4	5	6	23
g	11	14	15	16	33

A $g = 5a + 6$
B $g = a - 10$
C $g = a + 10$
D $g = 10a$
E $g = 5a - 10$

b

m	1	2	3	4	11
t	2	5	8	11	32

A $t = 3m$
B $t = 3m - 1$
C $t = m + 5$
D $t = 3m + 1$
E $t = 3m - 2$

5B

4 Look carefully at the pattern in each of the following tables. Then complete the formula for each table by inserting a number into the gap.

a

x	1	4	7	8	23
y	5	32	59	68	203

$y = 9x -$ ___

b

p	0	3	5	6	12
q	6	15	21	24	42

$q =$ ___ $p + 6$

5B

5 Use the pronumerals given to write a formula for each table.

a

x	0	2	3	4	7
y	6	8	9	10	13

b

c	5	6	8	12	13
d	0	1	3	7	8

c

g	3	4	6	9	23
h	17	22	32	47	117

d

m	1	4	5	6	11
n	3	15	19	23	43

5C

6 Complete the tables below by substituting each of the input numbers (x) into the rule.

a $y = x - 1$

x	1	3	4	7	105
y					

b $d = 3x$

x	1	2	6	7	12
d					

c $h = 6 - x$

x	0	2	4	5	6
h					

d $n = 11x + 3$

x	0	1	3	4	7
n					

5D

7 Otto works in a warehouse, packing boxes of CDs for distribution to music stores. An order from Sanity's city store fills 35 boxes with 18 CDs left over.

a Write an expression to represent the Sanity order, using the pronumeral x to represent a full box of CDs.

b If each full box contains 30 CDs, use your expression to calculate how many CDs the store ordered.

5D

8 Write expressions for the following:

a the difference between M and C

b money earned by selling B cakes for \$3 each

c the product of X and Y

d 15 more than G

e 1 more than D

f the cost of 12 bananas at H cents each

g T multiplied by 5.

5E

9 Simplify the following expressions by adding or subtracting like terms.

a $3g + 4g$ **b** $8y - 2y$ **c** $4h + 5h$ **d** $7ag - 2ag$

e $6gy - 3yg$ **f** $8r - 8r$ **g** $6y - 2y + y$ **h** $4t + 6 + 3t$

i $12gh + 6hg$ **j** $8t - 2m + 3t$ **k** $3m + m$ **l** $7g + 8g + 8 + 4$

m $7h + 4t - 3h$ **n** $2b + 7c + 8b$ **o** $11axy - 3axy$

5E

10 Simplify the following expressions first, then find the value of the expression if $x = 5$.

a $7x + 3x$ **b** $2x + 3x - 4$ **c** $11x + 12x$

d $x + 2x$ **e** $4x - x$ **f** $3x - 2x + 16$

g $21x - 13x + 7$ **h** $11 + 2x + 5x$ **i** $7x - 4x + 3x$

5F

11 The metal ceiling rafters in a school classroom consist of a series of triangles connected end to end as shown below. The number of triangle side lengths required in each rafter depends on the total number of triangles. However, adjacent triangles share a common side length.

1 triangle, 3 side lengths

2 triangles, 5 side lengths

3 triangles, 7 side lengths

How many side lengths will be needed to build a single rafter if each rafter contains 25 triangles? Follow the steps below:

a Draw up a table with columns headed T (the number of triangles), and S (the number of side lengths needed). Write in the numbers from the diagrams above.

b Complete a diagram showing the number of side lengths needed for a rafter containing 4 triangles. Add these values to your table.

c Look at the table to determine the formula which connects S and T. Write the formula in the form $S = \ldots$

d Substitute $T = 25$ into the formula to find the number of side lengths needed for each ceiling rafter.

test yourself CHAPTER 5

6 Equations

Walking to help the Bush Nursing hospital

	Sponsor's name	Address	Suburb	Amount per km
1	James McLennan	12 Thomas St	Hampton	\$2.00
2	Gillian McLennan	12 Thomas St	Hampton	\$2.00
3	Tania Smith	18 Thomas St	Hampton	\$3.00
4	Tony Intihar	51 Rochester St	Richmond	\$1.50
5	Nadia Intihar	51 Rochester St	Richmond	\$1.00
6	Emma Wu	7 Smith Ave	Northcote	\$2.50
7	Trent Taylor	25 Perth St	Eltham	\$3.00

To raise money for the Bush Nursing hospital, your school has organised a walk-a-thon. A copy of the sponsor form for your friend, Alex is shown on the left. If he raised \$75.00, how far did he walk?

By the end of this chapter, you will be able to answer this easily by writing the information as an equation that can be solved.

Are you ready?

Try the questions below. If you have difficulty with any of them, extra help can be obtained by completing the matching SkillSHEET. Either click on the SkillSHEET icon next to the question on the *Maths Quest 7* CD-ROM or ask your teacher for a copy.

SkillSHEET 6.1

Completing number sentences

1 Complete the following number sentences by calculating the missing values.

a $5 + \square = 8$ **b** $\square - 7 = 4$
c $\square \times 2 = 30$ **d** $\square \div 3 = 9$

SkillSHEET 6.2

Writing number sentences from written information

2 Construct a number sentence, using $\square$ for the unknown number, to match each problem described below.

a I am thinking of a number, so that when I add 12, the answer is 30.
b I am thinking of a number, so that when I subtract 9, the answer is 6.
c I am thinking of a number, so that when I multiply by 4, the answer is 48.
d I am thinking of a number, so that when I divide by 8, the answer is 7.

SkillSHEET 6.3

Applying the four operations

3 Use 8 as your starting number and find the result after applying the following operations.

a $+3, \times 2$ **b** $\times 2, -7$ **c** $-6, \div 2$
d $\div 4, +5$ **e** $+7, \div 3$ **f** $-4, \times 3$

SkillSHEET 6.5

Inverse operations

4 Write the inverse operation for each of the following.

a $+2$ **b** -5 **c** $\times 7$ **d** $\div 3$

SkillSHEET 6.7

Combining like terms

5 Simplify each of the following expressions by combining like terms.

a $2x + 3x + 6y$ **b** $3x + 2y + 4x + 7y$
c $x + 7 + 9x - 3$ **d** $8x - 5x + 6 + x$

SkillSHEET 6.8

Checking solutions by substitution

6 Answer true or false for each of the following.

a The solution to the number sentence $\square + 3 = 8$ is 5.
b The solution to the number sentence $\square \times 4 = 12$ is 8.
c The solution to the number sentence $\square - 1 = 4$ is 3.
d The solution to the number sentence $\square \div 10 = 2$ is 20.
e The solution to the number sentence $\square \times 2 + 4 = 18$ is 6.
f The solution to the number sentence $\square \div 6 - 6 = 0$ is 36.

Using inverse operations

Here is a simple number puzzle:
I am thinking of a number. When I add 5 to the number my answer is 18. What number am I thinking of?

We can write this puzzle as an equation (an expression containing the 'equals' sign), using a pronumeral to stand for the unknown number. In this case, using x to stand for the number, we would write: $x + 5 = 18$.

To solve this equation we must find the value of x. You can probably see that x must be 13 so we say that '$x = 13$ is the solution of the equation'. This is called solving the equation by inspection, and it works for very easy equations. For harder equations we will need to find better methods.

Flow charts

We will use diagrams called **flow charts** to help us organise harder equations. Here is an example of a flow chart:

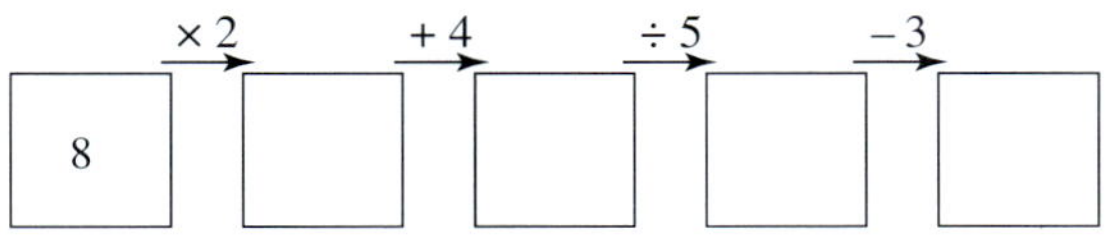

The 8 at the front is called the *input number*, and by carrying out each operation we can move through the flow chart until we reach the last number, called the *output number*. If we fill in this flow chart, we see that the output number is 1.

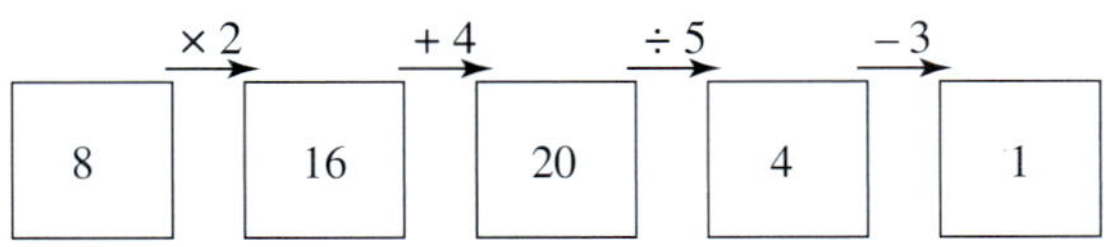

Backtracking

It is possible to travel in either direction through a flow chart. Working backwards, against the arrows, is called **backtracking**. For the flow chart above, we can backtrack by showing the inverse operations.

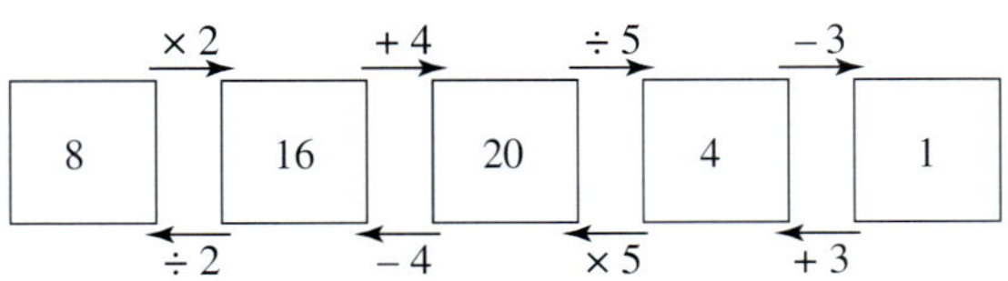

Each arrow carries an operation, for example − 3 or ÷ 5, and when we backtrack we must 'undo' the original operation; that is, carry out the **inverse operation**.

Inverse operations are undoing operations so:

+ 3 is the inverse of − 3
× 5 is the inverse of ÷ 5
− 4 is the inverse of + 4
÷ 2 is the inverse of × 2

WORKED Example 1

Write an equation to represent each of these puzzles. I am thinking of a number.

a **When I multiply the number by 8 the answer is 24.**
b **When I divide the number by 5 the answer is 7.**
c **When I divide 60 by the number, the answer is 10.**
d **When I subtract 7 from the number, the answer is 25.**
e **When I subtract the number from 72, the answer is 52.**
f **When I square the number, the result is 36.**

THINK	WRITE
a 1 Use a pronumeral to describe the number.	**a** Let m be the number.
2 Multiply the number by 8.	$m \times 8 = 8m$
3 Write the equation.	$8m = 24$
b 1 Use a pronumeral to describe the number.	**b** Let t be the number.
2 Divide the number by 5.	$t \div 5 = \frac{t}{5}$
3 Write the equation.	$\frac{t}{5} = 7$
c 1 Use a pronumeral to describe the number.	**c** Let s be the number.
2 Divide 60 by the number.	$60 \div s = \frac{60}{s}$
3 Write the equation.	$\frac{60}{s} = 10$
d 1 Use a pronumeral to describe the number.	**d** Let l be the number.
2 Subtract 7 from the number (that is, take 7 away from the number).	$l - 7$
3 Write the equation.	$l - 7 = 25$
e 1 Use a pronumeral to describe the number.	**e** Let a be the number.
2 Subtract the number from 72 (that is, take the number away from 72).	$72 - a$
3 Write the equation.	$72 - a = 52$
f 1 Use a pronumeral to describe the number.	**f** Let z be the number.
2 Square the number (that is, multiply the number by itself).	$z \times z = z^2$
3 Write the equation.	$z^2 = 36$

Sometimes the solution to an equation is obvious after a close inspection of the equation.

WORKED Example 2

Solve the following equations by inspection.

a $\frac{w}{3} = 4$

b $h - 9 = 10$

THINK	WRITE
a 1 Write the equation.	**a** $\frac{w}{3} = 4$
2 Think of a number which when divided by 3 gives 4. Try 12.	$12 \div 3 = 4$
3 So w must be 12.	$w = 12$
b 1 Write the equation.	**b** $h - 9 = 10$
2 Think of a number which equals 10 when 9 is subtracted from it. Try 19.	$19 - 9 = 10$
3 So h must be 19.	$h = 19$

WORKED Example 3

Use backtracking and inverse operations to find the input number in this flow chart.

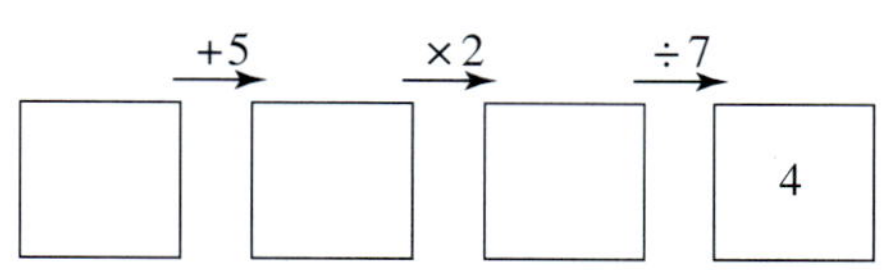

THINK	WRITE
1 Fill in the numbers as you backtrack. The inverse of ÷ 7 is × 7 ($4 \times 7 = 28$).	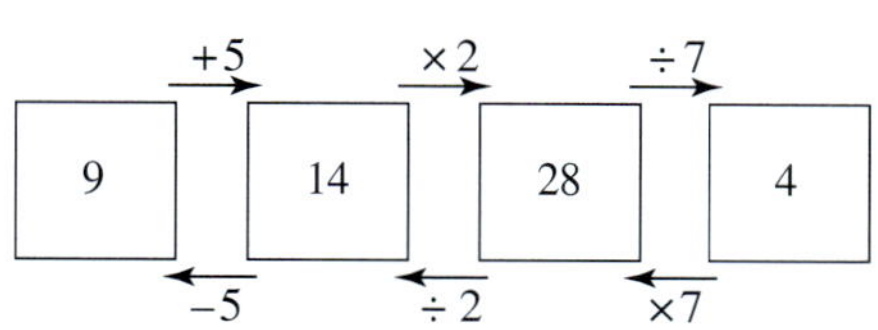
2 The inverse of × 2 is ÷ 2 ($28 \div 2 = 14$).	
3 The inverse of + 5 is − 5 ($14 - 5 = 9$).	

remember

1. *Flow charts* help us to organise harder equations.
2. We can use *backtracking* and *inverse operations* to solve equations.

EXERCISE 6A Using inverse operations

Note: If you would like to fill in the flow charts in this exercise, select the pages and print them out from the interactive *Maths Quest 7* CD-ROM.

SkillSHEET 6.1 Completing number sentences

SkillSHEET 6.2 Writing number sentences from written information

WORKED Example 1

1 Write an equation to represent each of these puzzles.
I am thinking of a number:

- **a** when I add 7, the answer is 11
- **b** when I add 3, the answer is 5
- **c** when I add 12, the answer is 12
- **d** when I add 5, the answer is 56
- **e** when I subtract 7, the answer is 1
- **f** when I subtract 11, the answer is 11
- **g** when I subtract 4, the answer is 7
- **h** when I subtract 8, the answer is 0
- **i** when I multiply by 2, the answer is 12
- **j** when I multiply by 6, the answer is 30
- **k** when I multiply by 5, the answer is 30
- **l** when I multiply by 6, the answer is 12
- **m** when I divide by 7, the answer is 1
- **n** when I divide by 3, the answer is 100
- **o** when I divide by 5, the answer is 2
- **p** when I divide by 7, the answer is 0
- **q** when I subtract the number from 15, the answer is 2
- **r** when I subtract the number from 52, the answer is 8
- **s** when I divide 21 by the number, the answer is 7
- **t** when I square the number, the answer is 100.

SkillSHEET 6.3 Applying the four operations

WORKED Example 2

2 Solve the following equations by inspection.

a $x + 7 = 18$ **b** $y - 8 = 1$ **c** $3m = 15$

d $\frac{m}{10} = 3$ **e** $w + 25 = 26$ **f** $m - 1 = 273$

g $4w = 28$ **h** $\frac{k}{5} = 0$ **i** $b + 15 = 22$

j $\frac{w}{3} = 6$ **k** $5k = 20$ **l** $h - 14 = 11$

m $b - 2.1 = 6.7$ **n** $\frac{c}{3} = 1.4$ **o** $5x = 14$

SkillSHEET 6.4 Flow charts

3 Complete the following flow charts to find the output number.

a i 5 $\xrightarrow{\times 3}$ □ $\xrightarrow{+1}$ □

ii 5 $\xrightarrow{+1}$ □ $\xrightarrow{\times 3}$ □

b i 15 $\xrightarrow{\div 5}$ ☐ $\xrightarrow{+10}$ ☐

ii 15 $\xrightarrow{+10}$ ☐ $\xrightarrow{\div 5}$ ☐

c i 7 $\xrightarrow{-5}$ ☐ $\xrightarrow{\times 2}$ ☐

ii 7 $\xrightarrow{\times 2}$ ☐ $\xrightarrow{-5}$ ☐

d i 4 $\xrightarrow{\div 2}$ ☐ $\xrightarrow{+6}$ ☐

ii 4 $\xrightarrow{+6}$ ☐ $\xrightarrow{\div 2}$ ☐

e i 30 $\xrightarrow{\div 3}$ ☐ $\xrightarrow{-9}$ ☐

ii 30 $\xrightarrow{-9}$ ☐ $\xrightarrow{\div 3}$ ☐

f i 0 $\xrightarrow{+6}$ ☐ $\xrightarrow{\div 3}$ ☐

ii 0 $\xrightarrow{\div 3}$ ☐ $\xrightarrow{+6}$ ☐

g i 7 $\xrightarrow{+3}$ ☐ $\xrightarrow{\times 5}$ ☐

ii 7 $\xrightarrow{\times 5}$ ☐ $\xrightarrow{+3}$ ☐

h i 15 $\xrightarrow{-5}$ ☐ $\xrightarrow{\times 10}$ ☐

ii 15 $\xrightarrow{\times 10}$ ☐ $\xrightarrow{-5}$ ☐

4 **a** In question **3**, what did you notice about each pair of flow charts?
b Does changing the order of operations affect the end result?

5 **a** Complete the two statements below:
Adding and ______________ are inverse operations.
______________ and dividing are inverse operations.
b Can you think of any other pairs of inverse operations?

WORKED Example 3

6 Use backtracking and inverse operations to find the input number in each of these flow charts.

6.5 SkillSHEET Inverse operations

EXCEL Spreadsheet Backtracking

a ☐ $\xrightarrow{+5}$ ☐ $\xrightarrow{\times 2}$ 12

b ☐ $\xrightarrow{-3}$ ☐ $\xrightarrow{\div 2}$ 2

c ☐ $\xrightarrow{\div 5}$ ☐ $\xrightarrow{-3}$ 0

d ☐ $\xrightarrow{\times 2}$ ☐ $\xrightarrow{\div 7}$ 4

e $\square \xrightarrow{\times 7} \square \xrightarrow{-3} \boxed{32}$

f $\square \xrightarrow{-7} \square \xrightarrow{\times 8} \boxed{24}$

g $\square \xrightarrow{+3} \square \xrightarrow{\times 2} \square \xrightarrow{-8} \boxed{6}$

h $\square \xrightarrow{\div 3} \square \xrightarrow{+2} \square \xrightarrow{\times 10} \square \xrightarrow{-6} \boxed{44}$

i $\square \xrightarrow{+4} \square \xrightarrow{\div 9} \square \xrightarrow{\times 8} \square \xrightarrow{-7} \boxed{1}$

j $\square \xrightarrow{\times 8} \square \xrightarrow{+6} \square \xrightarrow{-4} \square \xrightarrow{\times 10} \boxed{20}$

k $\square \xrightarrow{\div 2.17} \square \xrightarrow{-3.41} \boxed{3.25}$

l $\square \xrightarrow{\times \frac{2}{3}} \square \xrightarrow{\times \frac{3}{4}} \boxed{2}$

m $\square \xrightarrow{\times 28} \square \xrightarrow{-56} \square \xrightarrow{\times 15} \boxed{420}$

n $\square \xrightarrow{\times 9} \square \xrightarrow{-152} \square \xrightarrow{\times 19} \square \xrightarrow{+53} \boxed{72}$

o $\square \xrightarrow{\div 1.4} \square \xrightarrow{+2.31} \square \xrightarrow{\times 6.5} \square \xrightarrow{-0.04} \boxed{27}$

p $\square \xrightarrow{\times 78} \square \xrightarrow{+2268} \square \xrightarrow{\div 12} \square \xrightarrow{-2605} \boxed{1042}$

GAME time
Equations — 001

Joke time

Solve the equations below to find the puzzle answer code.

What tools do we use in arithmetic?

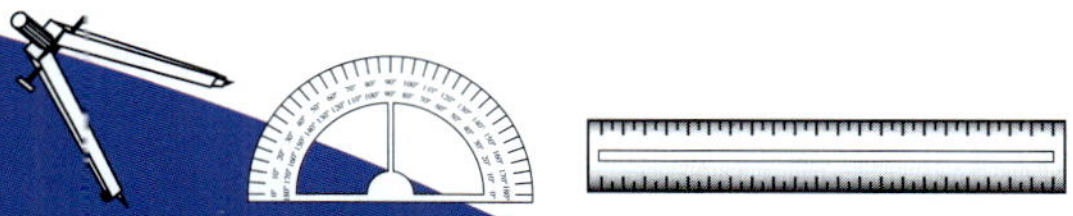

9	14	17	16	19	12	17	19	8	6	10

Equation	Answer	Equation	Answer
$A + 7 = 18$	A =	$G - 12 = 3$	G =
$12 + B = 15$	B =	$H - 7 = 0$	H =
$27 = C + 9$	C =	$16 = I - 3$	I =
$5 = 4 + D$	D =	$11 = K - 2$	K =
$E + 38 = 46$	E =	$L - 13 = 4$	L =

Why was Christopher Columbus a crook?

7	8	1	5	14	3	17	8	18	6	5	10	10
8	1	16	7	8	11	16	17	11	4	16	19	18

Equation	Answer	Equation	Answer
$3M = 27$	M =	$S \div 5 = 2$	S =
$5N = 20$	N =	$\frac{T}{4} = 4$	T =
$12O = 60$	O =	$\frac{U}{2} = 7$	U =
$7P = 84$	P =	$W \div 4 = 5$	W =
$9R = 54$	R =	$\frac{Y}{2} = 1$	Y =

What walks on its head all day?

11	1	6	11	20	19	4	15	12	19	4	10	16
14	18	13	19	4	2	5	14	6	10	7	5	8

Building up expressions

We can use flow charts to construct algebraic expressions. Here is an example, with an input number equal to m.

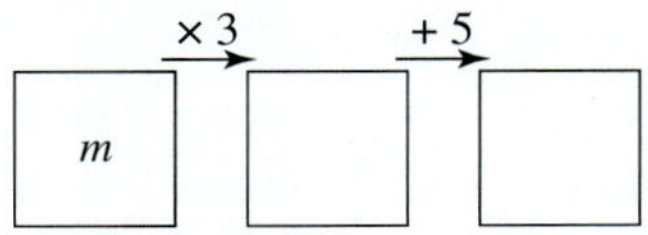

Multiplying m by 3 gives us $3m$, and then adding 5 gives $3m + 5$ which is the output number.

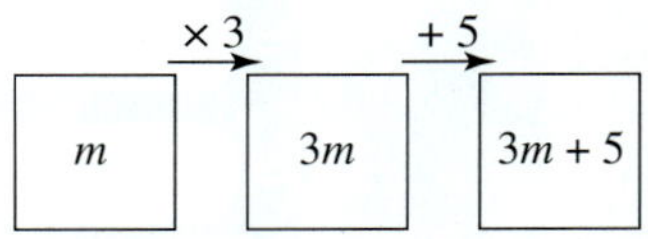

Compare it with this flow chart.

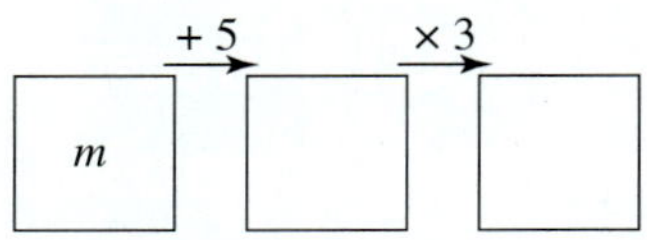

Adding 5 to m gives us $m + 5$, but to show that we are multiplying all of $m + 5$ by 3 we need to place $m + 5$ inside a pair of brackets. Now, $(m + 5) \times 3$ is written as $3(m + 5)$. A pair of brackets is a type of grouping device. It groups $m + 5$ together.

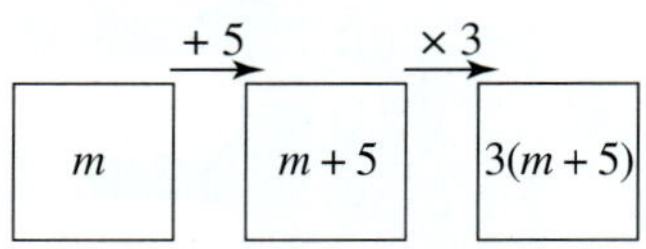

Remember that there is a difference between $2 + 5 \times 3$ and $(2 + 5) \times 3$.
So the flow charts which gave us the expressions $3m + 5$ and $3(m + 5)$ contained the same operations, but they were carried out in a different order.

A different grouping device

Compare these two flow charts.

Flow chart A

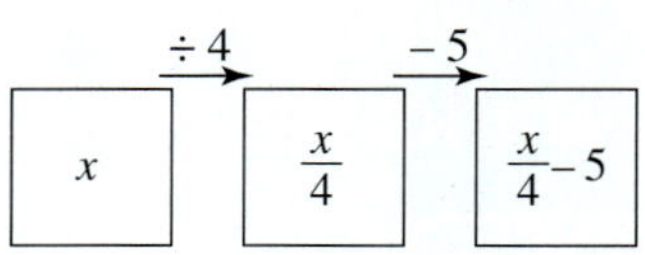

Flow chart B

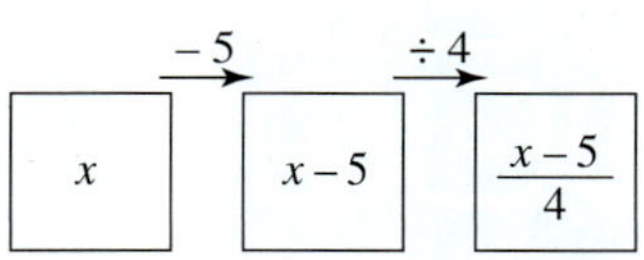

Can you see the grouping device in flow chart B? It is the division line (called a vinculum) drawn beneath $x - 5$ and we must take care that it runs beneath all of $x - 5$.

WORKED Example 4

Draw a flow chart whose input number is m and whose output number is given by the expressions: a $2m - 11$ b $\frac{m+9}{5}$ c $4\left(\frac{m}{3} + 2\right)$

THINK	WRITE
a 1 The first step is to obtain $2m$; that is, multiply m by 2.	**a** 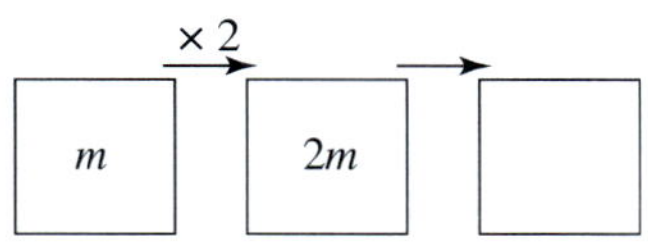
2 This is followed by subtracting 11.	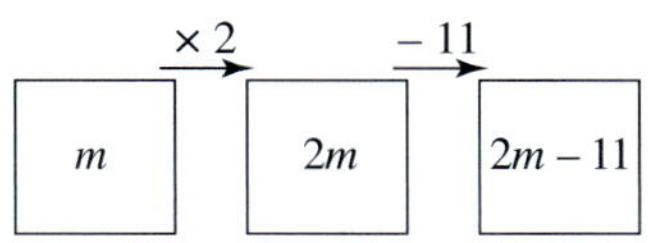
b 1 The expression $m + 9$ is grouped as in a pair of brackets, so we must obtain this part first. Therefore add 9 to m.	**b** 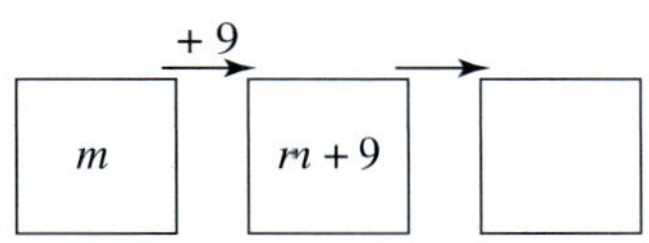
2 Then the whole expression is divided by 5.	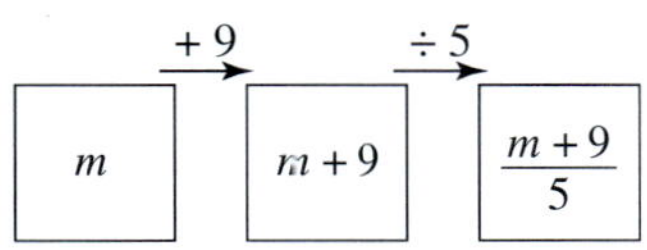
c 1 The pair of brackets indicates we must first work from within the brackets: (a) divide m by 3 (b) then add 2 to this result.	**c** 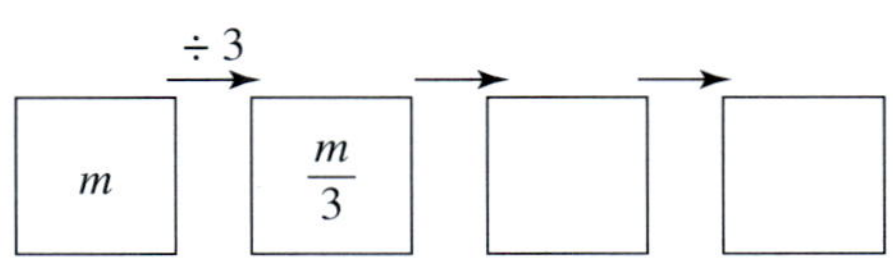 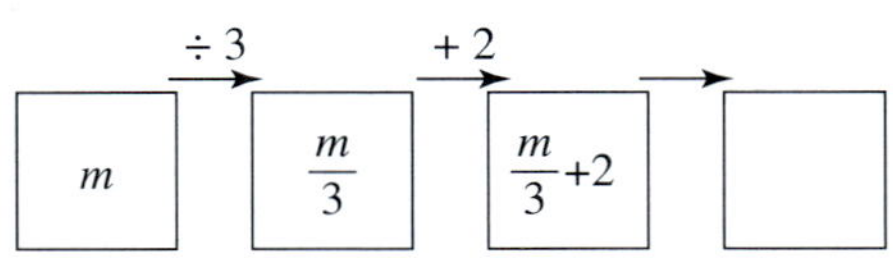
2 Multiply the final result obtained in step 1 by 4.	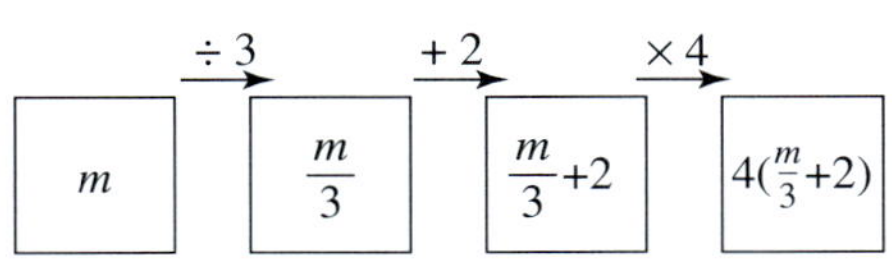

As well as constructing an algebraic expression using a flow chart, we can backtrack to our input number using inverse operations.

WORKED Example 5

Complete the flow chart at right by writing in the operations which must be carried out in order to backtrack to x.

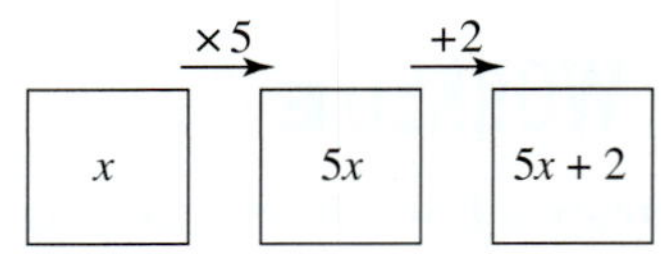

THINK / **WRITE**

1. Copy the flow chart and look at the operations that have been performed.

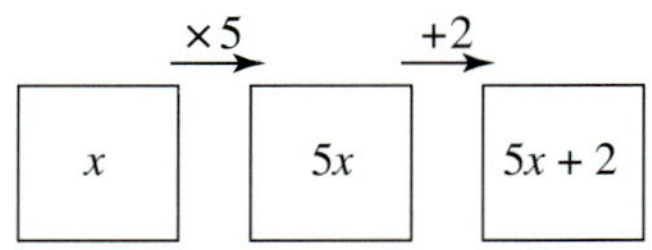

2. The inverse operation of + 2 is – 2. Show this on the flow chart.

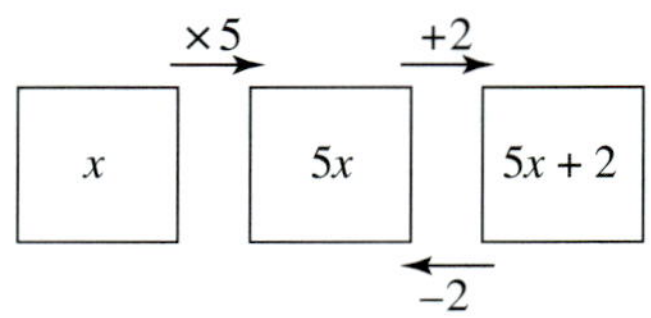

3. The inverse operation of × 5 is ÷ 5. Show this on the flow chart.

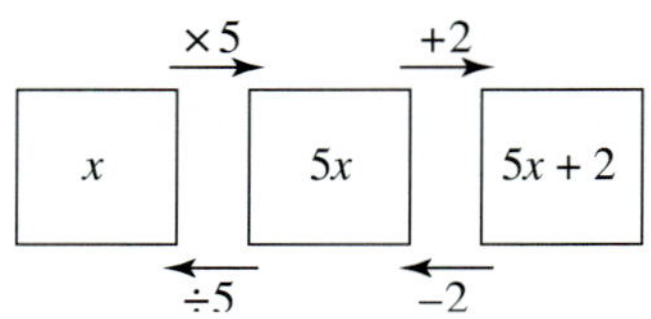

remember

1. We can use flow charts to construct algebraic expressions.
2. As well as constructing an algebraic expression using a flow chart, we can backtrack to our input number using inverse operations.

EXERCISE 6B Building up expressions

Note: If you would like to fill in the flow charts in this exercise, select the pages and print them out from the interactive *Maths Quest 7* CD-ROM.

SkillSHEET 6.6 Building expressions

1 Build up an expression by following the instructions on the flow charts.

a
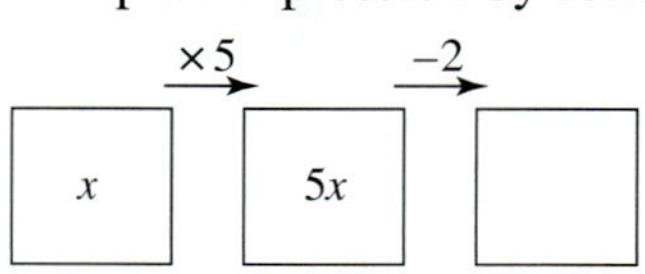

b
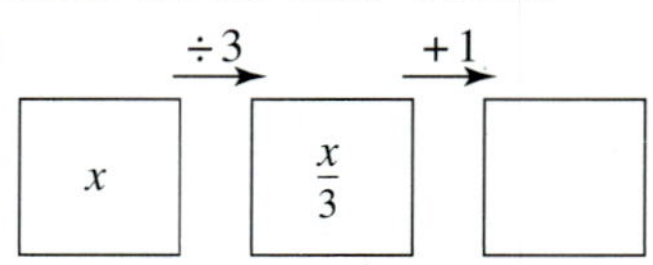

c
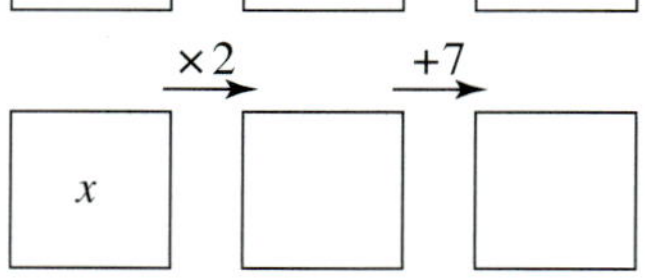

d
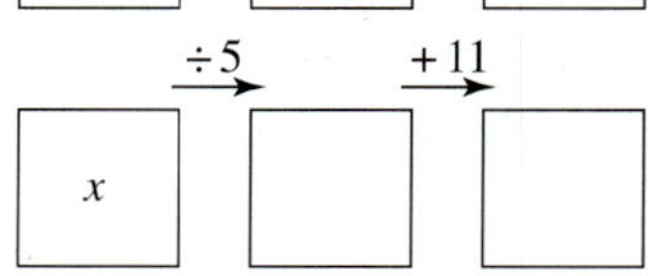

e
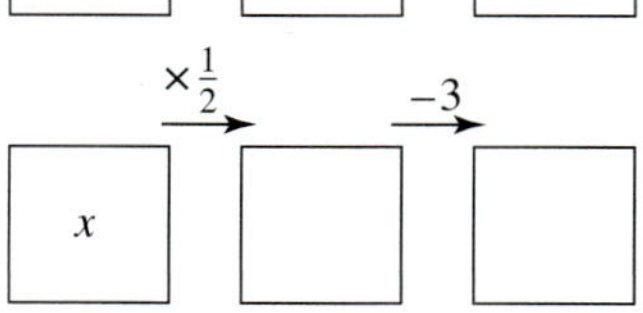

f
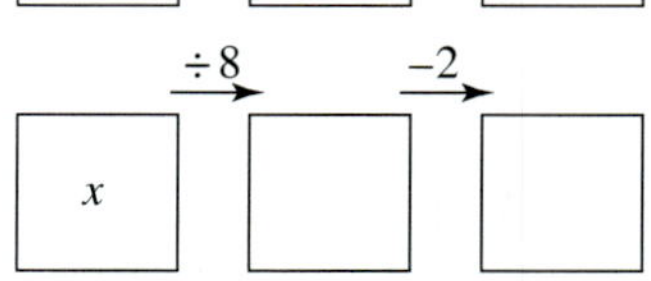

g $\boxed{x} \xrightarrow{\div 7} \boxed{} \xrightarrow{+9} \boxed{}$

h $\boxed{x} \xrightarrow{\times 3.1} \boxed{} \xrightarrow{+1.8} \boxed{}$

2 Build up an expression by following the instructions on the flow charts. Use a grouping device, such as a pair of brackets or a vinculum; for example $2(x + 3)$ or $\dfrac{x-5}{4}$.

a $\boxed{x} \xrightarrow{+5} \boxed{} \xrightarrow{\times 4} \boxed{}$

b $\boxed{x} \xrightarrow{+10} \boxed{} \xrightarrow{\div 3} \boxed{}$

c $\boxed{x} \xrightarrow{-2} \boxed{} \xrightarrow{\div 7} \boxed{}$

d $\boxed{x} \xrightarrow{+3} \boxed{} \xrightarrow{\times 9} \boxed{}$

e $\boxed{x} \xrightarrow{-2.1} \boxed{} \xrightarrow{\times 3} \boxed{}$

f $\boxed{x} \xrightarrow{+4.9} \boxed{} \xrightarrow{\div 5} \boxed{}$

g $\boxed{x} \xrightarrow{-\frac{1}{2}} \boxed{} \xrightarrow{\times 7} \boxed{}$

h $\boxed{x} \xrightarrow{-3.1} \boxed{} \xrightarrow{\div 1.8} \boxed{}$

3 Copy and complete the following flow charts by filling in the missing expressions.

a $\boxed{x} \xrightarrow{+2} \boxed{} \xrightarrow{\times 6} \boxed{}$

b $\boxed{x} \xrightarrow{\times 3} \boxed{} \xrightarrow{-2} \boxed{}$

c $\boxed{x} \xrightarrow{-8} \boxed{} \xrightarrow{\div 12} \boxed{}$

d $\boxed{x} \xrightarrow{\times 4} \boxed{} \xrightarrow{+1} \boxed{}$

e $\boxed{x} \xrightarrow{\div 7} \boxed{} \xrightarrow{+3} \boxed{}$

f $\boxed{x} \xrightarrow{-3} \boxed{} \xrightarrow{\div 5} \boxed{}$

g $\boxed{x} \xrightarrow{+5} \boxed{} \xrightarrow{\times 6} \boxed{} \xrightarrow{-2} \boxed{}$

h $\boxed{x} \xrightarrow{\times 3} \boxed{} \xrightarrow{+4} \boxed{} \xrightarrow{\div 7} \boxed{}$

i $\boxed{x} \xrightarrow{-8} \boxed{} \xrightarrow{\div 5} \boxed{} \xrightarrow{+9} \boxed{}$

j $\boxed{x} \xrightarrow{\div 3} \boxed{} \xrightarrow{+4} \boxed{} \xrightarrow{\times 6} \boxed{}$

k $\boxed{x} \xrightarrow{\times 11} \boxed{} \xrightarrow{-2} \boxed{} \xrightarrow{\times 4} \boxed{}$

l $\boxed{x} \xrightarrow{\times 3} \boxed{} \xrightarrow{\div 4} \boxed{} \xrightarrow{+2} \boxed{}$

WORKED Example 4

4 Draw a flow chart whose input number is x and whose output is given by the following expressions.

a $5x + 9$ b $2(x + 1)$ c $\frac{x}{6} + 4$

d $\frac{x - 8}{7}$ e $12(x - 7)$ f $\frac{x}{5} - 2$

g $7x - 12$ h $\frac{x + 6}{3}$ i $3(x + 7) - 5$

j $\frac{3x + 7}{2}$ k $4(3x + 1)$ l $3\left(\frac{x}{5} + 6\right)$

WORKED Example 5

5 Complete the following flow charts by writing in the operations which must be carried out in order to backtrack to x.

a
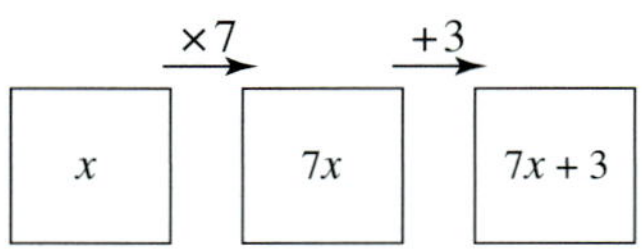

b
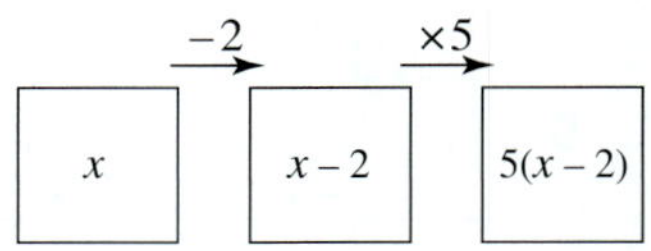

c
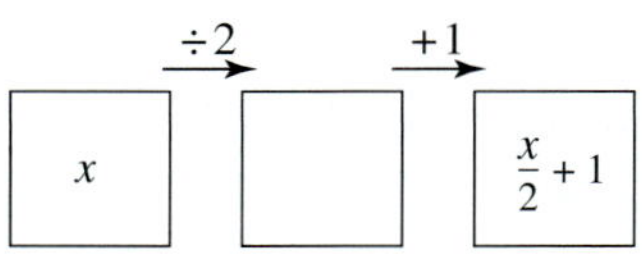

d
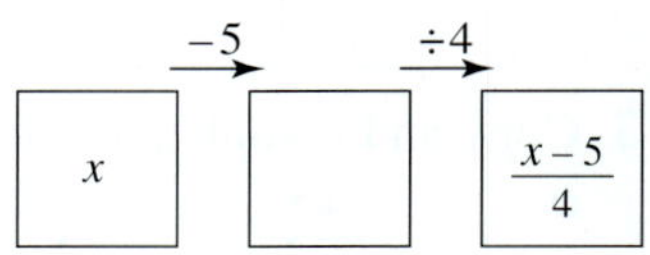

e
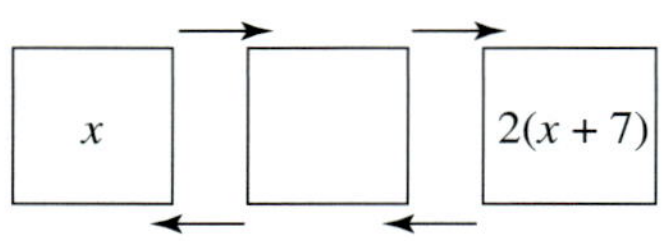

f
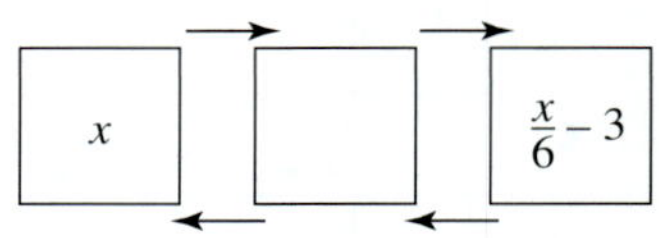

g
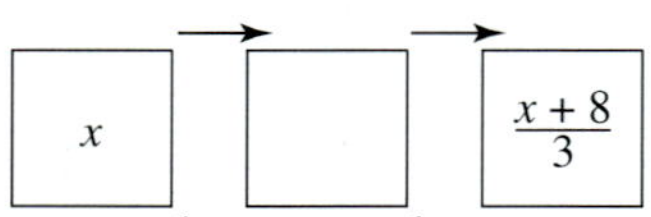

h
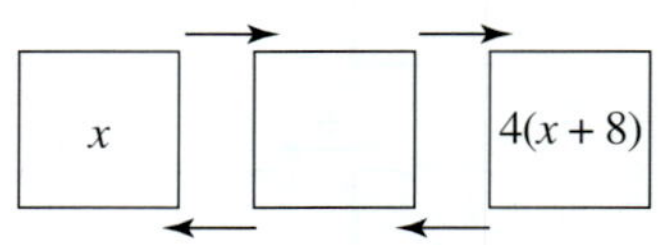

i
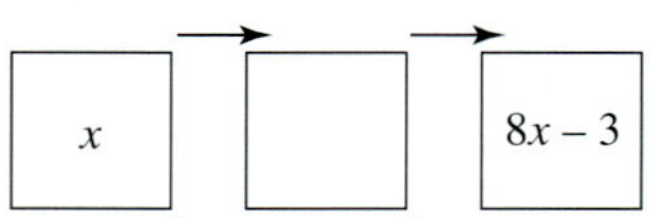

j
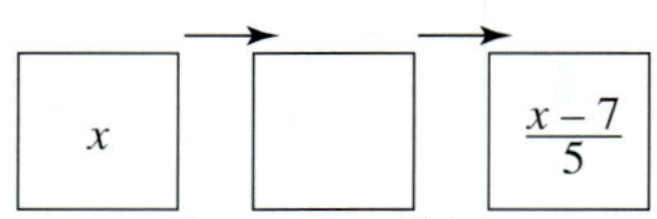

k
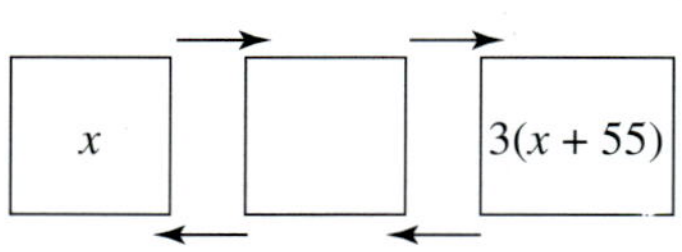

l
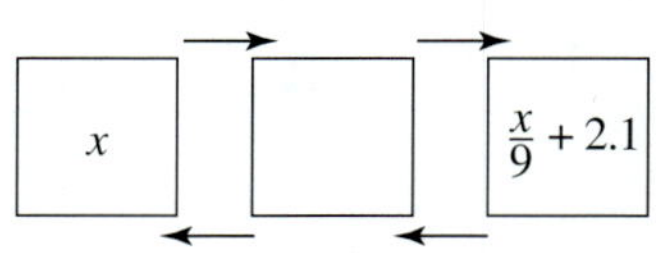

m
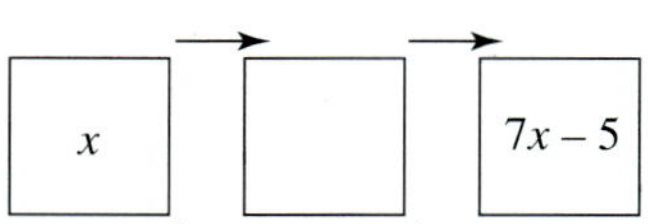

n
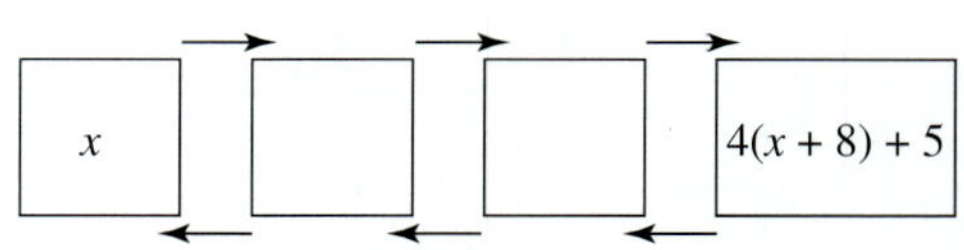

History of mathematics

ROBERT RECORDE (1510–1558)

During his life:
The astronomer Copernicus proposed that the planets revolve around the sun, the Portuguese navigator Vasco da Gama discovered the sea route from Western Europe to the East, and Michelangelo painted the ceiling of the Sistine Chapel.

Robert Recorde was born in Tenby, Wales and was educated at Oxford and Cambridge. He became the physician to King Edward VI and Queen Mary.

Recorde was principally responsible for establishing the English School of Mathematics and first introduced algebra to England. He is best known for inventing the 'equals' symbol '=' which appears in his book *The Whetstone of Witte* (1557). He thought that the symbol with two parallel line segments was the most appropriate as, in his words, 'noe 2 thynges can be moare equalle'. The symbol '=' was not immediately popular. The symbol '||' was used by some and 'ae' from the word 'aequalis' meaning equal, was also widely used until the 1700s.

Robert Recorde wrote many other successful maths textbooks. Another one was called *The Grounde of Artes* (1540) in which he discusses operations with Arabic numerals, computation with counters, proportion and fractions.

In 1549 Recorde was made controller of the Bristol mint. When he refused to divert money to support young King Edward VI's army he was accused of treason by Sir William Herbert. He regained favour in 1551 and was appointed surveyor of mines and monies in Ireland. Silver was mined and minted into coins including the first English coin to have a date written in Arabic rather than Roman numerals. After Edward's death, Queen Mary took the throne, but Recorde crossed paths again with Sir William Herbert, who was powerful in Mary's court as the Earl of Pembroke. In 1556 Recorde charged Pembroke with misconduct but the earl retaliated. Recorde was ordered to pay £1000 to the earl for libel. Recorde did not pay the fine.

Recorde died in King's Bench prison in Southwark, England in 1558 where he was sentenced and imprisoned for the debt.

Questions

1. What symbol did Robert Recorde invent?
2. Where does 'ae' come from and what does it mean?
3. How old was Robert Recorde when he died?

Solving equations using backtracking

We can now solve harder puzzles and equations by drawing a flow chart and then backtracking. For example, the following problem can be expressed on a flow chart.

I am thinking of a number. If I multiply it by 5 and then add 6 my answer is 21. What is the number? I will call the number m.

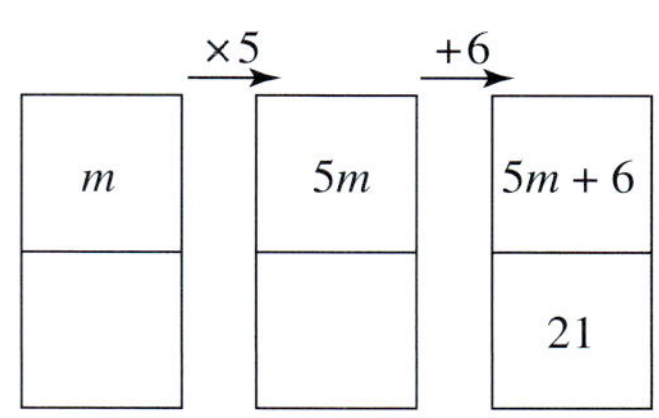

By backtracking we can see that m (my number) must be 3.

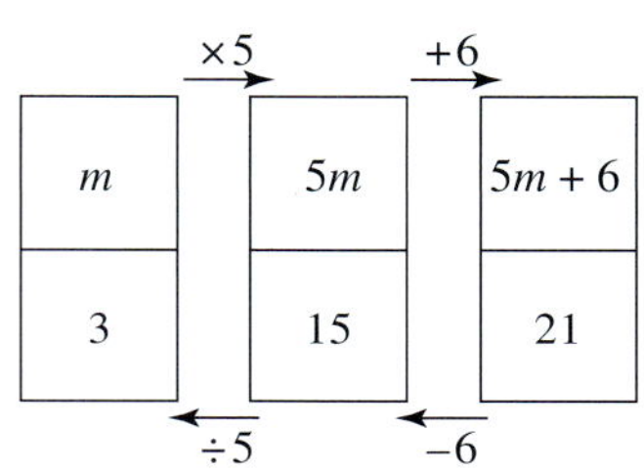

WORKED Example 6

Draw a flow chart to represent the following puzzle and then solve it by backtracking. I am thinking of a number. When I multiply it by 4 and then add 2 the answer is 14.

THINK

1. Build an expression using x to represent the number.
Start with x, multiply by 4 and add 2.
The output number is 14.
2. Backtrack to find the value of x.
The inverse operation of + 2 is −2 (14 − 2 = 12).
The inverse operation of × 4 is ÷ 4 (12 ÷ 4 = 3).
3. State the answer.

WRITE

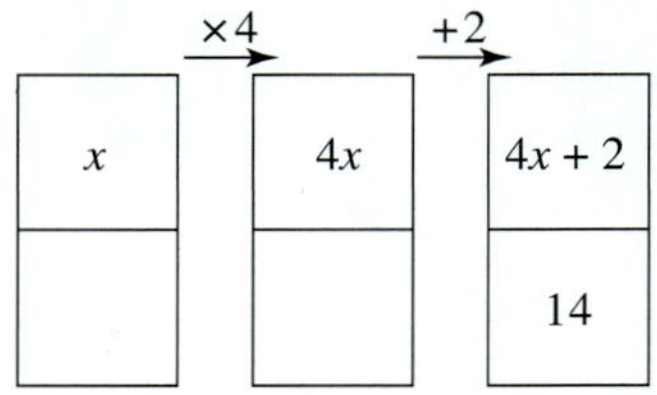

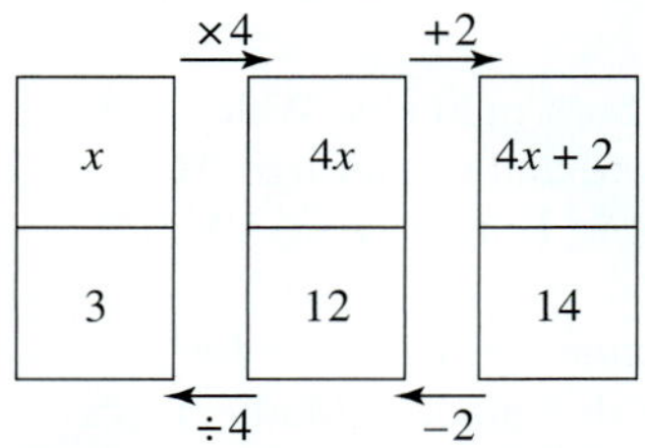

So $x = 3$. The number is 3.

WORKED Example 7

Solve the following equations by backtracking. **a** $3(x + 7) = 24$ **b** $\frac{x}{3} + 5 = 6$

THINK

a
1. Build the expression on the left-hand side of the equation.
Start with x, add 7 and then multiply by 3.
The output number is 24.
2. Backtrack to find x.
The inverse operation of × 3 is ÷ 3 (24 ÷ 3 = 8).
The inverse operation of + 7 is − 7 (8 − 7 = 1).
3. State the answer.

b
1. Build the expression on the left-hand side of the equation.
Start with x, divide by 3 and then add 5.
The output number is 6.

WRITE

a

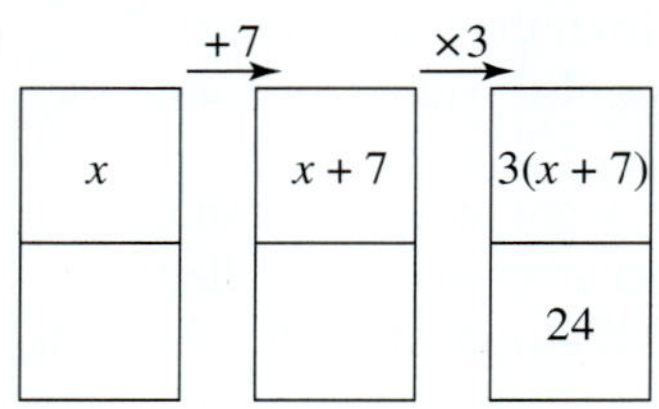

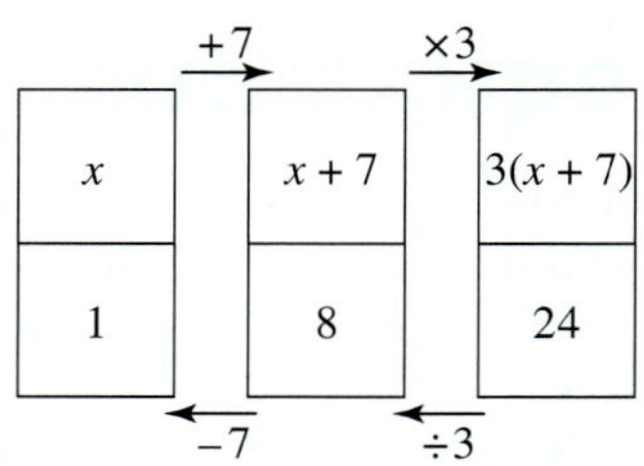

$x = 1$

b

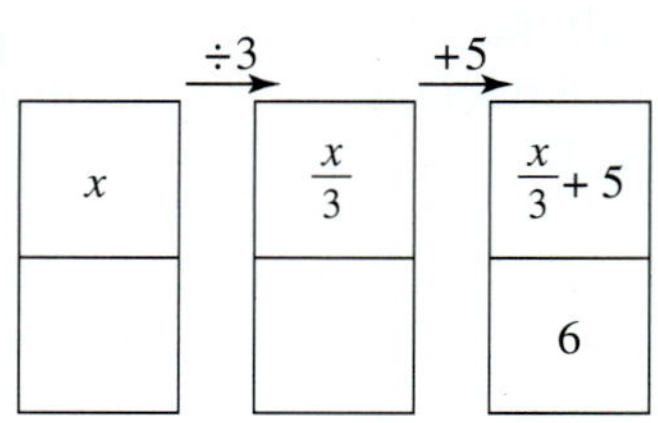

THINK | **WRITE**

2 Backtrack to solve for x.
The inverse operation of + 5 is − 5 ($6 - 5 = 1$).
The inverse operation of ÷ 3 is × 3 ($1 \times 3 = 3$).

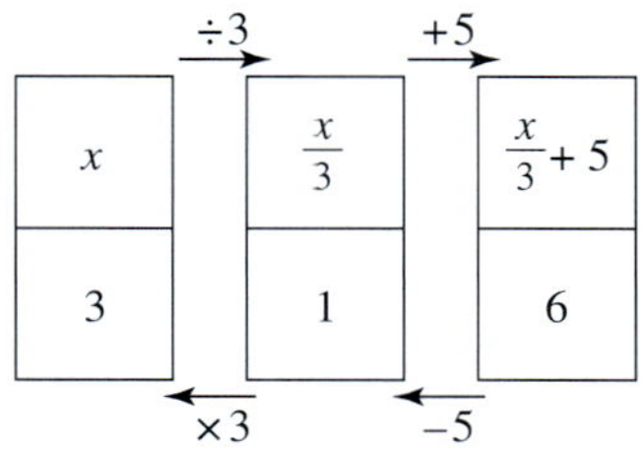

3 State the answer. $x = 3$

WORKED Example 8

Simplify and then solve the following equation by backtracking.
$5x + 13 + 2x - 4 = 23$

THINK | **WRITE**

1 Simplify by adding the like terms together on the left-hand side of the equation.
$5x + 2x = 7x$, $13 - 4 = 9$

$5x + 13 + 2x - 4 = 23$
$7x + 9 = 23$

2 Draw a flow chart and build the expression $7x + 9$.
Start with x, multiply by 7 and add 9.
The output number is 23.

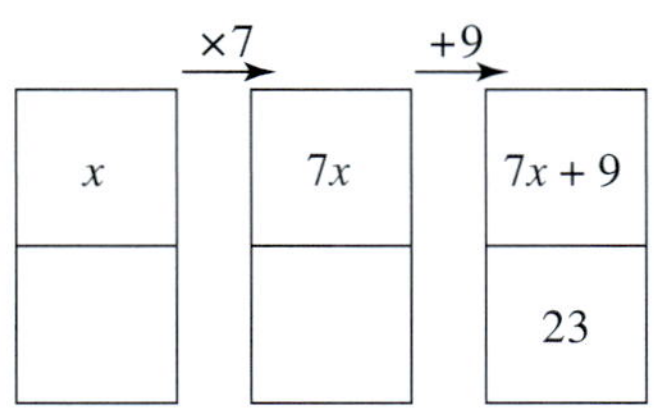

3 Backtrack to solve for x.
The inverse operation of + 9 is − 9 ($23 - 9 = 14$).
The inverse operation of × 7 is ÷ 7 ($14 \div 7 = 2$).

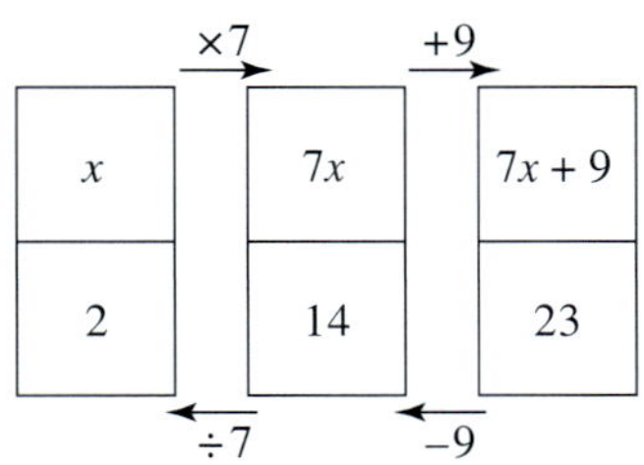

4 State the answer. $x = 2$

Note: You can select and print out questions to fill in the flow charts when using the interactive *Maths Quest 7* CD-ROM.

remember

We can solve equations by drawing a flow chart and backtracking.

EXERCISE 6C Solving equations using backtracking

WORKED Example 6

EXCEL Spreadsheet Backtracking

1 Draw a flow chart to represent each of the following puzzles and then solve them by backtracking.

I am thinking of a number.

a When I multiply it by 2 and then add 7 the answer is 11.
b When I add 3 to it and then multiply by 5 the answer is 35.
c When I divide it by 4 and then add 12 the answer is 14.
d When I add 5 to it and then divide by 3 the answer is 6.
e When I subtract 7 from it and then multiply by 6 the answer is 18.
f When I multiply it by 8 and then subtract 11 the answer is 45.
g When I subtract 4 from it and then divide by 9 the answer is 7.
h When I divide it by 11 and then subtract 8 the answer is 0.
i When I add 5 to it and then multiply by 2 the answer is 12.
j When I multiply it by 6 and then add 4 the answer is 34.
k When I multiply it by 5 and then subtract 10 the answer is 30.
l When I subtract 3.1 from it and then multiply by 6 the answer is 13.2.
m When I divide it by 8 and then subtract 0.26 the answer is 0.99.
n When I divide it by 3.7 and then add 1.93 the answer is 7.62.
o When I add $\frac{2}{5}$ to it and then divide by 6 the answer is $\frac{4}{5}$.
p When I subtract $\frac{3}{4}$ from it and then divide by $\frac{2}{3}$ the answer is $\frac{1}{6}$.

Mathcad Solving equations

2 Draw a flow chart and use backtracking to find the solution to the following equations.

a $5x + 7 = 22$ b $9y - 8 = 1$ c $3m - 7 = 11$
d $4x + 12 = 32$ e $8w + 2 = 26$ f $11m - 1 = 274$
g $4w + 5.2 = 28$ h $6b - \frac{5}{9} = \frac{2}{3}$ i $2a + \frac{1}{3} = \frac{3}{5}$

GC program–TI Solving equations

WORKED Example 7a

3 Solve the following equations by backtracking.

a $3(x + 7) = 24$ b $2(x - 7) = 22$ c $5(x - 15) = 40$
d $11(x + 5) = 99$ e $6(x + 9) = 72$ f $3(x - 11) = 3$
g $4(w + 5.2) = 26$ h $6(b - \frac{5}{9}) = \frac{2}{3}$ i $8(x - \frac{1}{4}) = \frac{4}{5}$

GC program–Casio Solving equations

WORKED Example 7b

EXCEL Spreadsheet Solving equations

4 Solve the following equations by backtracking.

a $\frac{x}{3} + 5 = 6$ b $\frac{x}{9} - 2 = 3$ c $\frac{x}{4} + 7 = 10$
d $\frac{x}{2} - 11 = 6$ e $\frac{x}{7} - 5 = 6$ f $\frac{x}{3} + 1 = 1$
g $\frac{x}{5} + 2.3 = 4.9$ h $\frac{x}{4} - \frac{3}{11} = \frac{4}{11}$ i $\frac{x}{2} + \frac{1}{9} = \frac{4}{9}$

5 Solve the following equations by backtracking.

a $\dfrac{x+4}{3}=6$ b $\dfrac{x-8}{7}=3$ c $\dfrac{x-8}{7}=10$

d $\dfrac{x+11}{2}=6$ e $\dfrac{x-5}{7}=0$ f $\dfrac{x+100}{17}=23$

g $\dfrac{x+2.21}{1.4}=4.9$ h $\dfrac{x-1}{5}=\dfrac{4}{7}$ i $\dfrac{x+1}{4}=\dfrac{3}{8}$

6 Use backtracking to find the solution to the following equations.

a $3x-7=23$ b $4(x+7)=40$ c $\dfrac{x+6}{9}=6$

d $\dfrac{x}{5}-2=8$ e $5(x-3)=15$ f $\dfrac{x+3}{8}=3$

g $6(x-4)=18$ h $\dfrac{x}{3}+10=12$ i $\dfrac{x}{2.1}-1.7=3.6$

j $4x+\frac{3}{5}=1$ k $\dfrac{x+5}{3}-3=7$ l $3(2x+5)=21$

m $4(x-2)+5=21$ n $3\left(\dfrac{x}{2}+1\right)=15$ o $2(3x+4)-5=15$

WORKED Example 8

7 Simplify and then solve the following equations by backtracking.

a $2x+7+3x+5=27$ b $x+x+1+x+2=18$

c $3x+9+x-4=17$ d $3x+5x+2x=40$

e $6x+6-x-4=37$ f $3x-11+4x=17$

g $5x-2x+5-x=19$ h $2x+3x+4x+5=7$

i $7x-4x+8-x=10$

6.7 SkillSHEET Combining like terms

8 **multiple choice**

The correct flow chart required to solve the equation $6x+5=8$ is:

A
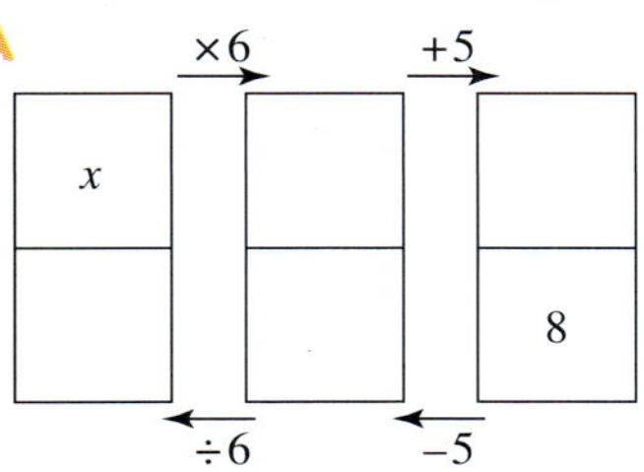

B
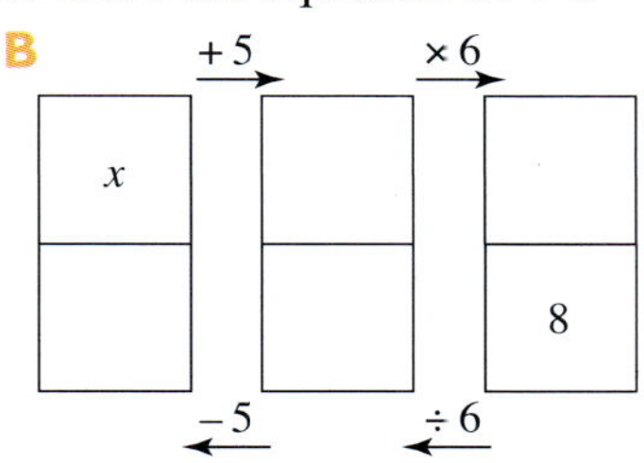

C
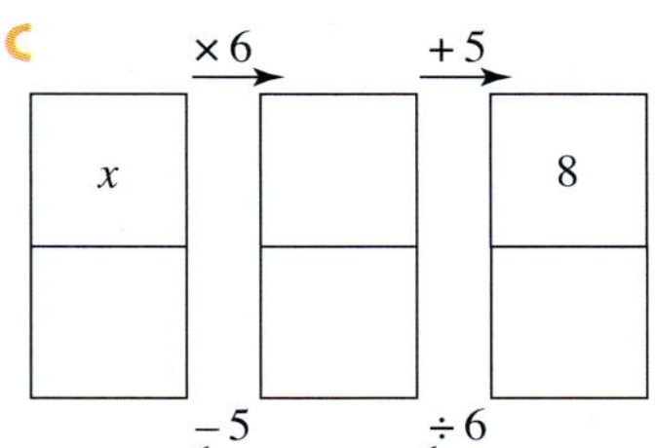

D
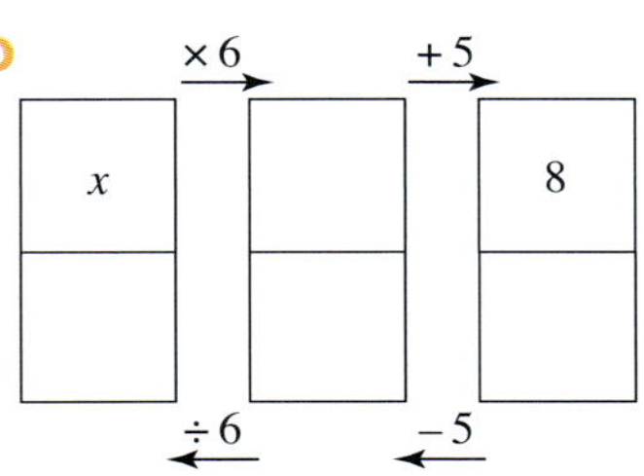

E
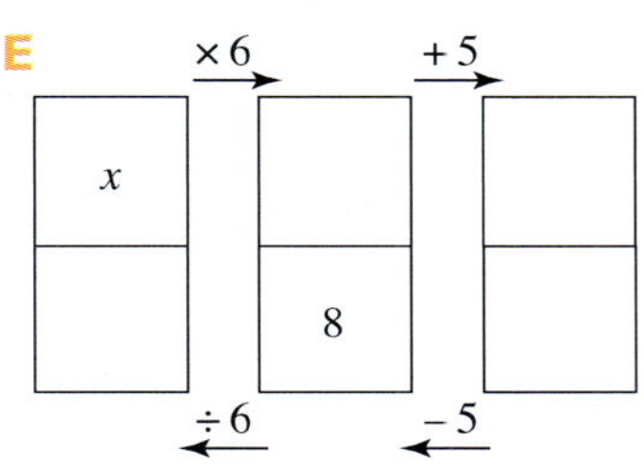

WorkSHEET 6.1

DESIGN

Mystical powers

Mel loves playing 'Think of a number' with friends. Here's an example of one of her puzzles.

- Think of a number.
- Double it.
- Add 10.
- Divide by 2.
- Take away the number you first thought of.
- Your answer is … 5!

Let's investigate to see why the answer is always 5, whatever number you first thought of. We can form expressions for each of the steps, using a variable as the starting value.

- Think of a number .... n
- Double it .... $n \times 2 = 2n$
- Add 10........ $2n + 10$
- Divide by 2 .... $(2n + 10) \div 2 = n + 5$
- Take away the number you first thought of........ $n + 5 - n = 5$
- Your answer is 5

So, your answer will always be 5, for any starting number.

1 Write expressions for each step in the following, showing that you can determine the answer in each case.

Puzzle 1
- Take the year in which you were born.
- Subtract 500.
- Multiply by 2.
- Add 1000.
- Divide by 2.
- Your answer is … your birth year!

Puzzle 2
- Take your age (in years).
- Add 4.
- Multiply by 10.
- Subtract 10.
- Divide by 5.
- Subtract your age.
- Take away 6.
- Your answer will be … your age!

Puzzle 3
- Think of a number.
- Divide it by 2.
- Add 2.
- Multiply by 4.
- Take away your original number.
- Subtract your original number again.
- Your answer should be … 8.

2 Write some 'Think of a number' puzzles yourself. Try them out on friends. They will marvel at your mystical powers!

10 QUICK QUESTIONS 1

Note: If you would like to fill in the flow charts for these questions, select the page and print it out from the interactive *Maths Quest 7* CD-ROM.

1 Write an equation to represent the following puzzle.
I am thinking of a number. When I add 11 to it the answer is 15.

2 Solve the equation $\frac{k}{8} = 4$ by inspection.

3 Complete the following flow chart to find the output number.

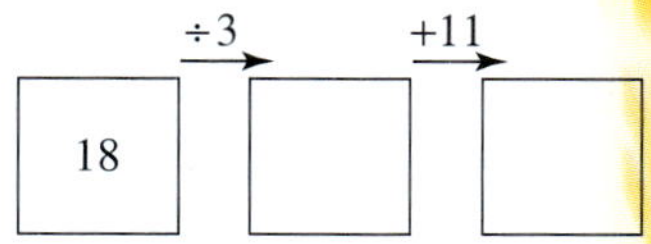

4 Use backtracking and inverse operations to find the input number in this flow chart.

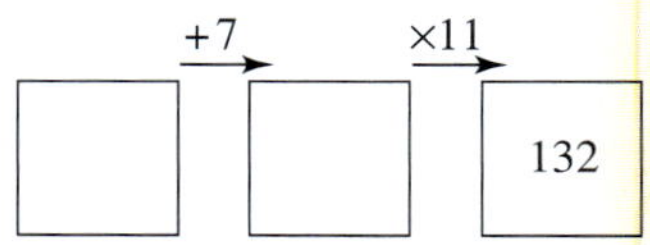

5 Build up an expression by following the instructions on the flow chart.

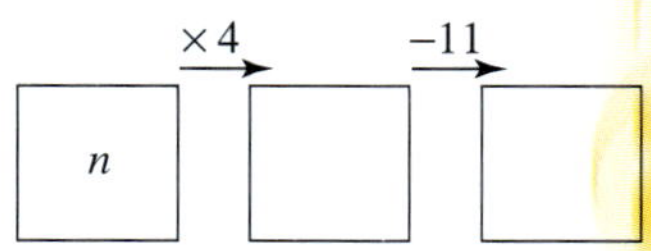

6 Build up an expression by following the instructions on the flow chart.

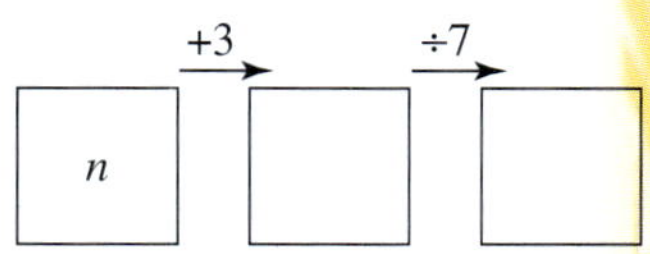

7 Draw the flow chart whose input number is m and whose output is given by the expression $3m + 2$.

8 Draw the flow chart whose input number is k and whose output is given by the expression $\frac{2k - 18}{6}$.

9 Complete this flow chart by writing in the operations that must be carried out in order to backtrack to a.

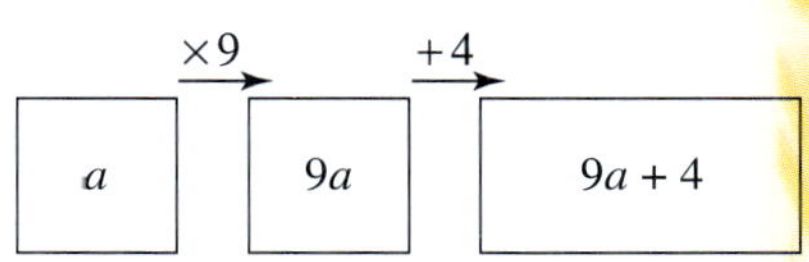

10 Complete this flow chart by writing in the operations that must be carried out in order to build the expression and backtrack to t.

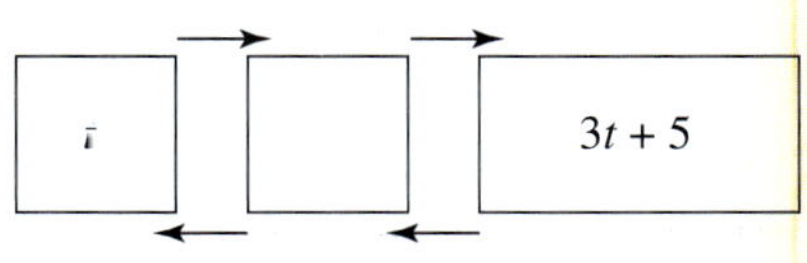

Checking solutions

It is always wise to check to see whether the **solution** to an equation is correct.

Hamish and Helen disagree about the solution to the equation $2x + 7 = 13$. Hamish says that the solution is $x = 10$, but Helen thinks that it is $x = 3$. They decide to check their solutions by looking at the value of $2x + 7$ and using substitution.

If $x = 10$, then $2x + 7 = 2 \times 10 + 7$
$= 27$.

But if $x = 3$, then $2x + 7 = 2 \times 3 + 7$
$= 13$.

The statement $2x + 7 = 13$ is true when $x = 3$. This means that $x = 3$ is the solution to the equation and so Helen is correct.

Sometimes we may encounter a situation where we are given the solution and asked to determine whether it is correct. The procedure is outlined below:

1. Substitute the possible solution into the equation (both sides if necessary).
2. Compare the value of the left-hand side with the value of the right-hand side.
3. If these two values are equal, the solution is correct.
4. If these two values are not equal, the solution is not correct.

WORKED Example 9

For each of the following, determine whether $x = 7$ is the solution to the equation.

a $\dfrac{x+5}{3} = 4$ **b** $2x - 8 = 10$

THINK	WRITE
a 1 Write the equation.	**a** $\dfrac{x+5}{3} = 4$
2 Write the left-hand side (LHS) of the equation and substitute $x = 7$.	If $x = 7$, LHS $= \dfrac{x+5}{3}$
3 Perform the calculation.	$= \dfrac{7+5}{3}$ $= \frac{12}{3}$ $= 4$
4 Compare this with the right-hand side (RHS) of the equation.	RHS $= 4$
5 Comment on the result.	The solution is $x = 7$, since LHS = RHS.
b 1 Write the equation.	**b** $2x - 8 = 10$
2 Write the left-hand side of the equation and substitute $x = 7$.	If $x = 7$, LHS $= 2x - 8$ $= 2(7) - 8$
3 Perform the calculation.	$= 14 - 8$ $= 6$
4 Compare this with the right-hand side of the equation.	RHS $= 10$
5 Comment on the result.	Therefore, $x = 7$ is not the solution since LHS $\neq$ RHS.

WORKED Example 10

For each equation below there is a solution given. Is the solution correct?

a $\frac{x+2}{3} = 2x - 12, x = 10$ **b** $3x - 7 = 2x + 3, x = 10$

THINK	WRITE
a 1 Write the equation.	**a** $\frac{x+2}{3} = 2x - 12$
2 Write the left-hand side of the equation and substitute $x = 10$.	If $x = 10$, LHS $= \frac{x+2}{3}$
3 Perform the calculation.	$= \frac{10+2}{3}$ $= \frac{12}{3}$ $= 4$
4 Write the right-hand side of the equation and substitute $x = 10$.	RHS $= 2x - 12$ $= 2(10) - 12$
5 Perform the calculation.	$= 20 - 12$ $= 8$
6 Comment on the results.	Therefore $x = 10$ is not the solution, since LHS $\neq$ RHS.
b 1 Write the equation.	**b** $3x - 7 = 2x + 3$
2 Write the left-hand side of the equation and substitute $x = 10$.	If $x = 10$, LHS $= 3x - 7$ $= 3(10) - 7$
3 Perform the calculation.	$= 30 - 7$ $= 23$
4 Write the right-hand side of the equation and substitute $x = 10$.	RHS $= 2x + 3$ $= 2(10) + 3$
5 Perform the calculation.	$= 20 + 3$ $= 23$
6 Comment on the results.	The solution is $x = 10$, since LHS = RHS.

Guess, check and improve

We can now try to guess the solutions to some equations, even very difficult ones, because we are able to check our solutions. After each guess we must stop to think and comment so that our guesses improve and lead us towards the correct solution. A table is very useful to help us organise our ideas.

WORKED Example 11

Use guess, check and improve to solve the equation $2x + 21 = 4x - 1$.

THINK

1. Set up a table with four columns displaying the value of x, the value of the left-hand and right-hand side equations (after substitution) and a comment on how these two values compare.
2. Substitute the first guess, say $x = 1$, into the LHS and RHS equations and comment on results.
3. Repeat step 2 for $x = 6$, $x = 10$ until the correct answer is obtained.
4. State the solution.

WRITE

Guess	Check		
x	$2x + 21$	$4x - 1$	Comment
1	23	3	$4x - 1$ is too small.
6	33	23	This is closer.
10	41	39	Very close
11	43	43	That's it!

The solution is $x = 11$.

So far we have dealt only with **solving equations** involving one variable. However, in problem solving you may encounter a situation which requires the use of two variables. The next worked example illustrates how the 'Guess, check and improve' method may again be applied to solve such a problem.

WORKED Example 12

Find two numbers whose sum is 31 and whose product is 150.

THINK

1. The numbers add up to 31 so guess two numbers that do this. Then check by finding their product.
2. Guess 1 and 30.
3. Guess 10 and 21.
   Try a number between 1 and 10 for the first number.
   Try a number between 5 and 10 for the first number.
   Try a number between 5 and 8 for the first number.
4. State the solution.

WRITE

Guess	Check	
Sum (small number first)	Product (P)	Comment
1, 30	$1 \times 30 = 30$	P is too low.
10, 21	$10 \times 21 = 210$	P is too high.
5, 26	$5 \times 26 = 130$	P is too low.
8, 23	$8 \times 23 = 184$	P is too high.
6, 25	$6 \times 25 = 150$	That's it!

The numbers are 6 and 25.

remember

We can check our solution to an equation by substituting the value for the pronumeral to see if it is true.

EXERCISE 6D Checking solutions

WORKED Example 9

6.8 SkillSHEET Checking solutions by substitution

1 For each of the following determine whether:

a $x = 3$ is the solution to the equation $x + 2 = 6$
b $x = 3$ is the solution to the equation $2x - 1 = 5$
c $x = 5$ is the solution to the equation $2x + 3 = 7$
d $x = 4$ is the solution to the equation $6x - 6 = 24$
e $x = 10$ is the solution to the equation $3x + 5 = 20$
f $x = 5$ is the solution to the equation $4(x - 3) = 8$
g $x = 7$ is the solution to the equation $3(x - 2) = 25$
h $x = 8$ is the solution to the equation $5(x + 1) = 90$
i $x = 12$ is the solution to the equation $6(x - 5) = 42$
j $x = 81$ is the solution to the equation $3x - 53 = 80$
k $x = 2.36$ is the solution to the equation $5x - 7 = 4.8$
l $x = 4.4$ is the solution to the equation $7x - 2.15 = 18.64$.

2 Complete the table below to find the value of $2x + 3$ when $x = 0, 1, 2, 3, 4$.

x	0	1	2	3	4
$2x + 3$					

a For what value of x does $2x + 3 = 11$?
b What is the solution (that is, the value of x) for $2x + 3 = 11$?
c What is the solution (that is, the value of x) for $2x + 3 = 5$?

3 Complete the table below to find the value of $5(x - 2)$ when $x = 2, 3, 4, 5, 6$.

x	2	3	4	5	6
$5(x - 2)$					

a What is the solution (that is, the value of x) to $5(x - 2) = 10$?
b What is the solution (that is, the value of x) to $5(x - 2) = 20$?
c What do you guess is the solution (that is, the value of x) to $5(x - 2) = 30$? Check your guess.

Mathcad Substitution

EXCEL Spreadsheet Substitution

4 a Copy and complete the table below.

x	$2x + 1$	$3x - 5$
3	7	
4		
5		10
6		
7		

b What is the solution to $2x + 1 = 3x - 5$?

5 **a** Copy and complete the table below.

x	$\frac{x+3}{2}$	$2x - 6$
3	3	
5		
7		8
9		
11		

b What is the solution to $\frac{x+3}{2} = 2x - 6$?

WORKED Example 10

6 For each equation below there is a solution given. Is the solution correct?

a $2x + 1 = 3x - 5$ $\quad x = 6$

b $5x + 1 = 2x - 7$ $\quad x = 8$

c $3x - 5 = x + 8$ $\quad x = 10$

d $5x = 2x + 12$ $\quad x = 4$

e $4x = 3x + 8$ $\quad x = 8$

f $3x = x + 20$ $\quad x = 15$

g $3x - 1.2 = x + 2.9$ $\quad x = 1.9$

h $6x + 1.5 = 2x + 41.5$ $\quad x = 10$

i $2.4(x + 1) = 9.6$ $\quad x = 3$

j $1.2(x + 1.65) = 0.2(x + 9.85)$ $\quad x = 0.45$

GC program—TI: Solving equations

GC program—Casio: Solving equations

EXCEL Spreadsheet: Solving equations

Mathcad: Solving equations

WORKED Example 11

7 Use guess, check and improve to solve the equation $3x + 11 = 5x - 1$. The process has been started for you.

Guess	Check		Comment
x	$3x + 11$	$5x - 1$	
1	14	4	$5x - 1$ is too small.
10	41	49	$5x - 1$ is too big.
5			

8 Use guess, check and improve to solve the following equations.

a $5x + 15 = x + 27$

b $2x + 12 = 3x - 2$

c $x + 20 = 3x$

d $12x - 18 = 10x$

e $10(x + 1) = 5x + 25$

f $x(x + 1) = 21x$

g $x(x + 7) = 12x$

h $6(x - 2) = 4x$

i $3(x + 4) = 5x + 4$

WORKED Example 12

9 Find two numbers whose sum is 21 and whose product is 98. The process has been started for you.

Guess	Check	
Sum (small number first)	Product (P)	Comment
1, 20	$1 \times 20 = 20$	P is too low.
10, 11	$10 \times 11 = 110$	P is too high.

10 Use guess, check and improve to find two numbers whose sum and product are given:

a sum = 26, product = 165
b sum = 27, product = 162
c sum = 54, product = 329
d sum = 45, product = 296
e sum = 178, product = 5712
f sum = 104, product = 2703
g sum = 153, product = 4662
h sum = 242, product = 14 065
i sum = 6.1, product = 8.58
j sum = 8, product = 14.79
k sum = 978, product = 218 957
l sum = 35, product = 274.89.

11 Copy and complete this table by substituting each x-value into $x^2 + 4$ and $4x + 1$. The first row has been completed for you. Use the table to find a solution to the equation $x^2 + 4 = 4x + 1$. (Remember, x^2 means $x \times x$.)

x	$x^2 + 4$	$4x + 1$
0	4	1
1		
2		9
3	13	
4		

WorkSHEET 6.2

MATHS QUEST CHALLENGE

1 **A football team won 4 more games than it lost. The team played 16 games. How many did it win?**

2 **Lily is half the age of Pedro. Ross is 6 years older than Lily and 6 years younger than Pedro. How old is Pedro?**

3 **A plumber cut a 20-metre pipe into two pieces. One of the pieces is three times as long as the other. What are the lengths of the two pieces of pipe?**

4 **Julie has the same number of sisters as brothers. Her brother Todd has twice as many sisters as brothers. How many children are in the family?**

10 QUICK QUESTIONS 2

1 Write the inverse operation for + 8.

2 Draw a flow chart to represent the following puzzle and then solve by backtracking. I am thinking of a number. When I multiply it by 2 and then add 9 the answer is 43.

3 Draw a flow chart and use backtracking to find the solution to the equation $12k - 8 = 124$.

4 Solve the equation $4(a - 7) = 48$ by backtracking.

5 Solve the equation $\frac{e}{8} + 2 = 10$ by backtracking.

6 First simplify, then solve the equation $2m + 9 + 4m - 4 = 41$.

7 Determine whether this statement is true or false: $w = 30$ is the solution to $\frac{w}{10} + 5 = 8$.

8 Complete the table below to find the value of $7(2a + 5)$ when $a = 0, 1, 2, 3, 4, 5$.

a	0	1	2	3	4	5
$7(2a + 5)$						

What is the solution (that is, the value of a) to $7(2a + 5) = 63$?

9 Complete this table.

a	$3a + 4$	$5a - 10$
5		
6		
7		
8		
9		

What is the solution to $3a + 4 = 5a - 10$?

10 Use guess, check and improve to solve the equation $9m + 8 = 11m - 30$.

Guess	Check		Comment
m	$9m + 8$	$11m - 30$	

Solving word problems

Equations can be used to solve many everyday puzzles and problems. The trick is to change word problems into equations. We have done this with 'I am thinking of a number' problems, but many other situations can be described by equations as well. Here is an example.

Luke and Stefan play cricket in the same team. Last week they scored a total of 53 runs, but Luke scored 11 more runs than Stefan. How many runs did Luke score?

Step 1 Read the question carefully.
The important points are these:
Luke scored 11 more runs than Stefan.
Luke's score plus Stefan's score equals 53.

Step 2 Turn the information into an equation.
We are asked to find Luke's score, so we will call it x. We usually say 'Let x equal Luke's score'. Stefan scored 11 runs fewer than Luke so his score will be $x - 11$. Since their scores total 53 we know that $x + x - 11 = 53$.

Step 3 Simplify and then solve the equation $x + x - 11 = 53$. This is the same as $2x - 11 = 53$.
We can backtrack or solve by inspection to find $x = 32$.

Step 4 Check your solution. If Luke scored 32 runs, then Stefan scored 21 runs and their total is $32 + 21 = 53$. This fits the question.

Step 5 Write your answer in words. Luke scored 32 runs.

WORKED Example 13

If x is a whole number, write the value of:
a the number 10 more than x
b the number one-third of x
c the number of minutes in x hours.

THINK	WRITE
a To get 10 more means to add 10.	a $x + 10$
b To find one-third means to divide by 3.	b $x \div 3 = \frac{x}{3}$
c To change hours to minutes, multiply by 60. Remember there are 60 minutes in 1 hour.	c $60x$

WORKED Example 14

a Simone earns \$5 per week more than Jessica. If Jessica earns \$$x$, find their total earnings per week.
b If x is the smallest of 2 consecutive numbers, find the sum of the two numbers.

THINK	WRITE
a 1 State the amount Jessica earns.	a Jessica: x
2 State the amount Simone earns.	Simone: $x + 5$
3 State the total amount they earn.	Together: $x + x + 5$
4 Simplify the expression.	$x + x + 5 = 2x + 5$
5 State the answer.	Their total earnings are \$$(2x + 5)$.

Continued over page

THINK	WRITE
b 1 Use the information given to write an expression regarding the second number. *Note:* If two numbers are consecutive they follow each other, for example 3 and 4.	**b** If x equals the smallest number, the next consecutive number is equal to $x + 1$.
2 Write an expression for the sum of the two consecutive numbers. Remember the sum means to add.	$x + x + 1$
3 Simplify the expression.	$= 2x + 1$

WORKED Example 15

Anh is 2 years older than his sister Phuong and the sum of their ages is 30 years. How old is Phuong?

THINK	WRITE
1 Identify what is needed and represent this with x. We need to find Phuong's age.	Let x equal Phuong's age.
2 Form an expression using the information that Anh is 2 years older.	Anh's age is $x + 2$.
3 Form an equation showing that their ages add up to 30.	Phuong + Anh = 30 $x + x + 2 = 30$
4 Simplify the equation.	$2x + 2 = 30$
5 Backtrack to solve for x.	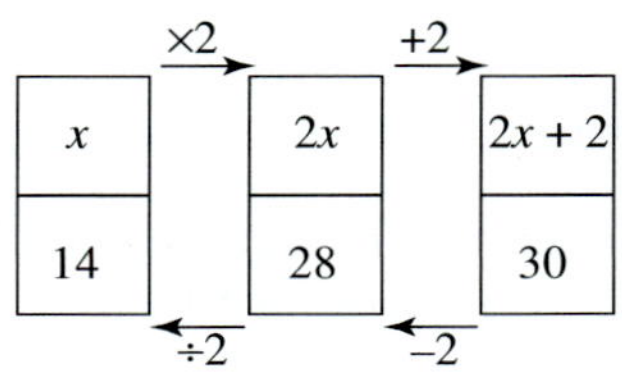 $x = 14$
6 Check the solution.	Check: Anh is 16, and 16 + 14 = 30.
7 Write the answer in words.	Phuong is 14 years old.

remember

Solving word problems involves five steps.

1. Read the question carefully.
2. Turn the information into an equation.
3. Simplify and then solve the equation.
4. Write your answer in words.
5. Check your solution.

EXERCISE 6E Solving word problems

WORKED Example 13

1 If x is a whole number, write down the value of:

a the number 5 more than x
b the number 8 less than x
c the number twice as large as x
d the number one quarter of x
e the next largest number after x
f the number of cents in x dollars
g the number of metres in x centimetres
h the number of days in x weeks
i the (mean) average of x and 20
j how much it costs to rent a tennis court for x hours at \$10 per hour
k the average of 5, 18 and k.

6.9 SkillSHEET Finding an average

2 **multiple choice**

a John is x years old. In 5 years his age will be:
A $5x$ B 17 C $x - 5$ D $x + 5$ E 10

b Chris is 2 years older than Jenny. If Jenny is x years old then Chris's age is:
A $x - 2$ B $2x$ C $x + 2$ D 2 E $\frac{x}{2}$

c Duy earns \$5 per week more than Ben. If Duy earns \$$x$, then Ben earns:
A $x + 5$ B $5x$ C $x - 5$ D $5 + x$ E $5 - x$

d Assume x is an even number. The next largest even number is:
A $x + 1$ B $2x + 2$ C 4 D $2x$ E $x + 2$

e Assume x is an odd number. The next largest odd number is:
A $x + 1$ B $x + 2$ C 3 D $2x + 1$ E $2x + 2$

f Gian is half Carla's age. If Gian is x years old, then Carla's age is:
A $\frac{x}{2}$ B $x + \frac{1}{2}$ C $2x$ D $\frac{1}{2}x$ E $x - \frac{1}{2}$

6.10 SkillSHEET Consecutive numbers

WORKED Example 14

3 a Ali earns \$7 more per week than Halit. If Halit earns \$$x$, find their total earnings per week.
b Sasha is twice as old as Kapila. If Kapila is x years old, find their total age.
c Frank has had 3 more birthdays than James. If James is $2x$ years old, find their total age.
d If x is the smaller of 3 consecutive whole numbers, find the sum of the 3 numbers.
e If x is the smallest of 3 consecutive odd numbers, find the sum of the 3 numbers.
f If x is the smallest of 3 consecutive even numbers, find the sum of the 3 numbers.

Mathcad Solving equations

GC program—TI Solving equations

Use equations to solve questions 4 to 14.

WORKED Example 15

4 Imraan is 5 years older than his brother Gareth and the sum of their ages is 31 years. How old is Gareth? (Let x represent Gareth's age.)

GC program—Casio Solving equations

5 In three basketball games Karina has averaged 12 points each game. In the first game she scored 11 points, in the second she scored 17 points, and in the third game she scored x points.
a From the given information, what is the average of 11, 17 and x?
b Write an equation using the answer to part a.
c Solve the equation.
d How many points did Karina score in the third game?

EXCEL Spreadsheet Solving equations

6 Melanie and Callie went tenpin bowling together. Melanie scored 15 more pins than Callie, and their total score was 207. What did Callie score?

7 The sum of three consecutive whole numbers is 51. Find the numbers. (*Hint:* Let the smallest number equal x.)

8 The sum of three consecutive odd numbers is 27. What are the numbers?

9 Three consecutive multiples of 5 add up to 90. What are the 3 numbers?

10 David is 5 years younger than his twin brothers. If the sum of their ages is 52, then how old is David?

11 In the high jump event Chris leapt 12 centimetres higher than Tim, but their two jumps made a total of 3 metres. How high did Chris jump?

12 Daniel and Travis are twins, but their sister Phillipa is 3 years older. If the sum of their three ages is 36, how old are the twins?

13 a Write a rule to describe how you would calculate the total cost of 3 items, two hamburgers and one drink.
b Let h represent the cost of a hamburger. If the drink costs \$3.75 and the total cost is \$7.25, write an equation which links all this information.
c Solve the equation to find the cost of a hamburger.

14 

a How much will it cost to hire a windsurfer for:
i 1 hour? ii 2 hours? iii 3 hours?
b Write a rule that could be used to calculate the cost of hiring a windsurfer for n hours.
c Use the rule to calculate the cost of hiring a windsurfer for 8 hours.
d You have \$100 to spend. Write an equation to help you work out the number of hours for which you could hire a windsurfer.
e Solve the equation to find the number of hours for which you could hire the windsurfer.
f Work out how much money (if any) you would have left over when you pay the hire charge.
g After a great day's windsurfing, you have returned the windsurfer (with no damage). Would you have enough money to buy fish and chips on the way home?

GAME time Equations — 002

WorkSHEET 6.3

THINKING

Bush Nursing walk-a-thon

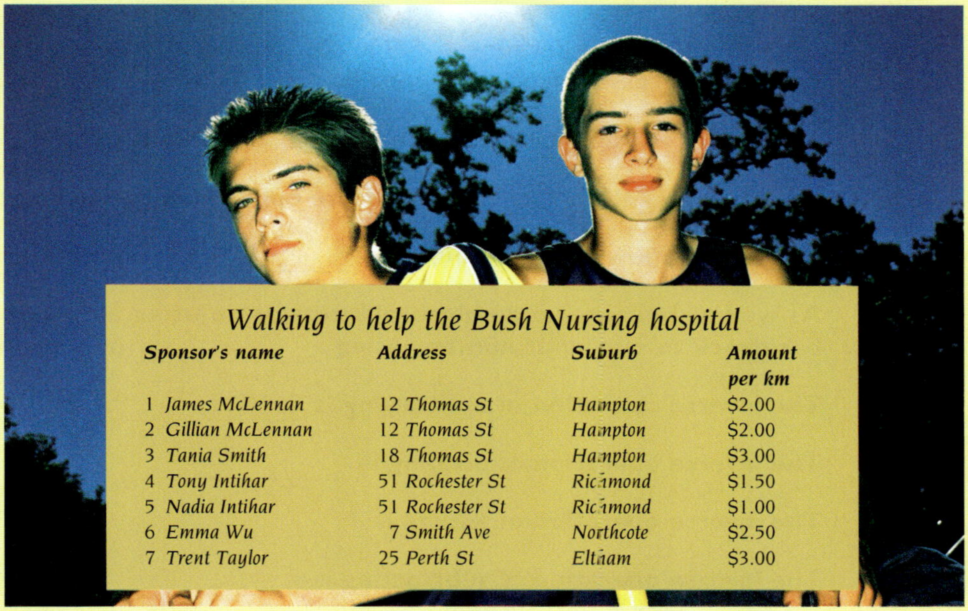

Walking to help the Bush Nursing hospital

	Sponsor's name	Address	Suburb	Amount per km
1	James McLennan	12 Thomas St	Hampton	$2.00
2	Gillian McLennan	12 Thomas St	Hampton	$2.00
3	Tania Smith	18 Thomas St	Hampton	$3.00
4	Tony Intihar	51 Rochester St	Richmond	$1.50
5	Nadia Intihar	51 Rochester St	Richmond	$1.00
6	Emma Wu	7 Smith Ave	Northcote	$2.50
7	Trent Taylor	25 Perth St	Eltham	$3.00

Your friend Alex has raised $75.00 for the Bush Nursing hospital. Can you work out how far he has walked? Look at his sponsor form above.

1 Let k represent the number of kilometres Alex has walked. Write an expression for the amount of money James McLennan would need to pay him.

2 In a similar way, write an expression for the amount of money each person listed on his sponsor form would need to pay.

3 Write a rule showing the total amount of money raised by Alex if he walked k kilometres in the walk-a-thon.

4 If the total amount of money raised by Alex is $75.00, write an equation which could be solved to find the value of k.

5 Solve the equation to find k and write a sentence to state your answer.

6 If instead of raising $75.00, Alex has raised $120.00 from his sponsors, how far would he have walked?

Another friend, Daniel had the following sponsor form.

Walking to help the Bush Nursing hospital

	Sponsor's name	Address	Suburb	Amount per km
1	Megan Talty	26 Toy Rd	Strathfield	$1.50
2	Michael Lamb	121 Holston St	Homebush	$3.50
3	Jessica Bacash	74 Mason Dr	Flemington	$2.00
4	Dinesh Jones	5 Eagle St	Croydon	$2.50

7 Daniel is not sure whether he should collect $82.50 or $85.50 from his sponsors. Using equations, work out which amount he should collect.

8 Find the total number of kilometres Daniel walked in the walk-a-thon.

9 If Daniel wanted to raise over $100 for the Bush Nursing hospital, how far would he have needed to walk? (Give the answer to the nearest kilometre.)

summary

Copy the following sentences. Fill in the gaps by choosing the correct word or expression from the word list that follows.

1 ____________ help us to organise harder equations.

2 We can use ____________ and inverse operations to solve equations.

3 We can use flow charts to build up ____________ expressions.

4 As well as building up an algebraic expression using a flow chart, we can backtrack to our input number using ____________ operations.

5 The inverse operation of multiplying is ____________.

6 The inverse operation of dividing is ____________.

7 The inverse operation of adding is ____________.

8 The inverse operation of subtracting is ____________.

9 We can ____________ equations by drawing a flow chart and back-tracking.

10 We can check our solution to an equation by ____________ the value for the pronumeral to see if it is true.

11 Changing ____________ problems into equations involves the 5 steps shown below.

Step 1 ____________ the question carefully.

Step 2 Turn the information into an ____________.

Step 3 ____________ and then solve the equation.

Step 4 ____________ your solution.

Step 5 Write your ____________ in words.

WORD LIST

substituting	flow charts	subtracting	answer
inverse	multiplying	backtracking	read
dividing	check	simplify	algebraic
equation	solve	adding	word

CHAPTER review

Note: If you would like to fill in the flow charts for this review, select the pages and print them out from the interactive *Maths Quest 7* CD-ROM.

1 Solve these equations by inspection. 6A

a $m + 7 = 12$ **b** $5h = 30$ **c** $s - 12 = 7$ **d** $\frac{d}{5} = 4$

2 Complete these flow charts to find the output number. 6A

a $\boxed{4} \xrightarrow{\times 5} \square \xrightarrow{-9} \square$

b $\boxed{5} \xrightarrow{+3} \square \xrightarrow{\div 4} \square$

c $\boxed{10} \xrightarrow{\div 2} \square \xrightarrow{+11} \square$

d $\boxed{8} \xrightarrow{-3} \square \xrightarrow{\times 7} \square$

3 Use backtracking and inverse operations to find the input number in each of these flow charts. 6A

a $\square \xrightarrow{\div 2} \square \xrightarrow{+10} \boxed{18}$

b $\square \xrightarrow{-7} \square \xrightarrow{\div 5} \boxed{7}$

c $\square \xrightarrow{-6} \square \xrightarrow{\times 7} \boxed{35}$

d $\square \xrightarrow{\times 2} \square \xrightarrow{+2} \boxed{18}$

4 Build up an expression by following the instructions on the flow chart. 6B

a $\boxed{x} \xrightarrow{\times 7} \square \xrightarrow{+8} \square$

b $\boxed{x} \xrightarrow{\div 3} \square \xrightarrow{-5} \square$

c $\boxed{x} \xrightarrow{+2} \square \xrightarrow{\times 6} \square$

d $\boxed{x} \xrightarrow{-7} \square \xrightarrow{\div 5} \square$

e $\boxed{x} \xrightarrow{+3} \square \xrightarrow{\times 5} \square \xrightarrow{-9} \square$

f $\boxed{x} \xrightarrow{\times 4} \square \xrightarrow{+11} \square \xrightarrow{\div 8} \square$

5 Draw the flow chart whose input is x and whose output is given by the expression. 6B

a $5(x + 7)$ **b** $\frac{x}{4} - 3$ **c** $6x - 14$ **d** $\frac{x + 2}{5}$

6B **6** Complete these flow charts by writing in the operations which must be carried out in order to backtrack to x.

a

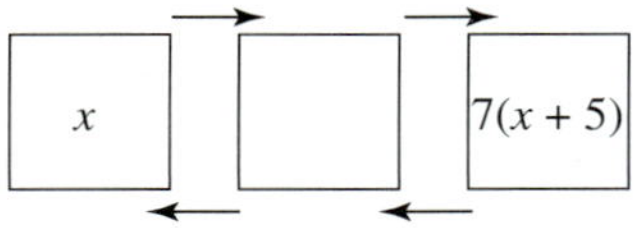

b

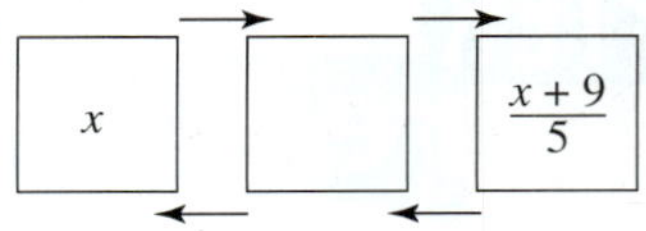

c

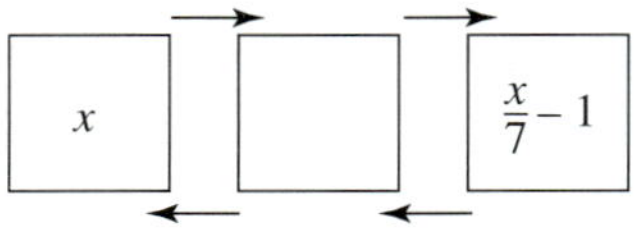

d

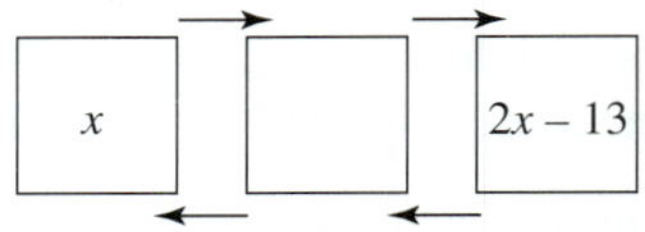

6C **7** Draw a flow chart and use backtracking to find the solution to the following equations.

a $7x + 6 = 20$ **b** $9(y - 8) = 18$ **c** $\frac{m}{5} - 3 = 9$ **d** $\frac{s+7}{5} = 5$

6C **8** Use backtracking to find the solution to these equations.

a $3(d + 1) = 15$ **b** $\frac{t}{4} - 11 = 14$ **c** $6d - 3 = 15$ **d** $\frac{a+6}{4} = 3$

6C **9** Simplify the expression and then solve the equation for each of the following.

a $7v + 3 + 3v + 4 = 37$ **b** $6c + 15 - 5c - 8 = 19$

6D **10** For each equation below there is a solution given. Is the solution correct?

a $5x - 7 = 2x + 2$ $x = 3$ **b** $\frac{x+9}{2} = 2x - 7$ $x = 5$

6D **11** Use guess, check and improve to find two numbers whose sum and product are given.

a Sum = 83, product = 1632 **b** Sum = 86, product = 1593

6E **12** Sophie and Jackie each have a collection of football cards. Jackie has 5 more cards than Sophie, and together they have 67 cards. By writing and solving an equation, find out how many cards Sophie owns.

6E **13** Andreas has completed 2 more pieces of homework than Richard who submitted x pieces of homework for the semester. If the total number of pieces of homework submitted by the two boys is 12, how many pieces of homework did Andreas submit?

Length and perimeter

7

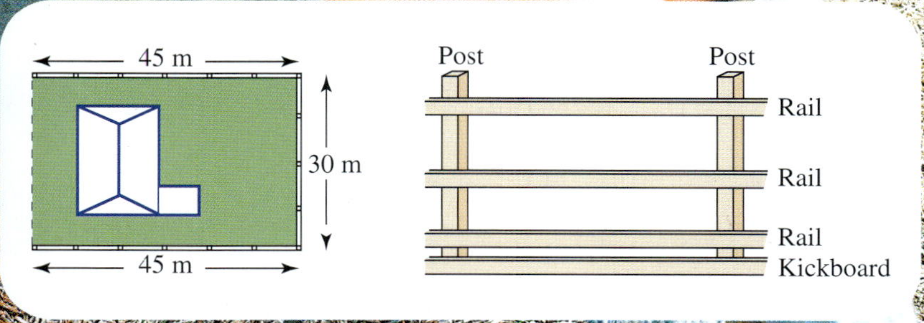

Marc and Cathy are seeking quotes on the cost of building a timber fence on three sides of their property.

The fence requires three rails and a kickboard as well as posts and palings. What length of timber will be needed for the kickboard?

If the timber for the rails costs $2.25 per metre, what will be the total cost of the timber for the railings?

These are just some of the calculations needed to produce a quote.

By the end of this chapter, you will have developed and practised further skills in measurement to complete an estimate of the total costs involved.

Are you ready?

Try the questions below. If you have difficulty with any of them, extra help can be obtained by completing the matching SkillSHEET. Either click on the SkillSHEET icon next to the question on the *Maths Quest 7* CD-ROM or ask your teacher for a copy.

SkillSHEET 7.1

Metric units of length

1 a For each metric unit of length, match the name with the correct abbreviation.

i	centimetre	m
ii	kilometre	mm
iii	metre	cm
iv	millimetre	km

b Give an example to describe the length of each unit in part **a**.

SkillSHEET 7.2

Reading scales (how much is each interval worth?)

2 For the scales shown, how much is each interval worth?

a 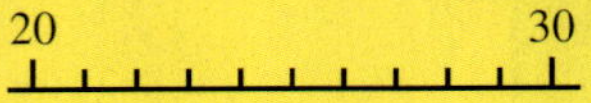

b 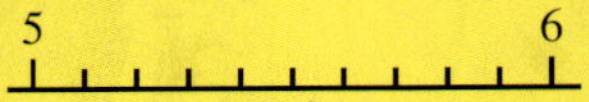

SkillSHEET 7.3

Measuring the length of a line

3 a Measure the length of this line to the nearest mm.

b Measure the length of this line in cm.

SkillSHEET 7.4

Relationship between unit size and the number of units used

4 The distance from your home to school is measured in both metres and kilometres. Which unit would you have more of for this measurement?

SkillSHEET 7.5

Converting units (metres and kilometres)

5 a Convert 18 km to m. **b** Convert 6000 m to km.

SkillSHEET 7.6

Converting units to compare lengths or distances

6 Arrange the following in order from smallest to largest.

a 25 mm, 2.3 cm, 0.2 m **b** 1.8 m, 1700 mm, 185 cm

SkillSHEET 7.7

Perimeter

7 Calculate the perimeter of each shape shown below. (*Note:* The figures are not drawn to scale.)

a 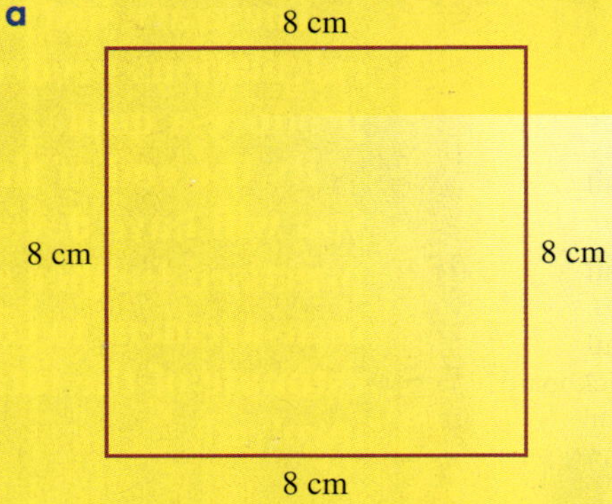

b 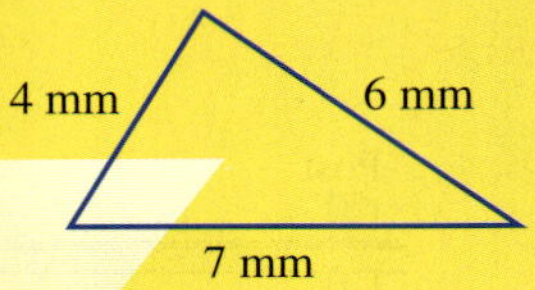

Metric units of length

The **metric** system was developed during the French Revolution (1789–99) to provide a standardised, easy-to-use system based on the number 10. The base unit of this system, the metre, derives its name from the Greek word *metron* meaning 'measure'. The metre was defined to be $\frac{1}{10\,000\,000}$ of the distance from the North Pole to the equator along the Earth's surface. This definition has changed a number of times. In 1983 it was defined as the distance travelled by a beam of light in a vacuum in $\frac{1}{299\,792\,458}$ of a second. The other units of length in the metric system are related to the metre by powers of 10.

The most commonly used units of length (in Australia, and most other countries) with their abbreviations (shortened versions of unit names) and some approximate examples are shown below.

1. Kilometre (km)

The distance travelled by a car travelling at the **speed** limit of 60 kilometres per hour in one minute

2. Metre (m)

The length of an adult's stride

3. Centimetre (cm)

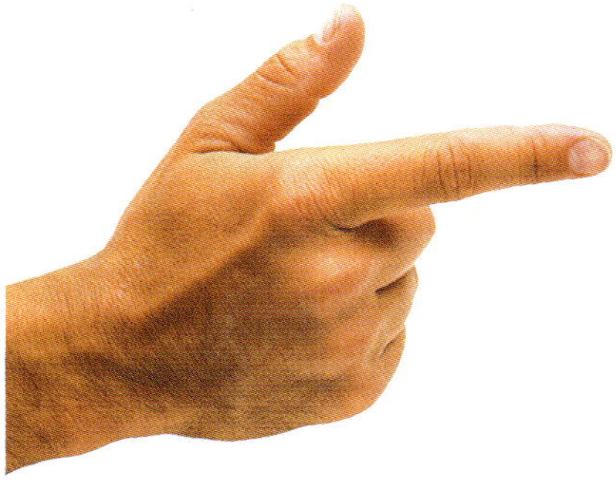

The width of each of your fingers

4. Millimetre (mm)

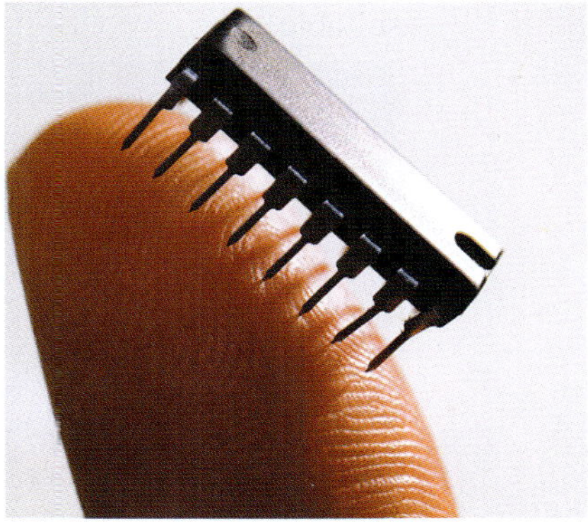

The width of a wire in this computer chip

Measuring lengths is easiest if we use units that give us convenient values; that is, values that are not too large or too small. For example, you'd be more likely to measure the height of a blade of grass using millimetres, while the distance from Melbourne to Wonthaggi would be best measured in kilometres.

WORKED Example 1

You have been given the task of measuring the length of a piece of rope to be used around the boundary of a cricket ground. Which metric units of length would you use?

THINK	WRITE
Estimate the length involved. In this case, it is in the order of 100s of metres. Would other units give a more manageable figure? (Would using centimetres, for example, 10 000 cm, be more convenient? No! What about kilometres? No — the length would be less than 1 kilometre.)	metres (or m)

remember

1. The most commonly used units to measure length are kilometres (km), metres (m), centimetres (cm) and millimetres (mm).
2. When measuring length, it is easiest to use units that give us convenient values (not too large and not too small).

EXERCISE 7A Metric units of length

SkillSHEET 7.1 Metric units of length

WORKED Example 1

1 Which units (mm, cm, m or km) would be most suitable for measuring the real lengths marked in each photograph? (You may give two different units if you think both would be suitable.)

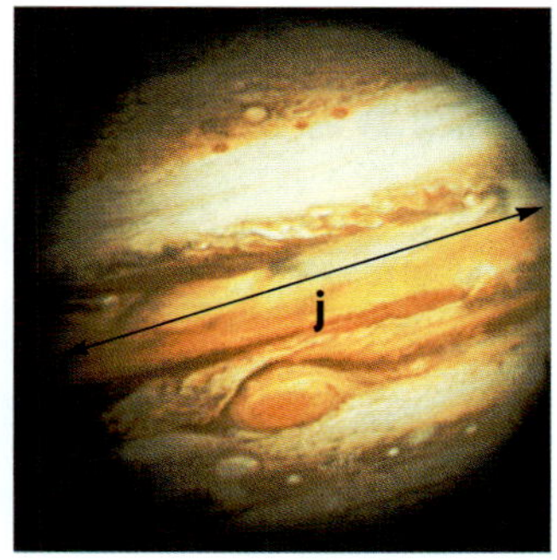

2 Which metric units of length would you use for measuring the following (you may give two different units if you think both could be used)?
- a The length of a netball court
- b The diameter of a netball
- c A photo frame's width
- d The thickness of a slice of bread
- e The length of a worm
- f The distance between Melbourne and Sydney
- g The depth of an Olympic swimming pool
- h The depth of a bowl of soup
- i The length of a mathematics textbook
- j The thickness of a magazine
- k The distance around the shores of a reservoir
- l The length of a cut-out shape on a GeoMat mathematics template

3 **multiple choice**

The thicknesses of two different brands of chocolate biscuit would be best compared by measuring in:
- A millimetres
- B kilometres
- C metres
- D centimetres
- E kilometres or metres

4 **multiple choice**

Girouk wishes to measure the length of his mathematics folder to see what size paper he could fit into it. Which of the following would be the best measuring instrument to use?
- A A science department metre ruler
- B A 10-cm mini ruler
- C A 30-cm plastic ruler
- D A 20-m flexible tape measure
- E A one-metre trundle wheel

5 Give three examples (not mentioned earlier in this exercise) of something that would be best measured in:
- a metres
- b millimetres
- c kilometres
- d centimetres.

6 Give an example of a length which each of the following people might measure in their jobs. (For example, a carpet layer would measure the length of a room.)
- a veterinary surgeon
- b cartographer (map maker)
- c costume designer
- d carpenter
- e doctor
- f landscape gardener
- g plumber
- h computer technician
- i track and field athlete
- j weather reporter

History of mathematics

JOSEPH-LOUIS LAGRANGE (1736–1813)

Joseph-Louis Lagrange — the father of the metric system

During his life: Captain James Cook landed in Botany Bay, Ludwig van Beethoven wrote his first symphony, America declared independence, Napoleon invaded Russia, and Joseph Michel and Jacques Etienne Montgolfier became the first people known to fly when they developed the hot air balloon.

Joseph-Louis Lagrange was one of the most brilliant, courteous and generous mathematicians of the eighteenth century. He helped develop the metric system — in fact some refer to him as the father of the metric system. He also made many valuable contributions to science including physics, in which he laid some of the groundwork for Einstein's theory of relativity.

Lagrange's wealthy father lost his fortune in the early 1700s, forcing Joseph-Louis to pursue a career in mathematics as he no longer had an inheritance! Although he initially found mathematics boring, Lagrange soon began to succeed, and became a professor of mathematics before he was eighteen. He later made contributions in calculus, astronomy, sound, mechanics and probability.

In 1764, Lagrange won the first of many Grand Prixs (not the race, but a mathematical award of the French Academy) for explaining why the same part of the surface of the moon is always visible from Earth.

In 1766, Lagrange replaced Leonhard Euler (another very famous mathematician) as court mathematician to Frederick the Great in the Berlin Academy.

Lagrange's first marriage was to a younger cousin. When she died at a young age, Lagrange threw himself into his work.

Eventually, overwork broke Lagrange's spirit. By 1783, he had become so depressed that he could no longer work in mathematics. He was sacked from the Berlin Academy in 1786, and accepted a job with the French Academy, but could still not shake his depression. It was said that when an important piece of his mathematical work was published in 1788, he left it unopened on his desk for two years, preferring to concentrate on other subjects.

The French Revolution began in 1789. Many scholars fled to avoid the terror inflicted on certain individuals during the Revolution, but Lagrange did not.

In 1792, at the age of 56, Lagrange was snapped out of his sadness by the daughter of an astronomer friend of his. Despite the difference in their ages, they married.

In 1797, Lagrange was appointed to a position at the Ecole Polytechnique, where he was required to supervise the development of the metric system. Some mathematicians wanted to base the system on the number 12, but Lagrange insisted that 10 be used. At one stage, in order to confuse his opponents, he argued that 11 be used! Eventually, Lagrange got his way, and the metric system was based on the number 10.

The famous general, Napoleon, is said to have thought very highly of Lagrange, who died in 1813, aged 77.

Questions

1. In which year was Lagrange born?
2. List some areas of study (besides mathematics) to which Lagrange contributed.
3. What was Lagrange required to do in his position at Ecole Polytechnique?
4. Why do you think 10 was used as the base for the metric system and not 11 or 12?
5. Can you think of any measurement systems that use 12 as a base?
6. Draw a timeline showing the major events in Lagrange's life.

Reading scales and measuring length

A scale is a series of marks used to measure length, mass, temperature or any other quantity. It can be used to measure different amounts of the same type of quantity. For example, the distances obtained by members of the long jump team at a school sports meeting can all be measured using a measuring tape.

When reading scales and measuring length:
1. **check that the scale starts at zero**
2. **check the value of each small division by counting along the scale to the next major mark**
3. **always give units (for example, centimetres) with your answer.**

It is important to realise that all measurements are approximations. For example, when measuring a child's height, we take our reading to the nearest millimetre mark on the measuring tape; yet children grow continuously — not in millimetre increments.

Clearly, we are approximating when we take a measurement.

WORKED Example 2

State the reading indicated by the arrow in each case below giving answers in:
(i) decimal form
(ii) fractional form where appropriate.

a

b

THINK	WRITE
a 1 Check that the line starts at 0. It does.	**a**
2 Note the units printed on the ruler (cm).	
3 Read the last centimetre mark (5).	
4 Does the line go past the last centimetre mark? No.	
5 Write the value with units. *Note:* There is no need to go any further as the value is a whole number.	5 cm

Continued over page

THINK	WRITE
b i 1 Check that the line starts at 0. It does.	**b i**
2 Note the units printed on the ruler (cm).	
3 Read the last centimetre mark (7).	
4 Consider how many smaller intervals there are between the numbered divisions. There are 10, so each smaller division represents 0.1 of a unit.	
5 Count the number of small divisions past the last centimetre mark (3).	
6 Write the value with units.	7.3 cm
ii Express the answer obtained in **b i** as a fraction. *Note:* The value of the first place to the right of the decimal point is tenths, so the digit 3 is 3 tenths.	**ii** $7\frac{3}{10}$ cm

WORKED Example 3

Use the given length of the climber's lower leg to estimate the labelled length of the rope on the diagram at right.

THINK	WRITE
1 Refer to the 0.5-m length indicated on the diagram and determine how many of these lengths correspond to the unknown length of the rope above the climber. *Note:* There are approximately $3\frac{1}{2}$ or 3.5 of these lengths in the rope.	 From the above diagram there are approximately $3\frac{1}{2}$ (or 3.5) of these lengths in the rope.
2 Multiply the number of lengths by 0.5 m.	Length = 3.5×0.5 = 1.75
3 Write the answer.	The length of the rope is approximately 1.75 m.

remember

1. The most commonly used units to measure length are kilometres (km), metres (m), centimetres (cm) and millimetres (mm).
2. When measuring length, it is easiest to use units that give us convenient values (not too large and not too small)
3. When reading scales and measuring length:
   (a) check that the scale starts at zero
   (b) check the value of each small division by counting along the scale to the next major mark
   (c) always give units (for example, centimetres) with your answer.

EXERCISE 7B Reading scales and measuring length

WORKED Example 2a

7.2 SkillSHEET Reading scales (how much is each interval worth?)

1 State the reading indicated by the arrow in each case below giving answers in:
i decimal form
ii fractional form where appropriate.

a

b

c

WORKED Example 2b

d

e
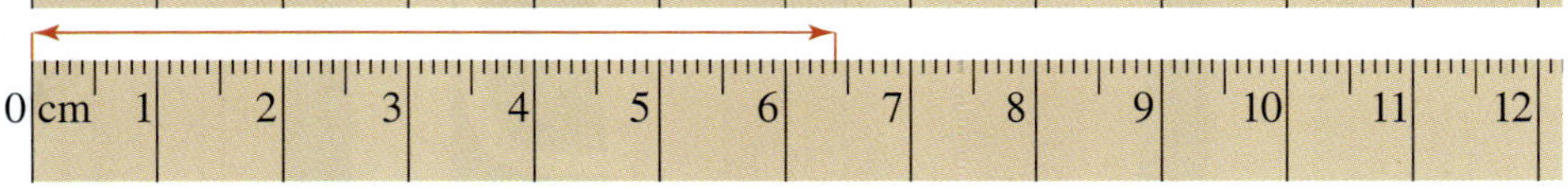

f
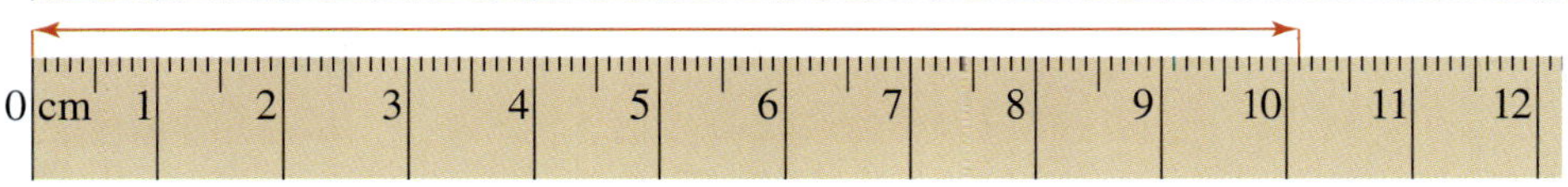

g
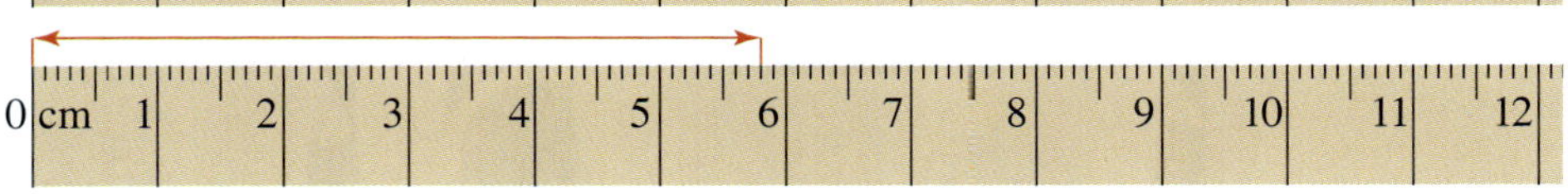

h
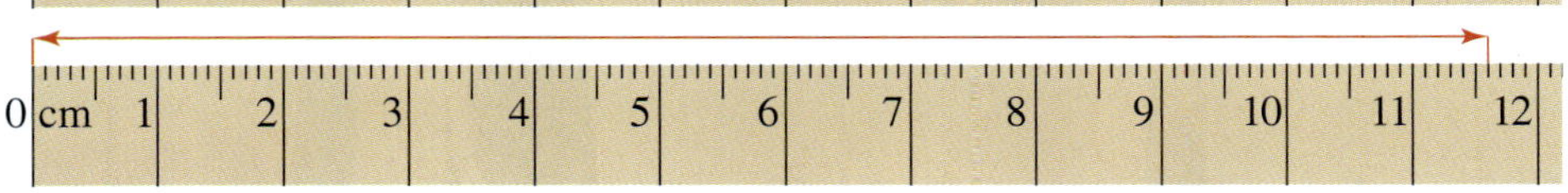

i
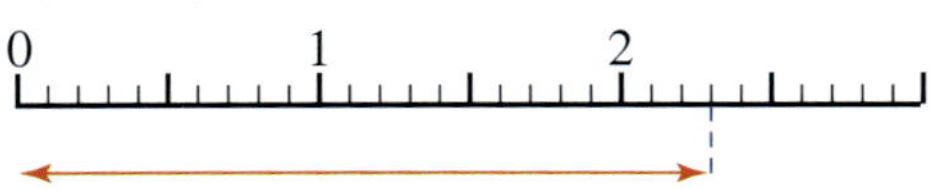

j
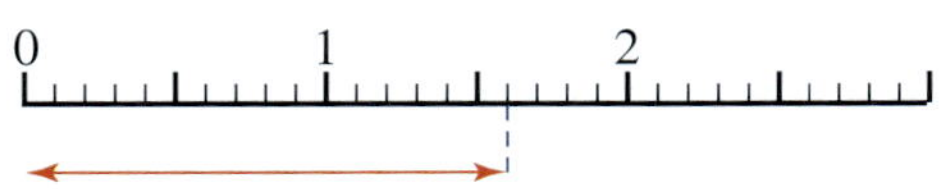

2 Read the values shown on the following scales. For parts **a** to **d** and **g** read from the arrow.

a

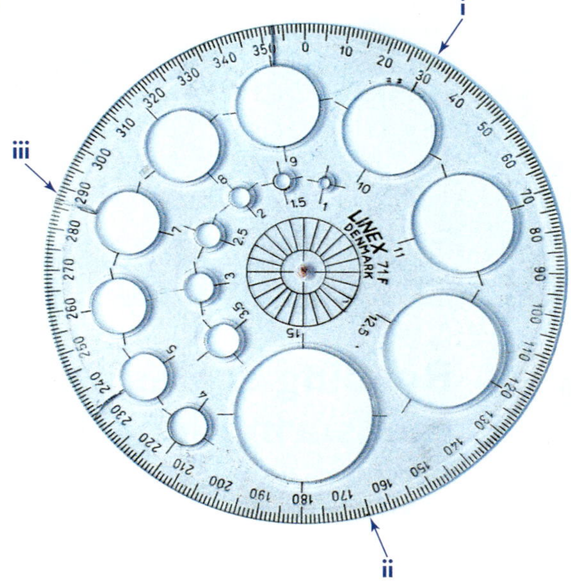

b

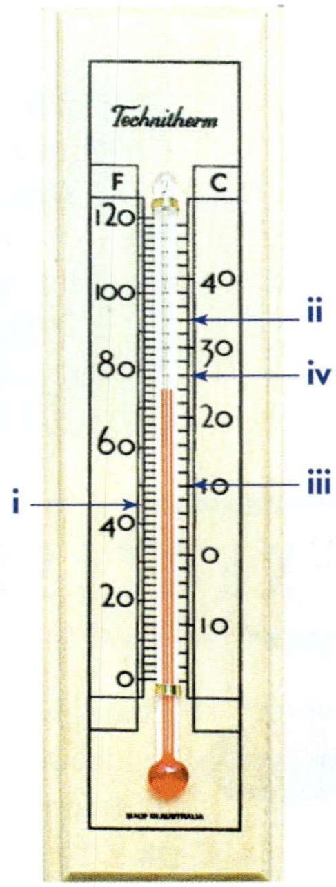

c

d

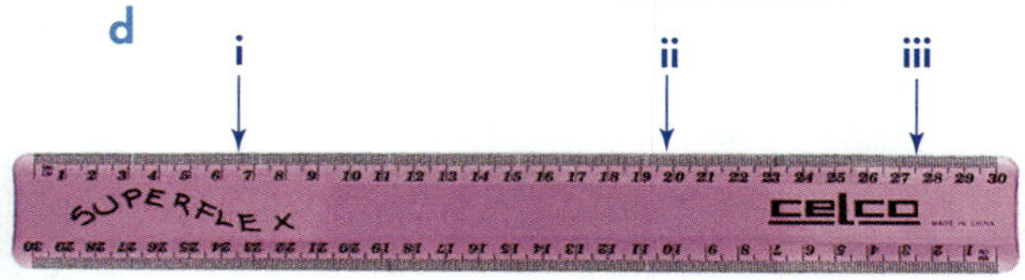

e i

ii

iii

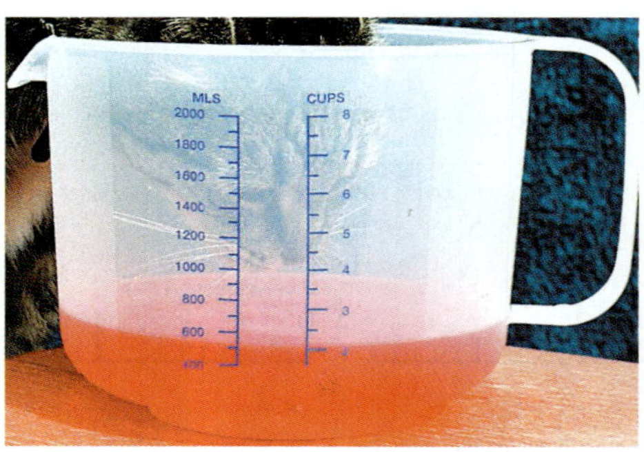

iv

f 

g 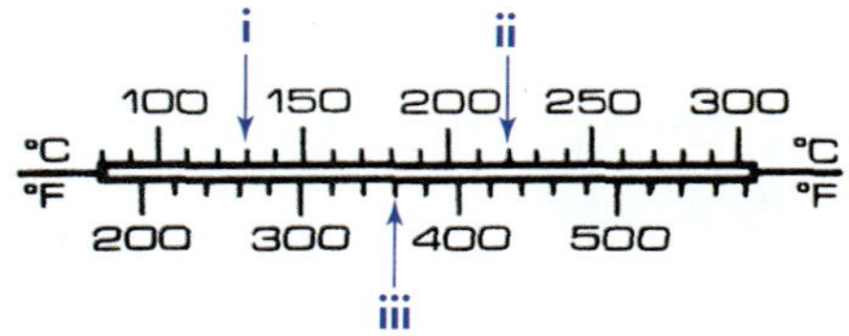

3 multiple choice

The reading on the scale below is:

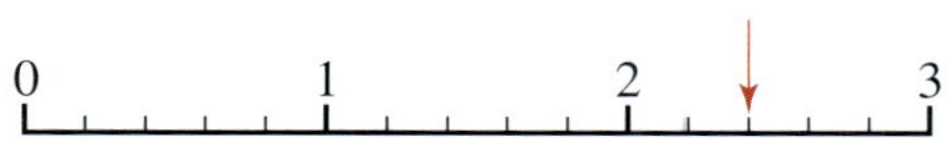

A 2.2 **B** 2.4 **C** 2.6 **D** 2.8 **E** 3.3

4 multiple choice

The reading on the scale below is:

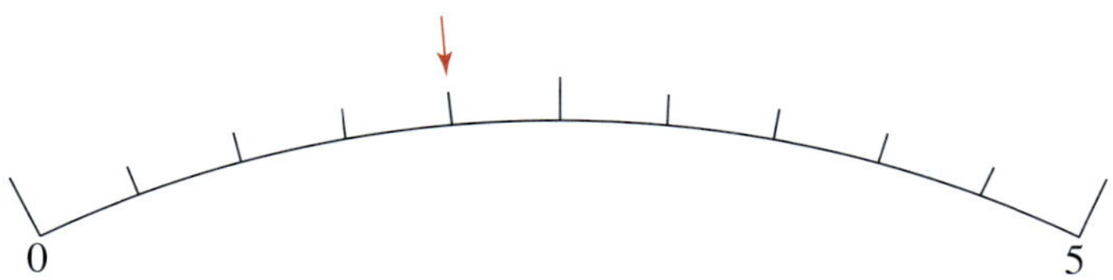

A $\frac{4}{5}$ **B** 4 **C** $\frac{2}{5}$ **D** $\frac{2}{10}$ **E** $\frac{1}{2}$

5 Measure each of the following lines in centimetres.

7.3 SkillSHEET Measuring the length of a line

a

b

c

d

e

f

g

h

6 Measure each of the following lines to the nearest millimetre.

a

b

c

d

e

f

g

h

7 **Convert** each of the lengths obtained in question **6** to centimetres in fraction form.

WORKED Example 3

8 In each of the following illustrations, use the given length to estimate the other lengths mentioned.

a The distance from the centre of the Earth to the centre of the Sun

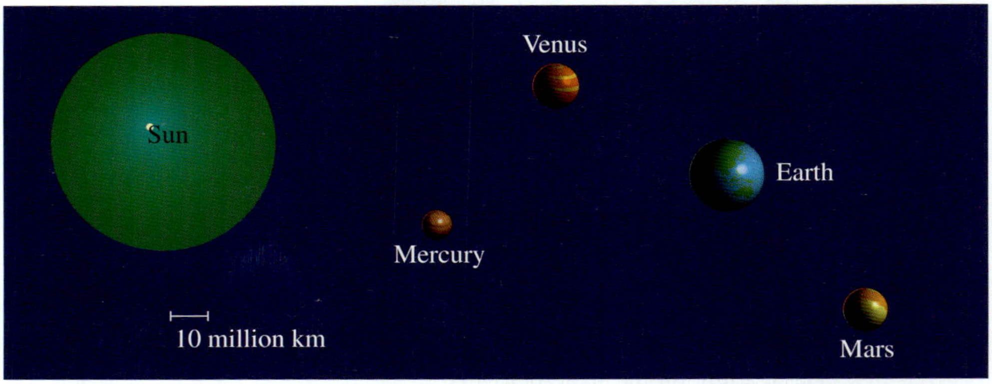

b The length of the train if it has four carriages

c The height of the giraffe

d The height of the window (to the inside centre of the frame)

e The distance in a straight line from:
i Cann River to Falls Creek **ii** Melbourne to Underbool **iii** Geelong to Canberra

f The height of the building in the foreground

g The height (x) of the pole near the utility

GAME time
Length and perimeter — 001

WorkSHEET 7.1

THINKING

Measuring lengths

Estimate and measure

Select several items whose length you can conveniently measure. Estimate the length before you measure, and present your results in a table similar to the one that follows.

Item	Estimated length	Measured length	Error
Desk width/length			
Door width			
Room length/width			
Length of walking pace			

(*Note:* Error is found by calculating the difference between the estimated length and the measured length.)

Paper thickness

Explain how you could use a normal ruler to find the thickness of a sheet of paper in this book.

Calculate the thickness of a page in this and one other book.

Pace lengths

You will need a long tape measure (preferably 30 m) or trundle wheel.
Devise a way of measuring the length of a student's pace during walking, jogging and running.

MATHS QUEST CHALLENGE

You will need 12 matches (or you can draw 12 lines on paper) for the following activity. Draw diagrams to show your results.

1 **Use the 12 matches to form a large square containing 4 small squares (a total of 5 squares). Use this arrangement for the following questions.**

2 **Show how you can remove 4 matches to leave exactly 2 squares.**

3 **Show how you can remove 4 matches to leave exactly 1 square.**

4 **Show how you can remove 2 matches to leave exactly 2 squares.**

5 **Show how you can move 3 matches to new positions so that there are exactly 3 squares.**

10 QUICK QUESTIONS 1

1 Which metric unit of length would be best for measuring the length of your fingernail?

2 Which metric unit of length would be best for measuring the length of a netball court?

3 How long is the snail shown below?

4 What is the reading on the dial shown below?

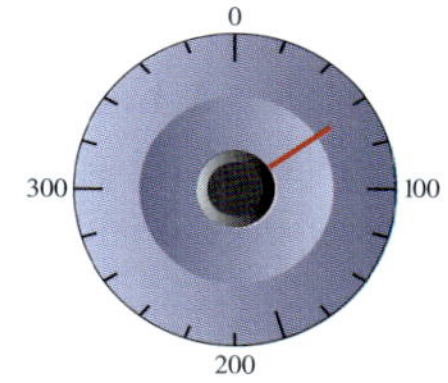

5 What is the length of this line segment?

6 Use the height of the woman to estimate the height of the building indicated.

7 True or false? The height of a student would be best measured in kilometres.

8 Arrange the following in ascending order according to the size of the unit: metres, kilometres, millimetres, centimetres.

9 Draw a line measuring 8.2 cm.

10 Draw a line measuring 23 mm.

Converting units of length

The main metric units of length are related as follows:

1 km = 1000 m 1 m = 100 cm 1 cm = 10 mm.

In chart form, the mathematical relationship can be shown in a way that is easy to remember. The numbers next to each arrow are called *conversion factors*.

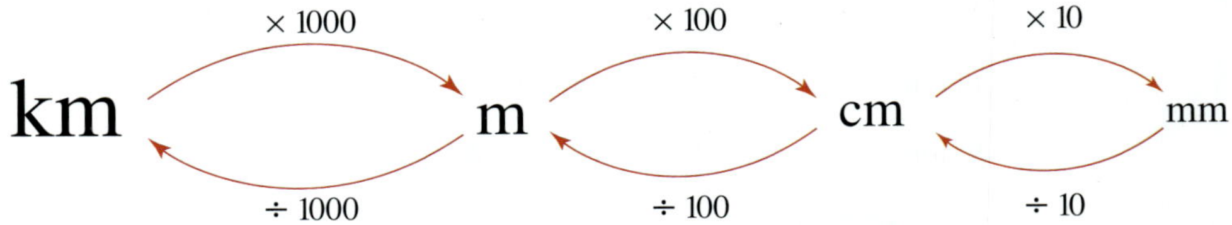

Notice that when converting to a smaller unit, we need to multiply by the conversion factor and, when converting to a larger unit, we need to divide by the conversion factor.

WORKED Example 4

Complete each of the following metric conversions.

a **0.3285 km = ______________ m** **b** **560 m = ______________ mm**

c **480 cm = ______________ km** **d** **$2\frac{3}{5}$ m = ______________ cm**

THINK	WRITE
a Refer to the metric conversion chart above. To convert kilometres to metres, multiply by 1000. (Move the decimal point 3 places to the right.)	**a** 0.3285 km = 0.3285 × 1000 m = 328.5 km
b This conversion involves two steps in the chart. First **convert** metres to centimetres by multiplying by 100 (since there is no decimal point place two zeros after the final digit). Then convert centimetres to millimetres by multiplying by 10 (place one zero after the final digit). *Note:* Overall, we need to multiply by 100 × 10 or 1000.	**b** 560 m = 560 × 100 cm = 56 000 cm = 56 000 × 10 mm = 560 000 mm
c To convert 480 centimetres to kilometres involves two division steps. From centimetres to metres, divide by 100 (move the decimal point 2 places to the left). From metres to kilometres, divide by 1000 (move the decimal point 3 places to the left).	**c** 480 cm = 480 ÷ 100 m = 4.8 ÷ 1000 km = 4.8 ÷ 1000 km = 0.004 8 km

THINK	WRITE
d 1 Convert $2\frac{3}{5}$ m to an improper fraction.	**d** $2\frac{3}{5}\text{ m} = \frac{2 \times 5 + 3}{5}\text{ m}$
	$= \frac{13}{5}\text{ m}$
2 Convert metres to centimetres by multiplying by 100. *Note:* Remember when multiplying fractions by a whole number, the number multiplies the numerator.	$= \frac{13}{5} \times 100\text{ cm}$
	$= \frac{1300}{5}\text{ cm}$
3 Simplify the fraction.	$= 260\text{ cm}$

remember

When converting units of length, it is useful to note the points below.

1. Imagine a decimal point at the right hand end of any whole numbers; for example, the number 35 can be thought of as having a decimal point after the 5 (35.).
2. Look at or draw a conversion chart.

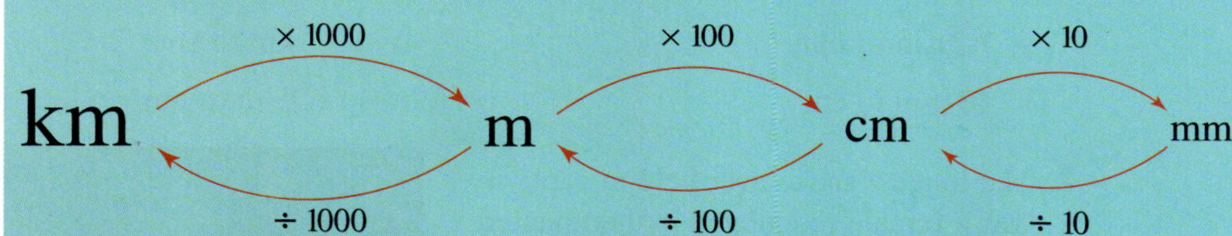

3. When converting to a smaller unit, multiply by the conversion factor.
4. When converting to a larger unit, divide by the conversion factor.

7.4 SkillSHEET Relationship between unit size and the number of units used

EXERCISE 7C Converting units of length

WORKED Example 4a

7.5 SkillSHEET Converting units (metres to kilometres)

1 Complete each of the following metric conversions.

a 2.0 km = ______________ m
b 7.0 km = ______________ m
c 65.0 km = ______________ m
d 5.3 km = ______________ m
e 0.66 km = ______________ m
f 9.0 m = ______________ cm
g 0.25 m = ______________ cm
h 28.0 cm = ______________ mm
i 200.0 cm = ______________ mm
j 700.0 m = ______________ cm

EXCEL Spreadsheet Length conversions

2 Convert to the units indicated.

a 8000 m = ______________ km
b 6500 m = ______________ km
c 700 m = ______________ km
d 50 m = ______________ km
e 10.5 m = ______________ km
f 6000 cm = ______________ m
g 300 cm = ______________ m
h 57 cm = ______________ m
i 9 cm = ______________ m
j 45 mm = ______________ cm
k 835 mm = ______________ cm
l 25 600 mm = ______________ cm

Mathcad Converting units

WORKED Example 4b, c, d

3 Copy and complete.

a 8 km = ______________ cm
b 5400 mm = ______________ m
c 101 m = ______________ mm
d 7800 cm = ______________ km
e $60\frac{1}{4}$ km = ______________ mm
f $112\frac{4}{5}$ cm = ______________ m
g 72.33 m = ______________ mm
h 40.5 mm = ______________ cm
i 5.05 km = ______________ m
j $30\frac{7}{20}$ mm = ______________ m

4 **multiple choice**

A distance of 6.25 km is the same as:

A 625 m **B** 0.006 25 m **C** 625 000 mm **D** 625 000 cm **E** 62.50 cm

5 **multiple choice**

A distance of 7 860 000 cm is equal to:

A 78.6 km **B** 786 m **C** 786 km **D** 786 000 mm **E** 78.6 km

6 Convert to the units indicated.

a $67\frac{7}{10}$ km to m
b $45\frac{1}{5}$ m to km
c 560 mm to m
d 450 000 cm to km
e $8\frac{3}{4}$ cm to mm
f 0.0006 km to cm
g 7.21 m to mm
h $3\frac{9}{100}$ km to cm
i 48 mm to cm
j 11.655 m to mm

7 The longest snake ever held in captivity was a female reticulated python named 'Colossus'. It was measured to be 8.68 m long. Convert this to centimetres and compare it to the adult length of the shortest species of snake, the West Indian *Leptoyphylops bilineata*, which grows to only 108 mm.

8 The world's largest mountains are Everest (8863 m) and 'K2' in Pakistan (8607 m). Convert these heights to kilometres.

9 The tallest tomato plant on record reached 16.3 m. How many centimetres is this?

10 The tallest Rialto Tower (shown to the left of the photograph of the Melbourne skyline at right) stands 0.448 km high. How many metres is this?

11 Arrange the following in order from smallest to largest.

a 12.5 m, 150 cm, 0.02 km
b 350 cm, 0.445 m, 3000 mm
c 50 km, 500 m, 50 000 mm
d 1700 cm, 1.7 m, 0.17 km
e 0.052 cm, 0.0052 mm, 0.000 052 m
f 990 cm, 0.909 m, 9000 mm

7.6 SkillSHEET Converting units to compare lengths or distances

12 Add the following lengths, giving your answer in either unit.

a 75 cm and 3 m
b 2700 m and 7.5 km
c 3800 mm and 52 cm
d 1.66 m and 58.2 cm
e 208 cm and 1.83 m
f 5500 mm and 20.1 m
g 40 km and 7800 m
h 0.000 675 km and 87.8 cm

13 Find the difference between each of the following lengths, giving your answer in either unit.

a 72 km and 5600 m
b 418 000 mm and 7.6 m
c $34\frac{3}{5}$ cm and $\frac{9}{20}$ m
d $2\frac{4}{5}$ km and 450 000 cm

14 A particular brand of computer CD is 1.2 mm thick. How high (in centimetres) would a pile of 23 such CDs be?

15 A builder needs to build a wall 3.5 m high. If each layer of bricks adds 8 cm of height, how many layers high will the wall be?

16 Deanne buys a length of rope and cuts it into three smaller sections, each of length 7200 cm. How many metres long was the original piece of rope?

17 Norbert is 1.53 m tall in bare feet. If he wears shoes with 6.3 cm thick soles, how high would he stand then? Give your answer in metres.

18 Adrian is driving a truck with a rooftop 3.4 m above road level, when he approaches an overpass bridge which has a clearance of 376 cm. Will Adrian's truck get under the bridge? If so, how much room does he have to spare? Give your answer in centimetres.

19 A childcare centre has three large cardboard boxes which the children stack up in various combinations. What stack heights are possible for a stack if the boxes' individual heights are 600 mm, 45 cm and 1.1 m? Give your answer in centimetres.

20 Finita attaches a trellis that is 0.6 m high to the top of her 180 cm high fence. How high is it to the top of the new trellis? Give your answer in metres.

21 Zvenglo is stacking identical boxes of height 330 mm. How high would a stack be if it contained six boxes? Give your answer in centimetres.

GAME time Length and perimeter — 002

22 Waldo's noticeboard is 1.5 m long and 1.2 m wide. If he pins a calendar of length 70 cm and width 60 cm exactly in the middle of the board, how much space is there above and below the calendar? (*Hint:* Draw a diagram of the situation.)

WorkSHEET 7.2

23 A licorice strap machine takes 3.75 m lengths of licorice, and chops them into 10 cm long pieces. How many pieces does each 3.75 m length produce?

ICT A metric unit converter

EXCEL Spreadsheet
Length conversions

The instructions below show one way of using an Excel spreadsheet to convert metric units of length. Alternatively, you may wish to use a pre-written spreadsheet 'Length conversions' on the *Maths Quest* CD-ROM to perform some conversions. The screen on the right shows how your spreadsheet should look once you have followed the steps below and completed the suggested task.

	A	B	C	D	E	F
1						
2	A metric unit converter					
3						
4						
5						
6		kilometre converter				
7		4	km =	4	km	
8				4000	m	
9				400000	cm	
10				40000000	mm	
11						
12						

1 In cell B6, type: kilometre converter
2 In cell B7, type: 4
3 In cell C7, type: km =
4 In cell D7, type: = B7*1
5 In cell E7, type: km

The steps above 'convert' kilometres to kilometres. They let us check that our unit converter is working as it should.

6 In cell E8, type: m
7 In cell E9, type: cm
8 In cell E10, type: mm
9 In cell D8, type: = B7*1000
10 In cell D9, type: = B7*100 000
11 In cell D10, type: = B7*10 000 000

Can you see the conversions? Now try entering a different value in cell B7. Does your spreadsheet update?

Now try to complete a new section on your spreadsheet to convert metres into the other units. You will need to use a forward slash (/) as a division sign.

10 QUICK QUESTIONS 2

1 Name the unit most suitable for measuring the length of a skateboard.

2 Give two examples of things that would be best measured in millimetres.

3 Convert 4.3 km to m.

4 Convert 87 000 mm to m.

5 Convert 2.1 km to cm.

6 Arrange the following in ascending order: 67 m, 2000 cm, 1345 mm, 0.04 km, 9 m.

7 Add 56 cm and 23 mm, giving your answer in the smallest unit.

8 In 2003 Ronnie was 182 cm tall. In 2006 his height was 2.1 m. How much did Ronnie grow in 3 years? Express your answer in centimetres.

9 A pile of cardboard to be recycled has become quite high. If there are 945 boxes and each is 32 mm thick when flattened, how high is the pile of boxes? Express your answer in two different units.

10 Asako is getting ready to go to the Melbourne Cup. If she is 167 cm tall and wears high heels which measure 8 cm and a hat which is 16 cm tall, find Asako's total height in metres.

U.S. Open winner 1997 and 1998. Which sport?

240	0.24	0.78	92	32	56	78	7.8	3.2	0.078	9.2	560	320	0.32	2.4	920	5.6	0.56	24

A 3200 m ____ km

F 7800 cm ____ km

N 56 mm ____ cm

S 24 million mm ____ km

T 32 cm ____ m

N $9\frac{1}{5}$ m ____ cm

C $\frac{14}{25}$ m ____ cm

I 0.000 32 km ____ cm

P 24 cm ____ mm

T 78 cm ____ m

E 0.000 56 km ____ mm

R 78 mm ____ cm

T 0.92 cm ____ mm

E 2400 mm ____ m

K $\frac{39}{500}$ km ____ m

R $\frac{8}{25}$ m ____ mm

A 240 m ____ km

I 0.000 005 6 km ____ cm

R $\frac{23}{250}$ km ____ m

Perimeter

A perimeter is the distance around the outside (border) of a shape. Fences go around the perimeter of many sporting fields.

Cyclists like Anna Mears at right race around the perimeter of a velodrome.

To calculate the perimeter, change all lengths to the same unit before adding them.

WORKED Example 5

Find the perimeter of each shape below.

a

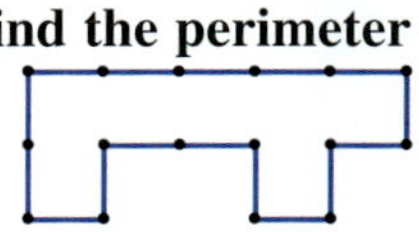

Note: The dots are 1 unit apart.

b

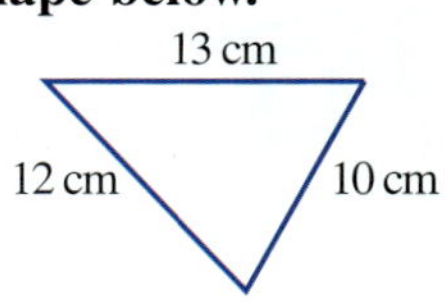

c

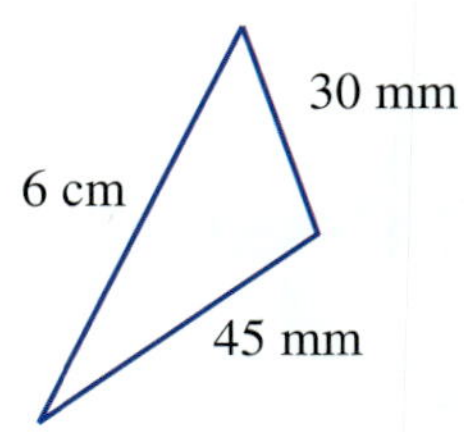

THINK	WRITE
a 1 Count the number of unit intervals around the outside of the shape (16). 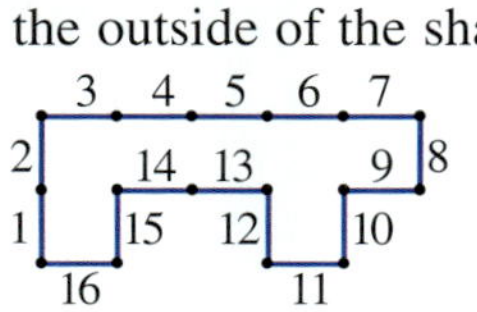	**a**
2 Write the number.	16 units

THINK

b 1 Check that the measurements are in the same units. They are.
2 Add the measurements.
3 Write the total, with units.

c 1 Notice the measurements are not all the same. Convert to the smaller unit (6 cm = 60 mm).
2 Add the measurements that now have the same unit (mm).
3 Write the total using the smaller unit.

WRITE

b
$$\begin{array}{r} 12 \\ 13 \\ +\ 10 \\ \hline 35 \text{ cm} \\ \hline \end{array}$$

c 6 cm = 60 mm
$$\begin{array}{r} 60 \\ 30 \\ +\ 45 \\ \hline 135 \text{ mm} \\ \hline \end{array}$$

Many problems involving finding the **perimeter** provide the description of a shape in words, but do not include its picture. In such cases, it is always a good idea to draw a diagram before attempting to find the perimeter.

WORKED Example 6

Find the perimeter of:

a a rectangular print which is 49.3 cm long and 22.0 cm wide

b a square print whose side length is 28 cm.

THINK

a 1 Draw a diagram of the print and write in the measurements.
2 The perimeter is the distance around the print, so add all the distances together.
3 Write the worded answer, with the correct unit.

b 1 Draw a diagram of the print and write in the measurements.
2 The perimeter is the distance around the print, so add all the distances together.
3 Write the worded answer, with the correct unit.

WRITE

a

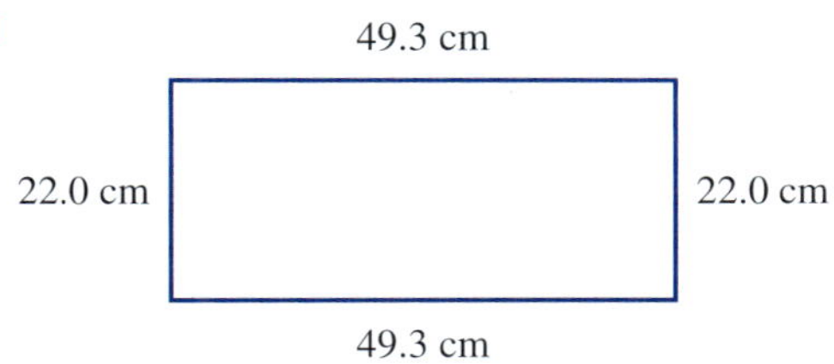

$P = 49.3 + 22.0 + 49.3 + 22.0$
$= 142.6$
The perimeter of the print is 142.6 cm.

b

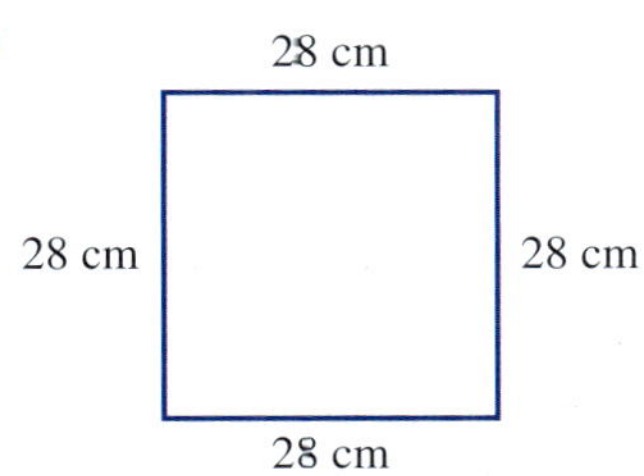

$P = 28 + 28 + 28 + 28$
$= 112$
The perimeter of the print is 112 cm.

Note: There is an alternative way to find the perimeter in the previous worked example. Observe that when finding the perimeter in worked example 6a, we had to count each of the two measurements (length and width) twice. So instead we could add the length and the width and then multiply the result by 2.

Check these calculations: $P = 2 \times (49.3 + 22.0)$
$= 2 \times 71.3$
$= 142.6$ cm (as before)

Similarly, the perimeter in worked example 6b could be obtained by multiplying the side length by 4.

Check these calculations: $P = 4 \times 28$
$= 112$ cm (as before)

These observations could be generalised as follows.

The perimeter of a rectangle, P, can be found using the formula $P = 2(l + w)$, where l is the length of the rectangle and w is its width.
The perimeter of a square, P, can be found using the formula $P = 4l$, where l is the side length of the square.

remember

1. The perimeter is the distance around the outside of a shape.
2. To calculate the perimeter, change all lengths to the same unit before adding them.
3. The perimeter of a rectangle is given by the rule $P = 2(l + w)$.
4. The perimeter of a square is given by the rule $P = 4l$.

EXERCISE 7D Perimeter

SkillSHEET 7.7 Perimeter

WORKED Example 5a

1 Find the perimeter of each shape below. The dots are 1 unit apart.

a
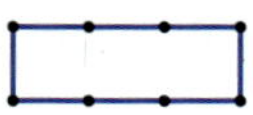
b
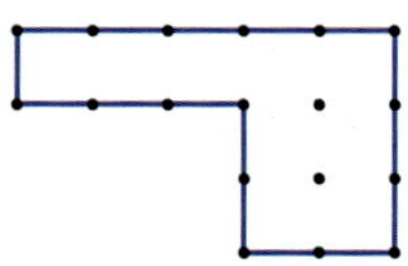
c
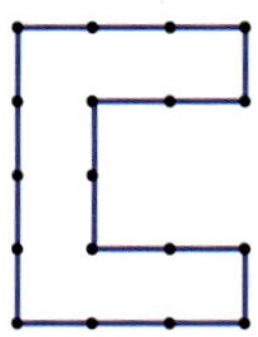
d
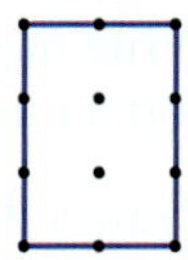
e
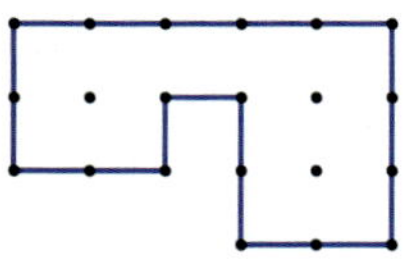
f
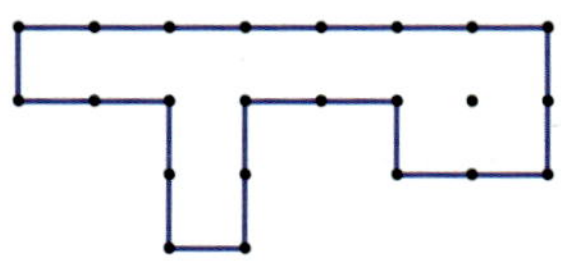
g
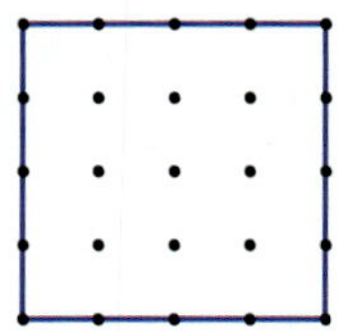
h
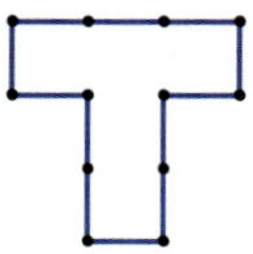
i
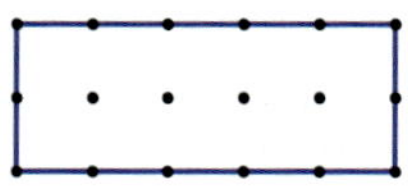
j
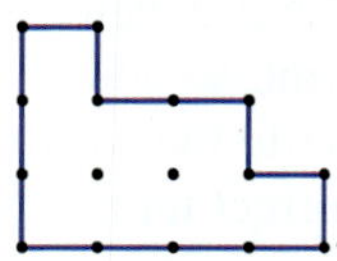

WORKED Example 5b

Cabri Geometry — Perimeter of a rectangle

2 Find the perimeter of each of the following.

a

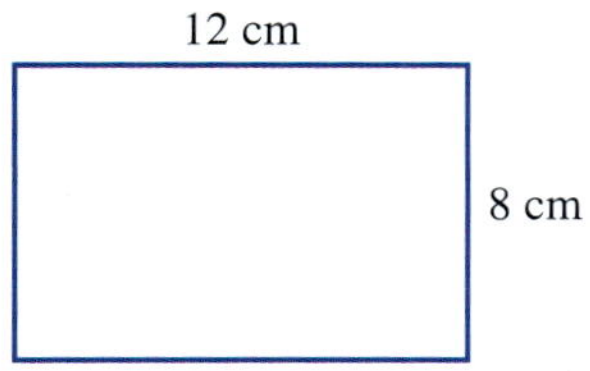

b

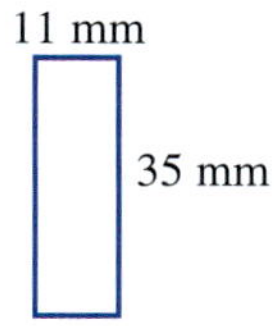

c

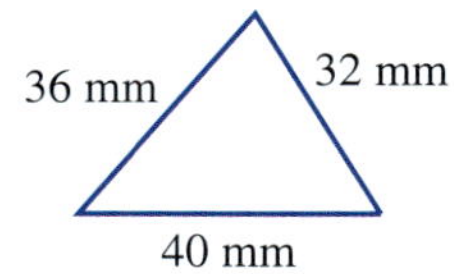

d

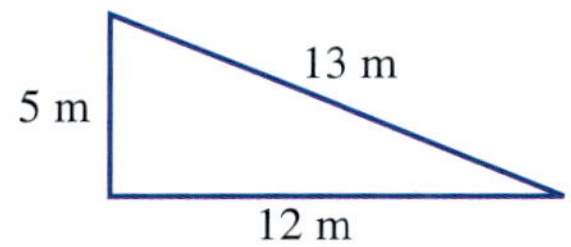

e

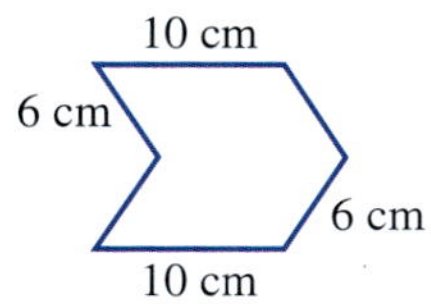

f

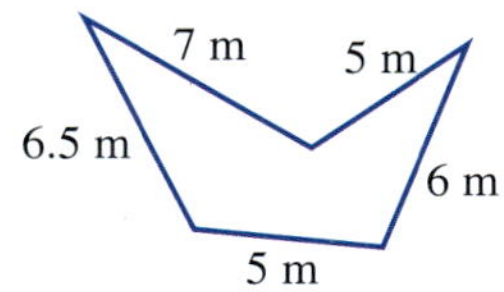

g

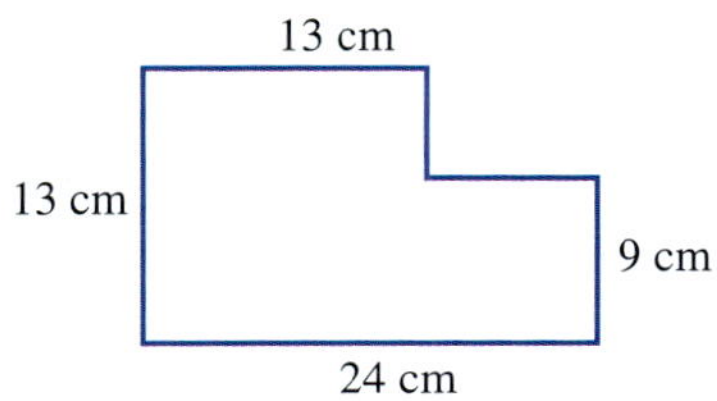

h

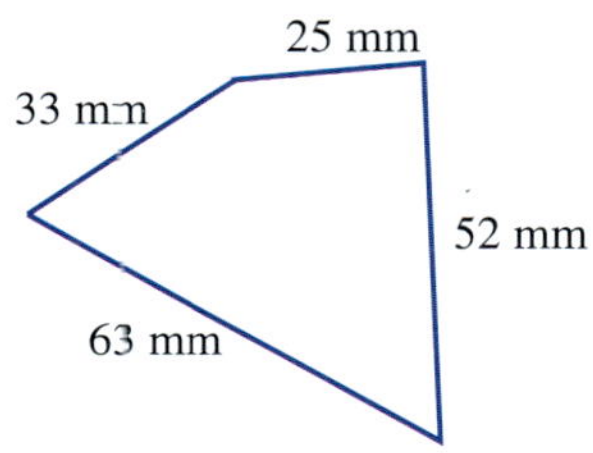

i

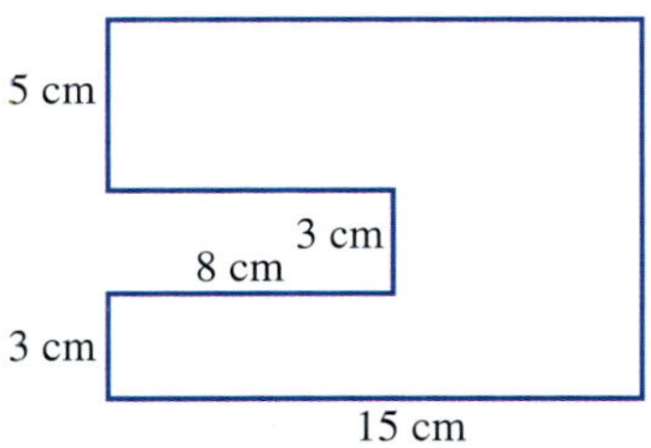

j

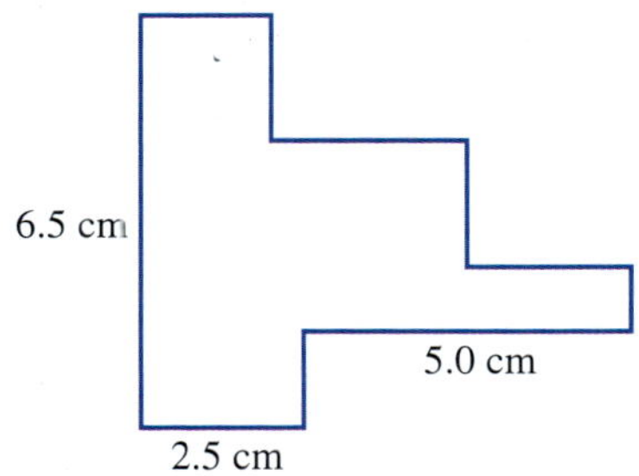

3 Measure the perimeter of each of the following, giving your answers in centimetres.

a

b

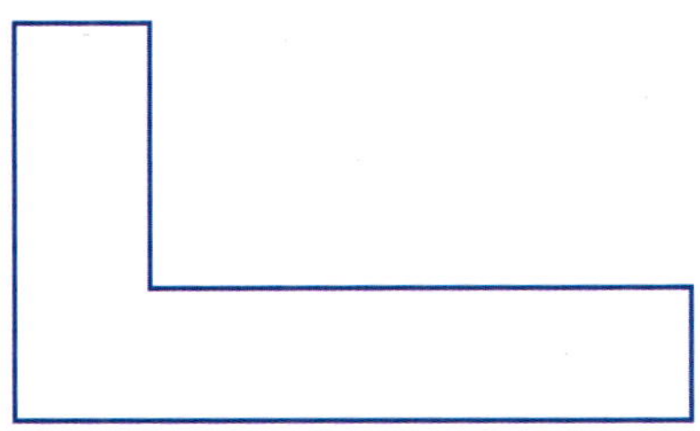

c

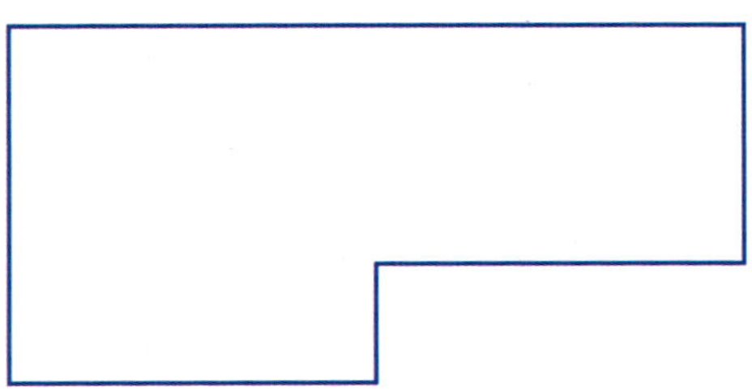

d

e

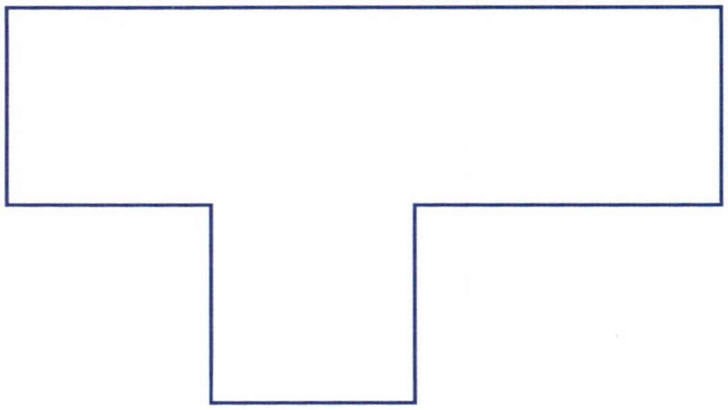

f

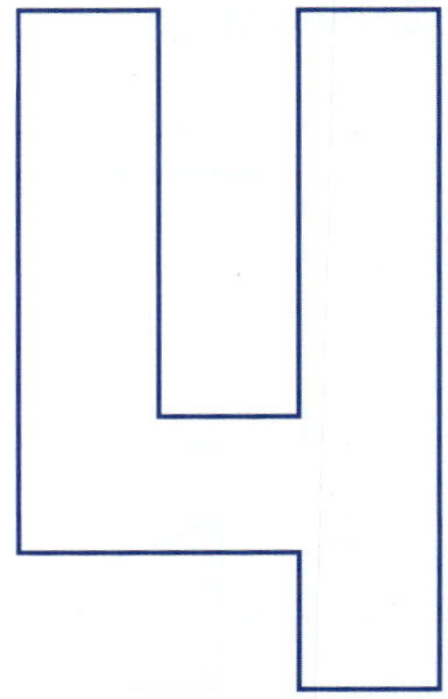

g

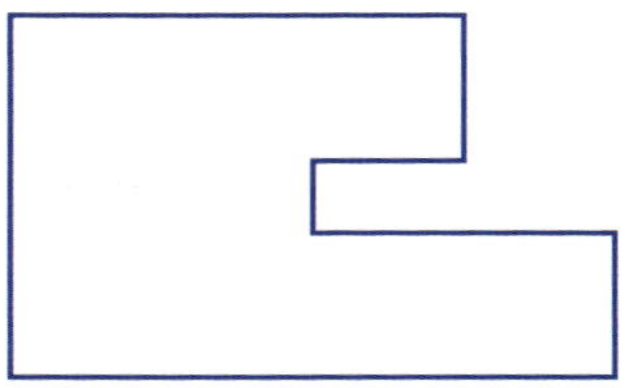

h

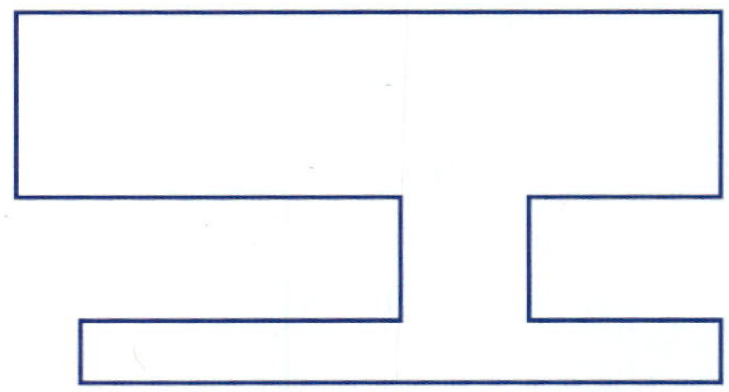

i

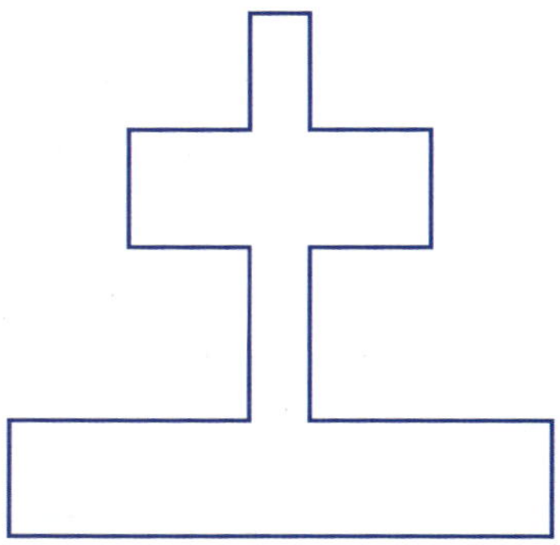

j

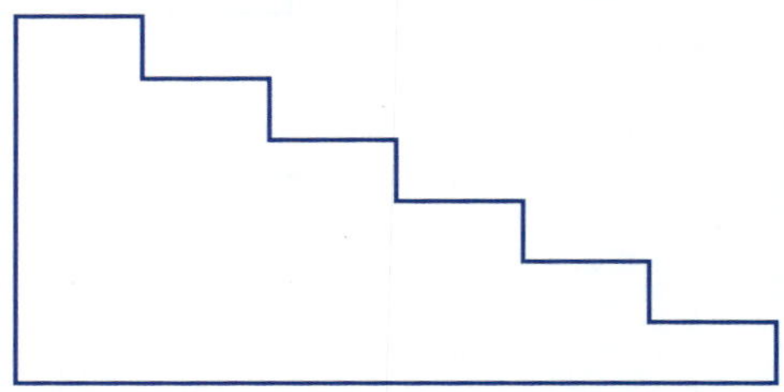

WORKED Example 5c

4 Find the perimeter of each shape below, giving answers in the smaller unit in each case.

a

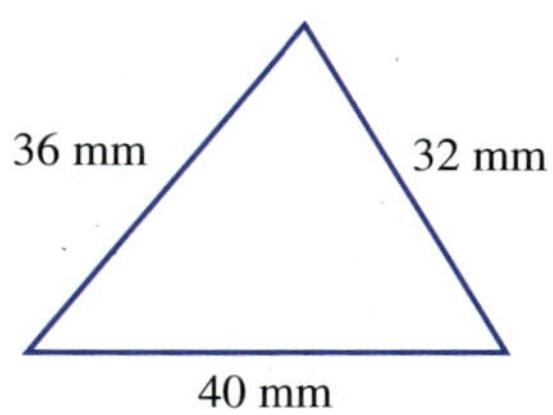

b

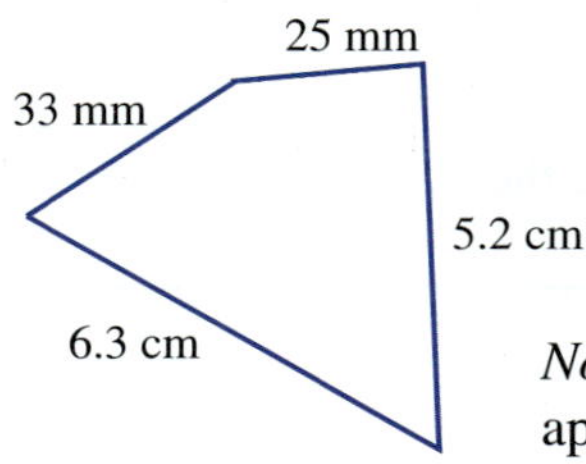

c

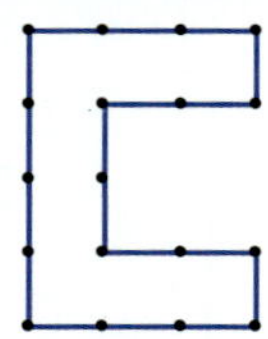

Note: The dots are 1 unit apart.

d

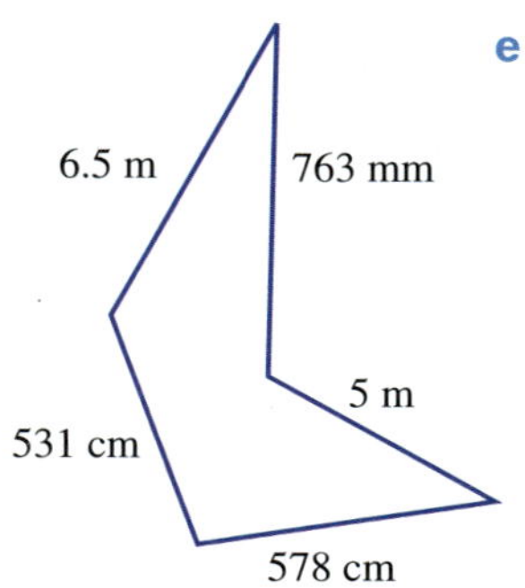

e

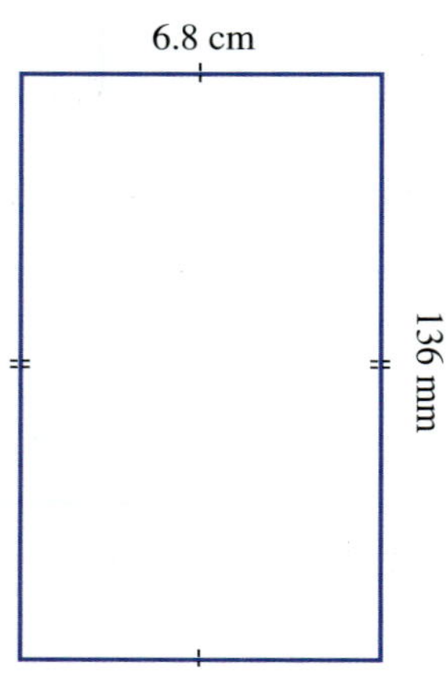

f

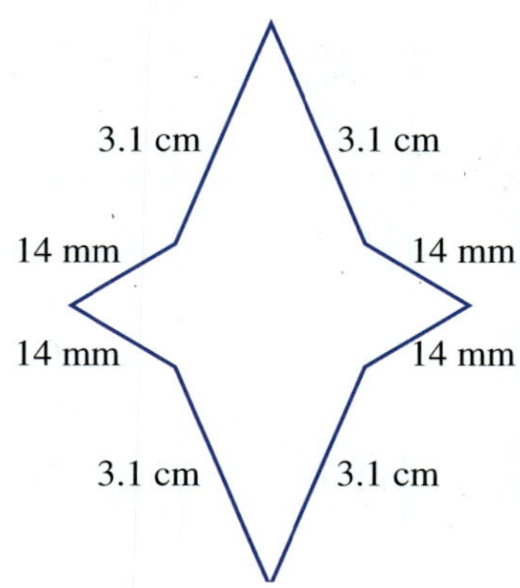

g

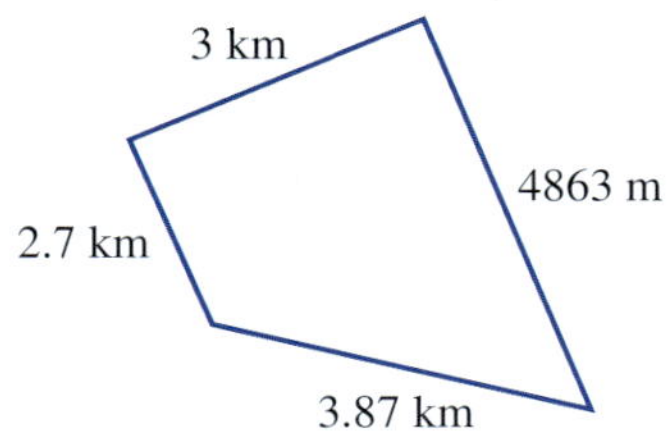

h

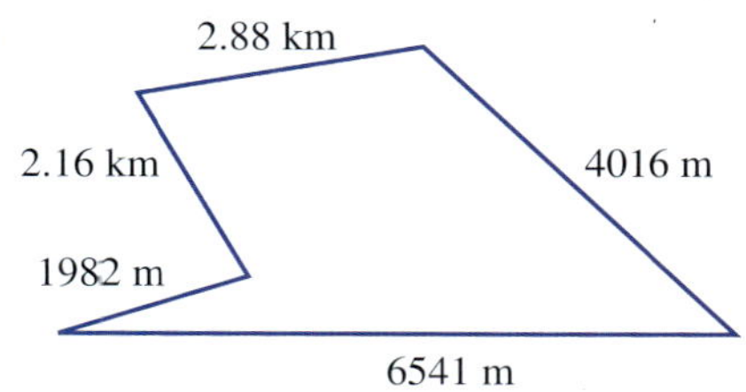

5 **multiple choice**

The perimeter of the shape at right is:

A 9 units **B** 18 units **C** 16 units
D 14 units **E** 12 units

Note: The dots are 1 unit apart.

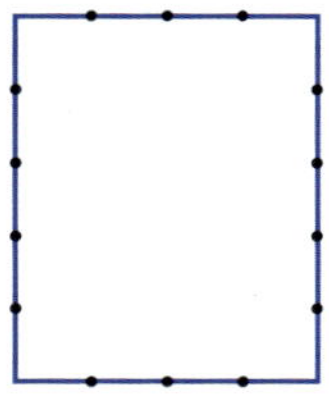

6 **multiple choice**

The shape at right has a perimeter of:

A 12 cm
B 4 cm
C 30 cm
D 29 cm
E 60 cm

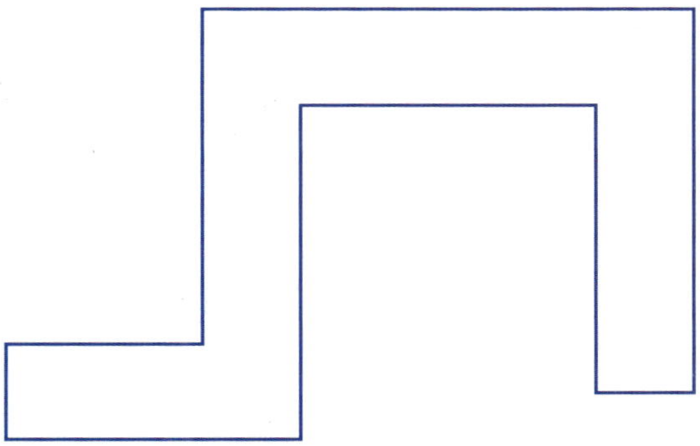

7 **multiple choice**

Measuring to the nearest half centimetre, the perimeter of the shape at right is:

A 24 cm **B** 24.5 cm
C 26 cm **D** 27 cm
E 28 cm

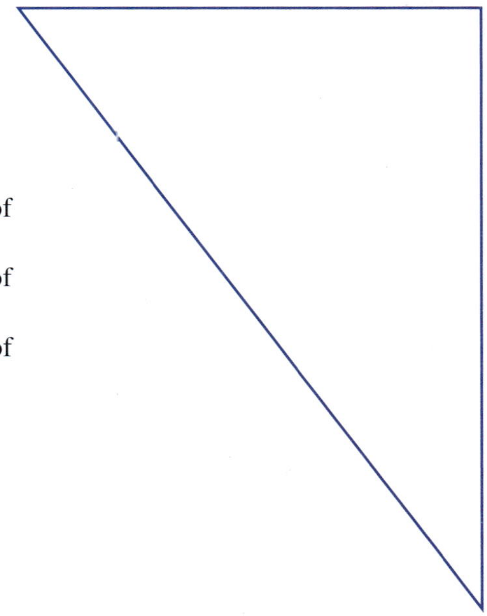

WORKED Example 6a

8 **a** Calculate the perimeter of a rectangle of length 45 cm and width 33 cm.
b Calculate the perimeter of a rectangle of length 2.8 m and width 52.1 cm.
c Calculate the perimeter of a rectangle of length $4\frac{1}{4}$ m and width $2\frac{1}{5}$ m.

WORKED Example 6b

9 **a** Calculate the perimeter of a square of side length 3.7 cm.
b Calculate the perimeter of a square of side length 2.14 m.
c Calculate the perimeter of a square of side length $8\frac{1}{5}$ mm.

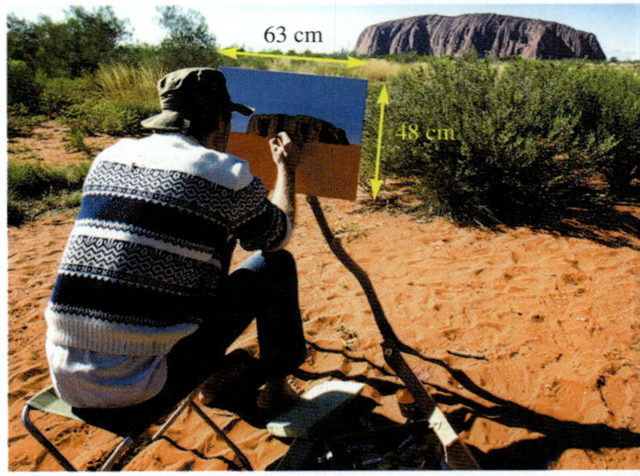

10 What length of party lights is needed to decorate the perimeter of a rectangular tent with dimensions 15.5 m by 8.75 m?

11 Allowing an extra 30 cm for waste, what length of picture frame is required to frame the artwork shown at right?

12 Phang's property boundary dimensions are shown on the right.

a How many metres of fencing will he need to fence all but the front boundary?

b If the fencing costs $19 per metre, what will the total cost be?

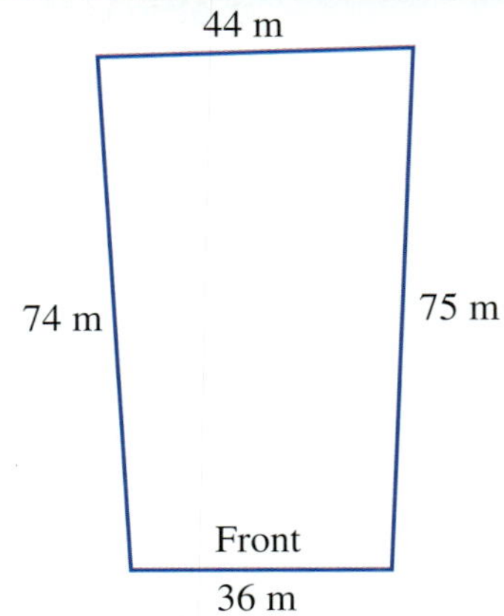

13 Lucille has a quote of $37 per metre for new tennis court fencing. A tennis court is 23.77 m long and 10.97 m wide. There should be a 3.6-m space between the sides of the court and the fence, and a 6.4-m gap between the ends of the court and the fence.

a Draw a diagram showing all given measurements.

b How many metres of fencing are needed?

c What will the total cost of the quote be?

14 Zedken wishes to install three strands of barbed wire at the top of the fences around a rectangular work site. The length of the site is 34.5 m, and its width is 19.8 m. What length of wire will Zedken need?

WorkSHEET 7.3

15 A new game, 'Bop-ball' is played on a triangular field, where each side of the triangle measures 46.6 m. A greenkeeper is marking the field's perimeter using a chalk dispensing trundle wheel. How far will the greenkeeper walk in marking the field?

MATHS QUEST CHALLENGE

1 On planet Mathematico, the main units of length are the yot, the blot and the snop. There are 13 yots in a blot and 17 blots in a snop. How many yots are there in 23 snops?

2 A square garden is enclosed by a fence. Each side of the fence has 10 posts. How many posts are there? (The answer is not 40.)

3 If a stock of food can feed 20 chickens for 24 days, how long would the same food stock last if there were 16 chickens?

4 Look for a pattern in the following letters.
A B C D E F G A B C D E F G A B C D E F G …
What will be the 500th letter in this pattern?

COMMUNICATION

Measuring curves

Each figure below consists of a curved line.

1 Estimate the length of each line.
2 Work out a method for measuring the length of each line and describe how it works.
3 Measure each line and compare your results with your estimates and those of a classmate. Comment on your findings.
4 For which of the figures is the length of the line also called the perimeter? Explain your answer.

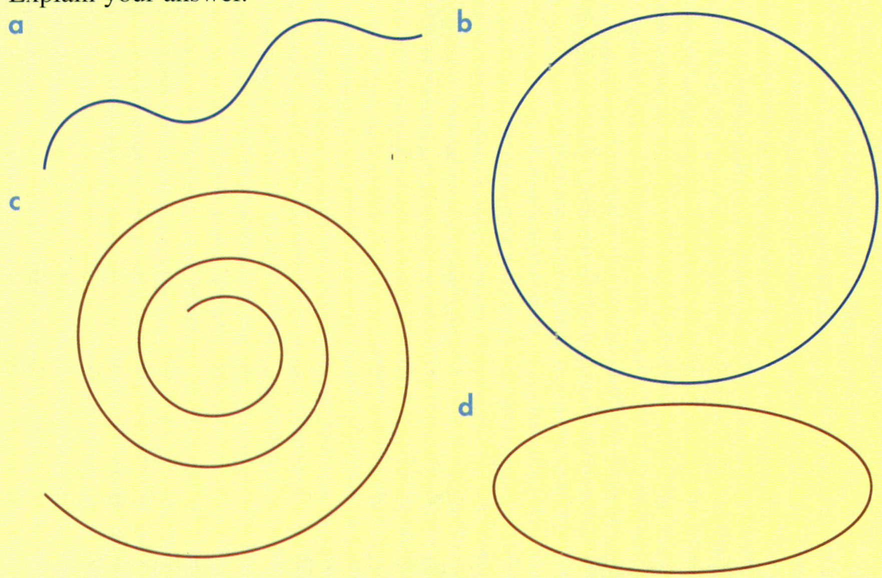

ICT

Maximise the perimeter

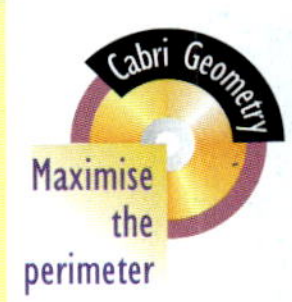

Open the Cabri Geometry file 'Maximise the perimeter', or start a new file and construct a circle with four points on it joined by line segments. Measure the lengths of each segment, and calculate the total of the segment lengths. You could even try this activity without a computer, using a pencil and paper. Drag the points on the circle around until you think you've found the maximum possible perimeter of the four-sided shape. Check with your teacher about how close you were to finding the maximum.

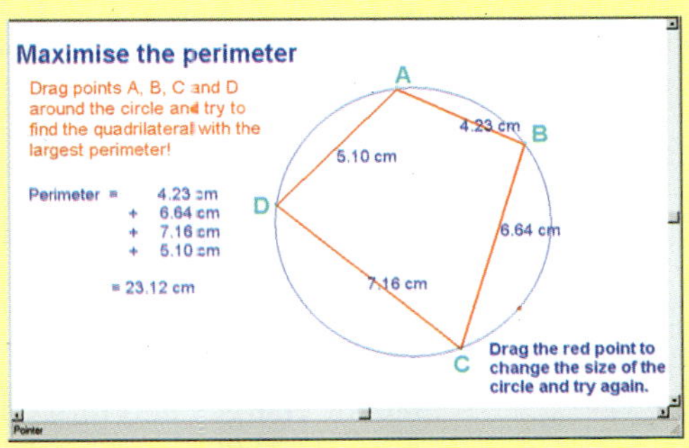

THINKING

Cost of a new fence

Marc and Cathy are seeking quotes on the cost of building a fence on three sides of their property. They want to calculate approximate costs for each item to decide whether the quotes supplied sound reasonable.

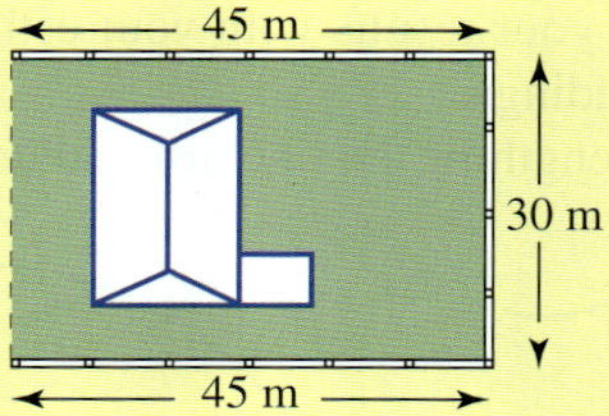

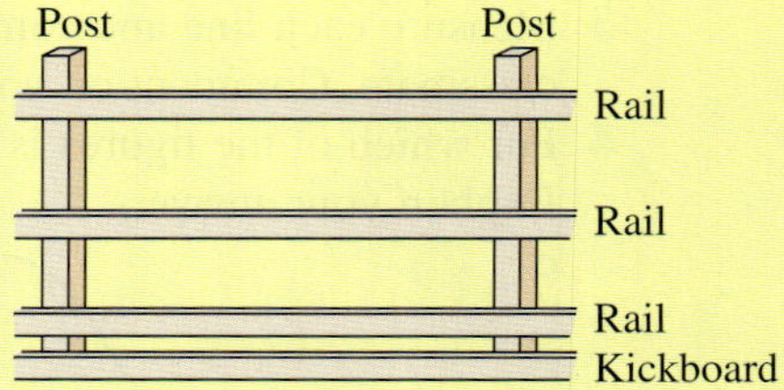

The new fence requires three rails, a kickboard, posts and palings.

1 What length of timber will be needed for the kickboard?

2 Calculate the cost of the kickboard if the timber required for this costs $1.90 per metre.

3 If the timber for the rails costs $2.25 per metre, what will be the total cost of the timber for the railings?

4 How many posts will be needed for the new fence if each post is to be 5 metres apart and there needs to be a post at the end of each straight section of fence?

5 Calculate the cost of the posts if the price of each post is $13.65.

Challenge

Palings are 9 cm wide and are nailed so they overlap each other by 15 mm on each side.

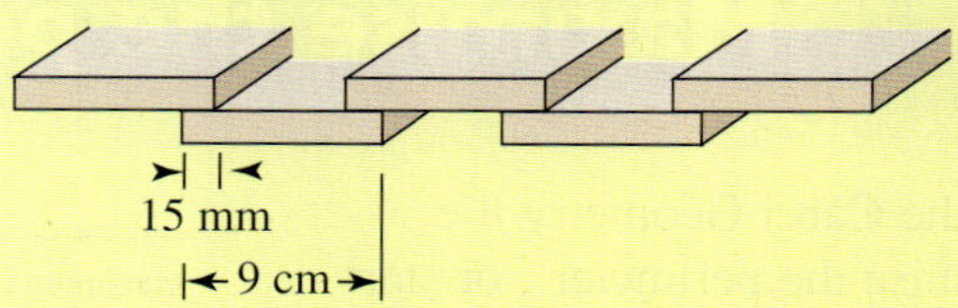

6 Calculate the approximate number of palings needed for the fence.

7 Palings cost $1.05 each. How much money should be allowed for the total cost of the palings?

8 Write an itemised list of all the costs involved. Include an amount to cover labour charges and miscellaneous items like the cost of nails. This amount is around $1000 (two people for two days at approximately $30.00 per hour for an eight-hour day). Estimate the cost of the new fence. This will provide Marc and Cathy with information to use when comparing builders' quotes.

summary

Copy the sentences below. Fill in the gaps by choosing the correct word or expression from the word list that follows.

1 The most appropriate unit for measuring an athletics track is ____________.

2 When using a ruler to measure a line, the ruler's ____________ mark should be on one end of the line.

3 The abbreviation km is short for ____________.

4 The abbreviation m is short for ____________.

5 The abbreviation ____________ is short for centimetres.

6 The abbreviation ____________ is short for millimetres.

7 Copy and complete:

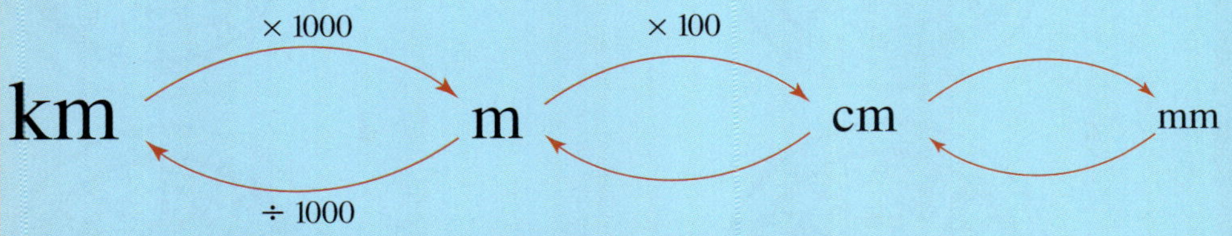

8 To convert from kilometres to metres, you need to ____________ by 1000.

9 To convert from metres to kilometres, you need to ____________ by 1000.

10 To convert from kilometres to millimetres, you need to multiply by ____________.

11 To convert from centimetres to metres, you need to divide by ____________.

12 The ____________ around the ____________ of a shape is called the perimeter.

13 When adding lengths to find a perimeter, all lengths should be in the ____________ units.

WORD LIST

distance	zero	outside	multiply
divide	same	cm	mm
metre	× 10	1 000 000	÷ 10
100	÷ 100	kilometre	metres

CHAPTER review

7A **1** Which metric unit would be most suitable for measuring:

a the distance walked by a teacher at school during a week?
b the length of a piece of spaghetti?
c the width of a pencil?

7B **2** What is the reading in kilograms on these bathroom scales? *Note:* The scale for kilograms is the inner circle

a

b

c

d

7B **3** Measure each of the following line segments in centimetres, giving the answers in:

i decimal form
ii fractional form.

a

b

c

7B **4** Estimate the height of the tree on the right, given that the person in the photograph at right is 1.7 m tall.

7C **5** Copy and complete the following conversions.

a 560 mm = ______________ m
b 2300 cm = ______________ km
c 17 m = ______________ cm

d $\frac{3}{4}$ km = ______________ mm **e** 2.09 m = ______________ mm

f $6\frac{4}{5}$ cm = ______________ m **g** 22.5 mm = ______________ cm

h $\frac{63}{10\,000}$ km = ______________ m **i** 82 000 000 m = ______________ km

j $5\frac{9}{10}$ mm = ______________ cm

6 Arrange from smallest to largest: 44.5 m, 455 cm, 455 000 mm, 0.004 45 km. 7C

7 **a** Add 45.6 km to 5600 m. 7C
b Find the difference between 80 m and 4300 cm.

8 During a rescue operation in calm seas, a 16.5 m rope is dangled from a helicopter hovering 20 m above sea level. A 175 cm tall man standing on the deck of a boat reaches 50 cm above his head for the rope. By how much does he fail to reach the rope if the deck is 1 m above sea level? 7C

9 The Amazing Ape-Impersonating Gymnastic Troupe consists of two people who are 150 cm tall, and three people who are 1.8 m tall to their shoulders. When these people form a tower five people high by standing on each other's shoulders, how high is the tower? 7C
Note: The head of the top person is 30 cm high.

10 Find the perimeter of each shape below. 7D

a

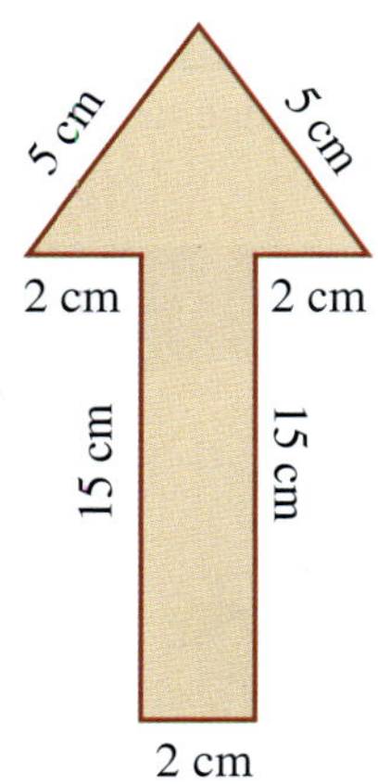

b

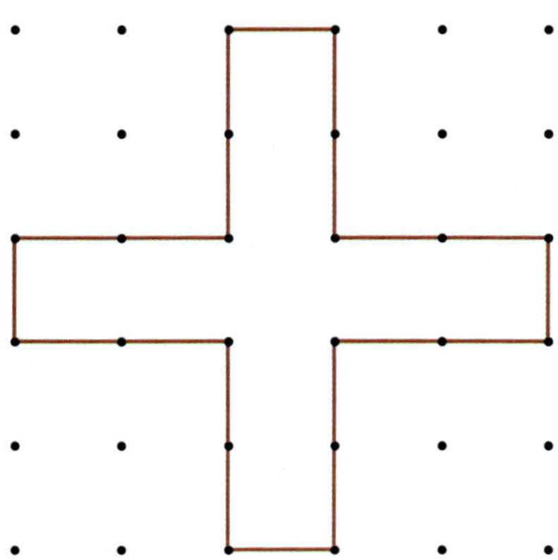

c

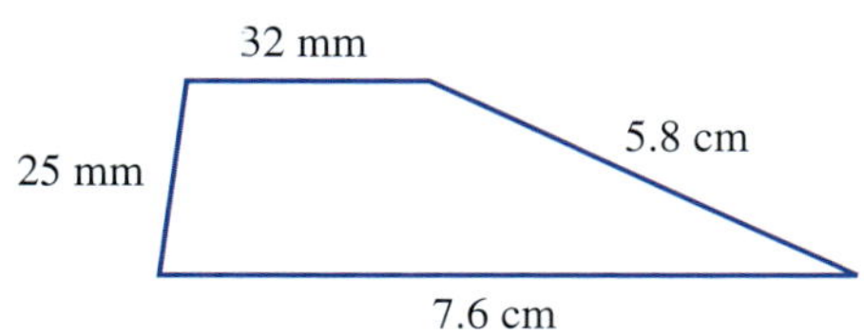

d

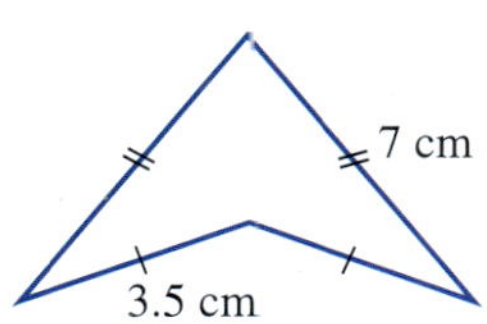

Note: The dots at left are 1 cm apart.

7D

11 Michelle rides three laps of the dirt-bike track shown at right. How far does she ride altogether?

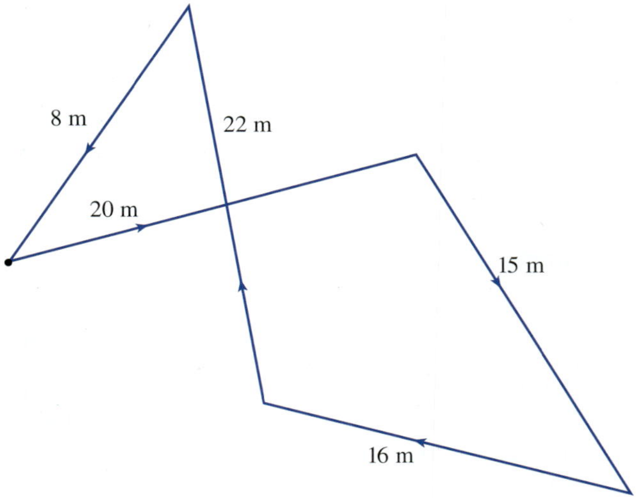

7D

12 A woven rectangular rug is 160 cm wide and 230 cm long. Find the perimeter of the rug.

7D

13 **a** What length of concrete edging will be required to form an edge around the lawn area shown at right?

b What will be the cost of the job if edging costs \$8.45 per metre?

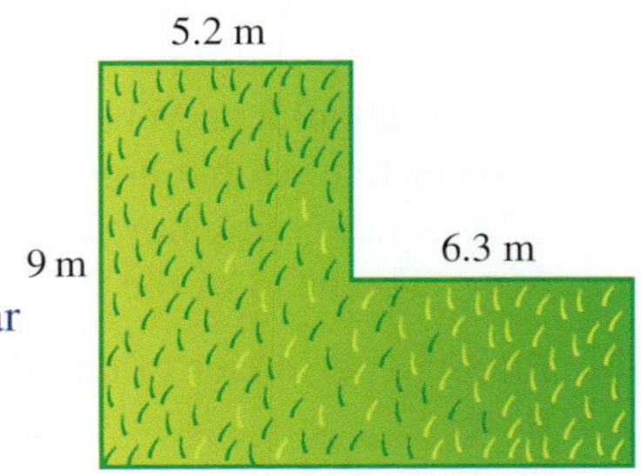

7D

14 What length of satin ribbon is required to edge a rectangular blanket on all four sides if the blanket is 240 cm long and 195 cm wide? (Assume there is no overlap.)

7D

15 Flavio runs five times around the baseball field shown below. What distance has he run in total?

8 Area, volume and capacity

How much would it cost to carpet the lounge room if you chose a carpet that cost $28 per metre fully laid? You need to calculate the floor area of the lounge room.

In this chapter, you will learn how to calculate the area of rectangles and triangles as well as other measurements such as volume and capacity.

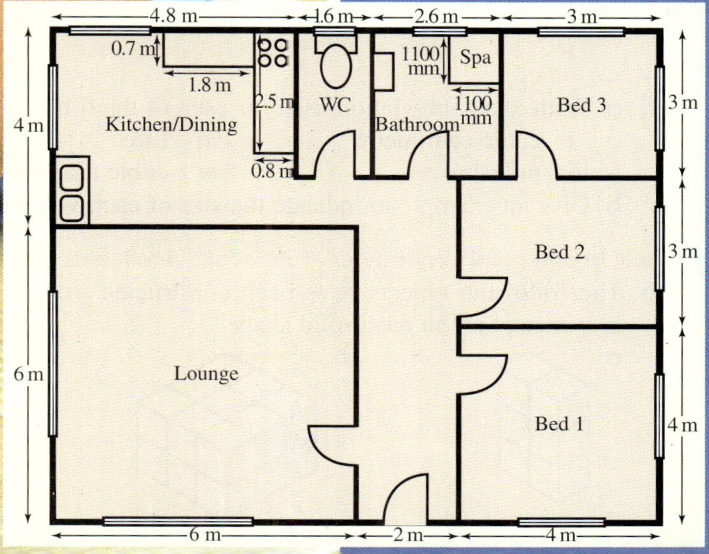

Are you ready?

Try the questions below. If you have difficulty with any of them, extra help can be obtained by completing the matching SkillSHEET. Either click on the SkillSHEET icon next to the question on the *Maths Quest 7* CD-ROM or ask your teacher for a copy.

SkillSHEET 8.1

Area units

1 **a** Write the abbreviation used for each of the following area units.

- **i** square metre
- **ii** square centimetre
- **iii** square kilometre
- **iv** hectare

b Give an example to indicate the size of each area unit.

SkillSHEET 8.2

Area of figures drawn on centimetre grid paper

2 The figures below are drawn on centimetre grid paper. What is the area of each one?

a

b

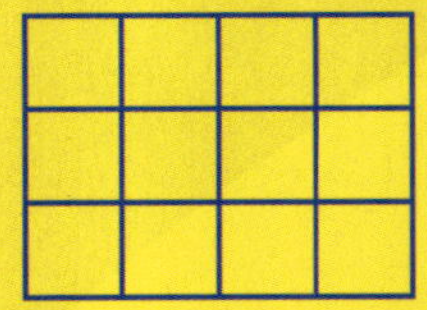

c

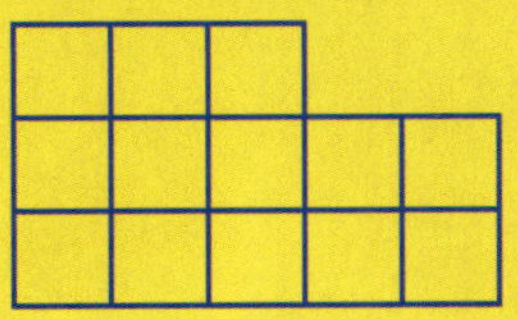

SkillSHEET 8.3

Area of rectangles

3 **a** Explain how to find the area of a rectangle using its length and width.

b Find the area of this rectangle.

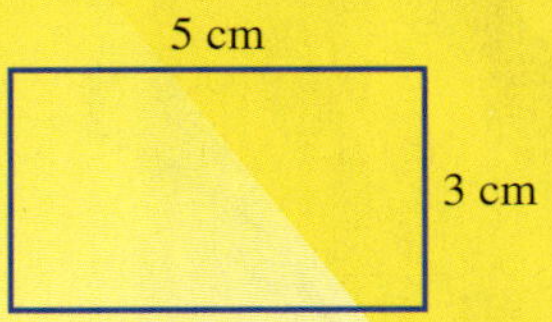

SkillSHEET 8.4

Area of triangles

4 **a** Explain how to find the area of a triangle using its base and height.

b Find the area of this triangle.

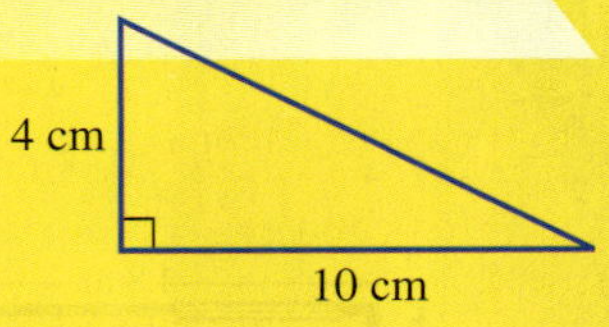

SkillSHEET 8.5

Volume units

5 **a** Write the abbreviation used for each of the following volume units.

- **i** cubic centimetre
- **ii** litre
- **iii** millilitre
- **iv** cubic metre

b Give an example to indicate the size of each volume unit.

SkillSHEET 8.6

Volume of a solid constructed from cubic-centimetre blocks

6 The following objects have been constructed using cubic centimetre blocks. How many cubic centimetres are in each solid shape?

a

b

Area

The **area** of a shape is the amount of flat surface enclosed by the shape. The area of this chessboard is covered with black and white tiles.

The perimeter of a sheet of paper is the distance around its edges. The area is the amount of space it covers. Similarly, the frame of a window goes around its perimeter, while the glass represents the area. The grass in the middle of a sports oval covers the area of the field.

Metric units of area

Commonly used metric units (for measuring areas) with their abbreviations and some examples are shown below:

1. Square kilometre (km^2)

The area of a country or a large city like Sydney is given in square kilometres.

2. Hectare (ha)

The area of a farm or a large city park is measured in hectares.

3. Square metre (m^2)

Square metres are used to measure the area of a classroom floor, blackboard or this window.

4. Square centimetre (cm^2)

Small areas, such as the area of a sheet of A4 paper or a book cover, are measured in square centimetres.

5. Square millimetre (mm^2)

Very small areas, such as the area of a button or a postage stamp, are measured in square millimetres.

For some shapes it is possible to measure the area easily because we can imagine it to be made up of a number of squares. A square with a side measure of 1 cm has an area of 1 square centimetre (1 cm^2).

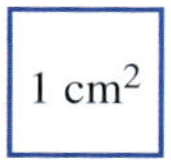

WORKED Example 1

The figures below are drawn on centimetre grid paper. Find the area of each one. (*Note:* The figures are not drawn to scale.)

a

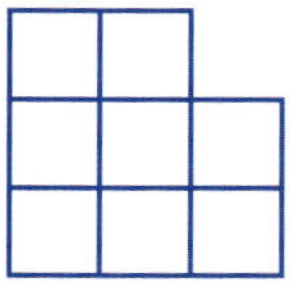

b

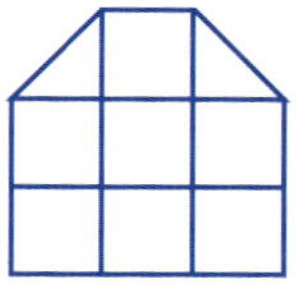

THINK	WRITE
a Count the squares. Remember to include the correct unit (cm^2) in the answer.	**a** 8 cm^2
b Some of the squares are cut in half by the diagonal line. It takes two of these to make up one square centimetre. Count the squares. Remember to include the correct unit (cm^2) in the answer.	**b** 8 cm^2

If a square is not completely enclosed by the shape, use the following rule to obtain an estimate of the area. If more than half the square is covered, count it. If less than half of the square is covered, don't count it. Tick the squares as each one is counted.

WORKED Example 2

Estimate the shaded area in the diagram at right, if the figure is drawn on centimetre grid paper.

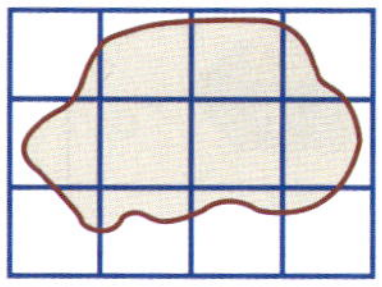

THINK	WRITE
1 Tick the squares that are more than half covered and count them.	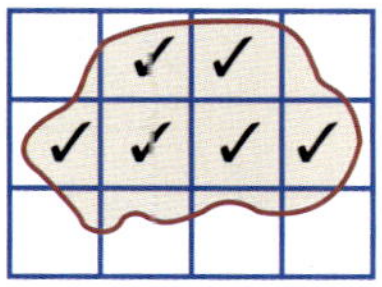
2 State the answer.	6 cm^2

remember

1. Area may be found by counting the number of squares needed to cover a surface.
2. A square with a side measure of 1 cm has an area of 1 square centimetre (1 cm^2).

EXERCISE 8A Area

SkillSHEET 8.1 Area units

1 Which unit would be most suitable to measure the following areas? Choose from mm^2, cm^2, m^2, ha or km^2.

a A computer screen
b A sheet of A3 paper
c Melbourne Cricket Ground
d A shirt button
e The Melbourne metropolitan area
f A compact disc
g A house block
h A room
i Alaska
j A basketball court
k A dairy farm
l The area inside the letter D

SkillSHEET 8.2 Area of figures drawn on centimetre grid paper

WORKED Example 1a

2 The figures below are drawn on centimetre grid paper. Find the area of each one.

a 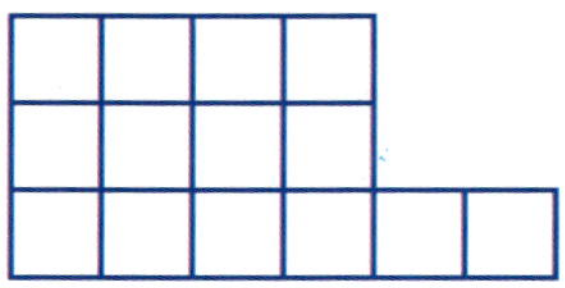

b 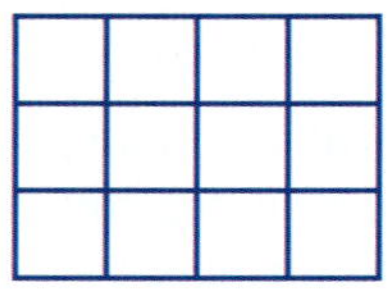

c 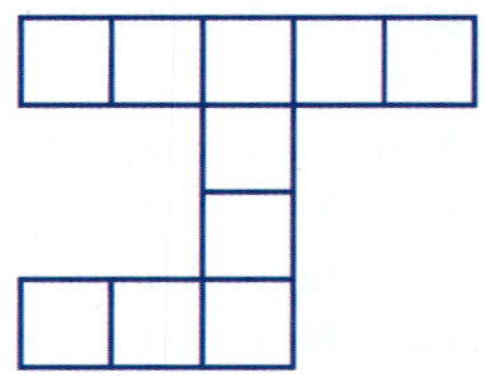

d 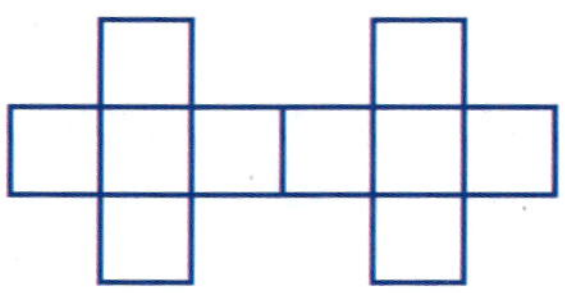

e 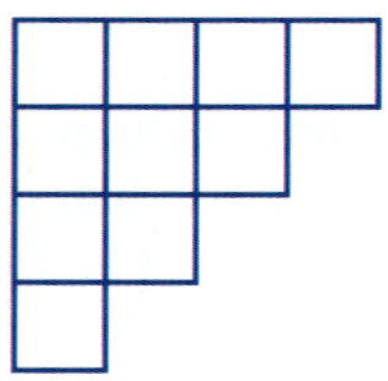

f 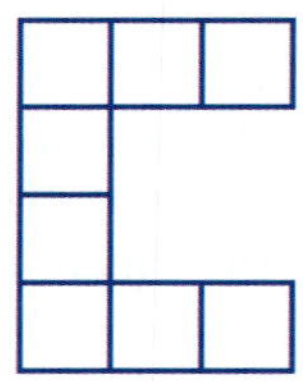

WORKED Example 1b

3 Find the area of the figures below, which are drawn on centimetre grid paper.

a 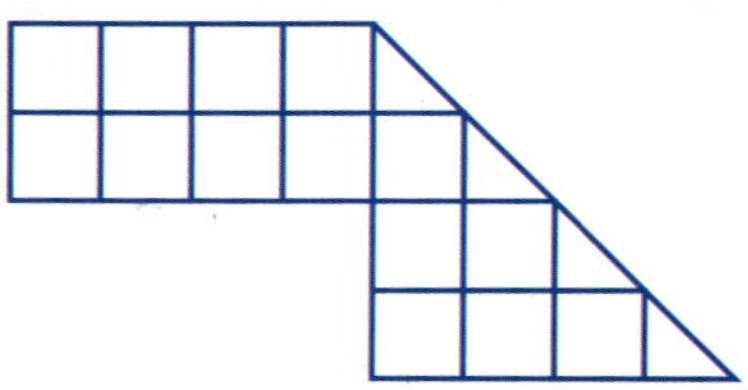

b 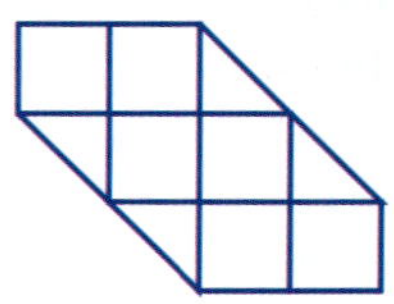

c 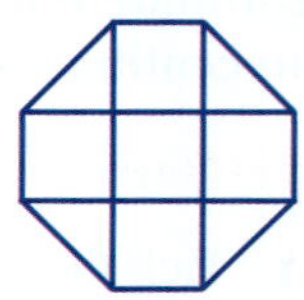

d 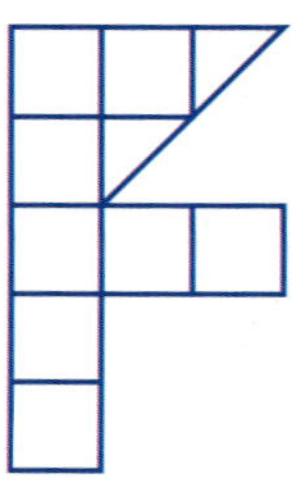

e 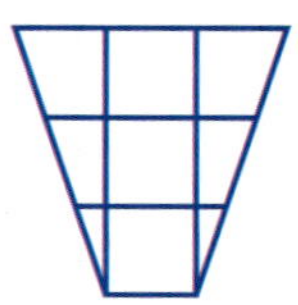

f 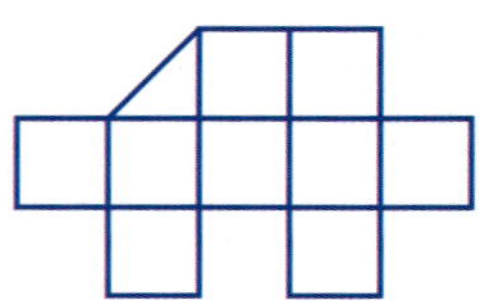

g 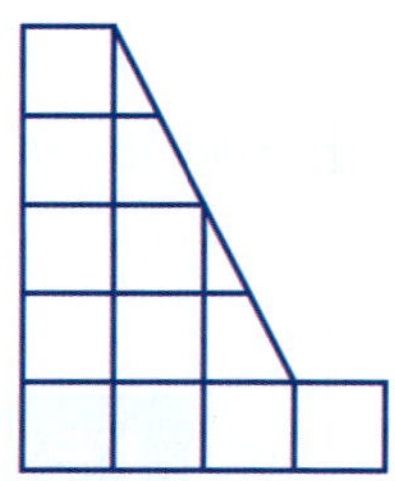

h 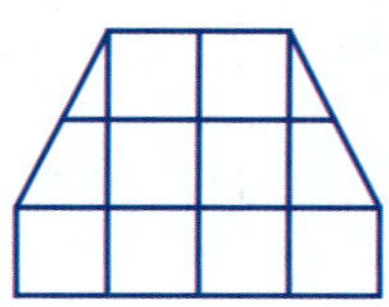

WORKED Example 2

4 Estimate the shaded area of each of the following figures if each is drawn on centimetre grid paper.

a

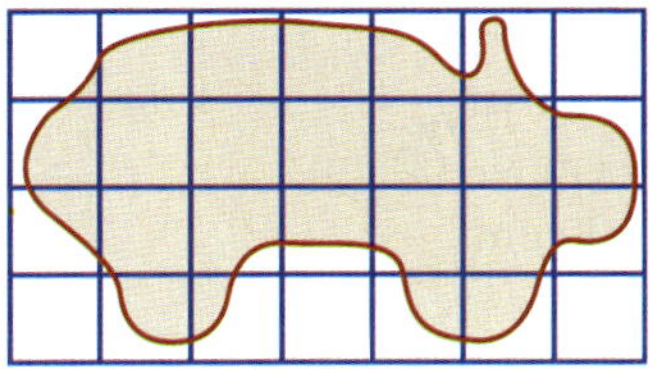

b

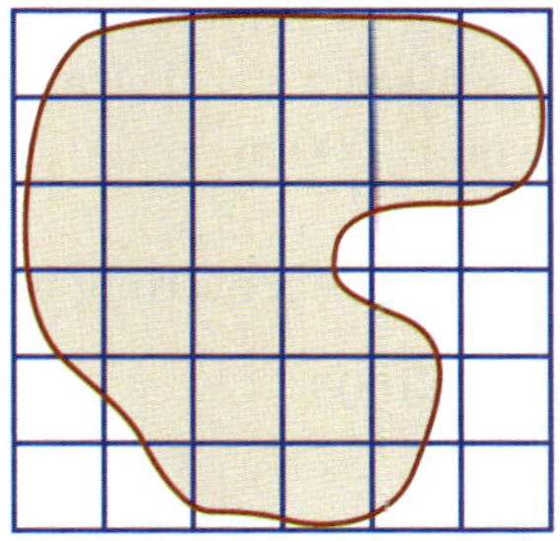

c

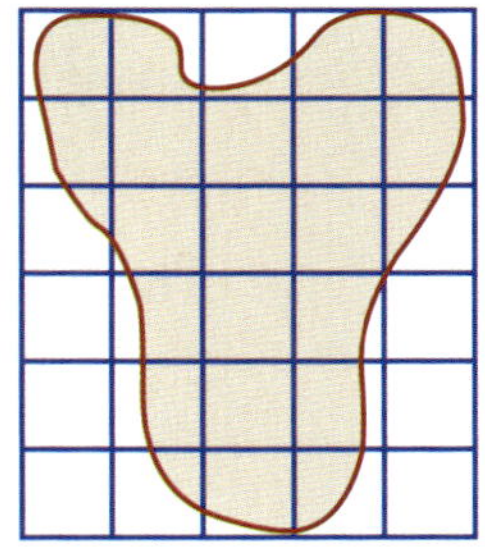

5 a Estimate the area of the sails of this yacht if a metre grid is over it as shown.

b Estimate the area of this butterfly if a centimetre grid is over it as shown.

c Estimate the area of Australia if a 470 km grid is over it as shown. *Note:* The area of each square of the grid is 470 km × 470 km or 220 900 km^2.

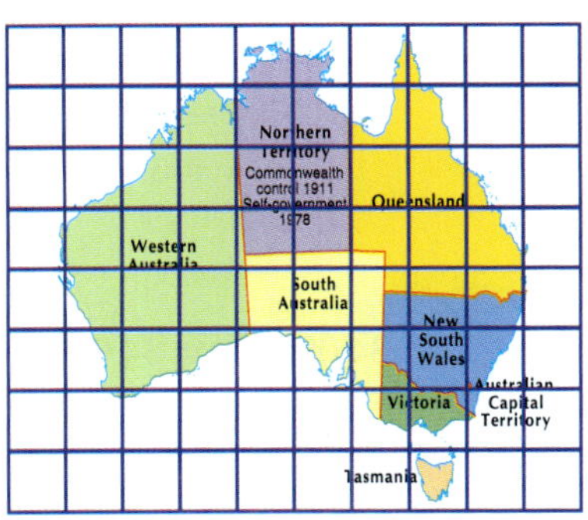

d Estimate the area of the chocolate scoop in this picture of an ice-cream if a centimetre grid is over it as shown.

Converting units of area

The main units of area are related as follows:

$1\text{ km}^2 = 1\text{ km} \times 1\text{ km}$
$\quad = 1000\text{ m} \times 1000\text{ m}$
$\quad = 1\,000\,000\text{ m}^2$

$1\text{ m}^2 = 1\text{ m} \times 1\text{ m}$
$\quad = 100\text{ cm} \times 100\text{ cm}$
$\quad = 10\,000\text{ cm}^2$

$1\text{ cm}^2 = 1\text{ cm} \times 1\text{ cm}$
$\quad = 10\text{ mm} \times 10\text{ mm}$
$\quad = 100\text{ mm}^2$

The chart below makes converting units simpler.

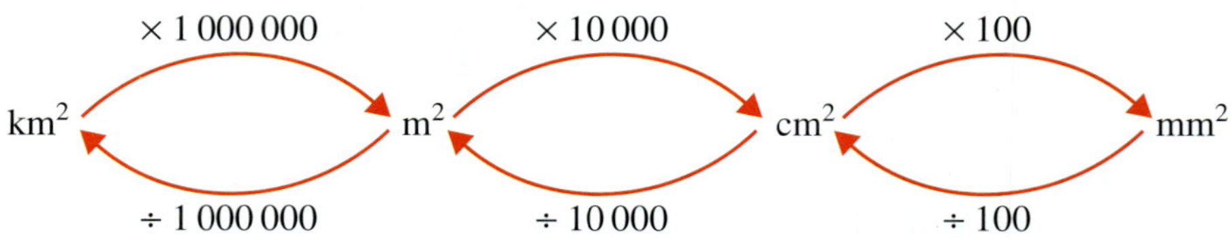

To convert to smaller units, multiply by the conversion factor.
To convert to larger units, divide by the conversion factor.

Hectares

The area represented by the diagram at right is equal to 1 hectare (1 ha). This unit of area is often used in real estate to measure the area of land, particularly for large properties.

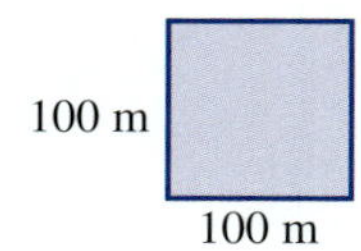

Can you explain why $1\text{ ha} = 10\,000\text{ m}^2$?

WORKED Example 3

Complete the following metric conversions.

a $75\text{ m}^2 =$ ________ cm^2
b $680\text{ mm}^2 =$ ________ cm^2
c $34\,000\text{ m}^2 =$ ________ ha
d $3\frac{3}{4}\text{ cm}^2 =$ ________ mm^2

THINK / **WRITE**

a 1 Look at the metric conversion chart. To convert square metres to square centimetres, multiply by 10 000.
2 Multiply by 10 000. Since 75 is a whole number, we do this by adding four zeros to the end of the number.

a $75\text{ m}^2 = 75 \times 10\,000\text{ cm}^2$
$\quad = 750\,000\text{ cm}^2$

b 1 Look at the metric conversion chart. To convert square millimetres to square centimetres, divide by 100.
2 Divide by 100; that is, move the decimal point two places to the left. Since 680 is a whole number, start from the end.

b $680\text{ mm}^2 = 680 \div 100\text{ cm}^2$
$\quad = 6.8\text{ cm}^2$

c 1 Recall that $1\text{ ha} = 10\,000\text{ m}^2$.
2 To convert square metres to hectares, divide by 10 000. To do this, move the decimal point four places to the left. Since 34 000 is a whole number, start from the end.

c $34\,000\text{ m}^2 = 34\,000 \div 10\,000\text{ ha}$
$\quad = 3.4\text{ ha}$

THINK	WRITE
d 1 Look at the metric conversion chart. To convert square centimetres to square millimetres, multiply by 100.	**d** $3\frac{3}{4}\text{ cm}^2 = 3\frac{3}{4} \times 100\text{ mm}^2$
2 Convert $3\frac{3}{4}$ to an improper fraction.	$= \frac{3 \times 4 + 3}{4} \times 100\text{ mm}^2$ $= \frac{15}{4}\text{ mm}^2$
3 Multiply the numerator by 100. Since 15 is a whole number, we do this by adding two zeros to the end of the number.	$= \frac{15}{4} \times 100\text{ mm}^2$ $= \frac{1500}{4}\text{ mm}^2$
4 Simplify the fraction.	$= 375\text{ mm}^2$

remember

1. When converting units of area, use a conversion chart or remember the conversion factors.

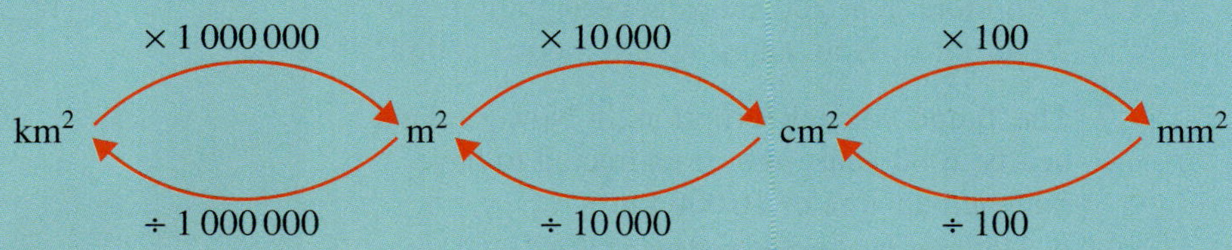

2. To convert to smaller units, multiply by the conversion factor.
3. To convert to larger units, divide by the conversion factor.

EXERCISE 8B Converting units of area

WORKED Example 3a, b, d

1 Complete the following metric conversions.

a $3\text{ m}^2 =$ ________ cm^2 **b** $3\frac{1}{2}\text{ cm}^2 =$ ________ mm^2
c $9.2\text{ km}^2 =$ ________ m^2 **d** $12\frac{17}{50}\text{ m}^2 =$ ________ cm^2
e $6\frac{3}{5}\text{ cm}^2 =$ ________ mm^2 **f** $2.4\text{ cm}^2 =$ ________ mm^2

Mathcad — Converting units of area

2 Copy and complete the following.

a $23\,000\text{ cm}^2 =$ ________ m^2 **b** $235\text{ mm}^2 =$ ________ cm^2
c $74\,000\,000\text{ m}^2 =$ ________ km^2 **d** $0.089\text{ km}^2 =$ ________ m^2
e $11.67\text{ m}^2 =$ ________ cm^2 **f** $\frac{1}{5}\text{ km}^2 =$ ________ m^2

EXCEL Spreadsheet — Area units converter

WORKED Example 3c

3 Copy and complete the following conversions.

a $53\,000\text{ m}^2 =$ ________ ha **b** $786\text{ m}^2 =$ ________ ha
c $1\,000\,000\text{ cm}^2 =$ ________ m^2 **d** $10\,646\text{ m}^2 =$ ________ ha
e $35\,000\,000\text{ m}^2 =$ ________ km^2 **f** $0.0006\text{ km}^2 =$ ________ m^2
g 56 ha = ________ m^2 **h** 1.48 ha = ________ m^2

EXCEL Spreadsheet — Area units converter (DIY)

4 **multiple choice**

An area of 23 000 000 m^2 is equal to:

A 23 km^2 B 2.3 km^2 C 23 ha D 23 000 ha E 230 ha

5 **multiple choice**

An area of 0.034 m^2 is equal to:

A 3.4 cm^2 B 34 cm^2 C 340 cm^2 D 3400 cm^2 E 34 000 cm^2

6 Arrange the following in order from smallest to largest.

a 12 500 m^2, 1.3 ha, 0.3 km^2

b 0.2 ha, 250 000 m^2, 1 km^2

c 200 000 cm^2, 2 m^2, 0.2 m^2

d 0.002 m^2, 500 mm^2, 100 cm^2

7 Add the following areas, giving your answer in either unit.

a 240 mm^2 and 3 cm^2

b 4 m^2 and 63 000 cm^2

c 7.26 cm^2 and 1480 mm^2

d 22 000 m^2 and 34 ha

e 0.003 km^2 and 5400 m^2

f 2 million square metres and 4.6 km^2

8 One of the largest open-air shopping centres is the Ala Moana centre in Honolulu, Hawaii, USA with more than 200 shops covering an area of 20 hectares. How many square metres is this?

9 The Taupo volcanic eruption, in New Zealand nearly 2000 years ago, is estimated to have flattened an area of 16 000 km^2.

a How many square metres is this?

b How many hectares is this?

COMMUNICATION

The area of a rectangle

Draw four different rectangles on a sheet of graph paper or centimetre grid paper. Make the length and width of each rectangle a whole number of centimetres. Calculate the area of each rectangle by counting the squares. Copy and complete the table below.

Rectangle	Length (cm)	Width (cm)	Area (cm^2)
1			
2			
3			
4			

Look at the relationship between length, width and area. Write a sentence describing how to find the area of a rectangle using its length and width.

Finding the area of a rectangle

Using the pronumerals A for area, l for length and w for width, the simple rule you discovered in the previous investigation can be written as the formula below.

$$A = l \times w \quad \text{or} \quad A = lw$$

A square is a rectangle whose length and width are equal. The formula for the area of a square with side length l can be written:

$$A = l \times l \quad \text{or} \quad A = l^2$$

WORKED Example 4

Find the area of the following shapes.

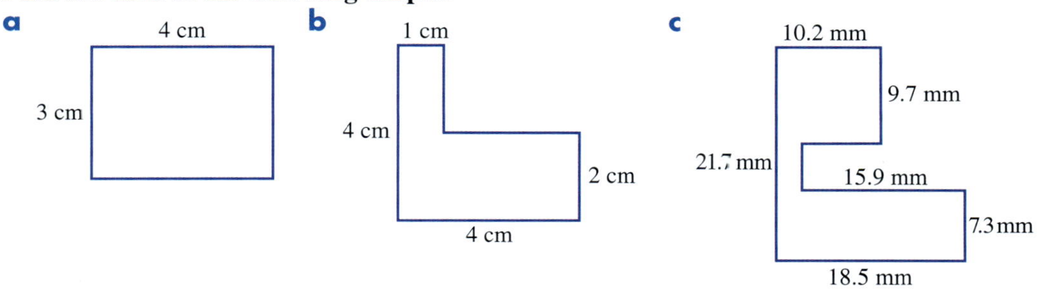

THINK	WRITE
a 1 Write the formula for the area of a rectangle.	**a** $A = lw$
2 Substitute the value 4 for l and 3 for w.	$A = 4 \times 3$
3 Calculate the result. Remember to answer in the correct unit (cm^2).	$= 12\ cm^2$
b 1 Divide the shape into two rectangles.	**b** 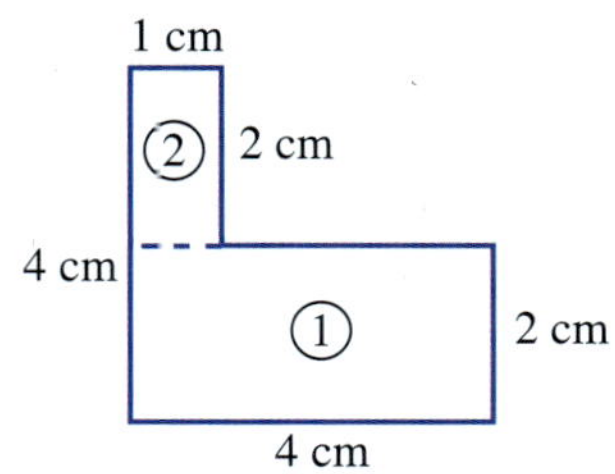
2 Calculate the area of each rectangle separately by substituting the correct values of l and w into the formula $A = lw$.	Area of rectangle 1 $= l \times w$ $= 4 \times 2$ $= 8\ cm^2$ Area of rectangle 2 $= l \times w$ $= 2 \times 1$ $= 2\ cm^2$
3 Add the two areas. Remember to answer in the correct unit (cm^2).	Area of shape = area of rectangle 1 + area of rectangle 2 $= 8\ cm^2 + 2\ cm^2$ $= 10\ cm^2$

Continued over page

THINK | **WRITE**

c 1 Divide the shape into three rectangles.

c

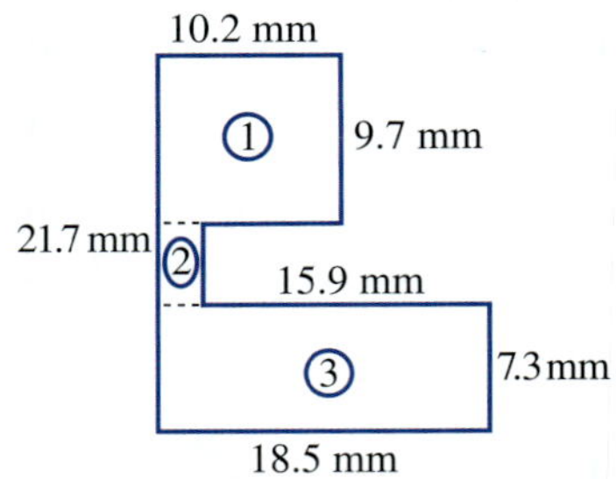

2 Calculate the area of each rectangle separately by substituting the correct values of l and w into the formula $A = lw$.

Area of rectangle 1 $= l \times w$
$= 10.2 \times 9.7$
$= 98.94 \text{ mm}^2$

Area of rectangle 2 $= l \times w$
$= (21.7 - 9.7 - 7.3) \times (18.5 - 15.9)$
$= 4.7 \times 2.6$
$= 12.22 \text{ mm}^2$

Area of rectangle 3 $= l \times w$
$= 18.5 \times 7.3$
$= 135.05 \text{ mm}^2$

3 Add the three areas. Remember to answer in the correct unit (mm^2).

Area of shape = Area 1 + Area 2 + Area 3
$= 98.94 \text{ mm}^2 + 12.22 \text{ mm}^2 + 135.05 \text{ mm}^2$
$= 246.21 \text{ mm}^2$

remember

1. The area of a rectangle is found by using the formula $A = lw$, where A is the area of the rectangle, l is its length and w is its width.

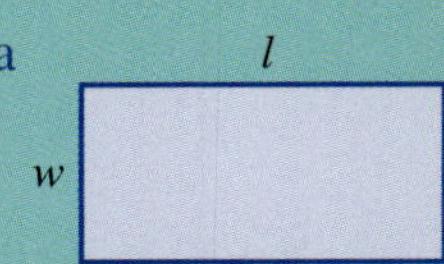

2. A square is a rectangle whose length and width are equal. The area of a square is found by using the formula $A = l^2$, where l is its side length.

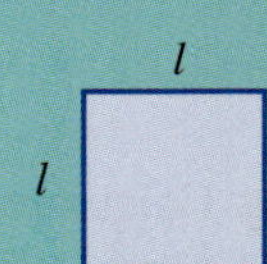

EXERCISE 8C Finding the area of a rectangle

1 Four rectangles are drawn on centimetre grids that follow.
- i State the length and width of each rectangle.
- ii Use your answers to part i to calculate the area of each rectangle.
- iii Count the number of 1-cm^2 squares to find the area of each rectangle.
- iv Compare your answers to parts ii and iii.

a 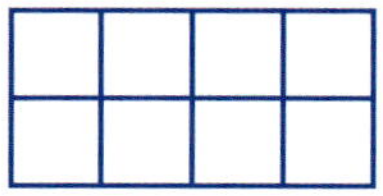

b 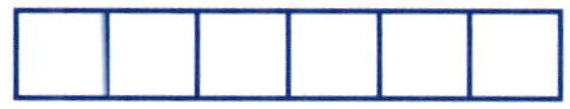

c 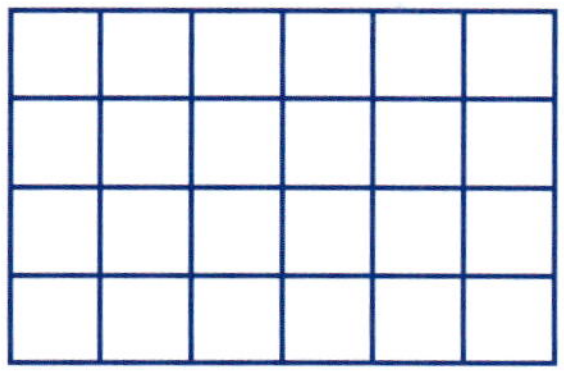

d 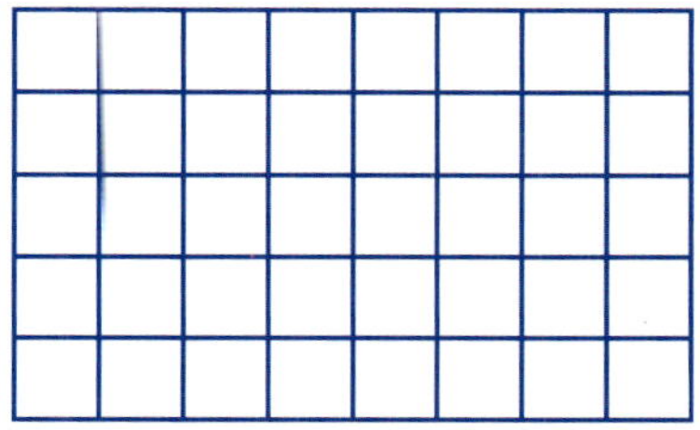

8.3 SkillSHEET — Area of rectangles

Mathcad — Area of a rectangle

EXCEL Spreadsheet — Area of a rectangle

Cabri Geometry — Area of a rectangle

WORKED Example 4a

2 Find the area of the following rectangles. (*Hint:* Use the formula $A = lw$.)

a 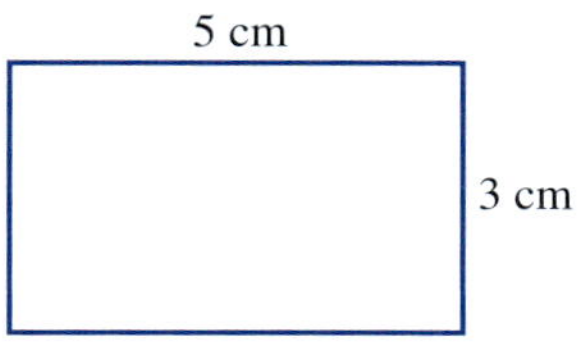

b 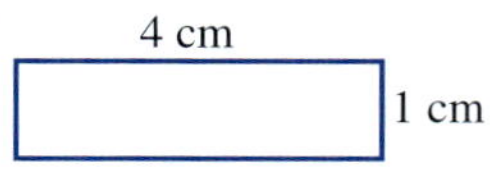

c 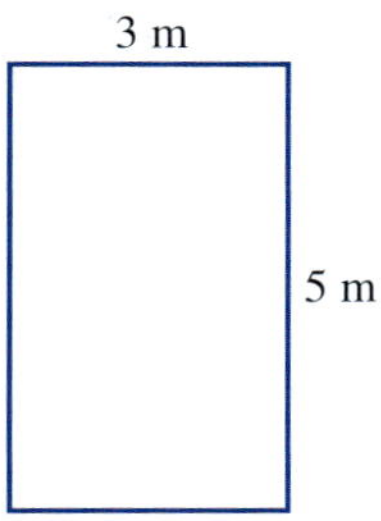

d 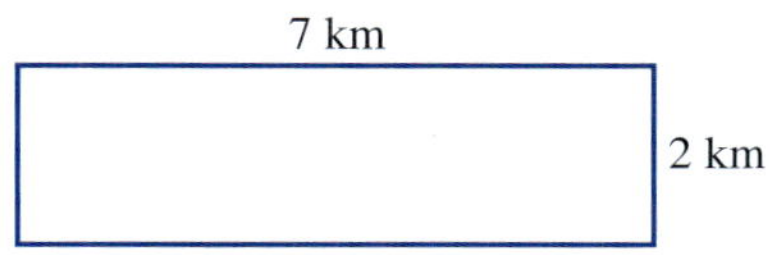

e 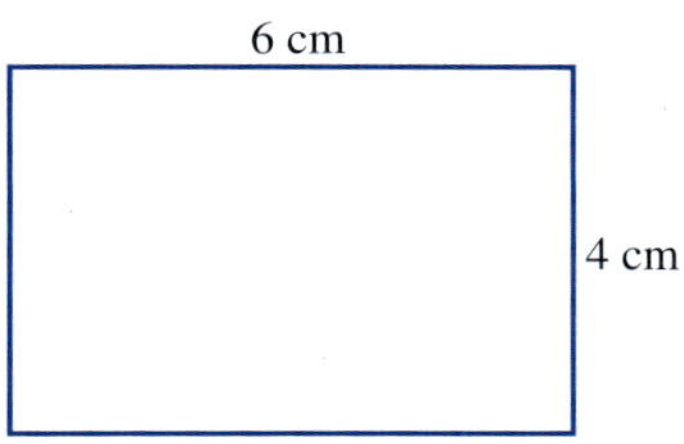

f 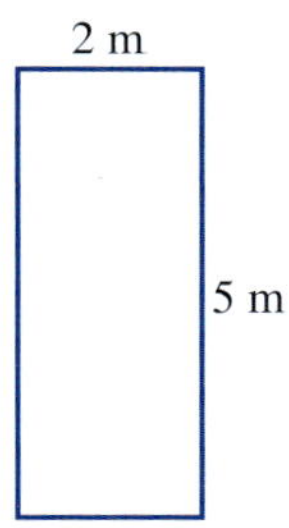

3 Accurately measure the length and width of each rectangle in the units specified by the bracket. Hence, calculate the area of the rectangle.

a 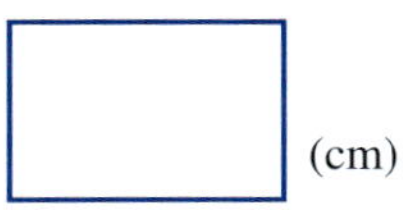

b 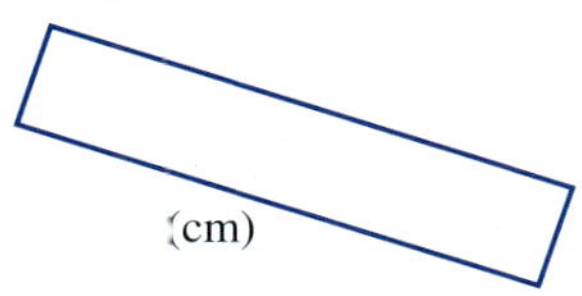

c 

d 

e 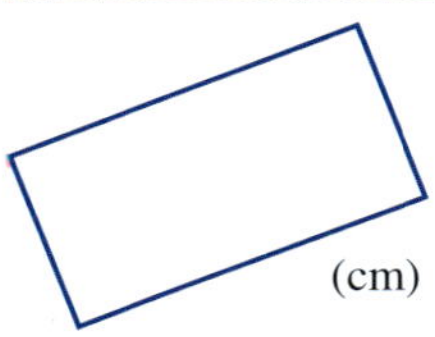

f 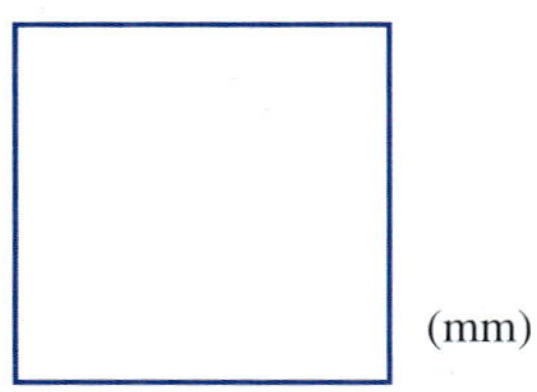

WORKED Example 4b

4 Find the area of the following shapes. (*Hint:* Divide the shapes into rectangles and squares before using the formula $A = lw$. Subtract the area of any sections 'cut out' of the main shape.)

a

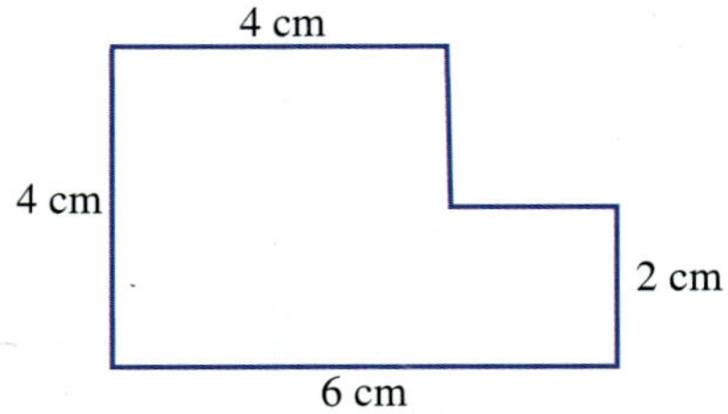

b

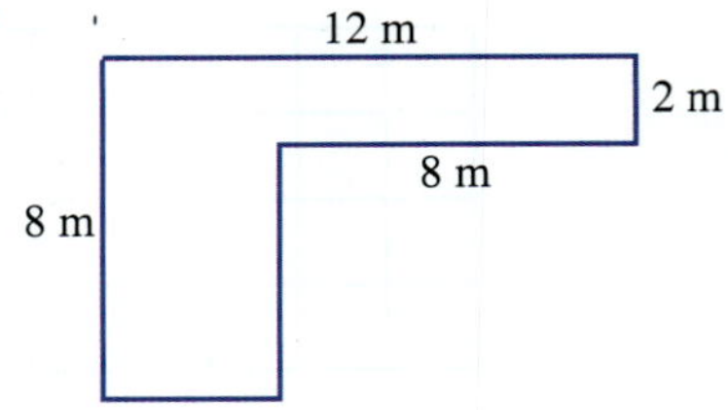

c

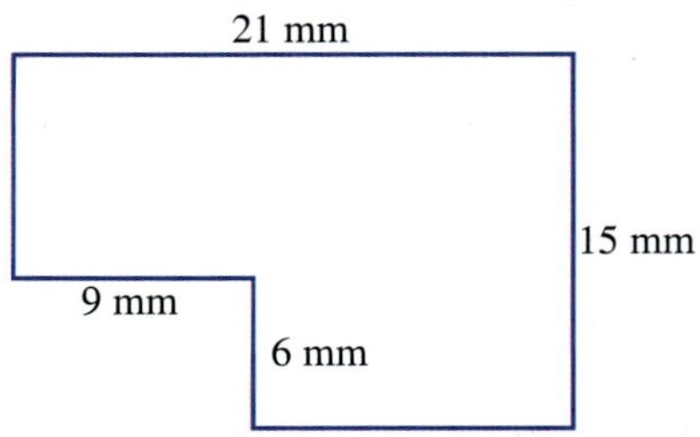

d

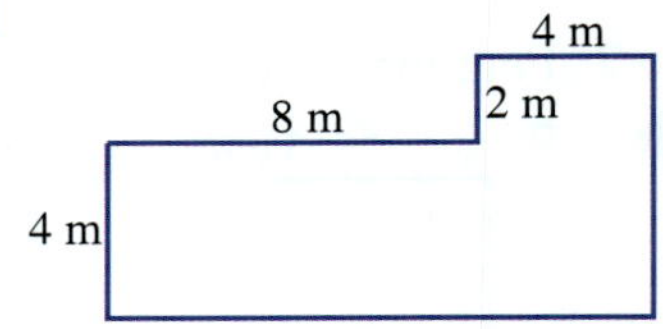

WORKED Example 4c

e

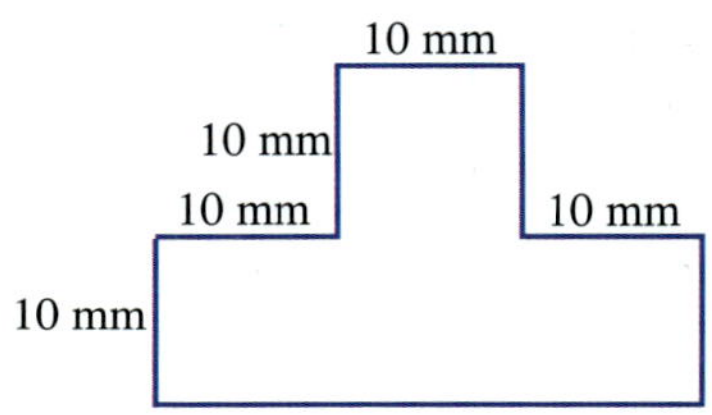

f

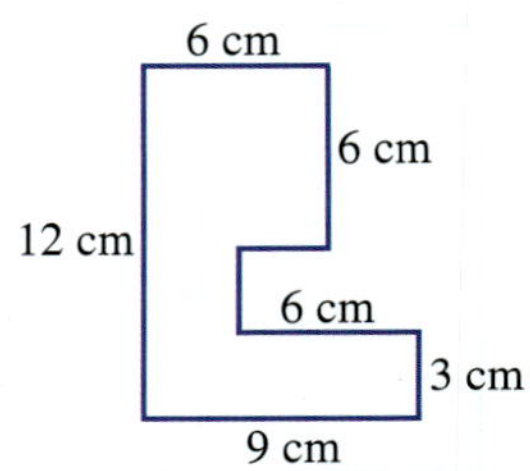

g

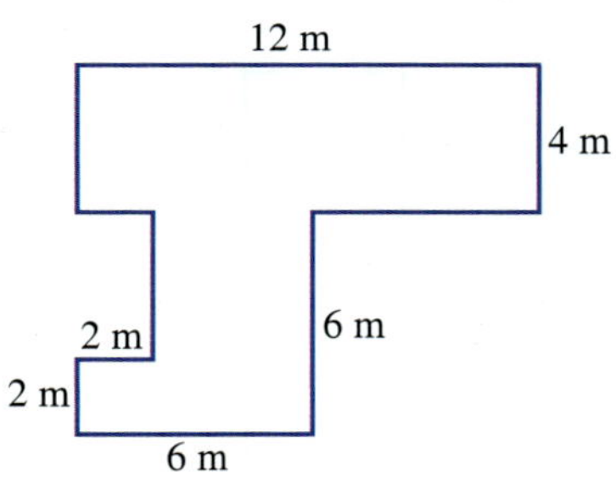

h

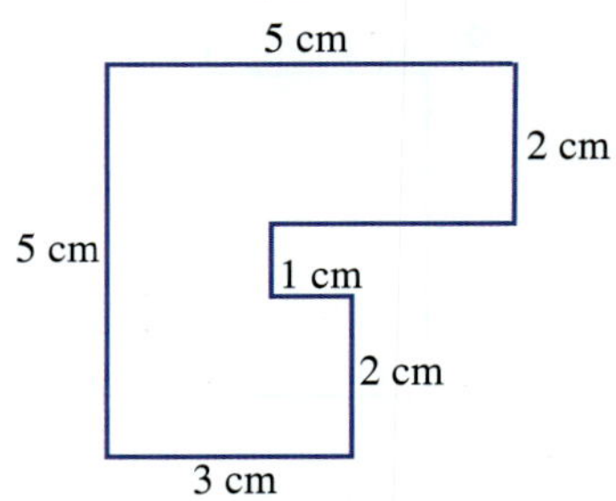

i

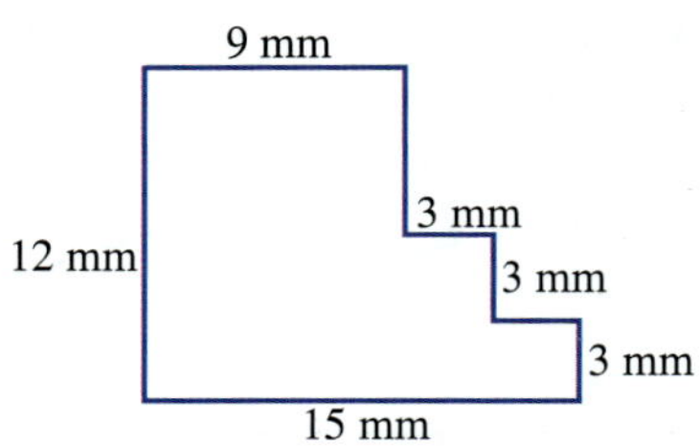

j

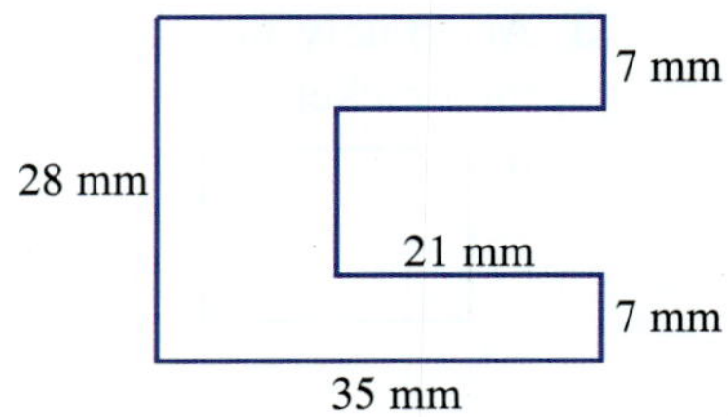

k

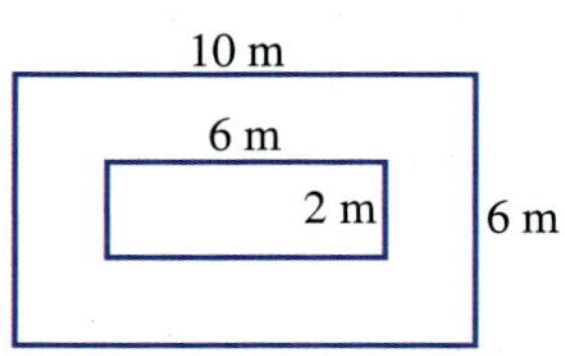

l

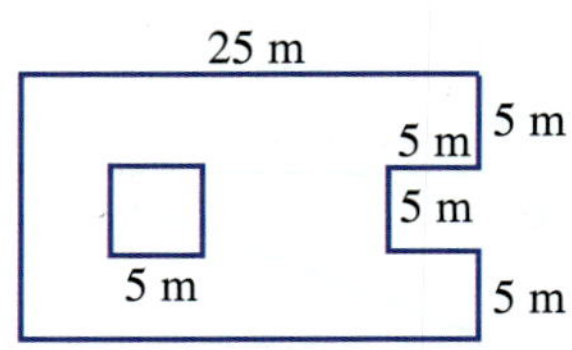

5 How many square metres of carpet are needed to carpet a rectangular room of length 5 m and width 3.5 m?

6 Find the area of material (cm^2) needed to make a rectangular table napkin 32 cm long and 25 cm wide.

7 Find the area of material needed to make a rectangular rug which is 4.2 m long and 230 cm wide. (Give your answer in square metres.)
(*Hint:* Convert the width measurement into metres first.)

8 Toby is tiling his bathroom with ceramic tiles that cost $35.20 per box.

a How many square metres of tiles will he need, if the rectangular room has a width of 2.5 m and a length of 3 m?

b How many boxes of tiles should he order if each box contains enough tiles to cover 0.5 m^2?

c What will be the cost of the tiles?

9 Jane is a landscape gardener who is laying a new lawn. The rectangular lawn is 13 m long and 8 m wide. How many square metres of turf should Jane order? What is the total cost of the turf if it costs $12.50 per square metre?

10 Members of the Lee family want to pave the area around their new swimming pool. The pool is set into the corner of the yard as shown at right. Find the area of paving (in square metres) required to cover the yard around the pool (shaded in the diagram).

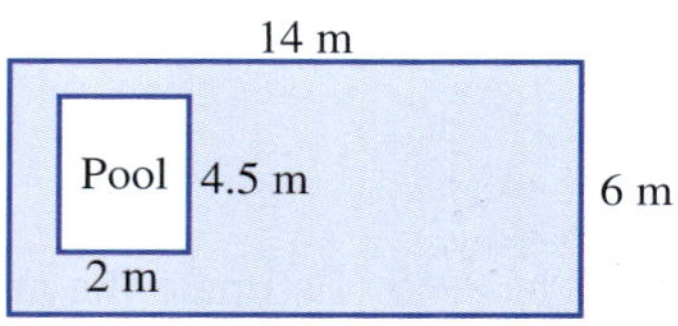

11 Alana, who works for Fast Glass Replacements, has been asked for a quote to replace three windows. Each window is 1.8 m long and 0.8 m wide. What price should Alana quote if the glass costs $27 per square metre? (Include a delivery cost of $25 in the quote.)

12 Find the total area of the cattle station shown at right, which has 3 large paddocks. (Give your answer in square kilometres.)

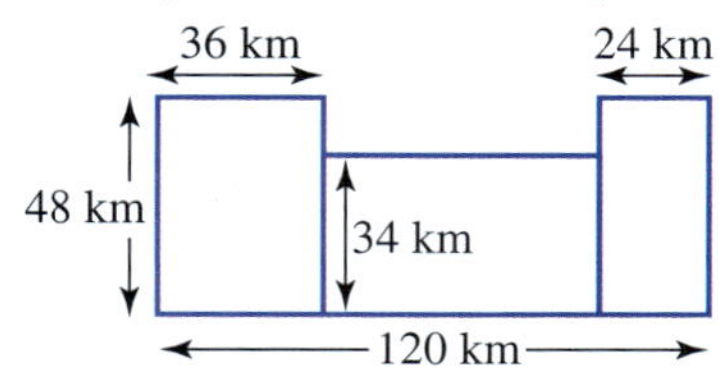

13 Rectangular laminate sheets are 1.8 m long and 0.9 m wide. How many sheets would be needed to cover 4.8 m^2 of bench space in a kitchen, if none is wasted?

14 Calculate the total floor area of a concrete slab for a house as shown at right.

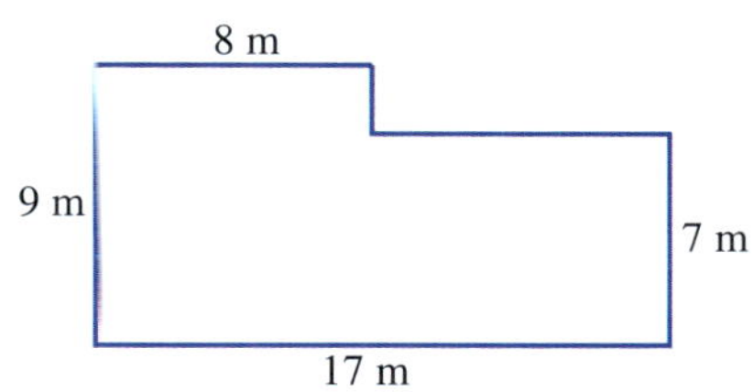

15 In order to construct an advertising sign, the letters below are cut from plastic sheets. Find the total area of plastic required to make all 3 letters.

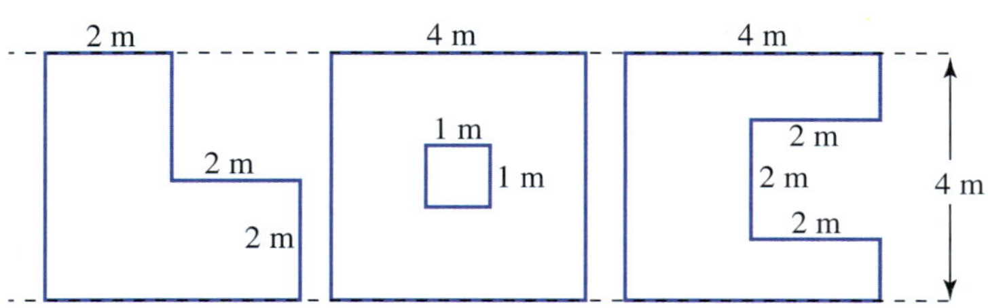

16 How many square centimetres of cardboard are needed to make the cube with an edge length of 12 cm shown at right?

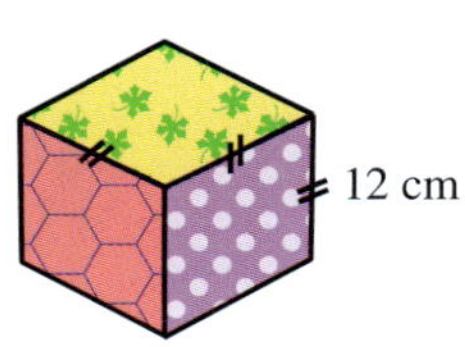

10 QUICK QUESTIONS 1

1 Select the unit (mm², cm², m², ha or km²) which would be the most appropriate to measure the area of a chessboard.

2 Select the most appropriate unit to measure the area of a netball court.

3 Find the area of the shape at right, which is drawn on centimetre grid paper.

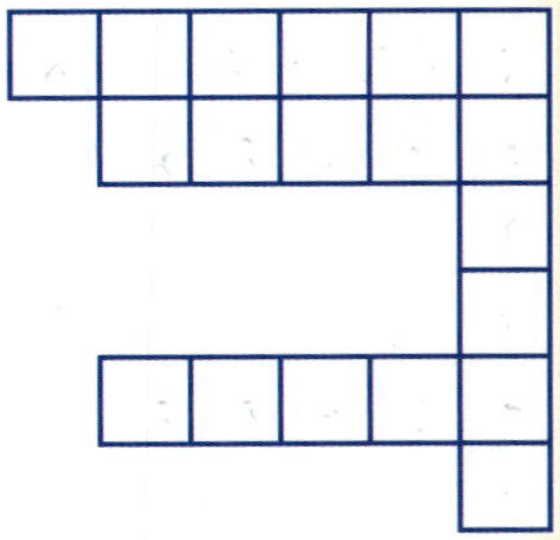

4 Find the perimeter of the shape shown in question 3.

5 Estimate the area covered by the paint blob if it is on centimetre grid paper.

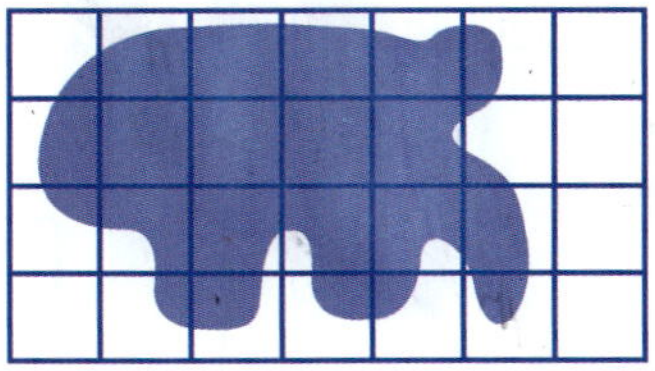

6 Estimate the area of water on a road if a metre grid is shown over it.

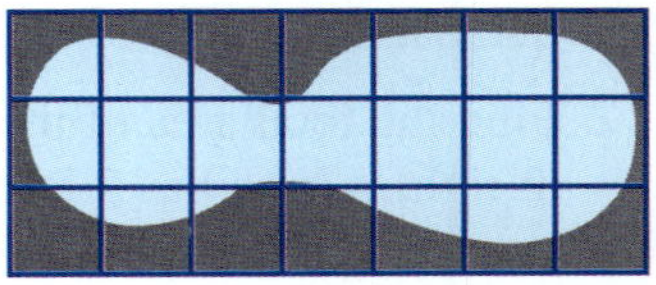

7 Convert 13 000 mm^2 to cm^2.

8 Convert 4.2 km^2 to m^2.

9 Find the area of the following rectangle.

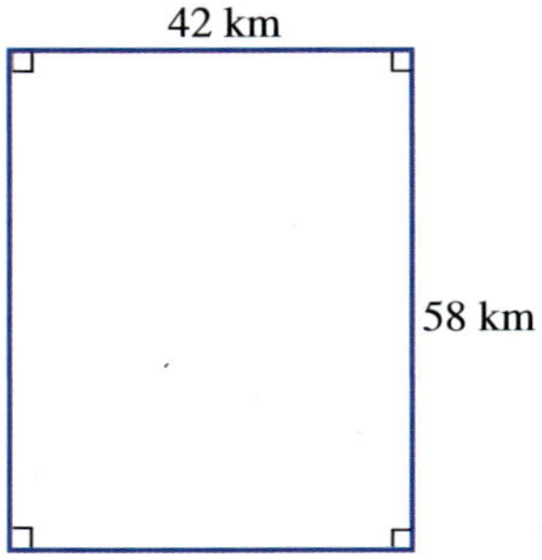

10 Find the area of the following shape.

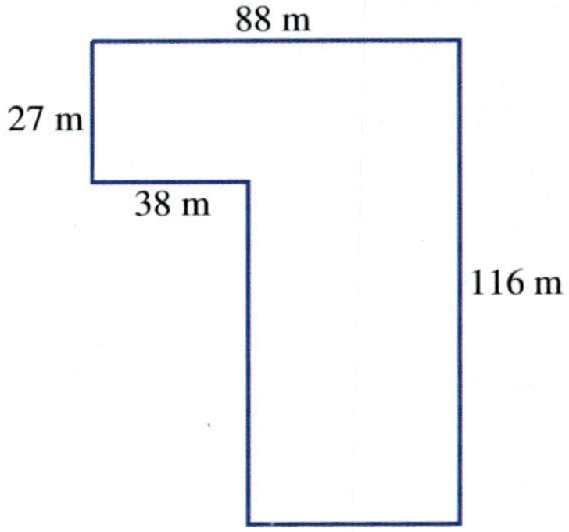

Finding the area of a triangle

Look at the triangles below, which have been formed by cutting rectangles in half.

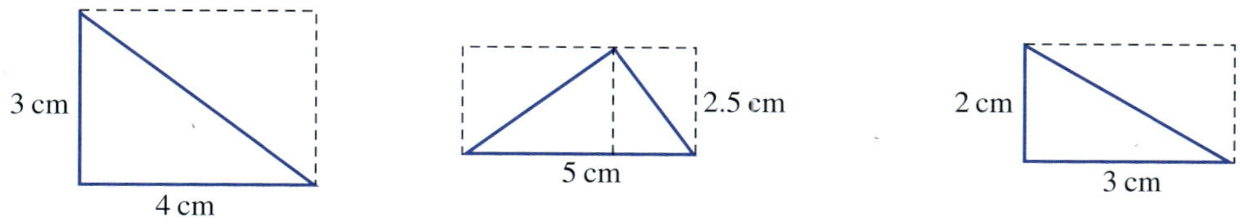

The area of a triangle is equal to half the area of a rectangle of the same base length and height.

The area of a triangle, A, is given by the formula: $A = \frac{1}{2}bh$, where b is the base and h is the height of the triangle. The base and the height of the triangle are perpendicular (at right angles) to each other.

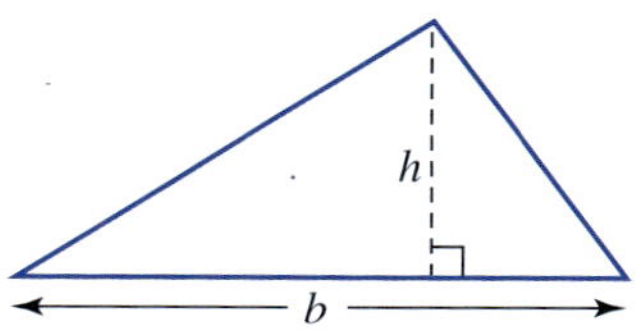

WORKED Example 5

Find the area of the shape at right.

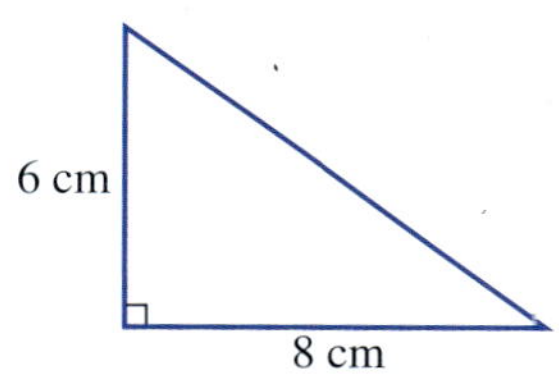

THINK

1. Draw an imaginary rectangle that contains the triangle. The rectangle should have the same base length and height as the triangle. Notice that the triangle forms half of the rectangle.
2. Use the formula $A = lw$ to find the area of this rectangle.
3. Halve the area of the rectangle to get the area of the triangle. Remember to answer in the correct unit (cm^2).
   Note: Formula $A = \frac{1}{2}bh$ could be used.

WRITE

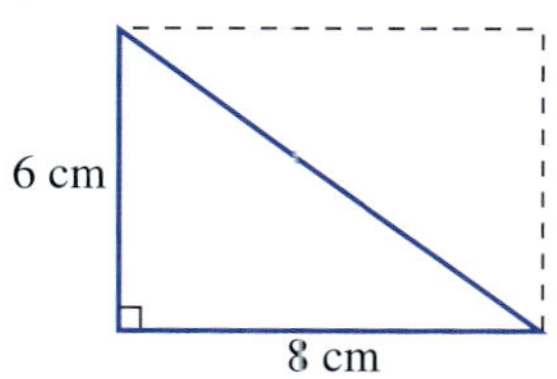

$A = lw$
$= 8 \text{ cm} \times 6 \text{ cm}$
$= 48 \text{ cm}^2$

Area of triangle = $\frac{1}{2}$ of area of rectangle
Area of triangle = $\frac{1}{2} \times 48$
$= 24 \text{ cm}^2$

WORKED Example 6

Find the area of each of these triangles.

a

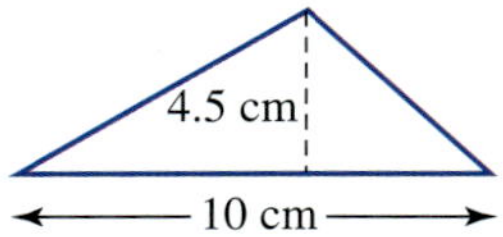

b

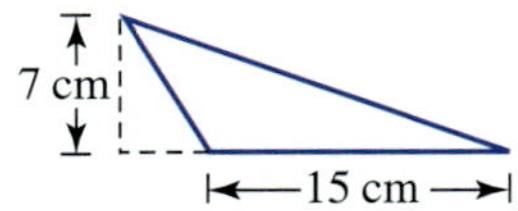

THINK	WRITE
a 1 Write the formula for the area of a triangle.	**a** $A = \frac{1}{2}bh$
2 Identify the values of b and h.	$b = 10, h = 4.5$
3 Substitute the values of b and h into the formula.	$A = \frac{1}{2} \times 10 \times 4.5$
4 Evaluate. (Since one of the values is even, halve it first to make calculations easier.) Remember to include the correct unit (cm^2).	$= 5 \times 4.5$ $= 22.5\ cm^2$
b 1 Write the formula for the area of a triangle.	**b** $A = \frac{1}{2}bh$
2 Identify the values of b and h.	$b = 15, h = 7$
3 Substitute the values of b and h into the formula.	$A = \frac{1}{2} \times 15 \times 7$
4 Evaluate. (Since neither value is even, multiply 15 and 7 first, and then divide by 2.) Remember to include the correct unit (cm^2).	$= \frac{1}{2} \times 105$ $= 52.5\ cm^2$

remember

1. The area of a triangle is equal to half that of a rectangle with the same base and height.
2. The area of a triangle may be found using the formula $A = \frac{1}{2}bh$, where b is the base and h is the height of the triangle.

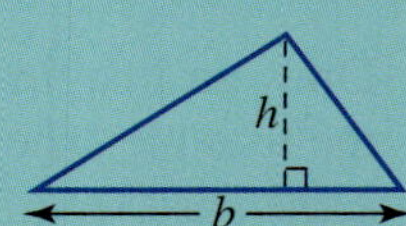

EXERCISE 8D Finding the area of a triangle

SkillSHEET 8.4 Area of triangles

WORKED Example 5

1 Find the area of the following triangles.

a

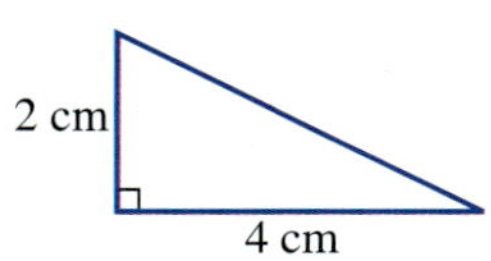

b

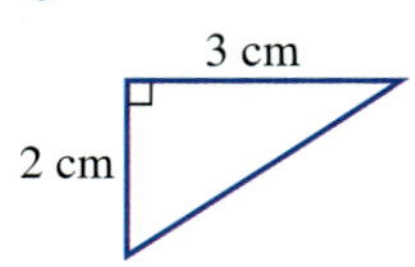

c

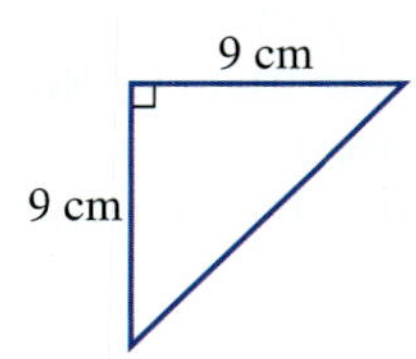

d

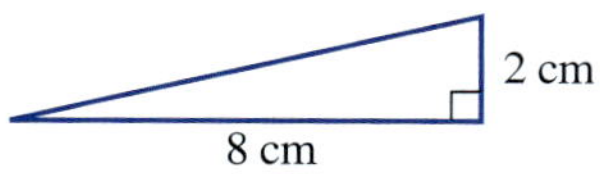

e

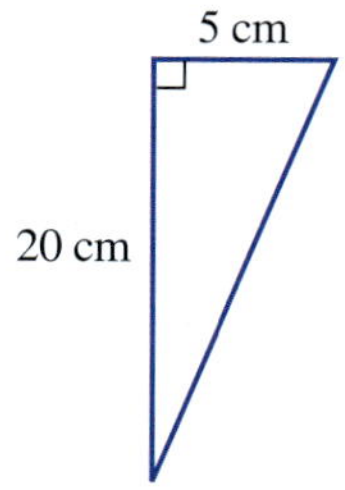

f

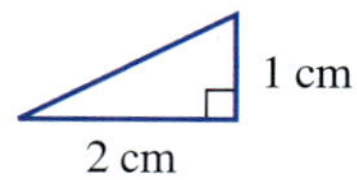

2 **multiple choice**

a In the diagram shown at right, the height of the triangle is:

A 60 mm
B 30 mm
C 27 mm
D 53 mm
E 49 mm

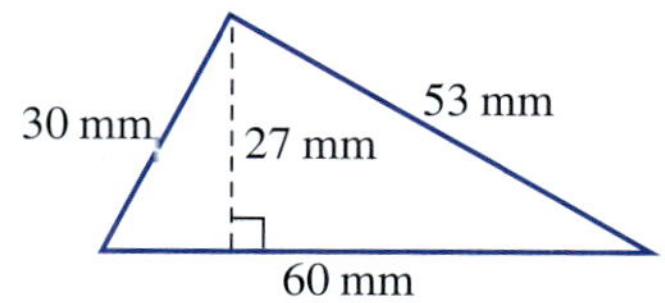

b In the diagram shown at right, the height of the triangle is:

A 2.6 cm
B 4.0 cm
C 3.0 cm
D 4.2 cm
E 3.8 cm

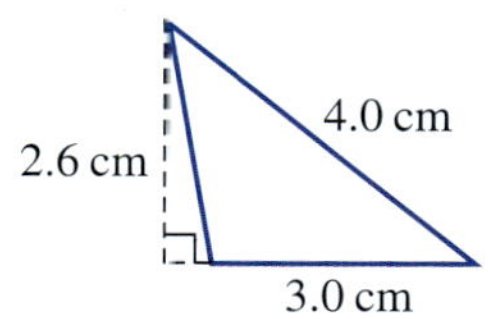

c In the diagram shown at right, the base length of the triangle is:

A 4.1 cm
B 2.9 cm
C 5.3 cm
D 7.2 cm
E 7.0 cm

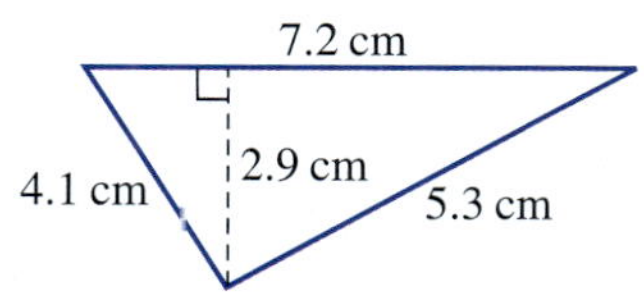

d In the diagram shown at right, the base length of the triangle is:

A 39 cm
B 58 cm
C 29 cm
D 25 cm
E 43 cm

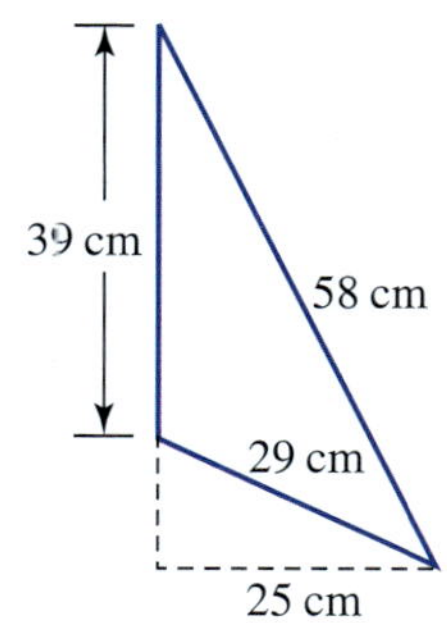

e The area of the triangle shown at right is:

A 19.2 cm^2
B 9.6 cm^2
C 12 cm^2
D 15 cm^2
E 24 cm^2

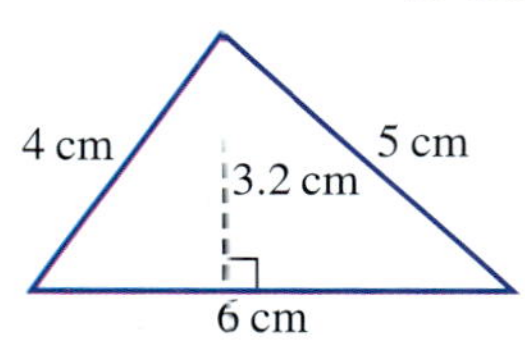

WORKED Example 6

3 Find the area of each of these triangles.

a
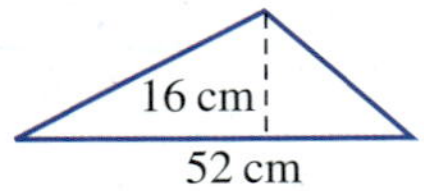

b
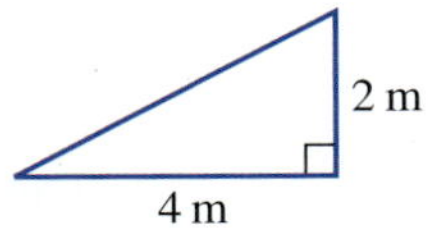

c
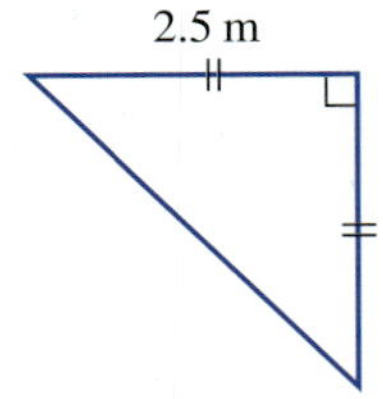

d
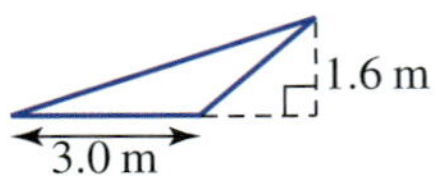

e
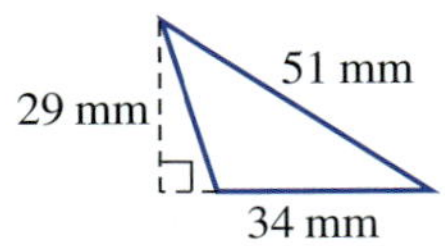

f
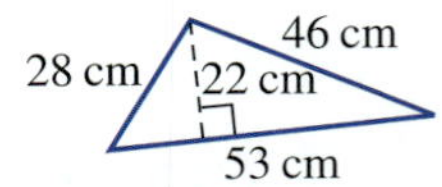

g
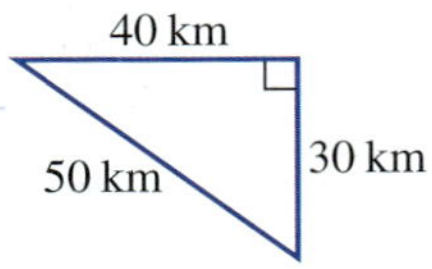

h
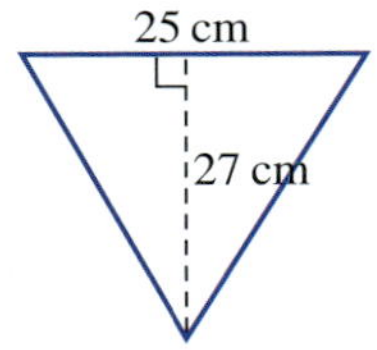

i
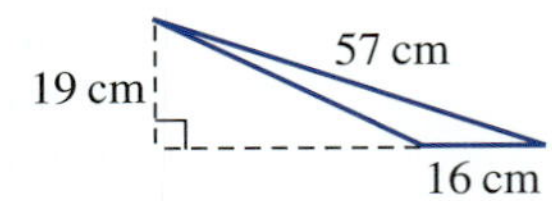

4 Calculate the approximate area of the triangular mainsail when the sail is flat. Use the dimensions shown on the yacht at right.

5 What area of cloth would be required to make one of the pennants shown below?

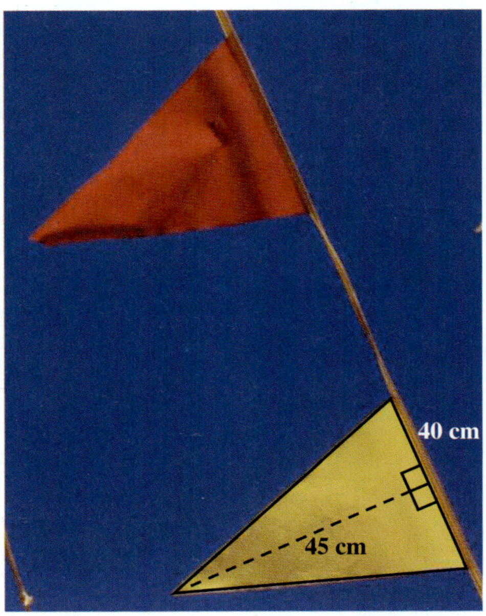

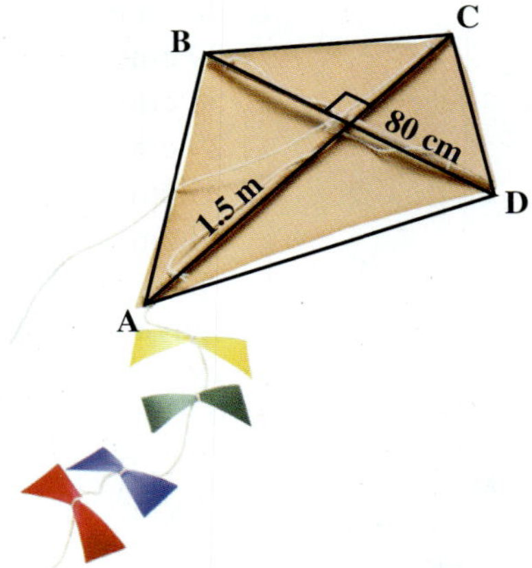

6 Find the area of cloth required to make the kite shown, given that the length BD is 80 cm and the length AC is 1.5 m.

7 A church spire has six identical triangular faces that have the dimensions shown at right. What area of copper roofing would be required to cover all six faces of the spire?

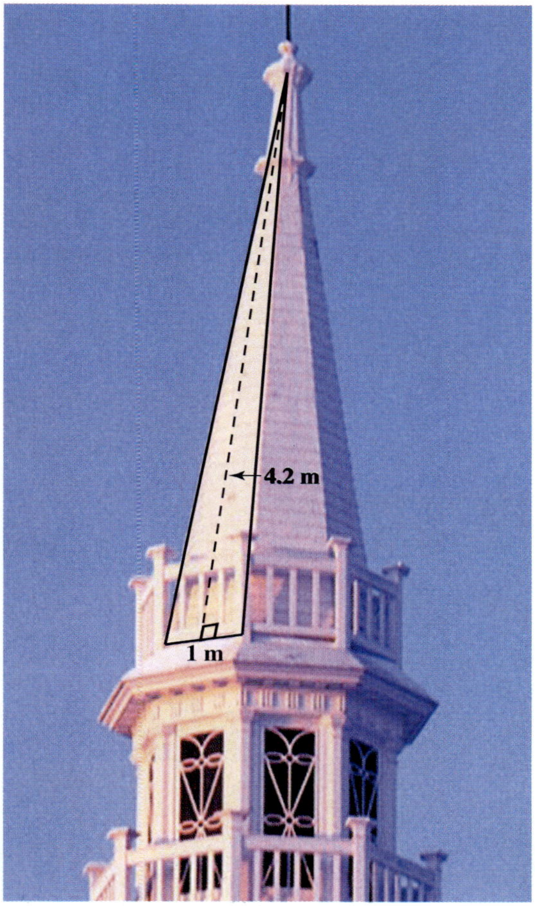

8 Find the total wing area of the delta winged jet aircraft shown below.

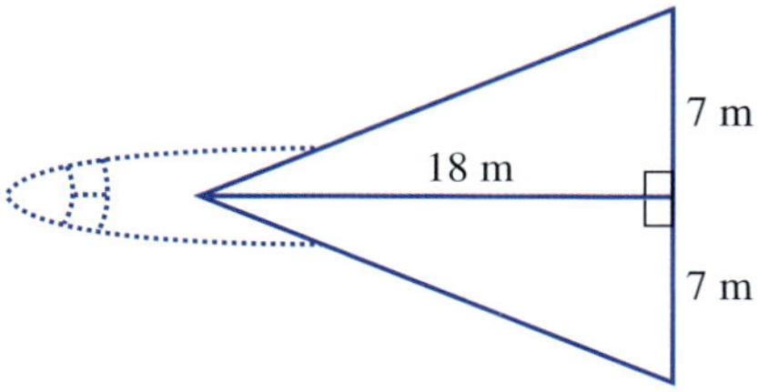

9 The triangle shown below can be used to make an estimate of the area of the state of Victoria.

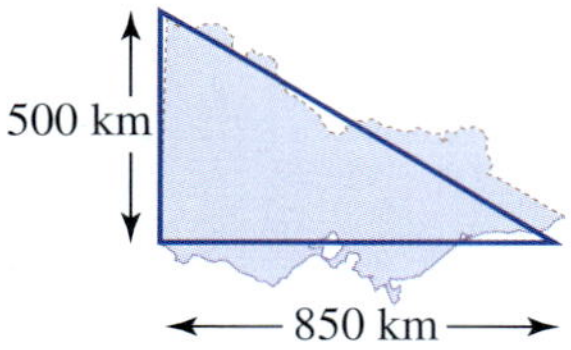

a Find the area of the triangle shown.

b Compare your estimate with the correct value from your atlas.

WorkSHEET 8.1

MATHS QUEST CHALLENGE

1 **A floor tiler charged $640.00 to tile a rectangular room. Her next job is to tile the floor of a rectangular room twice as long and twice as wide. How much should she charge for the larger room? (The answer is not $1280.00.)**

2 **A rectangle has an area of 36 cm^2 and a perimeter of 26 cm. Each side of the rectangle is a whole number of centimetres. What is the length and width of this rectangle?**

3 **What is the perimeter of a square that has the same area as a rectangle with length 18 mm and width 8 mm?**

4 **In a quiz, you start with 100 points. Ten points are added for each correct answer and four points are subtracted for each incorrect answer. If you answer 40 questions, what is the least number of questions you must answer correctly to have a score over 300?**

Have your eyes ever been checked?

Calculate the area of each figure below to discover the puzzle answer code.

28 cm²	49 m²	12 m²	54 m²	45 cm²	48 cm²			
14 m²	63 m²	36 m²	40 cm²	36 cm²	18 m²	70 cm²	12 cm²	22 cm²
60 m²	24 cm²	6 m²	39 m²	28 m²	18 cm²	15 cm²	32 cm²	8 m²

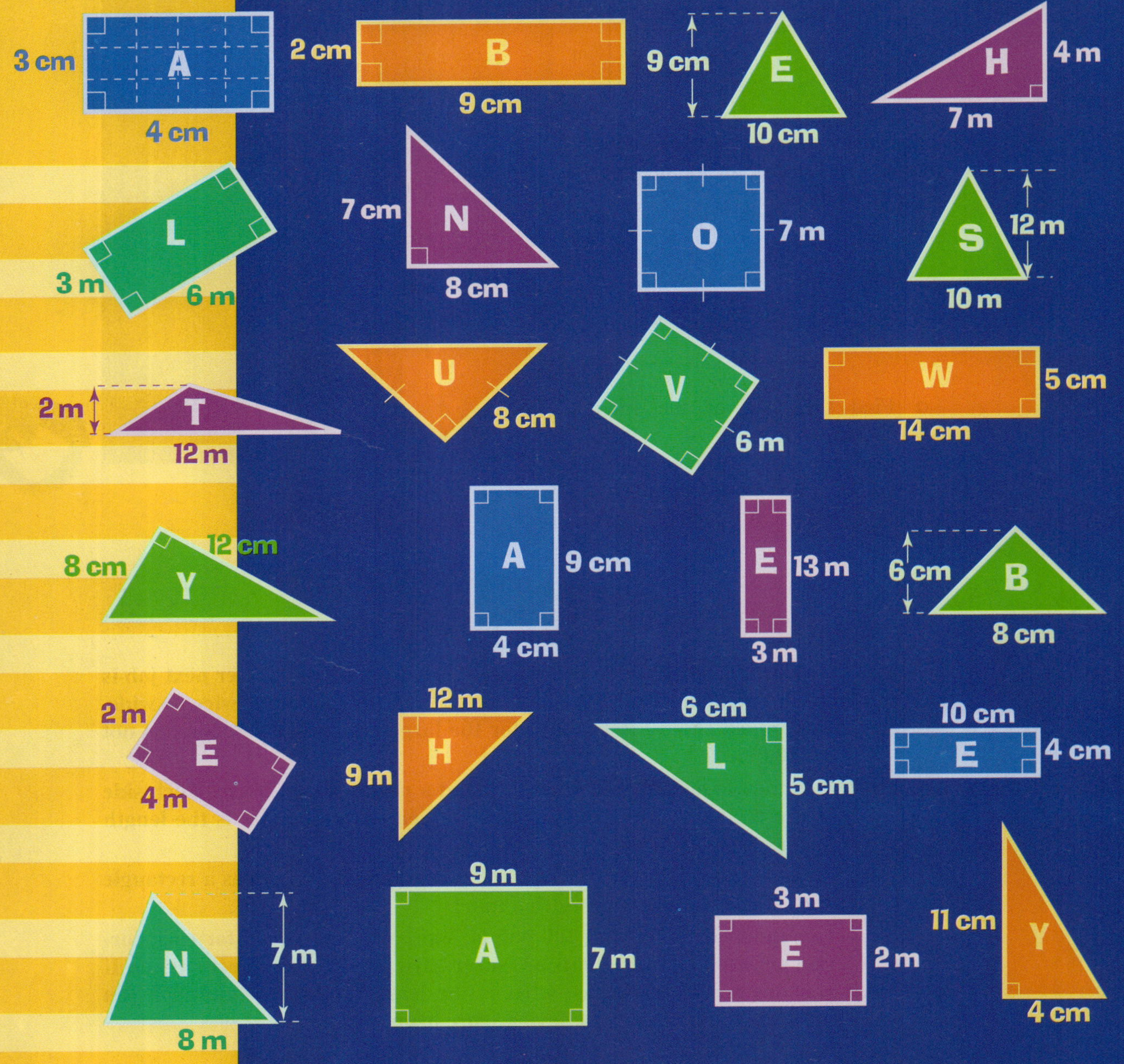

COMMUNICATION

What area does your hand cover?

You will need a sheet of graph paper.

1 Trace an outline of your hand on the graph paper.
2 Look at the 'larger' squares on the graph paper; that is, the squares made by the darker coloured lines. What is the area of one of these squares?
3 Estimate the number of 'larger' squares inside the outline of your hand.
4 Using your estimate from part 3, calculate the area covered by your hand.
5 Now look at the smaller squares on the graph paper; that is, the squares made by the fainter or lighter coloured lines. What is the area of one of these squares?
6 Estimate the number of 'smaller' squares inside the outline of your hand.
7 Using your estimate from part 6, calculate the area covered by your hand.
8 Compare the two area values you obtained by converting both to the same unit. What is the difference between the two values?
9 Which value do you think is more accurate? Explain why in a sentence.
10 Compare your results with another class member. Is the difference between your two area values smaller or larger than his or her values?
11 You may wish to extend this investigation and determine what area your body covers.

DESIGN

The size of tangram pieces

Resources: A4 paper, scissors

To begin this activity, we need to first make a tangram.

- Cut a piece of *square* paper from a sheet of unlined A4 paper.
- Fold the square sheet in half along its diagonal. Unfold and cut along the crease.
- You should now have two triangles. Take one of these triangles, fold it in half and cut along the crease line.
- Take your second triangle from above and lightly crease to find the midpoint of the longest side. Fold so that the vertex of the right angle touches that midpoint and cut along the crease. This forms a trapezium and a small triangle.
- Take the trapezium, fold it in half and cut along the crease line. You should now have two smaller trapeziums.
- Fold the acute base angle of one of the trapeziums to the adjacent right base angle and cut along the crease. This should result in a square and a small triangle.

- Take your other trapezium and fold its right base angle to the opposite obtuse angle. Cut along the crease. This should result in a parallelogram and a small triangle.
- You should now have 7 shapes — 5 triangles, 1 square and 1 parallelogram. These pieces together make up a tangram.
- See if you can now assemble the pieces to form the original square from which it was cut.

You can now use the tangram you have constructed for further activities.

1 Take your 7 shapes. Use the smallest triangle as the basic unit of area. Arrange your pieces in increasing order of area. Give the area of each piece in terms of 'small triangular units'. Some shapes may have the same area. Copy and complete the following table:

Shape	Sketch of shape	Area (such as 3 'triangular units')
1		
2		
3		
4		
5		
6		
7		

2 Your pieces can be fitted together in different combinations to form squares of various sizes. Experiment with your pieces to see whether you can form a *square* using the following number of pieces. Copy and complete the table below.

Number of pieces	Sketch
1	
2	
3	
4	
5	
6	
7	

3 The shape of various objects (animals, letters, numbers, ...) can be made using a combination of these 7 tangram pieces. Experiment making a collection of different figures. Draw diagrams showing the positioning of the pieces.

Area of composite shapes

Sometimes it is not possible to use a single formula to find the area of a particular shape. However, the area may be found by dividing the shape into parts and using the known formula for each part separately. For example, the area of the shape shown below can be found by dividing it into a triangle and a rectangle (simple shapes, for which we know the formula).

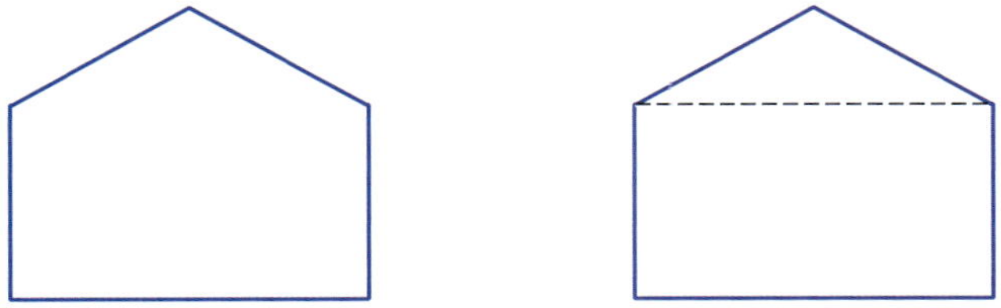

WORKED Example 7

Find the area of the shape at right.

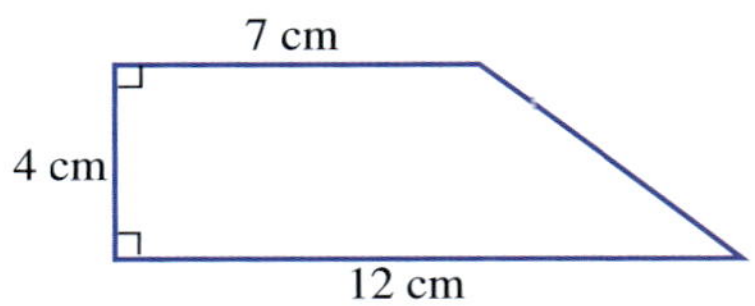

THINK

WRITE

1. Divide the shape into a rectangle and a triangle.

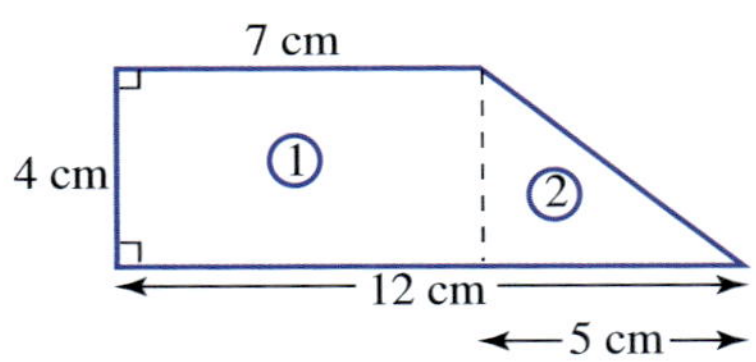

2. Use the formula $A = lw$ to find the area of the rectangle.

$$\text{Area 1} = lw = 7 \times 4 = 28 \text{ cm}^2$$

3. Use the formula $A = \frac{1}{2}bh$ to find the area of the triangle.

$$\text{Area 2} = \tfrac{1}{2}bh = \tfrac{1}{2} \times 5 \times 4 = 10 \text{ cm}^2$$

4. Add the two areas to get the total area.

$$\text{Total area} = A_1 + A_2 = 28 + 10 = 38 \text{ cm}^2$$

remember

1. Many problems such as finding a shaded area can be solved by subtracting one area from another.
2. Many composite shapes can be divided into two or more 'pieces' which have a simple area formula.

EXERCISE 8E Area of composite shapes

EXCEL Spreadsheet Area

GC program—TI Measurement

GC program—Casio Measurement

WORKED Example 7

1 Find the area of the following **composite shapes**.

a

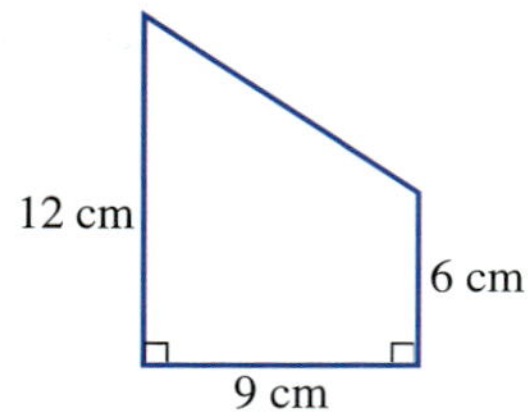

b

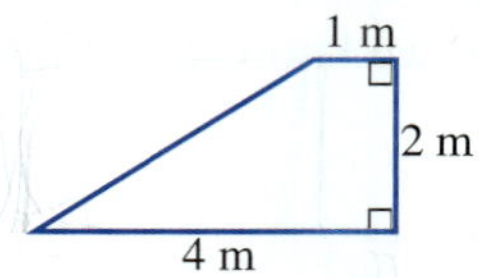

c

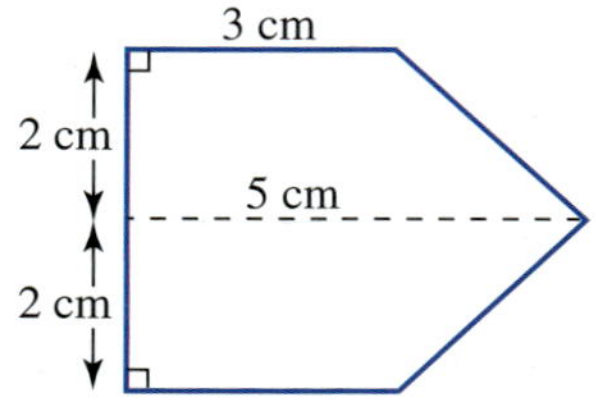

d

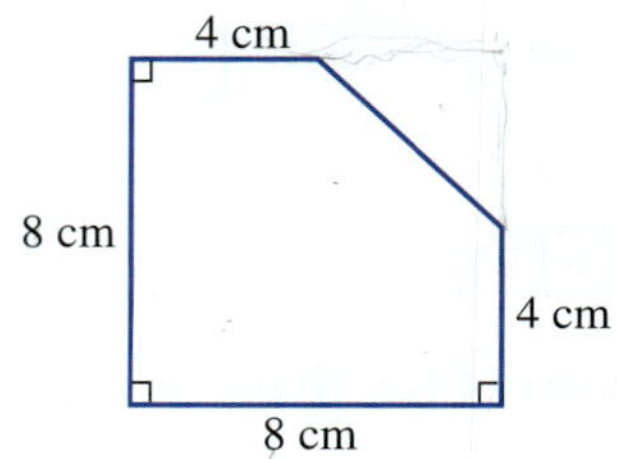

e

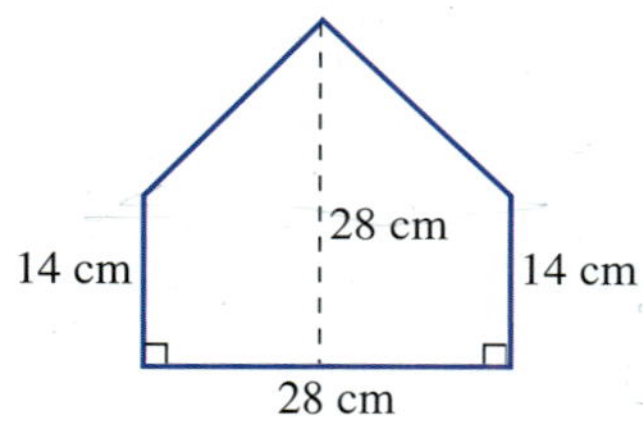

f

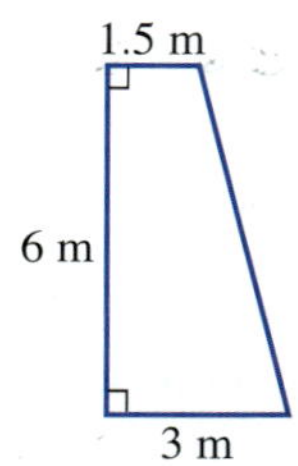

2 The sail whose measurements are shown at right has been constructed for the sailing vessel 'Nautilus'.

a Find the area of cloth required to construct the sail.

b What is the cost of the sail if the cloth costs $12.00 per square metre?

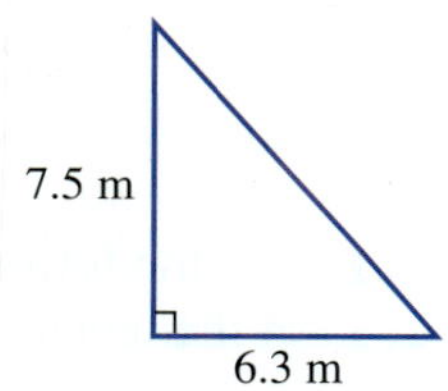

3 **a** Find the area of carpet needed for the room at right.

b What would the carpet for the room cost if it is sold for $25 per square metre?

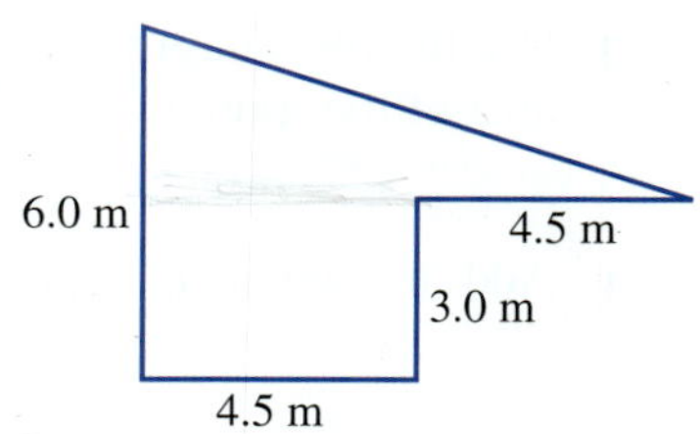

4 Find the area of the garage wall shown below.

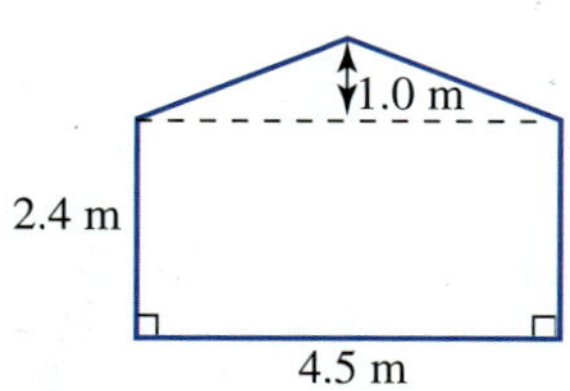

5 Determine the area of bricks needed to cover the courtyard shown below.

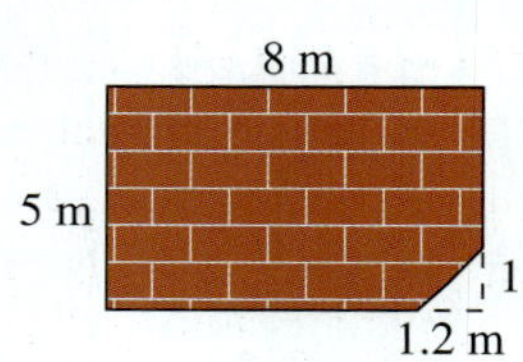

6 A sheet of A4 paper is approximately 30 cm long and 21 cm wide. Find the area of the triangle formed by cutting an A4 sheet in half diagonally.

7 Find the area of the entrance foyer shown on the right.

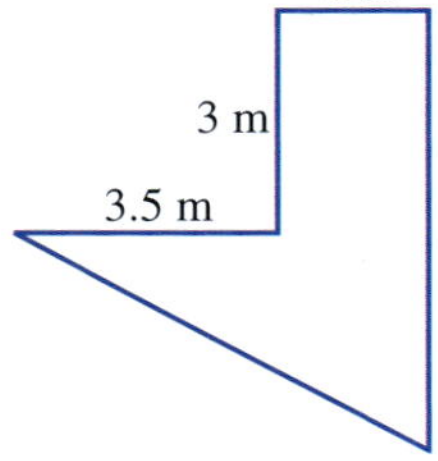

8 Michael is paving a rectangular yard, which is 15.5 m long and 8.7 m wide. A square fishpond with side lengths of 3.4 m is in the centre of the yard. What will be the cost of paving Michael's yard, if the paving material costs $17.50 per square metre?

GAME time
Area, volume and capacity — 001

Volume

The **volume** of a 3-dimensional object is the amount of space it occupies. To find the volume of a solid shape we must compare it with a standard volume unit. The commonly used units of volume are shown below.

1. Cubic centimetre (cm^3)

A cubic centimetre is the space occupied by a cube with sides of 1 cm.

A sugar cube has a volume of about 1 cm^3.

The volume of a discman is about 30 cm^3.

2. Cubic metre (m^3)

A cubic metre is the space occupied by a cube with sides of 1 m.

A bathtub can hold about $\frac{1}{2}$ m^3 of water and each of these recycling bins would hold about 1 m^3.

WORKED Example 8

How many cubic centimetres are in each solid shape? (Each small cube represents 1 cm³.)

a

b

THINK	WRITE
a 1 Count the cubes. There are 10 cubes in each layer. There are three layers altogether.	**a** Volume $= 10 \text{ cm}^3 \times 3$
2 Give the answer in cubic centimetres.	$= 30 \text{ cm}^3$
b 1 Count the cubes. There are 12 cubes in the first layer. There are 6 cubes in the second layer. There are 3 cubes in the third layer.	**b** Volume $= 12 \text{ cm}^3 + 6 \text{ cm}^3 + 3 \text{ cm}^3$
2 Give the answer in cubic centimetres.	$= 21 \text{ cm}^3$

remember

1. The volume of a 3-dimensional object is the amount of space it occupies.
2. Cubic centimetres and cubic metres are units that are commonly used to measure volume.

EXERCISE 8F Volume

SkillSHEET 8.5 Volume units

SkillSHEET 8.6 Volume of a solid constructed from cubic-centimetre blocks

WORKED Example 8a

1 How many cubic centimetres are in each solid shape? (Each small cube represents 1 cm³.)

a

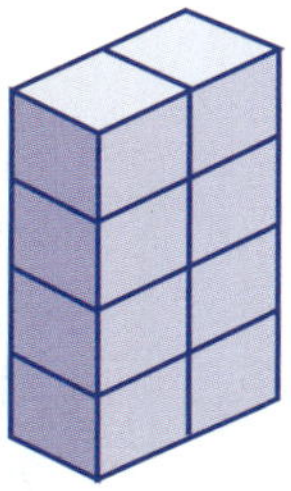

b

c

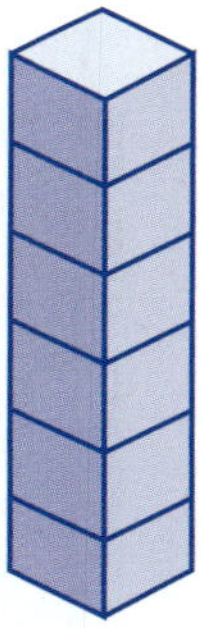

d

e

f

2 **multiple choice**

The volume of the prism shown at right is:

A 4 cm^3 B 12 cm^3 C 1 cm^3
D 8 cm^3 E 19 cm^3

WORKED Example 8b

3 Find the volume of the following solids. (Each small cube represents 1 cm^3.)

a

b

c

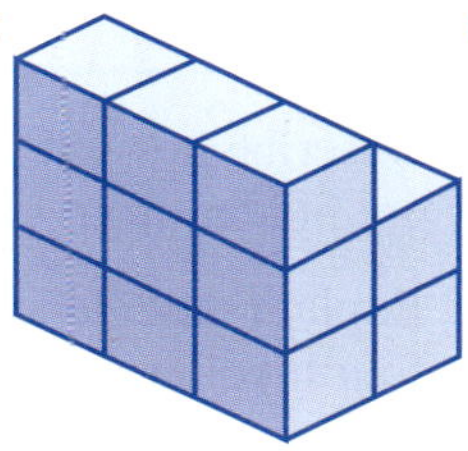

d

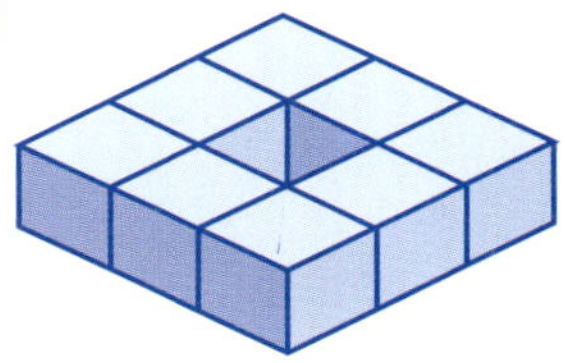

e

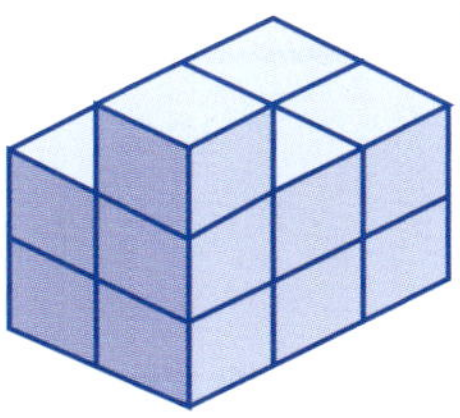

f

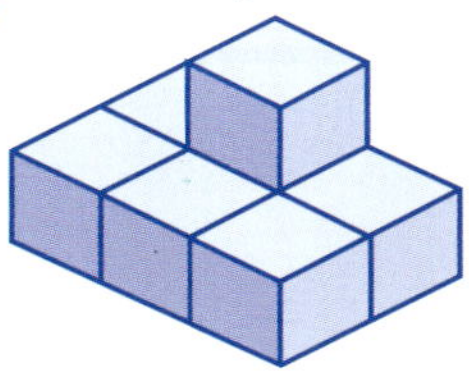

g

h

i

4 **multiple choice**

The volume of the prism on the right is:

A 20 cm^3 B 16 cm^3 C 10 cm^3
D 22 cm^3 E 3 cm^3

Note: Each cube has a volume of 1 cm^3.

THINKING Volume of a rectangular prism

The solid shapes in question 1 of exercise 8F are all rectangular prisms. A prism is a solid with identical opposite ends joined by straight sides.

1 Complete the table by looking carefully at each of the solids in exercise 8F, question 1.

Shape	Number of cubes in one layer	Number of layers	Volume (cm^3)
a			
b			
c			
d			
e			
f			

2 Describe how the number of cubes in one layer can be found even when some are hidden from view.

3 Explain the relationship between columns two and three of the table and the fourth column.

4 If you have a volume of 36 cm^3 made up of 9 layers, how many 1 cm^3 cubes are there in each layer? Explain your answer.

Finding the volume of a rectangular prism

The dimensions of a rectangular **prism** are its length (l), width (w) and height (h). The number of cubes in one layer of a rectangular prism is also equal to the area of the base of the prism. The number of layers is the height of the prism. The rule for finding the volume (V) of a rectangular prism can be found as shown below:

$$\text{Volume of a rectangular prism} = \text{Area of base} \times \text{height}$$
$$= A \times h$$
$$= l \times w \times h$$

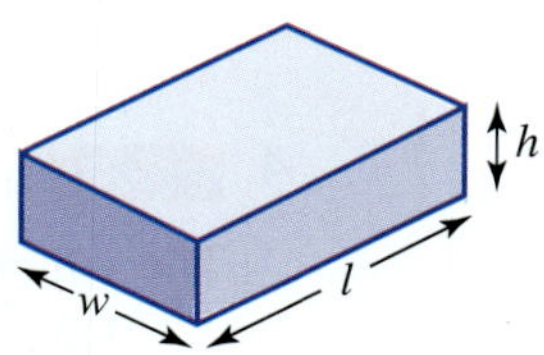

We write the formula without showing the multiplication signs as shown below.

$$V = lwh \quad \text{or} \quad V = Ah$$

WORKED Example 9

Use the formula $V = lwh$ to calculate the volume of the rectangular prism shown at right.

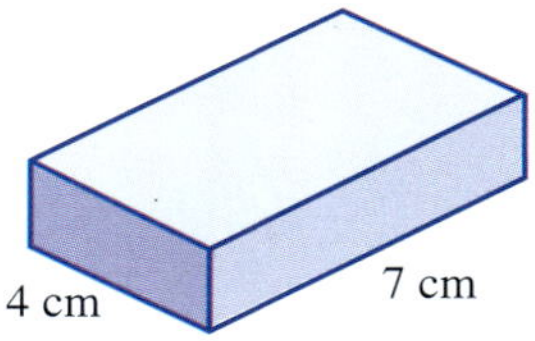

THINK	WRITE
1 Write the formula for volume.	$V = lwh$
2 Identify the length ($l = 7$ cm), width ($w = 4$ cm), and height ($h = 2$ cm) of the prism and substitute the values into the formula.	$= 7\text{ cm} \times 4\text{ cm} \times 2\text{ cm}$
3 Calculate the result. Remember to write the answer using the correct unit (cm^3).	$= 56\text{ cm}^3$

remember

Volume (V) of a rectangular prism: $V = lwh$ or $V = Ah$
where l is the length, w is the width, h is the height and A is the area of the base of the prism.

EXERCISE 8G Finding the volume of a rectangular prism

WORKED Example 9

EXCEL Spreadsheet: Volume of a rectangular prism

Mathcad: Volume of a rectangular prism

1 Use the formula $V = lwh$ to calculate the volume of the following rectangular prisms.

a

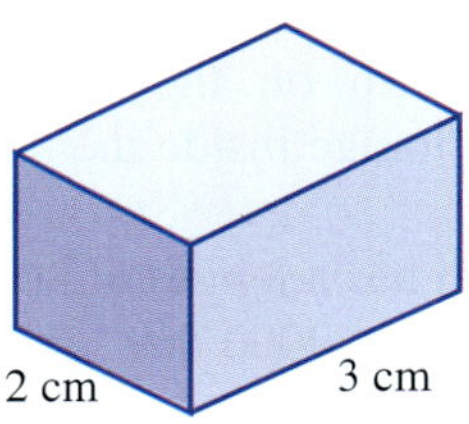

b

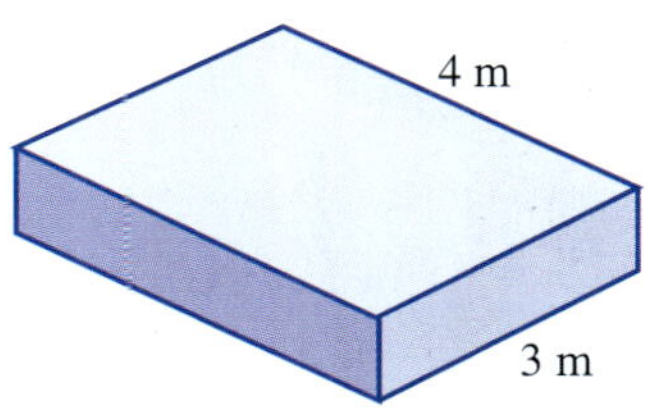

c

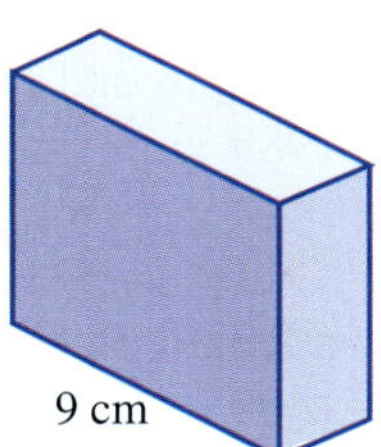

d

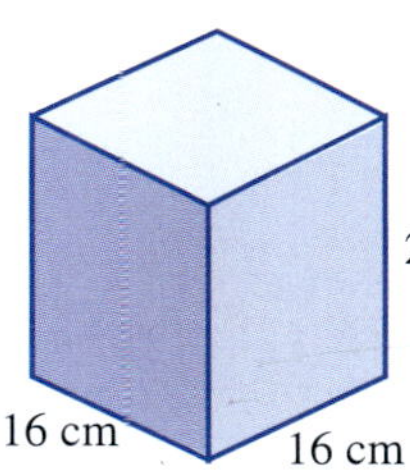

e

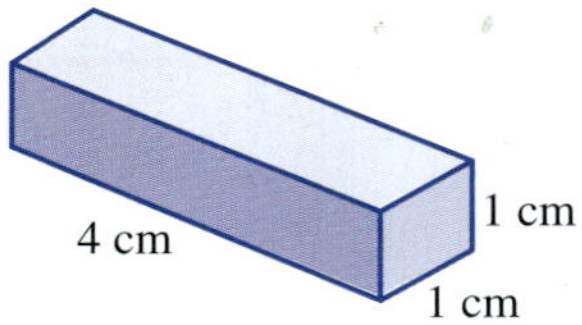

f

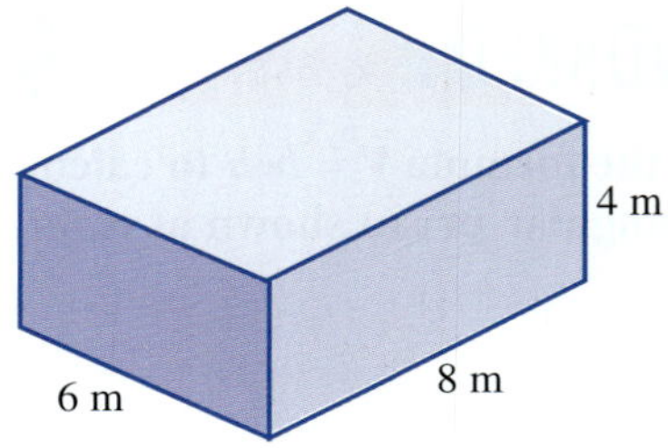

g

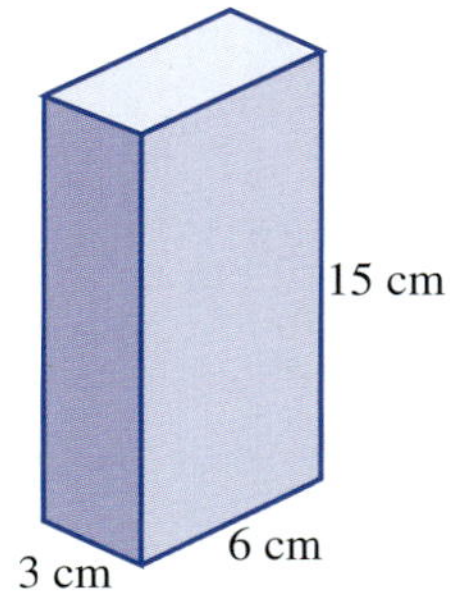

h

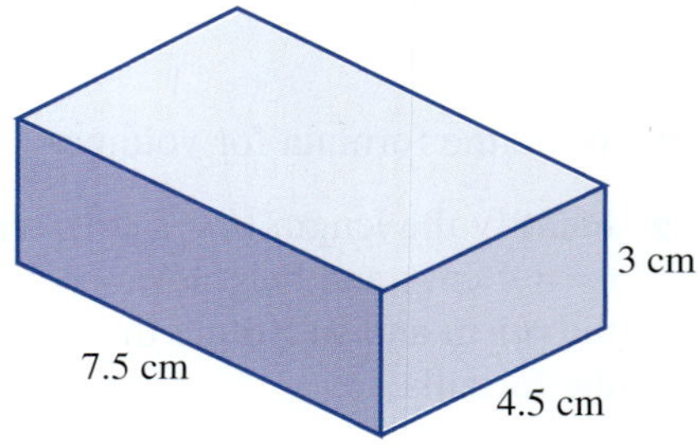

2 Find the volume of the shoe box shown at right. (Give your answer in cubic centimetres, cm^3.)

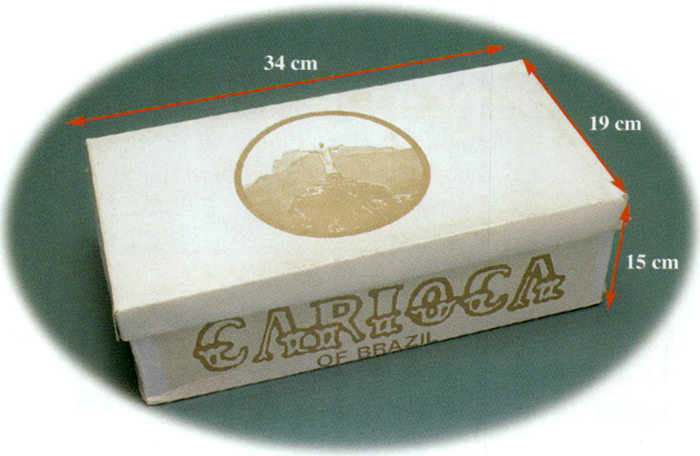

3 The inside dimensions of the 'Workinghorse' frost-free refrigerator are shown on the left. Find the volume available for food storage inside the refrigerator.

4 A rectangular prism has a length of 30 cm and a width of 15 cm. If its volume is 9000 cm^3, find the height of the rectangular prism.

5 A rectangular prism has a length of 13 cm and a width of 17 cm. If its volume is 1989 cm^3, find the height of the rectangular prism.

6 A rectangular prism has a length of 10 cm and a width of 15 cm. If its volume is 1800 cm^3, find the height of the rectangular prism.

7 The lunch box at right is a rectangular prism.

a Find the volume of the lunch box in cubic centimetres.

b Change each measurement to millimetres and hence find the volume of the lunch box in cubic millimetres.

8 Find the volume of concrete (in cubic metres) which would be needed to make the base for a garage (at right) that is 6.5 m wide and 3 m long. The concrete base is 0.25 m deep.

9 **a** Find the volume of the matchbox that has been converted to a cockroach hotel, shown at right (in cubic centimetres).

b Change each measurement to millimetres and hence find the volume of the matchbox in cubic millimetres.

c Matches are rectangular prisms of length 44 mm, width 2 mm and height 2 mm. Find the volume of a match (in cubic millimetres). (Ignore the red substance on the end of each match.)

d If a matchbox contains 50 matches, how much space is left in a full box?

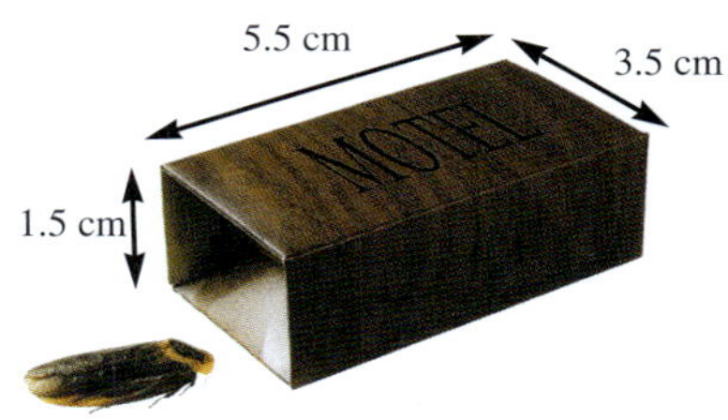

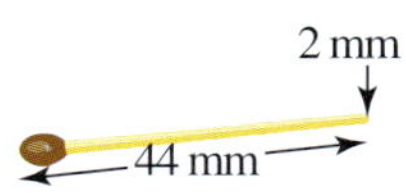

10 How many cubic metres of water would be needed to fill a diving pool which has a length of 16 m, a width of 12 m and a depth of 4 m?

11 Heather wishes to cover a rectangular lawn with topsoil to a depth of 0.1 m. If the lawn is 24 m long and 17 m wide, what volume of soil (in cubic metres) should she order?

12 The fruitcake shown at right is to be divided equally among 100 people at a wedding reception. What volume of cake (in cubic centimetres) will each guest receive?

13 Find the volume of clay in a house brick which has a length of 23 cm, a width of 11 cm and a height of 9 cm.

WorkSHEET 8.2

10 QUICK QUESTIONS 2

1 Construct a rectangle with an area of 30 cm^2.

2 Construct another rectangle that looks different from the one in question **1**, but still has an area of 30 cm^2.

3 Which rectangle (from questions **1** and **2**) has the larger perimeter?

4 Calculate the area of a triangle with base length of 12.5 cm and height of 7 cm.

5 Find the area of the shape at right.

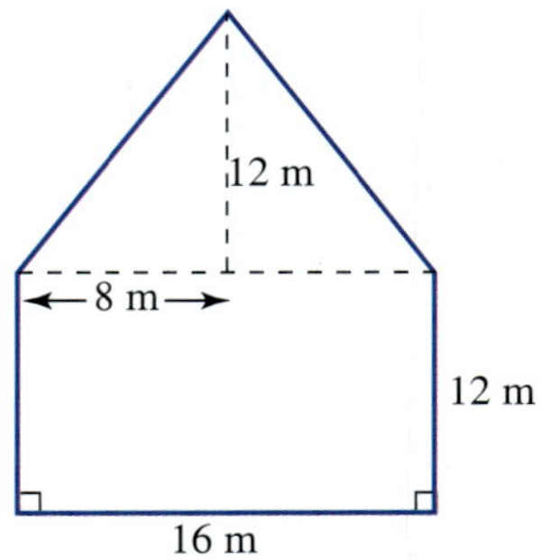

6 Find the total area of the shape below.

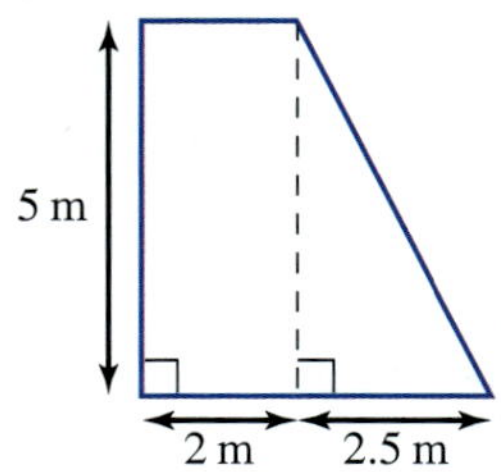

7 Find the shaded area in the figure below.

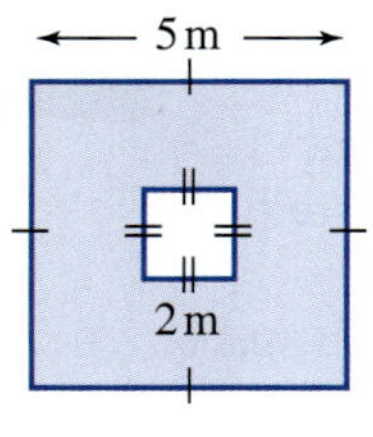

8 Find the volume of the prism at right by counting the cubes which are 1 mm^3 each.

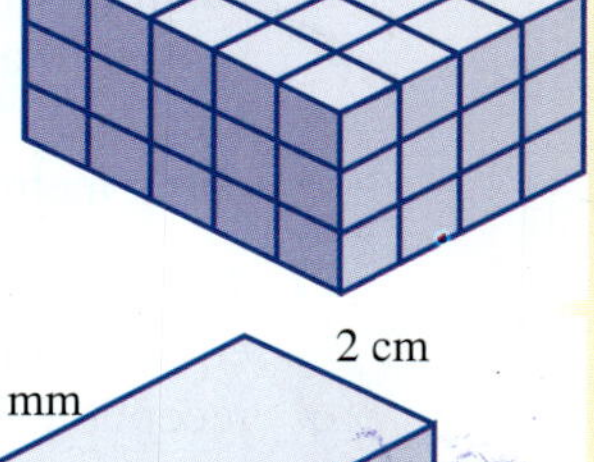

9 Find the volume of the rectangular prism at right by using the formula $V = l \times w \times h$. (Give your answer in cm^3.)

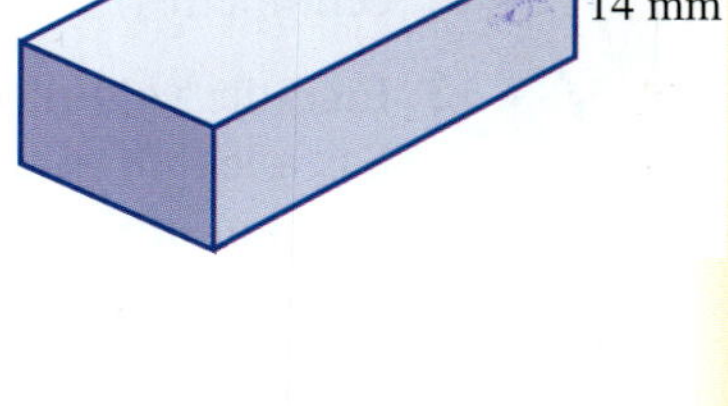

10 A new basketball stadium has just been built. It is 43 m wide, 125 m long and 12 m high. Find the volume of the new basketball stadium.

Finding the volume of other types of prisms and shapes

The rule: 'Volume = Area of base × height' which we used to find the volume of rectangular prisms, can be applied to any prism. In the diagram at right, the base of the prism has an area of 5 cm^2. The height of the prism is 3 cm. Notice that by counting cubes, the volume of the prism is 15 cm^3.

WORKED Example 10

Find the volume of the prism shown at right if each cube has a volume of 1 cm^3.

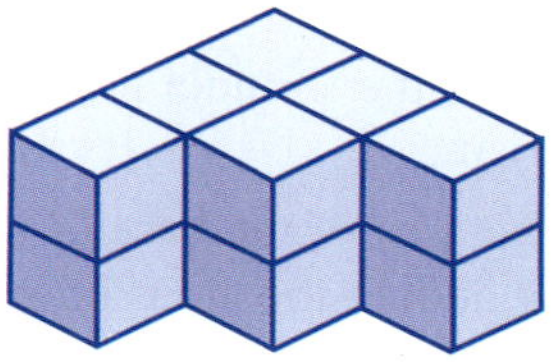

THINK	WRITE
1 Write the formula for the volume of a prism.	$V = Ah$
2 The area of the base (A) is the same as the area of the top layer which is 6 cm^2. The height is 2 cm.	$A = 6\text{ cm}^2$, $h = 2\text{ cm}$
3 Substitute the values of A and h into the formula $V = Ah$.	$V = 6\text{ cm}^2 \times 2\text{ cm}$
4 Calculate the answer and give the units in cubic centimetres.	$= 12\text{ cm}^3$

The formula $V = Ah$ works even if the prism has a base shape that is not a rectangle or square. In fact, it works whatever the shape of each end is, as long as the object has a uniform **cross-sectional area**.

WORKED Example 11

Calculate the volume of this shape.

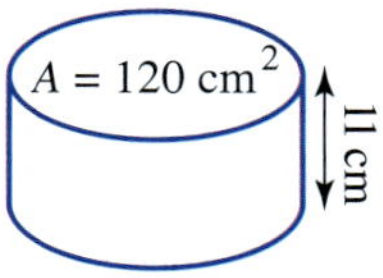

THINK	WRITE
1 Write the formula for the volume of the shape.	$V = Ah$
2 The area of the base, A, is given (120 cm^2), the height is 11 cm.	$A = 120\text{ cm}^2$ and $h = 11\text{ cm}$
3 Substitute the values of A and h into the formula $V = Ah$.	$V = 120\text{ cm}^2 \times 11\text{ cm}$
4 Calculate the answer and give the units in cubic centimetres.	$= 1320\text{ cm}^3$

WORKED Example 12

Use the information given in the diagram shown on the right to calculate the volume of the prism.

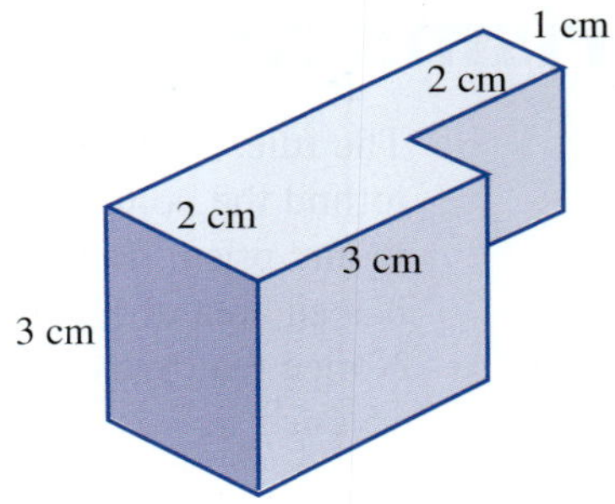

THINK	WRITE
1 Write the formula for the volume of a prism.	$V = Ah$
2 The area of the base (A) can be found by adding the areas of the 2 rectangles on the top surface. The height is 3 cm.	Area of large rectangle $= lw$ $= 3$ cm × 2 cm $= 6$ cm^2 Area of small rectangle $= lw$ $= 2$ cm × 1 cm $= 2$ cm^2 Total area of base $= 6$ cm^2 + 2 cm^2 $= 8$ cm^2 So $A = 8$ cm^2 and $h = 3$ cm.
3 Substitute the values of A and h into the formula $V = Ah$.	$V = 8$ cm^2 × 3 cm
4 Calculate the answer and give the units in cubic centimetres.	$= 24$ cm^3

remember

Volume of a prism (V) = Area of base (A) × height (h) or $V = Ah$.

EXERCISE 8H Finding the volume of other types of prisms and shapes

WORKED Example 10

1 Find the volume of each of the following prisms by copying and completing the following table. (Each cube has a volume of 1 cm^3.)

a

b

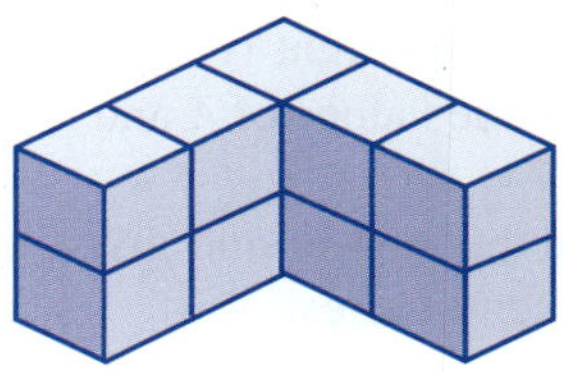

c

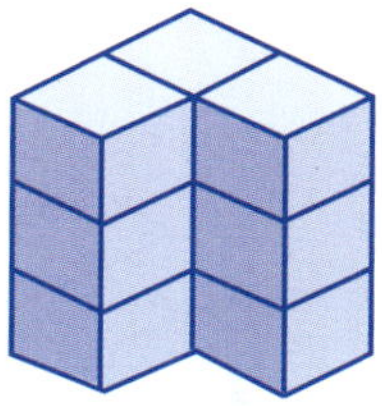

d

e

f

g

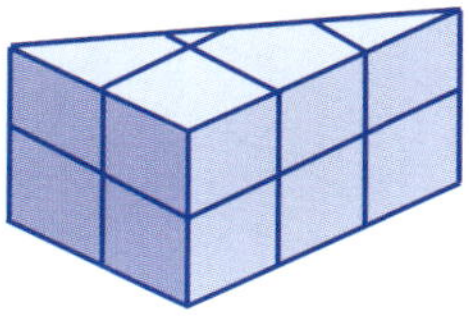

h

i

j

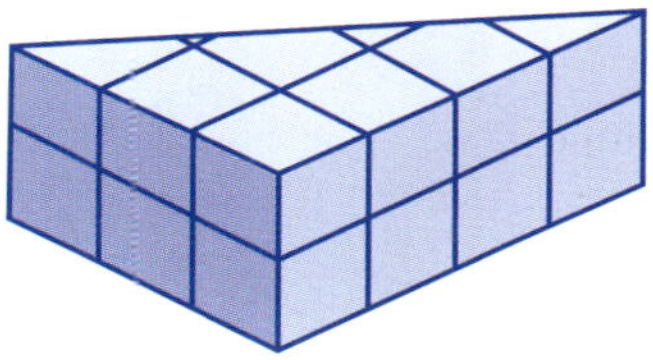

Prism	Area of base (cm^2)	Height (cm)	Volume (cm^3)
a			
b			
c			
d			
e			
f			
g			
h			
i			
j			

2 **multiple choice**

The volume of the prism shown at right (assuming each cube has a volume of 1 cm^3) is:

A 10 cm^3 **B** 20 cm^3 **C** 30 cm^3
D 40 cm^3 **E** 80 cm^3

3 **multiple choice**

The volume of the prism shown at right is:

A 16 cm^3 **B** 48 cm^3 **C** 32 cm^3
D 40 cm^3 **E** 160 cm^3

Each cube has a volume of 1 cm^3.

WORKED Example 11

4 Calculate the volume of each of the following shapes.

a

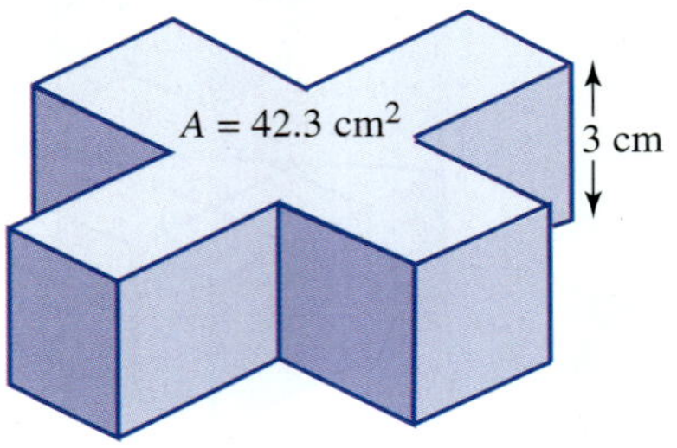

b

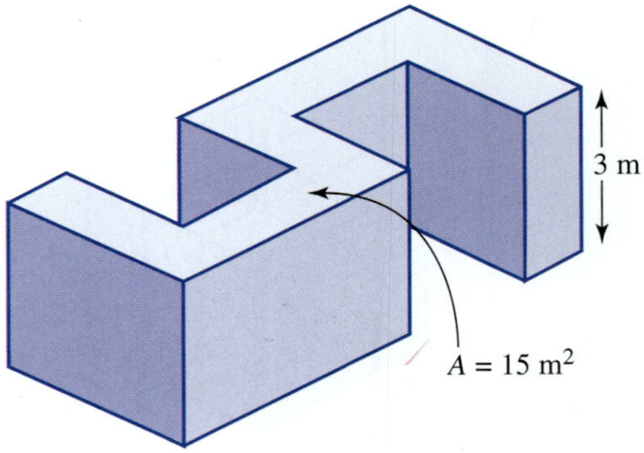

c

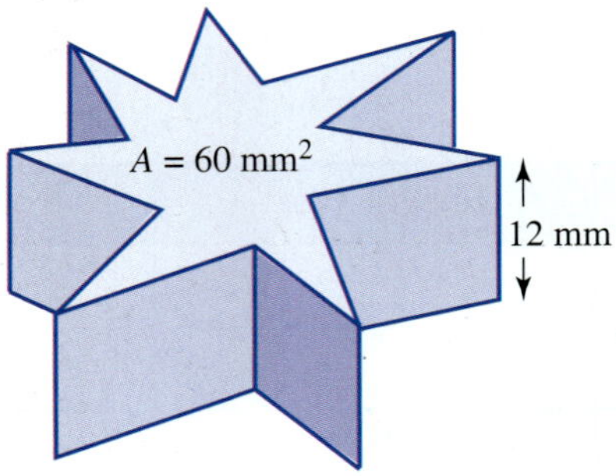

d

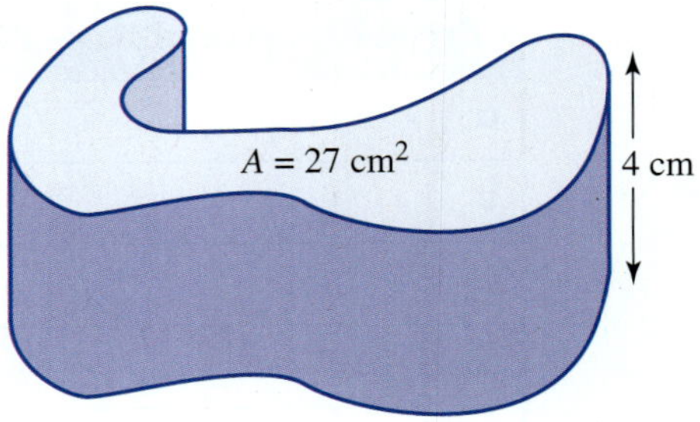

e

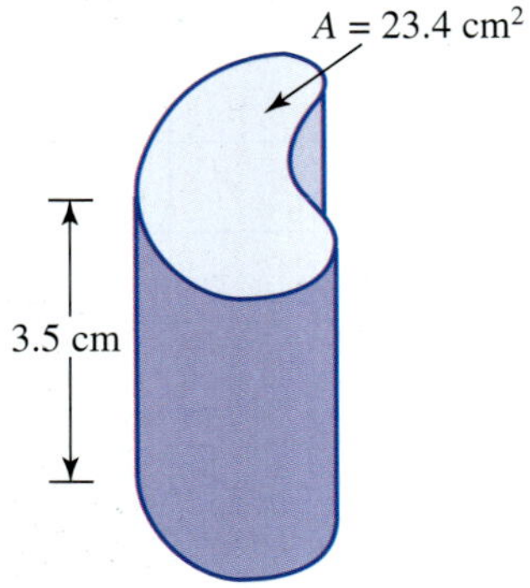

f

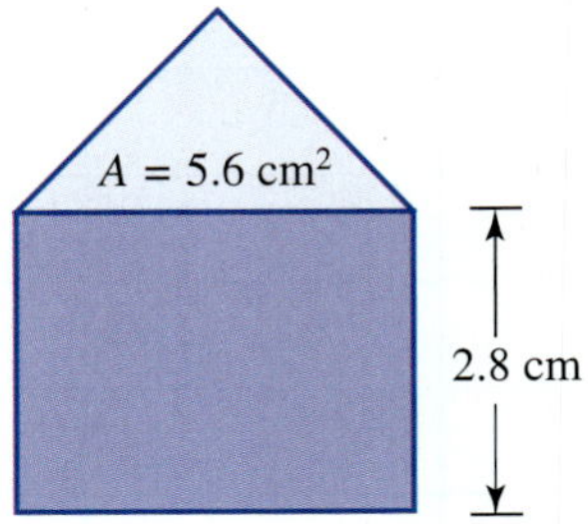

WORKED Example 12

5 Calculate the volume of the following prisms.

a

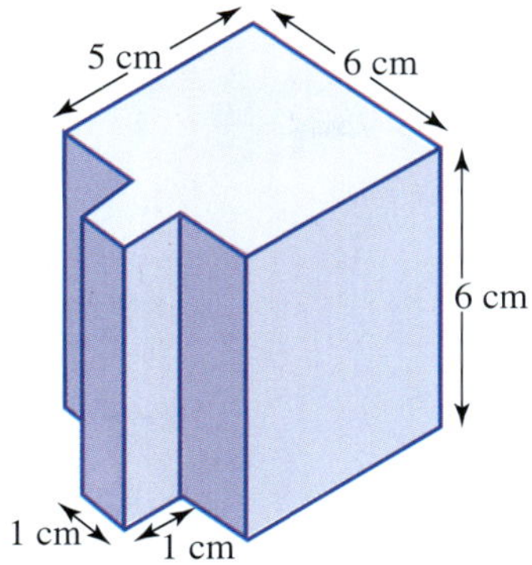

b

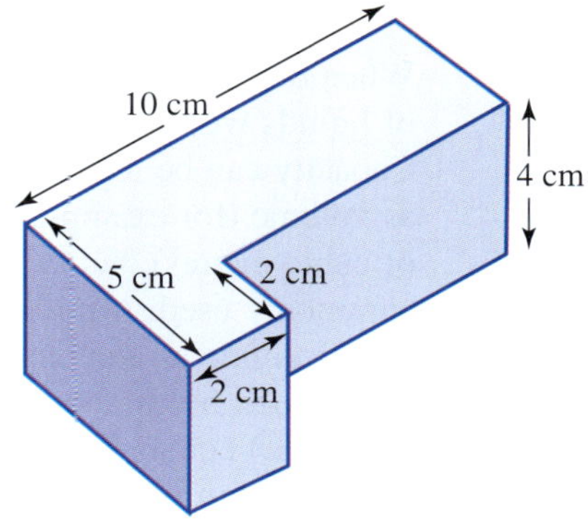

c

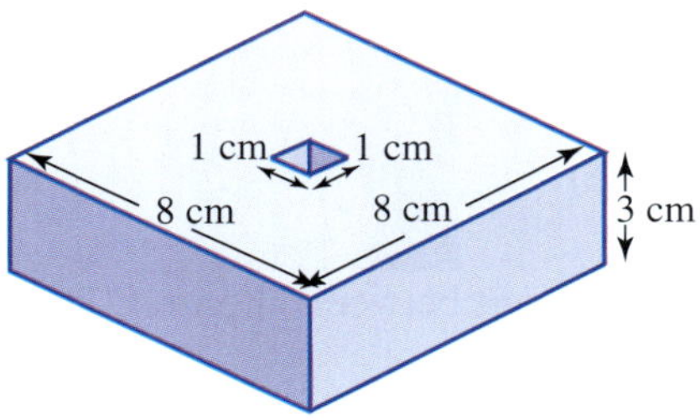

d

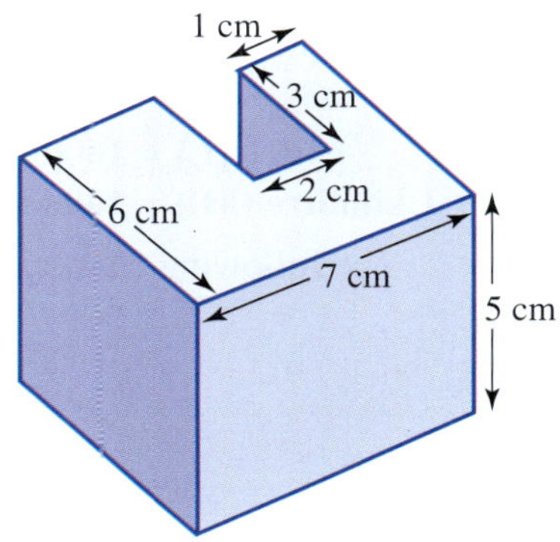

6 A swimming pool is rectangular and has a width exactly half its length. What volume of water would be needed to fill it if the swimming pool is 50 metres long and has a constant depth of 2 metres?

7 The photograph at right shows a 'round' of cheese that has a base of area 320 cm^2 when whole. If the height of the cheese is 6 cm, find the following:

a the volume of the whole cheese round

b the number of equal segments that could be obtained if the round is cut to produce segments of 160 cm^3 in volume.

Area, volume and capacity — 002

Capacity

When describing the volume of containers of liquids we often use the term **capacity**. Capacity can be measured in the same units as volume (for example, cubic centimetres or cubic metres), but two other units are commonly used. A millilitre (mL) is the same volume as a cubic centimetre (cm^3). A teaspoon holds about 5 millilitres of liquid. A litre (L) equals 1000 millilitres. A standard milk carton holds one litre. Most large plastic fruit juice containers hold 2 litres. For large volumes, the kilolitre (kL) may also be used.

1 litre (L) = 1000 millilitres (mL)
1 kilolitre (kL) = 1000 litres (L)

The following chart can be used to convert between kilolitres, litres and millilitres.

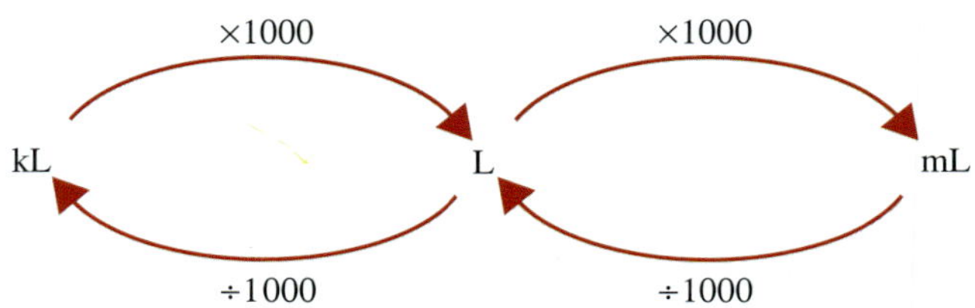

The following conversions may also come in handy.

$\mathbf{1\ cm^3 = 1\ mL}$ $\mathbf{1\,000\,000\ cm^3 = 1\ m^3}$

$\mathbf{1000\ cm^3 = 1\ L}$ $\mathbf{1\ m^3 = 1\ kL}$

WORKED Example 13

Copy and complete the following unit conversions.
a 6 L = ________ mL **b** 700 mL = ________ L **c** 0.45 L = ________ cm^3

THINK	WRITE
a Check the conversion chart. To convert litres to millilitres, multiply by 1000. Since 6 is a whole number, we do this by adding three zeros to the end of the number.	**a** $6\text{ L} = 6 \times 1000\text{ mL}$ $= 6000\text{ mL}$
b Check the conversion chart. To convert millilitres to litres, divide by 1000. To do this, move the decimal point three places to the left.	**b** $700\text{ mL} = 700 \div 1000\text{ L}$ $= 0.7\text{ L}$
c Check the conversion chart. To convert litres to millilitres, multiply by 1000. To do this, move the decimal point three places to the right. Note that $1\text{ mL} = 1\text{ cm}^3$.	**c** $0.45\text{ L} = 0.45 \times 1000\text{ mL}$ $= 450\text{ mL}$ $= 450\text{ cm}^3$

remember

1. Capacity is a term that is commonly used to describe the volume of liquids.
2. A millilitre is equivalent to 1 cm^3.
3. 1 litre (L) = 1000 millilitres (mL). 1 kilolitre (kL) = 1000 litres (L).

EXERCISE 8I Capacity

EXCEL Spreadsheet: Capacity unit conversions

Mathcad: Capacity unit conversions

WORKED Example 13a, b

1 Copy and complete the following unit conversions.

a 2 L = ________ mL
b 3000 mL = ________ L
c 13 kL = ________ L
d 7000 mL = ________ L
e 5500 mL = ________ L
f 260 L = ________ kL
g $2\frac{1}{2}$ L = ________ mL
h 32 000 mL = ________ L
i 55 mL = ________ L
j 0.035 L = ________ mL
k 420 L = ________ mL
l 0.99 kL = ________ L
m $1\frac{87}{100}$ L = ________ mL
n 22 500 mL = ________ L
o 350 mL = ________ L
p $\frac{1}{10}$ L = ________ mL
q 25 L = ________ kL
r 0.007 L = ________ mL

WORKED Example 13c

2 Copy and complete the following unit conversions.

a 750 cm^3 = _____ mL
b 2500 m^3 = _____ kL
c 800 cm^3 = _____ mL = _____ L
d 40 000 cm^3 = _____ mL = _____ L
e 6 L = _____ mL = _____ cm^3
f 5200 L = _____ kL = _____ m^3
g 2.45 L = _____ cm^3
h 78 000 cm^3 = _____ L

3 **multiple choice**

A capacity of 25 L is equal to:

A 0.025 mL B 250 mL C 0.25 mL D 25 000 mL E 2500 mL

4 **multiple choice**

A capacity of 35 400 mL is equal to:

A 35 400 000 L B 0.354 L C 3.5400 L D 35.4 L E 35 L

5 Arrange in order from smallest to largest:

a $2\frac{1}{2}$ L, 25 000 mL, $\frac{1}{4}$ L, 2.45 L

b 760 mL, 0.765 L, 7.65 mL, 7.60 L

c 110 mL, 0.1 L, 0.011 L, 1.1 L

6 A water bottle has a capacity of 2 litres. How many 125-mL bottlefuls are required to fill it?

7 A bottle contains 250 mL of orange juice concentrate. How much water should be added to make up 2 L of juice from the concentrate?

8 A scientist dilutes ('waters down') an acid solution by adding 120 mL of acid to 1.5 L of water. How much of the diluted solution will this make?

9 Most sports drinks are sold in 500-mL bottles. How many litres of sports drink are there in one dozen such bottles?

10 A medicine bottle contains 125 mL of cough syrup. How many 2.5-mL doses could be administered from this bottle, assuming that none is spilt?

11 Anthea runs a market stall selling detergent. How many 200-mL bottles of detergent could she fill from a 45-L bulk container?

12 A 185-mL container of 'Shine' hair conditioner is sold at the special price of \$3.70. A 0.5-L container of the same conditioner costs \$11.00. Which is the better buy? (*Hint:* Find the cost of 1 mL of hair conditioner in each case.)

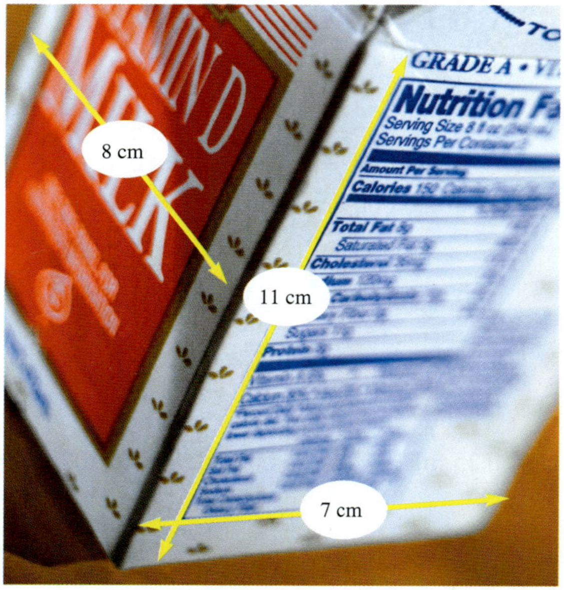

13 A milk bar sells 55 small bottles of lemon drinks in one week. How many litres of drink is sold if each bottle contains 180 mL?

14 Petrov is working as a school laboratory technician. How many litres of salt solution should Petrov prepare for an experiment in which there are 12 groups of students if each group requires 400 mL of solution?

15 How many millilitres of milk does the container on the left hold?

16 Claire has made her favourite green cordial with 0.25 L of water and 30 mL of cordial. Will a glass with a capacity of 250 mL hold this volume?

17 Sharmila has made a fresh juice with 0.3 L of apple juice and 40 mL of carrot juice. Will a glass with a capacity of 350 mL hold this volume?

18 Water is to be poured into this fish tank. How many litres of water are needed to fill the tank to a depth of 28 cm?

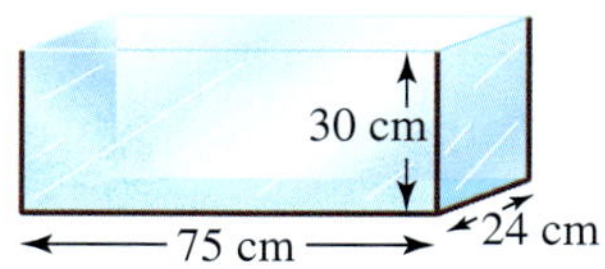

19 Estimate how many litres of water each sink will hold if filled to the top. Use the measurements shown. (*Hint:* First convert the measurements to centimetres.)

20 To achieve a world record, a 16.4-kL strawberry milkshake was made in the UK in 1996. To understand how large this is, consider the size of a rectangular prism that would have this capacity. Suggest three possible sets of measurements for this container.

21 One litre of paint covers an area of 20 square metres. How many millilitres of paint are required to cover an area of 6 square metres?

WorkSHEET 8.3

MATHS QUEST CHALLENGE

1 What is the perimeter of a square that has the same area as a rectangle with length 18 mm and width 8 mm?

2 Ivan cut up a rectangular slab of fudge by first making 4 equally spaced parallel vertical straight cuts and then making 4 equally spaced horizontal parallel straight cuts. How many pieces of fudge were produced?

3 A wooden cube with the length, width and height all measuring 3 cm is painted blue on all of its six faces. It is then cut into 27 separate cubes each of length, width and height of 1 cm.

a How many small cubes have 3 faces painted blue?

b How many small cubes have 2 faces painted blue?

c How many small cubes have 1 face painted blue?

d How many small cubes have no paint on any of their faces?

THINKING

Around the house

At the start of the chapter, you were asked to consider the cost of carpeting a lounge room using the measurements on the house plan shown below. Many other useful measurements can be calculated from the plan.

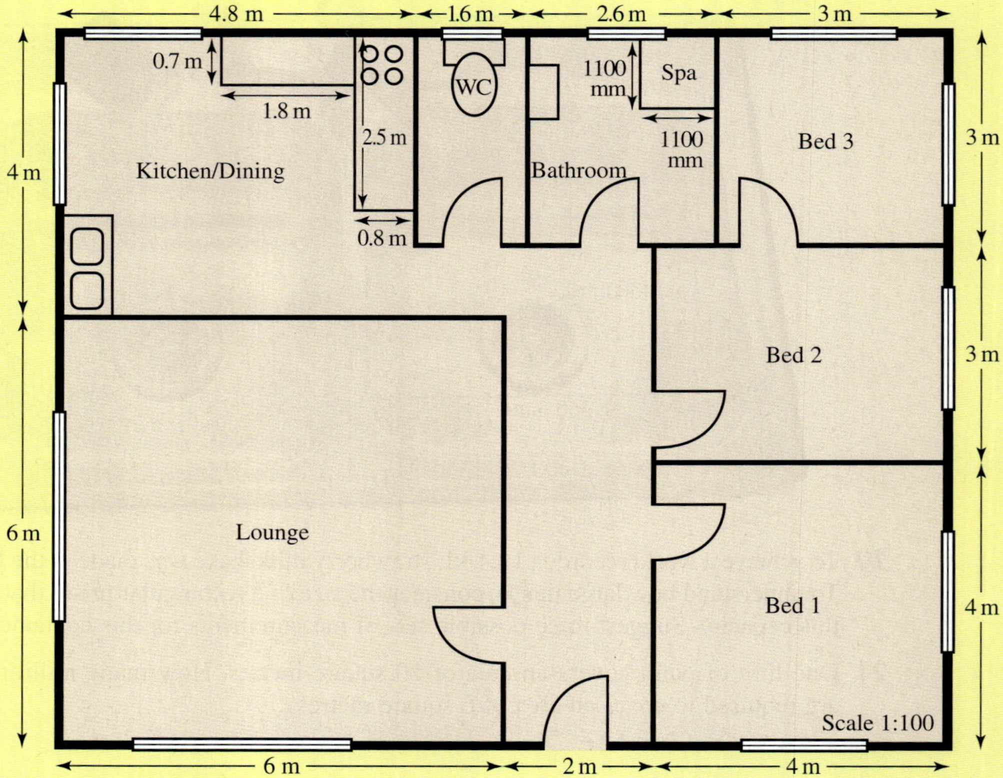

1 Calculate the area of the lounge room from the plan.

2 How much will it cost to have the lounge room carpeted if the cost of the carpet is $29 per square metre fully laid?

3 If the bedroom floors are also to be laid with carpet, calculate the total cost for the job.

4 The remaining floor surface is to be covered with square timber tiles which come in packs of 20. Each tile has a length of 12 cm. Estimate how many packs of tiles will be needed to cover the remaining floor surface.

5 The kitchen benches are to be laid with a laminated material. What area of material will be needed to cover the whole surface including the area where the hot plates will be located?

6 The house is to be built on a concrete slab. If the slab is to be 45 cm thick, calculate the volume of concrete that will be needed.

7 A spa bath is to be located in the bathroom. If it can be filled with water to a depth of 35 cm, how many litres of water will it hold?

summary

Copy the sentences below. Fill in the gaps by choosing the correct word or expression from the word list that follows.

1 ____________ is the amount of flat surface enclosed by a shape.

2 Two metric units of area are ____________ and ____________.

3 The formula $A = lw$ is used to find the area of a ____________.

4 To convert square kilometres to square metres you need to ____________ by ____________.

5 To convert square centimetres to square metres you need to ____________ by ____________.

6 One ____________ is equal to ten thousand square metres.

7 The area of a triangle is ____________ that of a rectangle with the same base and height.

8 The ____________ of a three-dimensional object is the amount of space that it occupies.

9 Two metric units of volume are ____________ and ____________.

10 The volume of a ____________ prism is given by the formula $V = lwh$ or $V = Ah$.

11 The volume of any prism can be found by multiplying the area of its ____________ by its ____________.

12 ____________ is the term used when describing the volume of liquids.

13 Three metric units of capacity are ____________, ____________ and ____________.

WORD LIST

height	base	divide
10 000	half	millilitres
multiply	volume	rectangle
cubic metres	square centimetres	1 000 000
litres	rectangular	hectare
area	cubic centimetres	kilolitres
square metres	capacity	

CHAPTER review

8A **1** Which unit of area would be most appropriate for measuring the following? (Choose from mm^2, cm^2, m^2, ha or km^2.)

- **a** The floor area of your classroom
- **b** The area of the state of Victoria
- **c** A pin head
- **d** A market garden
- **e** A floor tile
- **f** A bench top
- **g** The floor area in a house
- **h** The area of the continent of Antarctica

8A **2** On a sheet of graph paper, draw five different figures each having an area of $12\ cm^2$.

8A **3** Find the area of the following figures, which are drawn on centimetre grid paper.

a

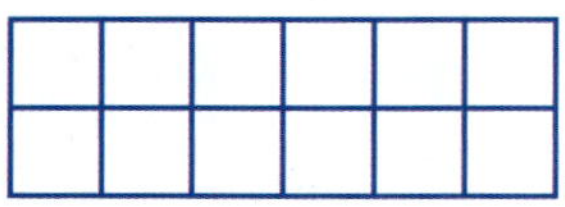

b

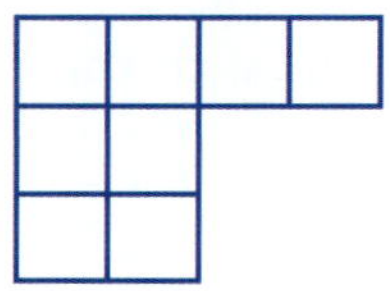

c

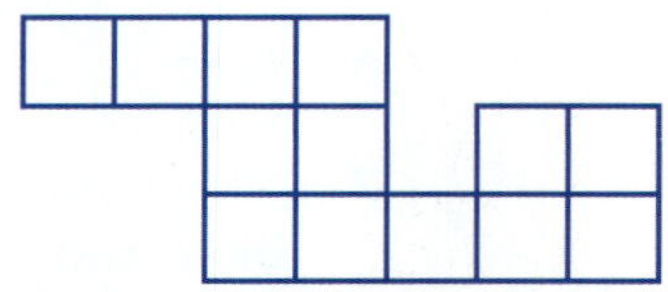

d

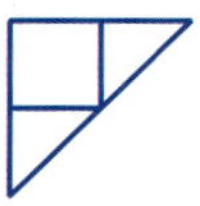

e

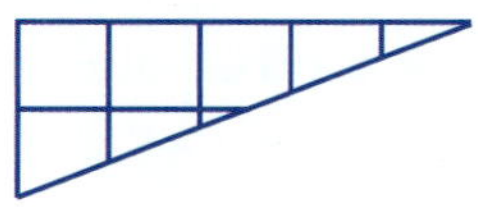

f

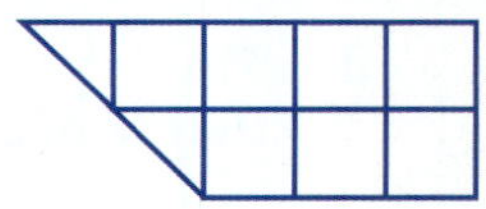

g

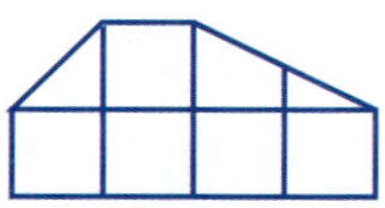

h

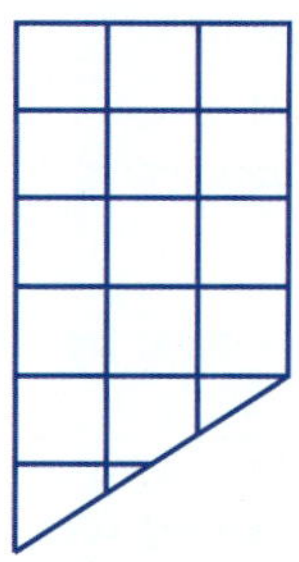

8A **4** Estimate the shaded area in each of the figures below if each is drawn on centimetre grid paper. *Note:* Each square on the grid has an area of $1\ cm^2$.

a

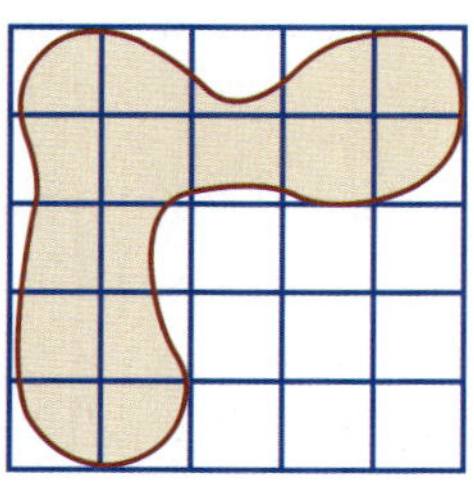

b

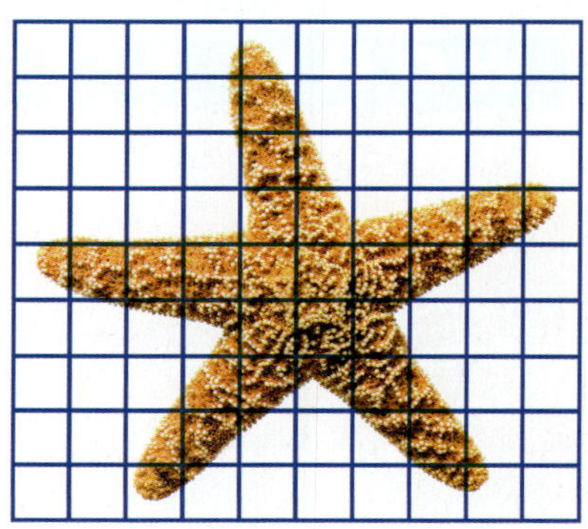

5 Complete the following unit conversions. 8B

a $56\ 000\ \text{cm}^2$ = ________ m^2 **b** $50\ 000\ \text{mm}^2$ = ________ cm^2

c $0.3\ \text{ha}$ = ________ m^2 **d** $25\ 300\ \text{m}^2$ = ________ ha

e $84\ 000\ 000\ \text{m}^2$ = ________ km^2 **f** $5.14\ \text{cm}^2$ = ________ mm^2

g $0.0034\ \text{m}^2$ = ________ cm^2 **h** $1.25\ \text{ha}$ = ________ m^2

6 Use the formula $A = lw$ to find the area of the following rectangles. 8C

a

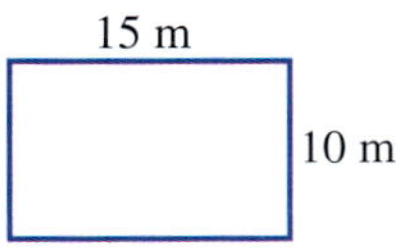

b

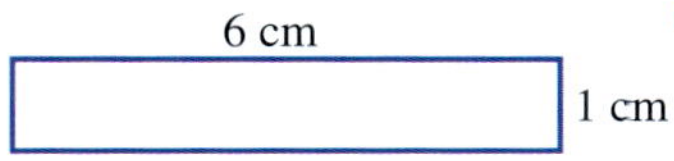

c

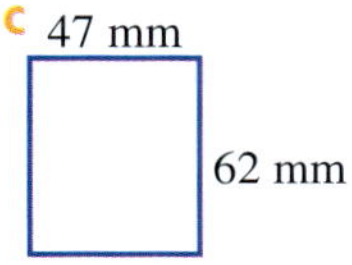

7 Measure the lengths of the sides in centimetres and hence calculate the area of each rectangle. 8C

a

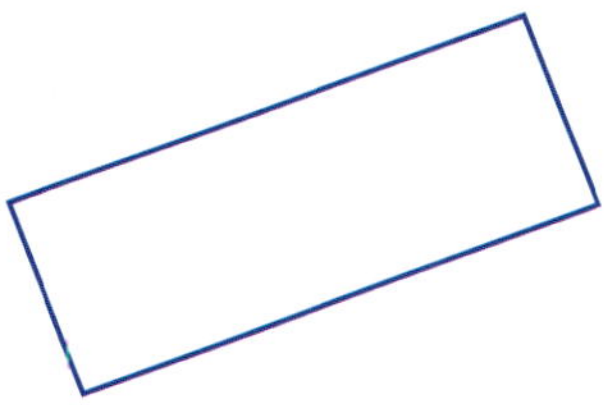

b

c

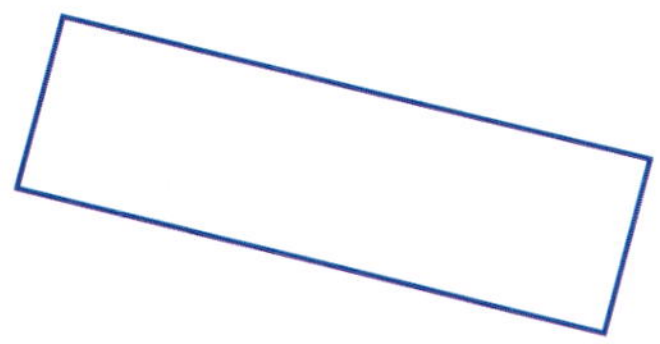

8 Find the area of the following shapes by first dividing them into rectangles. 8C

a

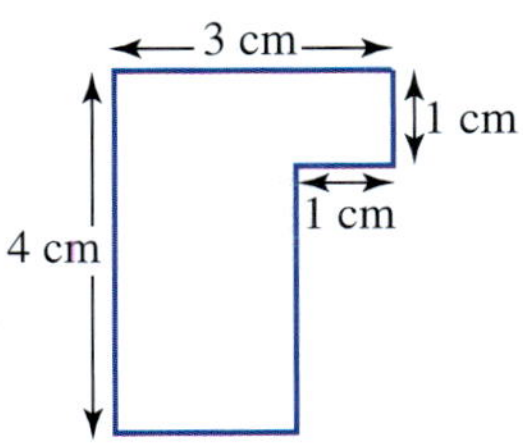

b

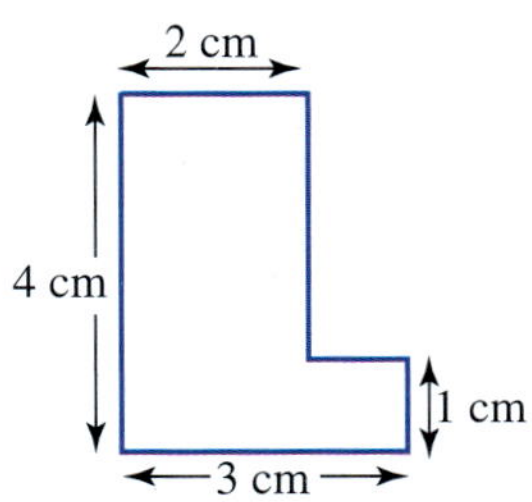

c

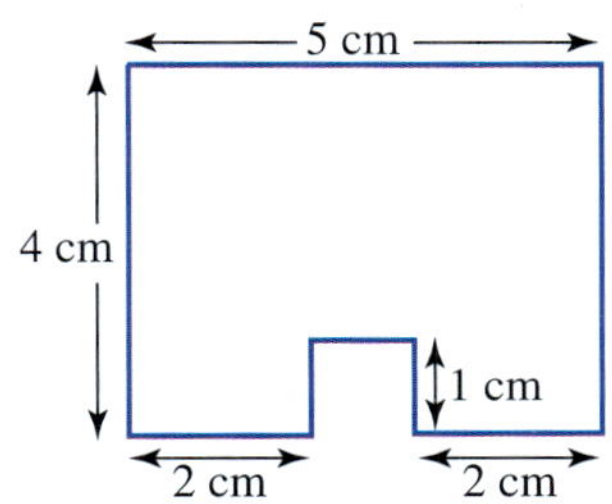

d

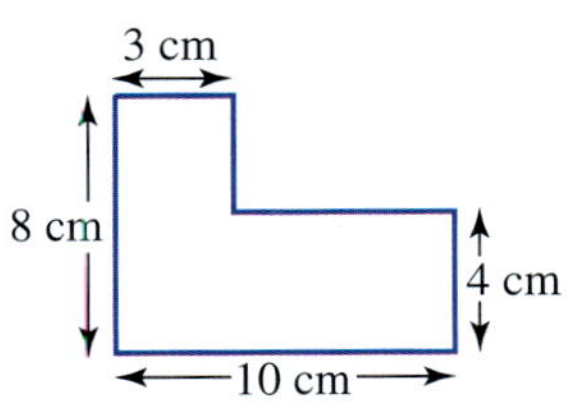

e

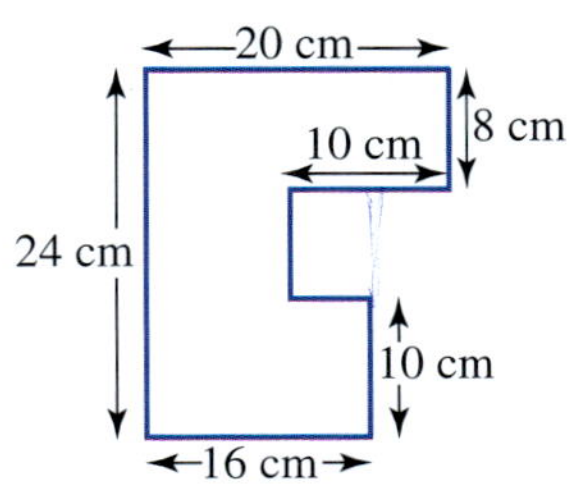

f

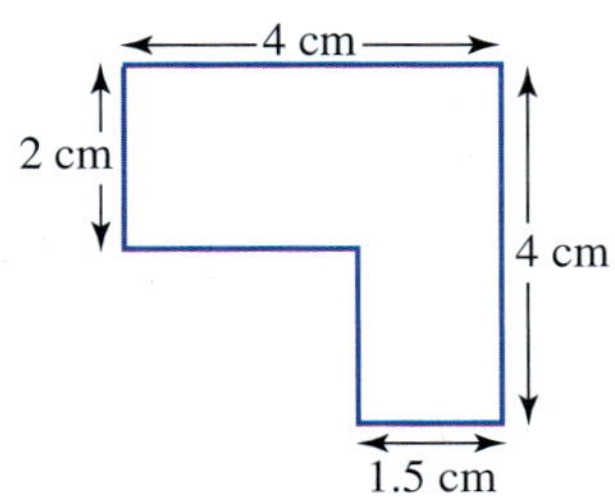

9 Find the area of a rectangular newspaper sheet which is 41 cm long and 30 cm wide. 8C

10 Find the cost of paving a rectangular courtyard which is 6.5 m long and 3.2 m wide. The courtyard is to be paved with concrete paving blocks, which cost \$28 per square metre. 8C

8D **11** Find the area of card needed to make the triangular display sign shown below and to the left.

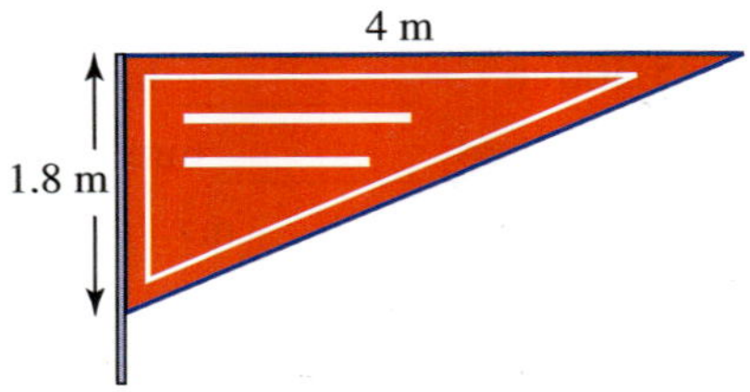

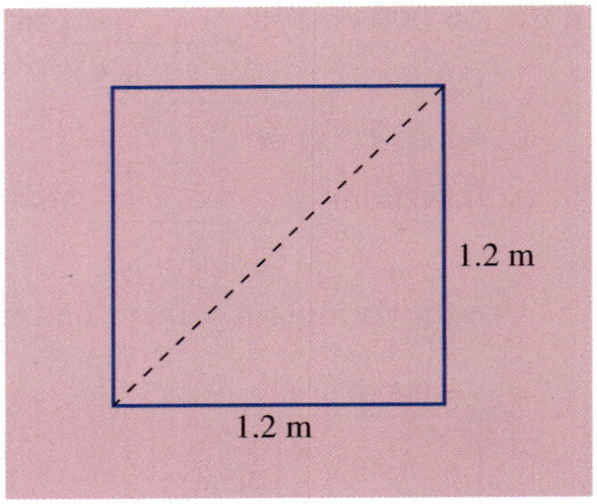

8D **12** When making a box kite, Katie cuts a square piece of material diagonally, as shown by the dotted line. What is the area of each triangular piece of material?

8E **13** Find the area of the following triangles.

a 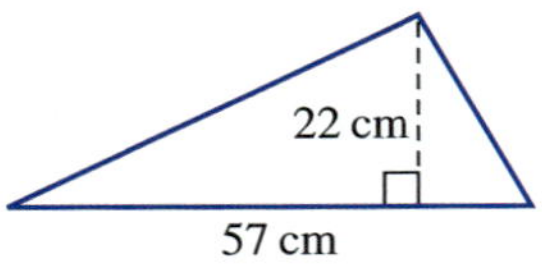

b 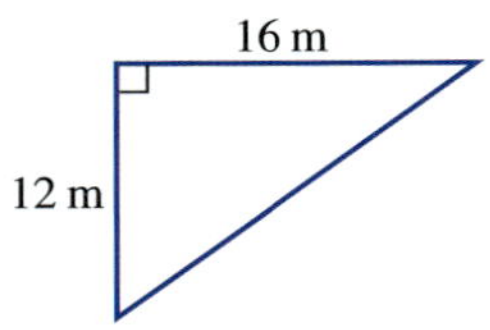

c 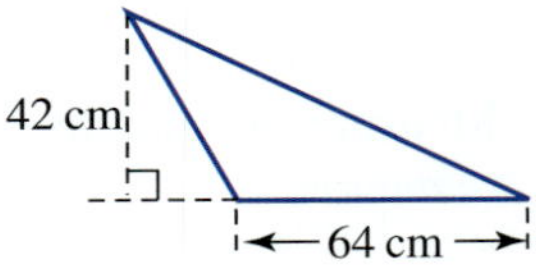

8E **14** Find the area of the following shapes.

a 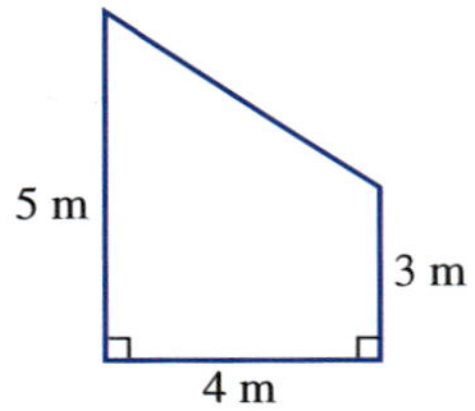

b 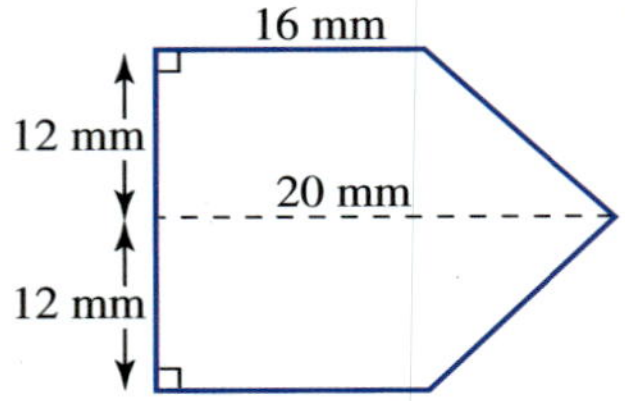

c 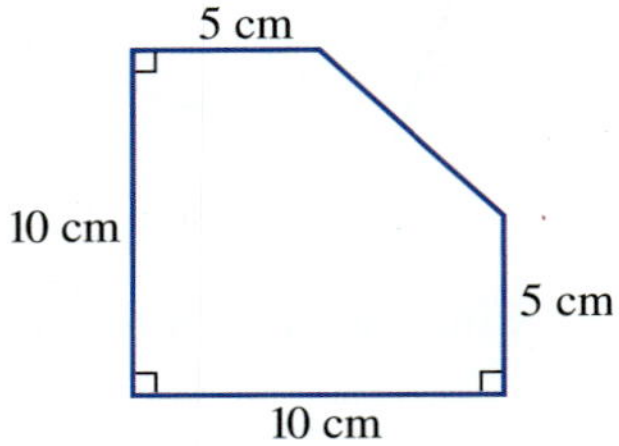

8F **15** How many cubic centimetres (cm^3) are in each shape below? Each cube has a volume of 1 cm^3.

a 

b 

c 

d 

e 

f 

16 The tip truck at right is filled with sand, level with the top of the tray. How many cubic metres of sand does the truck hold? 8G

17 Sandro wishes to cover a rectangular lawn with topsoil to a depth of 0.2 m. If the lawn is 30 m long and 24 m wide, what volume of soil (in cubic metres) should he order? 8G

18 A rectangular prism has a length of 40 cm and a width of 26 cm. If its volume is 29 120 cm^3, find the height of the rectangular prism. 8G

19 A rectangular prism has a length of 30 cm and a width of 15 cm. If its volume is 3825 cm^3, find the height of the rectangular prism. 8G

20 Use the rule 'Volume of a prism = area of base × height' to find the volume of the prisms shown below. Each cube has a volume of 1 cm^3. 8H

a

b

c

d

e

f

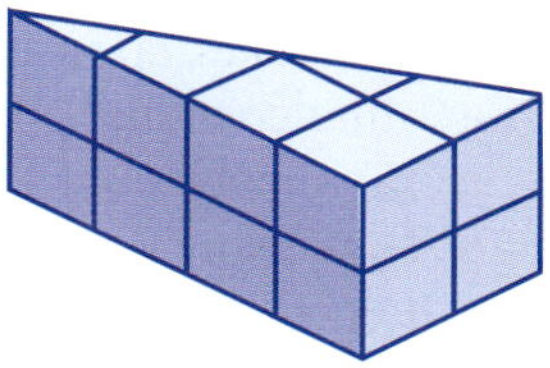

21 Use the information given in the diagrams below to find the volume of each of the following. 8H

a

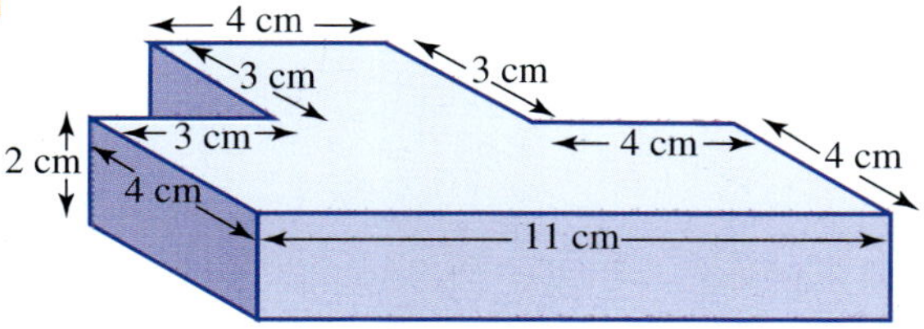

b

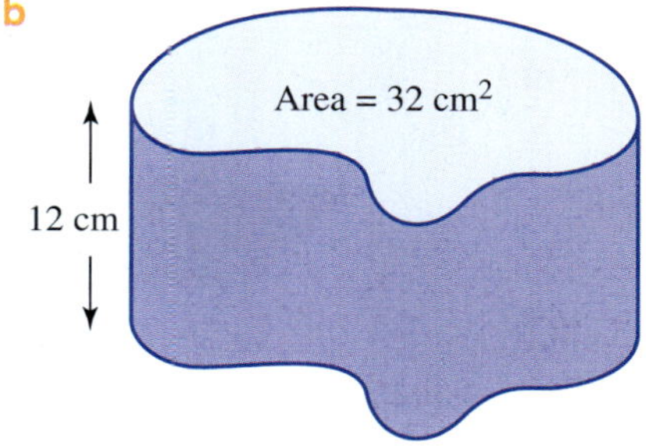

c

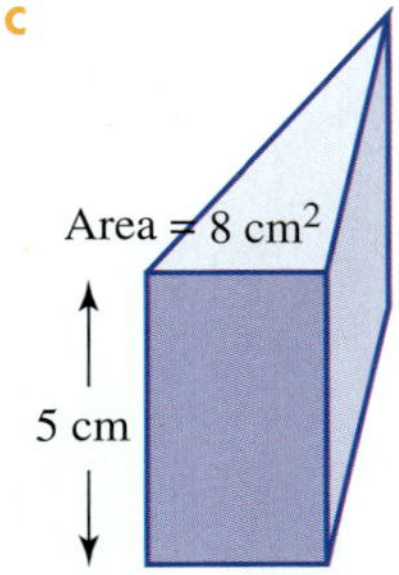

d

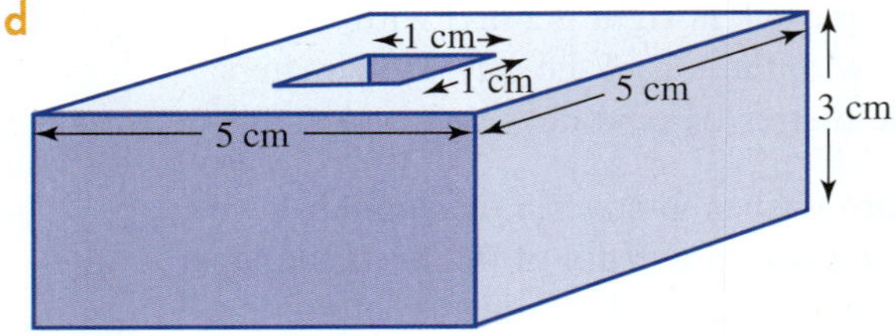

8H

22 If the base area of a compact disc case is 168 cm^2 and its height is 1 cm, find the volume of a stack of 95 compact disc cases.

8I

23 Copy and complete the following conversions.

a 8 L = ________ mL

b $\frac{21}{50}$ L = ________ mL

c 3300 mL = ________ L

d $1\frac{3}{250}$ L = ________ mL

e $4\frac{3}{10}$ kL = ________ L

f 0.0034 kL = ________ mL

g 4755 L = ________ kL

h 432 mL = ________ L

8I

24 In a food technology class, each of 14 groups of students uses 350 mL of fresh milk to make pancakes. How many litres of milk should be ordered to provide for the whole class?

8I

25 Mario makes up raspberry cordial by pouring 275 mL of concentrate into a 2-litre container and filling the container with cold water.

a How much cold water does Mario add?

b How many 250-mL glasses of cordial will Mario be able to pour from the large container?

8I

26 Estimate the number of litres of water that a goldfish bowl, similar to the one shown below, would hold.

Time, mass and temperature

9

Heather, James and Matt are looking forward to travelling with their parents to Los Angeles, USA, for a Disneyland holiday. Their flight schedule shows that they leave Melbourne at 1355 on 17 September and arrive in Los Angeles at 0925 on 17 September. What does a time of 1355 mean? How can it be possible to arrive before they leave?

In this chapter, you will learn to describe time and apply time calculations to solving practical problems like this.

Are you ready?

Try the questions below. If you have difficulty with any of them, extra help can be obtained by completing the matching SkillSHEET. Either click on the SkillSHEET icon next to the question on the *Maths Quest 7* CD-ROM or ask your teacher for a copy.

SkillSHEET 9.1

Reading the time (from analogue clocks)

1 What is the time shown on each of the following analogue clocks?

a

b

SkillSHEET 9.2

Comparing times on digital and analogue watches

2 Match the following times shown on a digital watch to that read from an analogue watch.

	Digital watch	Analogue watch
a	8.10	half past eight
b	8.45	twenty to nine
c	8.30	ten past eight
d	8.40	a quarter to nine

SkillSHEET 9.3

Calculating the time between two events

3 Calculate the time that has elapsed between these events.

a I woke at 7.30 am and arrived at school at 8.15 am.

b It started raining at 9.00 pm and stopped at 6.00 am the next day.

SkillSHEET 9.4

24-hour time

4 Match the following 12-hour times to the corresponding 24-hour times.

	12-hour time	24-hour time
a	9.00 pm	1000
b	10.00 am	2100
c	2.00 pm	1400

SkillSHEET 9.5

Drawing timelines

5 **a** Draw a timeline using an appropriate scale to show events that may occur between 8.00 am and 8.00 pm.

b Indicate on your timeline when you ate anything on the previous day.

SkillSHEET 9.6

Units of mass

6 **a** Write the abbreviation for each of the following mass units.

i kilogram **ii** gram

iii tonne **iv** milligram

b Give an example of an object that would be best measured in each of the units in part **a**.

Time — an introduction

Time is one of our most useful measurements. We can use it to work out how long we have been doing something, or how long we have until we must do something. We can use it to make future arrangements. It might be used to relate events in the past, present or future. If there were no time measurement, you would not know when to go to school, when it was lunchtime or, more importantly, when you were allowed to go home. Time is used to measure periods such as the number of weeks until your birthday, the number of weeks until the holidays, or the number of months since your last party. Time is something we all use, every day.

Time is divided into units. There are:

60 seconds in 1 minute
60 minutes in 1 hour
24 hours in 1 day
7 days in 1 week
2 weeks in 1 fortnight
about 4 weeks in 1 month
12 months in 1 year
about 365 days in 1 year
10 years in 1 decade
100 years in 1 century
1000 years in 1 millennium.

We also use the word **time** to refer to an instant (for example, 3 o'clock) rather than a period. A clock or watch can display the time in one of two ways. Some clocks display the time in analogue form using hour, minute and second hands that move continuously as they point to numbers on the clock face.

Other clocks and watches display a set of digits that change in steps as time passes.

Time calculations

WORKED Example 1

How many minutes are there in $4\frac{1}{4}$ hours?

THINK	WRITE
1 Convert the mixed number to an improper fraction.	$4\frac{1}{4}$ hours $= \frac{4 \times 4 + 1}{4}$
	$= \frac{17}{4}$ hours
2 Multiply the improper fraction by the number of minutes in 1 hour.	$\frac{17}{4}$ hours $= \frac{17}{4} \times 60$ minutes
	$= \frac{1020}{4}$ minutes
3 Evaluate.	$= 255$ minutes

WORKED Example 2

How many minutes are there in 5 days?

THINK

1. Convert the number of days to hours, that is, multiply 5 by 24 hours.
2. Evaluate.
3. Convert the number of hours to minutes, that is, multiply 120 by 60 minutes.
4. Evaluate.

WRITE

5 days = 5×24 hours
= 120 hours
120 hours = 120×60
= 7200 minutes

WORKED Example 3

Change the following into hours and minutes.
a 300 minutes b 425 minutes

THINK

a 1. Convert minutes to hours, that is, divide 300 by 60.
2. Evaluate.

b 1. Convert minutes to hours, that is, divide 425 by 60.
2. Evaluate.
3. Write the answer in hours and minutes.

WRITE

a 300 minutes = $300 \div 60$
= 5 hours

b 425 minutes = $425 \div 60$
$= \frac{425}{60}$
= 7 remainder 5
425 minutes = 7 hours 5 minutes.

The suffixes am and pm are used to indicate morning and afternoon, and are derived from 'ante meridiem' (meaning before midday), and 'post meridiem' (after midday).

12 noon may be written as 12.00 pm, and 12 midnight may be written as 12.00 am.

One way to find time differences is to mentally calculate how long it is to the next hour, then to the next 12.00 noon or 12.00 midnight. The following worked example illustrates this technique.

WORKED Example 4

Find the time difference between:

a 7.15 pm and 10.25 pm **b 2.50 am and 8.20 pm**

c 3.40 am on Tuesday and 5.10 pm the following Wednesday.

THINK

a 1 Construct a vertical column starting at 7.15 pm and ending at 10.25 pm. Set up two columns with the headings 'hours' and 'minutes' next to the first time.

2 Write 'key' times between the start and finish times.

3 Calculate the times added to get each key time, and write them under the hours and minutes headings as shown.

4 Add the time differences to find the total time difference.

5 Convert any minute values over 60 to hours and minutes and adjust the answer.

6 Answer the question.

b Repeat steps 1 to 6 in part **a**.

c Repeat steps 1 to 6 in part **a**.

WRITE

a

	Hours	Minutes
7.15 pm		
8.00 pm		45
10.25 pm	2	25
Total time	2	70
	= 3	10

The time difference is 3 hours 10 minutes.

b

	Hours	Minutes
2.50 am		
3.00 am		10
12.00 noon	9	00
8.20 pm	8	20
Total time	17	30

The time difference is 17 hours 30 minutes.

c

	Hours	Minutes
3.40 am Tuesday		
4.00 am Tuesday		20
4.00 am Wednesday	24	00
12.00 noon Wednesday	8	00
5.10 pm Wednesday	5	10
Total time	37	30

The time difference is 37 hours 30 minutes.

remember

When finding time differences:

1. use a vertical column
2. work to the next key time (exact hours, 12.00 am and 12.00 pm) until the final time is reached
3. 12 noon is 12 pm and 12 midnight is 12 am.

EXERCISE 9A Time calculations

1 Match the following activities with the most likely amount of time.

	Activity	Time
a	a cricket test match	1 year
b	writing your name	10 seconds
c	eating breakfast	15 minutes
d	building a house	5 days
e	flying time from Melbourne to Hong Kong	7 hours
f	being in Year 7	6 months

2 What are the times shown on each of the following analogue clocks?

SkillSHEET 9.1 Reading the time (from analogue clocks)

a

b

c

d

e

f

3 For each of the following, draw a 12-hour clock face and show the time.

a 8.20 am	b 8.45 pm	c 10.50 am
d 12.00 midnight	e 11.05 pm	f 5.11 pm
g 7.32 pm	h 9.24 am	i 11.16 am

9.2 SkillSHEET Comparing times on digital watches

4 Today we use clocks and watches to tell the time. Name two other devices that have been used in the past to measure time.

5 Explain the difference between 2 hours 25 minutes and 2.25 hours.

WORKED Example 1, 2

6 How many minutes are there in each of the following periods?

a 2 hours	b $2\frac{1}{2}$ hours	c $3\frac{1}{4}$ hours	d $\frac{1}{2}$ hour
e $\frac{3}{4}$ hour	f $7\frac{3}{4}$ hours	g 1 day	h 9 days

Mathcad Converting units of time

WORKED Example 3

7 Change the following to hours and minutes.

a 200 minutes	b 185 minutes	c 160 minutes
d 230 minutes	e 405 minutes	f 95 minutes
g 610 minutes	h 72 minutes	i 305 minutes

8 Change the following to minutes.

a 1 hour 15 minutes	b 2 hours 10 minutes	c 1 hour 50 minutes
d 4 hours 25 minutes	e 7 hours 35 minutes	f 3 hours 12 minutes
g $16\frac{3}{4}$ hours	h $5\frac{1}{2}$ hours	i $6\frac{1}{4}$ hours

9 How many:

a days in a week?	b weeks in a year?	c seconds in a minute?
d seconds in an hour?	e seconds in a day?	f hours in a day?
g hours in a year?	h minutes in a day?	i minutes in a year?

Note: Assume 1 year = 365 days

10 What is the time:

a 1 hour after 4.00 pm?	b 1 hour before 4.00 pm?
c 1 hour after 5.30 am?	d 1 hour 20 minutes after half past 10 am?
e 1 hour 20 minutes before 10.00 am?	f 3 hours 18 minutes after 2.00 pm?
g 2 hours 30 minutes before 4.30 am?	h 4 hours 35 minutes after quarter to 10 am?
i 4 hours 35 minutes before 10.05 pm?	j 5 hours 27 minutes after 1.08 am?
k 5 hours 27 minutes before 1.08 am?	l 1 hour 10 minutes before 11.30 am?

WORKED Example 4a, b

11 Find the time difference between:

a 8.20 pm and 8.35 pm	b 4.15 pm and 5.20 pm	c 7.15 am and 8.28 am
d 5.17 am and 6.32 am	e 9.15 pm and 10.08 pm	f 3.16 pm and 5.09 pm
g 11.28 pm and 12 midnight	h 9.21 am and 1.06 pm	i 11.10 am and 4.25 pm
j 6.05 am and 6.05 pm	k half past 6 pm amd quarter to 3 am.	

9.3 SkillSHEET Calculating the time between two events

WORKED Example 4c

12 Find the time difference between:

a 7.20 am on Monday and 6.30 pm the following Tuesday

b 4.38 am on Saturday and 1.25 pm the following Sunday

c 8.45 pm on Wednesday and 10.16 am the following Thursday

d 1.20 pm on Wednesday and 9.09 am the following Friday.

EXCEL Spreadsheet Time differences

13 **multiple choice**

225 minutes is the same as:

A 2 hours 25 minutes B 2 hours 15 minutes C 3 hours 45 minutes
D 3 hours 25 minutes E 2 hours 45 minutes

14 **multiple choice**

If the time was 8.45 pm, what would be the time 5 hours 20 minutes later?

A 1.05 am B 1.05 pm C 2.05 pm D 2.05 am E 1.10 pm

15 **multiple choice**

If a train takes 2 hours and 18 minutes to arrive at its destination and it arrives at 6.03 pm, at what time did it leave?

A 3.45 pm B 8.21 pm C 4.15 pm D 3.21 pm E 3.45 am

16 If it takes Joanne 16 minutes to write one page of a letter, how long would it take her to write a letter of three pages?

17 Mathew spends half an hour doing homework each week night. How many hours of homework has he done after 3 weeks?

18 If the time is now 7.55 am in Melbourne, what is the time in Adelaide if South Australian time is half an hour behind Victorian time?

19 If the time is now 8.30 am in Melbourne, what is the time in Perth if Western Australia is two hours behind Victoria?

20 If the time is 9.45 pm in Perth, what is the time in Melbourne? (See question **19**.)

21 If the time is 10.05 pm in Adelaide, what is the time in Melbourne? (See question **18**.)

22 If Susan takes 25 minutes to walk to the station, waits for the train for 7 minutes, travels on the train for 12 minutes and then takes 8 minutes to walk to her friend's house, how long has she taken altogether?

GAME time
Time, mass and temperature — 001

23 If James spends 35 minutes on Friday, 2 hours and 12 minutes on Saturday and $1\frac{1}{4}$ hours on Sunday to complete his assignment, how much time did he take altogether?

MATHS QUEST CHALLENGE

1 **In a mirror reflection, the time on an analogue clock face appears to be 25 minutes to 2. What time is it?**

2 **When would an analogue clock and its mirror image appear exactly the same?**

3 **Georgina works as a joiner. She needs to cut planks of wood that are 6 metres long into 2-metre lengths. Each saw cut takes 4 minutes. If the whole job is completed in 1 hour and 20 minutes, how many planks were cut?**

24-hour clock

Sometimes the time is given using a 24-hour system, or 24-hour clock. The hours are numbered from 1 to 24, beginning at midnight. The times from midnight to 12.59 pm look very similar to the 12-hour time. At 1.00 pm you need to add 12 hours, making the time 1300 hours.

Times in the 24-hour system are written as 'fourteen hundred hours' (1400) or 'twenty-three thirty hours' (2330). Notice that in 24-hour time, we always use 4 digits. Some examples are given in the table below.

12-hour time	24-hour time
6.00 am	0600 hours
10.00 am	1000 hours
12.30 pm	1230 hours
2.00 pm	1400 hours (add 12 hours to 2.00)
8.00 pm	2000 hours (add 12 hours to 8.00)
11.30 pm	2330 hours (add 12 hours to 11.30)

WORKED Example 5

Find the difference in hours and minutes between the following 24-hour times:
a 0635 and 2150 b 1055 and 1543.

THINK / **WRITE**

a
1. Set up two columns with the headings 'hours' and 'minutes'.
2. Write the values, putting the highest hour value first.
3. Calculate by subtraction.
4. Write as hours and minutes.

Hours	Minutes
21	50
06	35
15	15

15 hours 15 minutes

b
1. Set up two columns with the headings 'hours' and 'minutes'.
2. Write the values, putting the highest hour value first.

Hours	Minutes
15	43
10	55

3. You cannot take 55 minutes from 43 minutes. Convert 1 hour into 60 minutes and add it to the minutes column. Then rewrite 15 43 as 14 103.
(Using 14 103 helps us to do the subtraction, but it is not a real time.)
4. Calculate by subtraction.
5. Write as hours and minutes.

Hours	Minutes
14	103
10	55
4	48

4 hours 48 minutes

remember

1. 24-hour time is expressed as four-digit numbers, for example, 0300, 2300, 1530.
2. To convert 'pm' times between 1.00 pm and 12.00 midnight to 24-hour time, add 12 hours, and write the four digits.

EXERCISE 9B 24-hour clock

SkillSHEET 9.4 24-hour time

Mathcad 12-h/24-h time

EXCEL Spreadsheet 12-h → 24-h time (DIY)

EXCEL Spreadsheet 24-h → 12-h time (DIY)

WORKED Example 5

1 Write each time from the first column and match it with the 24-hour time from the second column.

10.12 am	1212
12.12 pm	1200
12 midnight	1012
12 midday	1120
11.20 pm	0000
11.20 am	2320

2 Write each of the following times using the 24-hour clock.
a 10.20 am **b** 11.30 am **c** 5.10 am **d** 4.15 am
e 5.15 pm **f** 6.30 pm **g** 8.30 am **h** 8.40 pm
i 12.00 midday **j** 11.30 pm **k** 4.35 am **l** 2.30 pm

3 Convert each of the following 24-hour times to 12-hour time.
a 2315 **b** 1310 **c** 0815 **d** 0115
e 1818 **f** 1220 **g** 0005 **h** 1005
i 2005 **j** 1135 **k** 1520 **l** 1414

4 Find the difference between each of the following times. (The first time is the earlier time in each case.)
a 1023 and 2312 **b** 1000 and 1215 **c** 1430 and 1615
d 1530 and 1615 **e** 1005 and 2315 **f** 1135 and 1440
g 0820 and 1550 **h** 0712 and 2008 **i** 1455 and 0015

5 Mary-Jane always arrived at school at 0855 and left at 1526. How long was she at school?

6 Peter wished to record a movie on video. He had a 180-minute tape and a 240-minute tape. The film he wished to record started at 2030 and finished at 2345.
a How long was the movie?
b Which tape should Peter use?

7 An aircraft left the airport at 0920 and arrived at its destination at 1305. How long was the flight?

8 **multiple choice**

A clock shows the time as 1543. What is the correct time if it is known that the clock is 33 minutes slow?
A 1576 **B** 1616 **C** 1510 **D** 1516 **E** 1606

9 **multiple choice**

A clock shows the time as 2345. What is the correct time if it is known that the clock is 27 minutes slow?

A 2318 **B** 2372 **C** 2412 **D** 0042 **E** 0012

10 **multiple choice**

A clock shows the time 0857. What is the correct time if it is known that the clock is 48 minutes fast?

A 0809 **B** 0811 **C** 0945 **D** 0805 **E** 0905

11 **multiple choice**

A clock shows the time 1004. What is the correct time if it is known that the clock is 31 minutes fast?

A 1035 **B** 0935 **C** 0973 **D** 1013 **E** 0933

12 Jamie has made a schedule for football training. He will start exercising at 0615 and stop at 0705. He will then continue training from 1600 until 1810. For how long will Jamie train each day?

13 A chart supplied with your bread-making machine shows the program process and the time it takes at each stage.

Program process	Basic light	Basic medium	Basic dark
First knead	5 min	5 min	5 min
Second knead	20 min	20 min	20 min
Dispenser activates at	2 h 33 s	2 h 43 s	2 h 53 s
First rise	39 min	39 min	39 min
First punch down	10 s	10 s	10 s
Second rise	25 min 50 s	25 min 50 s	25 min 50 s
Shape (use pause button for reshaping dough)	15 s	15 s	15 s
Third rise	49 min 45 s	49 min 45 s	49 min 45 s
Bake	30 min	40 min	50 min
Total cycle time			

GAME time: Time, mass and temperature — 002

a Calculate the total cycle time for the 3 types of loaves shown.

b You can preset the timer on your bread-making machine so that the bread will be ready at a specific time. The timer uses the 24-hour clock format. You decide to bake a basic light loaf so that it will be ready when you arrive home from school at 4.30 pm. At what time should the machine start the process? To what value should the timer be preset to begin the process?

WorkSHEET 9.1

THINKING Using 24-hour time

The 24-hour clock format is used in many everyday situations and jobs. Can you think of reasons why it is important that:

1 the armed forces (navy, army, airforce) use this system?

2 arrival and departure times at airports are in 24-hour time?

Suggest other situations or jobs where the 24-hour system is used.

10 QUICK QUESTIONS 1

1 How many minutes are there in 3 hours?

2 Convert 4 hours 13 minutes to minutes.

3 Find the time difference between 6.25 am and 7.32 pm.

4 Show 4.56 on an analogue clock face.

5 What is the time 2 hours 13 minutes after 7.09 am?

6 As his daily exercise, Peter runs 4 laps of a block around his house. If it takes him 17 minutes and 18 seconds to run one lap, how long does he spend running altogether?

7 Write 3.47 pm in 24-hour time.

8 Convert 2222 to 12-hour time.

9 If the school day begins at 0832 and ends at 1520, how many hours and minutes are spent at school each day?

10 A clock shows the time as 1543. What is the correct time if the clock is 32 minutes slow?

THINKING Up, up and away!

Heather, James and Matt are flying from Melbourne at 1355 on 17 September and arriving in Los Angeles at 0925 on 17 September.

1 What does a time of 1355 mean? Why is it shown like this?

2 Explain why they appear to arrive in Los Angeles at a time and date earlier than when they leave Melbourne.

3 Melbourne is 18 hours ahead of Los Angeles standard time. When they arrive in LA, what is the time and date in Melbourne?

4 How much time did it take them to fly from Melbourne to Los Angeles?

5 If their return flight leaves Los Angeles at 2230 on 21 October and takes 14 hours and 40 minutes, what will be the time and date when it arrives in Melbourne?

The calendar

We have a different calendar each year. This is because January 1 does not always start on the same day. Also, February has 28 days in some years and 29 each leap year (every fourth year). As you know, your birthday is not on the same day of the week this year as it was last year.

Early Christians considered the birth of Christ to be the most outstanding event of history. So one way of describing history was to divide it into two eras, BC and AD. The abbreviation AD stands for *anno Domini*, which means 'in the year of our Lord', and this is the beginning of the era. The AD period starts at AD 1. Each year after this time is increased by 1. If the year is 2000 AD it means 2000 years after Christ was born. To help us describe the time before this date we use BC (before Christ) and count backwards. If an event occurred at a time 350 years before Christ was born then we call this time 350 BC. It is generally accepted that there is no year zero between 1 BC and 1 AD. Because not all people are Christians, CE (of the Common Era) is often used instead of AD: and BP (Before Present) can be used instead of BC.

Each year is in turn split into 12 months, each with about 30 days, but inconveniently, some months have 28, 29 or 31 days!

Here are two very different methods of remembering the number of days in a particular month.

The first method involves remembering the following rhyme:

Thirty days have September,
April, June and November.
All the rest have 31
Except for February alone,
Which has 28 days clear
And 29 in each leap year.

The other method is a practical approach involving your knuckles as illustrated by the diagram below, and the following steps.

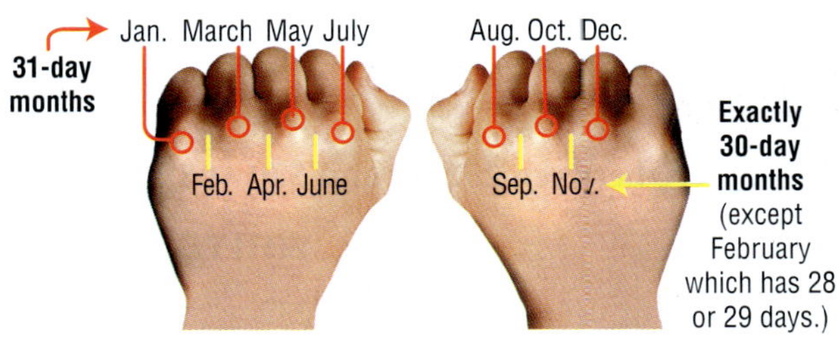

Make two fists and place them next to each other. Beginning with the knuckle of the little finger on your left hand, start reciting the months. January represents the first knuckle, February represents the first dip, March represents the next knuckle, and so on.

Each knuckle represents months with exactly 31 days, and each dip represents a month with exactly 30 days (except for February which has either 28 or 29 in a leap year).

The next example discusses leap years in more detail.

WORKED Example 6

How much time is there between:
a December 1925 and December 1835?
b 12 April 1978 and 15 April 1983 (without calculating for leap years)?

THINK	WRITE
a 1 Set out the years as for subtraction.	**a** 1925 − 1835
2 Calculate.	90
3 Write the answer in years.	90 years difference
b 1 Calculate how many years there are between 12 April 1978 and 15 April 1983.	**b** 5 years
2 Calculate how many days are left over.	3 days
3 Write the answer in years and days.	There is a difference of 5 years and 3 days.

WORKED Example 7

How much time is there between 232 AD and 530 AD?

THINK	WRITE
1 As both years are AD, write as a subtraction.	530 − 232
2 Calculate.	298
3 Write the answer in years.	There is a difference of 298 years.

WORKED Example 8

How much time is there between 232 BC and 530 AD?

THINK	WRITE
1 Calculate the number of years between 232 BC and year 1. *Note:* There are 232 years before year 1.	232
2 Calculate the number of years between year 1 and 530 BC. *Note:* There are 530 years after year 1.	530
3 Add the values obtained in parts **a** and **b**.	762
4 Write the answer in years.	There are 762 years between 232 BC and 530 AD.

History of mathematics

THE CALENDAR

Why was nobody born in Spain from 5 October 1582 to 14 October 1582? William Shakespeare died in England on *Tuesday*, 23 April 1616 and Miguel Cervantes died in Spain on *Saturday*, 23 April 1616. How did this occur?

At present there are about 40 different calendars being used around the world. Most calendars are based upon the number of days it takes for one complete cycle of the Earth around the Sun, but the number of days it takes to get around is not an exact integer such as 365, but rather 365.242 189 67 (to the nearest second, approximately).

Three main types of calendars are in use now.

1. A solar calendar is based upon the time it takes for the Earth to go around the Sun.
2. A lunar calendar is based upon the phases of the moon. The Islamic calendar is an example of a lunar calendar.
3. A lunisolar calendar is also based upon the movement of the moon with corrections of extra months to make it stay in cycle with the solar calendar. The Hebrew and Chinese calendars are examples of lunisolar calendars.

Calendars are generally either based upon some starting point in time such as the birth of Christ or on cycles such as the birth and death of rulers; for example, the Chinese calendar. As calendars developed separately over the centuries in different countries, there were problems in getting them to work together. The most common calendar system that we now use is the Gregorian calendar.

The Julian calendar, introduced by Julius Caesar in 45 BC, was a solar calendar with months of specific lengths and a leap day every 4 years. This was used by most of Europe until the Gregorian calender was developed. The Julian calendar had several months based upon numbers, such as September (7), October (8), November (9) and December (10), but these are no longer in the same place because of the addition of January and February.

In the sixth century a scholar, Dionysius Exiguus, tried to work out the times for Easter. To work this out he needed a consistent calendar system which he based upon the birth of Christ. He designated this as *anno Domini* which became AD with 1 AD being the birth date of Christ. However, it is now believed that Christ was born about 4 BC.

The Gregorian calendar was specified by an act of the British Parliament in 1751 as the calendar to be used by England and all of its colonies. At the time North America and India were British colonies and thus this calendar has become the one used by default around most of the world.

The Gregorian calendar was originally developed by the Catholic Church under Pope Gregory (the calendar system is named after him) to enable the dates of certain religious celebrations to be determined. Some dates such as Christmas on December 25 were fixed but others, including Easter, varied each year. Over the centuries the date of Easter was determined by calculations relating to the date of the first full moon after the equinox.

This date had moved about 10 days over the centuries so Pope Gregory declared that 4 October 1582 would be followed by 15 October 1582 so there were 10 'lost' days. This was proclaimed by the Roman Catholic Church so all of its people and countries (for example, Spain, Portugal and the Italian states) had to follow this new system. At first the Protestant church in England (and other countries) did not want to conform, meaning that an individual day occurred at different times in different countries.

The Hebrew calendar was based upon the Era Mundi (Era of Creation) which is equivalent to the Gregorian date of 7 October 3760 BC. Each year normally has 12 or 13 months of 29 or 30 days. The day is considered to start at 6 pm which would be 0 hours. Each hour is divided into 1080 *halakim*. A *halakim* is equivalent to $3\frac{1}{3}$ seconds. The Hebrew calendar was formalised about 4119 AM (*anno mundi*), which is the equivalent of 359 AD.

The Islamic calendar started at the Era of the Hijra when the Prophet went from Mecca to Medina. This is known as 1 AH (*anno higerae*) or Muharram 1, which is equivalent to 622 AD on the Gregorian scale. The day starts at sunset. For the purpose of religious ceremonies the month starts when the first crescent moon is seen after the new moon. If the moon cannot be seen because of the weather then a new month can be declared 30 days after the start of the last month.

Questions

1. What calendar do we currently use in Australia? After whom is it named?
2. Which calendar was named after Julius Caesar?
3. Why was nobody born in Spain from 5 October 1582 to 14 October 1582?
4. William Shakespeare died in England on Tuesday, 23 April 1616 and Miguel Cervantes died in Spain on Saturday, 23 April 1616. How did this inconsistency in days occur?

remember

1. Use either of the following methods to recall the number of days in a month.
   (a) Thirty days have September,
   April, June and November.
   All the rest have 31,
   Except for February alone,
   Which has 28 days clear
   And 29 in each leap year.
   (b)

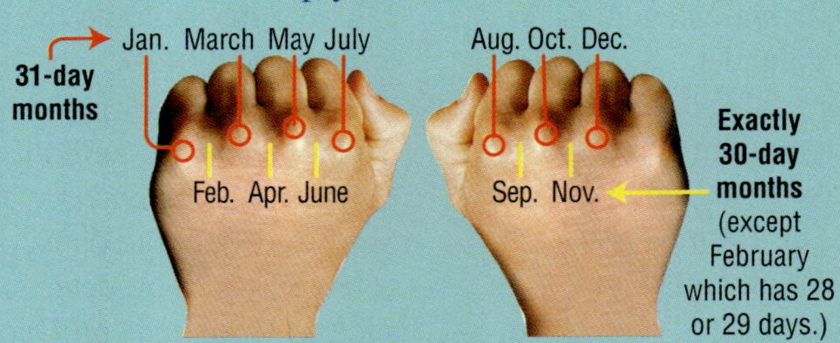

2. BC (before Christ) refers to the period before the birth of Christ, starting from 1 BC.
3. AD (*anno Domini*) means after the birth of Christ, starting from 1 AD. This year is an AD year.

EXERCISE 9C The calendar

1. a How many months have exactly 30 days? Name them.
   b How many months have 31 days? Name them.
   c How many months have less than 30 days? Name them.
   d How many days are there in most years?
   e How many days are there in a fortnight?
   f How many months in a year?
   g What is a leap year?
   h What months occur during summer?
   i What months occur during winter?
   j What months occur during autumn?
   k What months occur during spring?

2. Using the 2010 calendar shown, answer the following questions.

JANUARY

S	M	T	W	T	F	S
					1	2
3	4	5	6	7	8	9
10	11	12	13	14	15	16
17	18	19	20	21	22	23
24	25	26	27	28	29	30
31						

FEBRUARY

S	M	T	W	T	F	S
	1	2	3	4	5	6
7	8	9	10	11	12	13
14	15	16	17	18	19	20
21	22	23	24	25	26	27
28						

MARCH

S	M	T	W	T	F	S
	1	2	3	4	5	6
7	8	9	10	11	12	13
14	15	16	17	18	19	20
21	22	23	24	25	26	27
28	29	30	31			

APRIL

S	M	T	W	T	F	S
				1	2	3
4	5	6	7	8	9	10
11	12	13	14	15	16	17
18	19	20	21	22	23	24
25	26	27	28	29	30	

MAY

S	M	T	W	T	F	S
						1
2	3	4	5	6	7	8
9	10	11	12	13	14	15
16	17	18	19	20	21	22
23	24	25	26	27	28	29
30	31					

JUNE

S	M	T	W	T	F	S
		1	2	3	4	5
6	7	8	9	10	11	12
13	14	15	16	17	18	19
20	21	22	23	24	25	26
27	28	29	30			

JULY

S	M	T	W	T	F	S
				1	2	3
4	5	6	7	8	9	10
11	12	13	14	15	16	17
18	19	20	21	22	23	24
25	26	27	28	29	30	31

AUGUST

S	M	T	W	T	F	S
1	2	3	4	5	6	7
8	9	10	11	12	13	14
15	16	17	18	19	20	21
22	23	24	25	26	27	28
29	30	31				

SEPTEMBER

S	M	T	W	T	F	S
			1	2	3	4
5	6	7	8	9	10	11
12	13	14	15	16	17	18
19	20	21	22	23	24	25
26	27	28	29	30		

OCTOBER

S	M	T	W	T	F	S
					1	2
3	4	5	6	7	8	9
10	11	12	13	14	15	16
17	18	19	20	21	22	23
24	25	26	27	28	29	30
31						

NOVEMBER

S	M	T	W	T	F	S
	1	2	3	4	5	6
7	8	9	10	11	12	13
14	15	16	17	18	19	20
21	22	23	24	25	26	27
28	29	30				

DECEMBER

S	M	T	W	T	F	S
			1	2	3	4
5	6	7	8	9	10	11
12	13	14	15	16	17	18
19	20	21	22	23	24	25
26	27	28	29	30	31	

a On which day of the week is Christmas Day in 2010?
b If Fathers' Day is celebrated on the first Sunday in September, on what date will it fall in 2010?
c Australia Day is 26 January. On what day does this fall in 2010?
d If the local tennis club meets every third Monday in the month, state each of the meeting dates.
e If Mary Victoria's birthday is on 28 November, on what day will this be in 2010?
f Is 2010 a leap year? How do you know?
g Graham planned a party on 17 June. If the date is changed to 3 weeks later, what is the new date of his party?
h If a curriculum day is planned for 11 November 2010, on which day of the week will it be held?

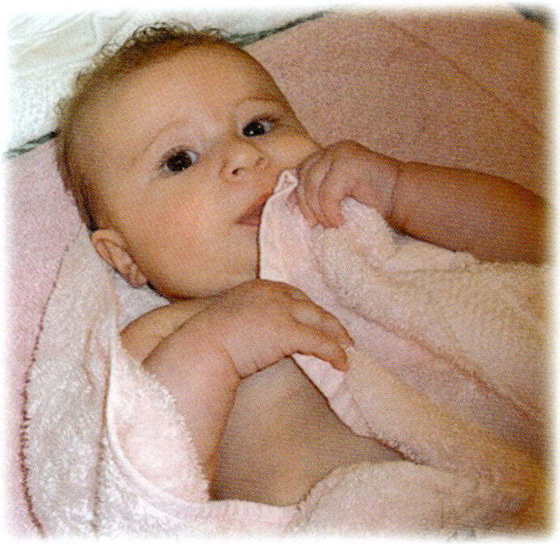

EXCEL Spreadsheet Time differences

WORKED Example 6

3 How much time is there between:
a June 1984 and June 1999?
b 1 April 2001 and 1 November 2001?
c 3 November 1998 and 3 June 1999?
d 5 January 2002 and 17 January 2002?
e 8 March 2010 and 12 May 2010?
f 16 October 1998 and 18 October 2004?
g 21 December 2001 and 30 January 2002?
h 18 January 2005 and 21 February 2005?
i 11 June 1945 and 12 June 1995?
j 14 May 1900 and 22 July 2010?

4 How many years are there between:
a 1910 and 1945? b 2002 and 2034? c 1989 and 2001?
d 2003 and 2200? e 1312 and 1610? f 1962 and 2010?

WORKED Example 7, 8

5 How much time is there between:
a 312 AD and 476 AD? b 234 AD and 560 AD? c 1100 BC and 20 BC?
d 2300 BC and 1100 BC? e 200 BC and 200 AD? f 450 BC and 1200 AD?
g 415 BC and 782 AD? h 1220 BC and 1220 AD? i 728 BC and 1342 AD?

6 If a leap year is exactly divisible by 4, how many leap years are there between 1923 and 1983?

7 If Joanne was born in 1985, what is her age on her birthday in 2004?

8 If Alex was born in 1982 and his younger brother, Simon, was born in 1994, how many years older than Simon is Alex?

9 Leap years are normally those years that can be divided by 4. However, in order to keep our calendars correct, some years that are divisible by 4 are not leap years. These are the years that end in '00'. If a year ends in '00', it must be divisible by 400 if it is to be a leap year. For example, the year 1900 is not divisible by 400 and is not a leap

year, whereas the year 2000 is divisible by 400 and so it is a leap year. Other years that are not leap years are 1700, 1800, 2100, 2200, 2300 and 2500. None of these can be divided by 400. The years 1600, 2000 and 2400 are leap years as these can be divided by 400.

EXCEL Spreadsheet Leap year calculator

Which of the following are leap years?

a 2008 b 2012 c 1999 d 1998
e 2000 f 2001 g 2003 h 2004
i 2100 j 2600 k 3000 l 3400
m 3800 n 4000 o 5000

EXCEL Spreadsheet Leap year calculator (DIY)

10 Penny was born on 29 February 1988. Considering leap years, how many 'birthdays' would she have had by the end of:

a 1990? b 1995? c 1999? d 2000?
e 2005? f 2025?

11 State whether the following statements are true or false.

a Christmas Day always falls on a Sunday.
b Good Friday always falls on a Friday.
c Easter is always in April.
d Summer always occurs over December, January and February.
e There are always 365 days in a year.
f There are always 30 days in a month.
g There are always 14 days in a fortnight.
h There are always 7 days in a week.
i The 1st of the month is always a Monday.
j All months have exactly the same number of days.
k Some years have less than 365 days.
l Every year is a leap year.
m Every year starts on 1 January.

12 **multiple choice**

If Kelly turned 13 on 3 February 2006, she was born in:

A 1993 B 1992 C 1994
D 2019 E 1991

13 **multiple choice**

If David turned 22 on 8 March 1990, he was born in

A 1967 B 1972 C 1968 D 1988 E 1978

14 **multiple choice**

If Michael turned 82 on 19 November 2005, he was born in:

A 1921 B 1922 C 1923 D 1924 E 1925

15 Andrea has a birthday on 12 March, and Jane has a birthday on 18 November. If both girls were born in the same year, how much older is Andrea than Jane?

16 Jake was born on 28 December 1985, and Luke was born on 15 January 1986. How much older is Jake?

Timelines

By using a **timeline** we can show events on a scale line. This means that we can see from our picture (timeline) how close or how far away some events are.

WORKED Example 9

Show the following events on a timeline.

7.30 am Amy got up.
7.35 am Amy had a shower.
7.45 am Amy got dressed.
8.00 am Amy had her breakfast.
8.20 am Amy cleaned her teeth and brushed her hair.
8.30 am Amy went to the bus stop.
8.40 am Amy caught the school bus.
8.55 am Amy arrived at school.

THINK	WRITE
1 Calculate the length of time that the timeline needs to represent.	Total time to be shown is from 7.30 am to 8.55 am. = 1 hour 25 minutes = 85 minutes
2 Decide on a suitable scale. Equally spaced marks of 1 cm representing every 10 minutes would be suitable.	Scale of 1 cm for every 10 minutes.
3 Rule a line about 9 to 10 cm long and mark off every 1 cm. Label each mark with a time starting from 7.30 am and increasing by 10 minutes.	
4 Show each event at the correct place on the scale.	

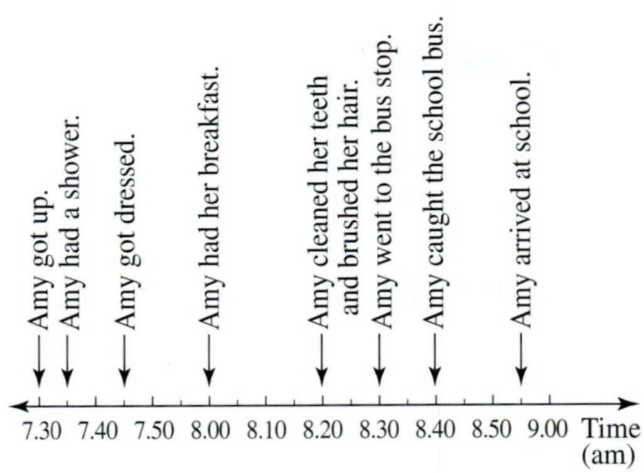

remember

When constructing timelines:
1. determine the total time to be spanned by the line
2. choose a suitable scale and mark equal divisions along the line.

EXERCISE 9D Timelines

WORKED Example 9

1 Carefully copy the following timeline into your workbook. Make sure that you use equal spacing to show equal amounts of time.

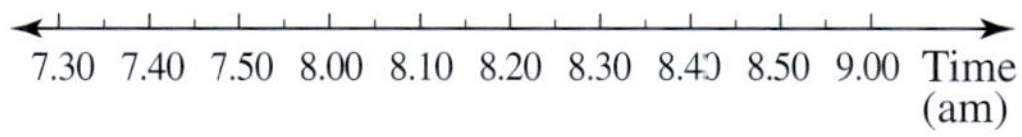

Using the following information, complete the given timeline.

7.50 am	Get out of bed
8.00 am	Start eating breakfast
8.15 am	Have a shower
8.25 am	Get dressed
8.30 am	Comb hair, clean teeth
8.35 am	Make lunch
8.45 am	Leave for school

2 Put the following information about milestones in Brett's life onto a timeline. Take care to use equal spacing to show equal amounts of time.

January 1994	Brett was born
April 1994	Rolled over by himself
May 1994	Had his first tooth
July 1994	Started sitting
September 1994	Began crawling
February 1995	Began walking
February 1998	Started preschool
January 1999	Started primary school
January 2006	Started secondary school

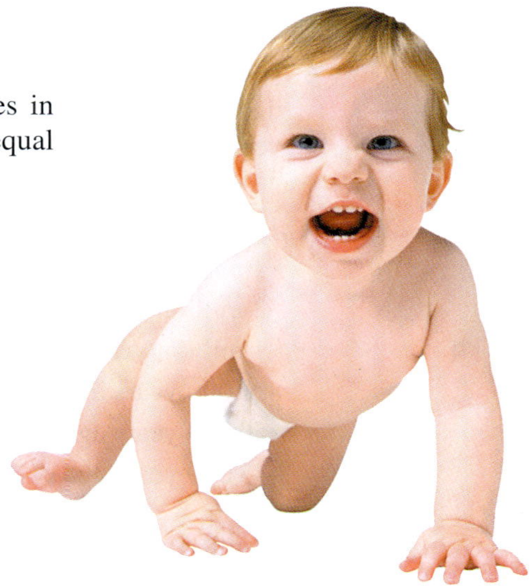

3 Carefully copy the following timeline and show the following inventions in the correct position.

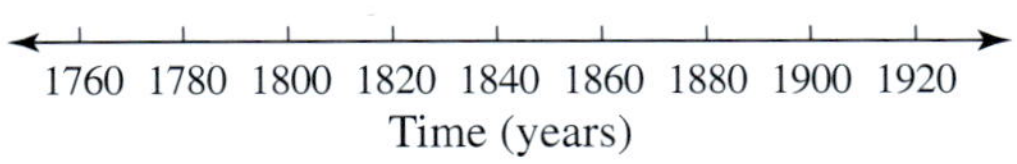

1769	First steam engine patented
1783	First person to fly in a hot-air balloon
1830	First railroad
1877	First recorded sound — phonograph
1885	First automobile
1903	First aeroplane
1908	First mass-produced car — Model T Ford

4 Show the following birth dates of musicians on a timeline.

1685 Johann Sebastian Bach
1685 George Handel
1756 Wolfgang Mozart
1770 Ludwig van Beethoven
1825 Johann Strauss
1840 Pyotr Tchaikovsky
1843 Edvard Grieg

5 Show each of the following events on a timeline.

10 500 BC The dog became domesticated.
8000 BC Other animals, such as goats, became domesticated.
7500 BC First town, Jericho, was formed.
3500 BC The wheel was invented.
3400 BC Egyptian numbers were invented.
3000 BC Beginning of a more complex written language.
2800 BC Papyrus, a writing material, was invented.
2680 BC First stone pyramid was built.
2500 BC First cats were domesticated.

6 Humans have developed over millions of years. At different stages of development their scientific name has changed. Show the following information on a timeline. Be very careful with your scale.

20 million years ago *Prosimian*
10 million years ago *Ramapithecus*
4 million years ago *Australopithecus*
2 million years ago *Homo habilis*
1 million years ago *Homo erectus*
800 000 years ago *Homo sapiens*
150 000 years ago *Homo sapiens neanderthalensis*
45 000 years ago *Homo sapiens sapiens*

7 At different times in the Earth's history, different creatures appeared. The following dates show the evolution of dinosaurs. Show this information on a timeline.

270 million years ago Dimetron
220 million years ago Coelophysis
210 million years ago Icarosaurus
180 million years ago Dimorphodon
150 million years ago Brachiosaurus
140 million years ago Archaeopteryx
130 million years ago Diplodocus
110 million years ago Kronasaurus
80 million years ago Triceratops
50 million years ago Basilosaurus

8 When we draw timelines, it is very important to have an accurate scale. Why?

9 What is the advantage of using a timeline over just writing out the information?

10 **a** Is the order of the information important when drawing a timeline?
b Explain why.

11 Think of some other topics that you could show on a timeline. Suggest 3 ideas that could be used.

Timetables

Trains

A **timetable** is a list which provides times of things that are scheduled to happen. A train timetable will tell you when and where the train should be each day. You can use a timetable (such as the one shown below) to help you plan your journey.

110

Melbourne to Mt Gambier

via Geelong, Terang, Warrnambool, Port Fairy and Heywood

Days of operation	Mon.–Thu.	Fri.	Mon.–Fri.	Mon.–Thu.	Fri.	Saturday			Sunday		
Train/Coach	Train	Train	Train	Train	Train	Train	Train	Train	Train	Train	Train
Melbourne (Spencer St) dep.	0848	0848	1240	WC 1747	WC 1747	0848	1220	WC 1840	0940	1600	WC 1900
Werribee									1007	1627	1927
Geelong arr.	0944	0944	1331	1854	1854	0944	1313	1933	1042	1702	1953
CHANGE SERVICE									COACH	COACH	
Geelong Stn dep.	0947	0947	1334	1857	1857	0947	1316	1936	1055	1715	1956
South Geelong				1900	1900						
Marshall	0955	0955	1342	1906	1906	0955	1324	1944			2004
Winchelsea Stn	1015	1015	1406	1926	1926	1015	1344	2003			2025
Winchelsea 1									1125	1745	
Birregurra									1142	1802	
Birregurra Stn			1420	1954	1955			2017			2039
Colac Stn arr.	1040	1040	1441	2006	2007	1040		2029			
Colac Stn dep.	1042	1042	1443	2009	2010	1043	1407	2031	1200	1820	2052
Pirron Yallock									1212	1832	
Camperdown Stn arr.	1109	1109		2036	2037	1109		2057			
Camperdown Stn dep.	1111	1111	1510	2038	2040	1111	1457	2059	1238	1858	2120
Terang Stn	1126	1126	1525	2054	2057	1126	1512	2114	1256	1916	2136
Deakin University T/o									1325	1945	
Warrnambool Stn arr.	1200	1200	1558	2129	2132	1201	1542	2146	1335	1955	2211
CHANGE SERVICE	COACH	COACH	COACH	COACH	COACH	COACH		COACH			COACH
Warrnambool Stn dep.	1215	1215	1620	2140	2145	1220		2155			2220
Illowa	1225	1225	1630	2150	2155	1230		2205			2230
Koroit arr.	1230		1635			1235					
Port Fairy arr.			1700								
Port Fairy dep.	1255	1243		2205	2210	1300		2218			2240
Yambuk	1310	1258		2217	2222	1315		2230			2255
Tyrendarra	1330	1318		2235	2240	1335		2250			2315
Narrawong	1335	1323		2242	2247	1340		2257			2322
Portland 1 arr.	1350	1338		2257	2302	1355		2312			2337
Portland 1 dep.	1355	1343		2300	2305	1400		2315			2340
Heywood 2	1415	1405		2320	2320	1420		2335			0001
Dartmoor	1445	1435			2325	1450					0030
Mount Gambier (CST) arr.	1455	1440			2359	1455					0035

WC — The Westcoaster

Source: West Coast Railway, Victoria

WORKED Example 10

If I wanted to be in Colac by 8:15 pm on a Friday, at what time would I catch the train from Melbourne?

THINK	WRITE
1 Read down the first column of the rail service timetable and find Colac Stn arr. to see when the train arrives in Colac.	
2 Read across until the last column for Friday and find the latest arrival time before 8.15 pm.	The train arrives in Colac at 2007 or 8.07 pm on Friday.
3 Read up this column to see when the same train departs Melbourne.	This train departs Melbourne at 1747 or 5.47 pm.

Tides

A tide timetable (like the one below) will tell you when the tides will be high or low. This is very useful for people like anglers and surfers.

TIDES FOR
WESTERN PORT (STONY POINT)
NOVEMBER 2005

Tue. 1		Wed. 2		Thu. 3		Fri. 4		Sat. 5		Sun. 6		Mon. 7		Tue. 8	
Time	Ht	Time	Ht	Time	Ht	Time	Ht	Time	Ht	Time	Ht	Time	Ht	Time	Ht
0524	1.19	0042	2.46	0135	2.68	0223	2.85	0308	2.95	0352	2.98	0434	2.96	0516	2.90
1136	2.20	0629	1.12	0721	1.03	0807	0.95	0849	0.89	0930	0.85	1012	0.84	1053	0.84
1744	0.80	1231	2.27	1321	2.34	1405	2.40	1446	2.45	1527	2.47	1607	2.48	1648	2.47
		1833	0.60	1918	0.42	2001	0.28	2044	0.19	2125	0.14	2207	0.14	2250	0.20

The tide heights (Ht) are in metres, red for low tide and blue for high tide.
Times stated are Australian Eastern Standard Time (24-hour clock). During Daylight Saving Time (when in force) one hour needs to be added to the times stated.

Source: Australian Government Bureau of Meteorology

WORKED Example 11

What was the height of the highest tide on Saturday 5 November? At what time did this occur?

THINK	WRITE
1 Read across the first row of the table to find 5 November.	
2 Read down to the column that gives the tide heights.	
3 Choose the largest number from the group that corresponds to a high water level. *Note:* High tides are in blue.	The height of the highest tide was 2.95 m.
4 Read the tide time from the tide time column. This is on the left side of tide height just found.	This occurred at 3.08 am.

remember

1. A timetable is a list that provides times of things that are scheduled to happen.
2. Headings and keys need to be checked carefully when reading timetables.

EXERCISE 9E Timetables

Use the West Coast Railway timetable (see page 363) to answer the following questions.

1 How many trains leave from Melbourne for Mt Gambier each Monday to Thursday?

WORKED Example 10

2 If I have a brief appointment in Melbourne on Wednesday at 9.00 am, but I want to be in Warrnambool by 4.00 pm, which train would I catch from Melbourne?

3 From what station in Melbourne does the train leave?

4 Do all trains finish their journey at Warrnambool?

5 What happens after Warrnambool?

6 At which towns do each of the joint-train/coach services finish?

7 What does WC (written just above the departure time of three of the train services from Melbourne) stand for?

8 Imagine you are travelling on the Westcoaster.
 a How long does it take you to travel from:
  i Melbourne to Geelong?
  ii Melbourne to Warrnambool?
  iii Colac to Terang?
  iv Colac to Warrnambool?
  v Port Fairy to Heywood?
  vi Port Fairy to Mount Gambier?
 b i How long does the train stay in Geelong?
  ii How long does the train stay in Colac?
  iii How long do the passengers have to wait at Warrnambool?
  iv At what other station(s) does the timetable show that the train or coach stays for a period of time?
  v On which day(s) does the 'Westcoaster' travel to Mount Gambier?

9 How many joint trains/coach services per day travel to Mount Gambier?

10 On what day of the week are there two services from Melbourne to Mount Gambier?

11 a Which train/coach service takes the shortest time to travel from Melbourne to Mount Gambier on a weekday?
 b How long is the total trip?

12 Why do you think it takes a little longer to travel from Melbourne to Heywood on the 'Westcoaster' on Friday than on other days of the week?

13 How many train/coach services travel to Port Fairy during the week? (Include those which also go past Port Fairy.)

Using the tides times given on page 364, answer the following questions.

14 How many low tides are there each day?

15 Do the high tides occur at the same time each day?

WORKED Example 11

16 a What is the height of the highest tide on Thursday 3 November 2005?
b At what time did this occur?

17 How much time is there between high tides on Monday 7 November 2005?

18 Is this time difference between high tides the same each day?

19 How much time is there between low tides on Monday 7 November 2005?

20 Is the time difference between low tides the same as the difference between high tides?

21 a What is the lowest low tide?
b When does this occur?

22 a What is the highest low tide?
b When does this occur?

Time zones and flight schedules

Time zones

EXCEL Spreadsheet Time zones

Time changes as we travel around the world. In Europe it is night time when in Australia it is day and it is day time in Europe when we have night. To help everyone know what time it is in different countries, the world is divided into 24 **time zones** (see opposite page for the map showing world time zones).

If we travel east, we need to set our watches ahead. If we travel west, we need to set our watches back.

Time in Perth	Time in Melbourne	Time in Auckland
3.00 pm	5.00 pm	7.00 pm
6.00 pm	8.00 pm	10.00 pm
10.00 am	12 midday	2.00 pm
10.00 pm	12 midnight	2.00 am next day

The time in Queensland, New South Wales, Victoria and Tasmania is called Eastern Standard Time or EST. In Western Australia the time is 2 hours behind EST. New Zealand is 2 hours ahead of EST.

The time in England is called Greenwich Mean Time or GMT. GMT is 10 hours behind EST.

Time in London (GMT)	Time in Perth	Time in Melbourne (EST)	Time in Auckland
8.00 am	4.00 pm	6.00 pm	8.00 pm
4.00 pm	12 midnight	2.00 am (next day)	4.00 am (next day)
12 midday	8.00 pm	10.00 pm	12 midnight
12 midnight	8.00 am	10.00 am	12 midday
3.00 am	11.00 am	1.00 pm	3.00 pm

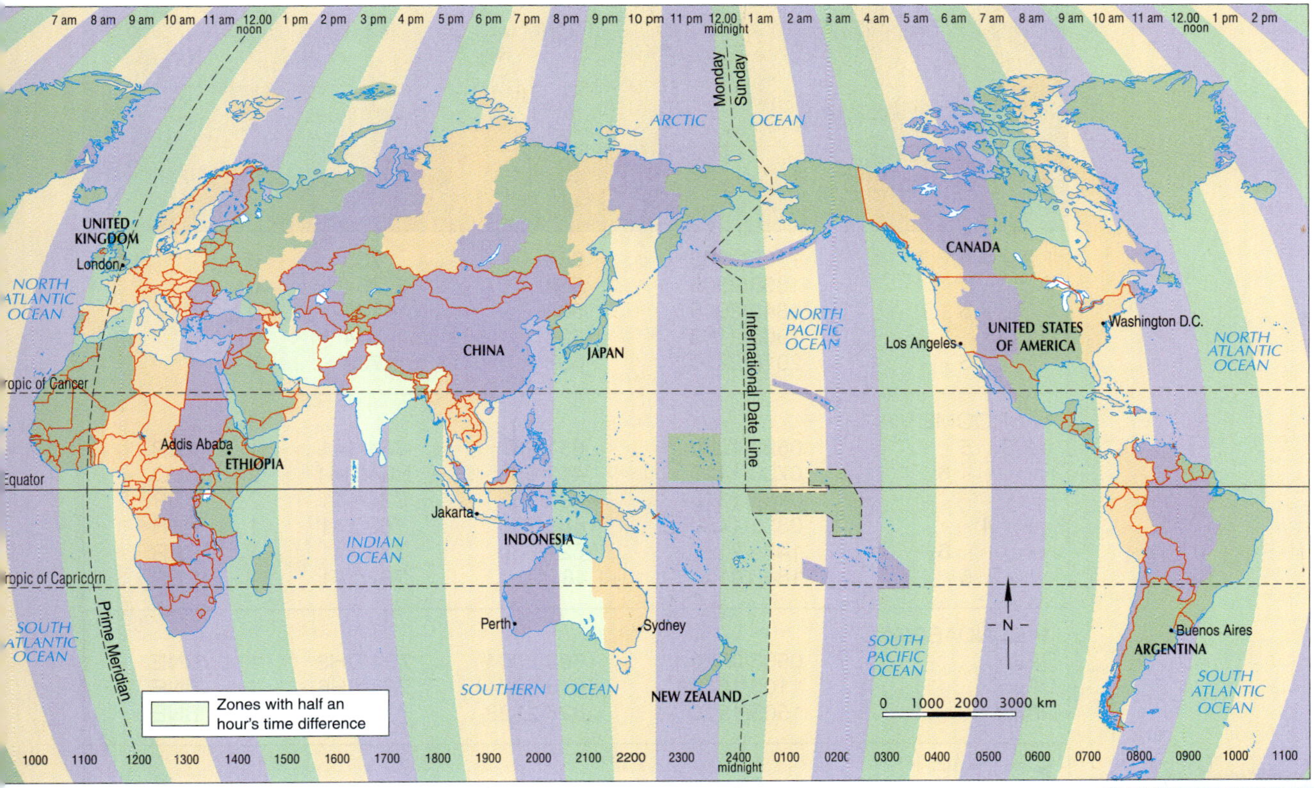

WORKED Example 12

a What is the time in Melbourne if the time in London (GMT) is 8.00 am?
b What is the time in London (GMT) if the time in Melbourne is 8.00 am?

THINK	WRITE
a 1 Melbourne is 10 hours ahead of London. Add 10 hours to the London time. 2 Calculate.	**a** London time: 8.00 am Melbourne time: 8.00 am + 10 hours It is 6.00 pm.
b 1 London is 10 hours behind Melbourne. Subtract 10 hours from the Melbourne time. 2 Calculate.	**b** Melbourne time: 8.00 am London time: 8.00 am − 10 hours It is 10.00 pm the previous day.

Flight schedules

A flight schedule is shown on the next page. A lot of information is given on this schedule. The days of the week on which the flight occurs are given in the first column. Note that numbers represent days: 1 represents Monday, 2 represents Tuesday, through to 7 for Sunday. If there is a flight every day it is shown as 'daily'. The departure and arrival times are given together in column 2, using 24-hour time. The flight number is given in column 3 and the number of stops is given in column 5.

Day	From To	Depart/ Arrive	Flight No.	Aircraft	Stops	Transfer at
FROM MELBOURNE (Cont'd)						
▶ BRISBANE (Cont'd)						
1 2 3 4 5 · 7		1705/1905	630	733	0	
DAILY		1805/2005	632	734	0	
· · · · 5 · ·		1845/2045	634	733	0	
DAILY		1905/2105	636	734	0	
· · · · 5 · ·		2010/2210	638	733	0	
▶ BROOME						
1 2 3 · 5 · ·		0620/1230	651/937	734/141	1	PER
· · · · · 6 ·		0730/1410	412/689	EQV/141	2	AYQ
· · · · · · 7		0755/1310	493/962	734/141	2	ASP
· 2 · 4 · · ·		0835/1430	485/956	762/141	1	PER
· · · · · 6 ·		1435/2120	621/960	733/141	1	PER
· · · · 5 · ·		1840/0120	+1 481/928	762/141	1	PER
▶ BUNDABERG						
DAILY		0605/0920	608/2304	734/DH8	1	BNE
DAILY		1105/1510	618/2324	733/SH6	1	BNE
DAILY		1505/1910	626/2328	734/DH8	1	BNE
▶ BURNIE						
1 2 3 4 5 · ·		0840/0940a	2731	DH8	0	
· · · · · 6 ·		1010/1110a	2733	DH8	0	
1 2 3 4 5 · 7		1515/1615a	2737	DH8	0	
DAILY		1830/1930a	2739	DH8	0	
▶ CAIRNS						
DAILY		0605/1050	608	EQV	1	
· 2 3 · · · ·		0605/1050	608	CHG	1	
· 2 · 4 · 6 7		0640/0955	644	EQV	0	
1 · 3 · 5 · ·		0645/1140	406	EQV	1	
DAILY		0705/1225	610	CHG	1	
1 2 · · · · ·		1030/1540	424	733	1	
· · · · 5 · ·		1845/2335	634	734	1	
DAILY		1005/1500	616/514	733/733	1	BNE
DAILY		1605/2055	628/538	733/734	1	BNE
DAILY		1805/2255	632/544	734/733	1	BNE
▶ CANBERRA						
1 2 3 4 5 · ·		0625/0720	496	733	0	
· · · · · 6 ·		0645/0800	978	DH8	0	
DAILY		0815/0910	474	734	0	
· · · · · · 7		1015/1130	980	DH8	0	
1 2 3 4 5 · 7		1140/1255	982	DH8	0	
1 2 3 4 5 · ·		1245/1340	480	734	0	
· · · · · 6 ·		1315/1430	984	DH8	0	
DAILY		1515/1610	482	733	0	
1 2 3 4 5 · 7		1710/1805	600	734	0	
DAILY		1925/2020	654	734	0	

Days of week →

1 = Monday 2 = Tuesday 3 = Wednesday 4 = Thursday 5 = Friday 6 = Saturday 7 = Sunday

remember

1. The world is divided into 24 one-hour time zones.
2. Places to the east are ahead of those to the west.

EXERCISE 9F Time zones and flight schedules

WORKED Example 12a

1 What is the time in Melbourne if the time in London (GMT) is as given below?

a 11.00 am b 12.30 pm c 10.20 am
d 12.00 midnight e 12.00 noon f 2.15 am
g 3.30 pm h 5.45 pm i 8.10 am
j 9.35 pm k 1.12 pm l 1.12 am?

WORKED Example 12b

2 What is the time in London (GMT) if the time in Melbourne is as given below?

a 3.30 am b 12 noon c 12 midnight
d 5.50 pm e 6.40 am f 7.20 pm
g 11.10 am h 1325 i 1550
j 0615 k 1855 l 1935?

3 **multiple choice**

If Suzie wanted to telephone her friend in England before he went to work for the day, what would be the best time for her to ring from Melbourne, Australia?

A 0700 B 1700 C 1900
D 2100 E 2300

4 If the time was 5.00 am in London, what would the time be in:
a Perth? b Melbourne? c Auckland?

5 If the time was 8.00 pm in Perth, what would the time be in:
a London? b Melbourne?
c Auckland?

6 If the time was 3.30 pm in Auckland, what would the time be in:
a London? b Perth?
c Melbourne?

Using the flight schedule on page 368, answer questions 7 to 23.

7 Flights from Melbourne to Brisbane leaving Melbourne at 1705 occur 6 days per week. On which day is there no flight from Melbourne to Brisbane at this time?

8 At what time(s) do flights leave daily for Brisbane?

9 How many flights are there from Melbourne to Brisbane on:
a Friday? b Monday? c Saturday?

10 How many flights are there from Melbourne to Bundaberg each day?

11 At what times do the flights to Bundaberg leave Sydney?

12 a Are there any stops when travelling from Melbourne to Bundaberg?
b If so: i how many? ii where are they?

13 The flight that goes to Bundaberg leaving at 0605 is flight 608. What is the flight that goes to Bundaberg leaving at 1505 hours?

14 The flight that goes from Melbourne to Cairns leaving at 0605 hours was the same flight, 608, as the flight from Melbourne to Bundaberg. Why do you think this is?

15 How many flights are there from Melbourne to Cairns (the second flight number indicates a transfer from another flight):
a every day? b on Friday? c on Sunday?

16 What day(s) of the week has (have) the most flights from Melbourne to Cairns?

17 How long does it take for flight 608 to travel from Melbourne to Cairns?

18 a How long does it take for flight 644 to travel from Melbourne to Cairns?
b Why do you think it takes so much longer to travel from Melbourne to Cairns on flight 608, than it does on flight 644?

19 How many daily flights are there from Melbourne to Canberra?

20 How many flights are there each day from Melbourne to Canberra on:
a Monday? b Saturday? c Sunday?

21 How long does it take to travel from Melbourne to Canberra on flight 600?

22 Do the flights from Melbourne to Canberra make any stops before they arrive in Canberra?

23 **multiple choice**

WorkSHEET 9.2

a What is the departure time for flight 2737 travelling from Melbourne to Burnie?
A 840 B 1010 C 1515 D 1830 E 1615

b How many stops are there on flight 493, travelling from Melbourne to Broome?
A 0 B 1 C 2 D 3 E 4

MATHS QUEST CHALLENGE

1 Halley's comet appeared in the years 1758, 1834, 1910, 1986. In what year will it next appear?

2 What is the date when the month of August is 20 000 minutes old?

3 Jessica and Connor have to catch the 8.20 am train. Jessica's watch is 10 minutes fast but she thinks that it is 5 minutes slow. Connor's watch is 10 minutes slow but he thinks that it is 10 minutes fast. Each leaves home expecting to arrive at the station just in time to catch the train. Who misses the train and by how many minutes?

10 QUICK QUESTIONS 2

1 How many minutes are there in 8 hours 15 minutes?

2 List the months of the year according to the number of days in each month.

3 How many years are there between 342 BC and 876 AD?

4 Is 2056 a leap year?

5 How many days are there between 12 March and 6 May?

6 Put the following list of winners of the Melbourne Cup onto a timeline:
1947, Hiraj; 1978, Arwon; 1980, Beldale Ball; 1943, Dark Felt; 1965, Light Fingers; 1977, Gold and Black; 1968, Rain Lover; 1954, Rising Fast.

Use the flight schedule for flights from Mildura to various destinations to answer questions **7** and **8**.

Frequency	Depart	Arrive	Flight No	Stops	Meals	From - To
From	**Mildura**					**MQL**
	Mildura 11km					
To	**Adelaide**					
M T W T F . .	**0630**	0935	***6201/61**	1	-/S	
Daily	**0955**	1320	***6203/63**	1	-/L	
. . . . F . .	**1320**	1640	***6205/195**	1	-/S	- Apr 12
. . . . . . S	**1320**	1640	***6205/195**	1	-/S	
M T W T F . S	**1640**	2110	***6207/99**	1	-/S	
To	**Brisbane**					
M T W T F . .	**0630**	1110	***6201/170**	1	-/S	
Daily	**0955**	1420	***6203/68**	1	-/L	
. . . . F . .	**1320**	1815	***6205/74**	1	-/S	- Apr 12
. . . . . . S	**1320**	1815	***6205/74**	1	-/S	
M T W T F . S	**1640**	2040	***6207/76**	1	-/D	
To	**Melbourne**					
M T W T F . .	**0630**	0740	***6201**	0		
Daily	**0955**	1105	***6203**	0		
. . . . F . .	**1320**	1430	***6205**	0		- Apr 12
. . . . . . S	**1320**	1430	***6205**	0		
M T W T F . S	**1640**	1750	***6207**	0		
. . . T . . .	**2000**	2110	***6209**	0		- Apr 12
. . . . F . .	**2000**	2110	***6209**	0		
To	**Perth**					
M T W T F . .	**0955**	1350	***6203/19**	1	-/L	
. . . . . . S	**0955**	1430	***6203/211**	1	-/L	
. . . . . S .	**0955**	1540	***6203/63/83**	2	-/L/L	
. . . . F . .	**1320**	2035	***6205/97**	1	-/D	- Apr 12
M T W T F . S	**1640**	2035	***6207/97**	1	-/D	
To	**Sydney**					
M T W T F . .	**0630**	0950	***6201/20**	1	-/B	
Daily	**0955**	1320	***6203/68**	1	-/L	
. . . . F . .	**1320**	1620	***6205/34**	1	-/S	- Apr 12
. . . . . . S	**1320**	1620	***6205/34**	1	-/S	
M T W T F . S	**1640**	1950	***6207/52**	1	-/D	

7 On what days of the week does the 6.30 am flight to Melbourne operate?

8 How long does it take to fly to Brisbane from Mildura if you travel on the 1.20 pm Sunday flight?

9 Rachel was flying on a direct route from Melbourne to London and she departed on an 0800 plane on 5 August. The flying time was exactly 25 hours. What was the time and date in Melbourne when she arrived in London?

10 What was the local time and date when Rachel (in question **9**) arrived?

Mass

Are **mass** and weight the same?

Mass describes how much matter makes up an object and is measured in kilograms (kg), whereas weight describes the gravitational force acting on an object and is measured in newtons (N).

A man of mass 60 kg on Earth has a weight of about 600 N. The same man on the Moon would still have a mass of 60 kg, but his weight would be about 10 N.

Units of mass

1. Milligram (mg)

A small fly or wasp's mass is approximately 4 milligrams.

2. Gram (g)

A scorpion has a mass of approximately 20 grams.

3. Kilogram (kg)

A full-grown female cheetah reaches a mass of 45 kilograms.

4. Tonne (t)

An elephant has a mass of approximately 1 tonne.

The relationship between these units of mass is as follows:

1000 mg = 1 g 1000 g = 1 kg 1000 kg = 1 t

When converting units of mass, if you start with a smaller unit and change to a larger unit, then you need to divide by the conversion factor. If you start with a larger unit and change to a smaller unit, then you need to multiply by the conversion factor.

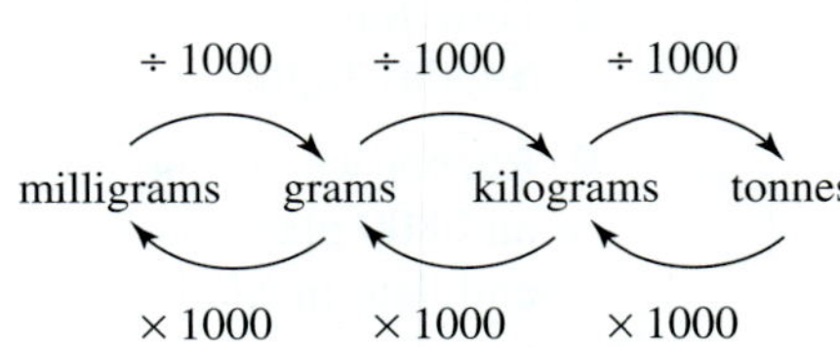

WORKED Example 13

Convert 2.8 kilograms to grams.

THINK	WRITE
1 Write the quantity.	2.8 kg
2 A kilogram is a larger unit than grams, so it is necessary to multiply. To convert kilograms to grams, multiply by 1000.	$= 2.8 \times 1000$ g
3 Calculate by moving the decimal place 3 units to the right.	$= 2800$ g

WORKED Example 14

Convert $21\frac{1}{4}$ tonnes to kilograms.

THINK	WRITE
1 Write the quantity.	$21\frac{1}{4}$ t
2 Change $21\frac{1}{4}$ tonnes into a decimal.	$= 21.25$ t
3 A tonne is a larger unit than a kilogram, so it is necessary to multiply. To convert tonnes to kilograms, multiply by 1000.	$= 21.25 \times 1000$ kg
4 Calculate by moving the decimal place 3 units to the right.	$= 21\,250$ kg

WORKED Example 15

Convert 950 kg to tonnes.

THINK	WRITE
1 Write the quantity.	950 kg
2 A kilogram is a smaller unit than a tonne so you will need to divide. To convert kilograms to tonnes, divide by 1000.	$= 950 \div 1000$ t
3 Calculate by moving the decimal point 3 units to the left.	$= 0.950$ t
4 Remove the last zero as there is no need for it here.	$= 0.95$ t

remember

1. To convert to a larger unit, you need to divide.
2. To convert to a smaller unit, you need to multiply.

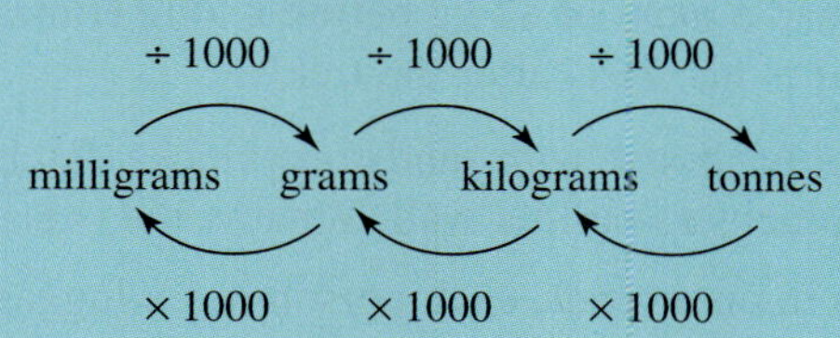

EXERCISE 9G Mass

SkillSHEET 9.6 Units of mass

1 Arrange the following in order of mass, from lightest to heaviest:
brick, feather, car, lunch box, egg, bag of concrete.

WORKED Example 13

EXCEL Spreadsheet Converting units of mass

2 Convert each of the following to grams.

a 1 kg b 2 kg c 2.5 kg
d 3.7 kg e 0.5 kg f 0.72 kg
g 6.7 kg h 109.02 kg i 14.08 kg
j 0.03 kg k 0.025 kg l 3.07 kg
m 5.12 kg n 1.004 kg o 10.04 kg

3 Convert each of the following to grams.

a $\frac{1}{2}$ kg b $\frac{1}{4}$ kg c $\frac{3}{4}$ kg
d $\frac{3}{5}$ kg e $3\frac{1}{2}$ kg f $22\frac{1}{4}$ kg
g $100\frac{17}{20}$ kg h $\frac{1}{100}$ kg i $\frac{17}{250}$ kg

Mathcad Converting units of mass

4 Convert the following to kilograms.

a 2 t b 2.7 t c 3.8 t
d 0.3 t e 0.04 t f 15.1 t
g 7.03 t h 18.005 t i 103.04 t

WORKED Example 14

5 Convert the following to kilograms.

a $\frac{1}{2}$ t b $\frac{1}{3}$ t c $\frac{1}{4}$ t
d $\frac{3}{4}$ t e $\frac{4}{5}$ t f $5\frac{1}{4}$ t
g $\frac{9}{100}$ t h $\frac{61}{500}$ t i $203\frac{3}{4}$ t

WORKED Example 13,14,15

6 Convert the following to the units specified.

a 2 g to mg b 250 kg to t c 1 kg to mg
d 820 mg to g e 8400 kg to t f 750 kg to t
g 3.2 kg to g h 2.5 t to kg i 7.03 kg to g
j 75 000 kg to t k 5 000 000 mg to g l 4500 t to kg
m 18 000 g to kg n 20 kg to t

7 If Peter goes to the supermarket and buys 1 kg of sugar, 500 g of butter, 200 g of yoghurt, 810 g of tinned tomatoes and 425 g of tinned apricots, what is the total mass of his purchases (in grams)?

8 There are 12 cans in a box. If each can has a mass of 425 g, what is the total mass held in the box (in kg)?

9 Graham had a mass of 82 kg before a long Himalayan trek. If he lost 8.3 kg during the trek, what is his mass after the trek?

10 Greg built his house with sandstone blocks. Each block has a mass of 20 kg, and Greg used 45 blocks to build a wall. What is the mass of the wall?

11 On a trip in an aircraft, each person has a luggage limit of 20 kg. If Peter's bag has a mass of only 18.7 kg, how much more mass is he entitled to take?

12 Megan wished to buy lots of items on her overseas trip. She took only 8.3 kg of luggage with her, but bought 3.3 kg of jumpers, 2.1 kg of gifts, 0.73 kg of books, 1.8 kg of shoes, 2.8 kg of jeans and 1.3 kg of souvenirs.

a What is the mass of her luggage on the return journey?

b Is she carrying too much and, if so, by how much?

13

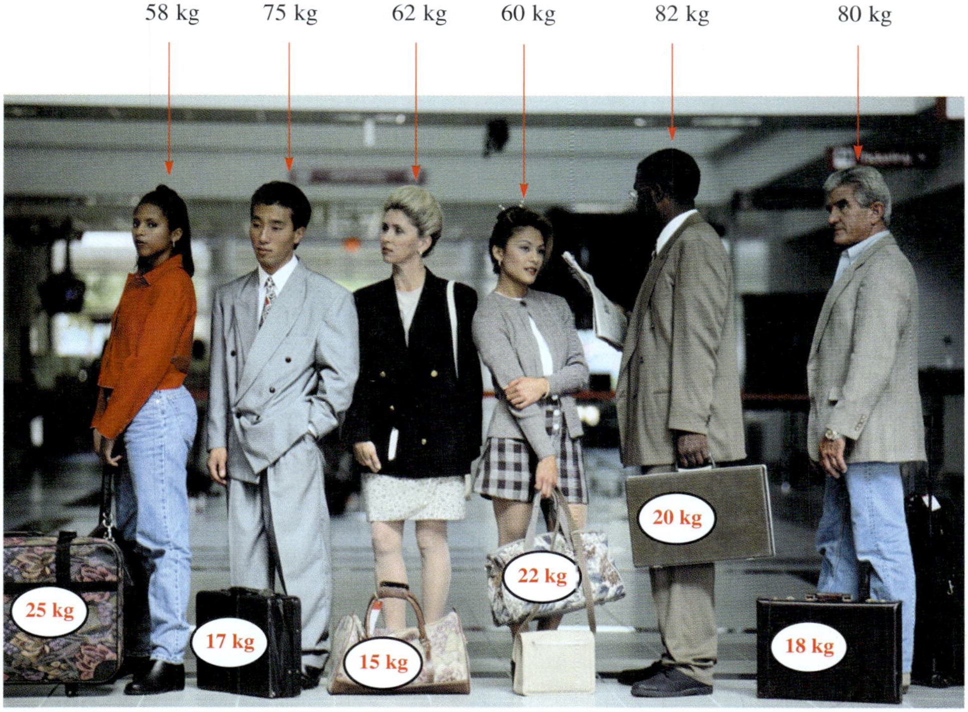

a Find the total mass of the 6 people.

b Find the total mass of their luggage. (The masses on their hand luggage includes the mass of their checked-in baggage.)

c If they travelled as a group, would they be within the luggage limit for 6 people? (Assume a limit of 20 kg per person.)

14 Estimate the mass of the following items:

a a basketball

b your textbook

c a compact disc

d a full can of soft drink

e a car.

15 Many products in the supermarket are labelled '500 g net'. What does the word 'net' mean?

16 A jar of honey has a net mass of 200 g and a gross weight of 245 g. Find the mass of the container.

WorkSHEET 9.3

17 A 250-g jar of coffee costs \$3.75 and a 500-g jar of the same coffee costs \$8. Which jar gives better value?

This happened around 400 BC

Join the dots next to equal masses with a straight line. Each line will pass through a number and letter giving the puzzle code.

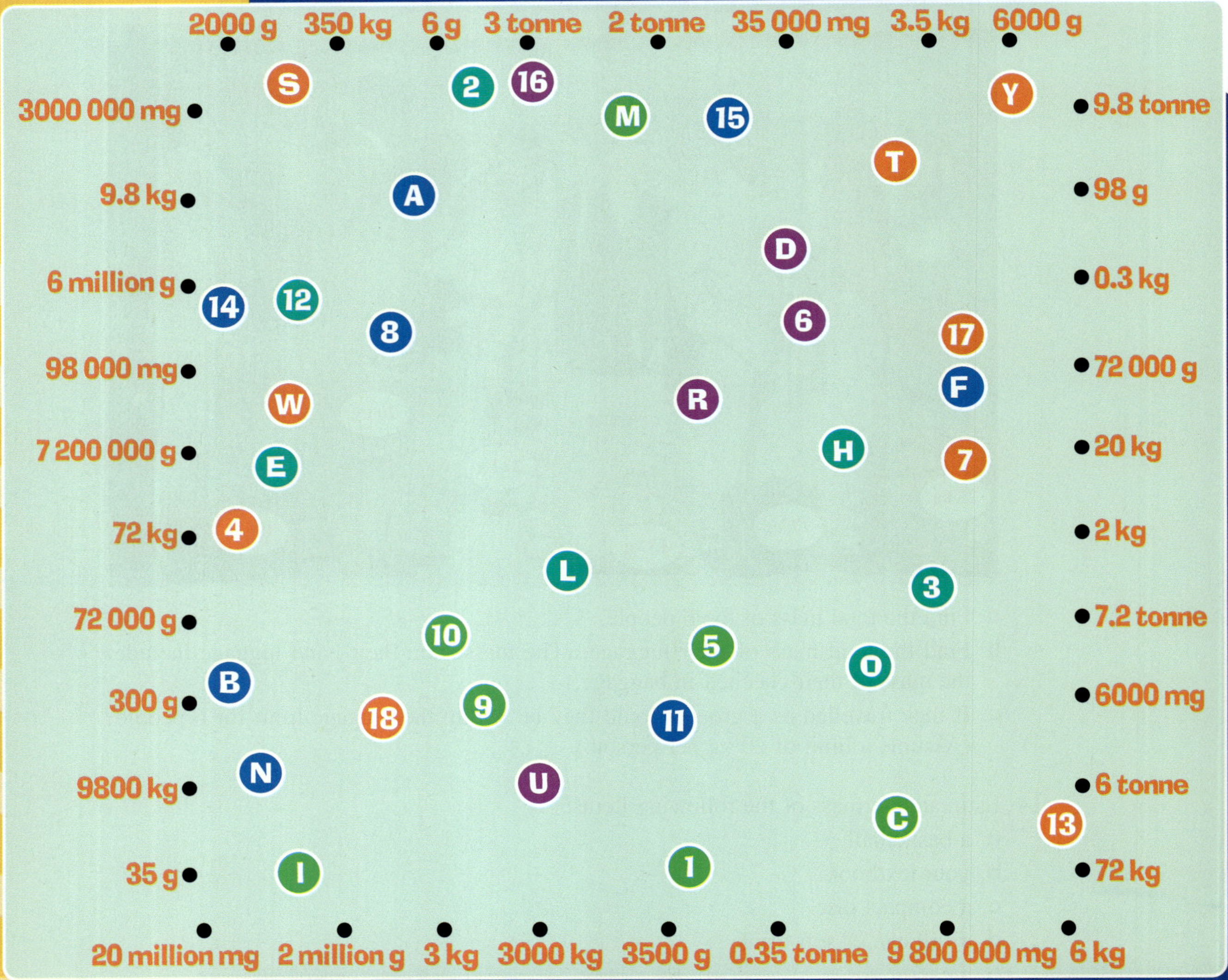

1	2	3	4	5	6	7	1	8	3	9	5	10	11	12
7	13	7	1	3	10	14	4	15	16	10	17	3	6	7
18	11	7	16	7	3	8	5	15	5	15	8	5	11	

THINKING Comparing mass

1 What is your mass?
2 What is the mass of 1 litre of water?
You can use your mass or the known mass of some object to compare with other objects. This helps you to understand how heavy or light an object may be.
3 The largest pure nugget of gold was found in Victoria in 1869. 'The Welcome Stranger' had a mass of 69.92 kg. Compare this to your own mass.
4 The smallest aircraft which could still carry one person and reach speeds up to 306 km/h had a mass of 179.6 kg. How many of 'you' would make up the same mass?
5 A chocolate Easter egg of mass 4.76 tonnes was made by the staff at Cadbury Ltd. Compare this to your own mass. How many of 'you' would make up the same mass?
6 In 1989, a 2.5-tonne hamburger was made in the USA. Estimate how many 'regular' hamburgers this would be equivalent to.
7 A world record was set for the largest meat pie in New York, USA. It was for a chicken pie of mass 10.06 tonnes. Choose an object that you could compare the mass with. How many of the chosen objects would be equivalent to the chicken pie?
8 Estimate how long it would take you to eat 1.5 kg of unmelted ice-cream. If you are not sure how much this would be, think of another object that you could compare it to (for example, a tub of margarine or butter).
In fact, Tony Dowderwell from the USA ate 1.53 kg of unmelted ice-cream in 31.67 s. Would you have come close?

Temperature

When is it cold? Would a hat and warm coat be appropriate clothing for a 32° day? How hot is the Sun?

Temperature measures how cold or warm an object is in comparison to a reference point. The units most commonly used for measuring temperature are **degrees Celsius (°C)**, and degrees Fahrenheit (°F). Both the Celsius and Fahrenheit scales use the freezing and boiling points of water as their reference points.

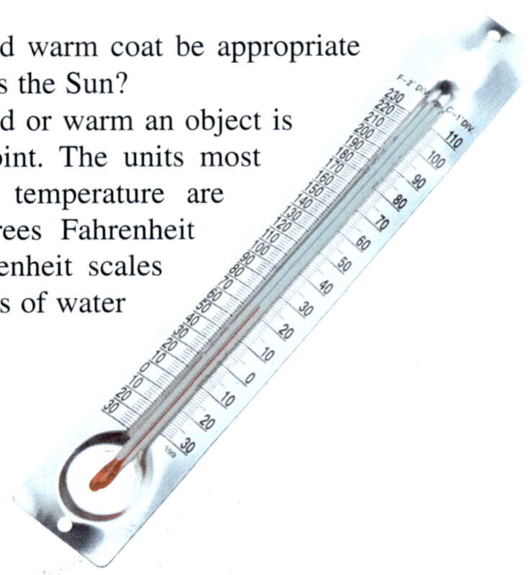

The scale most widely used today, including in Australia, is the Celsius (°C) scale.

The table below summarises the temperatures.

Scale	Freezing point of water	Boling point of water	Used since
Fahrenheit	32°F	212°F	1720s
Celsius	0°C	100°C	1960s

As for all measurements, the unit we are measuring with is very important; for example, a hat and warm coat would be appropriate for a day when the temperature was 32°F, but not for a 32°C day.

WORKED Example 16

Accurately read the following temperatures.

a

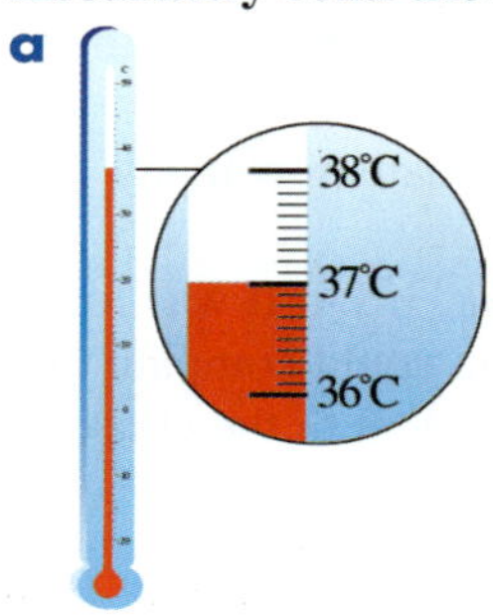

b

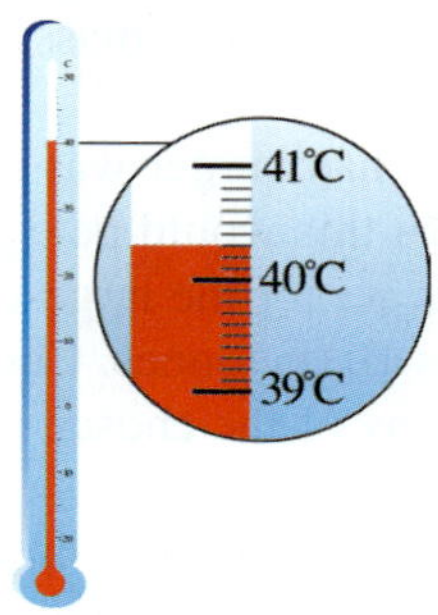

THINK	WRITE
a 1 Read the scale and determine which marking the red liquid lines up with. *Note:* The liquid is in line with the 37° marking.	**a**
2 Write the answer.	The temperature is 37° C.
b 1 Read the scale and determine which marking the red liquid lines up with. *Note:* The liquid has passed the 40° mark.	**b**
2 Consider how many smaller intervals there are between the larger divisions. There are ten intervals between the 40° and 41° marking, so each smaller interval is 0.1° of a unit.	
3 The liquid is in line with the third interval.	
4 Write the answer.	The temperature is 40.3° C.

Converting temperatures

Did you know that the body temperature of a healthy person is approximately 37ºC (98ºF), or the pizza oven's temperature is about 400ºC (750ºF): and the surface temperature of the Sun is around 5500ºC (9900ºF)?

It is important to be able to convert temperature from Celsius degrees to Fahrenheit degrees and vice versa. This can be done using the temperature conversion chart below.

ºC	0	10	20	30	36.9	40	50	60	70	80	90	100	200	300	400
ºF	32	50	68	86	98.4	104	122	140	158	176	194	212	392	572	752

WORKED Example 17

Using the conversion table as a guide, convert 15°C to °F.

THINK

1. The required temperature does not appear in the table; however 15°C is the average of (or halfway between) 10°C and 20°C. The table can therefore be used to obtain the corresponding Fahrenheit values of 10°C (50°F) and 20°C (68°F).
2. Calculate the average of 50°F and 68°F.
3. Write the answer.

WRITE

1. Average temperature (°C) $= \frac{10 + 20}{2}$
   $= \frac{30}{2}$
   $= 15$°C
2. Average temperature (°F) $= \frac{50 + 68}{2}$
   $= \frac{118}{2}$
   $= 59$°F
3. 15°C is equal to 59°F.

The temperature conversion table, although helpful, is very limited. What if we wanted to convert 7°C to °F or 182°F to °C? For this purpose the following formulas are extremely useful.

To convert Celsius to Fahrenheit we use $F = \frac{9}{5}C + 32$, where F is the temperature in °F and C is the temperature in °C.

To convert Fahrenheit to Celsius we use $C = \frac{5}{9}(F - 32)$.

WORKED Example 18

Use the appropriate conversion formula to:
a convert 7°C to °F
b convert 182°F to °C.

THINK

a
1. Write the appropriate formula for converting Celsius to Fahrenheit.
2. Substitute the value of 7 for C.
3. Perform the calculation.
4. Write the answer.

WRITE

a
1. $F = \frac{9}{5}C + 32$
2. $F = \frac{9}{5} \times 7 + 32$
   $= \frac{63}{5} + 32$
3. $= 44.6$
4. 7°C is equal to 44.6°F.

Continued over page

THINK	WRITE
b 1 Write the appropriate formula for converting Fahrenheit to Celsius.	**b** $C = \frac{5}{9}(F - 32)$
2 Substitute the value of 182 for *F*.	$C = \frac{5}{9}(182 - 32)$ $= \frac{5}{9}(150)$
3 Perform the calculation and round to 1 decimal place.	$= \frac{5}{9} \times 150$ $= \frac{750}{9}$ $= 83.333\,333\,33$ ≈ 83.3
4 Write the answer.	182° F is approximately equal to 83.3°C.

WORKED Example 19

The temperature of ice-cream in a bowl is 3°C and is increasing 1°C every minute. What is the temperature of the ice-cream:

a after 5 minutes?

b after 525 seconds?

THINK	WRITE
a 1 Determine how much the temperature has risen in 5 minutes. *Note:* The temperature rises 1°C for every minute, so in 5 minutes it rises 5°C.	**a** Temperature increase = 1°C × 5 = 5°C
2 Add the temperature of the ice-cream to the rise in temperature.	New temperature = 3°C + 5°C = 8°C
3 Answer the question.	The ice-cream's temperature after 5 minutes is 8°C.
b 1 Convert 525 seconds to minutes. *Note:* Divide 525 seconds by 60.	**b** 525 seconds = 525 ÷ 60 $= \frac{525}{60}$ = 8.75 minutes
2 Determine how much the temperature has risen in 8.75 minutes. *Note:* The temperature rises 8.75°C in 8.75 minutes.	Temperature increase = 1°C × 8.75 = 8.75°C
3 Add the temperature of the ice-cream to the rise in temperature.	New temperature = 3°C + 8.75°C = 11.75°C
4 Answer the question.	The ice-cream's temperature after 525 seconds is 11.75°C.

remember

1. Temperature measures how cold or warm an object is in comparison to a reference point.
2. The scale widely used in most parts of the world today, including in Australia, is the Celsius (°C) scale.
3. To convert Celsius (C) to Fahrenheit (F) we use $F = \frac{9}{5}C + 32$.
4. To convert Fahrenheit to Celsius we use $C = \frac{5}{9}(F - 32)$.

EXERCISE 9H Temperature

WORKED Example 16

1 Accurately read the following temperatures.

a
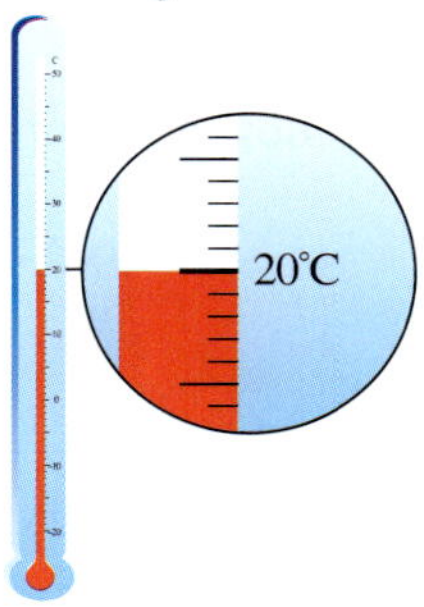

b
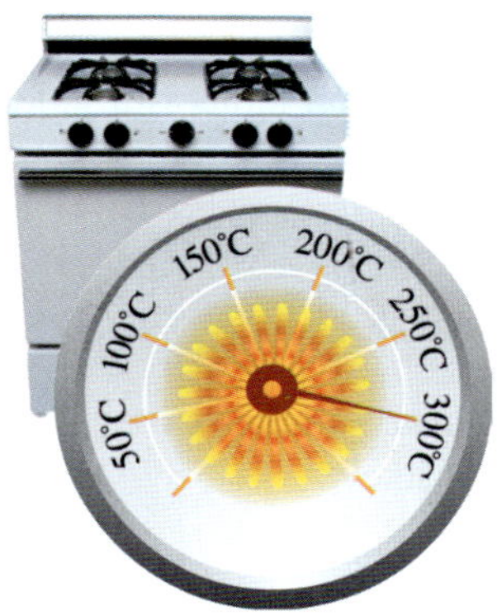

c
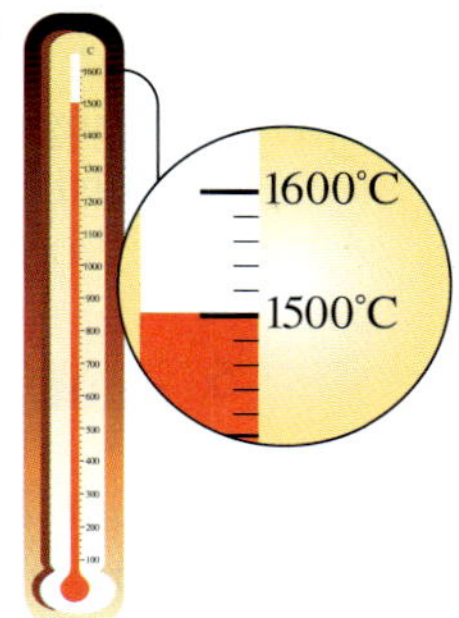

d
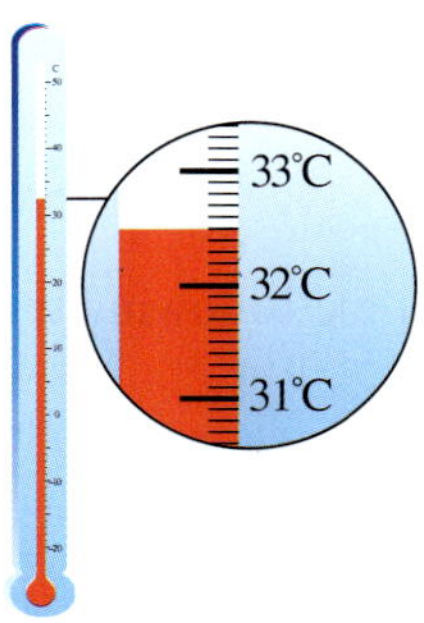

e
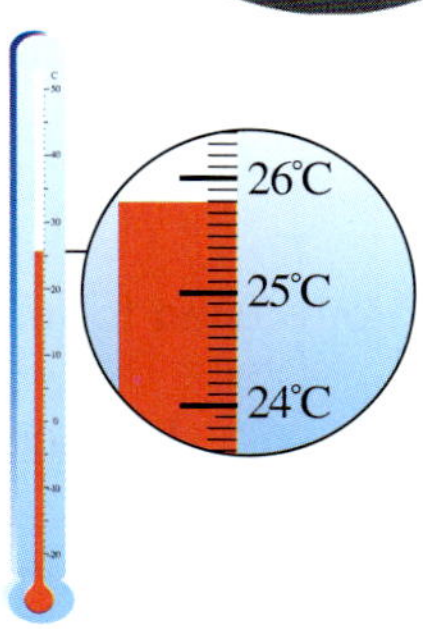

f

g
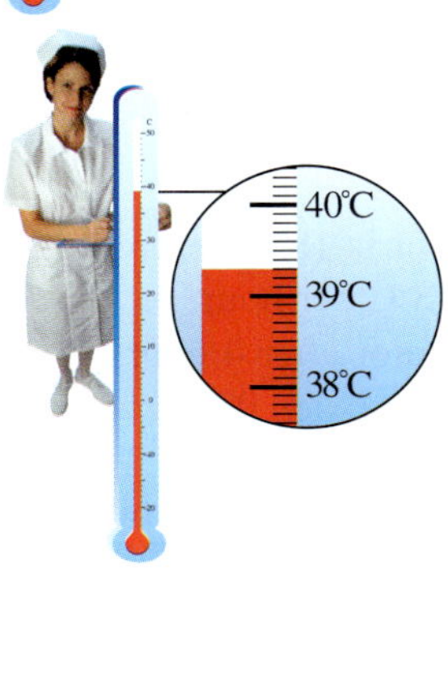

h

Note: In the Galileo thermometer at left, glass spheres with tags are sealed in a liquid-filled cylinder. Each sphere sinks at a different temperature (indicated by tag). Temperature is read from the sphere with the lowest marked temperature that is still floating.

2 Match the event with the possible temperature.

a A winter's morning in Melbourne	1535°C
b Cooking nachos in a microwave oven	30°C
c Melting iron	5500°C
d An air-conditioned office	5°C
e A winter's maximum temperature in Darwin	23°C
f The temperature of the Sun	250°C

Questions 3 and 4 refer to the following conversion table.

°C	0	10	20	30	36.9	40	50	60	70	80	90	100	200	300	400
°F	32	50	68	86	98.4	104	122	140	158	176	194	212	392	572	752

WORKED Example 17

3 Using the conversion table as a guide, convert the following temperatures to degrees Fahrenheit.
a 45°C b 150°C c 12°C

4 Using the conversion table as a guide, convert the following temperatures to degrees Celsius.
a 70°F b 41°F c 400°F

5 **multiple choice**

For a Melbourne summer's day, the weather was reported as very hot. The temperature was most likely:
A 38°F B 50°F C 50°C D 38°C E 25°C

WORKED Example 18a

6 Use the appropriate conversion formula to convert the following temperatures from °C to °F.
a 12°C b 37°C c 1235.6°C

WORKED Example 18b

7 Use the appropriate conversion formula to convert the following temperatures from °F to °C.
a 40°F b 2000°F c 89.95°F

WORKED Example 19

8 The temperature of a glass of water is 8°C and is increasing by 2°C every minute. What is the temperature of the water:
a after 3 minutes?
b after 230 seconds?

9 The temperature of the inside of a refrigerator is 18°C before it is switched on for the first time. If the refrigerator cools down by 3°C per hour:
a find the inside temperature of the refrigerator after 4 hours
b find how long it takes for the refrigerator to reach 0°C.

10 When initially warming up a mug of hot chocolate, the milk's temperature rises at 12.5°C per 30 seconds. If the initial temperature of the milk is 12°C, what is the temperature after 2 minutes?

ICT

How hot is my chocolate?

This activity investigates how quickly hot chocolate cools over a period of 10 minutes. Add 200 millilitres of hot water into a mug or cup containing your chocolate powder. Place a laboratory thermometer in the cup and take an initial reading. Continue to take readings at regular intervals of one minute for 10 minutes.

Copy and complete the following table. Or you can use the Excel file that has already been set up for you on the *Maths Quest 7* CD-ROM.

EXCEL Spreadsheet
Temperature table

	Temperature			
Time (min)	**200 mL water**	**400 mL water**	**Tap water**	**Ice cube**
0				
1				
2				
3				
4				
5				
6				
7				
8				
9				
10				

1 How much has the temperature dropped in the first 5 minutes?

2 How much has the temperature dropped in 10 minutes?

3 Does the temperature decrease by a constant amount? Explain.

4 Would twice as much water cool by the same amount? Investigate.

5 What effect would adding tap water or an ice cube have on how the hot chocolate cools? Investigate.

summary

Copy the sentences below. Fill in the gaps by choosing the correct word from the word list that follows.

1 For time calculations use a ____________ column.

2 In time calculations, work to the next key time (exact hours, 12.00 am and 12.00 pm) until the ____________ time is reached.

3 The time 12 ____________ is 12 pm and 12 midnight is 12 am.

4 24-hour time is expressed as ____________ digit numbers, for example, 0300, 2300, 1530.

5 To convert 'pm' times to 24-hour time, ____________ 12 hours, and write the four digits.

6 BC (before Christ) refers to ____________ the year 1 AD.

7 AD (*anno Domini*) means starting from and ____________ the year 1 AD. This year is an AD year.

8 Determine the ____________ time to be spanned by the line when drawing timelines.

9 Use ____________ divisions along the line.

10 The world is divided into ____________ one-hour time zones.

11 Places to the east are ____________ of those to the west.

12 When calculating mass, to convert to a smaller unit you need to ____________.

13 When calculating mass, to convert to a larger unit you need to ____________.

14 Fill in the gaps:

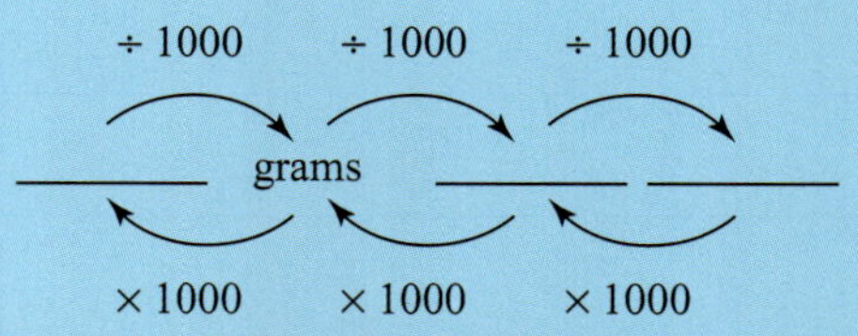

15 The Celsius (°C) scale is used to measure ____________ in Australia.

16 To convert ____________ to Fahrenheit we use ____________.

17 To convert ____________ to Celsius we use ____________.

WORD LIST

add	equal	multiply	24
after	final	$C = \frac{5}{9}(F - 32)$	vertical
ahead	4	noon	$F = \frac{9}{5}C + 32$
before	Fahrenheit	tonnes	
temperature	kilograms	Celsius	
divide	milligrams	total	

CHAPTER review

1 State how many minutes are in each of the following times. 9A

a 3 hours b $5\frac{1}{4}$ hours c $7\frac{1}{2}$ hours
d $1\frac{3}{4}$ hours e 1 day f 3 days

2 Change the following times to hours and minutes. 9A

a 165 minutes b 140 minutes c 210 minutes
d 220 minutes e 800 minutes f 75 minutes

3 Change the following times to minutes. 9A

a 1 hour 20 minutes b 2 hours 40 minutes c 3 hours 10 minutes
d 4 hours 18 minutes e 10 hours 35 minutes f 3 hours 42 minutes

4 Jamie and his family are preparing to visit relatives in Canberra. They begin their journey at 8.15 am and arrive at their destination at 4.30 pm. For how long did they travel? 9A

5 Write each of the following using 24-hour clock time. 9B

a 10.35 pm b 7.15 am c 3.20 am
d 4.42 pm e 9.50 pm f 1.05 am

6 Convert each of the following 24-hour times to 12-hour times. 9B

a 1240 b 0725 c 1550
d 0909 e 2121 f 1120

7 Find the difference between the following 24-hour times. 9B

a 1840 and 0920 b 2112 and 1115 c 2205 and 0627 d 1833 and 1158

8 How many years are there between these (AD) dates? 9C

a 1835 and 1942 b 2051 and 2077 c 2342 and 2105 d 1837 and 2252

9 Ignoring leap years, how much time is there (in years, months and days) between: 9C

a May 1997 and May 2006?
b 5 September 1995 and 10 September 2002?
c 28 December 2010 and 30 December 2112?
d 8 June 1922 and 21 January 1987?

10 Which of the following are leap years? 9C

a 2004 b 2005 c 2800 d 2804
e 2900 f 3204 g 3100 h 3900

11 Show the following information on a timeline. 9D

776 BC	First Olympic Games were held.
600 BC	First modern coins minted.
204 BC	Great Wall of China completed.
170 BC	First paved streets were built (in Rome).
110 BC	Romans began to cultivate oysters.
105 AD	First paper was made.
190 AD	The Chinese calculated pi (π) accurately to 5 decimal places.
271 AD	First simple compass was used by the Chinese.

9E **12** Use the Melbourne to Mt Gambier West Coast Railway timetable on page 363 to answer the following questions.
  a At what time did the 1042 from Colac leave the Spencer St station?
  b How long does it take to travel from Melbourne (Spencer St) to Warrnambool on a Tuesday morning?
  c How long does it take to travel from Geelong to Marshall?

9F **13** What is the time in Melbourne if the GMT (Greenwich Mean Time) is:
  a 11.20 am? **b** 3.30 pm? **c** 9.15 pm? **d** 6.23 am?

9F **14** What is the GMT (Greenwich Mean Time) when the time in Melbourne is:
  a 11.30 pm? **b** 11.30 am? **c** 1450 hours? **d** 0835 hours?

9F **15** Marian's grandmother lives in the UK and wants to ring Marian for her birthday. At what time (GMT) should Marian's grandmother ring if she wants to speak to Marian at 5.00 pm EST (Eastern Standard Time)?

9G **16** Convert the following units.
  a 3000 kg to tonnes **b** 25 kg to grams **c** 500 mg to grams **d** 20 000 g to kilograms

9G **17** Adrian was building a wall with bricks, each of a mass of 2.7 kg. What is the mass of the wall if he used 422 bricks?

9G **18** If Julia goes to the supermarket and buys 1 kg of flour, 250 g of rice, 100 g of sliced ham, 50 g of tinned baked beans and 675 g of fresh tomatoes, what is the total mass of her purchase (in grams)?

9G **19** As Mary Victoria grows, her mass increases from 2.73 kg to 5.48 kg. What is the increase in mass?

9H **20** Read the following temperatures.

a

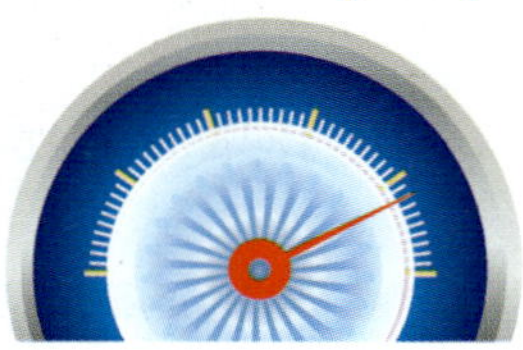

b

c

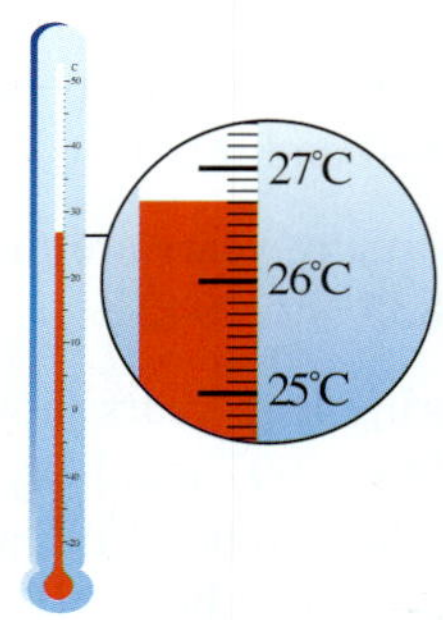

9H **21** Use the appropriate conversion formula to convert the following temperatures from °C to °F.
  a 55°C **b** 250°C **c** 18°C

9H **22** Use the appropriate conversion formula to convert the following temperatures from °F to °C.
  a 90°F **b** 180°F **c** 800°F

9H **23** The temperature of an ice cube of water is 0°C and is increasing by 3°C every hour. What is the temperature:
  a after 2 hours? **b** after 20 minutes?

9H **24** The temperature of a cup of tea is 60°C, if it cools at 8° per quarter hour, find the temperature after 1 hour.

test yourself CHAPTER 9

10 Angles

The accuracy of angles in the building trade is vital. Can you think of reasons for this? The timber framework or truss for a roof is made up of triangular shapes for strength.

By looking at the photograph, can you see how many angles in the framework are less than 90°? Are there any angles greater than 90°? If so, how many?

By the end of this chapter, you will be able to measure, classify and construct angles.

Are you ready?

Try the questions below. If you have difficulty with any of them, extra help can be obtained by completing the matching SkillSHEET. Either click on the SkillSHEET icon next to the question on the *Maths Quest 7* CD-ROM or ask your teacher for a copy.

SkillSHEET 10.1

Reading the scale on a protractor

1 What is the size of the angle shown in the diagram at right?

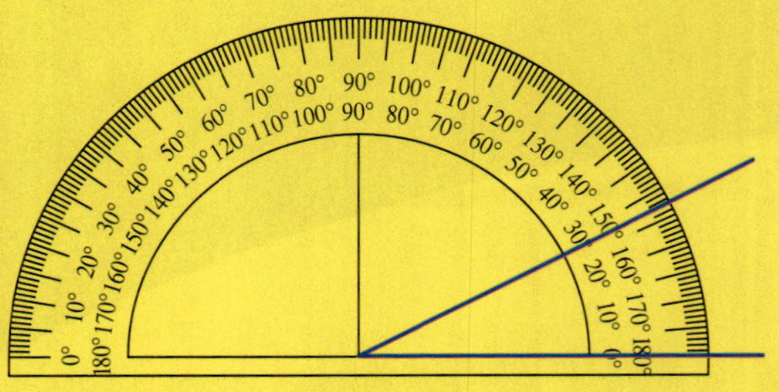

SkillSHEET 10.2

Comparing the sizes of angles

2 Consider the three angles shown.

A

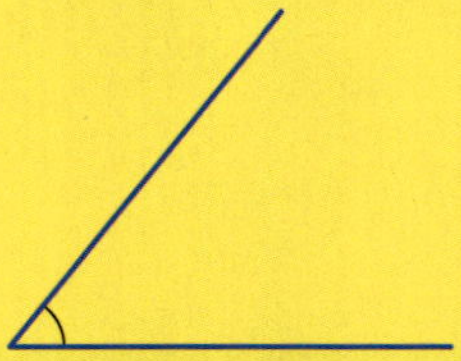

B

C

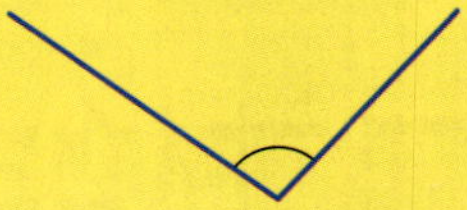

a Which diagram shows the largest angle? **b** Which diagram shows the smallest angle?

SkillSHEET 10.3

Estimating the size of an angle

3 Choose one of the following alternatives to complete the sentences below about the angles in question **2**.

i 10° and 20° **ii** 30° and 40° **iii** 45° and 55°
iv 70° and 80° **v** 90° and 100° **vi** 115° and 125°

a The size of the angle in diagram A can be estimated to be between ____________.
b The size of the angle in diagram B can be estimated to be between ____________.
c The size of the angle in diagram C can be estimated to be between ____________.

SkillSHEET 10.4

Measuring angles with a protractor

4 Measure the size of each angle in question **2** with a protractor.

SkillSHEET 10.5

Constructing angles with a protractor

5 Construct an angle of 50° using a protractor.

SkillSHEET 10.6

Classifying angles

6 Angles can be classified according to their size. Match the name of the angle to the given sizes.

a between 0° and 90°	**A** reflex
b 90°	**B** acute
c between 90° and 180°	**C** right
d 180°	**D** revolution
e between 180° and 360°	**E** straight
f 360°	**F** obtuse

SkillSHEET 10.7

Naming angles

7 Use three letters to name each of the following angles.

a

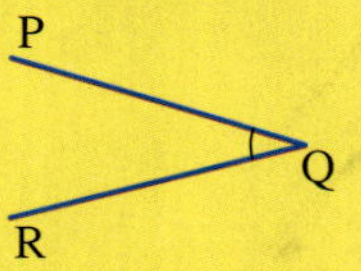

b

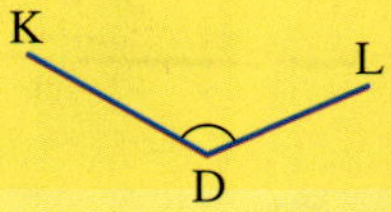

Introduction

Angles are used widely in everyday life. Builders use them in constructing houses; artists use them to help in sketching and painting; and architects and engineers use their knowledge and understanding of angles for design and construction work.

An **angle** is made when two lines meet at a point. The point where the two lines meet is called a **vertex** and the lines themselves are called *arms* of the angle.

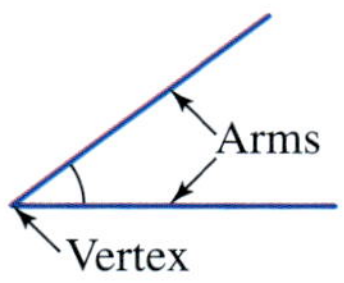

Measuring angles

Angles are measured in **degrees (°)**. The device used for measuring angles is called a **protractor**. There are two types of protractors: circular and semicircular.

The circular protractor

Circular protractors are marked from 0° to 360°.

To measure angles using a circular protractor:

1. place the centre of the protractor on the vertex of the angle
2. match the 0° horizontal centre line with one of the arms of the angle
3. read the size of the angle indicated by the other arm.

It is wise to estimate the size of an angle before measuring it. This gives an indication as to whether the measure is a reasonable one.

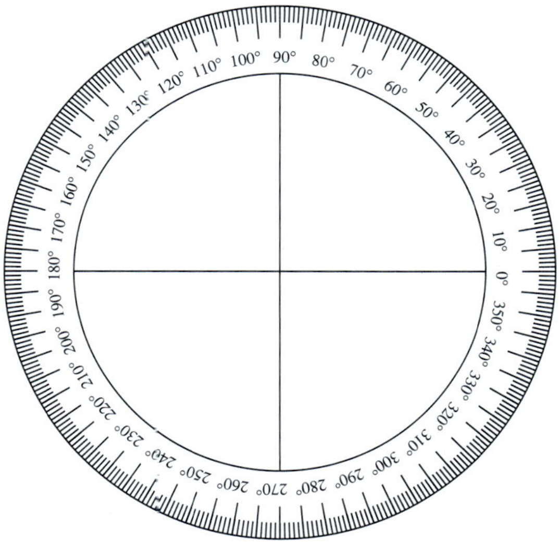

WORKED Example 1

First estimate and then measure the size of each of the following angles using a circular protractor. Comment on the accuracy of the estimate.

a

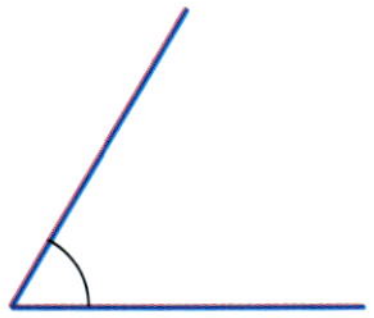

b

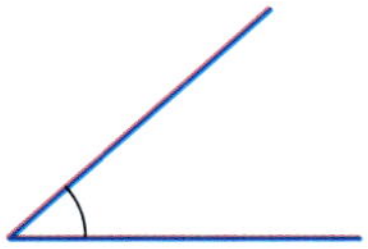

c

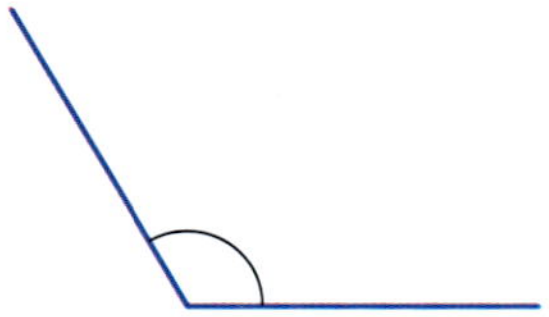

THINK	WRITE/MEASURE
a 1 Estimate the size of the angle. Write the estimate. *Note:* The angle is between 0° and 90° but closer to 90°.	**a** The estimate is approximately 60°.

Continued over page

THINK

WRITE/MEASURE

2 Place the centre of the protractor on the vertex of the angle.

3 Match the horizontal line that passes through the centre of the protractor and points to 0° with one of the arms of the angle.

4 Read the size of the angle indicated by the other arm. (You may need to extend the line so that you can read it on the protractor.)

a

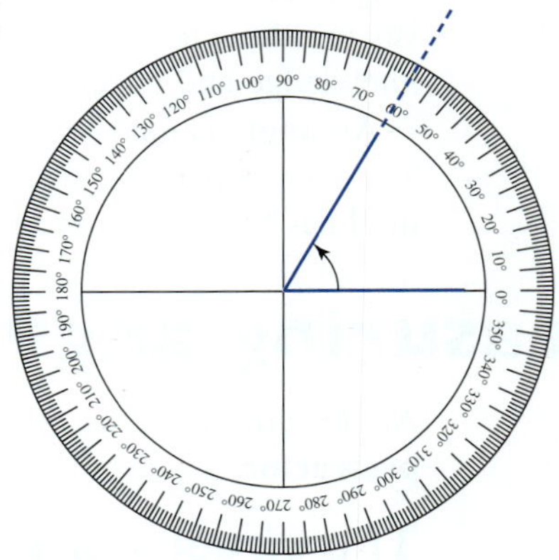

5 Comment on the accuracy of the estimate.

The size of this angle is 60°.
The estimate was an accurate one.

b 1 Estimate the size of the angle. Write the estimate. *Note:* The angle appears to be approximately halfway between 0° and 90°.

2 Repeat steps 2–5 from part **a**.

b The estimate is approximately 45°.

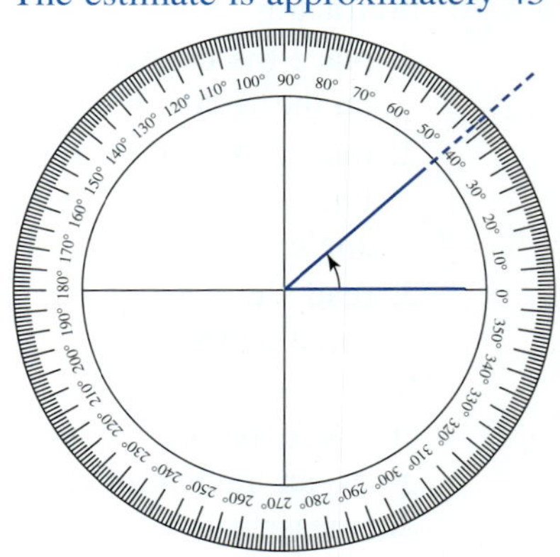

The size of this angle is 42°.
The estimate was a reasonable one.

c 1 Estimate the size of the angle. Write the estimate. *Note:* The angle is between 90° and 180° but closer to 90°.

2 Repeat steps 2–5 from part **a**.

c The estimate is approximately 115°.

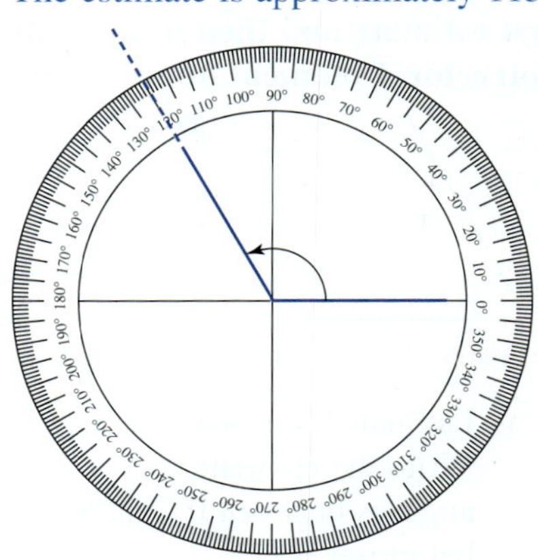

The size of this angle is 120°.
The estimate was a reasonable one.

The semicircular protractor

Semicircular protractors have two scales (inner and outer) which are marked from 0° to 180°. Either of these two scales can be used for measuring angles.

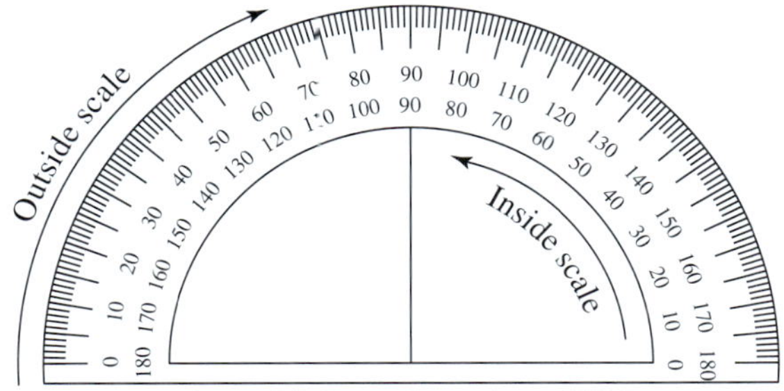

To measure angles which are less than 180°:

Step 1 Line up the centre of the protractor's baseline with the vertex (that is, where the vertical 90° line, intersects with the horizontal 180° baseline).

Step 2 Match the base of the protractor with one of the arms of the angle.

Step 3 Use the scale on the base starting at 0 to read the number of degrees spanned by the angle.

WORKED Example 2

First estimate and then measure the size of the angle shown at right using a semicircular protractor. Comment on the accuracy of the estimate.

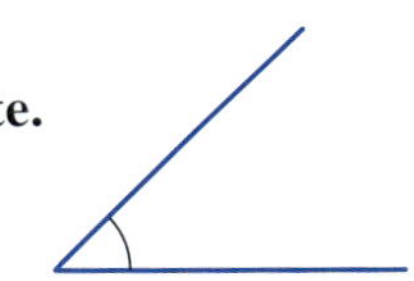

THINK	WRITE/MEASURE
1 Estimate the size of the angle. Write the estimate. *Note:* The angle appears to be approximately halfway between 0° and 90°.	The estimate is approximately 45°.
2 Line up the centre of the protractor's baseline with the vertex (that is, where the vertical, 90° line, intersects with the horizontal 180° baseline).	
3 Match the base of the protractor with one of the arms of the angle. (You may need to extend the line so that you can read it on the protractor.)	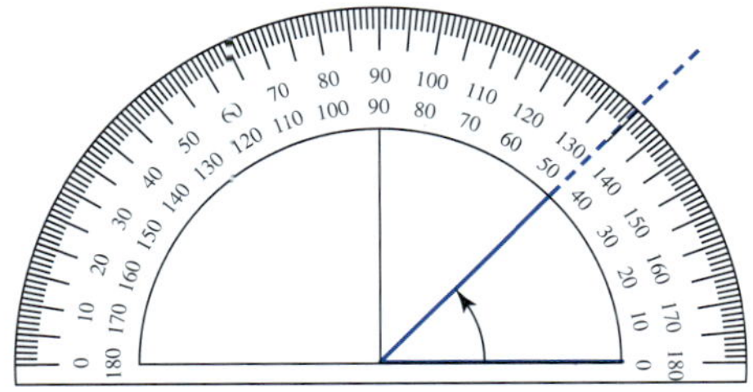
4 Use the scale on the base starting at 0° (in this case an inner scale) to read the number of degrees spanned by the angle.	The size of this angle is 45°.
5 Comment on the accuracy of the estimate.	The estimate was an accurate one.

To measure angles greater than 180°:

Step 1 Measure the angle enclosed by the straight lines (that is, the one that is smaller than 180°).

Step 2 Subtract the number obtained in step (1) from 360°.

WORKED Example 3

First estimate and then measure the size of the angle shown at right using a semicircular protractor. Comment on the accuracy of the estimate.

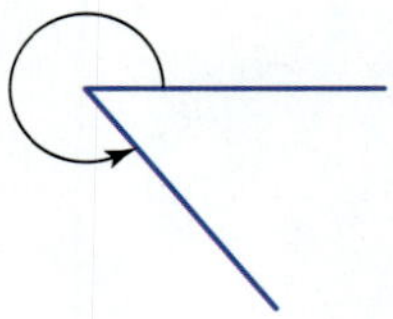

THINK	WRITE/MEASURE
1 Estimate the size of the angle. Write the estimate. *Note:* The angle appears to be approximately halfway between 270° and 360°.	The estimate is approximately 315°.
2 Measure the angle enclosed by the straight lines.	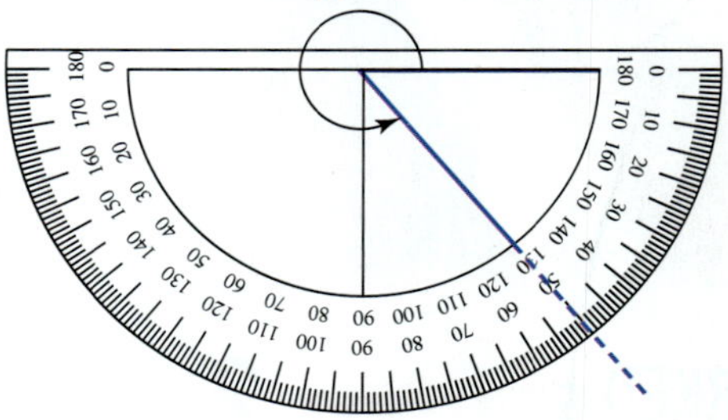 The smaller angle is 50°.
3 Subtract 50° from 360° to find the required angle.	Required angle: 360° – 50° = 310°
4 Comment on the accuracy of the estimate.	The estimate was a reasonable one.

Check the size of the angle in worked example 3 with a circular protractor.

remember

1. When two lines meet at a point, an *angle* is formed.
2. The point where the lines meet is called a *vertex* and the lines are called the *arms* of the angle.
3. All angles are measured in degrees using a protractor.
4. Circular protractors are marked from 0° to 360°. If the centre of the protractor is placed on the vertex and one of the arms points at 0°, the other arm will point to the angle being measured.
5. Semicircular protractors have two scales (inner and outer) which are marked from 0° to 180°. Either of these two scales can be used for measuring angles. Angles that are less than 180° can be measured in the same way as with circular protractors. To measure an angle that is larger than 180°, the angle enclosed by the arms (that is, the one that is less than 180°) is measured first and is then subtracted from 360°.
6. It is a good idea to estimate the size of an angle before measuring it.

EXERCISE 10A Measuring angles

WORKED Example 1, 2, 3

1 First estimate and then measure the size of each of the following angles using a protractor. Comment on the accuracy of the estimate.

Cabri Geometry
Angles

a
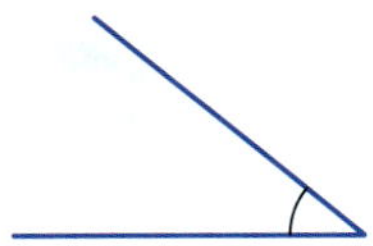

b
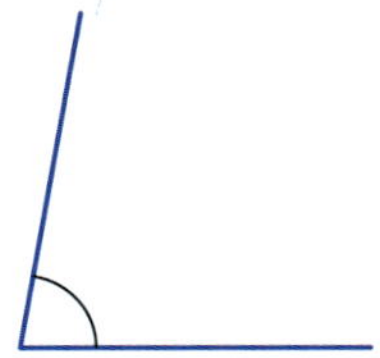

c
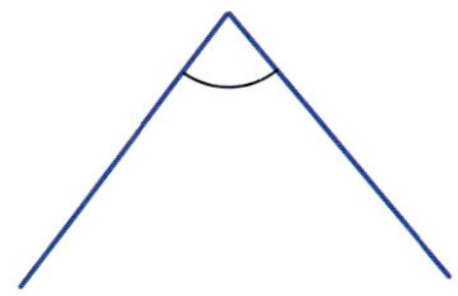

10.1 SkillSHEET
Reading the scale on a protractor

d
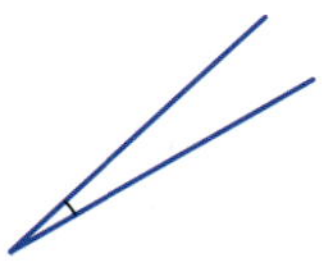

e

f
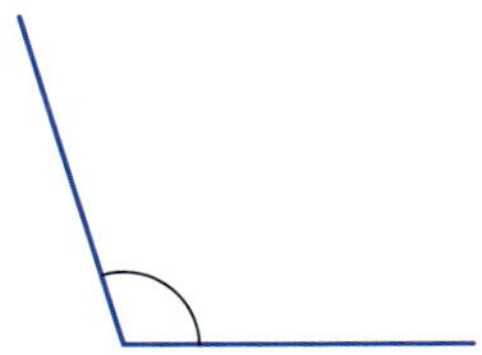

10.2 SkillSHEET
Comparing the sizes of angles

g
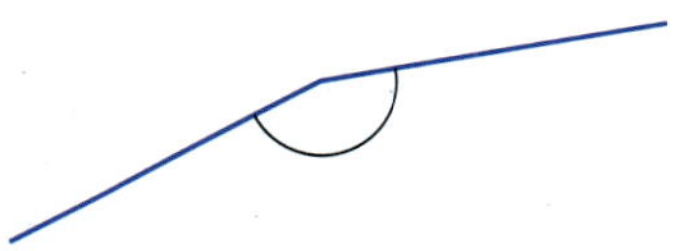

h
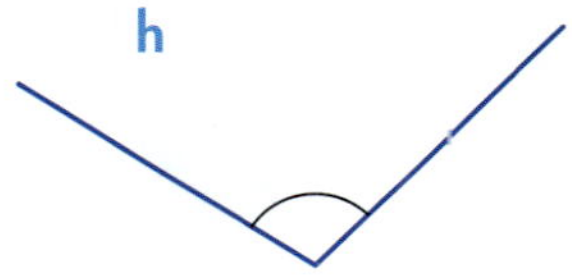

i
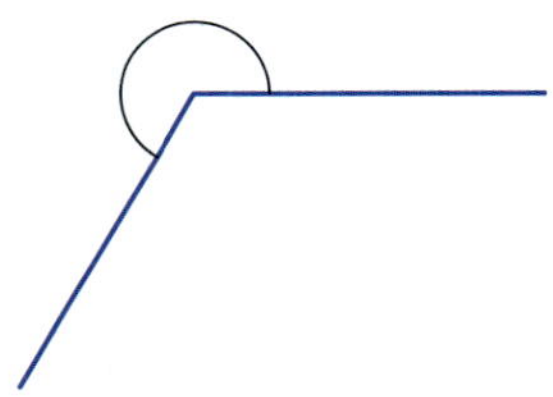

10.3 SkillSHEET
Estimating the size of an angle

j
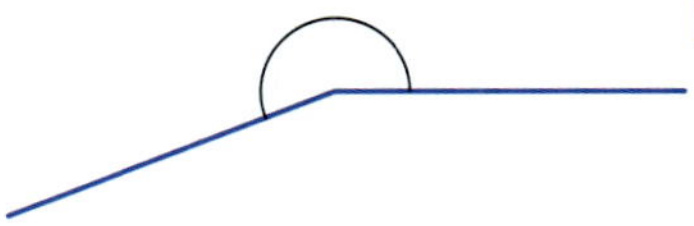

k
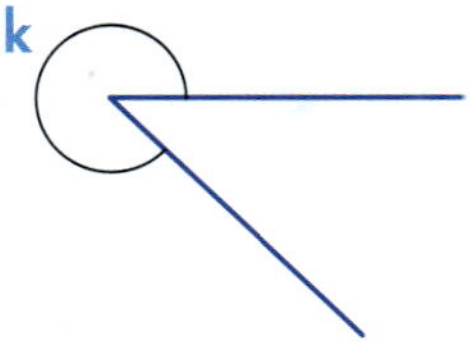

l
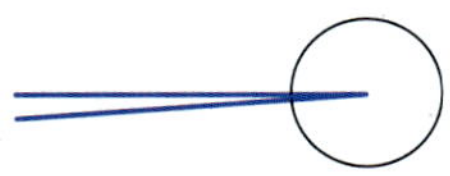

10.4 SkillSHEET
Measuring angles with a protractor

2 Comment on the accuracy of your measurement compared with the answer in the text.

3 **multiple choice**

a The size of the angle shown is closest to:

A 20° B 30° C 40°

D 150° E 140°

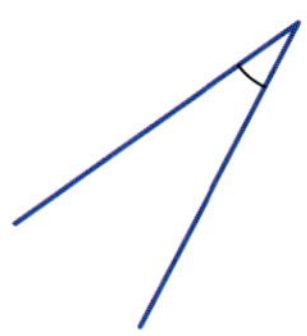

b The size of the angle shown is closest to:

A 42° B 138° C 142°

D 145° E 38°

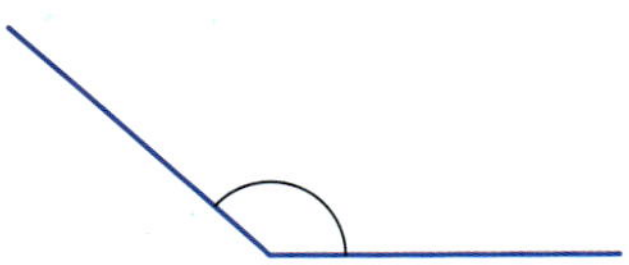

4 In each of the photographs, measure the indicated angle.

a

b

c

d

e

f

5 The recommended slope for wheelchair access is 4° from the horizontal. An angle greater than this makes it more difficult for the person in the wheelchair to travel up the ramp.

a Would you rate the slope of the bottom section of the ramp as 'difficult', 'as recommended' or 'easy'? Why?

b How would you rate the second part of the ramp?

GAME time
Angles — 001

Ten Olympic games sites

Use your protractor to measure the sizes of the angles in the 9 triangles below. The angles and letters or numbers give the answer code.

18° 72° 90° 23° 48° | 109° 109°
95° 90° 60° 25° 90° | 30° 85°
18° 25° 65° 132° 72° 25° | 28° 28°
20° 90° 18° 44° 90° 102° | 109° 28°
20° 23° 132° 34° 44° 81° | 42° 57°
130° 95° 48° 130° 132° 95° 130° | 40° 30°
20° 90° 132° 95° 10° 72° 130° 48° | 42° 30°
112° 130° 10° 44° 72° 48° 90° 132° 130° | 40° 57°
48° 90° 18° 130° 132° 52° 72° 48° 72° 18° | 109° 85°
20° 72° 16° 34° 44° 90° 44° 34° 95° 25° | 30° 109°

COMMUNICATION

Estimating the size of an angle

1 Copy the following table.

Diagram	Estimate of angle	Measured angle	Difference
1			
2			
3			
4			

2 Without using a protractor, estimate the size of each angle in the first four diagrams of the golf sequence above and complete the second column of the table.

3 Explain how you obtained an estimate. What steps did you follow?

4 Complete the third column by measuring each angle with a protractor.

5 Calculate the difference between your estimate and the actual value of each angle.

6 Describe how you could improve your skills in estimating the size of an angle.

7 Try these improvements for estimating the size of the angles in the remaining four diagrams. Copy and complete the following table.

Diagram	Estimate of angle	Measured angle	Difference
5			
6			
7			
8			

8 Has the difference between the estimated and actual value for each angle become smaller? Have your estimating skills improved?

Constructing angles with a protractor

Protractors can be used not only for measuring, but also for constructing angles. To construct an angle using a circular protractor, follow the steps given below.

Step 1 Draw a straight line.

Step 2 Put a dot at one end of the line. (The line represents one of the arms and the dot represents the vertex of the angle.)

Step 3 Place the protractor so that its centre is on the vertex and 0° is on the baseline.

Step 4 Find the required angle on the scale and mark a small dot at the edge of the protractor.

Step 5 Join the small dot with the vertex to form the second arm of the angle.

Step 6 Label the angle.

A semicircular protractor can also be used for constructing angles. If the angle to be constructed is between 0° and 180°, the steps are exactly the same as when using a circular protractor.

WORKED Example 4

Construct each of the following angles.
a 50° **b** 152.5°

THINK	WRITE/DRAW
a 1 Draw a baseline. Put a dot at one end. This is to be the vertex of the angle.	**a** 
2 Place the protractor so that its centre is on the vertex and 0° is on the baseline. 3 Find 50° on the scale and mark a small dot at the edge of the protractor.	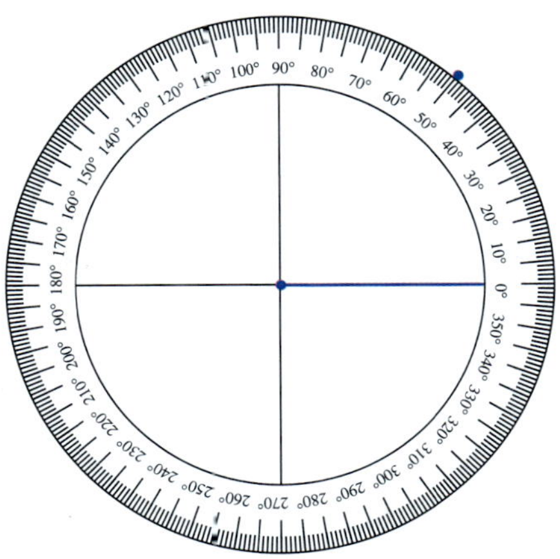
4 Join the small dot and the vertex with a straight line. 5 Label the angle.	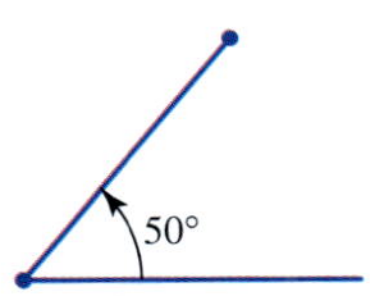

Continued over page

THINK

b 1 Repeat steps 1 and 2 from part **a**.

2 Find 152.5° on the scale and mark a small dot at the edge of the protractor.
Remember to start at 0° and, using the inside scale, move in an anticlockwise direction.

3 Join the vertex and the dot where you have marked the 152.5° with a straight line.

4 Label the angle.

WRITE/DRAW

b

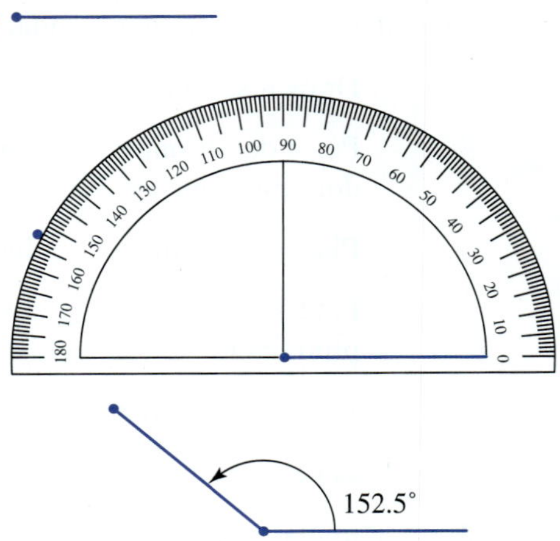

If the angle is larger than 180°, the procedure is different.

To construct an angle larger than 180° using a semicircular protractor:

Step 1 subtract the given angle from 360°

Step 2 construct the resulting angle (it will be less than 180°)

Step 3 label the required angle: it is 'outside' the one that has been constructed.

WORKED Example 5

Construct an angle of 210° using a semicircular protractor.

THINK

1 Subtract 210° from 360°.

2 Draw a baseline. Put a dot at one end. This will become the vertex of the angle.

3 Place the protractor upside down on the baseline so that its centre is on the vertex.

4 Start at 0° and, using the ouside scale, move in a clockwise direction; mark the position of 150° with a small dot.

5 Remove the protractor and join the vertex to the small dot representing 150°.

6 Mark the 'outside' angle as 210°.

WRITE/DRAW

$360° - 210° = 150°$

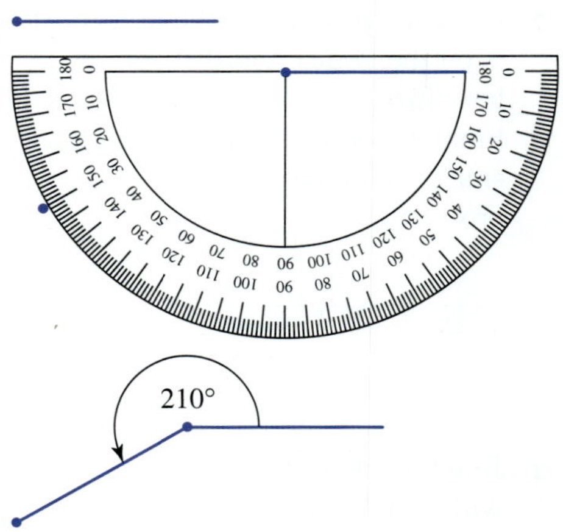

Note that it is not necessary to place the protractor upside down. The diagram on the right shows the angle of 210°, constructed by placing the protractor in the upright position.

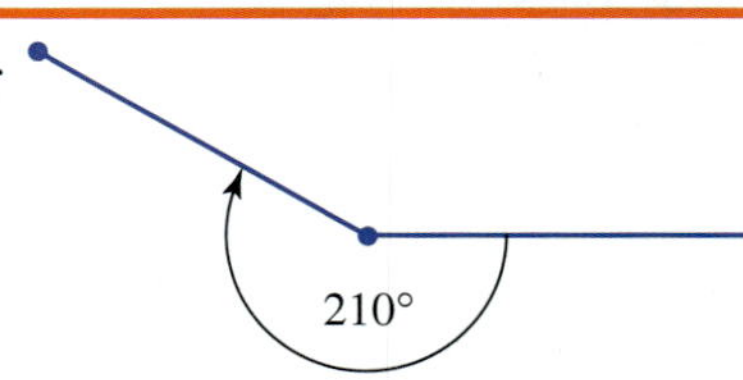

remember

1. To construct an angle, follow the steps given below.
   (a) Rule a straight line.
   (b) Put a dot at one end of the line. (The line represents one of the arms and the dot represents the vertex of the angle.)
   (c) Place the protractor so that its centre is on the vertex and 0° is on the baseline.
   (d) Find the required angle on the scale and mark a small dot at the edge of the protractor.
   (e) Join the small dot with the vertex to form the second arm of the angle.
   (f) Label the angle.
2. To construct an angle that is larger than 180°, using a semicircular protractor:
   (a) subtract the given angle from 360°
   (b) construct the resulting angle (it will be less than 180°)
   (c) label the required angle: it is 'outside' the one that has been constructed.

EXERCISE 10B Constructing angles with a protractor

10.5 SkillSHEET Constructing angles with a protractor

GC program–TI Constructing angles

GC program–Casio Constructing angles

EXCEL Spreadsheet Constructing angles

WORKED Example 4a

1 Construct each of the following angles.
a 15° b 9° c 53° d 75.5°
e 45° f 40.8° g 88° h 76°

WORKED Example 4b

2 Construct each of the following angles.
a 96° b 92° c 165.2° d 143°
e 140° f 156.4° g 127° h 149°

WORKED Example 5

3 Construct each of the following angles using a semicircular protractor.
a 185° b 210.5° c 235° d 260°
e 243° f 192° g 249.3° h 214°

4 **multiple choice**
To construct an angle of 212° using a semicircular protractor, we need to construct an angle less than 180° first and then label the 'outside' angle. The size of the first angle to be constructed is:
A 212° B 58° C 148° D 112° E none of these

5 Construct each of the following angles.
a 295° b 269° c 307° d 349° e 328° f 300° g 345° h 358°

Types of angles

Angles can be classified according to their size.

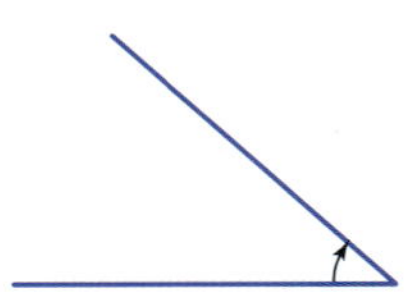

An **acute angle** is greater than 0°, but less than 90°.

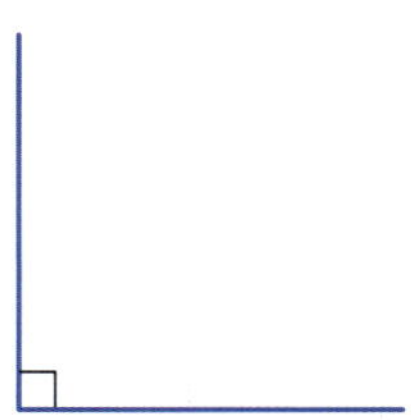

A **right angle** is an angle that equals exactly 90°.

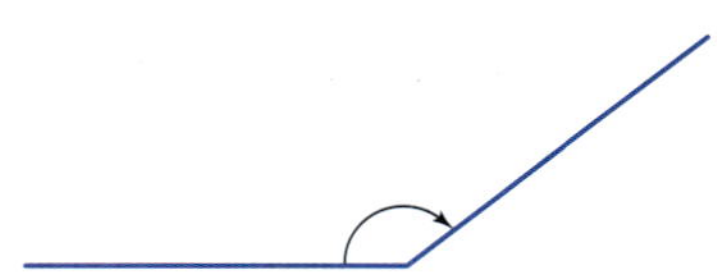

An **obtuse angle** is greater than 90° but less than 180°.

A **straight angle** equals exactly 180°.

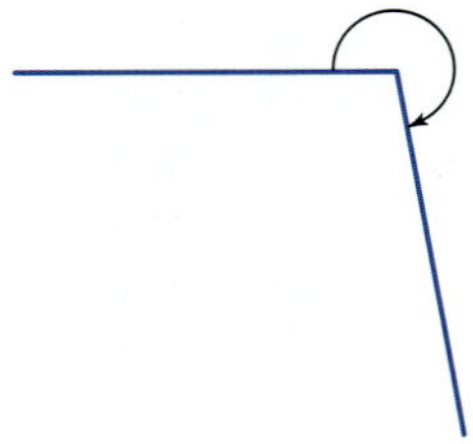

A **reflex angle** is greater than 180° but less than 360°.

A **revolution** or a **perigon** is an angle of 360° (a full circle).

WORKED Example 6

Classify each of the following angles according to their size.
a 115° b 27° c 300°

THINK	WRITE
a The given angle is larger than 90°, but smaller than 180°, so classify it accordingly.	**a** 115° is an obtuse angle.
b The given angle is between 0° and 90°, so classify it accordingly.	**b** 27° is an acute angle.
c The given angle is larger than 180°, but less than 360°, so classify it accordingly.	**c** 300° is a reflex angle.

remember

Angle size	Diagram	Name of angle
between 0° and 90°		Acute
90°		Right
between 90° and 180°		Obtuse

Angle size	Diagram	Name of angle
180°		Straight
between 180° and 360°		Reflex
360°		Revolution

EXERCISE 10C Types of angles

SkillSHEET 10.6 Classifying angles

WORKED Example 6

1 Classify each of the following angles according to its size.

a 12° b 215.3° c 98° d 156°
e 180° f 62° g 355.2° h 90°
i 4.8° j 360° k 100° l 45°
m 82° n 270° o $36\frac{1}{2}$° p 195°

2 The following list gives values of particular angles.

3°, 45°, 65°, 123°, 69°, 234°, 90°, 360°, 300°, 270°, 165°, 210°, 180°

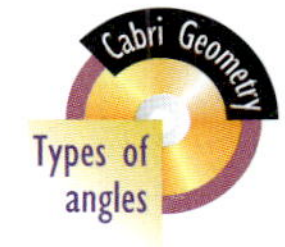

a How many of these are acute angles? List them.
b How many of these are obtuse angles? List them.
c Is there a right angle in the list?
d Is there a straight angle?
e How many reflex angles are there? List them.
f Is there a full revolution? What is its value?

3 Consider the diagrams below and write down the type of angle shown in each case.

a

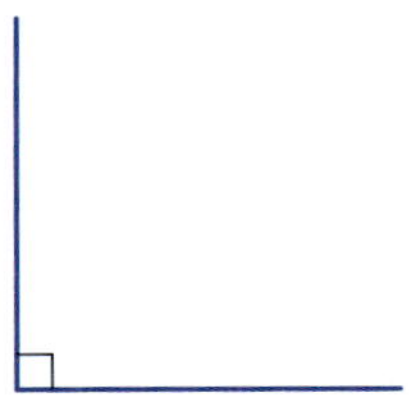

b

c

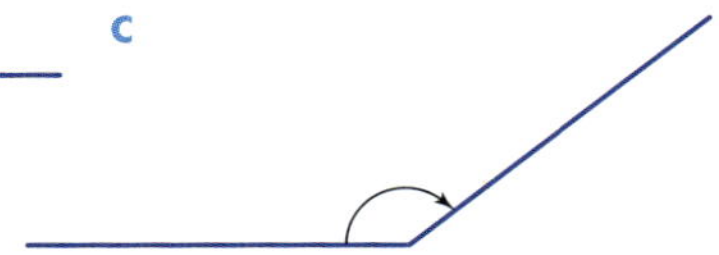

d

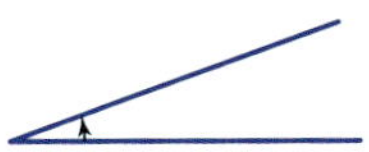

e

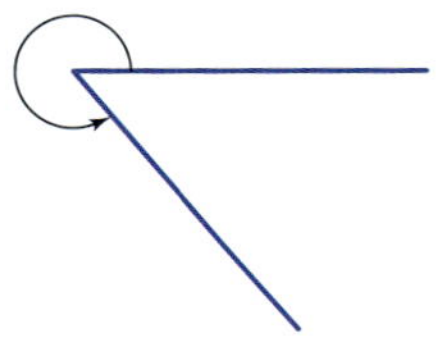

f

4 State the type of angle shown in each photograph.

a

b

c

d

5 State the type of angle that the minute hand of a clock sweeps while moving for the period of:

a 10 minutes
b 15 minutes
c 20 minutes
d 30 minutes
e 45 minutes
f 1 hour.

6 At various times of the day the hands of a clock form two angles: the 'inside' angle (the smaller one) and the 'outside' angle (the larger one).

a State the type of the inside angle, formed by the hands of the clock at:

i 1 o'clock ii 20 minutes to 12 iii 6 o'clock
iv 9 o'clock v quarter to 4 vi half past 10.

b Write two different times when the inside angle of the clock is:

i acute ii obtuse iii right.

7 Angles are very important in engineering and architecture. Three angles have been marked in the photograph shown.

a Classify each angle.

b Use a protractor to measure each angle.

c With a partner, locate as many other angles as you can that are the same size as the marked angles 1, 2 and 3.

GAME time Angles — 002

10 QUICK QUESTIONS 1

1 Use the protractor provided to find the size of the angle.

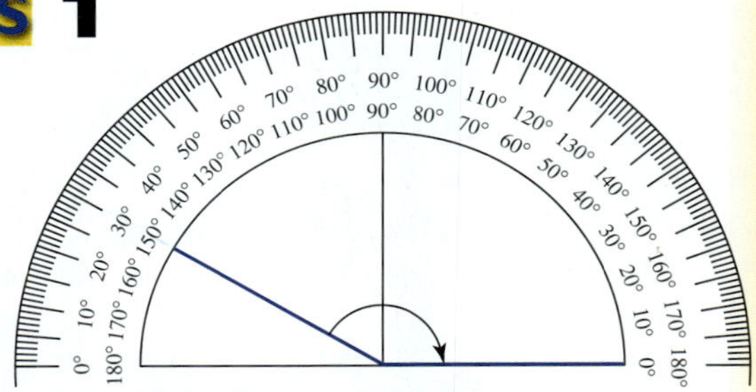

2 First estimate, then use a protractor to measure the size of the angle shown on the right. Comment on the accuracy of your estimate.

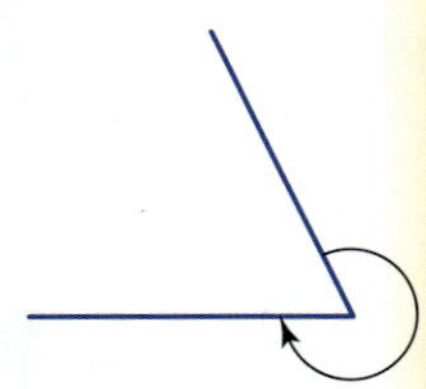

3 Use a protractor to draw an angle of 25°.

4 Use a protractor to draw an angle of 212°.

Name the following angles in questions 5 and 6 as either acute, right, obtuse, straight, reflex or a revolution.

5 156° 6 340°

Use the photo of a beach hut (on the right) to answer questions 7 to 10 for the white timbers.

7 How many acute angles are there?

8 How many right angles are there?

9 How many obtuse angles are there?

10 How many reflex angles are there?

Naming angles

Angles can be named using the capital letters of the English alphabet. A common way to name an angle is to use three letters: two letters to represent either arm of the angle and the third letter to represent its vertex. The letter, denoting the vertex is always placed in the middle (between the two other letters).

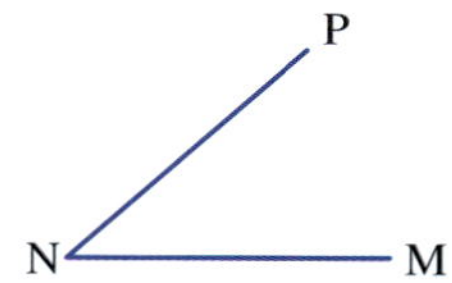

Instead of writing the word 'angle' we use the symbol ∠ before the letters. For example, the angle shown at right can either be named ∠MNP, or ∠PNM.

WORKED Example 7

Name each of the following angles.

a

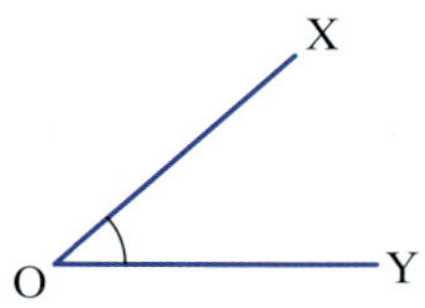

b

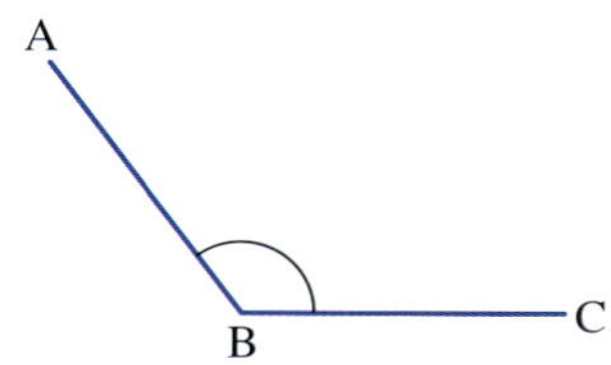

THINK

a Name the angle by starting either from X, or from Y, and placing O (the letter for the vertex) in the middle. Remember to place the angle symbol (∠) before the letters.

b Name the angle by starting either from A, or from C, and placing B (the letter representing the vertex) in the middle. Remember to place the angle symbol (∠) before the letters.

WRITE

a ∠XOY or ∠YOX

b ∠ABC or ∠CBA

WORKED Example 8

Draw the acute angle ∠ADG.

THINK

1. Construct any acute angle (unless specified otherwise).
2. Letter D is in the middle, so place it at the vertex.
3. Place letter A at one arm of the angle and letter G at the other. (Note that it does not matter which arm is represented by AD and GD.)

DRAW

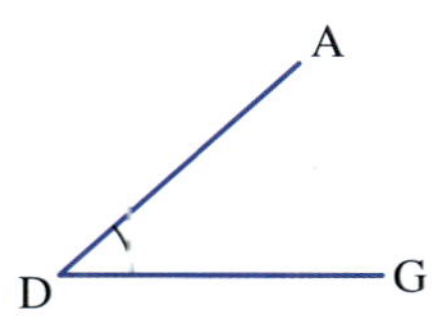

remember

1. An angle can be named using three capital letters from the English alphabet: two for the arms and one for the vertex of the angle.
2. The letter representing the vertex of the angle is always placed in the middle. Use the symbol ∠ instead of the word 'angle' before the three letters.

EXERCISE 10D Naming angles

SkillSHEET 10.7 Naming angles

Cabri Geometry Naming and measuring angles

WORKED Example 7

1 Name each of the following angles.

a 

b 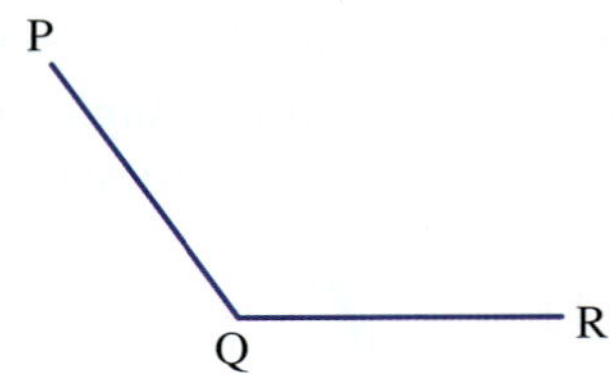

c 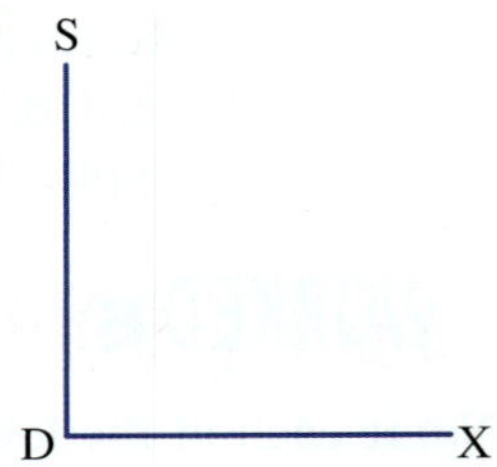

d 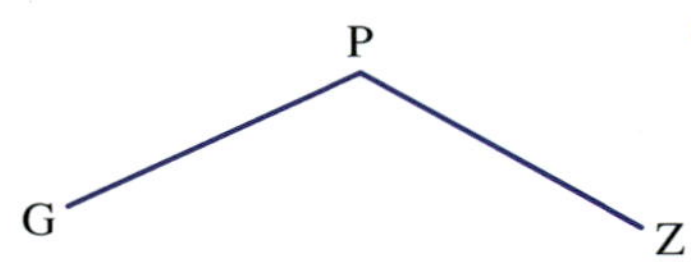

e 

f 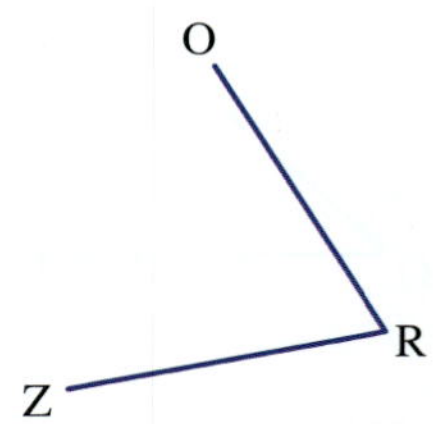

WORKED Example 8

2 Draw each of the following angles, to the type as specified in brackets.

a ∠CDE (acute)
b ∠TRE (obtuse)
c ∠QAS (straight)
d ∠FGH (reflex)
e ∠KJF (right)
f ∠NBV (acute)

3 Give the letter name of each acute and obtuse angle in the following diagrams.

a 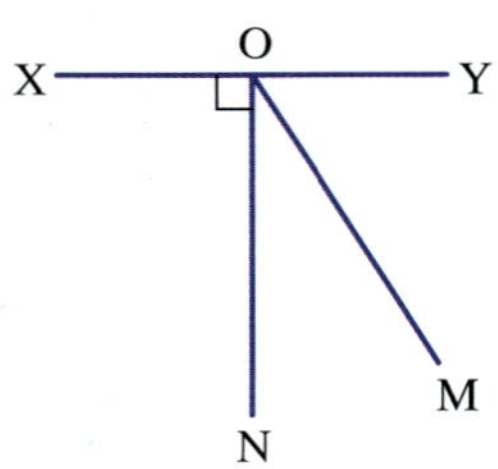

b 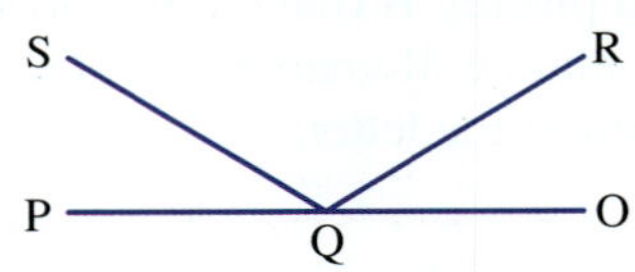

c 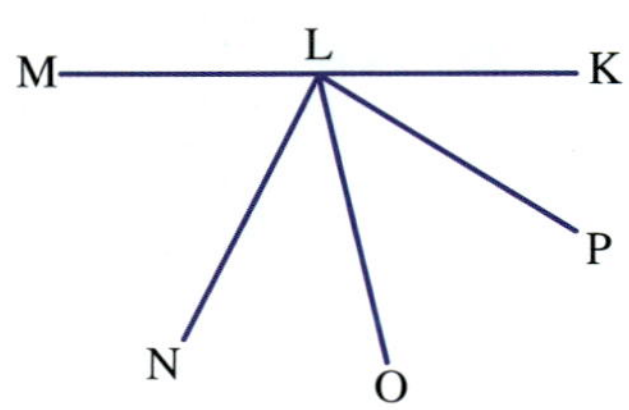

d 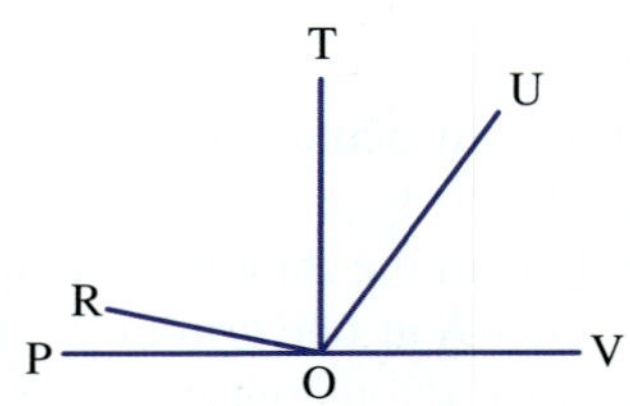

e 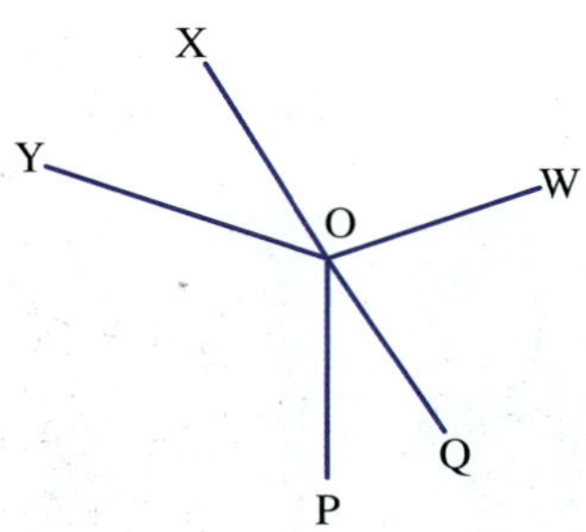

f 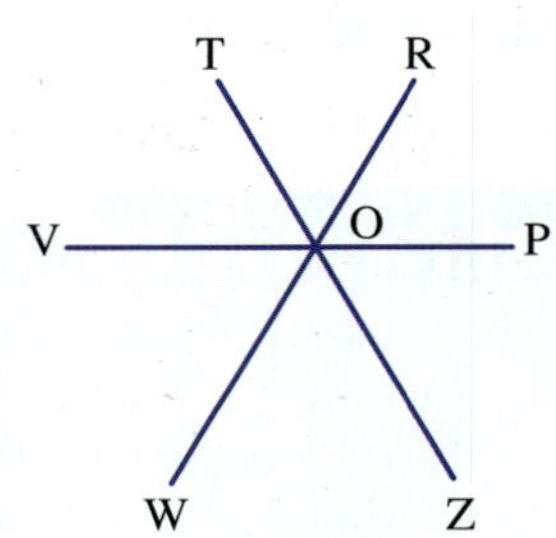

g

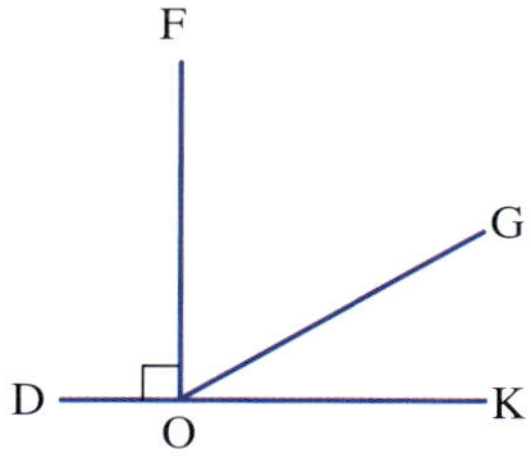

h

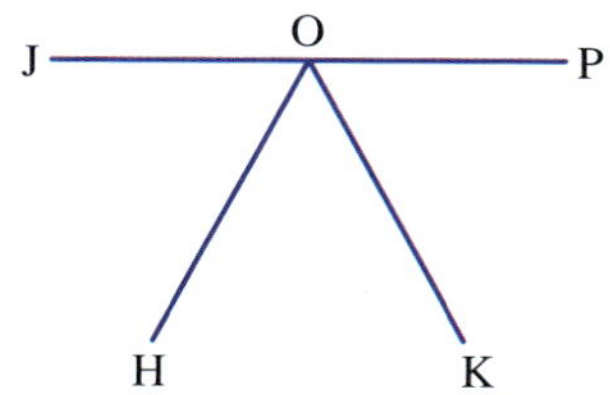

4 In each photograph below, acute, right-angled and obtuse angles have been used.

- **i** Name an acute angle.
- **ii** Name two right angles.
- **iii** Name an obtuse angle.
- **iv** Can you name a reflex angle?

a

b

Calculating the size of angles

Complementary and supplementary angles

Complementary angles are angles that add up to 90°.

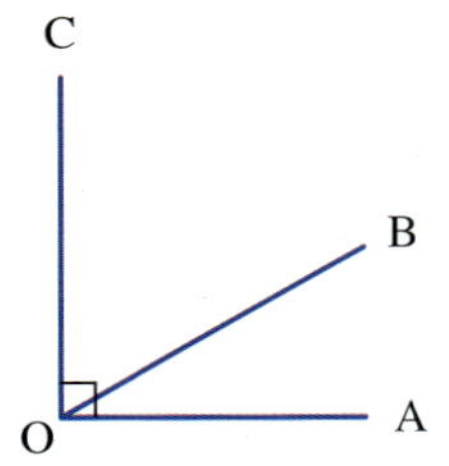

$\angle AOB + \angle BOC = 90°$
$\therefore$ $\angle OB$ and $\angle COB$ are complementary angles.

Supplementary angles are angles that add up to 180°.

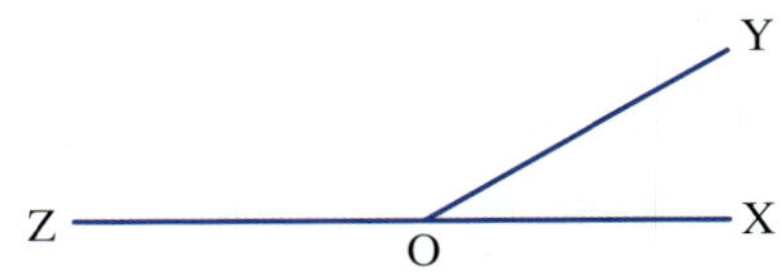

$\angle XOY + \angle YOZ = 180°$
$\therefore$ $\angle XOY$ and $\angle YOZ$ are supplementary angles.

WORKED Example 9

Find the complement of 60°.

THINK	WRITE
Complementary angles add up to 90°, so subtract 60° from 90°.	$90° - 60° = 30°$

WORKED Example 10

Find the supplement of 100°.

THINK	WRITE
Supplementary angles must add up to 180°, so subtract 100° from 180°.	$180° - 100° = 80°$

If the size of an angle is unknown, we can write a *pronumeral* to take its place. A pronumeral can be any letter of the alphabet or a symbol. It takes the place of a number. In the following example, the pronumeral x is used to indicate the size of the unknown angle in each case, but any pronumeral could have been used.

Each statement should be followed with a reason stating the geometrical property used.

WORKED Example 11

Find the value of x in each of these diagrams.

a

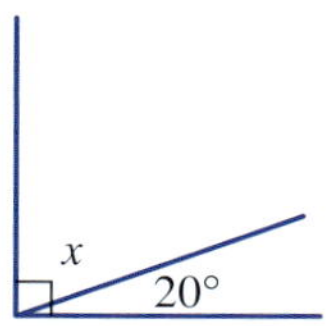

b

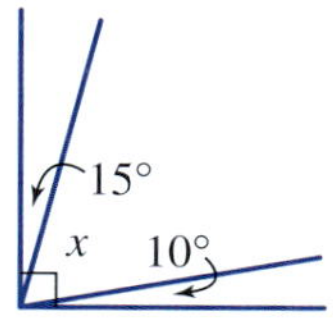

THINK

a 1 From the diagram we see that the right angle (with size 90°) is split into two smaller angles: x and 20°. Write this as an equation and give the reason.

2 Use inspection or backtracking to solve the equation for x.

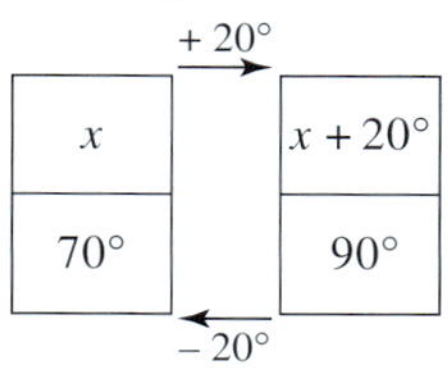

b 1 From the diagram we can see that three angles (x and 10° and 15°) form a right angle. State this as an equation.

2 Simplify by adding 10° and 15° together.

3 Solve the equation to find the value of x. Using backtracking:

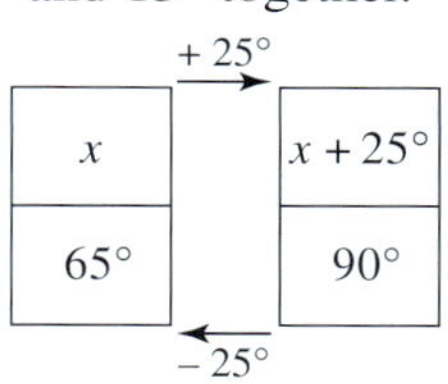

WRITE

a $x + 20° = 90°$
(Complementary angles)

$x = 90° - 20°$
$x = 70°$

b $x + 10° + 15° = 90°$
(Complementary angles)

$x + 25° = 90°$
$x = 90° - 25°$
$x = 65°$

WORKED Example 12

Find the value of y in each of the diagrams.

a

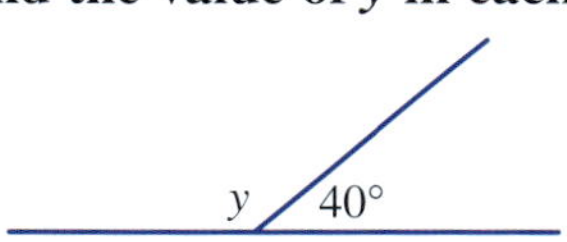

b

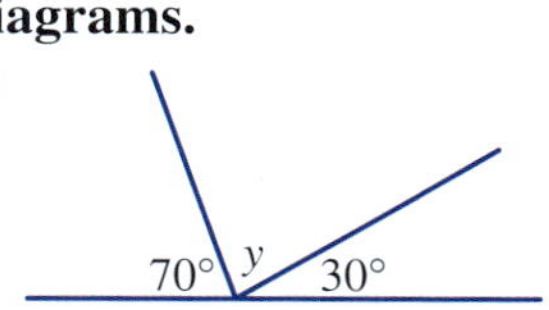

THINK

a 1 Angles y and 40° are supplementary and so they add to 180°. State this as an equation.

2 Solve the equation to find the value of y (that is, find the supplement of 40°).

b 1 The three angles y and 30° and 70° form a straight line angle. Write this as an equation.

2 Simplify by adding 30° and 70° together.

3 Solve the equation to find the value of y (that is, find the supplement of 100°).

WRITE

a $y + 40° = 180°$
(Supplementary angles)
$y = 180° - 40°$
$y = 140°$

b $y + 30° + 70° = 180°$
(Supplementary angles)
$y + 100° = 180°$
$y = 180° - 100°$
$y = 80°$

Angles forming a circle

There are 360° in a circle. So angles with the common vertex that form a circle must add up to 360°.

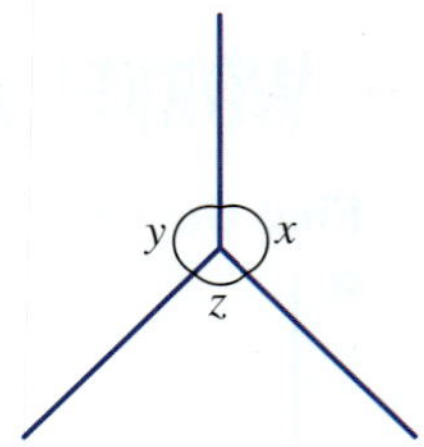

$y + x + z = 360°$

WORKED Example 13

Find the value of the pronumeral in each of the following.

a

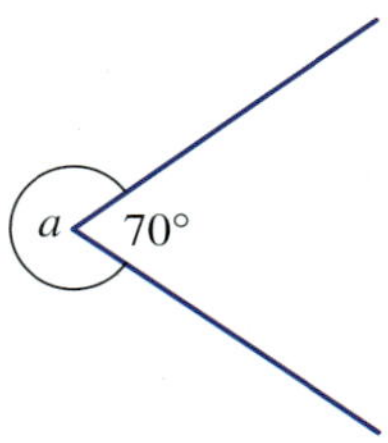

b

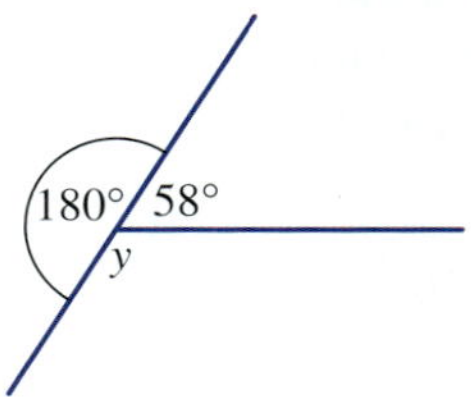

c

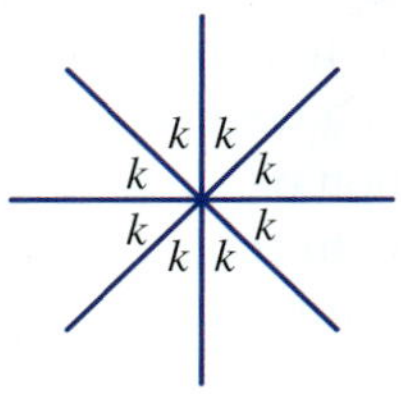

THINK	WRITE
a 1 Angles *a* and 70° form a circle and hence must add up to 360°. State this as an equation.	**a** $a + 70° = 360°$
2 Solve the equation to find the value of *a*.	$a = 360° - 70°$ $a = 290°$
b 1 Angles *y*, 58° and 180° form a circle and so must add up to 360°.	**b** $y + 58° + 180° = 360°$
2 Simplify by adding 58° and 180° together.	$y + 238° = 360°$
3 Solve the equation to find the value of *y*.	$y = 360° - 238°$ $y = 122°$
c 1 Form an equation by putting the sum of all angles on one side and 360° on the other.	**c** $k + k + k + k + k + k + k + k = 360°$
2 Simplify.	$8k = 360°$
3 Solve the equation to find the value of *k*.	$k = \dfrac{360°}{8}$ $k = 45°$

Vertically opposite angles

When two straight lines intersect, they form four angles with a common vertex. The angles that are opposite to each other are called **vertically opposite angles**. These angles are equal in size.

In the diagram at right there are two pairs of vertically opposite angles.
∠AOB = ∠DOC and ∠BOC = ∠AOD

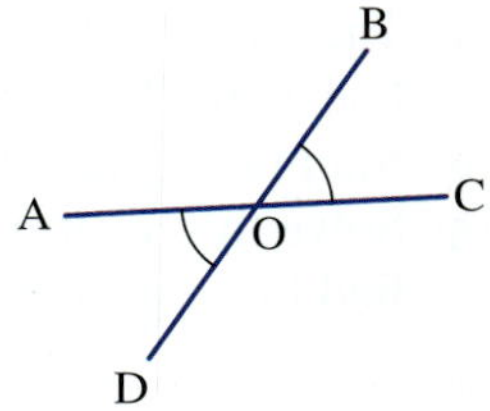

WORKED Example 14

Find the value of the pronumeral in the diagram at right.

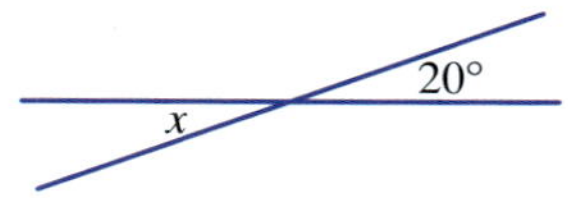

THINK

Angles shown on the diagram are vertically opposite and hence are equal in size.

WRITE

$x = 20°$

remember

1. *Complementary angles* add up to 90°.
2. *Supplementary angles* add up to 180°.
3. Angles with a common vertex that form a circle add up to 360°.
4. *Vertically opposite angles* are equal in size.

EXERCISE 10E Calculating the size of angles

10.8 SkillSHEET Complementary angles

10.9 SkillSHEET Supplementary angles

Mathcad Finding the complement

Mathcad Finding the supplement

Cabri Geometry Complementary angles

Cabri Geometry Supplementary angles

WORKED Example 9

1 Find the complement of each of the following angles.

a 12° b 32° c 88° d 45°
e 73° f 25° g 69° h 6°

WORKED Example 10

2 Find the supplement of each of the following angles.

a 75° b 6° c 112° d 48°
e 167° f 62° g 33° h 125°

3 Which of the following angles are complementary?

a 20° and 48° b 20° and 70° c 45° and 45°
d 40° and 60° e 50° and 130° f 30° and 60°

4 Which of the following angles are supplementary?

a 25° and 155° b 30° and 70° c 40° and 140°
d 35° and 80° e 50° and 100° f 120° and 60°

WORKED Example 11

5 Find the value of x in each of the following diagrams, giving reasons for your answers.

a
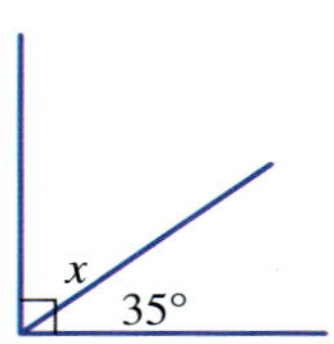

b
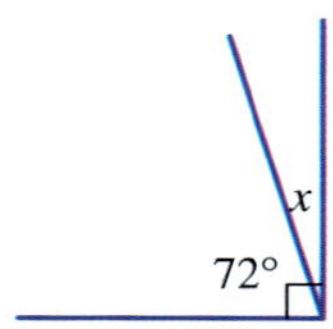

c
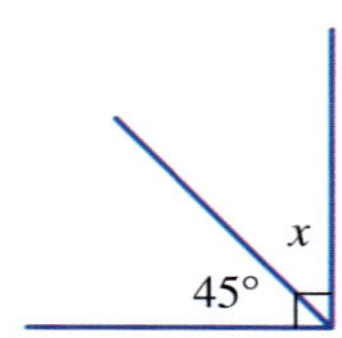

d
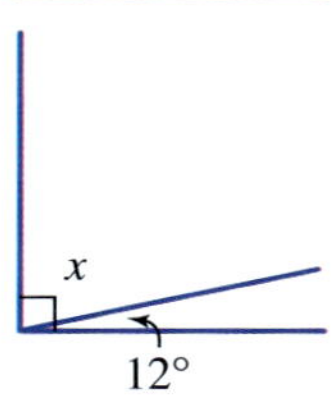

e
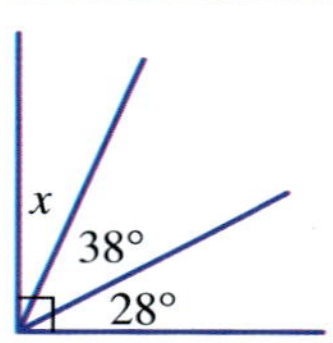

f
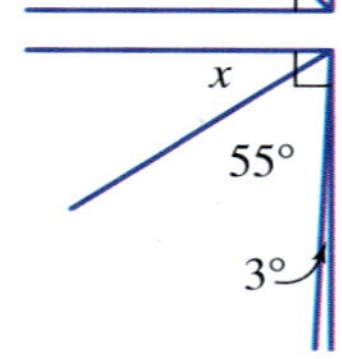

g

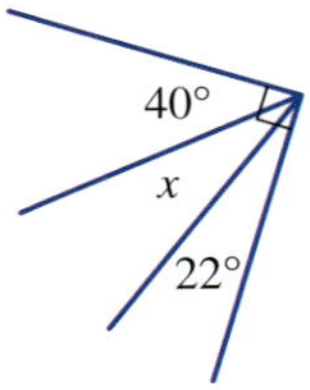

h

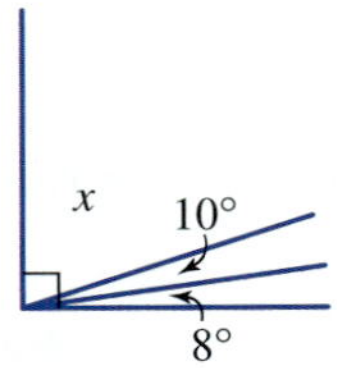

i

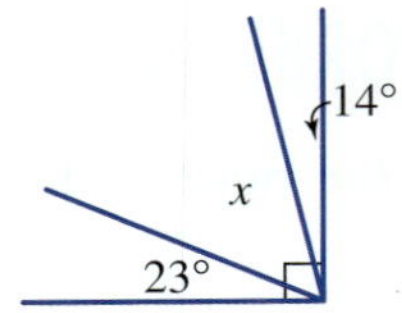

6 A farmer has a rectangular gate on his property. The frame of the gate is shown in the photograph below. Find the value of x.

WORKED Example 12

7 Find the value of y in each of the following diagrams, giving reasons.

a

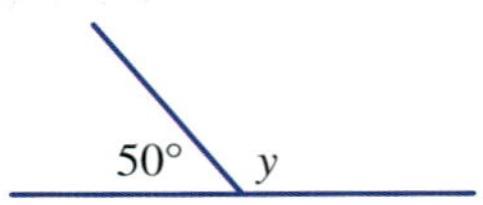

b

c

d

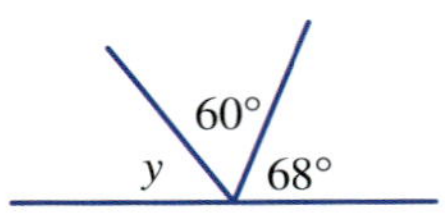

e

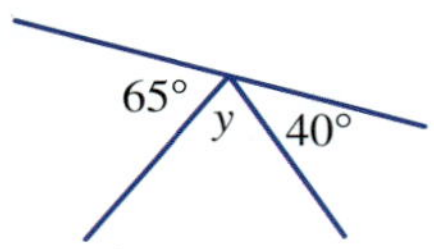

f

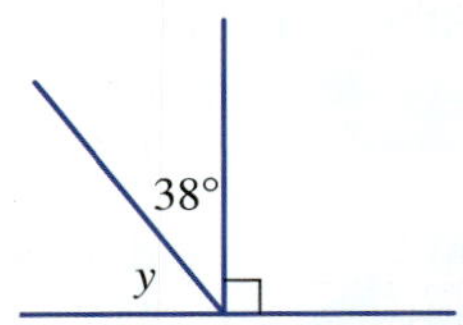

g

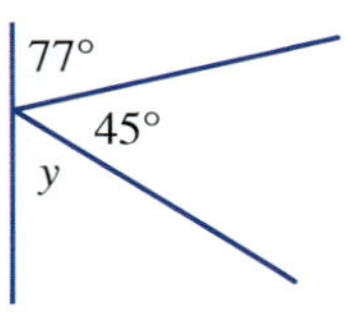

h

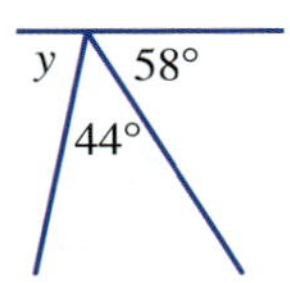

i

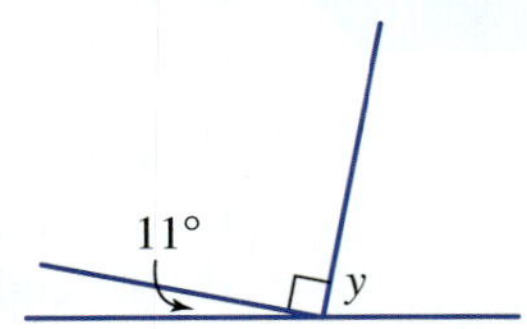

WORKED Example 13

8 Find the value of the pronumeral in each of the following, giving reasons.

a

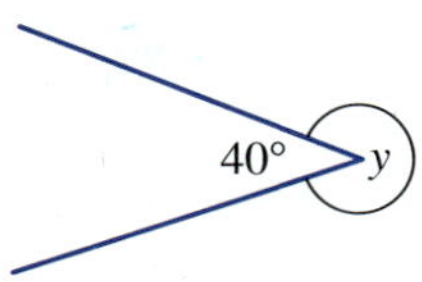

b

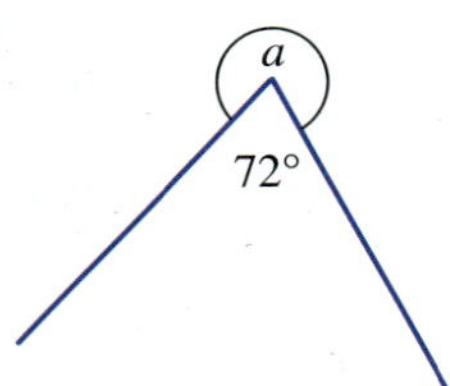

c

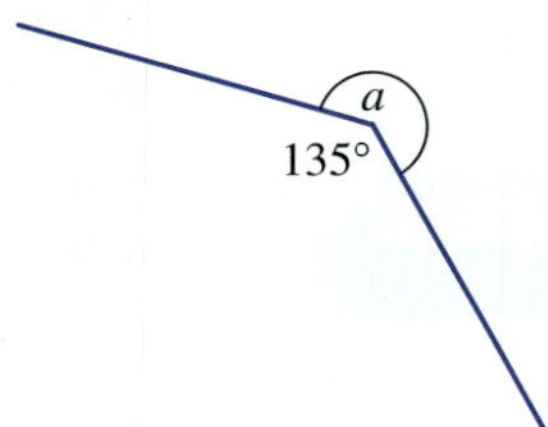

d

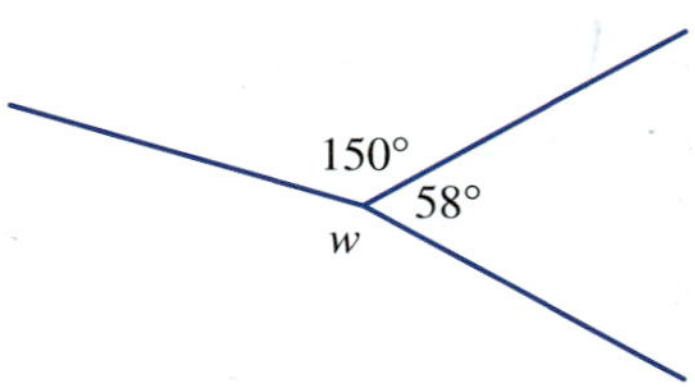

e

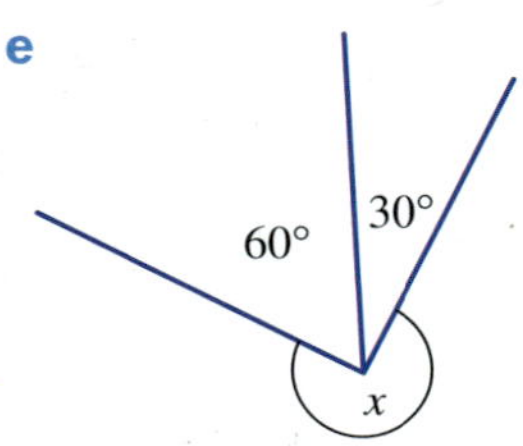

f

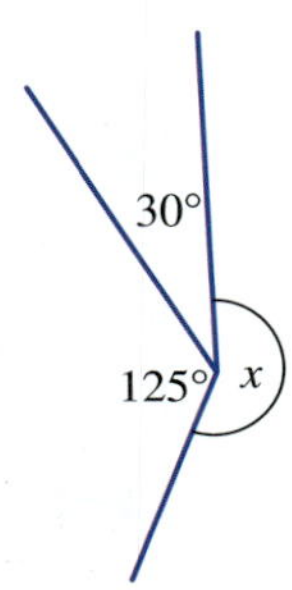

g

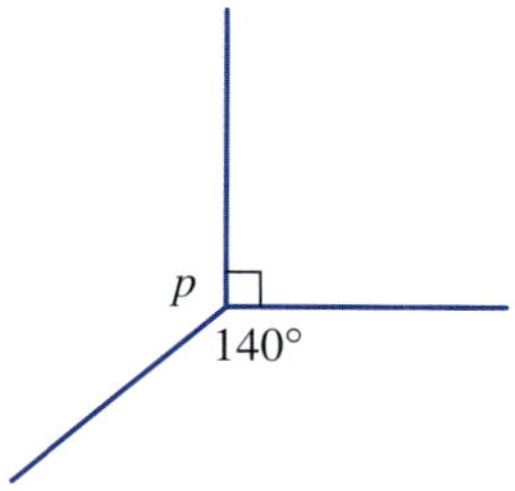

h

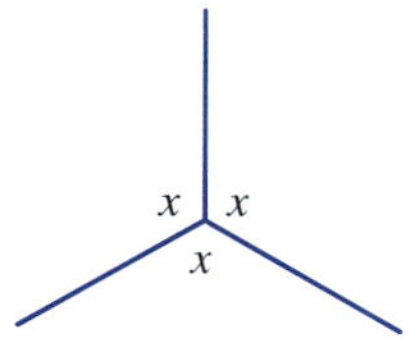

i

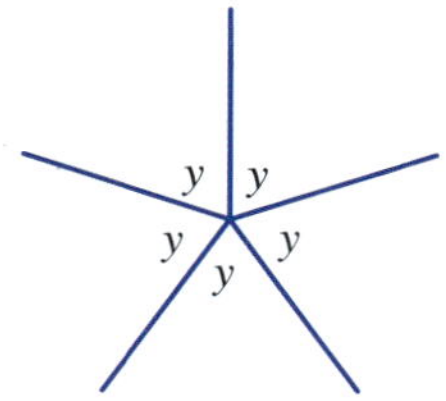

WORKED Example 14

9 Find the value of the pronumeral in each of the following, giving reasons for your answer.

a

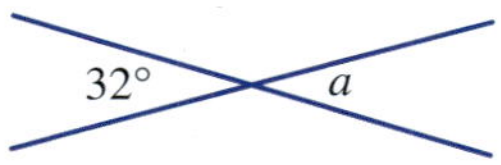

b

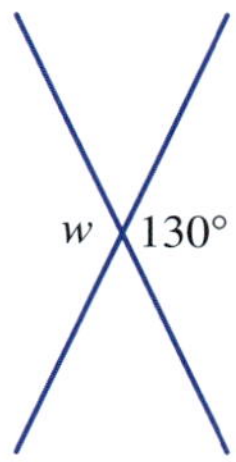

c

d

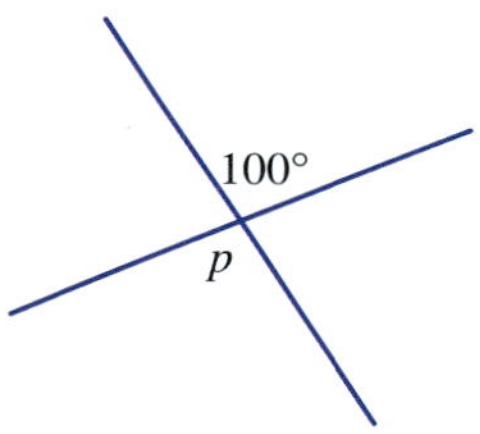

e

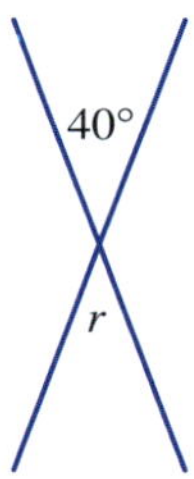

f

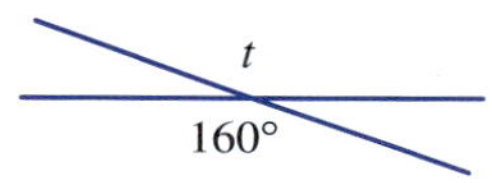

10 Find the value of the pronumeral in each of the following.

a

b

c

d

WorkSHEET 10.2

10 QUICK QUESTIONS 2

1 Use a protractor to measure the size of the angle shown.

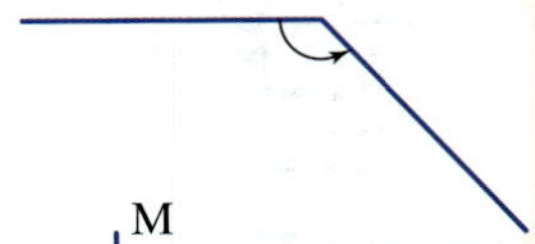

2 What type of angle is 94°?

3 Using the letters shown, name the angle.

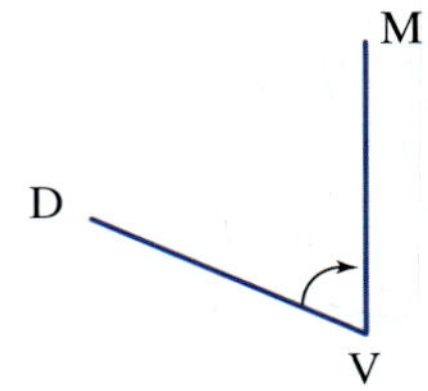

4 Give the letter name of the obtuse angle in the diagram at right.

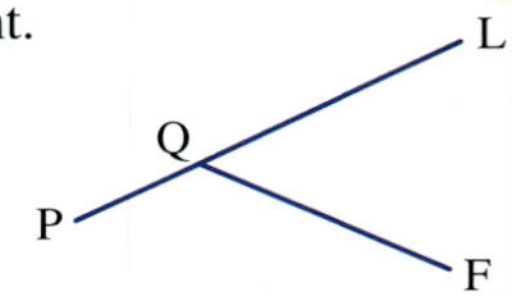

5 Find the complement of 34°.

6 Find the supplement of 21°.

For questions 7 to 10, find the value of the pronumeral, giving reasons for your answers.

7

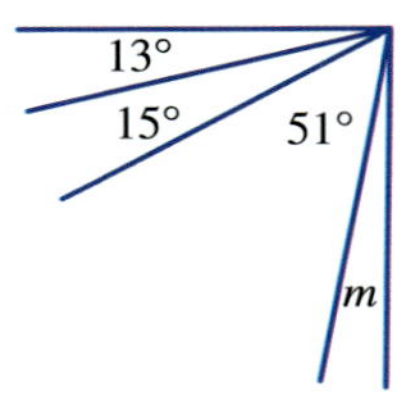

8

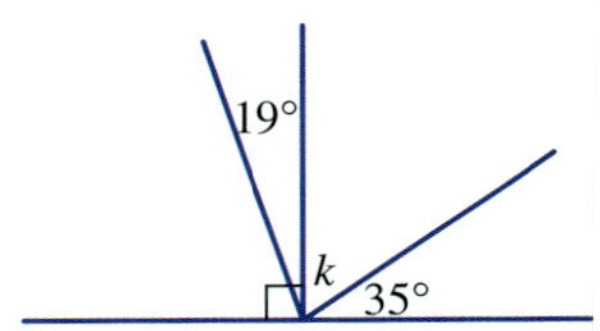

9

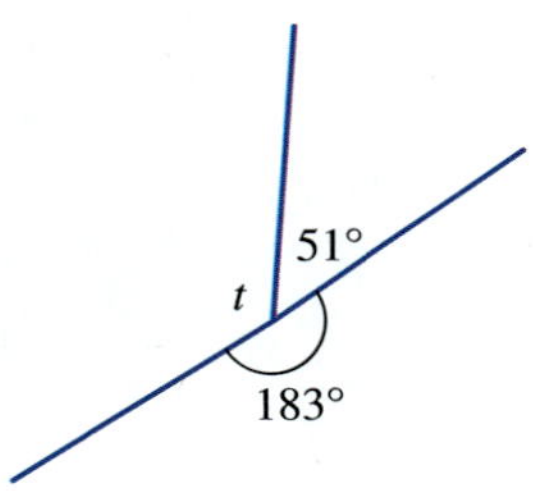

10

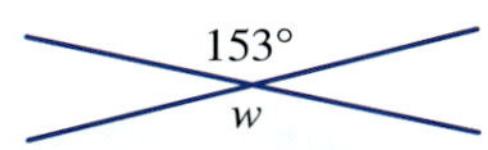

MATHS QUEST CHALLENGE

1 **A circular pizza is cut into 8 equal slices. What angle is formed at the centre of the pizza by each slice?**

2 **Julia is seated in a Ferris wheel which makes one full revolution in 1 minute. Through what angle does Julia travel in 25 seconds?**

3 **What is the smaller angle formed between the hour hand and the minute hand of a clock at:**

a **6 o'clock?** b **9 o'clock?** c **12 o'clock?**
d **1 o'clock** e **7 o'clock** f **half past 6?**

4 **What angle does the minute hand sweep through while moving for:**

a **45 minutes?** b **20 minutes?** c **55 minutes?**

Constructing angles with a pair of compasses

Constructing angles

Earlier in this chapter we discussed constructing angles with the aid of a protractor. Some angles can also be constructed using a pair of compasses and ruler.

Note: The correct term for this instrument is a **pair of compasses** (not to be confused with the compass used in navigation). For convenience, the pair of compasses will sometimes be referred to only as *compasses*.

WORKED Example 15

Construct a right angle using a pair of compasses and a ruler.

Cabri Geometry: Constructing a right angle (DIY)

THINK	DRAW
1 Rule a baseline AB (make it about 4 to 5 cm).	
2 Place the point of the compasses at A. Draw one small arc above and another one below the line.	Place the point of the compasses here and draw the two arcs. 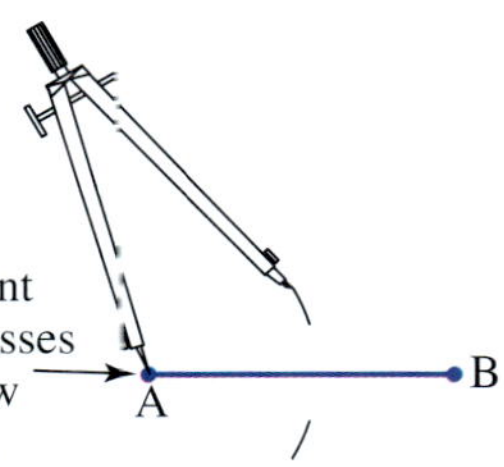
3 Place the point of the compasses at B. (Do not change the width of the compasses.) Draw two small arcs, which intersect the first two at points C and D.	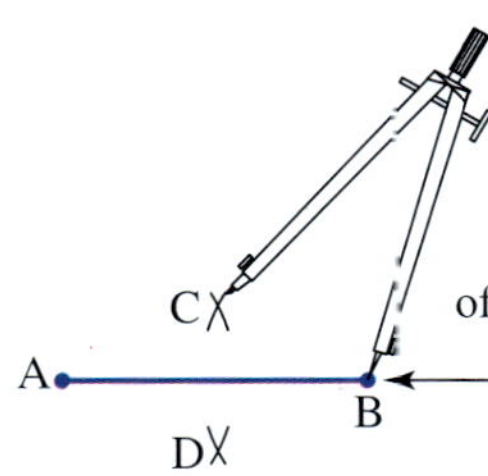 Place the point of the compasses here and draw two other arcs at C and D.
4 Join points C and D with a straight line. The line AB intersects (cuts) the line CD at point O. The four angles thus formed are: ∠COB, ∠BOC, ∠DOA and ∠AOC. Each of these angles is equal to 90°.	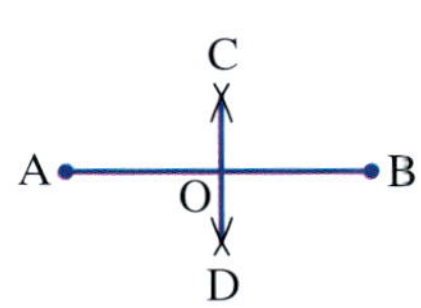

Cabri Geometry — Constructing an angle of 60° (DIY)

WORKED Example 16

Construct an angle of 60°.

THINK

1. Rule a baseline AB.
2. Place the point of the compasses at A and draw a large arc so that it intersects the line AB at C.
3. Place the point of the compasses at C (without changing its width) and draw a small arc which cuts the large arc at D.
4. Join points A and D. ∠DAC = 60°.

DRAW

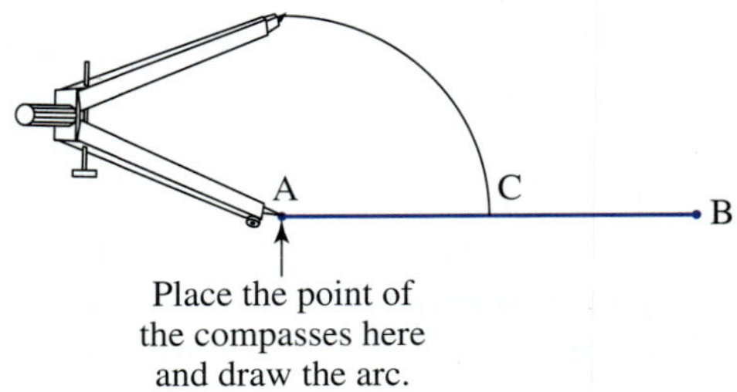

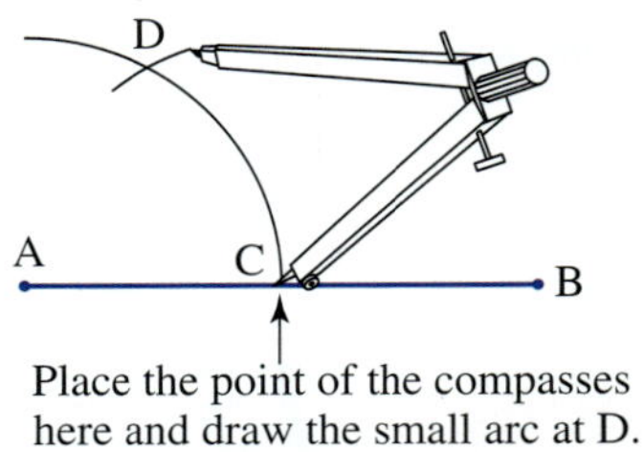

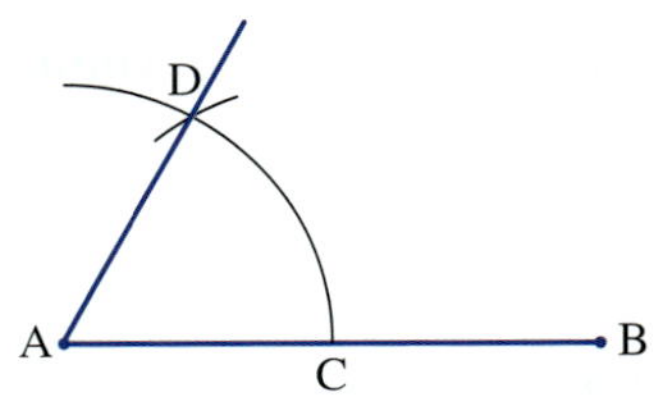

Bisecting angles

A pair of compasses can also be used to **bisect** (or halve) angles. The following worked example shows how this can be done.

Cabri Geometry — Bisecting an angle (DIY)

WORKED Example 17

Bisect the following angle.

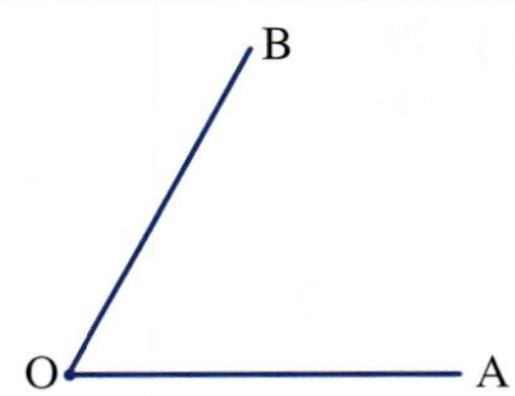

THINK

1. Place the point of the compasses at O and draw a large arc so that it intersects the arm OA at point X and the arm OB at point Y.

DRAW

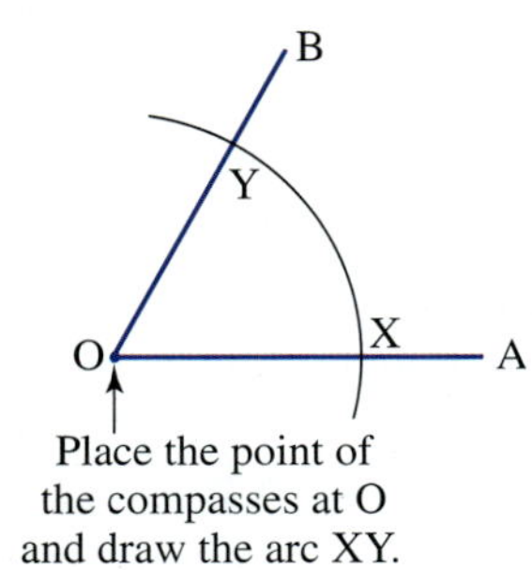

THINK	DRAW
2 Place the point of the compasses at X and draw a small arc.	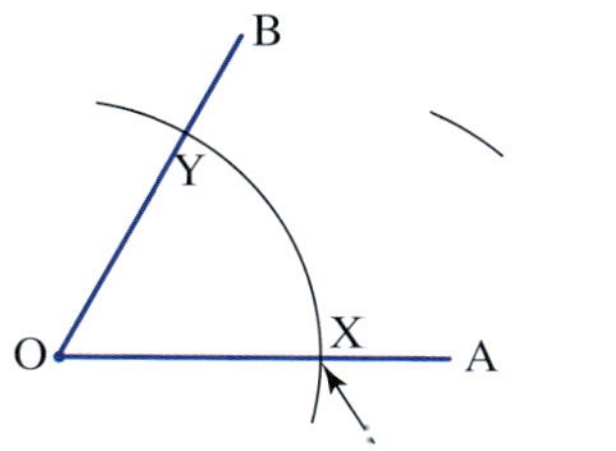
3 Place the point of the compasses at Y and draw another small arc so that it intersects the first one at D. (Do not change the width of the compasses.)	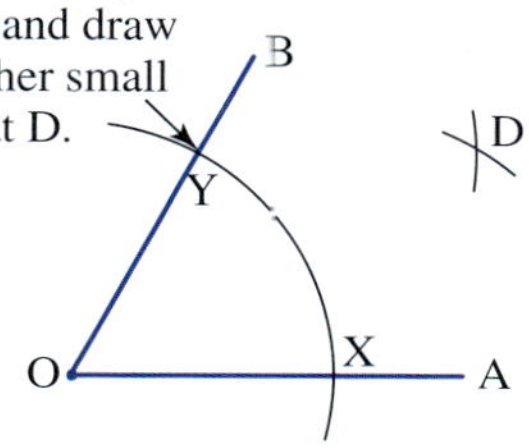
4 Join points D and O with a straight line. The angle is now bisected, that is: $\angle AOD = \angle DOB = \frac{1}{2}\angle AOB$.	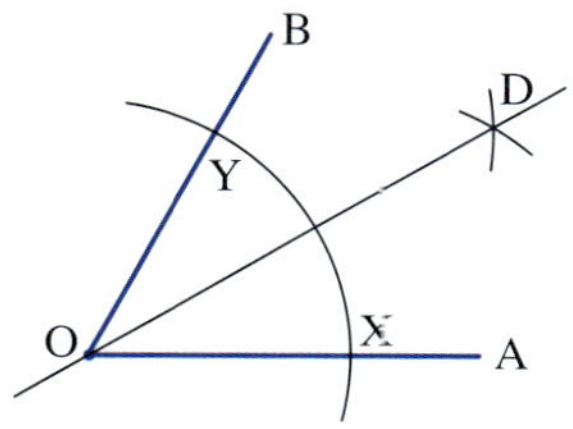

Note that the technique of bisecting angles, when combined with techniques of constructing 90° and/or 60° angles allows us to construct some other angles without using a protractor. For example, to construct a 45° angle we can construct a 90° angle first and then bisect it by following instructions outlined in the above examples.

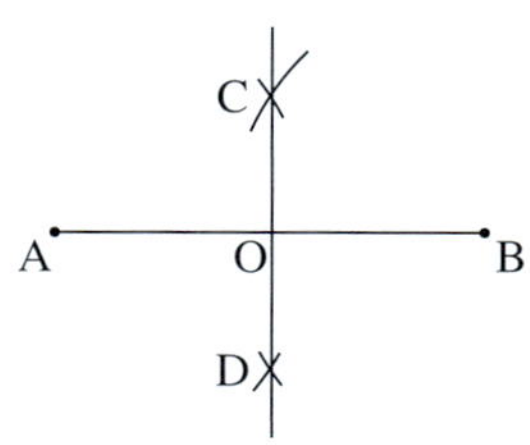

$\angle COB$

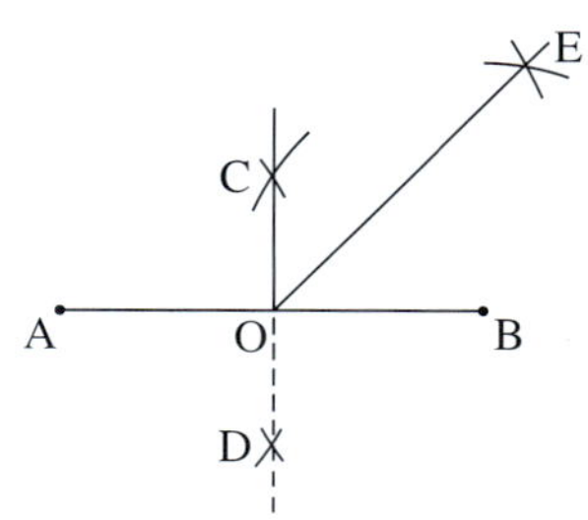

$\angle COE = \angle BOE = \frac{1}{2}\angle COB$

Copying angles

Suppose you need to construct an angle of exactly the same size as the one that is given; that is, to copy the given angle. One method of doing this is to measure the given angle using a protractor and then to construct exactly the same angle (again using the protractor). Alternatively, an angle can be copied using a pair of compasses as shown in the following example.

Cabri Geometry — Copying an angle (DIY)

When using the *Maths Quest 7* CD-ROM, you can use Cabri Geometry instructions from this icon.

WORKED Example 18

Copy the following angle exactly. (Do not use a protractor.)

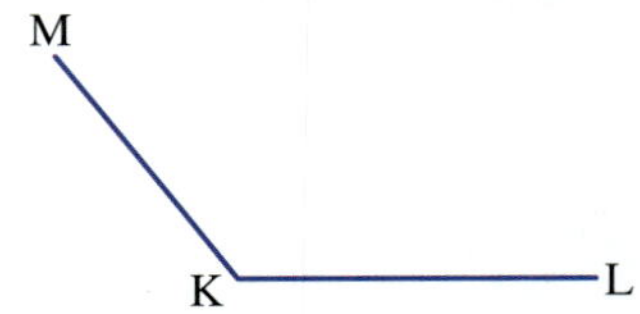

THINK

DRAW

1. Place the point of the compasses at K and draw a large arc so that it intersects KL at T and KM at P.

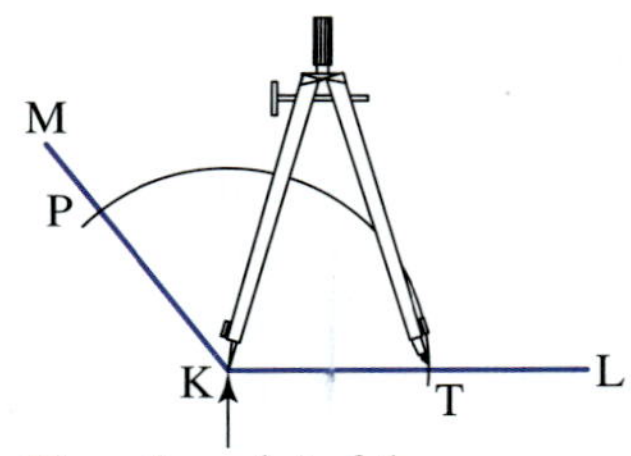

Place the point of the compasses at K and draw the arc PT.

2. Draw a new baseline K′L′ (for the new angle).
3. Place the point of the compasses at K′ and draw the arc of the same width as in step 1 so that it intersects K′L′ at T′.
4. With the compasses, 'measure' the distance between P and T in the original angle (that is, open the compasses to the width PT).

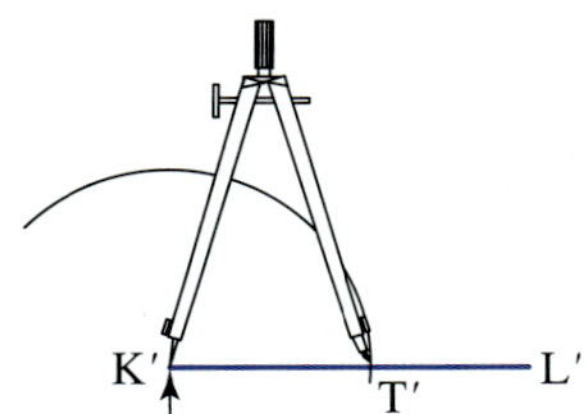

Place the point of the compasses at K′ and draw the arc.

5. With the compasses width unaltered, place the point of the compasses at T′ and draw a small arc so that it intersects the large arc at P′.

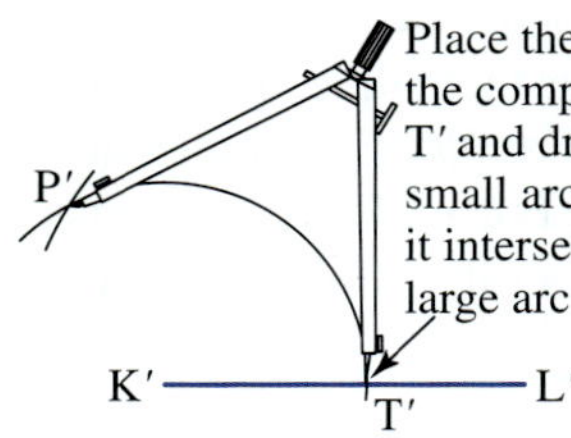

Place the point of the compasses at T′ and draw a small arc so that it intersects the large arc at P′.

6. Join the points K′ and P′. ∠L′K′M′ is an exact copy of ∠LKM.

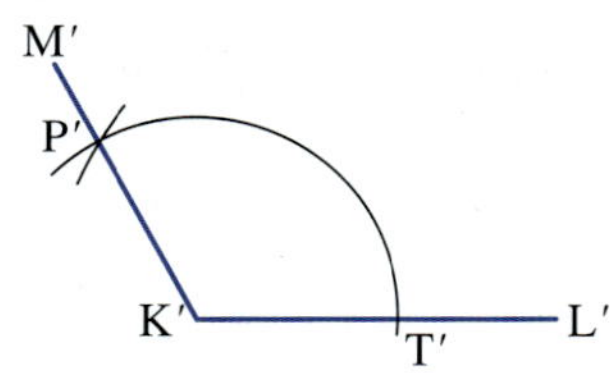

remember

A pair of compasses can be used for:
1. constructing certain angles
2. bisecting angles
3. copying angles.

EXERCISE 10F Constructing angles with a pair of compasses

For this exercise you will need to use a pair of compasses, a pencil and a ruler.

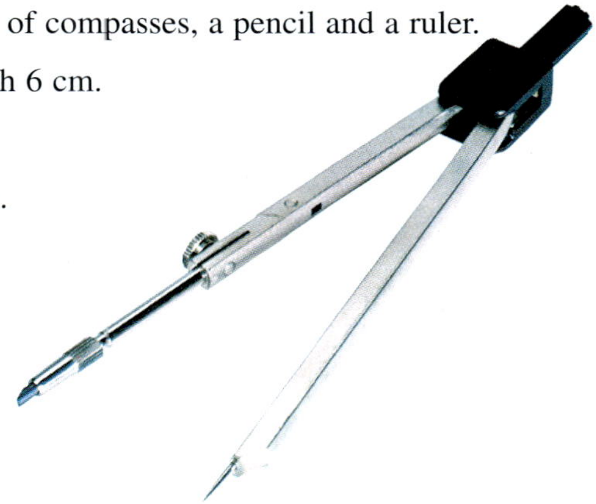

WORKED Example 15

1 Construct a 90° angle on a baseline of length 6 cm.

WORKED Example 16

2 Construct a 60° angle on a baseline of 5 cm.

WORKED Example 17

3 Construct a 90° angle and bisect it.

4 Construct a 30° angle.

5 Construct a 30° angle and bisect it.

6 Construct the following angles.

a 45° b 75° c 105° d 150° e 120°

WORKED Example 18

7 Copy the following angles.

a

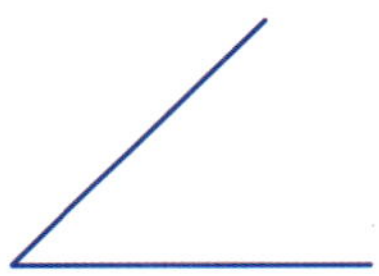

b

c

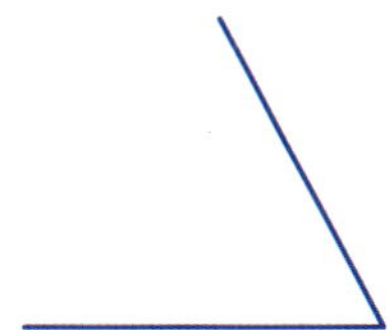

d

e

f

8 Bisect each of the angles you copied in question 7.

DESIGN Copying triangles

What minimum information is needed to copy a triangle exactly?

Consider the triangle ABC.

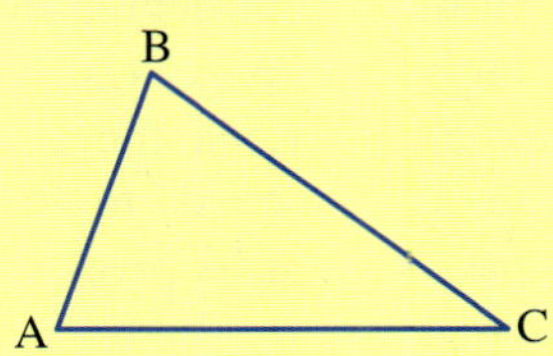

To copy this triangle, follow the steps outlined below.

1 Using a pair of compasses and a ruler, copy ∠BAC using the technique discussed in this section. *(continued)*

2 Measure the length of the side AC and find the position of point C′ on the corresponding arm of the angle, so that A′C′ = AC. (This can be done using either a ruler, or a pair of compasses.)

3 Measure the length of the side AB and find the position of point B′ on the corresponding arm of the angle, so that A′B′ = AB.

4 Join points C′ and B′ with a straight line to complete the triangle A′B′C′.

By comparing the sizes of the angles and the lengths of the sides of the two triangles, verify that triangle A′B′C′ is an exact copy of the triangle ABC.

Now consider triangle MNP.

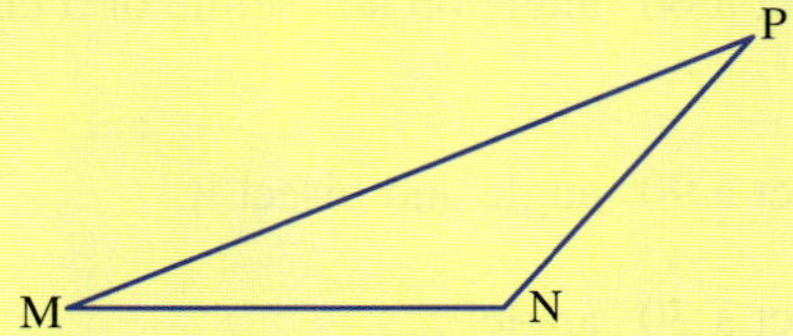

To copy this triangle, follow the steps outlined below.

1 Draw a straight line M′N′, so that M′N′ = MN.

2 Use a pair of compasses and a ruler to copy ∠NMP. (The new vertex is point M′ on the line M′N′.)

3 Use a pair of compasses and a ruler to copy ∠MNP. (The new vertex is point N′ on the line M′N′.)

4 Extend the arms of the copied angles until they meet at P′.

By comparing the sizes of the angles and the lengths of the sides of the two triangles, verify that triangle M′N′P′ is an exact copy of the triangle MNP.

Experiment with other triangles. Do the above methods work for any triangle? Are there any other methods of copying triangles?

Copy and complete the following sentence:

To copy a triangle, the minimum information needed is: . . .

Bearings

This section deals with a directional compass that is used to identify **bearings**.

A directional compass has the cardinal points north, east, south and west (N, E, S and W) and half-cardinal points (NE, SE, SW, and NW) marked in a circle on the compass. Many directional compasses also have internal numerical degree markings from 0° to 360°. The compass on page 420 has degree markings from 0° to 90° between the cardinal points. A magnetic needle is attached to the disc of the directional compass so that its pivot is at the centre of the circle.

To use a directional compass, hold it in front of you so that the point marked north is pointing away from you. You will then need to turn your body so that the needle points to the north.

True bearings

All true bearings are written as the number of degrees from north measured in a clockwise direction. They are written using three digits.

WORKED Example 19

Identify the true bearing of the point A in each of the following.

a

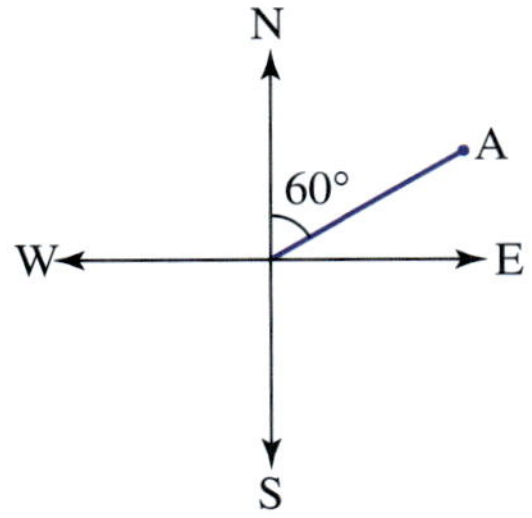

b

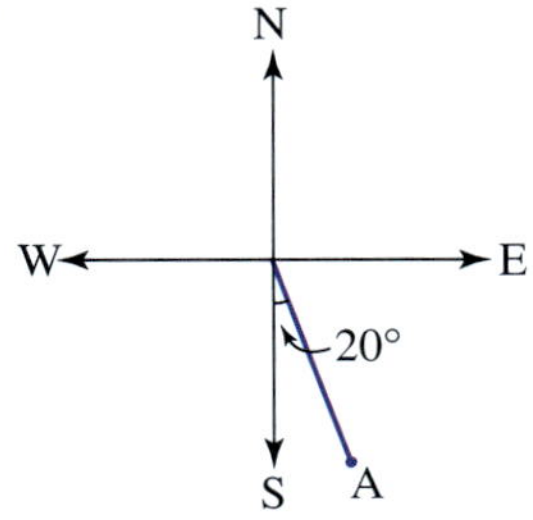

c

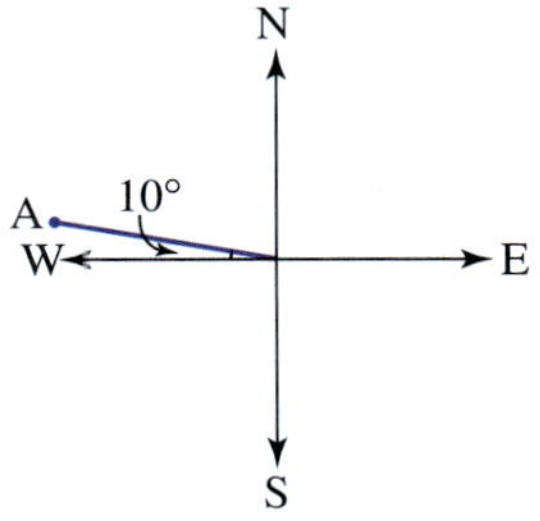

THINK

a 1 The true bearing is measured in a clockwise direction from north. On the diagram mark the angle to be found (that is, the angle between the north line and the line joining the centre of the compass with point A).

2 The angle representing the true bearing coincides with the angle that was given. So simply state the bearing of point A, using 3 digits.

b 1 Mark the angle that represents the true bearing of point A.

2 Cardinal point S corresponds to the 180° mark. From the diagram we can see that the angle to be found is 20° less than 180°. So subtract the given angle from 180° to find the true bearing of point A.

WRITE

a

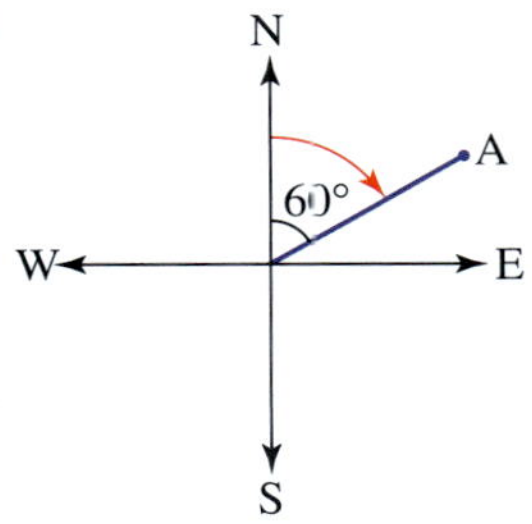

The bearing of point A is 060° true or 060° T.

b

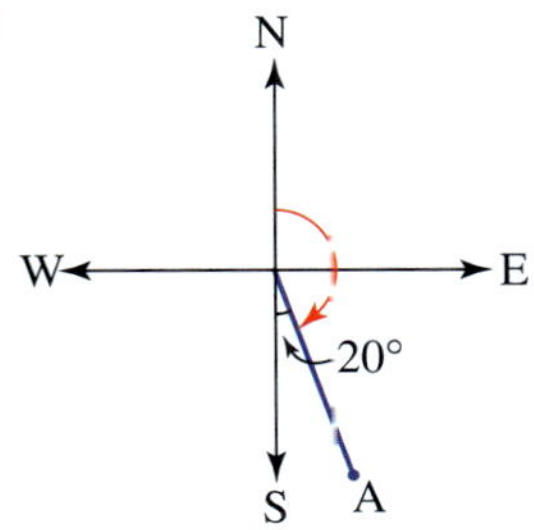

180° – 20° = 160°
So the bearing of point A is 160° true, or 160° T.

Continued over page

THINK	WRITE
c 1 On the diagram, mark the angle that represents the true bearing of point A.	**c** 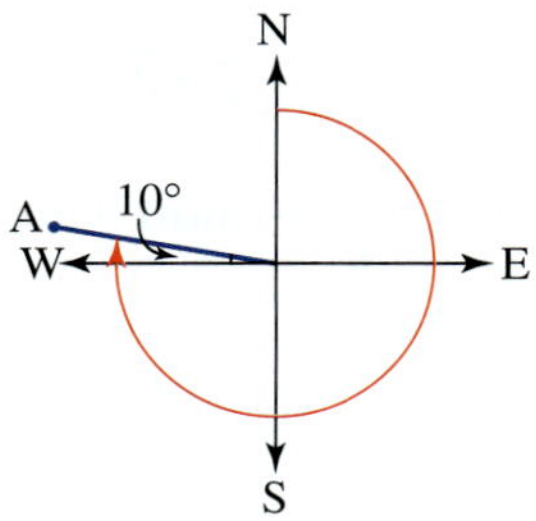
2 Cardinal point W corresponds to the 270° mark. From the diagram we can see that the required angle is 10° larger than 270°. So add the given angle to 270° to find the true bearing of point A.	$270° + 10° = 280°$ So the bearing of point A is 280° true, or 280° T.

Compass bearings

Compass bearings are measured in degrees from the north–south line in either a clockwise or anticlockwise direction.

To identify the **conventional compass bearing** of a point we need to state (in this order):

1. whether the angle is measured from north or south
2. how many degrees there are in the angle
3. whether the angle is measured in the direction of east or west.

For example, in the diagram at right, the angle that the north–south line makes with the line joining the centre of the compass with point A is 55°, measured from the south in the direction of west. The compass bearing of point A in this case is written as S 55° W.

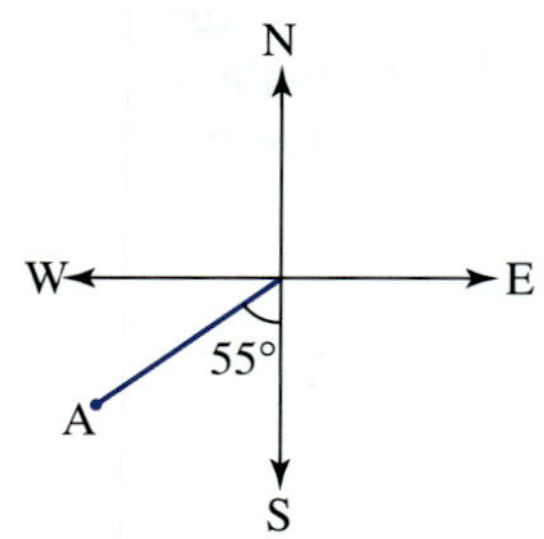

Sometimes the point is positioned on the actual direction line. In such cases, the bearing is written as *due north* (or south or east or west).

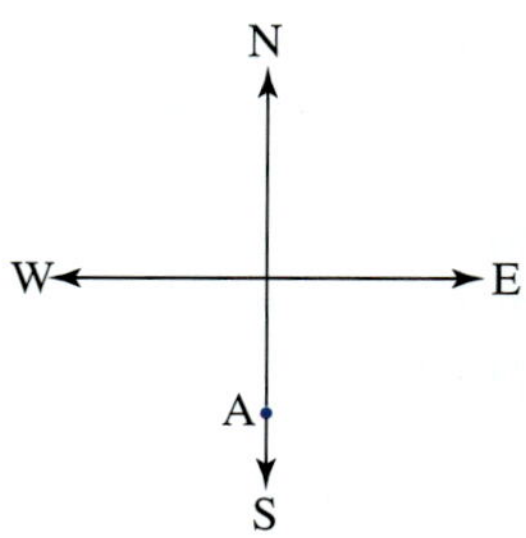

A is *due South.*

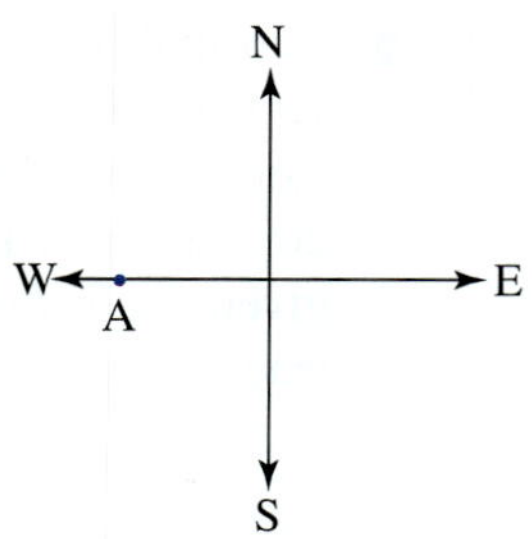

A is *due West.*

WORKED Example 20

Express each of the following as compass bearings.
a 050° true b 220° true c 270° true

THINK	WRITE
a 1 Draw a diagram to show the position of the point on a bearing of 050° true. *Note:* Measure an angle of 50° clockwise from the line pointing north.	**a** 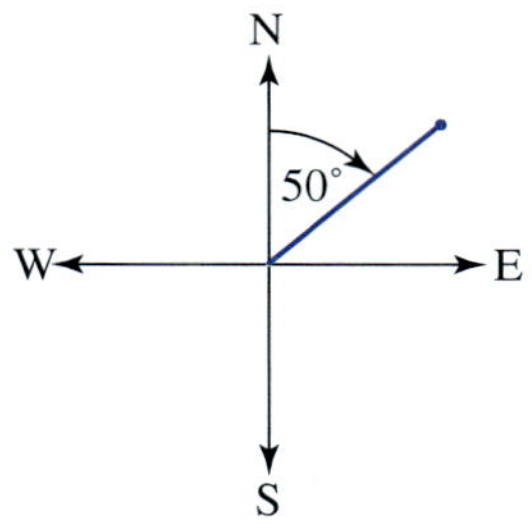
2 The angle between the vertical line and the line representing the direction of the point is measured from the north towards east and is 50°. Write this as a compass bearing.	The compass bearing is N 50° E.
b 1 Draw a diagram to show the position of the point on a bearing of 220° true. *Note:* Measure an angle of 220° clockwise from the line pointing north.	**b** 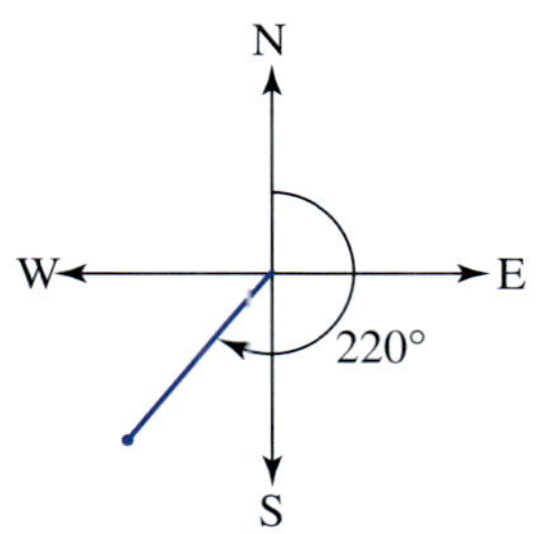
2 The angle is measured from south towards west. To find the size of the angle, subtract 180° from 220° (as point S corresponds to 180°).	220° – 180° = 40°
3 State the compass bearing.	The compass bearing is S 40° W.
c 1 Draw a diagram to show the position of the point on a bearing of 270° true.	**c** 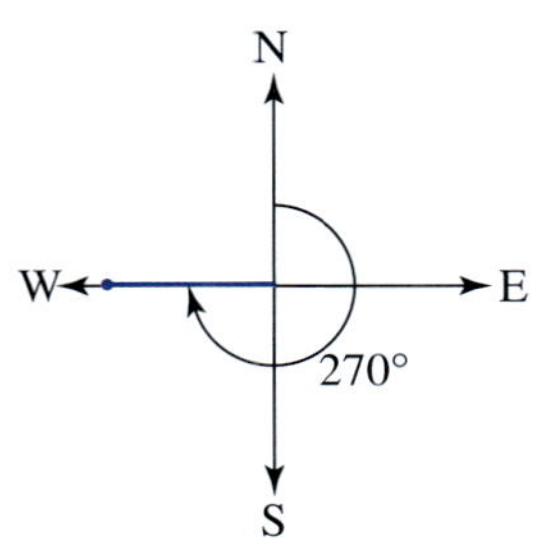
2 The point is exactly on the line directed west (as cardinal point W corresponds to 270°). So state its bearing.	The point is due west.

Cabri Geometry: True bearings

The Cabri geometry file 'True bearings' found on the *Maths Quest 7* CD-ROM allows you to view different true bearings. This can assist you in working out the equivalent compass bearing.

remember

1. The *true bearing* is given by the angle, measured from north in a clockwise direction. It is written using three digits.
2. The *compass bearing* is given by the angle measured from the north–south line in either a clockwise or anticlockwise direction.
3. To specify the compass bearing we need to state (in the following order):
   (a) whether the angle is being measured from north or south
   (b) the size of the angle
   (c) whether the angle is being measured in the direction of east or west.

EXERCISE 10G Bearings

Cabri Geometry: True bearings

WORKED Example 19

1 Identify the true bearing of point A in each of the following.

a
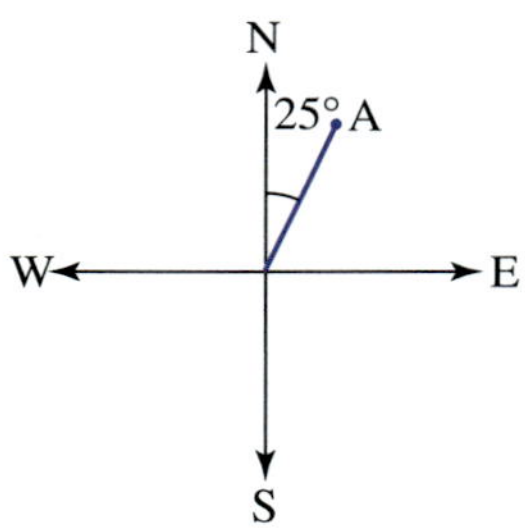

b
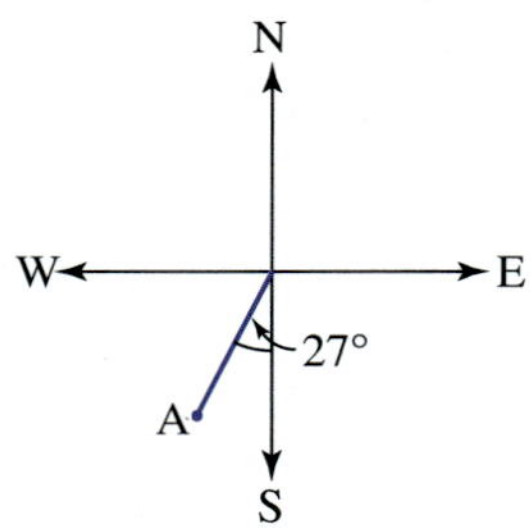

c
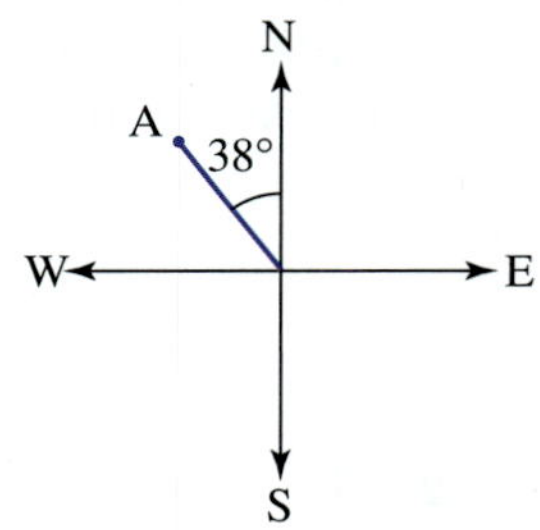

d
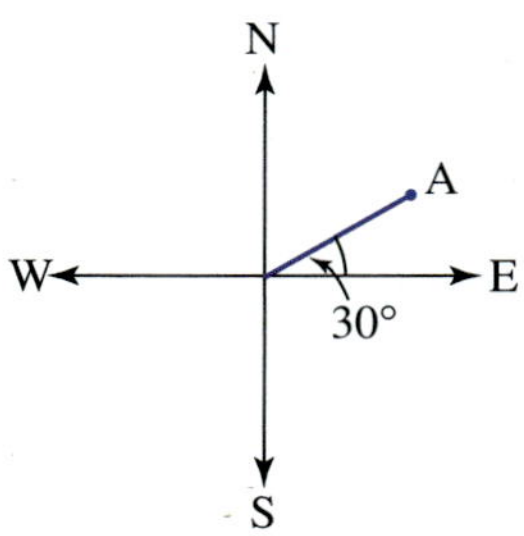

e
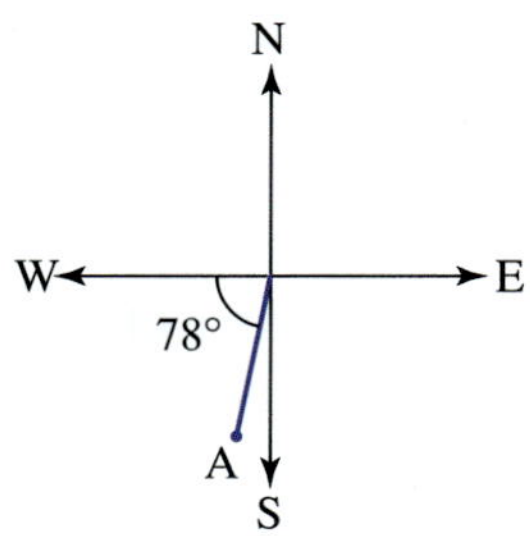

f
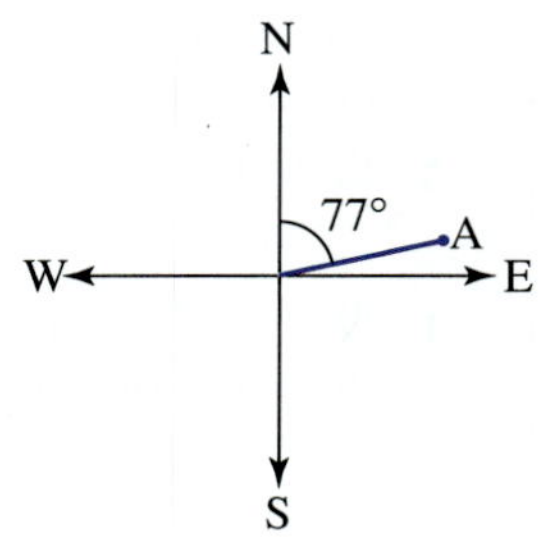

g
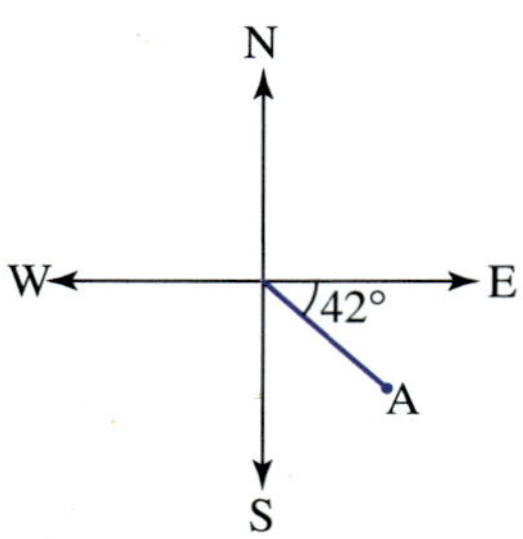

h
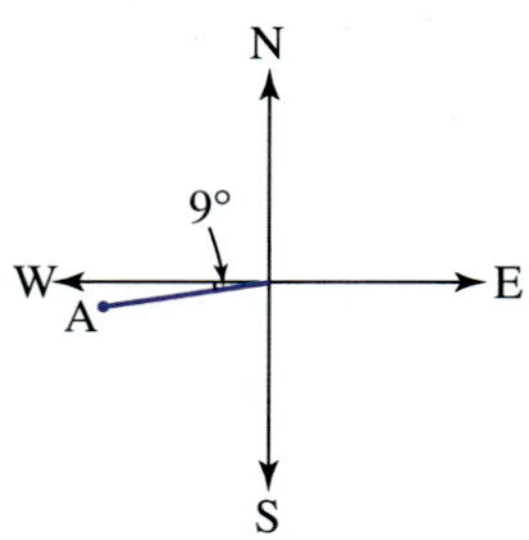

i
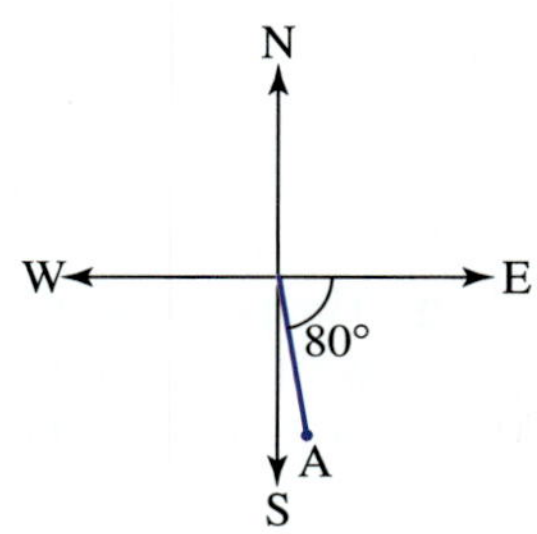

j

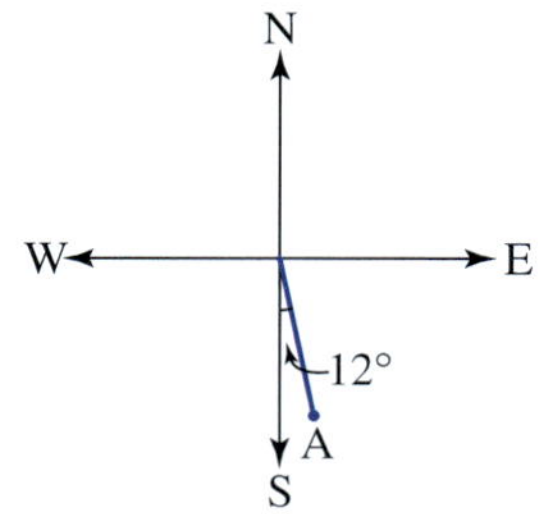

k

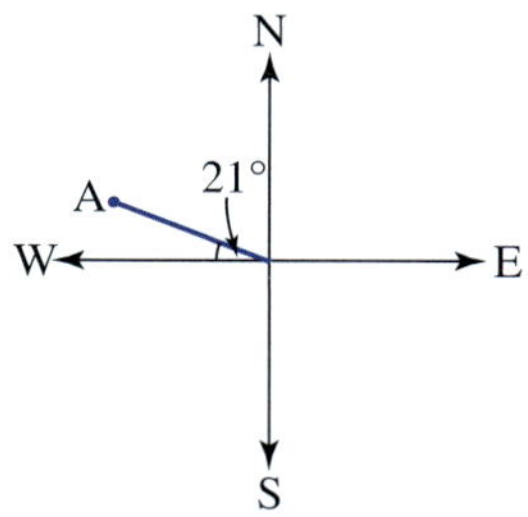

l

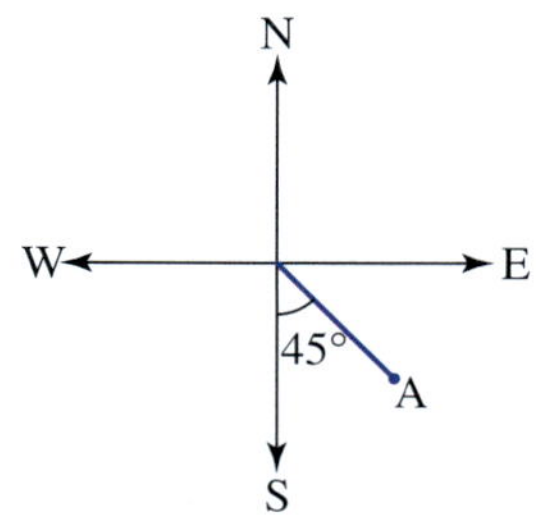

2 Identify the compass bearing of each point A in question **1**.

WORKED Example 20

3 Express each of the following as compass bearings.

EXCEL Spreadsheet True → compass bearings

a 023° true	**b** 144° true	**c** 198° true	**d** 225° true
e 270° true	**f** 315° true	**g** 358° true	**h** 070° true
i 090° true	**j** 042° true	**k** 180° true	**l** 205° true

4 Draw diagrams to show the position of the points with the following compass bearings.

a N 45° E	**b** S 50° E	**c** N 30° W	**d** S 60° W
e N 80° E	**f** S 20° E	**g** N 10° W	**h** S 40° W

5 Write each of the following as true bearings.

EXCEL Spreadsheet Compass → true bearings

a N 23° E	**b** N 15° W	**c** N 5° W	**d** N 88° E
e S 55° E	**f** S 11° W	**g** S 74° E	**h** S 48° W

6 Copy and complete the table below by writing both true and compass bearings of each of the points 1–12 represented on this diagram.

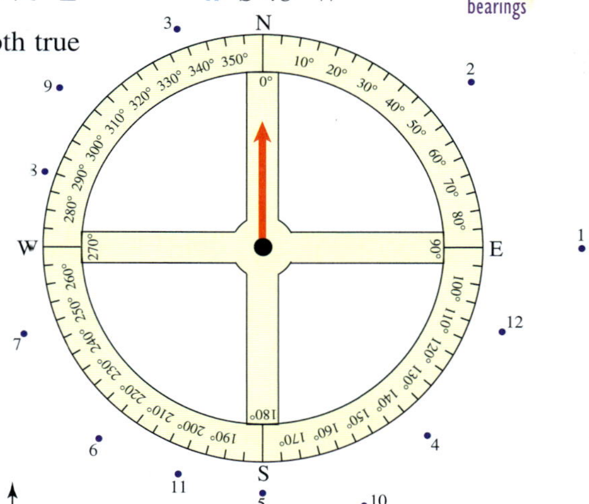

Point	True bearing	Compass bearing
1		
2		
3		
4		
5		
6		
7		

7 A group of Year 7 students went on the cross-country run shown at right, as a part of their physical education training.

From their starting point what is the true bearing of:

a the waterfall?
b the river?
c the mountain lookout?
d the finish line?

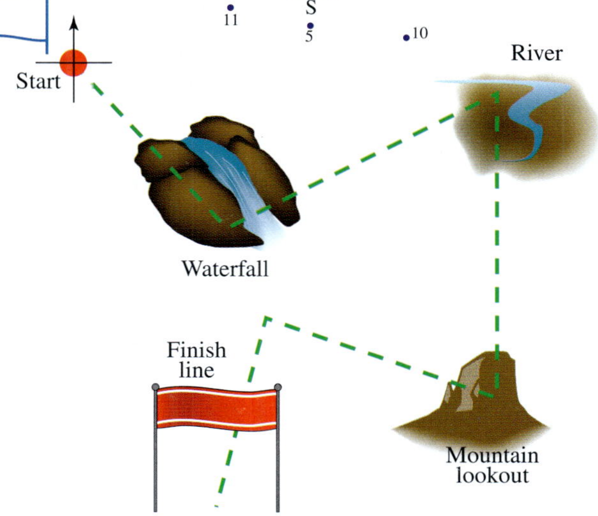

8 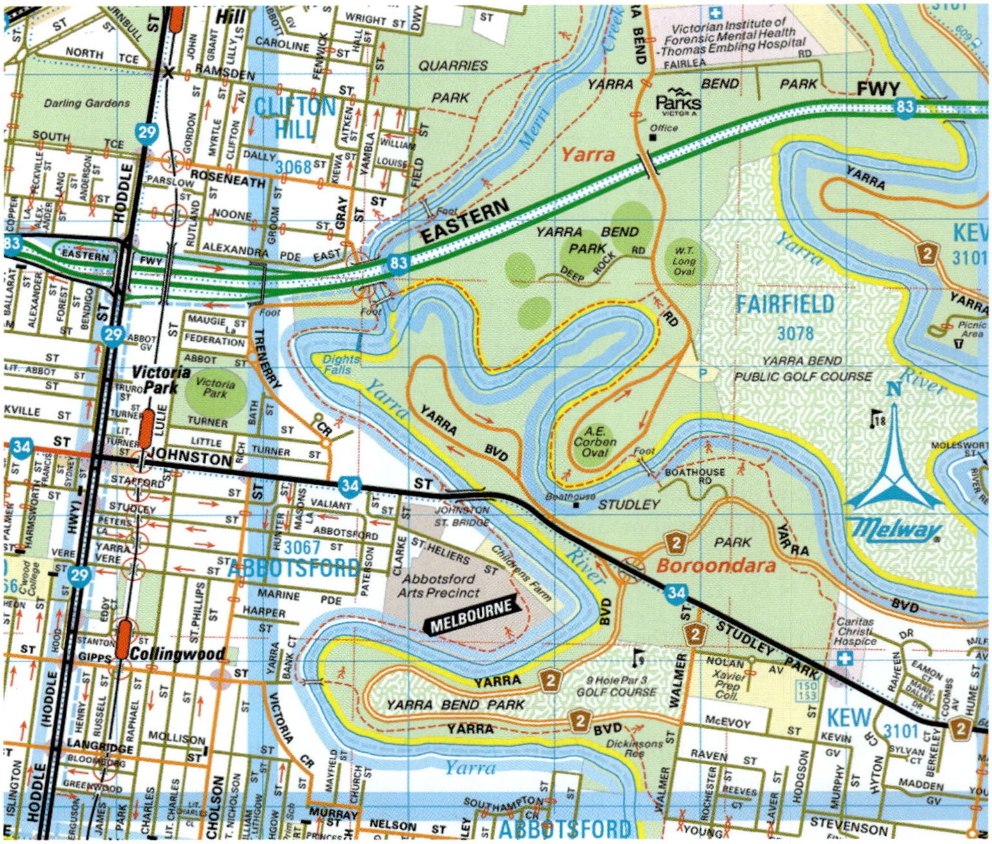

Using the map above, give the true bearings on the following places from the 18th hole of the Yarra Bend Public Golf course.

a Victoria Park oval

b Xavier Prep College

c Abbotsford Arts Precinct

WorkSHEET 10.3

MATHS QUEST CHALLENGE

1 **What distance will a car travelling at 60 km per hour travel in 1 second?**

2 **In a mail sorting centre, one electronic reader can sort 160 envelopes per minute. Another reader can sort 220 envelopes per minute. Using both readers, how long will it take to sort 95 000 envelopes?**

3 **Jordan was offered two jobs, one involves washing cars, the other mowing lawns. The job washing cars pays \$1 the first day, \$2 the second day, \$3 the third day, \$4 the fourth day and so on; the amount increasing by \$1 each successive day. The other job, mowing lawns, pays 1 cent the first day, 2 cents the second day, 4 cents the third day, 8 cents the fourth day and so on; the amount doubling each successive day. Each job is 16 days long. Which job pays more and by how much?**

DESIGN

Angles and more angles

The photograph shows a framework or truss for a roof with points labelled.

1 Copy and complete the table below by:
  a using a protractor to measure each angle
  b classifying each angle by type (for example, acute, obtuse, and so on).

Angle	Measurement	Classification
∠CAB		
∠ABC		
∠BCA		
∠BCD		
∠CDB		
∠DBC		
∠FAB		
∠ABF		
∠AFB		
∠BFE		
∠FBE		
∠FEB		

2 Use the measurements you have obtained in question 1 to accurately redraw the framework in your book. (The overall size of the framework is not important; however, the angles are.)

3 From the table, name two sets of complementary angles.

4 Name two sets of supplementary angles.

5 Find the sum of each set of three angles.
 a ∠CBD + ∠BCD + ∠CDB =
 b ∠FBE + ∠BEF + ∠EFB =
 c ∠BAF + ∠AFB + ∠ABF =
 d ∠CAB + ∠ABC + ∠ACB =

6 Each set of three angles forms a triangle. Can you see a pattern? Write a sentence describing what you have found.

7 For extra strength the builders decide to add an extra piece of timber from A to the strut BC (the timber between B and C) and another from A to the strut BF. These pieces of timber are to bisect the angles ∠CAB and ∠BAF. Use a compass and ruler to draw in these extra struts on your diagram.

8 Imagine that you are an ant walking along the frame from A to B while the framework is still lying flat on the ground. If the bearing of B from A is due south or 180° true, copy and complete the following table which shows the direction you travel between each point in your journey.

Section of journey	Compass bearing	True bearing
From A to B	Due south	180° true
From B to C		
From C to D		
From D to E		
From E to F		
From F to B		
From B to A		

summary

Copy the sentences below. Fill in the gaps by choosing the correct word or expression from the word list that follows.

1 When two ______________ meet at a point, an angle is formed.

2 The point where the lines meet is called ______________ and the lines are called ______________ of the angle.

3 Angles are measured in ______________ with the help of a protractor. It is wise to first ______________ the size of the angle.

4 To measure an angle that is larger than 180° using a semicircular protractor, first measure the small angle enclosed by the arms and then ______________.

5 To construct an angle that is larger than 180° using a semicircular protractor, subtract the given angle from 360° and construct the resulting angle. Label the required angle: it is ______________ the one that has been constructed.

6 An ______________ angle is an angle that is greater than 0° and less than 90°.

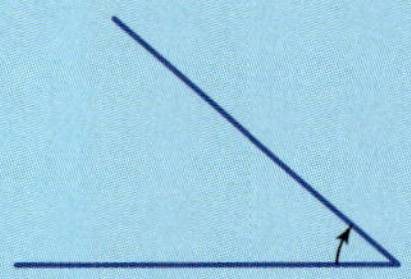

7 A 90° angle is called a ______________ angle.

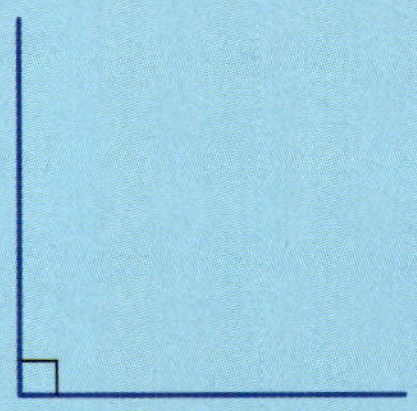

8 An ______________ angle is larger than 90°, but smaller than 180°.

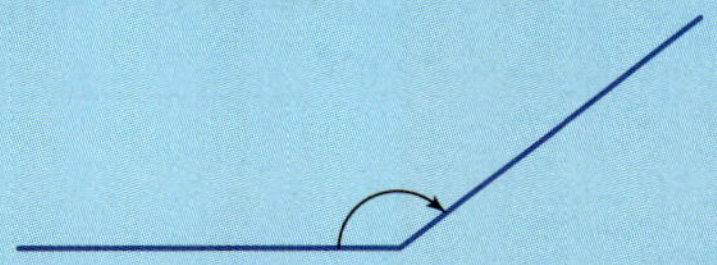

9 A straight angle equals ______________.

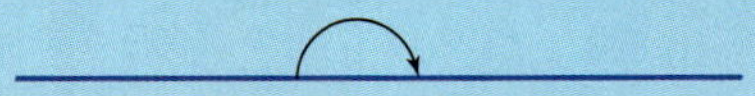

10 A ______________ angle is greater than 180°, but less than 360.°

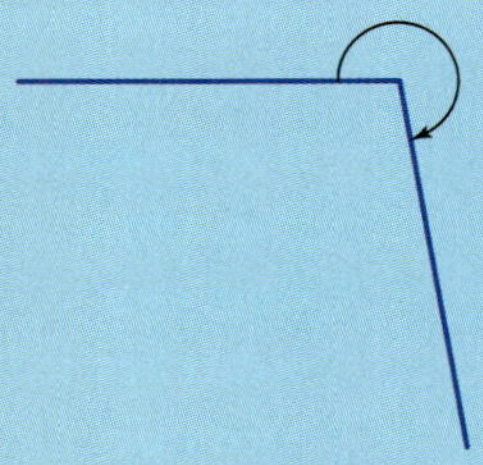

11 A 360° angle is called ______________.

12 An angle can be named using three capital letters of the English alphabet: two for the arms and one for the vertex of the angle. The letter representing the ____________ is always placed in the middle.

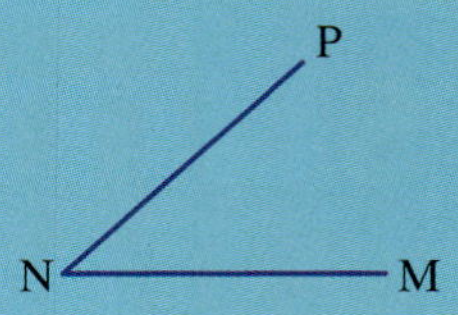

13 Instead of the word 'angle' the ____________ is used.

14 ____________ angles add up to 90°.

15 ____________ angles add up to 180°.

16 Angles with a common vertex that form a ____________ add up to 360°.

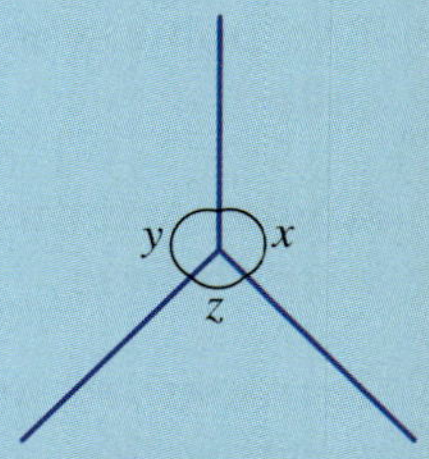

17 Vertically opposite angles are ____________ in size.

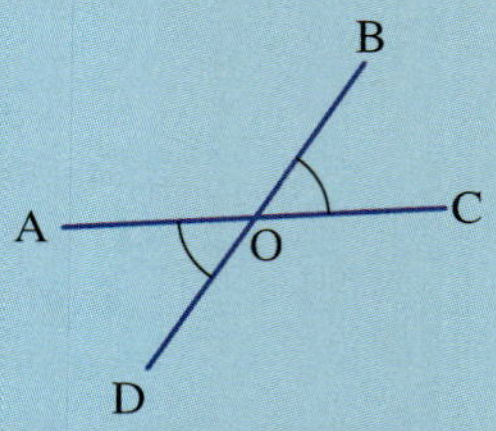

18 To bisect an angle means to ____________ it. This can be done with the help of a pair of compasses. A pair of compasses can also be used for constructing certain angles and for copying angles.

19 The true bearing is given by the angle measured ____________ in a clockwise direction. It is written using ____________ digits.

20 The conventional bearing is given by the angle measured from the ____________ in either a clockwise or anticlockwise direction. To specify the conventional bearing we need to state (in the following order) whether the angle is being measured from the north or the south, the size of the angle and whether the angle is being measured in the direction of ____________.

WORD LIST

complementary	180°	acute
a vertex	one revolution	halve
north–south line	east or west	from north
the arms	symbol ∠	right
degrees	supplementary	vertex of the angle
subtract it from 360°	equal	circle
'outside'	lines	estimate
obtuse	reflex	three

CHAPTER review

1 First estimate, then use a protractor to measure the magnitude of each of the following angles. Comment on the accuracy of your estimate. 10A

a

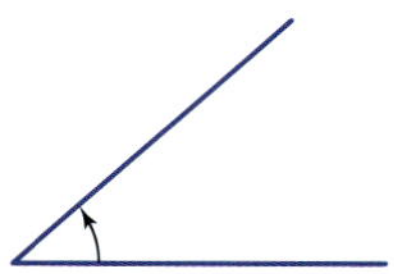

b

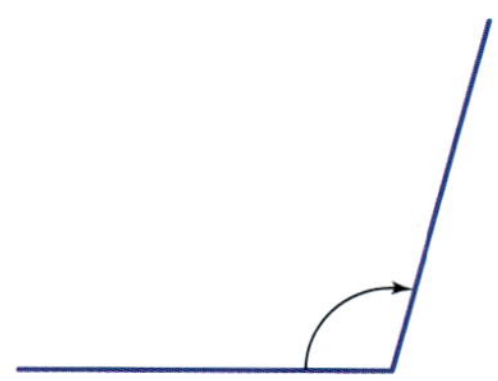

c

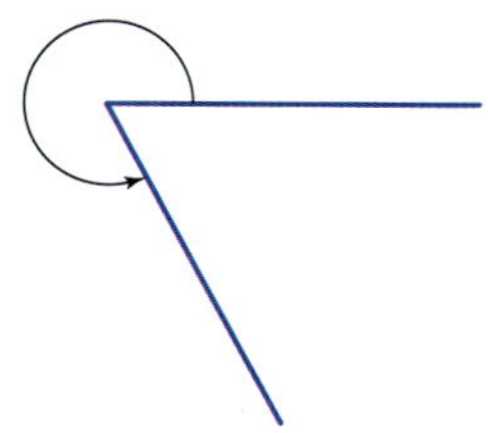

2 Use a protractor to construct each of the following angles. 10B

a 43° **b** 68° **c** 90° **d** 123° **e** 170°
f 226° **g** 270° **h** 285° **i** 300° **j** 310°

3 State the type of each of the angles shown below. 10C

a

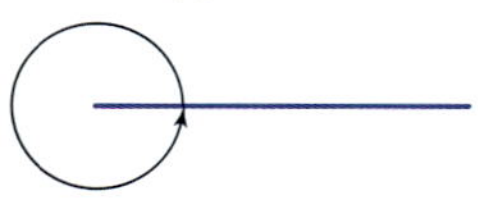

b

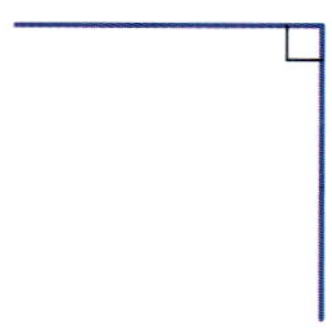

c

d

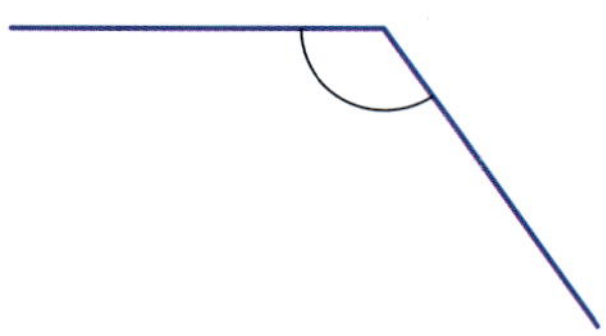

e

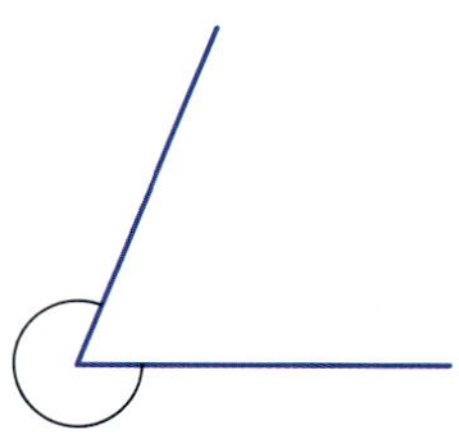

f

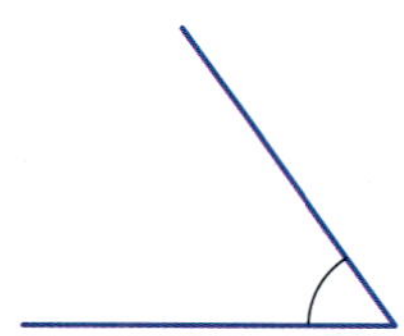

Questions **4** and **5** refer to the diagram at right:

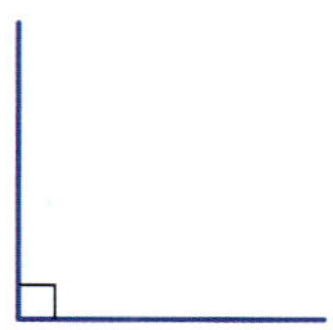

4 **multiple choice** 10C

The angle shown in the diagram has a magnitude of:

A 50° **B** 40° **C** 90° **D** 100° **E** 190°

5 **multiple choice** 10C

The angle shown is:

A a right angle **B** a straight angle **C** an acute angle
D an obtuse angle **E** a reflex angle

6 **multiple choice** 10C

An acute angle has a magnitude that is:

A more than 90° **B** less than 90° **C** exactly 90°
D more than 90° but less than 180° **E** less than 180°

10C **7** **multiple choice**

A straight angle has a magnitude of:

A 90° **B** 270° **C** 360° **D** 180° **E** 60°

10D **8** Name the angle shown.

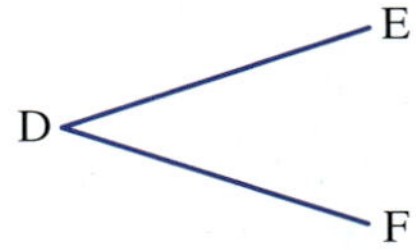

10E **9** Find the supplement of:

a 45° **b** 49° **c** 80° **d** 65° **e** 25°.

10E **10** Find the complement of:

a 12° **b** 40° **c** 56° **d** 38° **e** 70°.

10E **11** Find the value of the pronumeral in each of the following, giving reasons for your answers.

a

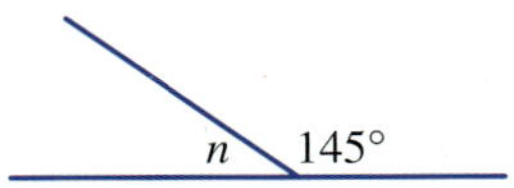

b

c

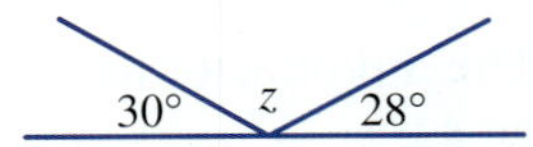

d

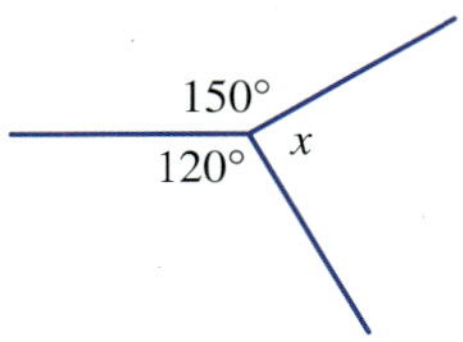

e

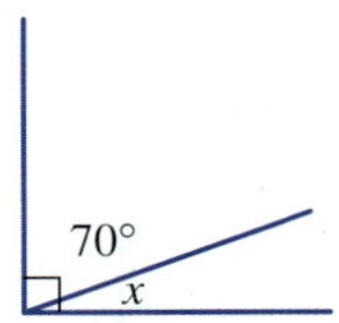

f

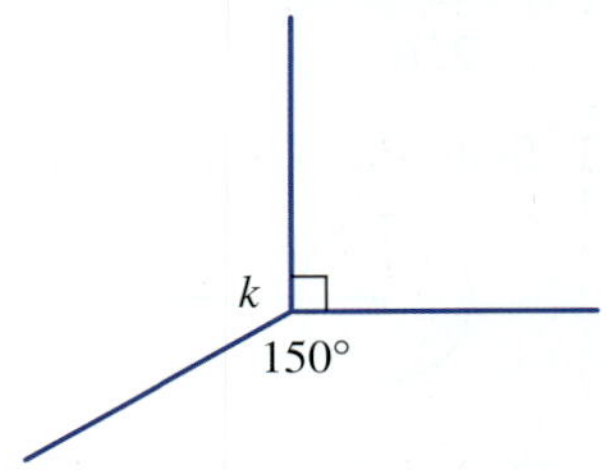

g

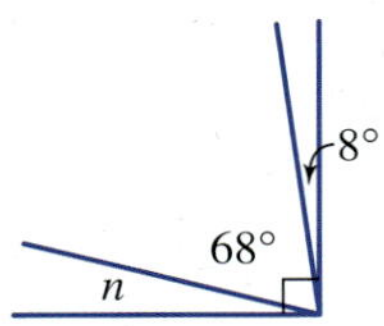

h

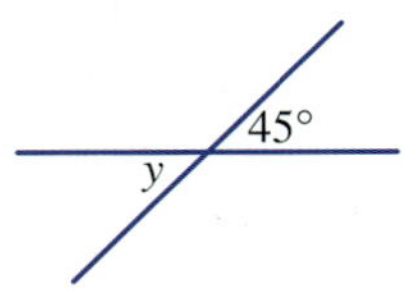

i

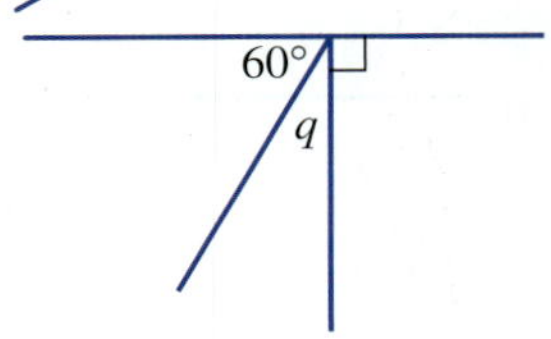

10F **12** Construct an angle of 90° and bisect it by using a pair of compasses and a ruler.

10F **13** Copy exactly the shown angle using a pair of compasses.

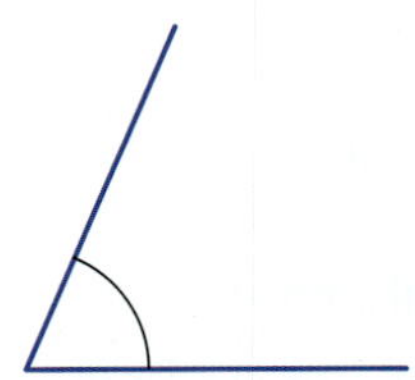

10F **14** Construct the following angles using a pair of compasses and a ruler.

a 45° **b** 60° **c** 30° **d** 120° **e** 240°

10G **15** Change each of the following to compass bearings.

a 078° true **b** 125° true **c** 189° true **d** 250° true **e** 295° true

10G **16** Change each of the following to true bearings.

a N 45° W **b** N 72° E **c** S 55° E **d** S 12° W **e** Due west

test yourself CHAPTER 10

Plane and solid shapes

11

Val and Peter want to replace their front gate with another of the same design. What shapes are formed by the metal bars of the gate? To have this gate made, they need to supply a diagram of it with all measurements and angles shown.

In this chapter, you will look at different shapes and their properties.

Are you ready?

Try the questions below. If you have difficulty with any of them, extra help can be obtained by completing the matching SkillSHEET. Either click on the SkillSHEET icon next to the question on the *Maths Quest 7* CD-ROM or ask your teacher for a copy.

SkillSHEET 11.1

Classifying triangles according to the lengths of their sides

1 A triangle can be classified as equilateral, isosceles or scalene. Classify each of the following triangles.

a

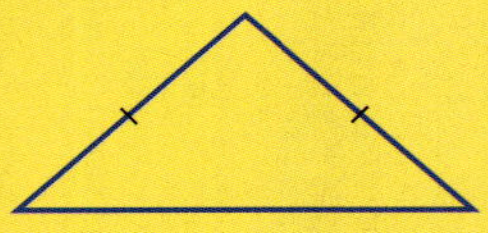

b

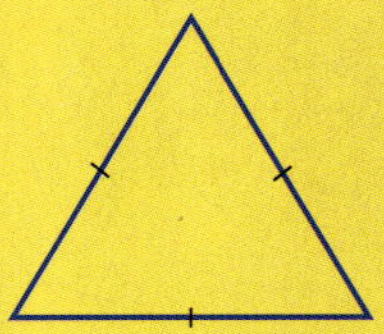

c

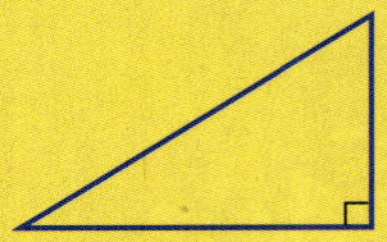

SkillSHEET 11.2

Classifying triangles according to the size of their angles

2 A triangle can also be classified according to the size of its angles. Match one of the following descriptions to each of the triangles in question **1**.

i acute-angled **ii** right-angled **iii** obtuse-angled

SkillSHEET 11.3

Naming polygons

3 Match the correct name with each polygon shown.
Choose from: triangle, octagon, hexagon, quadrilateral.

a

b

c

d

Use the grid shown at right for questions **4**, **5** and **6**.

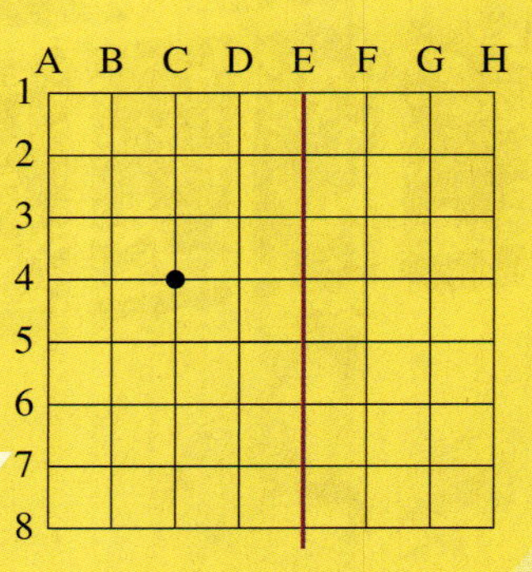

SkillSHEET 11.5

Translation of a point

4 In the grid shown, the point at position C4 is moved 3 units right and 2 units down. What is the final position of the point?

SkillSHEET 11.6

Reflection of a point

5 In the grid shown, the point at position C4 is reflected in the red line. What is the final position of the point?

SkillSHEET 11.7

Rotation of a point

6 In the grid shown, the point at position C4 is rotated about a point at position E4. It is rotated a quarter turn in a clockwise direction. What is the final position of the point?

Introduction

The world around you is filled with many different shapes and objects. The roof of a house in Falls Creek is shaped as a triangle when viewed from the front. This lets the winter snow slide off the roof.

Dice are shaped as cubes so that each of the six numbers is equally likely to appear on the uppermost face.

A soccer ball, which can be kicked and rolled along the ground, is made up of five-sided and six-sided shapes that form a sphere.

As you learn about the properties of shapes and objects you will understand how and why they are used in the world around you.

Triangles

Types of triangles

Every triangle has three angles and three sides.

Capital letters of the English alphabet at each **vertex** are used to identify triangles and the symbol △ is used in place of the word triangle.
For example, the triangle shown at right can be referred to as △ABC. However, it would be equally appropriate to name it △BCA, △CAB, △ACB, △BAC or △CBA.

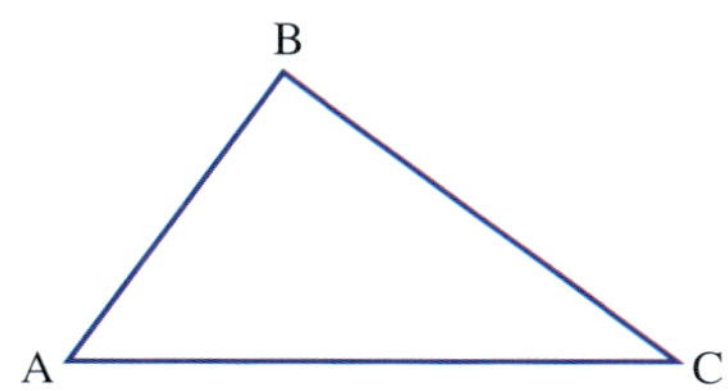

Triangles can be **classified** according to the length of their sides or the size of their angles.

Classifying triangles according to the length of their sides

An *equilateral triangle* has all sides equal in length.

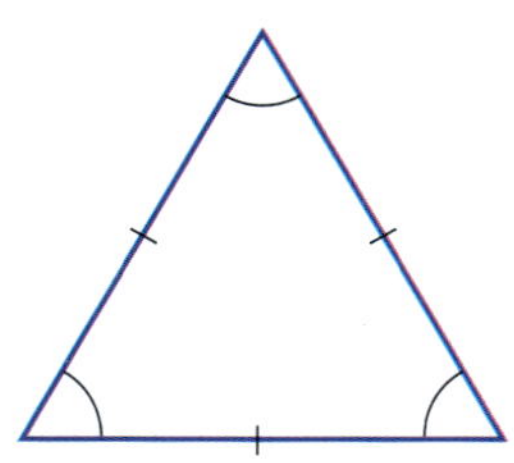

Note that identical marks on the sides of a triangle are used to indicate that the sides have the same length. The angles of an **equilateral triangle** are equal in size. This is shown by placing identical curves on each angle.

An *isosceles triangle* has exactly two sides of equal length.

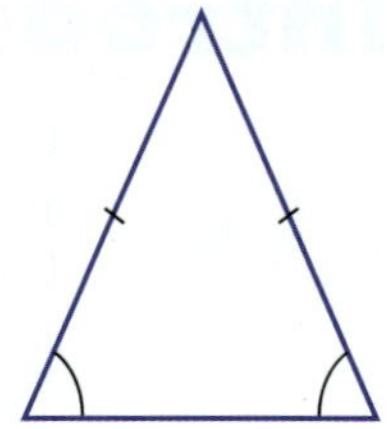

The side of the **isosceles triangle** that has a different length is often called the base of the triangle. The angles adjacent to the base of the isosceles triangle are equal in size. On the diagram at right, the side markings show the two sides that are equal and the angle markings show the two angles that are equal.

A *scalene triangle* has no equal sides.

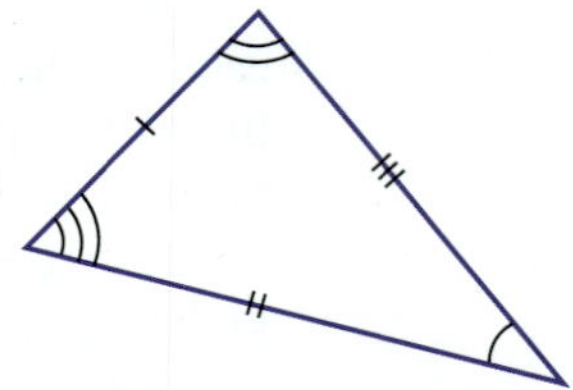

The different side markings on the diagram show that the three sides have different lengths. A **scalene triangle** has all three angles of different size. This is shown by using different angle markings.

WORKED Example 1

Classify each of these triangles according to the lengths of its sides.

a

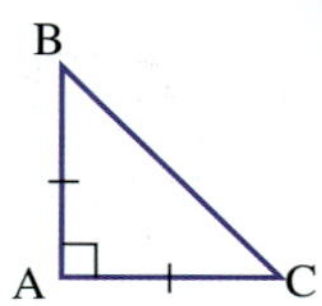

b

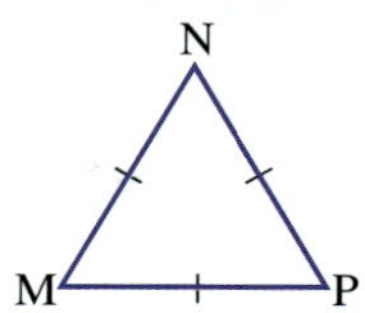

c

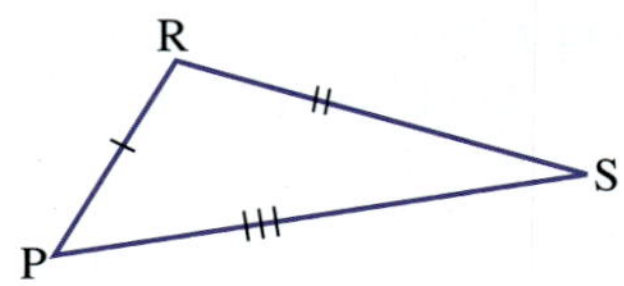

THINK	WRITE
a Sides AB and AC have identical markings on them, which indicate that they are of equal length. So △ABC has 2 equal sides. Classify it accordingly.	**a** △ABC is an isosceles triangle.
b The 3 sides of △MNP have identical markings on them, which means that all 3 sides are equal in length. Classify this triangle.	**b** △MNP is an equilateral triangle.
c All 3 sides of △PRS are marked differently. Therefore, no sides in this triangle are equal in length. Use this information to classify the triangle.	**c** △PRS is a scalene triangle.

Classifying triangles according to the size of their angles

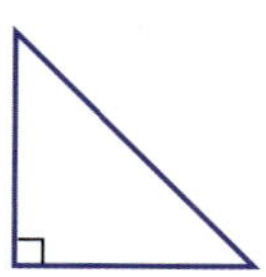

A right-angled triangle has one of its angles equal to 90° (one of its angles is a right angle). A small square in the corner marks the right angle.

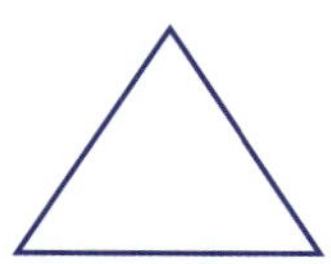

An acute-angled triangle has all angles smaller than 90° (all three angles are acute).

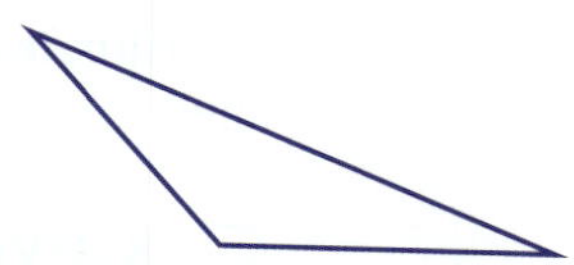

An obtuse-angled triangle has one angle greater than 90° (one angle is obtuse).

WORKED Example 2

Classify each of the triangles in worked example 1 according to the size of its angles.

THINK	WRITE
a In $\triangle ABC$, $\angle CAB$ is marked as the right angle, so classify it accordingly.	**a** $\triangle ABC$ is a right-angled triangle.
b In $\triangle MNP$ all angles are less than 90°, so classify this triangle.	**b** $\triangle MNP$ is an acute-angled triangle.
c In $\triangle PRS$, $\angle PRS$ is greater than 90°; that is, it is obtuse. Use this information to classify the triangle.	**c** $\triangle PRS$ is an obtuse-angled triangle.

Angles in a triangle

COMMUNICATION Sum of angles in a triangle

Using a ruler and a protractor, follow these steps.

1 Draw an acute-angled triangle in your workbook.

2 Use a protractor to measure each of the 3 angles.

3 Find the sum of the 3 angles.

4 Draw up a table like the one shown below and write in your results.

Triangle	First angle	Second angle	Third angle	Sum of angles
a Acute-angled **b** Obtuse-angled **c** Right-angled **d** Isosceles **e** Scalene				

5 Repeat steps 1–3 for the four other triangles in the table.

6 Write any patterns that you have observed in relation to the sum of the angles in a triangle.

7 Cut out each of your triangles. Label the three angles '*a*', '*b*' and '*c*'. Tear out these three angles and place together at a point.

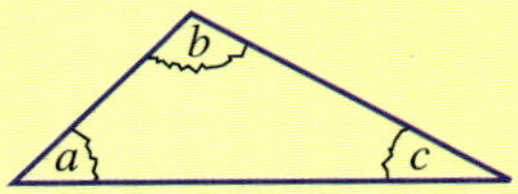

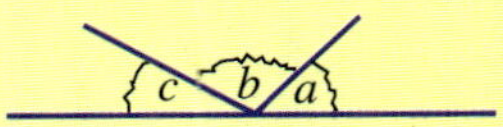

What do you conclude? Is it the same for all your triangles?

As a result of this investigation you should have discovered the rule that is stated below.

It can be shown that the sum of the 3 angles in any triangle is equal to 180°.
In the triangle at right, $a + b + c = 180°$ where the 3 angles of the triangle are a, b and c

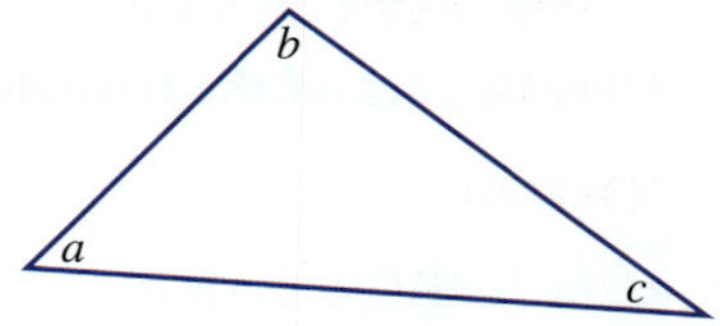

This rule can be used to find missing angles in triangles, as shown in the following examples.

WORKED Example 3

Find the value of the pronumeral in each of the following triangles.

a

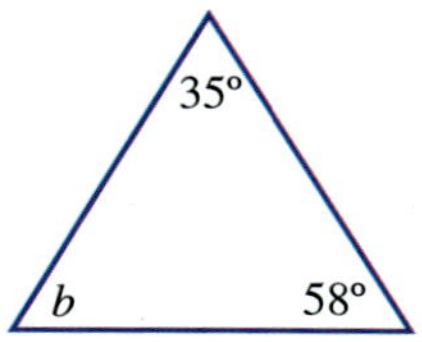

b

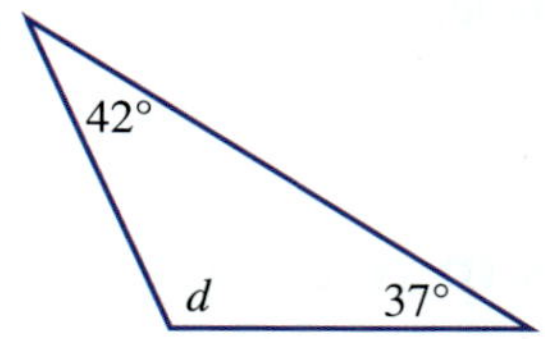

THINK

a 1. The sum of the three angles (b, 35° and 58°) must be 180°. Write this as an equation.
2. Simplify by adding 35° and 58° together.
3. Use inspection or backtracking to solve for b.

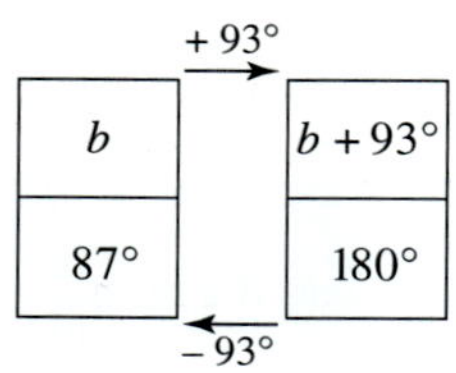

WRITE

a $b + 35° + 58° = 180°$

$b + 93° = 180°$

$b = 180° - 93°$

$b = 87°$

THINK

b 1. The sum of the three angles (d, 37° and 42°) must be 180°. Write this as an equation.
2. Simplify by adding 37° and 42° together.
3. Use inspection or backtracking to solve for d.

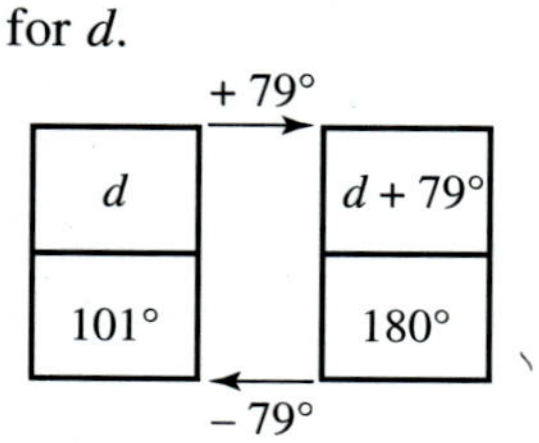

WRITE

b $d + 37° + 42° = 180°$

$d + 79° = 180°$

$d = 180° - 79°$

$d = 101°$

In the previous section it was discussed that the angles at the base of an isosceles triangle are equal in size. Worked examples 4 and 5 illustrate the use of this property.

WORKED Example 4

Find the value of the pronumeral in the following triangle.

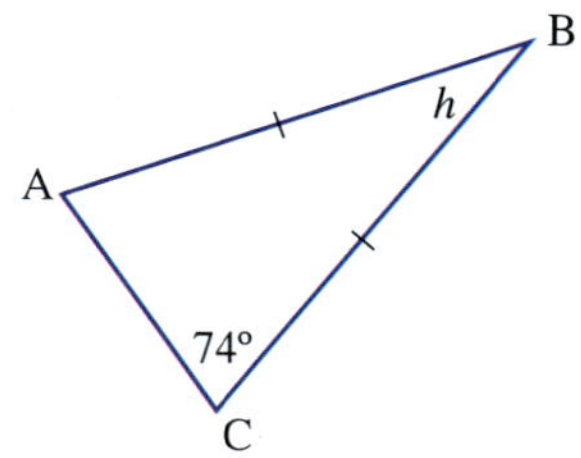

THINK

1. The markings on the diagram indicate that $\triangle$ABC is isosceles with AB = BC. Therefore, the angles at the base are equal in size; that is, $\angle$BCA = $\angle$BAC = 74°.
2. All 3 angles in a triangle must add up to 180°.
3. Simplify.
4. Solve for h, that is, subtract 148° from 180°.

WRITE

$\angle BCA = \angle BAC = 74°$

$$\angle ABC + \angle BAC + \angle BCA = 180°$$
$$h + 74° + 74° = 180°$$
$$h + 148° = 180°$$
$$h = 180° - 148°$$
$$h = 32°$$

WORKED Example 5

Find the value of the pronumeral in the following triangle.

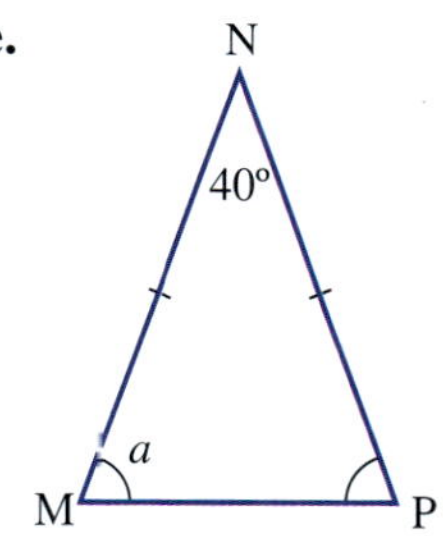

THINK

1. From the diagram we can see that $\triangle$MNP is isosceles with MN = NP. Hence, $\angle$NPM = $\angle$NMP = a.
2. Form an equation by putting the sum of the angles on one side and 180° on the other side of the equals sign.
3. Simplify by collecting like terms.
4. Use inspection or backtracking to solve for a.

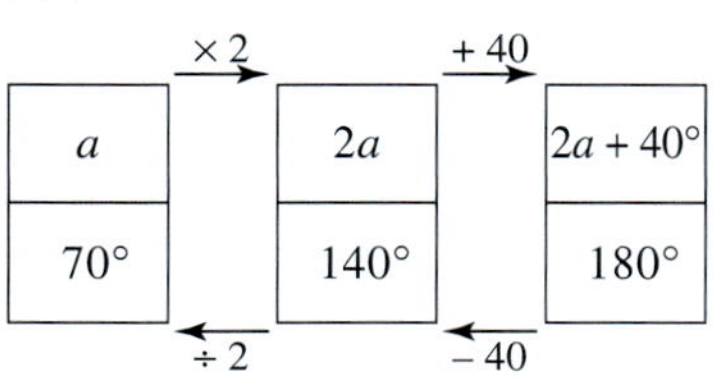

WRITE

$\angle NMP = \angle NPM = a$

$$\angle NMP + \angle NPM + \angle MNP = 180°$$
$$a + a + 40° = 180°$$

$$2a + 40° = 180°$$
$$2a = 180° - 40°$$
$$2a = 140°$$
$$a = \frac{140°}{2}$$
$$= 70°$$

Interior and exterior angles of a triangle

The angles inside a triangle are called **interior angles**. If any side of a triangle is extended outwards, the angle formed is called an exterior angle. The **exterior angle** and the interior angle adjacent (next) to it are supplementary and therefore add up to 180°.

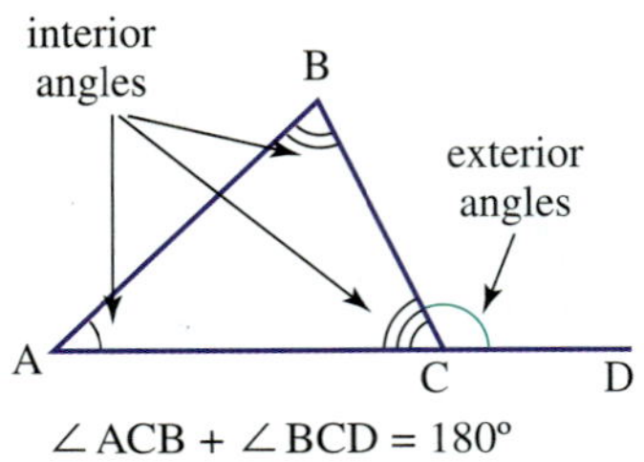

From the above illustration it can also be seen that:

$$\angle BAC + \angle ABC = \angle BCD.$$

Therefore the sum of the opposite interior angles is equal to the exterior angle.

WORKED Example 6

Find the value of the pronumerals in the diagram at right.

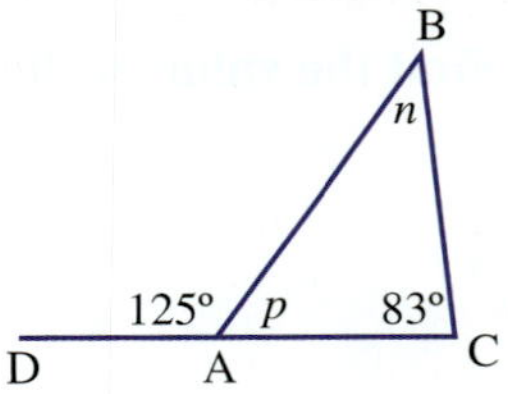

THINK	WRITE
1 ∠BAC (angle p) together with its adjacent exterior angle (∠DAB) add up to 180°. Furthermore, ∠DAB = 125°. So, form an equation.	$\angle BAC = p$; $\angle DAB = 125°$; $\angle BAC + \angle DAB = 180°$ $p + 125° = 180°$
2 Solve for p, that is subtract 125° from 180°.	$p = 180° - 125°$ $p = 55°$
3 The interior angles of △ABC add up to 180°. Identify the values of the angles and form an equation.	$\angle BCA + \angle BAC + \angle ABC = 180°$ $\angle BCA = 83°$ $\angle BAC = p = 55°$ $\angle ABC = n$ $83° + 55° + n = 180°$
4 Simplify by adding 83° and 55°.	$n + 138° = 180°$
5 Solve for n, that is subtract 42° from 180°.	$n = 180° - 138°$ $n = 42°$

remember

1. According to the lengths of the sides, a triangle can be classified as being:
   (a) *equilateral* (three equal sides)
   (b) *isosceles* (exactly two equal sides)
   (c) *scalene* (no equal sides).
2. A triangle can be classified according to the angle size, as being:
   (a) *acute-angled* (all three angles are acute)
   (b) *right-angled* (one angle is a right angle)
   (c) *obtuse-angled* (one angle is obtuse).
3. The sum of the interior angles in any triangle is equal to 180°.
4. The angles at the base of an isosceles triangle are equal in size.
5. An exterior angle of a triangle, and an interior angle adjacent (next) to it, are supplementary (that is, add up to 180°).
6. The sum of the two opposite interior angles is equal to the exterior angle.

EXERCISE 11A Triangles

WORKED Example 1

1 Name each of the following triangles using the capital letters, then classify each triangle according to the lengths of its sides.

a
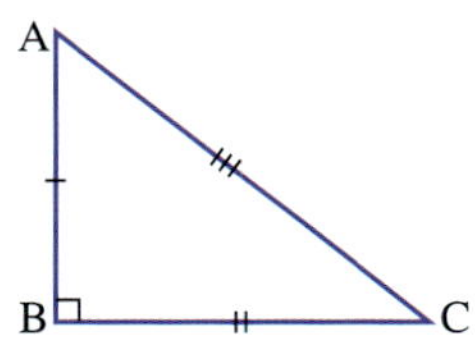

b
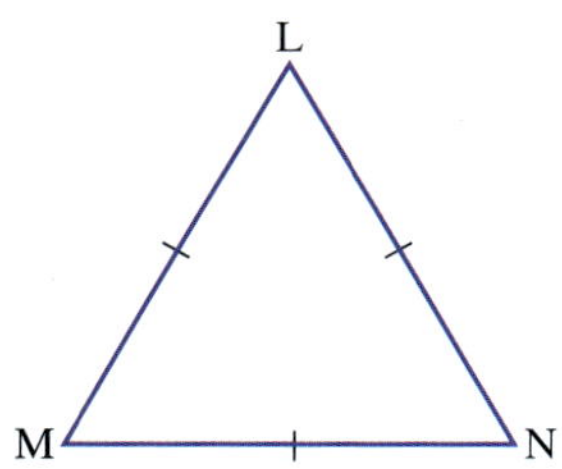

c
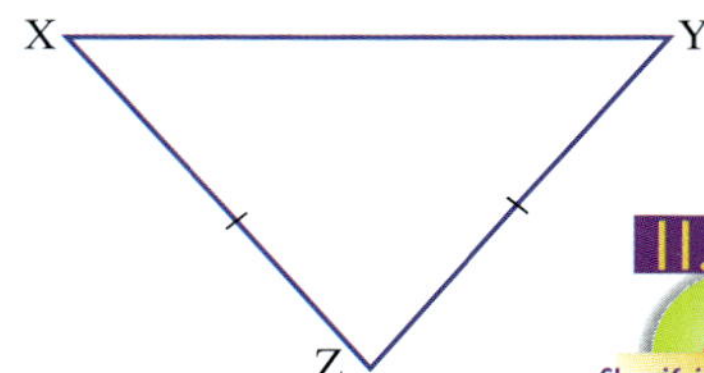

d
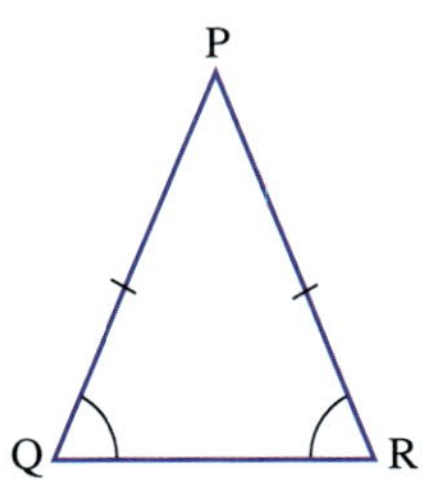

e
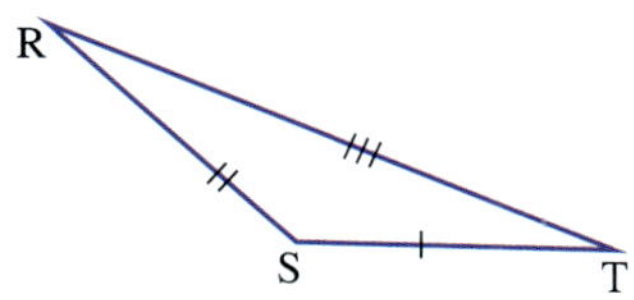

f
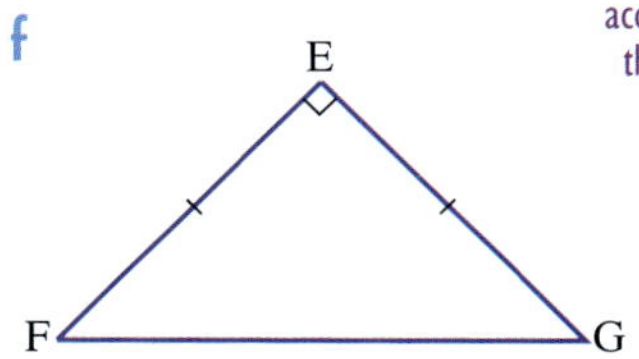

11.1 SkillSHEET Classifying triangles according to the lengths of their sides

Cabri Geometry Classifying triangles (sides)

WORKED Example 2

2 Classify each of the triangles in question **1** according to the size of its angles.

3 Add side and angle markings to these diagrams to show that:

11.2 SkillSHEET Classifying triangles according to the size of their angles

a
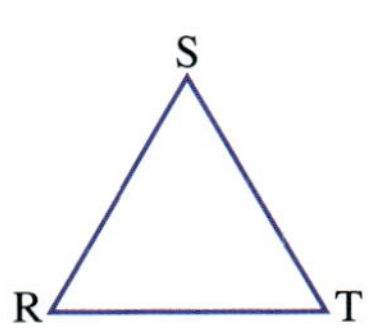

△RST is an equilateral triangle

b
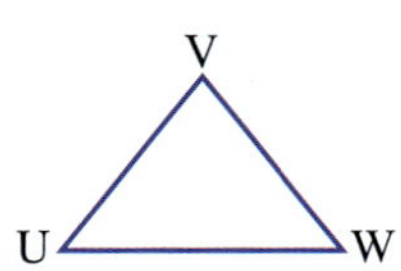

△UVW is an isosceles triangle

c
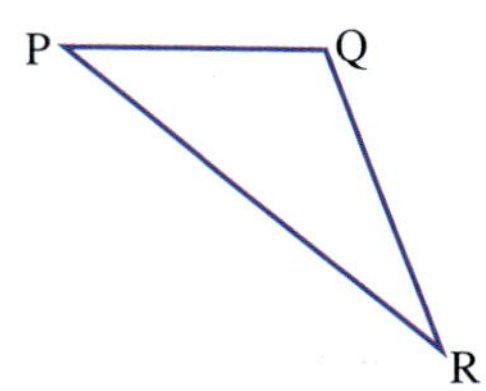

△PQR is a scalene triangle

Cabri Geometry — Classifying triangles (angles)

d

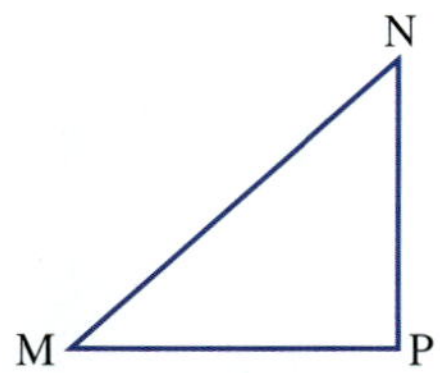

$\triangle$MNP is a right-angled triangle

e

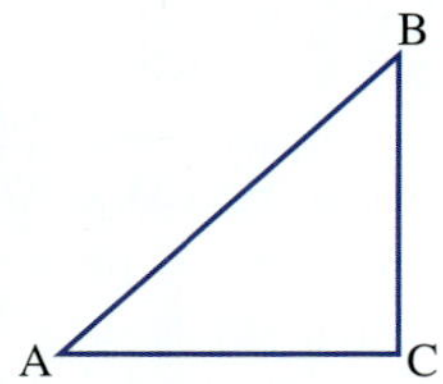

$\triangle$ABC is a right-angled and isosceles triangle

f

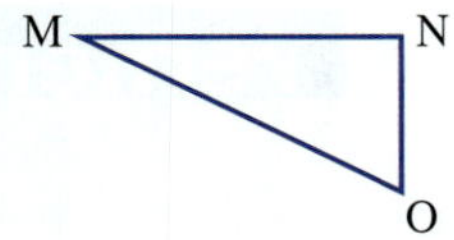

$\triangle$MNO is a right-angled and scalene triangle.

4 **multiple choice**

a Which of these triangles is an equilateral triangle?

A

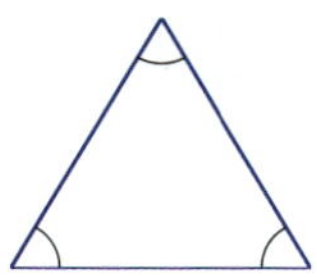

B

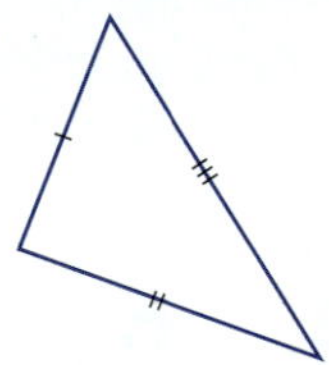

C

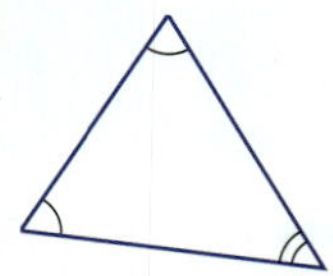

D

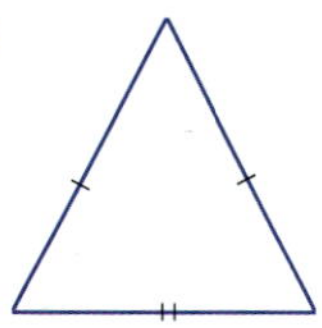

E

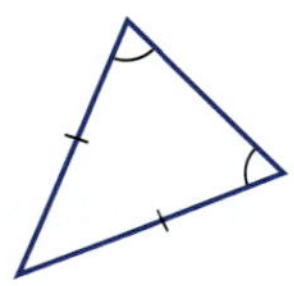

b Which of these triangles is not a scalene triangle?

A

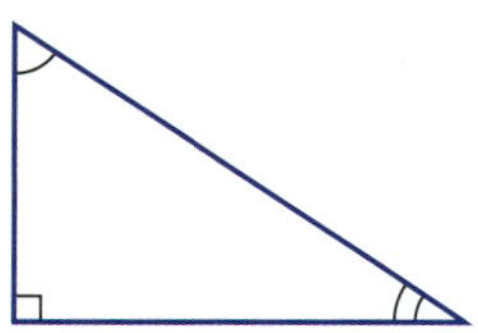

B

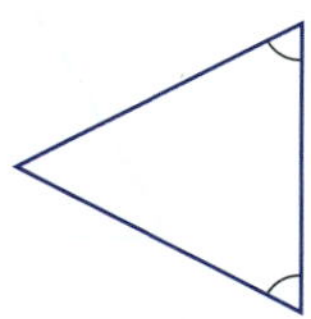

C

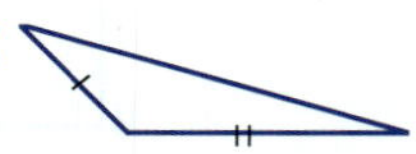

D

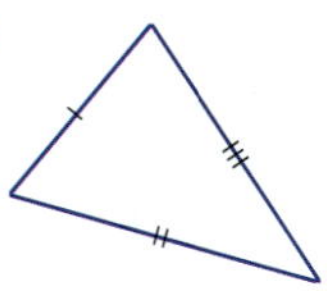

E

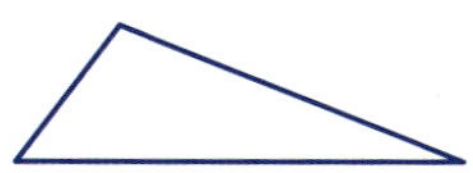

5 **multiple choice**

a Which of these triangles is both right-angled *and* scalene?

A

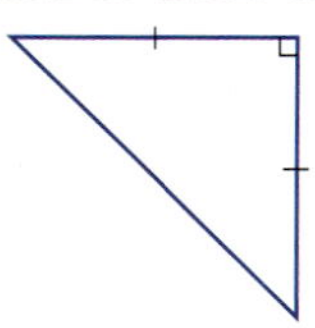

B

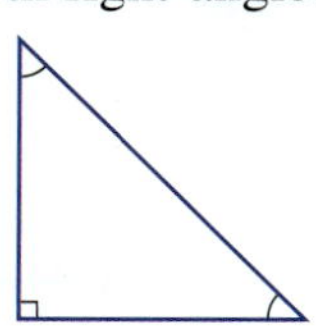

C

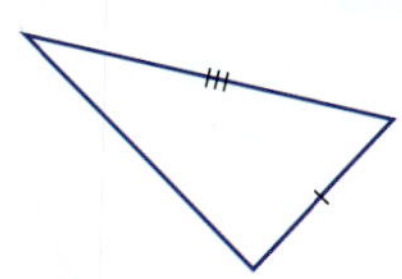

D

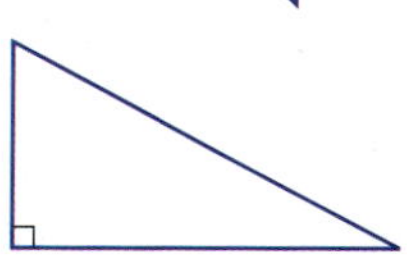

E

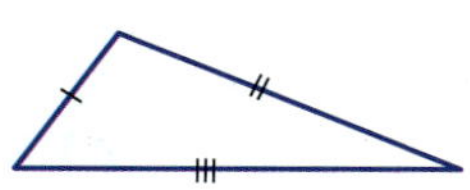

b Which of these triangles is both acute-angled *and* isosceles?

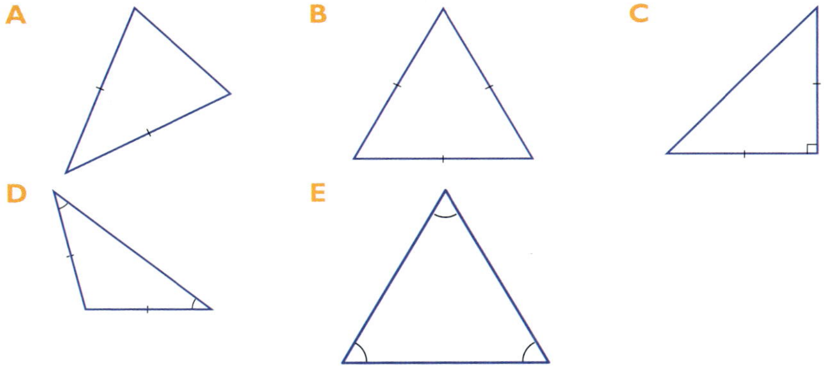

6 What types of triangles can you see in the sandwiches below?

7 Write down three acute triangles you can see around you.

8 Find one example in your classroom or home of each of the six types of triangles described in this chapter. Describe clearly where the triangle occurs, draw the triangle and classify it according to both side and angle types.

9 In the picture at right:

a how many equilateral triangles can you find?
b how many right-angled triangles can you find?
c how many isosceles triangles can you find?

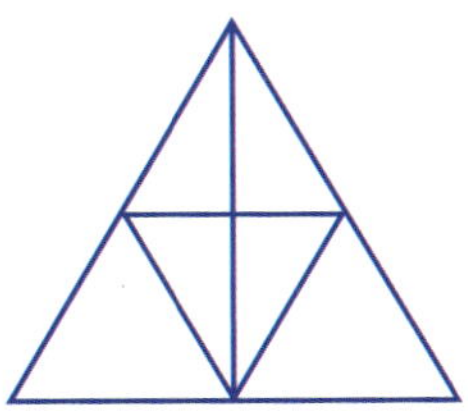

10 Use your ruler, pencil and protractor to accurately draw:

a an equilateral triangle with side lengths 6 cm and all angles 60°
b an isosceles triangle with two sides which are 6 cm each with a 40° angle between them
c a right-angled triangle whose two short sides are 6 cm and 8 cm. (How long is the longest side?)
d a scalene triangle with two of the sides measuring 4 cm and 5 cm and an angle of 70° between the two sides.

WORKED Example 3

Cabri Geometry — Angle sum of a triangle

Mathcad — Angles in a triangle

Cabri Geometry — Angles in right-angled triangles

11 Find the value of the pronumeral in each of the following triangles, giving coded reasons.

a
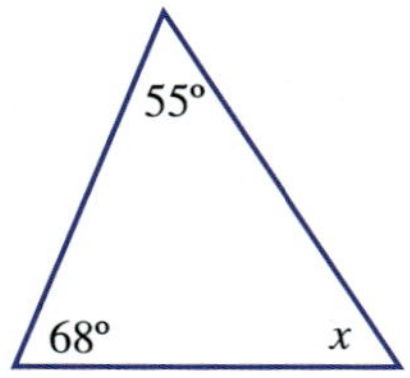

b
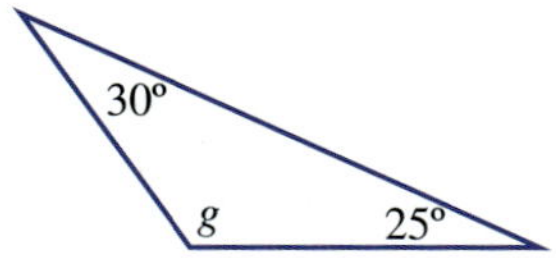

c
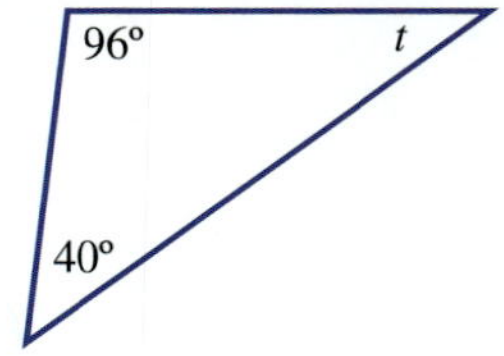

d
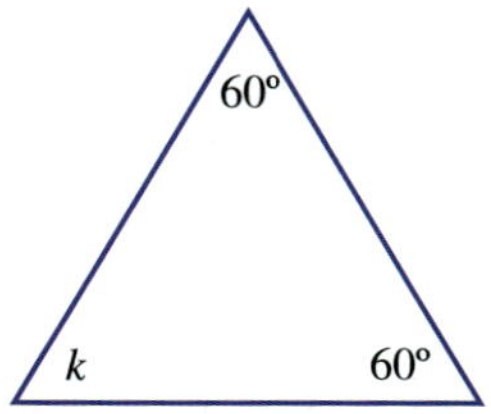

e
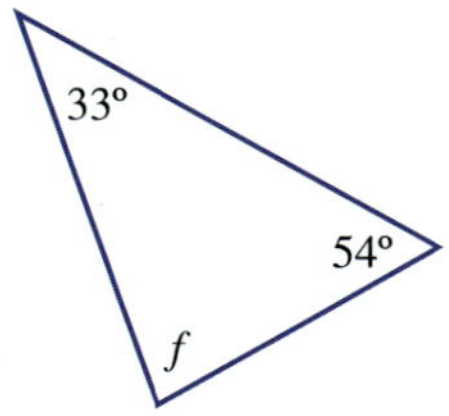

f
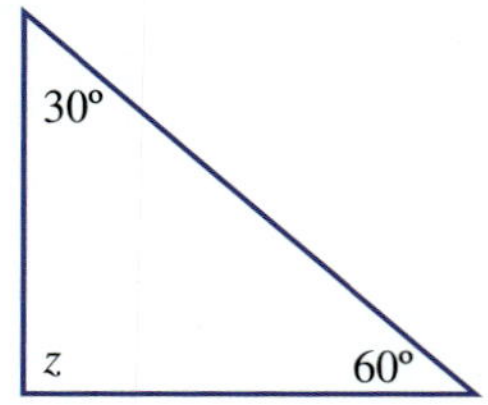

12 Find the value of the pronumeral in each of the following right-angled triangles, giving reasons.

a
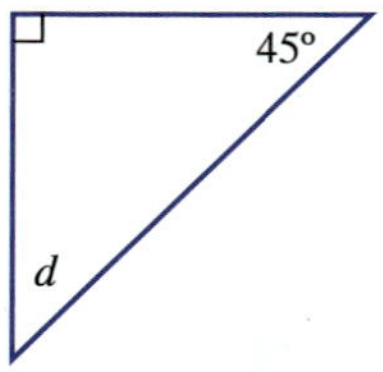

b
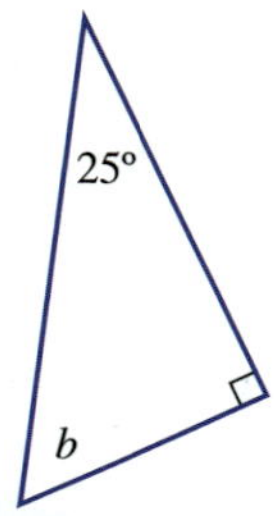

c
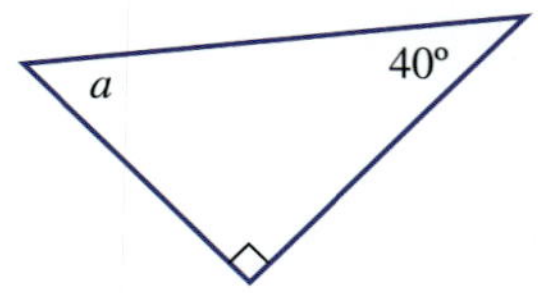

WORKED Example 4

13 Find the value of the pronumeral in each of the following triangles, giving reasons.

a
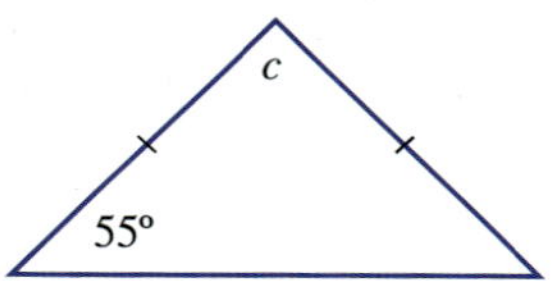

b
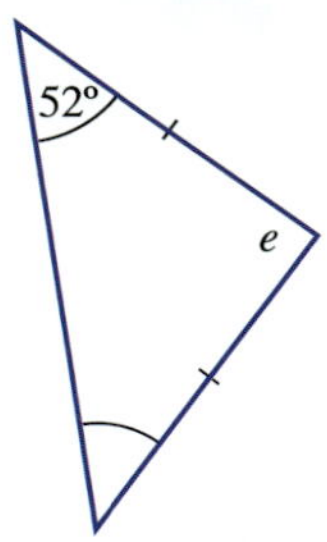

c
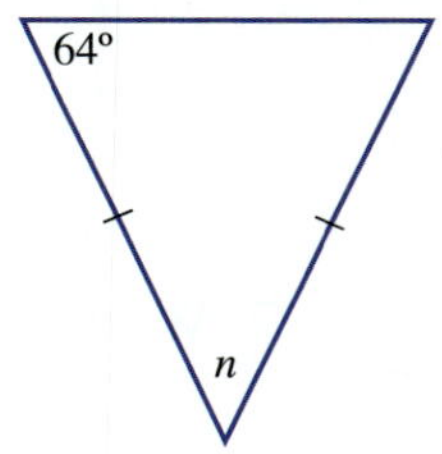

WORKED Example 5

14 Find the value of the pronumeral in each of the following triangles, providing reasons.

a
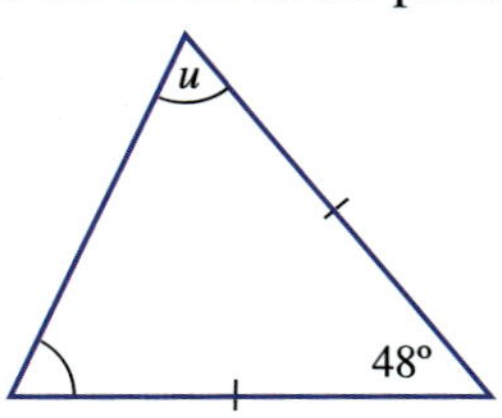

b
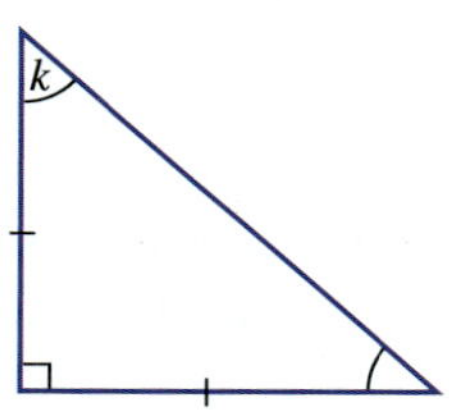

c
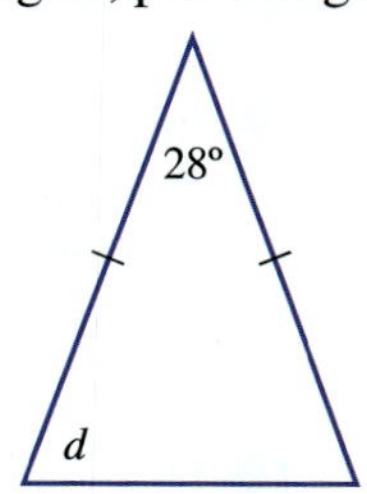

d
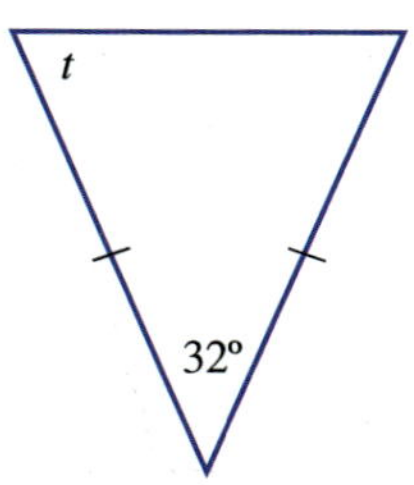

e
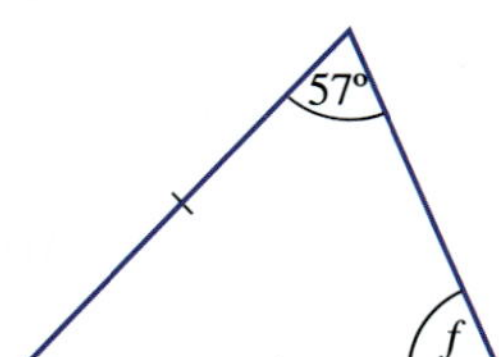

f
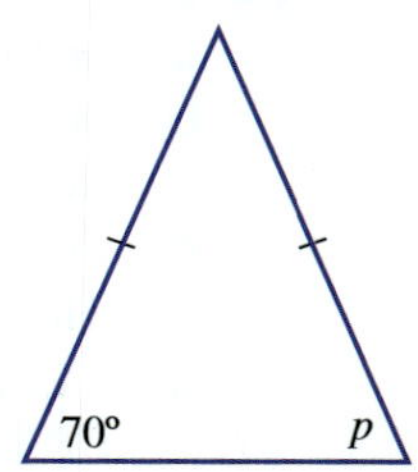

15 Find the missing angle in each of the triangles marked on the following photographs.

a

b

c

d

16 a An isosceles triangle has 2 angles of 55° each. Find the size of the third angle.

b An isosceles triangle has 2 angles of 12° each. Find the size of the third angle.

c Two angles of a triangle are 55° and 75° respectively. Find the third angle.

d Two angles of a triangle are 48° and 68° respectively. Find the third angle.

WORKED Example 6

17 Find the value of the pronumerals in each of the following diagrams.

a

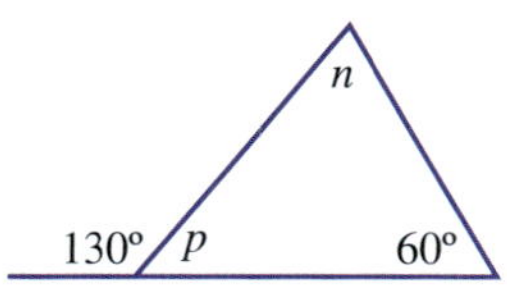

b

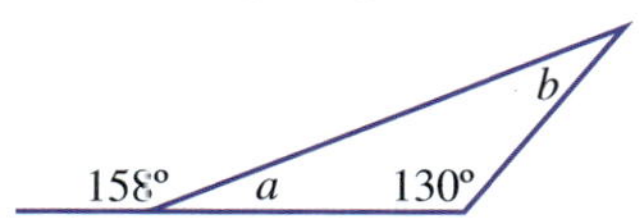

c

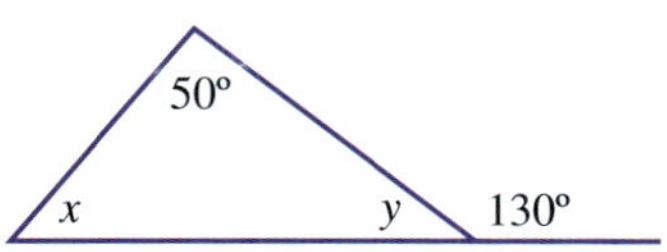

d

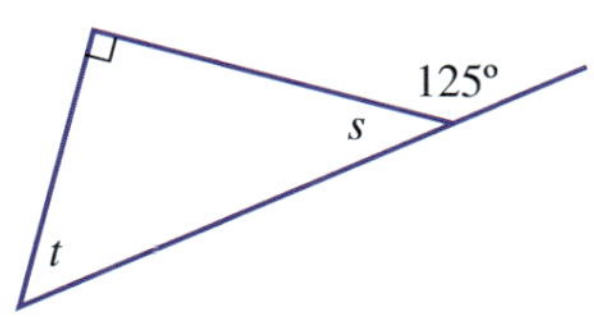

e

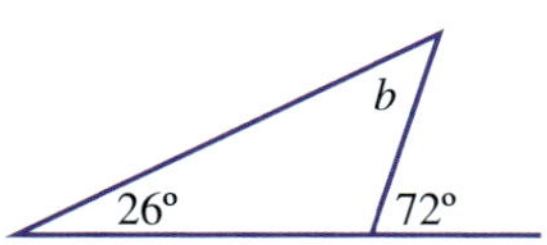

f

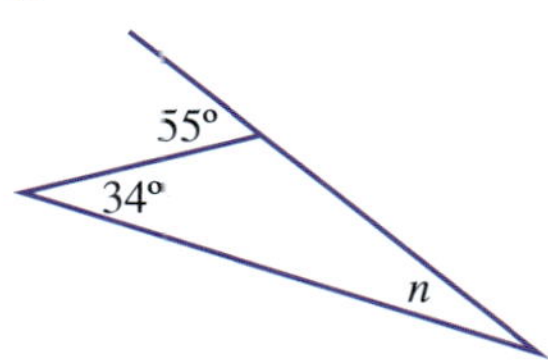

g

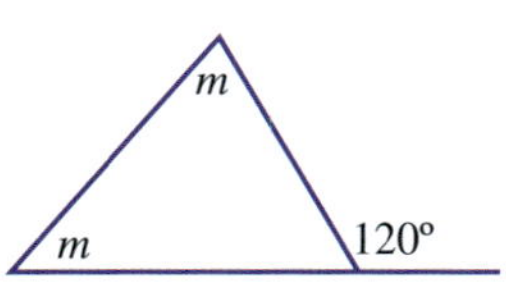

h

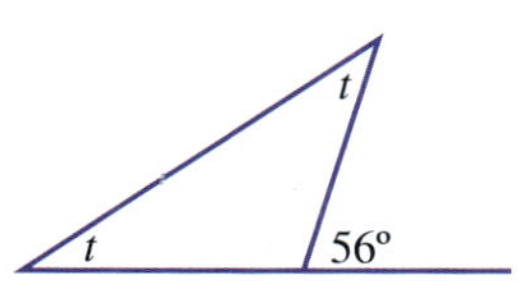

18 **a** Use a ruler and a protractor to construct each of the following triangles.

  i An isosceles triangle with a base of 4 cm and equal angles of 50° each.

  ii An isosceles triangle with two sides which are 5 cm each and two equal angles which are 45° each.

b On your diagrams label the size of each angle. Classify the triangles according to the size of their angles.

19 Below are sets of 3 angles. For each set state whether or not it is possible to construct a triangle with these angles. Give a reason for your answer.

a 40°, 40°, 100° **b** 45°, 60°, 70° **c** 45°, 55°, 85° **d** 111°, 34.5°, 34.5°

20 Explain in your own words why it is impossible to construct a triangle with 2 obtuse angles.

DESIGN

Properties of triangles

Cut some thin strips of paper, three of each of the following lengths:

3 cm, 4 cm, 5 cm, 6 cm, 8 cm and 10 cm.

1. Using a combination of three of these strips (without bending them), how many different triangles can you make?
2. What type of triangle is each one? How many are there of each kind?
3. Are there any similar triangles (same shape, but different size)?
4. Measure the size of the angles. What general relationship can you observe between the lengths of the sides of a triangle and the size of the angles?
5. Formulate a general statement which would enable you to determine whether it is possible to form a triangle from three given side lengths. Why is it not possible in some cases?
6. Create a poster, displaying the results of your investigation.

MATHS QUEST CHALLENGE

1 How many triangles can you find in these shapes?

a

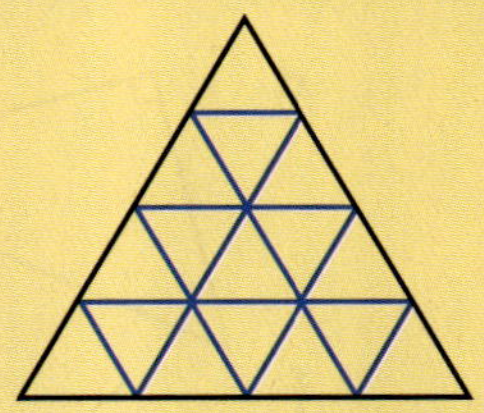

b

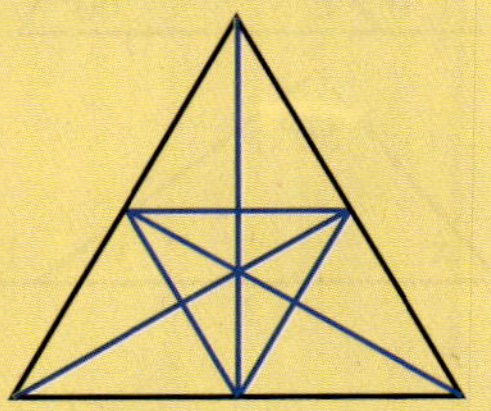

2 How many triangles can you find in these shapes?

a

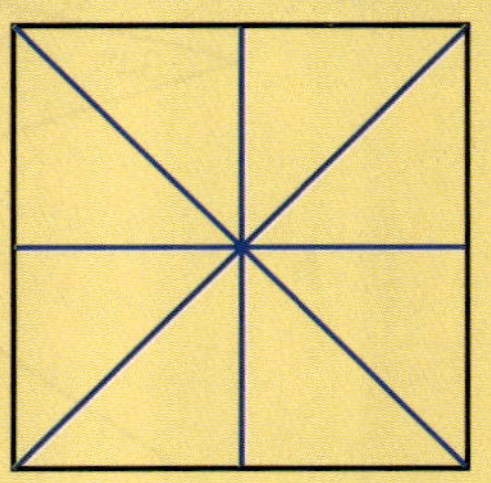

b

Quadrilaterals

Any 2-dimensional closed shape with four straight sides is called a **quadrilateral**. *Quad-* means four, as in quadruplets (four babies), or quadriplegic (paralysed in all four limbs). *Lateral* means sides, as in *lateral movement* (sideways movement) or *lateral thinking* (thinking sideways, or around, a problem).

All quadrilaterals can be divided into two major groups:

(a) parallelograms and

(b) other quadrilaterals.

Parallelograms are quadrilaterals with both pairs of opposite sides being parallel to each other. Parallelograms include:

(a) rectangles,

(b) squares

(c) rhombuses (diamonds).

The table below shows different parallelograms and their properties. Note that **parallel** sides are marked with identical arrows.

Parallelograms	Shape	Properties
Parallelogram	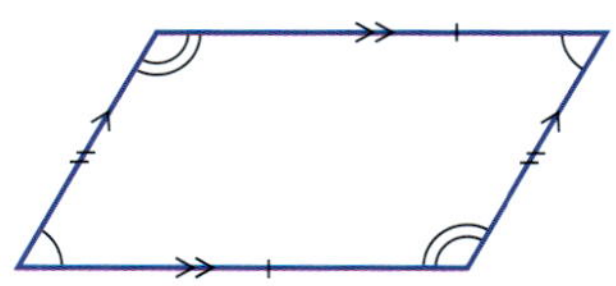	Opposite sides are equal in length. Opposite angles are equal in size.
Rectangle	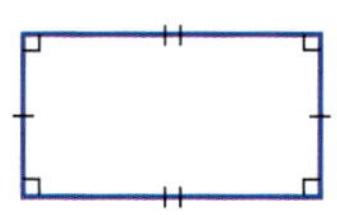	Opposite sides are equal in length. All angles are the same and equal 90°.
Rhombus	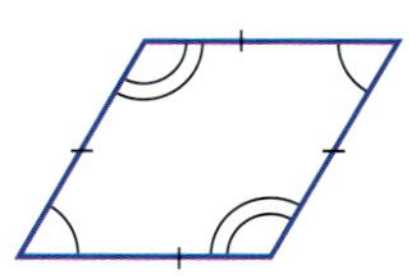	All sides are equal in length. Opposite angles are equal in size.
Square	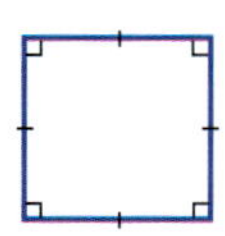	All sides are equal in length. All angles are the same and equal 90°.

Other quadrilaterals include **trapeziums**, **kites** and **irregular quadrilaterals**. The following table shows properties of these shapes.

Cabri Geometry: Trapeziums

Cabri Geometry: Kites

Other quadrilaterals	Shape	Properties
Trapezium	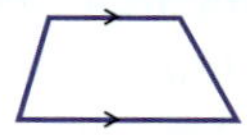	One pair of opposite sides is parallel.
Kite	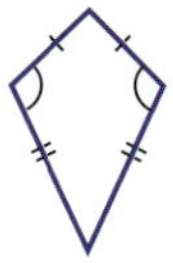	Two pairs of adjacent (next to each other) sides are equal in length. One pair of opposite angles (the ones that are between the sides of unequal length) are equal.
Irregular quadrilateral	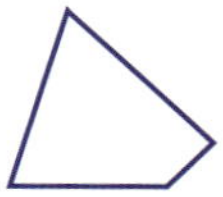	This shape does not have any special properties.

WORKED Example 7

Name the following quadrilaterals, giving reasons for your answers.

a

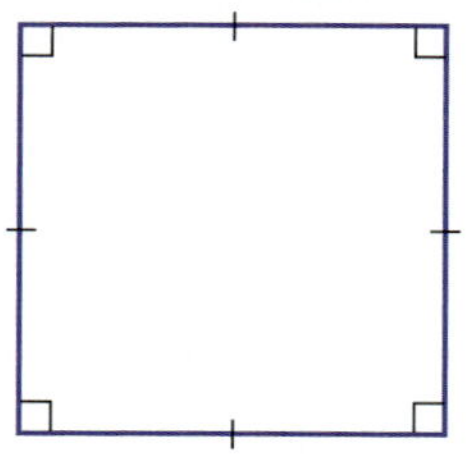

b

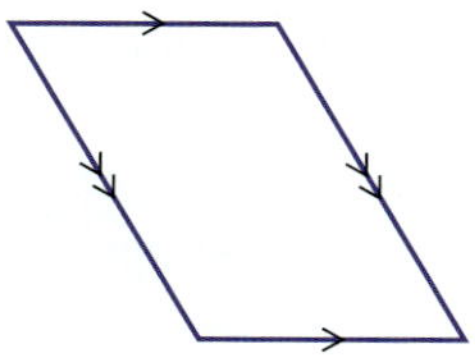

THINK

a The markings on this quadrilateral indicate that all sides are equal in length and all angles equal 90°. Classify the quadrilateral by finding the matching description in the table.

b The arrows on the sides of this quadrilateral indicate that there are two pairs of parallel sides and hence it is a parallelogram. Check the descriptions in the table to see if it is a particular type of parallelogram.

WRITE

a The given quadrilateral is a square, since all sides are equal and all angles are 90°.

b The given quadrilateral is a parallelogram, since it has two pairs of parallel sides.

Angles in a quadrilateral

DESIGN Sum of angles in a quadrilateral

Using a ruler and a protractor, follow these steps.

1 Draw an irregular quadrilateral in your workbook.
2 Measure each of the 4 angles using a protractor.
3 Find the sum of the 4 angles.
4 Record your results for the irregular quadrilateral into the following table.

Quadrilateral	First angle	Second angle	Third angle	Fourth angle	Sum of angles
Irregular quadrilateral					
Parallelogram					
Trapezium					
Kite					
Square					

5 Repeat steps 1 to 4 for each of the other quadrilaterals in the table.

6 Study your results and write any patterns that you have noticed, regarding the sum of angles in a quadrilateral.

7 Cut out each of your quadrilaterals. Label the three angles '*a*', '*b*', '*c*' and '*d*'. Tear out these four angles and place together at a point.

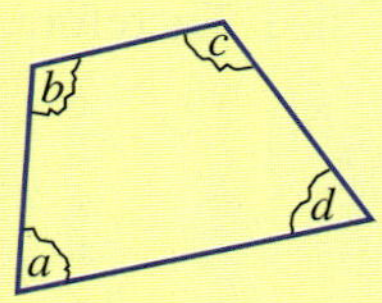

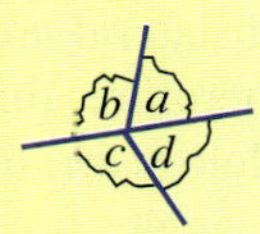

What do you conclude? Is it the same for all your quadrilaterals?

As a result of your investigation you should have discovered this rule.

The sum of the angles in any quadrilateral is 360°.
In the quadrilateral at right

$$a + b + c + d = 360°$$

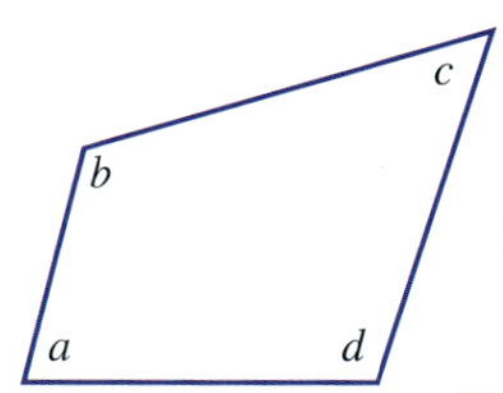

Alternatively from the diagram at right, it can be seen that a quadrilateral is comprised of two triangles. Therefore the sum of the angles in each triangle gives the angle sum of the quadrilateral.

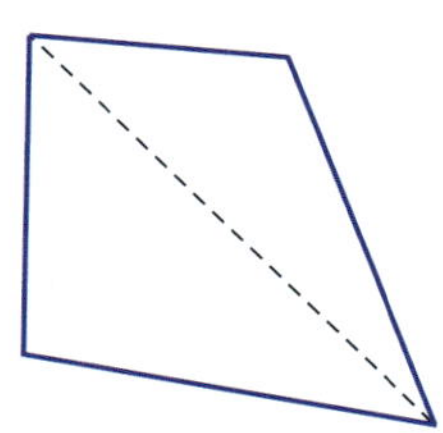

Angle sum of quadrilateral = 2 × 180°
= 360°

We can use this rule to find missing angles in quadrilaterals, as shown in the examples that follow.

WORKED Example 8

Find the value of the pronumeral in the diagram.

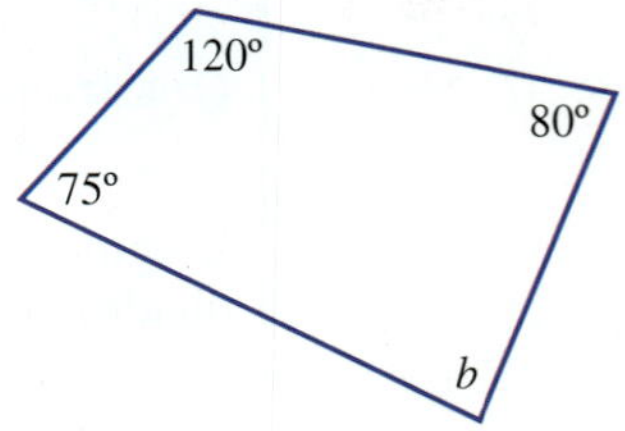

THINK

1. The sum of the angles in a quadrilateral is 360°. So, express this as an equation.
2. Simplify by adding 120°, 80° and 75°.
3. Solve to find the value of b, that is, subtract 275° from 360°.

WRITE

$b + 80° + 75° + 120° = 360°$

$b + 275° = 360°$

$b = 360° - 275°$

$b = 85°$

WORKED Example 9

Find the value of the pronumeral in the diagram, giving a reason for your answer.

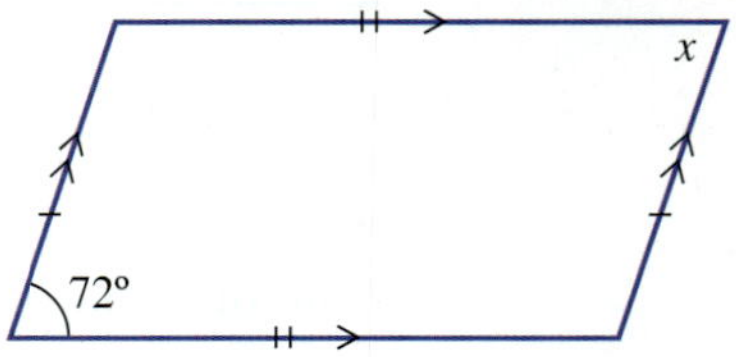

THINK

According to the markings, the opposite sides of the given quadrilateral are parallel and equal in length. Therefore, this quadrilateral is a parallelogram. In a parallelogram opposite angles are equal. So, state the value of the pronumeral.

WRITE

Opposite angles in a parallelogram are equal in size. Therefore, $x = 72°$.

WORKED Example 10

Find the value of the pronumerals in the diagram at right.

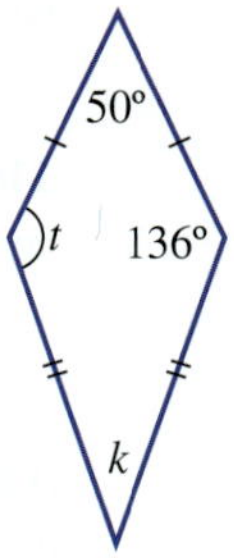

THINK

1. Form an equation by writing the sum of the angles on one side and 360° on the other side of an equals sign.
2. The quadrilateral shown in the diagram is a kite. Angle t and angle 136° are the angles between unequal sides and therefore must be equal in size.
3. Replace t in the equation with 136°.
4. Simplify.
5. Solve to find the value of k, that is, subtract 322° from 360°.

WRITE

$k + t + 50° + 136° = 360°$

$t = 136°$
(The figure is a kite.)

$k + 136° + 50° + 136° = 360°$

$k + 322° = 360°$

$k = 360° - 322°$

$k = 38°$

remember

1. A *quadrilateral* is a 2-dimensional closed shape with four straight sides.
2. All quadrilaterals can be divided into two major groups: parallelograms and other quadrilaterals.
3. *Parallelograms* have two pairs of parallel sides and include rectangles, squares and rhombuses.
4. Other quadrilaterals include trapeziums, kites and irregular quadrilaterals.
5. The sum of angles in any quadrilateral is equal to 360°.

EXERCISE 11B Quadrilaterals

WORKED Example 7

Cabri Geometry: Squares, Parallelograms, Rectangles, Rhombuses, Trapeziums, Kites

1 Name the following quadrilaterals, giving reasons for your answers.

a
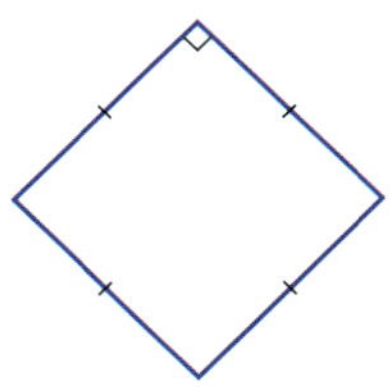

b
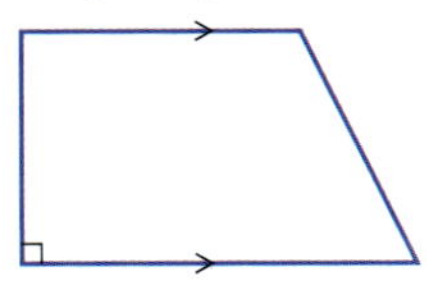

c
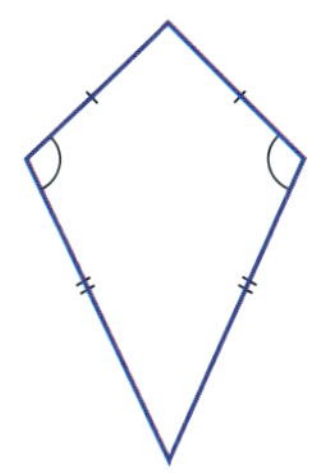

d
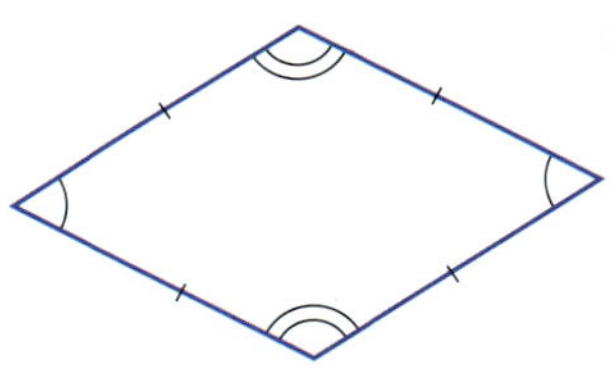

e
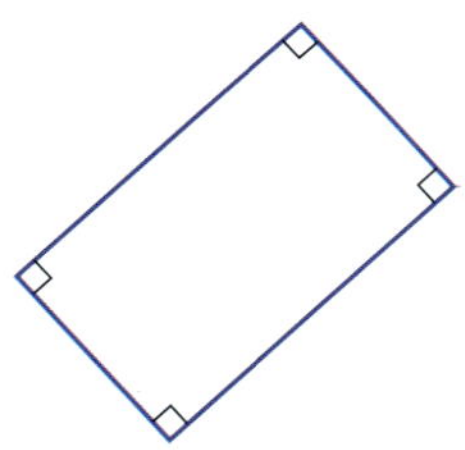

f

2 **multiple choice**

a
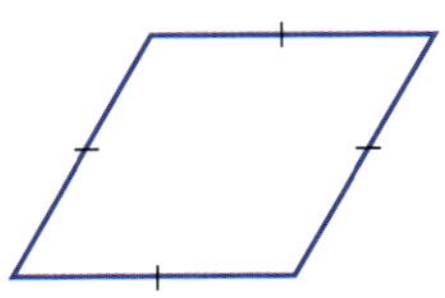

This quadrilateral is a:
- **A** square
- **B** rectangle
- **C** kite
- **D** parallelogram
- **E** rhombus

b
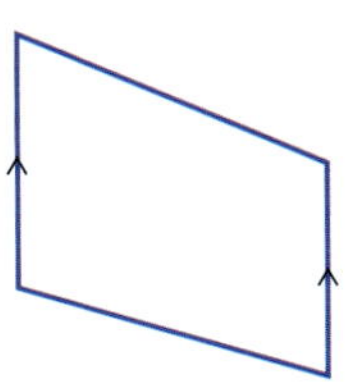

This quadrilateral is a:
- **A** trapezium
- **B** parallelogram
- **C** rhombus
- **D** kite
- **E** square

c
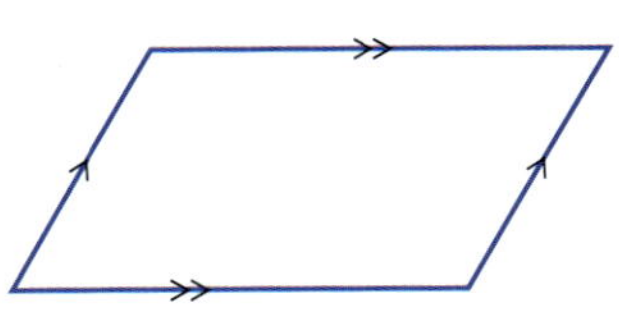

This quadrilateral is a:
- **A** trapezium
- **B** square
- **C** kite
- **D** rhombus
- **E** parallelogram

3 State whether each of the following statements is true or false.
- **a** All squares are rectangles.
- **b** All squares are rhombuses.
- **c** All rectangles are squares.
- **d** Any rhombus with at least one right angle is a square.
- **e** A rectangle is a parallelogram with at least one angle equal to 90°.

4 **multiple choice**

A rectangle is a quadrilateral because:
- **A** it has 4 right angles
- **B** it has 2 pairs of parallel sides
- **C** its opposite sides are equal in length
- **D** it has 4 straight sides
- **E** it has 2 pairs of parallel sides and 4 right angles

5 Draw 4 equilateral triangles with side lengths 4 cm and cut them out.
- **a** Use 2 of these triangles to make a rhombus. Draw your solution.
- **b** Use 3 of these triangles to make a trapezium. Draw your solution.
- **c** Use all 4 triangles to make a parallelogram. Draw your solution.

6 Copy and cut out the set of shapes shown at right. Arrange the shapes to form a square. Draw your solution.

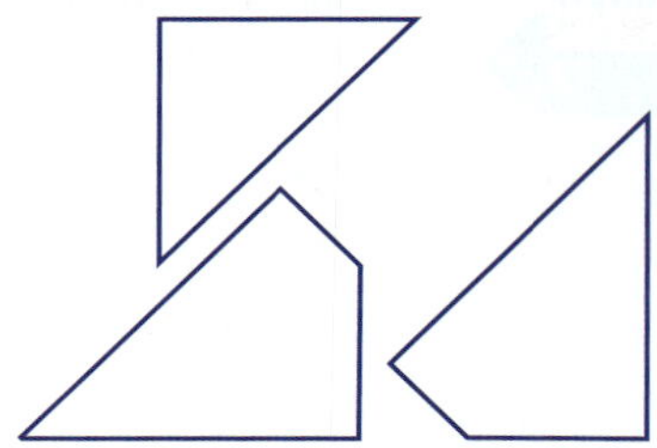

7 In your house, find an example of each type of quadrilateral discussed in this section. Write down the type of quadrilateral and where you found it.

In the following questions, remember to provide coded reasons for any relationships you form.

Cabri Geometry — Angle sum in a quadrilateral

Mathcad — Angles in a quadrilateral

WORKED Example 8

8 Find the value of the pronumeral in each of the following diagrams.

a

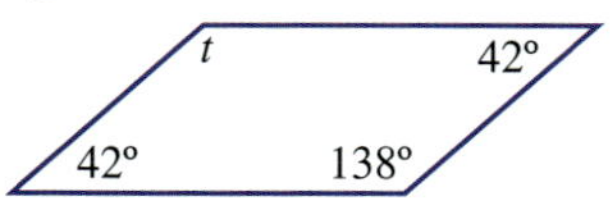

b

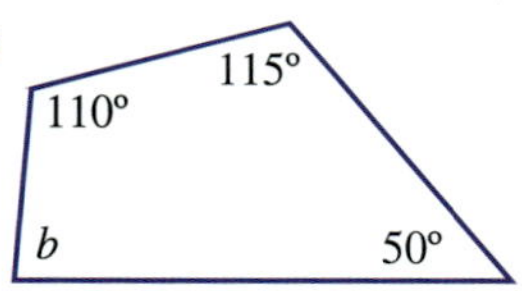

c

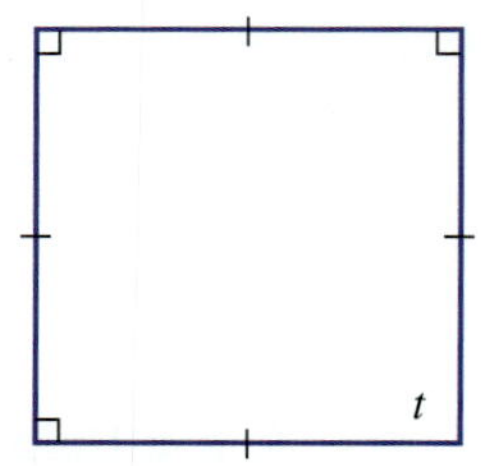

d

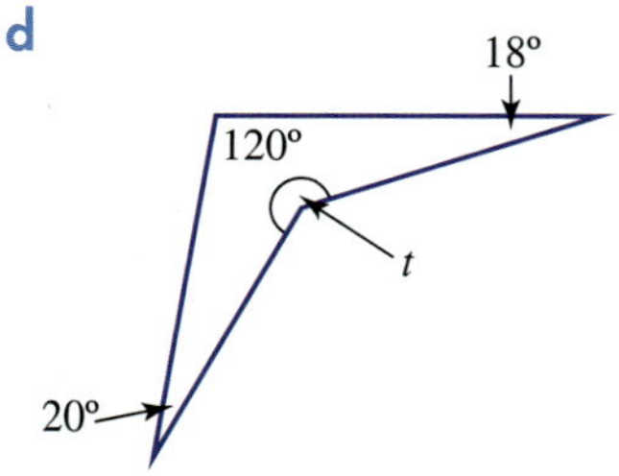

e

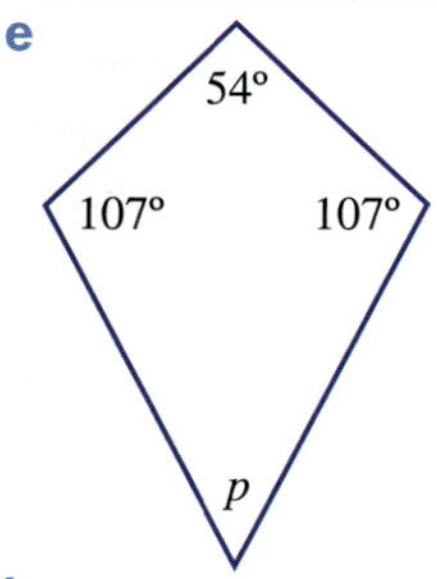

f

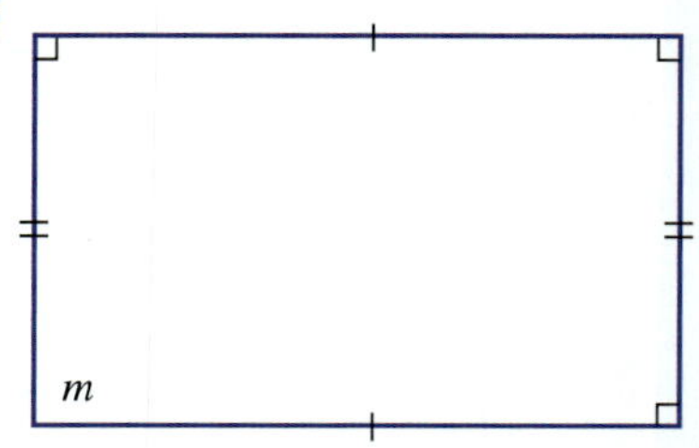

g

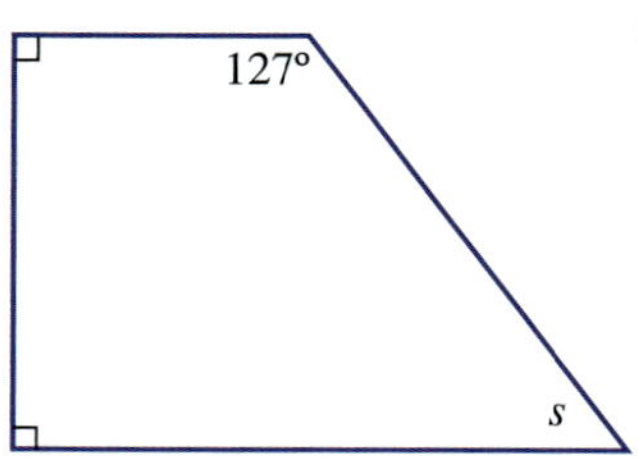

h

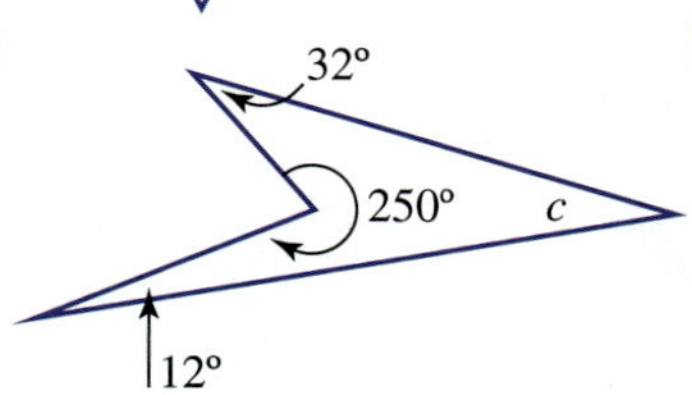

i

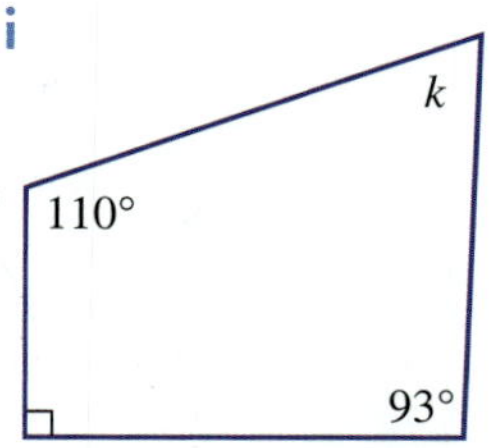

WORKED Example 9

9 Find the value of the pronumeral in each of the following diagrams, giving reasons for your answers.

a

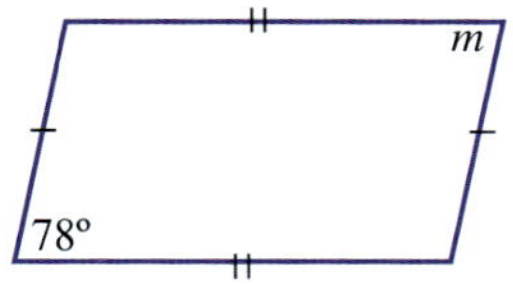

b

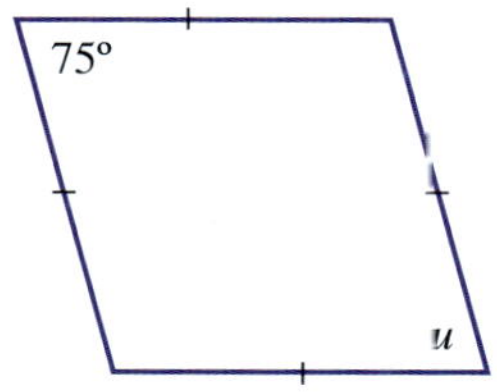

c

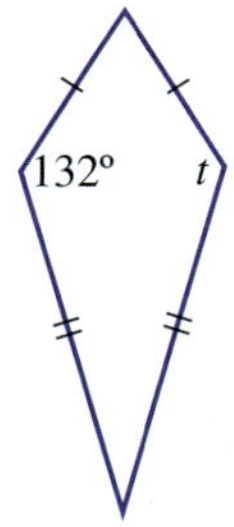

d

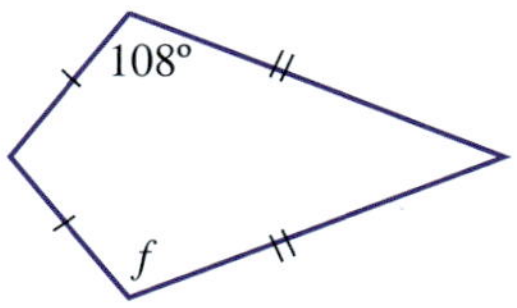

e

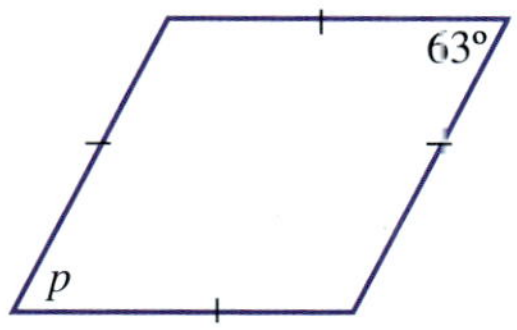

f

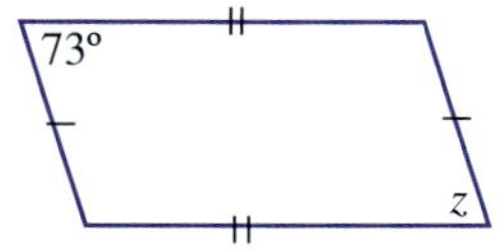

WORKED Example 10

10 Find the value of the pronumerals in each of the following diagrams.

a

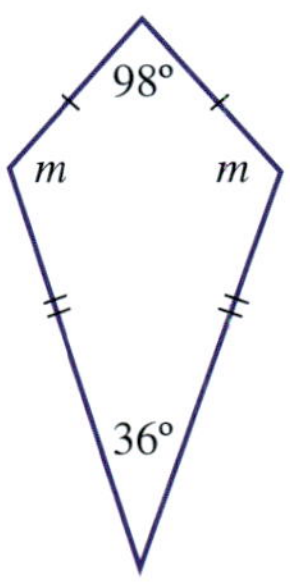

b

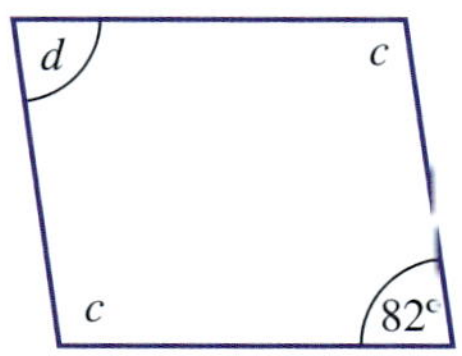

c

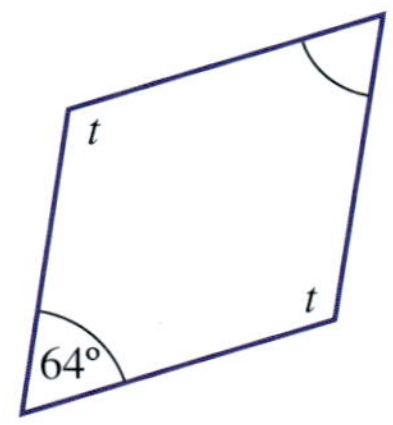

d

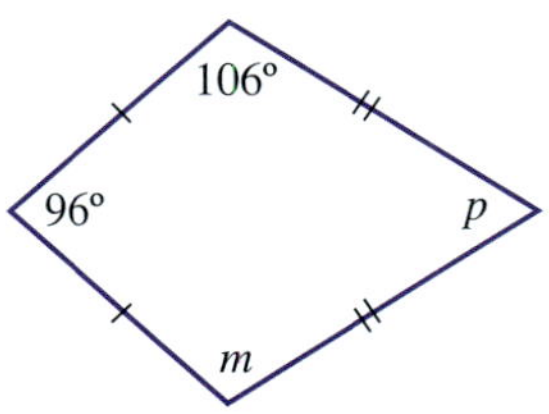

e

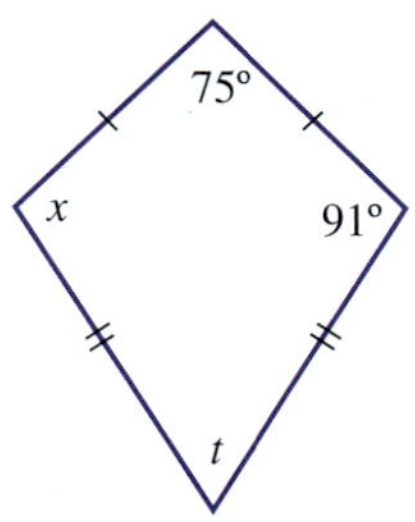

f

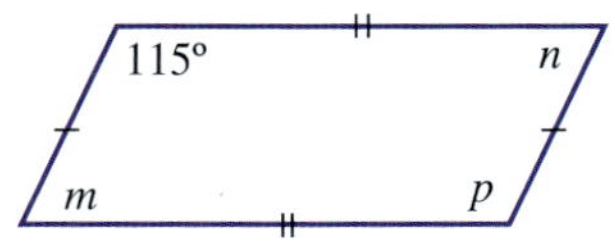

11 multiple choice

The value of t in the diagram at right is:

A 360° **B** 112° **C** 222°
D 138° **E** 180°

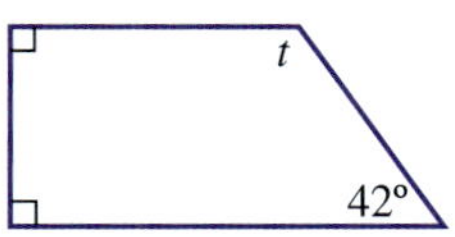

12 multiple choice

The value of r in the diagram at right is:

A 117° **B** 180° **C** 234°
D 126° **E** 63°

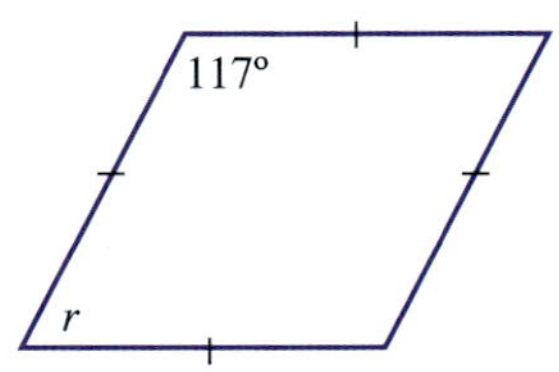

13

This photograph shows the roof of a fast food restaurant. Calculate the value of p.

14 Find the size of the obtuse angle in the kite shown at right.

15 Two angles in a parallelogram are 45° and 135°. Find the other 2 angles.

16 Tom measures 2 angles of a kite at 60° and 110°, but forgets which angle is which. Draw 3 different kites that Tom may have measured, showing the size of all angles in each diagram.

17 Below are sets of 4 angles. For each of the sets decide whether it is possible to construct a quadrilateral. Explain your answer.

a 25°, 95°, 140°, 100° b 40°, 80°, 99°, 51°

GAME time Plane and solid shapes — 001

18 Three angles of a quadrilateral are 60°, 70° and 100°.

a What is the size of the fourth angle of this quadrilateral?

b How many quadrilaterals with this set of angles are possible?

c Construct one quadrilateral with the given angle sizes in your book. (The choice of the length of the sides is yours.)

DESIGN

Design for a front gate

Val and Peter want to replace their front gate with another of the same design. To have this gate made, they need to supply a diagram of it with all measurements and angles shown.

1 There are 4 different shapes formed by the metal bars of the gate. How many different types of triangles are there? Can you name them?

2 How many types of quadrilaterals are there? Name them.

3 Draw a diagram of the gate showing the length measurements and the one angle that is given.

4 Use this angle to calculate all the remaining angles in the diagram.

5 Explain how you were able to achieve this.

6 Using a ruler and protractor, design a fence that is to be constructed using metal bars. Include different triangles and quadrilaterals to make your design as interesting as possible. Write a short report describing the shapes you have used and important angles which need to be marked on your design to assist in the construction of the fence.

What did Dorothy's 3 friends in the Wizard of Oz want from the wizard?

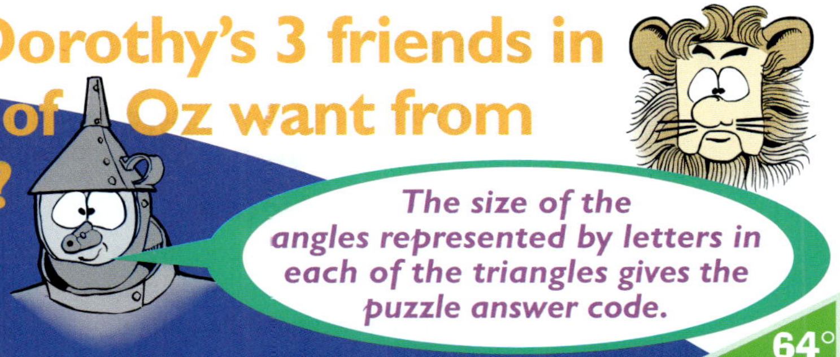

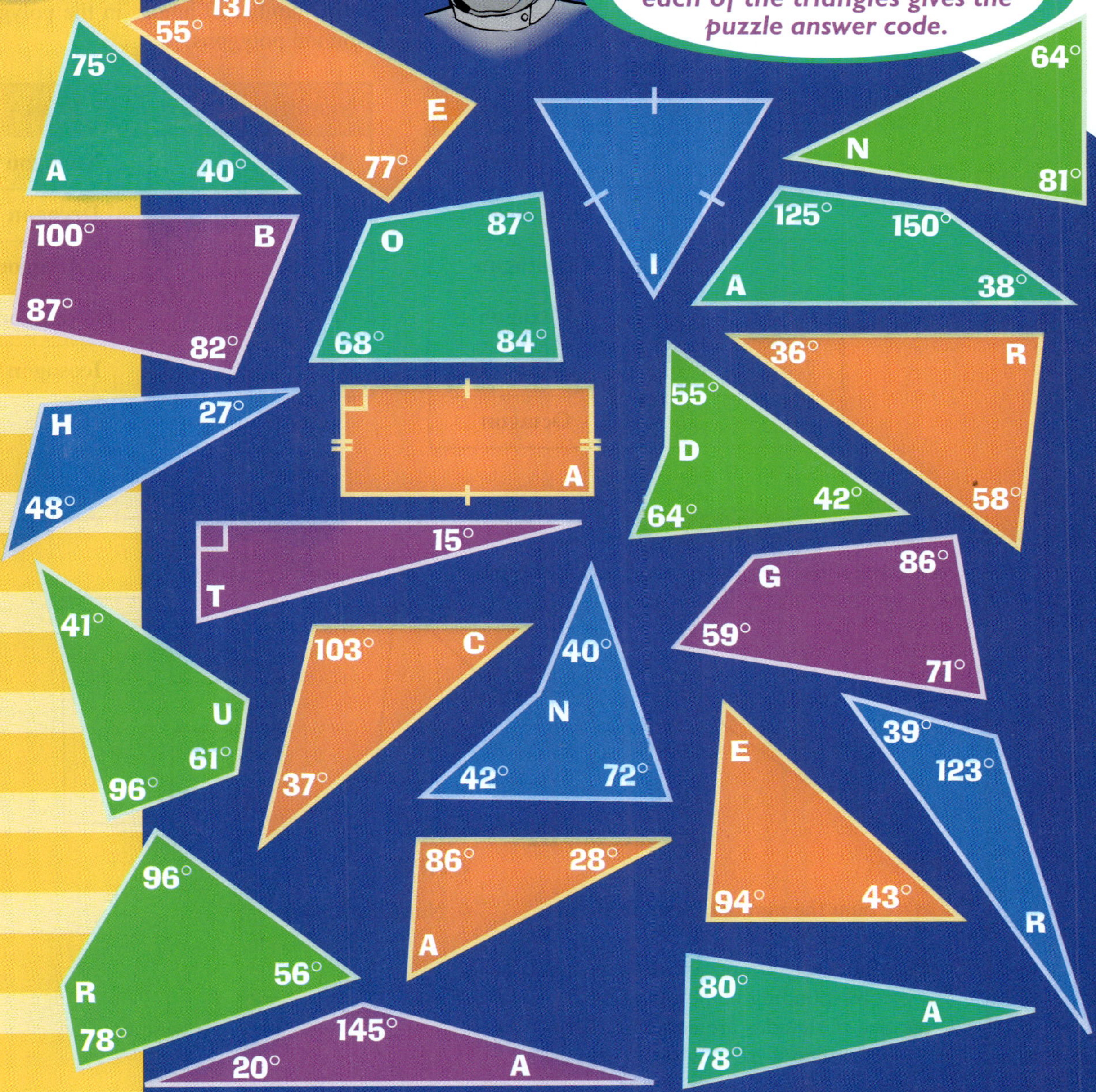

66°	105°	43°	47°	18°	75°	40°	121°	162°	86°	90°	144°	97°
22°	35°	199°	15°	91°	130°	65°	60°	206°				

Polygons

A **polygon** is any closed shape with three or more sides, each of which is a straight line. If the polygon has all sides and angles equal it is said to be a **regular polygon**.

Naming polygons

Polygons are named according to the number of sides or angles in the shape. (Note that the number of sides in any polygon is equal to the number of angles in the polygon.) The table below gives the names of the most common polygons.

Number of sides	Name
3	Triangle
4	Quadrilateral
5	**Pentagon**
6	**Hexagon**
7	**Heptagon**
8	**Octagon**

Number of sides	Name
9	**Nonagon**
10	**Decagon**
11	**Undecagon**
12	**Dodecagon**
20	**Icosagon**

WORKED Example 11

Name the following polygons. State if the polygon is regular.

a

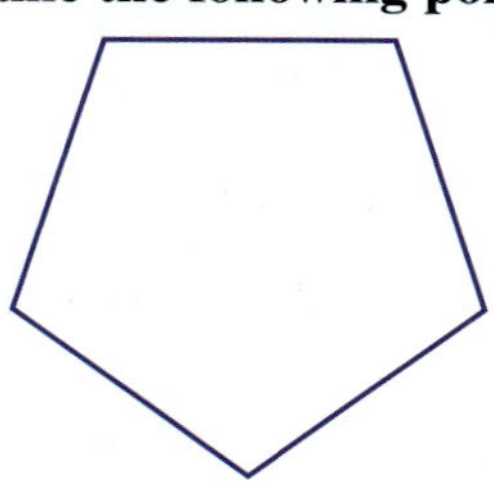

b

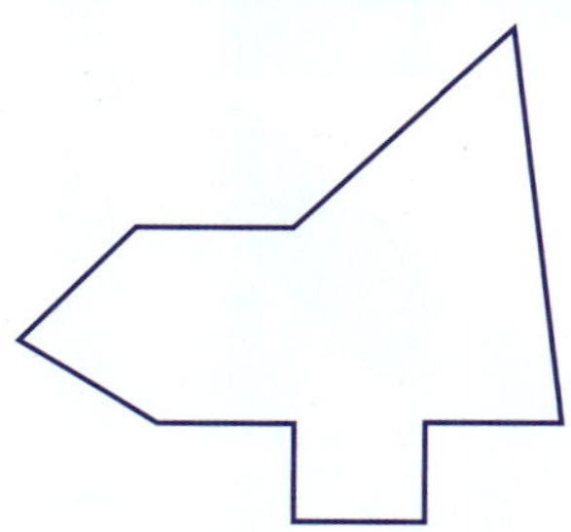

c

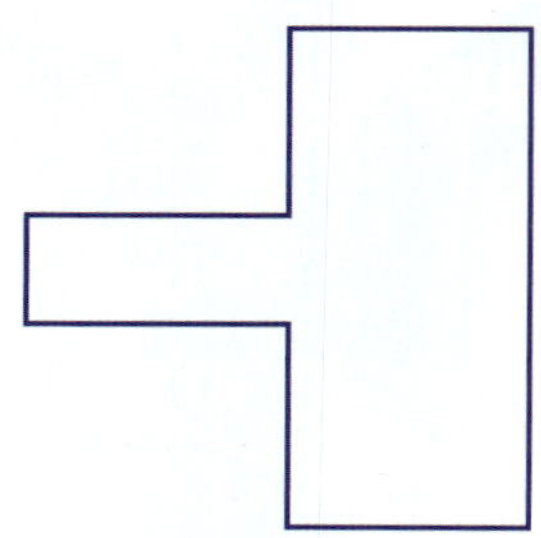

THINK

a 1 Count the number of sides in the polygon.
2 Match the number of sides with the corresponding name in the table.
3 See if all sides are equal to determine if it is regular.

b Repeat steps 1 to 3 as in part **a**.

c Repeat steps 1 to 3 as in part **a**.

WRITE

a Number of sides = 5
The polygon is a regular pentagon.

b Number of sides = 10
The polygon is a decagon, but is not regular.

c Number of sides = 8
The polygon is an octagon, but is not regular.

Can you see why this famous building in the USA is called the Pentagon?

A polygon that has an angle greater than 180° is called a *convex polygon*. The figure at right is a convex hexagon. The angles at B and at E are greater than 180°.

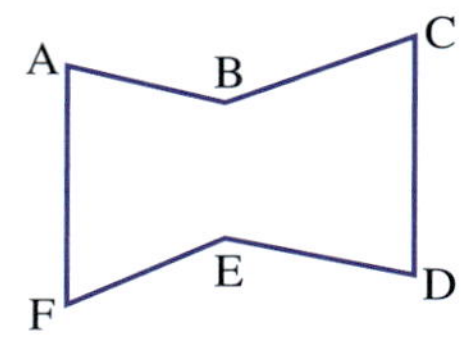

remember

1. A *polygon* is a closed shape with straight sides.
2. Polygons are named according to the number of sides or angles in the shape.
3. A polygon with equal sides and angles is called a *regular* polygon.
4. A polygon with an angle greater than 180° is called a *convex* polygon.

EXERCISE 11C Polygons

11.3 SkillSHEET Naming polygons

Cabri Geometry Polygons

Cabri Geometry Regular polygons

WORKED Example 11

1 Name the following polygons. State if the polygon is regular.

a
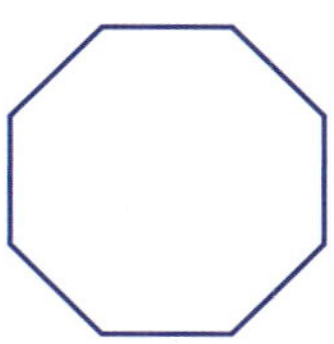

b
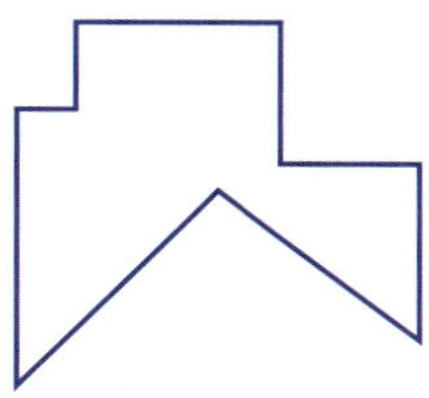

c

d
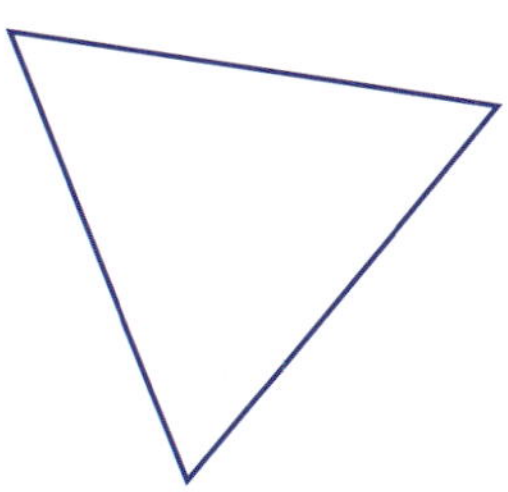

e

f
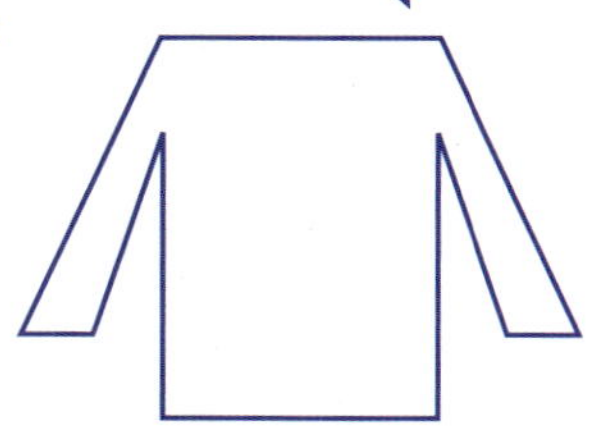

2 Which of the polygons in question 1 are convex?

3 Draw two different examples of each of the following polygons.

a hexagon
b quadrilateral
c nonagon
d pentagon
e octagon
f triangle

4 Count the number of sides on a 50-cent piece and name its shape.

5 **multiple choice**

a What shape is the following sign?

A hexagon
B quadrilateral
C pentagon
D heptagon
E octagon

b A horizontal cross-section of each of these pencils is a:

A quadrilateral
B hexagon
C octagon
D pentagon
E circle

6 These patterns are made up of different polygons. Can you name them all?

a

b

c

d

WorkSHEET 11.1

7 Name one place where you can find these polygons in your home or school.

a a triangle
b a quadrilateral
c a hexagon
d an octagon

DESIGN In search of polygons

1 Trace this shape, cut it out and cut along the dotted lines to make three pieces. Rearrange the pieces to make the following polygons. (Draw the solutions in your workbook.)

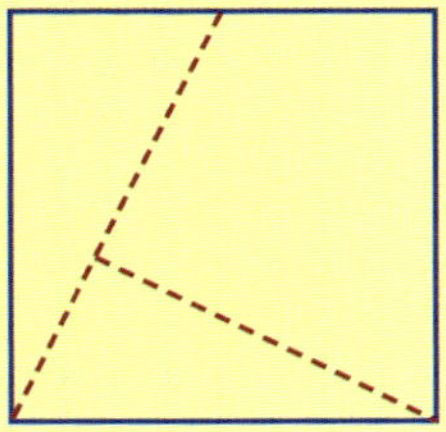

- a triangle
- b square
- c rectangle
- d trapezium
- e pentagon
- f parallelogram

2 The diagram at right is made up of equilateral triangles. How many rhombuses can you find in the picture? (One rhombus that is made up of two triangles is shown in red.)

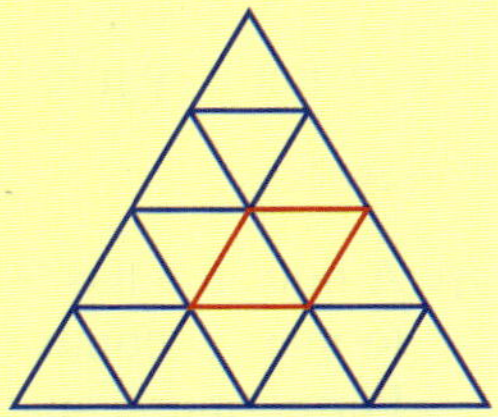

10 QUICK QUESTIONS 1

1 Classify the triangle at right according to its side lengths.

2 Classify the triangle in question 1 according to the size of the angles.

3 Draw an isosceles, right-angled triangle.

4 Name the quadrilateral drawn at right.

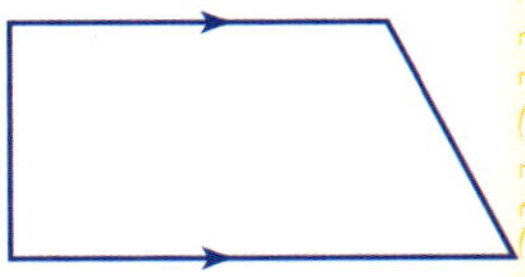

5 Draw a parallelogram.

6 Name the quadrilateral that has all four sides equal.

7 Name the quadrilateral that has the opposite angles equal.

8 Name the polygon drawn at right. State if it is regular.

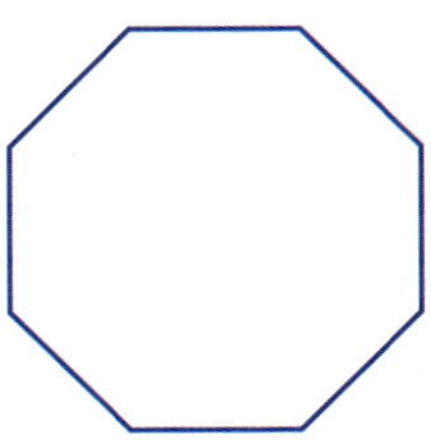

9 Draw an example of a convex pentagon.

10 What is the name given to a six-sided polygon?

Constructing polygons

In the preceding exercises there were some questions involving construction of triangles and quadrilaterals. In this section we will discuss further ways of constructing polygons using a ruler, pair of compasses and protractor.

A polygon with all sides equal in length and all angles equal in size is called a *regular polygon*.

A regular polygon can be constructed in a circle as shown in the following worked example.

WORKED Example 12

Construct a regular nonagon in a circle of radius 5 cm.

THINK

1. A nonagon has 9 vertices, so we need to mark 9 points on the circumference. Furthermore, since the nonagon is regular, the vertices must be equidistant from each other (that is, evenly spaced along the circumference). There are 360° in a circle, so divide 360° by 9, to find the distance between each point on the circumference.
2. Draw a circle of radius 5 cm.
3. Use a protractor to mark 9 points on the circle at 40° intervals. (These points are to become the vertices of the nonagon.)
4. Join the points with straight lines to construct a nonagon.

WRITE/DRAW

Number of sides = 9, so there are $360° \div 9 = 40°$ between each point.

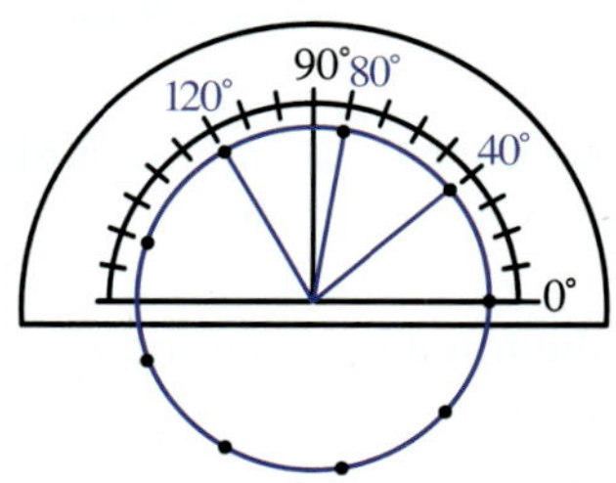

remember

1. A *regular polygon* has all sides of equal length and all angles of equal size.
2. To construct a regular polygon in a circle, first divide 360° by the number of sides. This gives you the angle between the vertices. Then mark the points on the circumference of the circle and join them together with straight lines.

EXERCISE 11D Constructing polygons

WORKED Example 12

11.4 SkillSHEET Constructing angles with a protractor

Mathcad Internal angle of a polygon

1 Construct a square in a circle of radius 5 cm.

2 Construct a regular pentagon in a circle of radius 5 cm.

3 Construct a regular hexagon in a circle of radius 5 cm.

4 Construct a regular octagon in a circle of radius 5 cm.

5 a Follow the instructions below to produce this design.
   i Draw a square of side 10 cm.
   ii Find the midpoint (middle) of each side.
   iii Join the midpoints to form a new square.
   iv Repeat steps ii and iii for your new square.
 b Repeat steps ii and iii to produce smaller and smaller squares. How many smaller squares can you make?
 c Colour in your final design.

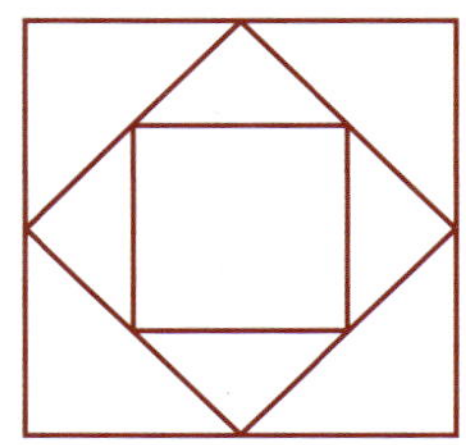

6 Draw your solutions for each part of this question in your book.

First draw a square with sides of length 5 cm.
 a With 1 straight line divide the square into:
   i 2 equal rectangles
   ii 2 equal right-angled triangles
   iii 2 equal trapeziums.
 b With 2 straight lines divide the square into:
   i 4 equal triangles
   ii 4 equal squares
   iii 3 equal rectangles
   iv 4 equal irregular quadrilaterals.
 c With 3 straight lines divide the square into:
   i 4 equal rectangles
   ii 6 equal rectangles.

7 Copy the shape at right and cut it out. Cut out along the dotted lines and rearrange the pieces to form a square. (Draw the solution in your workbook.)

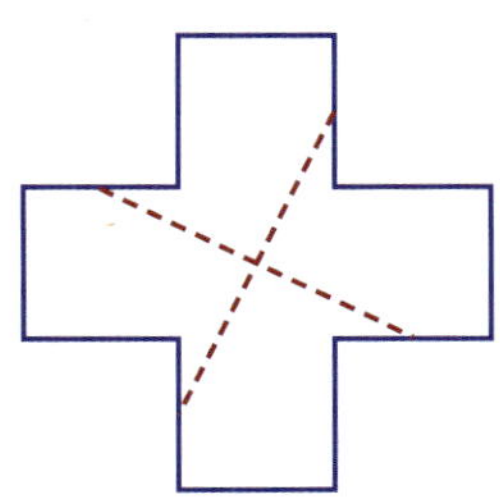

8 **multiple choice**

Which of the following is a regular quadrilateral?

A A rectangle B A parallelogram C A rhombus
D A square E All of the above

Plans and views

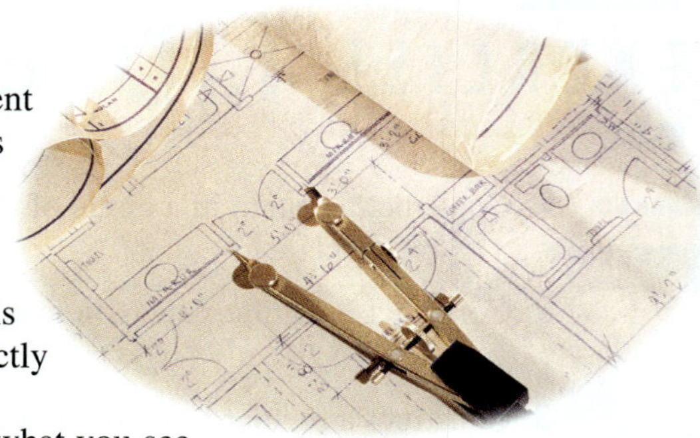

An object can be viewed from different angles. Architects and draftspersons often draw plans of building sites and various objects when viewed from the front, the side or the top.

The **front view**, or **front elevation**, is what you see if you are standing directly in front of an object.

The **side view**, or **side elevation**, is what you see if you are standing directly to one side of the object. You can draw the left view or the right view of an object.

The **top view**, or **bird's eye view**, is what you see if you are hovering directly over the top of the object looking down on it.

WORKED Example 13

The following object is made from 4 cubes. Draw plans of it showing:

a the front view b the right view c the top view.

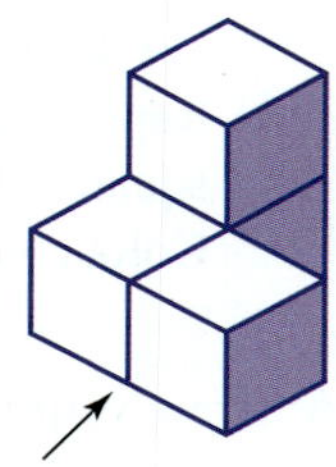

Front

THINK

DRAW

a Make this shape using cubes. Place the shape at a considerable distance and look at it from the front (this way you can see only the front face of each cube). Draw what you see. (Or simply imagine looking at the shape from the front and draw what you see.)

a

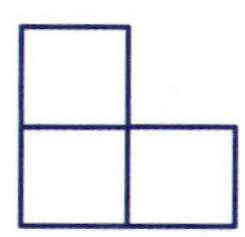

Front view

b Look at your model from the right, or imagine that you can see only the right face of each cube and draw what you see.

b

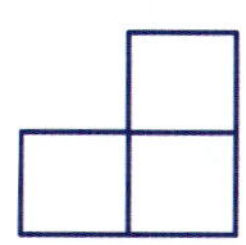

Right view

c Look at your model from the top, or imagine that you can see only the top face of each cube. Draw what you see.

c

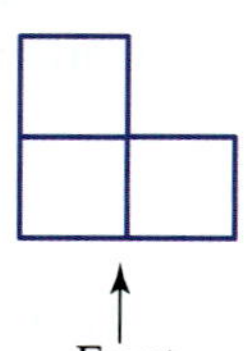

Top view

Front

Figures such as the one in worked example 13 can be drawn using isometric dot paper. This will help to give the 3-dimensional perspective of the object.

WORKED Example 14

Draw:

i the front view

ii the right view

iii the top view of this solid.

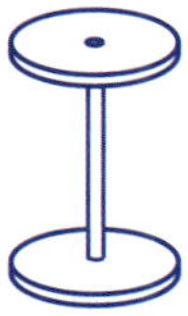

THINK

1. Find an object of similar shape, or visualise the object in your head.
2. Whether viewed from the front, or from the right of the object, the cylindrical shaft will appear as a long thin rectangle. The circular discs will also be seen as a pair of identical rectangles. So the front view and the right view are the same.
3. When the object is viewed from above, all we can see is the flat surface of the top disc; that is, a large and a small circle with the same centre. (Note that such circles are called concentric.)

DRAW

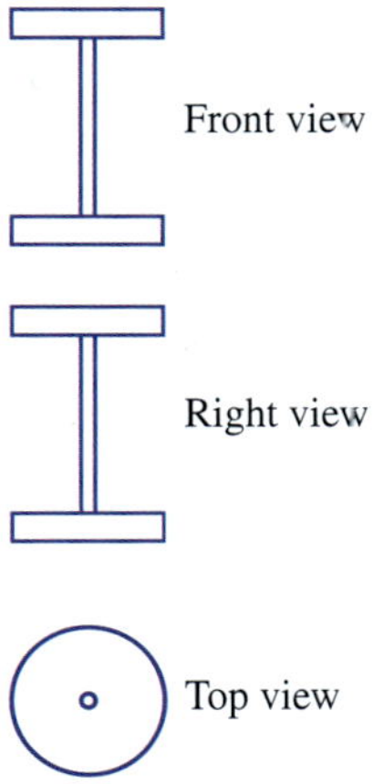

WORKED Example 15

The front, right and top views of a solid are shown. Use cubes to construct the solid.

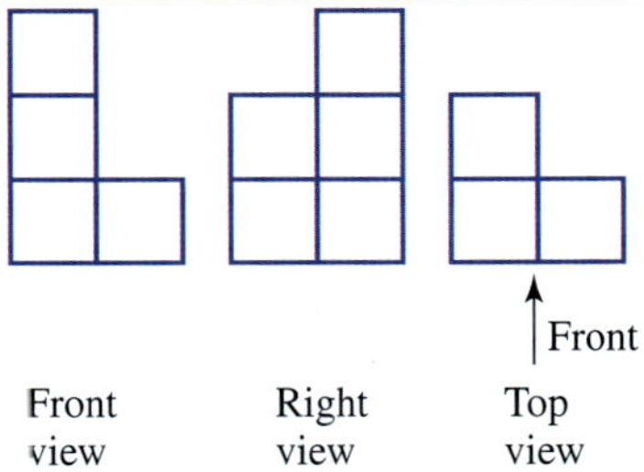

THINK

1. Use cubes to construct the solid.
2. Check carefully that your solid matches each of the 3 views you are given. Make adjustments if necessary.

CONSTRUCT

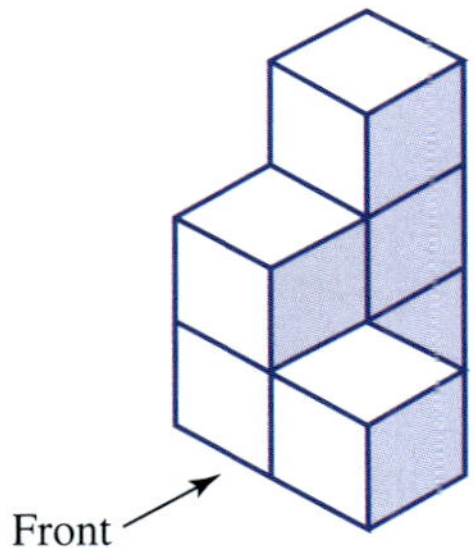

Perspective drawing

A **2-dimensional representation** of a **3-dimensional object** can be made using a single-point perspective. This shows the object as it appears to the eye in reality. When we look at a picture of a scene, the objects seem to converge or move towards a point in the distance. This point is called a vanishing point.

WORKED Example 16

Construct a perspective drawing of the rectangular prism shown.

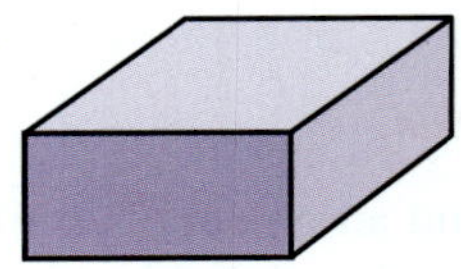

THINK | **WRITE**

1. Draw the front face of the prism as a rectangle.

2. Take a vanishing point in the distance behind the object and slightly to the right. Draw dotted lines from this point to the corners of the front face of the prism.

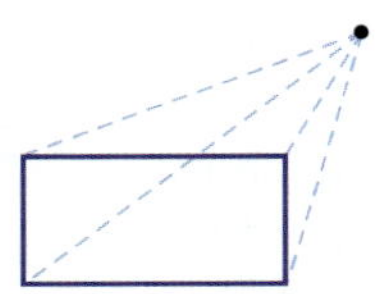

3. On these dotted lines, make the following markings. Draw the top edge of the back face parallel to the top edge of the front face. Also draw the right edge of the back face parallel to the right edge of the front face.

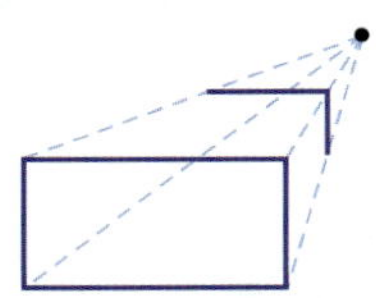

4. Join the top- and right-side vertices of the front face to the corresponding vertices of the back face.

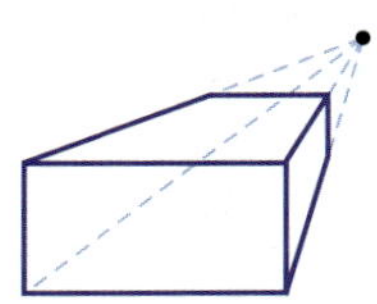

remember

1. The *front view*, or *front elevation*, is what you see if you are standing directly in front of an object.
2. The *side view*, or *side elevation*, is what you see if you are standing directly to one side of the object. You can draw the left view or the right view of an object.
3. The *top view*, or *bird's eye view*, is what you see if you are hovering directly over the top of the object looking down on it.
4. A 2-dimensional representation of a 3-dimensional object can be made using a single-point perspective.

EXERCISE 11E Plans and views

WORKED Example 13

1 The following objects are made from cubes. For every object draw the plans, showing the front view, the right view and the top view. (You may wish to use a set of cubes or building blocks to help you.)

a
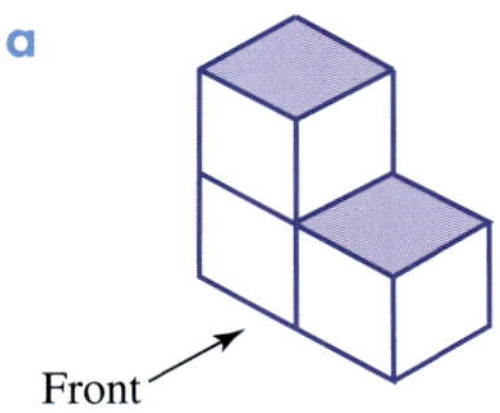

b
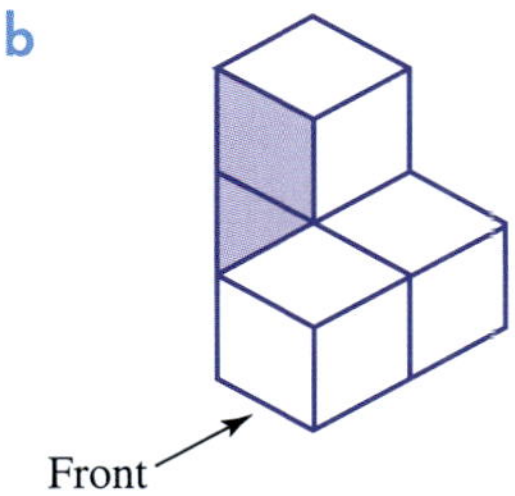

c
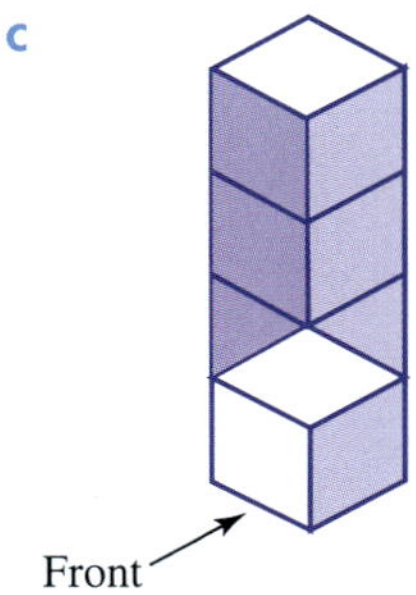

d
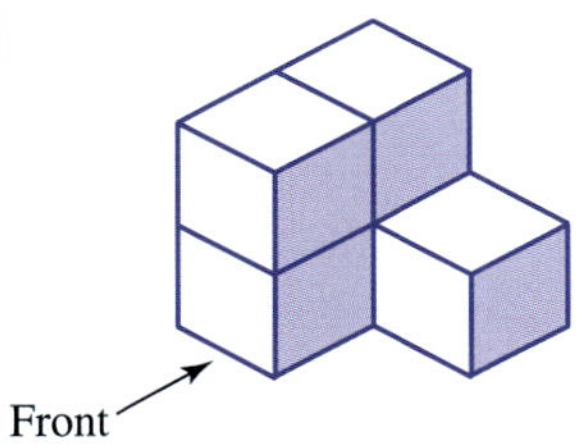

e
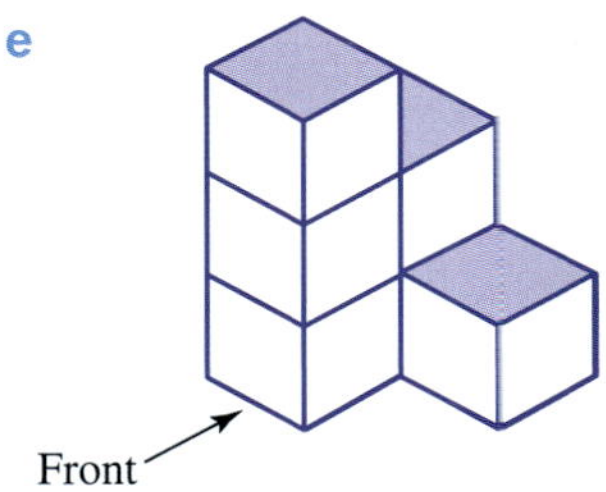

f
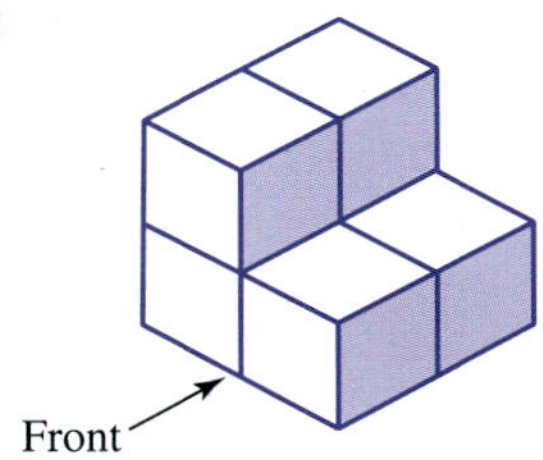

g
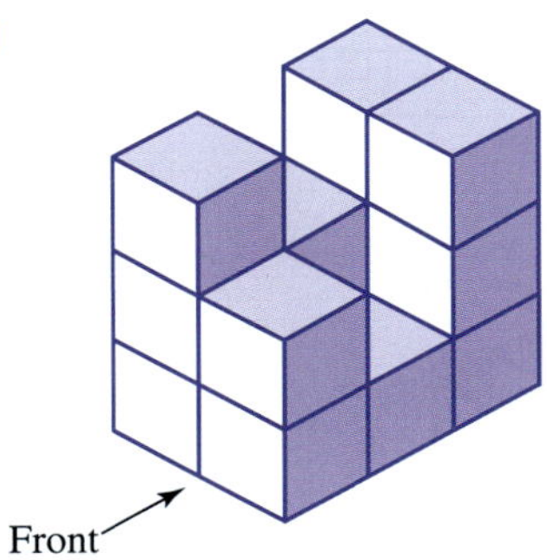

h
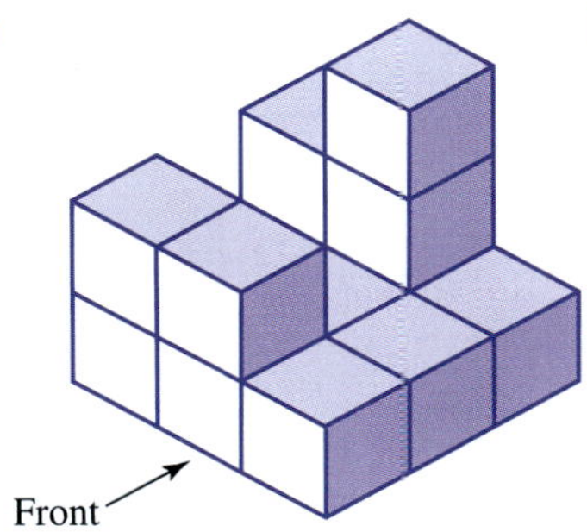

i
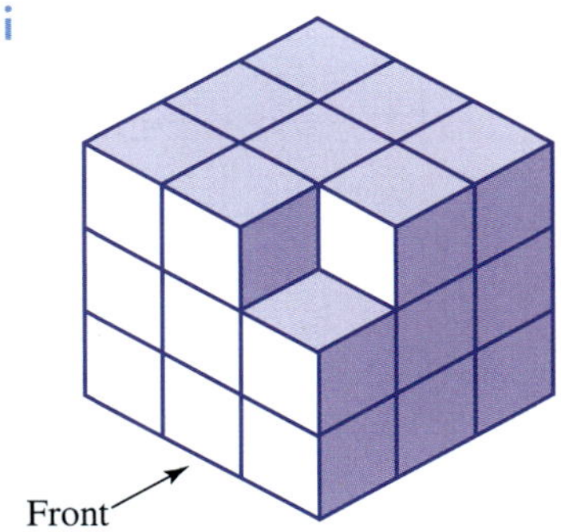

WORKED Example 14

2 Draw the front, right and top views of each solid shown.

a

b

c

d

WORKED Example 15

3 The front, right and top views of a solid are shown. In each case, use cubes to construct the solid.

a

Front view

Right view

Top view

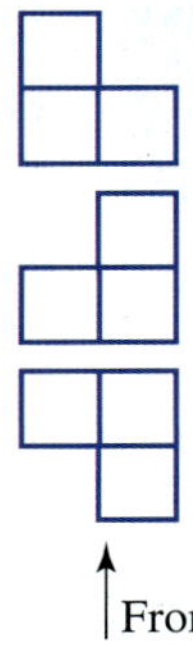

Front

b

Front view

Right view

Top view

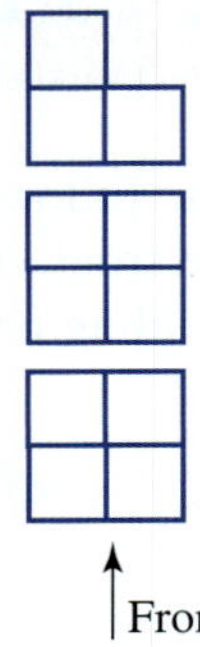

Front

c

Front view

Right view

Top view

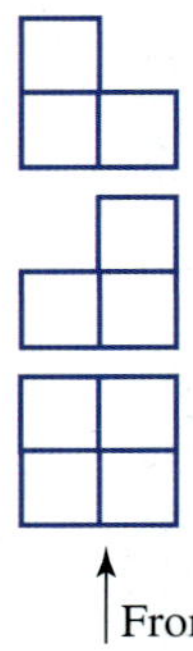

Front

d

Front view

Right view

Top view

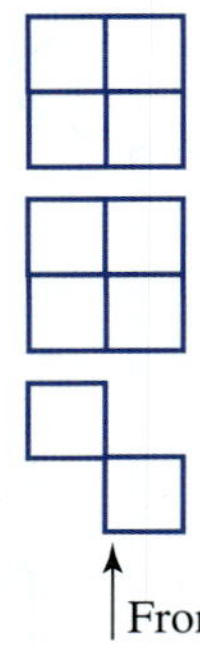

Front

4 **multiple choice**

The front, right and top views of a solid are shown. Which of the given drawings could represent the solid?

Front view 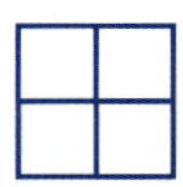 Right view 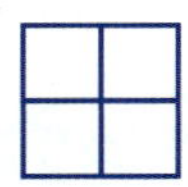 Top view 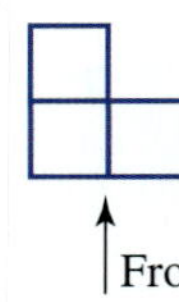

Front

A

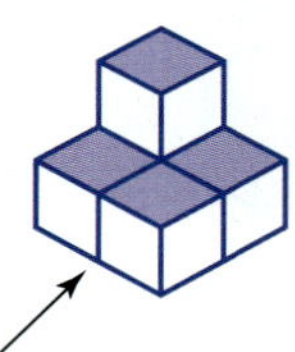

Front

B

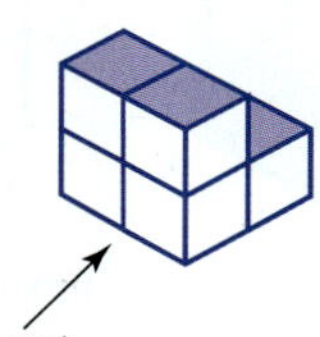

Front

C

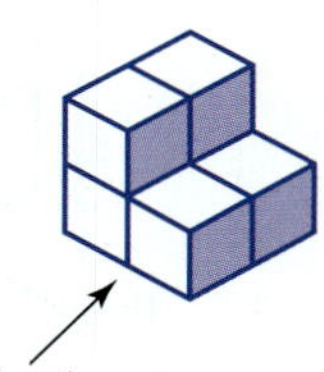

Front

D

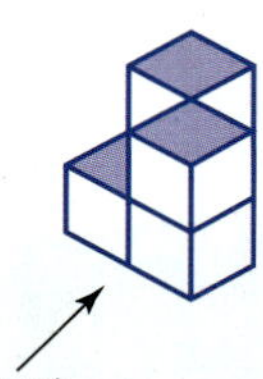

Front

E

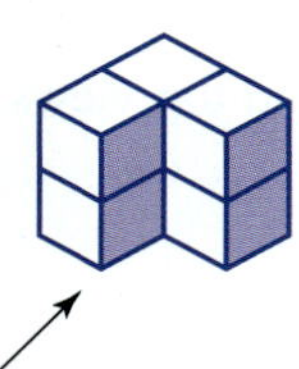

Front

5 **a** What shape is the top view of a telephone pole?
b What shape is the top view of the Melbourne Cricket Ground?
c What shape is the side view of a bucket?
d What shape is the top view of a car?

6 a Draw the side view of this pool table.
b Draw the front view of your house (seen from the street).
c Draw the side view of an electric kettle.
d Draw the top view of your television set.

7 A shape is made using only 4 cubes. Its front view, right view and top view are shown.

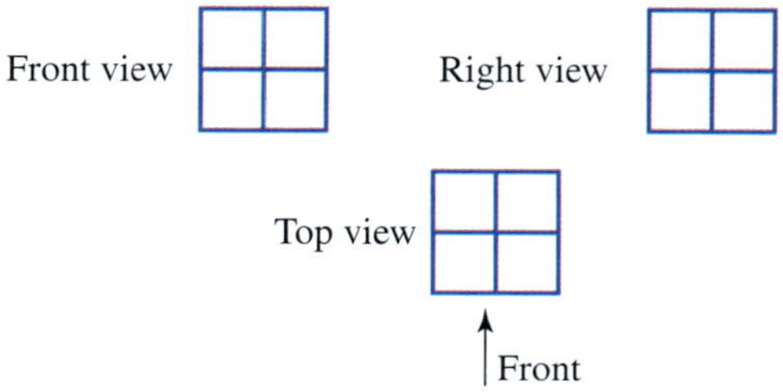

a Is it possible to construct this solid?
b Describe or draw what this solid would look like.

WORKED Example 16

8 Construct a perspective drawing of a cube.

Polyhedra, nets and Euler's Rule

A *polyhedron* is a 3-dimensional shape in which each flat surface is a polygon.

This photograph of an ancient Egyptian pyramid is an example of a polyhedron, since each of its 5 flat surfaces is a polygon.

The flat surfaces which make a **polyhedron** are called **faces**.
The lines where 2 faces of a polyhedron meet are called **edges**.
The points where 3 or more edges of a polyhedron meet are called **vertices**.

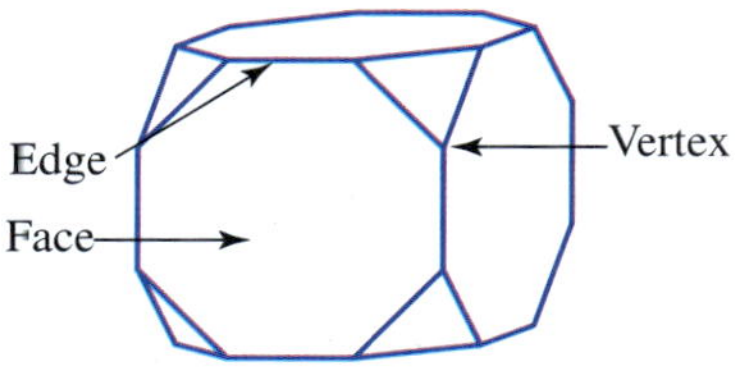

WORKED Example 17

For the polyhedron shown, write:

a the number of faces and the shape of each face

b the number of edges

c the number of vertices.

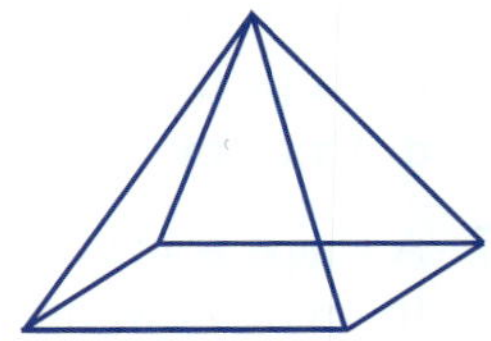

THINK

a The base of the polyhedron shown is a square and the other 4 faces are triangles.

b Four edges are formed where each of the triangles meets the square base. Another 4 edges are formed where the triangular faces meet each other.

c There are 4 vertices on the square base and 1 at the top.

WRITE

a Number of faces = 5
Shape of the faces: 1 square and 4 triangles

b Number of edges = 8

c Number of vertices = 5

Naming polyhedra

The table below gives the names of some common polyhedra that you might often see. The names are often associated with the number of faces of the polyhedron.

Polyhedron	Number of faces	Name
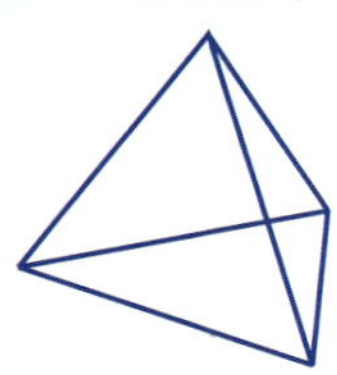	4	**tetrahedron**, or triangular-based pyramid
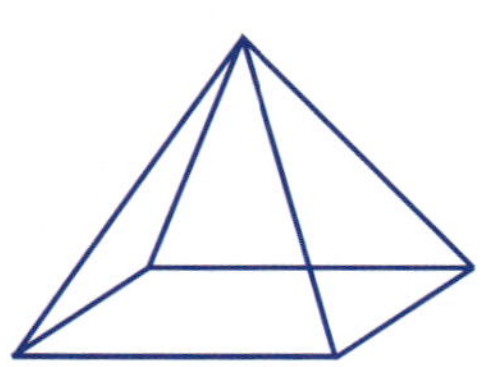	5	square-based pyramid
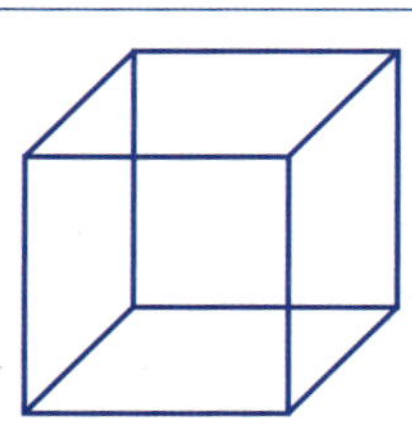	6	**cube**

Polyhedron	Number of faces	Name
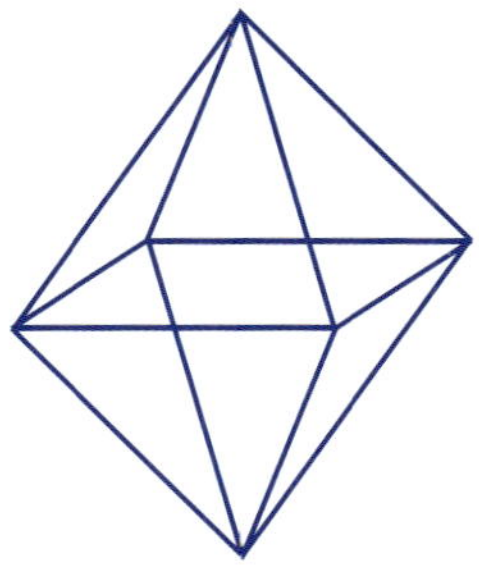	8	**octahedron**
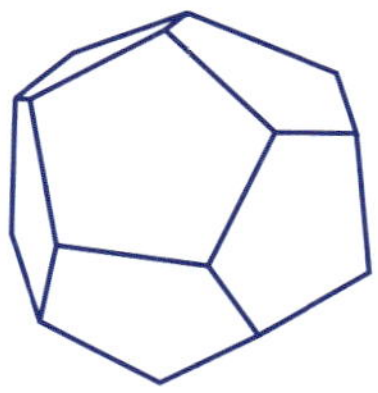	12	**dodecahedron**
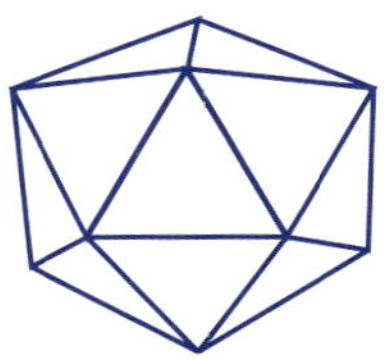	20	**icosahedron**

Nets of polyhedra

A 2-dimensional plan that can be folded to construct a 3-dimensional polyhedron is called a net of that polyhedron.

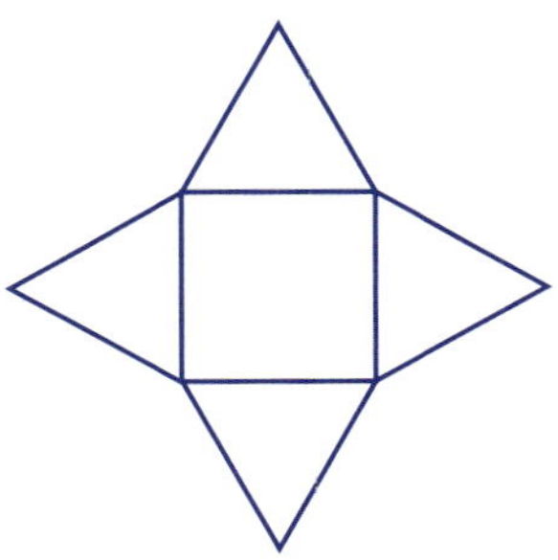

This is a net of a square-based pyramid: if you fold up the triangles and stick them together, the square-based pyramid shown at right will be formed.

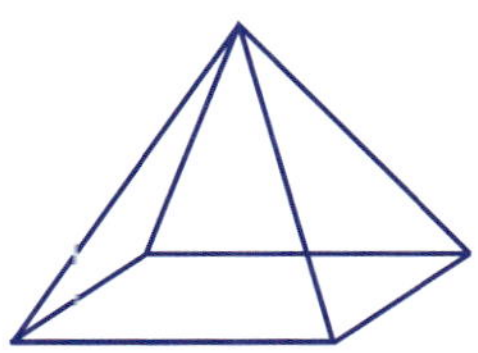

Square-based pyramid

A demonstration version of the program Poly is available on the *Maths Quest 7* CD-ROM. This program allows you to visualise polyhedra and their nets.

Click on the icon when using the *Maths Quest 7* CD-ROM for further instructions on how to use Poly.

remember

1. A *polyhedron* is a 3-dimensional shape in which every flat surface is a polygon.
2. The flat surfaces that form a polyhedron are called *faces*.
3. The lines where 2 faces of the polyhedron meet are called *edges*.
4. The points where 3 or more edges of the polyhedron meet are called *vertices*.
5. The *net* is a 2-dimensional plan, which can be folded to form a 3-dimensional object.

EXERCISE 11F Polyhedra, nets and Euler's Rule

Poly

WORKED Example 17

1 You may wish to use the program Poly to assist you in completing this exercise.

i 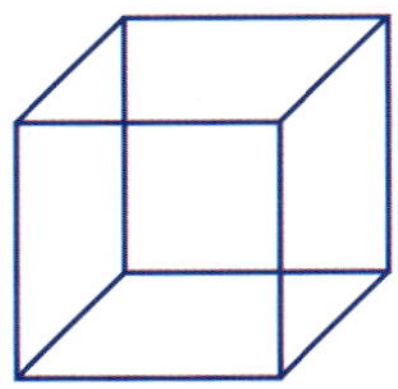 **ii** 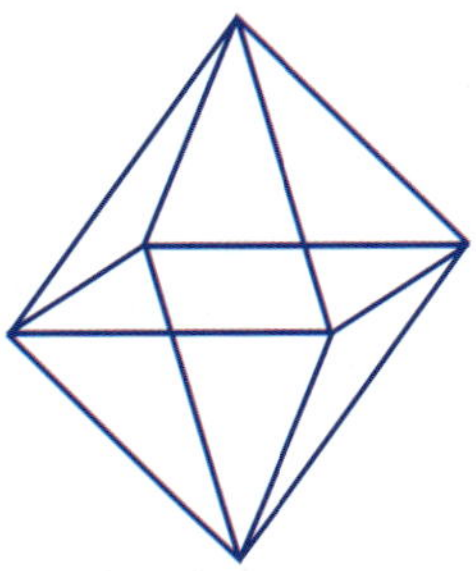 **iii** 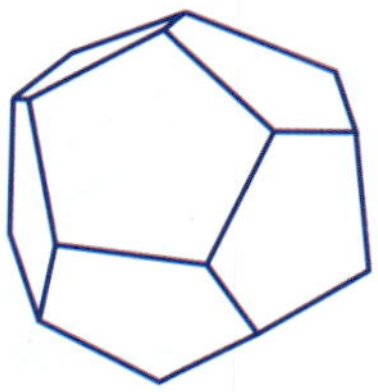

For each polyhedron shown, write down:

a the number of faces and the shape of each face

b the number of edges **c** the number of vertices.

2 Name each of the polyhedra in question **1**.

3 Copy this net of a tetrahedron. Tabs have been included to assist in the construction.
Cut out the net and fold it to construct the tetrahedron.
Look at the tetrahedron and write down:

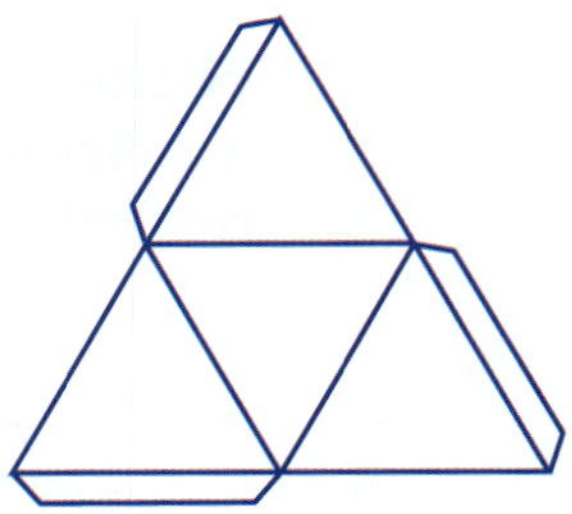

a the number of faces

b the number of vertices

c the number of edges.

4 This is the net of a *parallelepiped*. Each of its faces is a parallelogram. Copy the net, cut it out and fold it to construct the parallelepiped. Look at the parallelepiped and write down:

a the number of faces

b the number of vertices

c the number of edges.

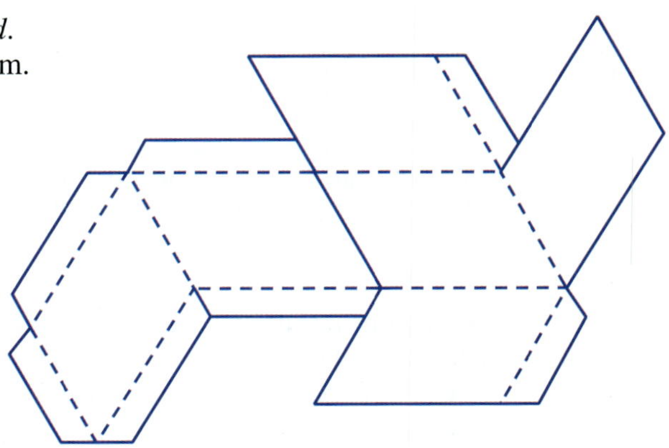

5 This is the net of a truncated tetrahedron. It is a tetrahedron with its corners cut off. Copy the net, cut it out and fold it to construct the truncated tetrahedron. Look at the truncated tetrahedron and write down:

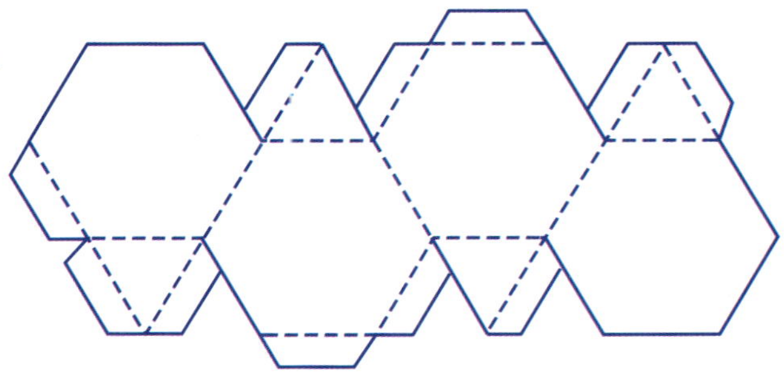

a the number of faces
b the number of vertices
c the number of edges.

6 Copy this net of an icosahedron.

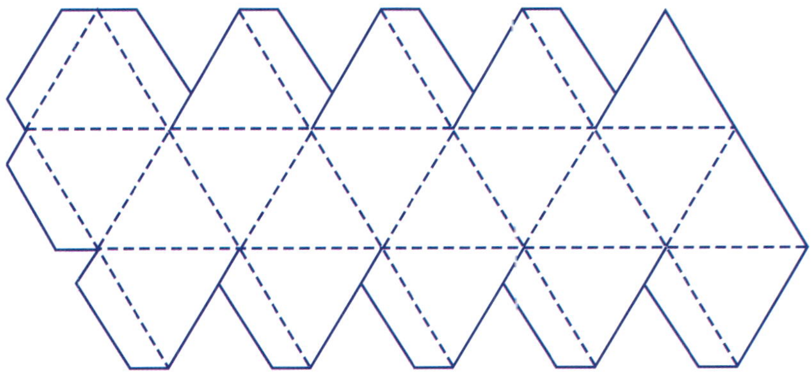

Cut out the net and fold it to construct the icosahedron. Look at the icosahedron and write down:

a the number of faces
b the number of vertices
c the number of edges.

7 a Copy this table into your workbook. Use the answers to questions 1, 3, 4, 5, 6 (or the program Poly) to complete the table.

Question number	Name of polyhedron	Number of faces (*F*)	Number of vertices (*V*)	Number of edges (*E*)	$F + V - E$
1 i	cube				
1 ii	octahedron				
1 iii	dodecahedron				
3	tetrahedron				
4	parallelepiped				
5	truncated tetrahedron				
6	icosahedron				

b State the pattern that you have found.

c The pattern that you have discovered is known as Euler's Rule. (Euler is pronounced 'Oiler'.) Copy the rule into your workbook for future reference.

Euler's Rule for polyhedra can be stated as:
for any polyhedron, $F + V - E = 2$,
where F is the number of faces, V is the number of vertices and E is the number of edges of the polyhedron.

In other words, the number of faces plus the number of vertices minus the number of edges equals 2.

Mathcad: Using Euler's rule

8 Use Euler's Rule, established in the previous question, to check whether it is possible for a polyhedron to have:
  a 6 faces, 8 vertices and 10 edges
  b 9 faces, 12 vertices and 19 edges.

9 Use Euler's Rule to answer each of the following questions.
  a A polyhedron has 10 faces and 16 vertices. How many edges does it have?
  b A polyhedron has 12 faces and 18 vertices. How many edges does it have?
  c A polyhedron has 10 edges and 6 vertices. How many faces does it have?
  d A polyhedron has 8 faces and 12 vertices. How many edges does it have?

10 a Use straws and plasticine (or any other suitable material) to construct these polyhedra:
  i a square-based pyramid

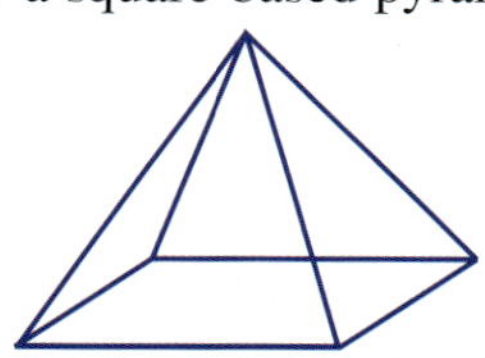

  ii a nonahedron

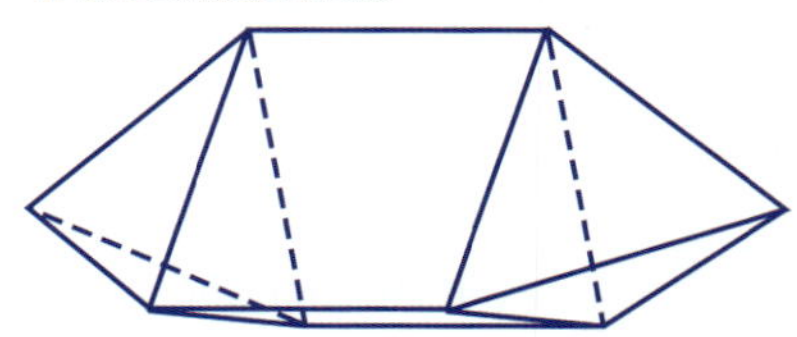

  b For each model count the number of faces, edges and vertices and hence verify Euler's rule.
  c Describe, in writing, how to build each shape.

WorkSHEET 11.2

THINKING Parallel, perpendicular, skew?

Consider the triangular prism at right.

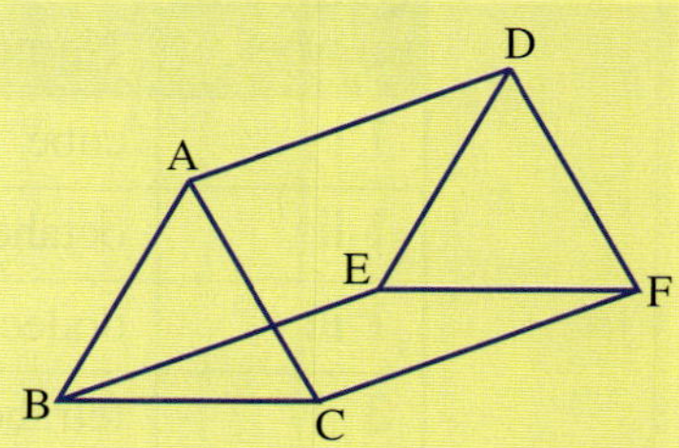

1 From the diagram we can see that the lines AC and DF are parallel. Name any other sets of parallel lines.

2 We can also see that the line BC meets the line CF at right angles. We can therefore say that BC and CF are perpendicular. List any other pairs of perpendicular lines that you can find.

3 Any pair of lines that are neither parallel nor perpendicular are said to be *skew;* that is, a pair of skew lines are not parallel but will never intersect. For example, in the diagram above, AC and DE are skew lines. List any other pairs of skew lines that you can find.

4 In the figures at right, list:
  a the pairs of parallel lines
  b the pairs of **perpendicular lines**
  c the pairs of skew lines.

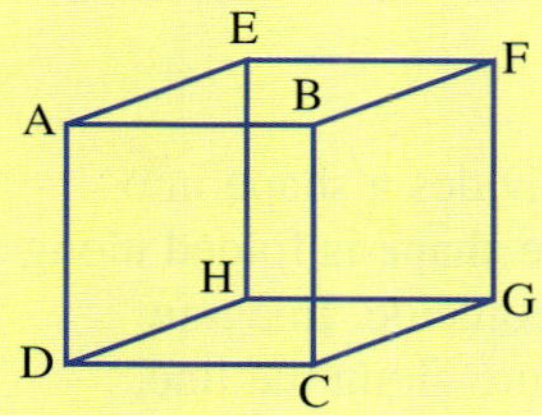

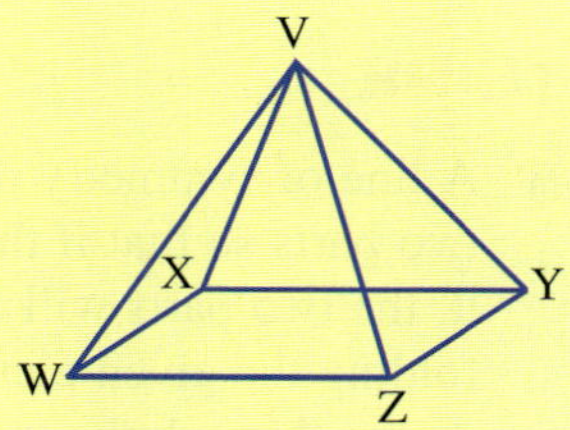

DESIGN

Making models of polyhedra

Use the program Poly to find the nets of 3 polyhedra. There are some suggestions below. Print out each net and trace onto coloured paper or card. (You may like to enlarge your net by using a photocopier first.) Cut and fold to form each of these polyhedra. Have fun!

- Hebesphenomegacorona (J89) (in Johnson Solids)
- Decagonal deltohedron (in Dipyramids and Deltohedrons)
- Hexakis octahedron (in Catalan Solids)
- Square anti-prism (in Prisms and Anti-prisms)
- Great rhombicosidodecahedron (in Archimedean Solids)

Poly

Axes of symmetry

If it is possible to draw a line which would divide a shape into two parts, so that when folded along that line the parts would coincide exactly, the shape is said to be symmetrical along that line. The line itself is called the *line (or the axis) of symmetry.*

If a mirror is placed along the **axis of symmetry**, the half of the shape that faces the mirror, together with its reflection, will form the whole shape (that is, the mirror reflection will look exactly like the other half of the shape). So we can think of the axis of symmetry as the line that acts like a mirror.

Many things in nature are symmetrical. Some objects have more than one axis of symmetry, while others have none.

WORKED Example 18

Decide whether the picture of the house shown at right is symmetrical and, if so, show where the line of symmetry is.

Continued over page

THINK

1. A line of symmetry divides a shape into two parts so that if the shape is folded along it, the two parts will coincide. Imagine folding the given picture along the line, drawn through the top point of the roof and the midpoint of the base of the house. The two halves will coincide exactly.
2. Show the line of symmetry using a dotted line.

WRITE

The shape of the house is symmetrical.

WORKED Example 19

How many axes of symmetry does this shape have?

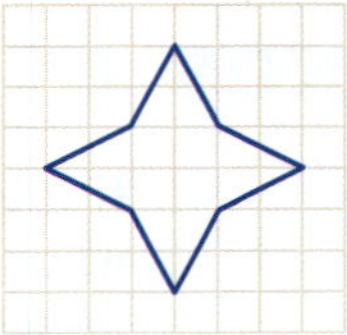

THINK

1. An axis of symmetry is a line that acts as a mirror. A vertical line through the top and bottom vertices of the given shape would divide it into two halves, which are mirror images of each other. So it is the axis of symmetry.
2. Look for other axes of symmetry. Indicate all the axes of symmetry with dotted lines.
3. State the number of axes of symmetry.

WRITE/DRAW

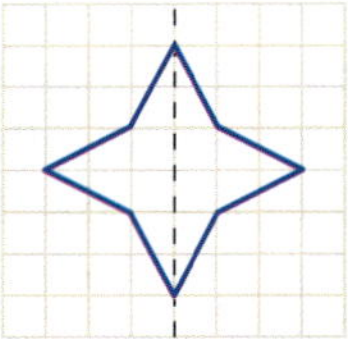

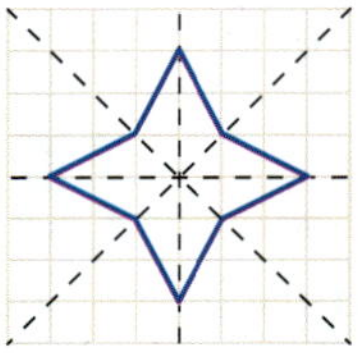

There are 4 axes of symmetry.

remember

1. *A line (or axis) of symmetry* is a line that can be drawn on a shape so that it acts like a mirror. One side of the object appears reflected in the line of symmetry.
2. If the shape is folded along the line of symmetry, the two sides will coincide exactly (that is, the shape will fold onto itself).
3. Some shapes have more than one axis of symmetry, while others do not have any.

EXERCISE 11G Axes of symmetry

WORKED Example 18

1 Decide whether each of the following are symmetrical and, if so, show where the line of symmetry is.

a

b

c

d

e

f

2 Copy each of the following shapes on to grid or squared paper. Carefully cut out each shape. Fold it to find the axes of symmetry. State the number of axes of symmetry for each shape. (Some shapes will have more than one axis of symmetry.)

a 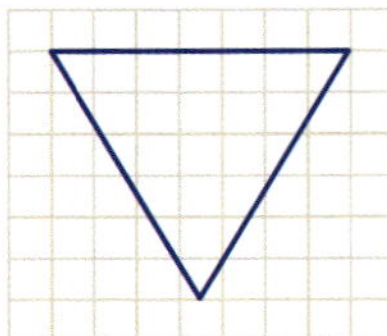
b 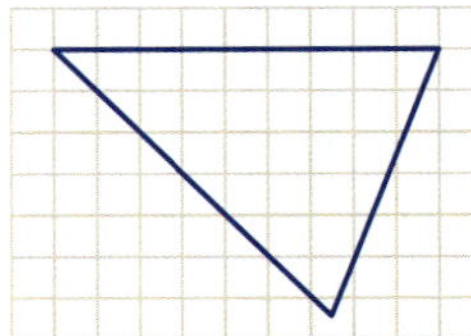
c 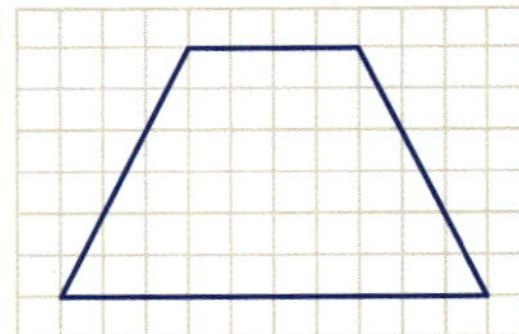
d 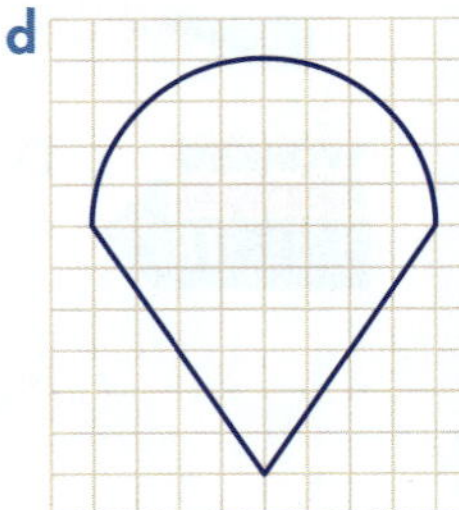
e 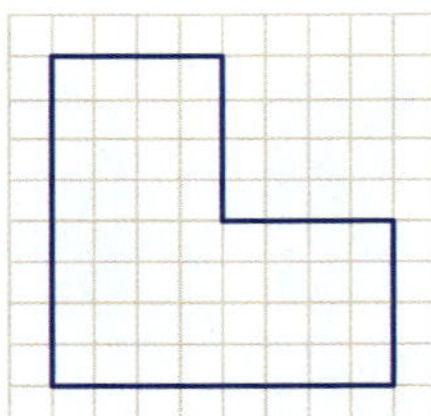
f 
g 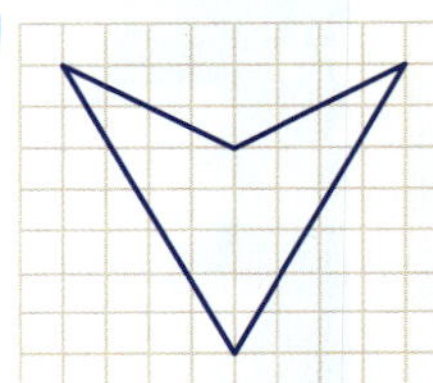
h 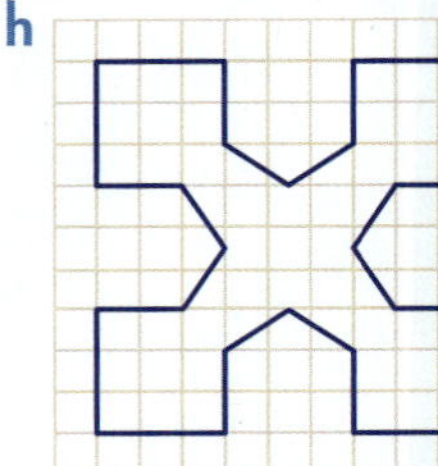

WORKED Example 19

3 How many axes of symmetry do each of the following shapes have?

a 
b 
c 
d 
e 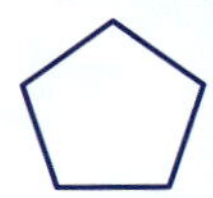
f 

4

A B C D E F G H I J K L M
N O P Q R S T U V W X Y Z

a b c d e f g h i j k l m
n o p q r s t u v w x y z

Copy out the letters and draw in the axes of symmetry for each if appropriate.

a Which upper case letters of the alphabet have a line of symmetry?
b Which lower case letters of the alphabet have a line of symmetry?
c Do any of the letters have more than one line of symmetry? If so, which ones?

5 **multiple choice**

The number of axes of symmetry in the shape at right is:

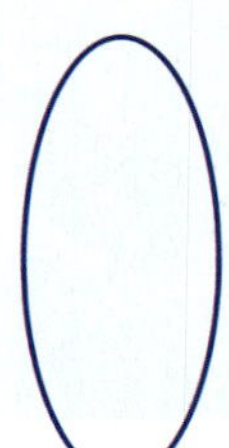

A 0 B 1 C 2
D 3 E 4

6 **multiple choice**

The number of axes of symmetry in the shape at right is:

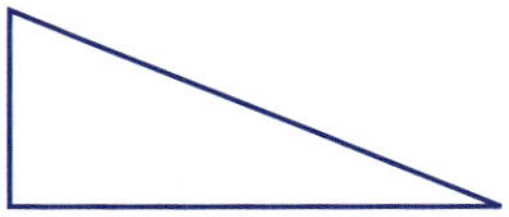

A 0 **B** 1 **C** 2 **D** 3 **E** 4

7 For each of the following diagrams, use the dotted line as an axis of symmetry to complete the picture.

a

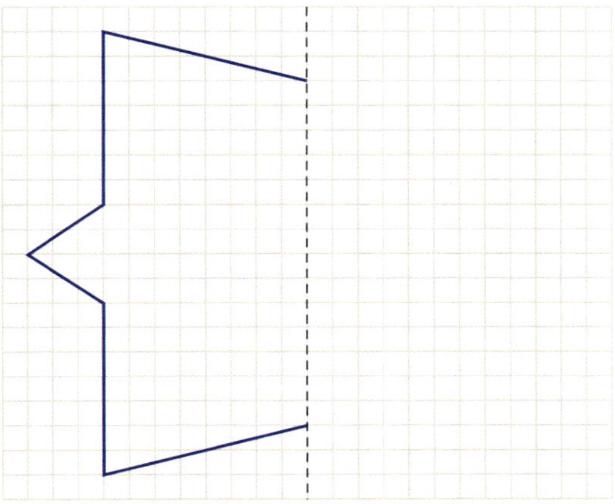

b

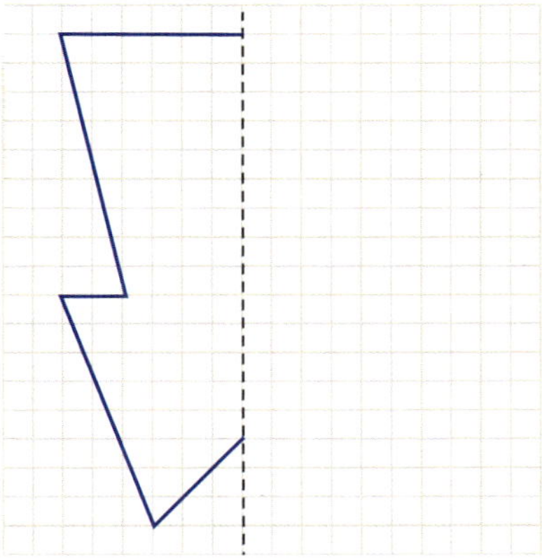

8 For each of the following, use the dotted line as an axis of symmetry to complete the picture.

a

b

MATHS QUEST CHALLENGE

1 **Ducks were walking in a line; one duck was behind two ducks, one duck was in front of two ducks and one duck was between two ducks. How many ducks were in the line?**

2 **This array of 9 nines contains 8 straight lines, each totalling 27 (3 across, 3 down, 2 diagonal). Show how to move 2 nines to new positions to make 10 straight lines totalling 27.**

9	9	9
9	9	9
9	9	9

3 **Thanh is looking at a photograph of a man and says, 'Uncles and aunts have I none, but that man's son is my son's father.' Who is the man in the photograph?**

10 QUICK QUESTIONS 2

1 Draw the top view of the solid drawn at right.

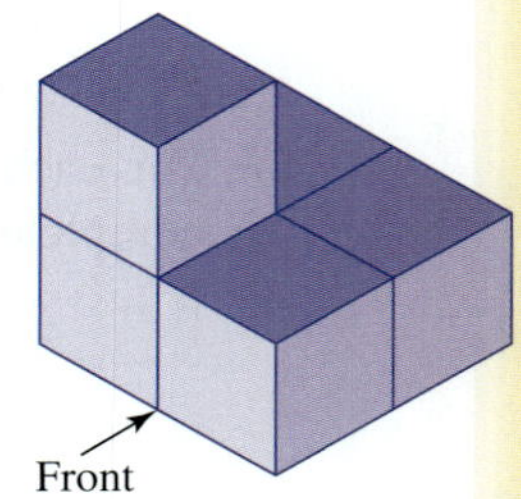

2 What shape is the front of a can of soft drink?

3 For the solid drawn at right, state the number of faces, vertices and edges.

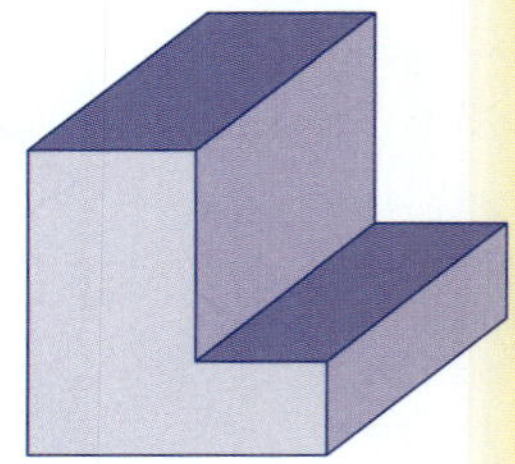

4 For the solid in question 3, verify Euler's Rule.

5 Name the solid shape drawn at right.

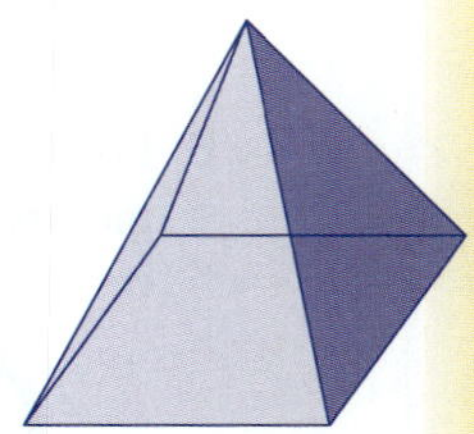

6 Draw the net of a cube.

7 Draw the solid shape to match the net shown at right.

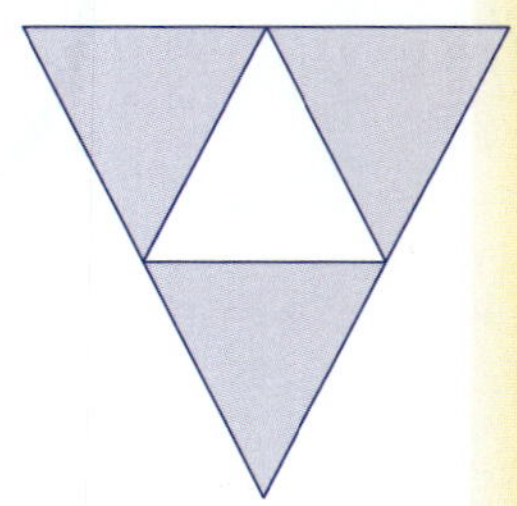

8 Is it possible to construct a solid shape with five faces, nine edges and ten vertices? Explain your answer.

9 How many axes of symmetry does this object have?

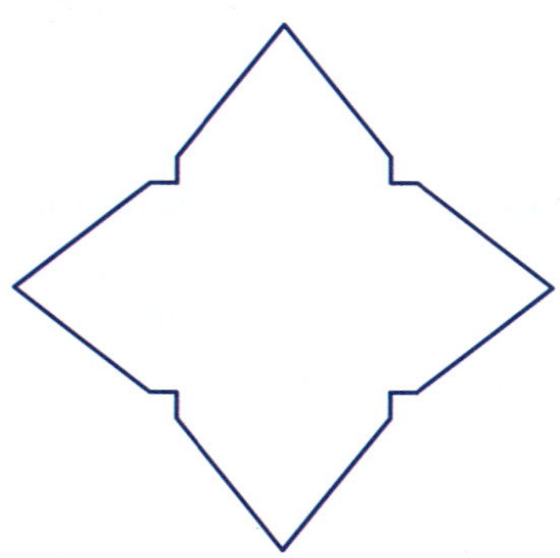

10 a Draw a shape that has no axes of symmetry.
   b Draw a shape that has one axis of symmetry.
   c Draw a shape that has two axes of symmetry.
   d Draw a shape that has exactly four axes of symmetry.

Transformations

Shapes can be moved, using an ordered set of moves, from one place to another. Such ordered sets of moves are called **transformations**. Prior to the transformation the shape is usually called *the original object*; after the transformation the shape is called *the image*. If after the transformation the image is exactly the same size and shape as the original object, the transformation is called **isometric**. The original object and its image are called congruent as they are identical.

In this chapter we will discuss the following isometric transformations: translations, reflections, and rotations.

When dealing with various transformations, it is customary to denote the image with the same letters as the original shape, but with an added symbol called a *prime*. For example, if the original object that is to undergo transformation is a triangle named ABC, then its image after the transformation is usually named A′B′C′.

Translations

If a shape is moved from one position to another, without turning, flipping over or changing size, then we can say the shape has been *translated*.

A shape can be translated up, down, right or left. A **translation** also results from a combination of these moves. It is just as if the object has moved from one place to the other.

WORKED Example 20

For each of the following shapes, specify the translation that took place.

a

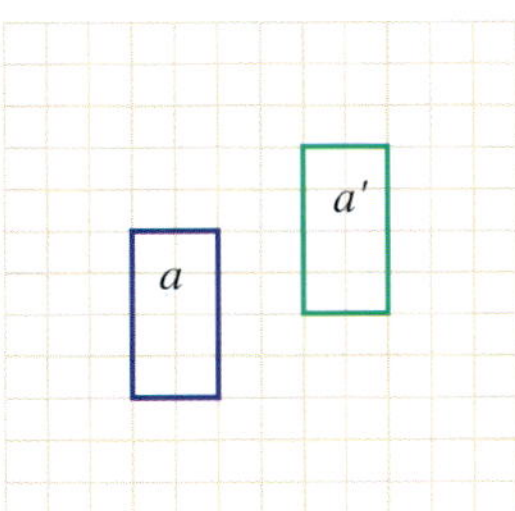

b

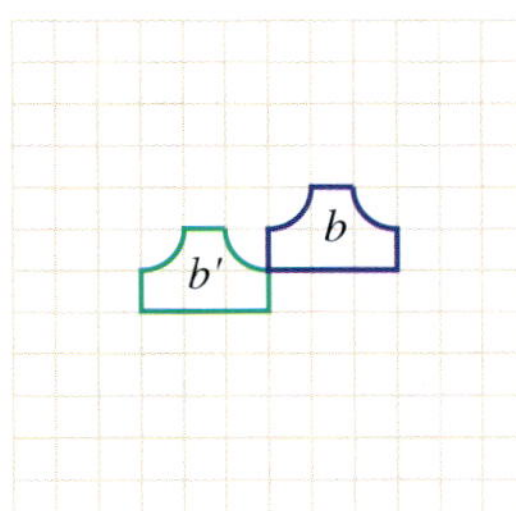

THINK

a 1 On the original shape (shape *a*), select any point (say right-hand-side bottom corner point) and call it P. Find the corresponding point on the image shape (shape *a*′) and call it P′. From point P draw:
(a) a horizontal line so that it ends directly below P′ and
(b) the vertical line to point P′.

2 Count and record the number of units the shape has been moved.

WRITE/DRAW

a

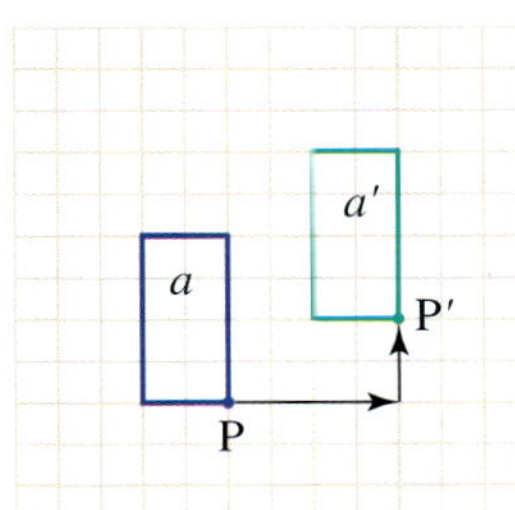

Translation: 4 units right, 2 units up

Continued over page

THINK

b 1 On the original shape (shape b), select any point and label it P. Find the corresponding point on the image shape (b') and label it P′. From point P draw a vertical line down until it levels with point P′, then draw a horizontal line to meet the image shape at P′.

2 Count the number of units the shape has been moved and hence record the translation that took place.

WRITE/DRAW

b

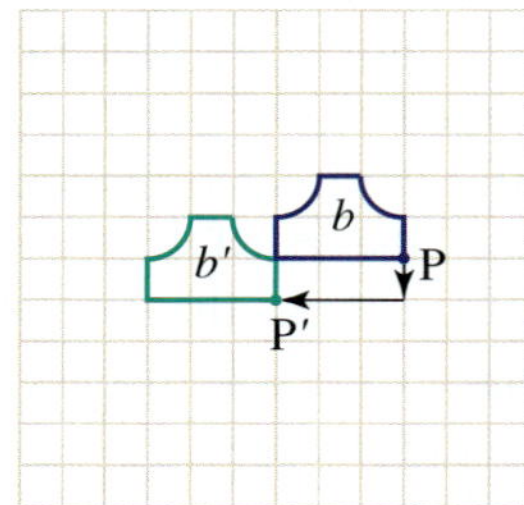

Translation: 1 unit down, 3 units left

WORKED Example 21

Given that a shape is translated 7 units up and 2 units left, use graph paper to show its original shape and position after translation.

THINK

Choose a point on the shape and imagine moving it 7 units up and 2 units to the left. Repeat this for other points. Draw the image of the shape and label it a'.

DRAW

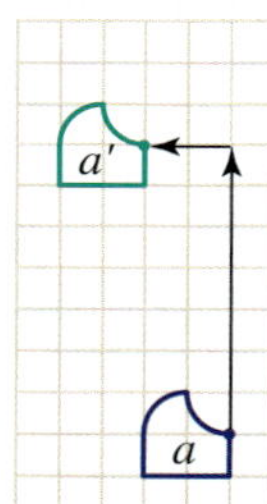

Objects that are more complex can be translated by following the steps given below.

Step 1 Select some key points first. (The corner points are usually the best to choose for this purpose.)

Step 2 Translate the selected points as required in the problem.

Step 3 Complete the image of the object. As a guide, use the points that have been translated earlier.

Reflections

A *reflection* is an exact image of an object, as seen in a mirror. A reflection is as far behind the mirror as the object is in front of it.

A **reflection** always has reversed orientation; that is, right appears left and left appears right. A mirror is useful for looking at reflections of various shapes.

WORKED Example 22

Draw reflections of each of the following objects using the dotted line as the 'mirror line'.

a

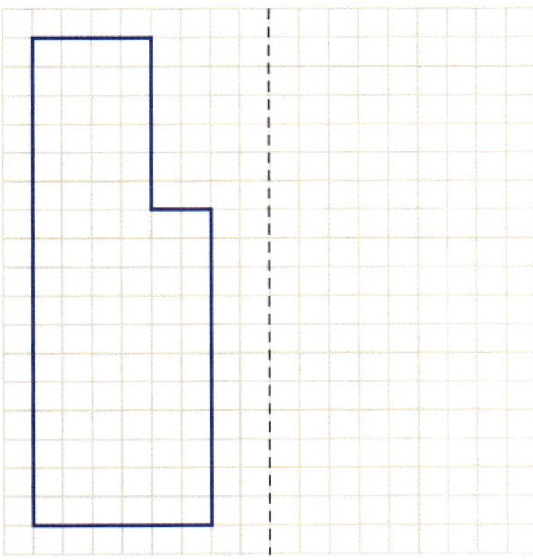

b

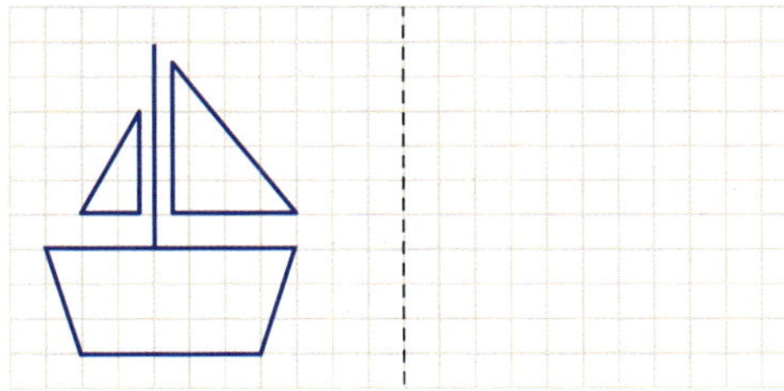

THINK

a A reflection is a reverse image of the shape that is as far behind the mirror as the shape is in front of it. Use a mirror to help you draw the reflection.

DRAW

a

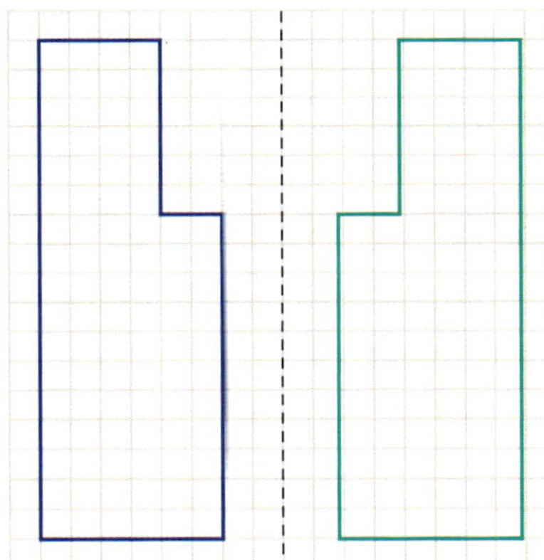

b Reflect the shape in the 'mirror' line. Make sure that all vertices of the image are as far from the line as the vertices of the original shape.

b

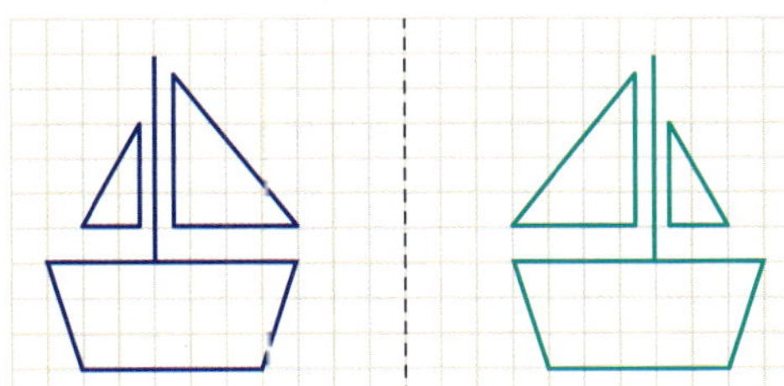

Some shapes are easier to reflect than others. If a mirror is not available to assist you in obtaining reflections of harder shapes, follow the steps outlined below.

Step 1 First select some key points. (The vertices are usually the most convenient to choose.)

Step 2 From the selected points draw perpendicular lines (that is, lines at right angles) to the 'mirror' line.

Step 3 Extend these lines beyond the 'mirror'.

Step 4 To find the reflection of any point, measure the distance from that point to the mirror along the line. The reflection is on the other side of the 'mirror', at the same distance from it, as the original point.

Step 5 Complete the reflection of the shape, using the points reflected earlier as a guide.

Rotations

An object turning about a certain point is called a **rotation**.

To specify the rotation we need to identify the centre of rotation (that is, the point about which the object is to be rotated). We also need to state the size of the angle of the rotation and whether the object is to be turned in a **clockwise** or **anticlockwise** direction.

Although an object can turn fully or partially, from a fraction of a circle to many turns of a circle, in this section we will consider only full-turn (360°), half-turn (180°) or quarter-turn (90°) rotations and their multiples.

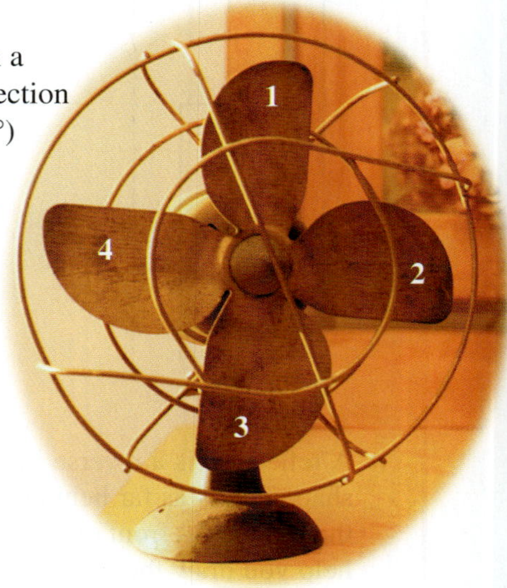

In the design of a fan, one blade is rotated to three different positions. In the photograph shown, the position of blade 2 is obtained by rotating blade 1 through 90° in a clockwise direction or 270° in an anticlockwise direction. The position of blade 3 is obtained by rotating blade 1 through 180° in either a clockwise or anticlockwise direction. Similarly, the position of blade 4 is obtained by rotating blade 1 through 270° in a clockwise direction or 90° in an anticlockwise direction.

WORKED Example 23

Specify the rotation that has taken place in the following figure. (Give two possible answers.)

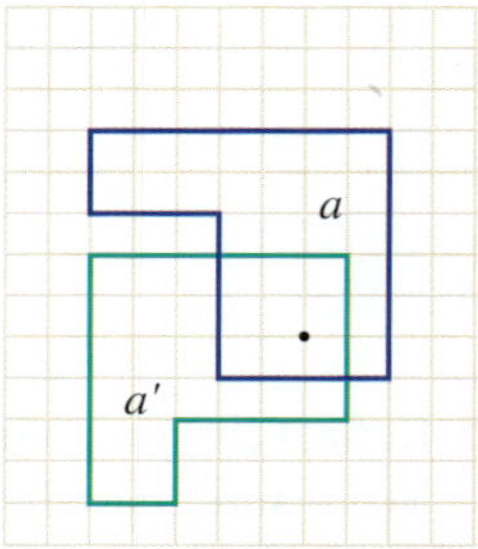

THINK

1. Select any point on the object (a vertex is always convenient to choose) and join it with the centre of rotation, using a straight-line segment. Find the corresponding point on the image and also join it with the centre of rotation.
2. If the object is rotated in an anticlockwise direction, the angle formed between the two lines is 90°. So specify the rotation.
3. If the object is rotated in a clockwise direction, the angle of rotation is $360° - 90° = 270°$ (or full turn $- \frac{1}{4}$ turn $= \frac{3}{4}$ turn). Specify the alternative rotation.

WRITE/DRAW

1. 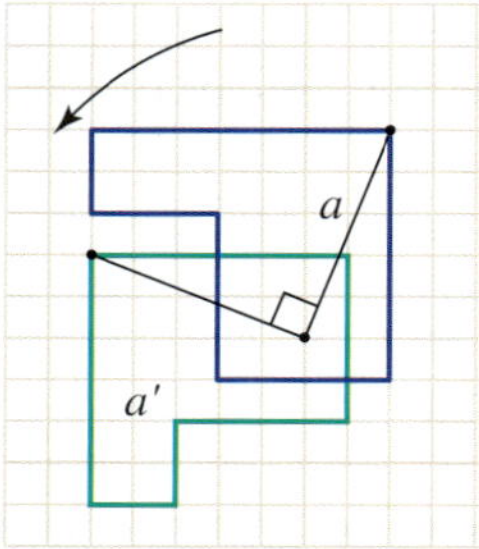

2. Anticlockwise, quarter turn

3. Clockwise, three-quarter turn

To perform a rotation of a shape, use the following steps.

Step 1 Select some key points on the object. (The vertices are usually convenient to choose.)

Step 2 Join the selected points with the centre of rotation, using straight-line segments.

Step 3 For each point, place the centre of your protractor at the centre of rotation, so that the zero mark is on the line, joining the centre with the point that is to be rotated. Make sure the protractor faces the specified direction (that is, clockwise or anticlockwise).

Step 4 Rotate each point by the specified number of degrees (the 'image' point is as far from the centre of rotation as the original point).

Complete the image using the points rotated earlier as a guide.

WORKED Example 24

Show the final position of the shape shown at right after a half-turn rotation in an anticlockwise direction about the dot.

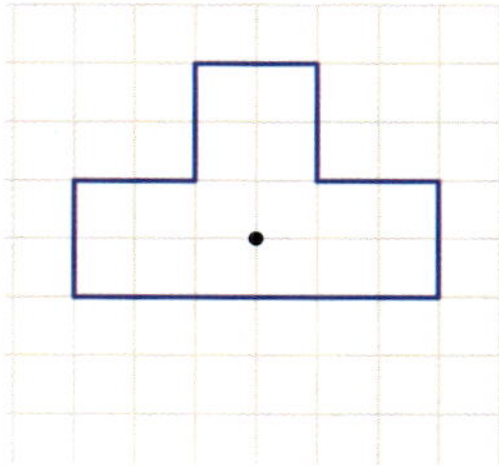

THINK

DRAW

1. Select a point on the shape (any vertex is convenient) and call it (say) P.
   (a) Join point P with the centre of rotation.
   (b) Place the centre of your protractor on the centre of rotation with its zero at the line, joining the centre with the point P. We need to rotate the shape in an anticlockwise direction by a half-turn (that is, 180°). So, measure out 180° and mark the point. (Make sure the protractor faces the right way; that is, anticlockwise.)
   (c) Join the 180° mark with the centre of rotation. Find the position of the 'image' point P′: it is as far from the centre of rotation as point P.

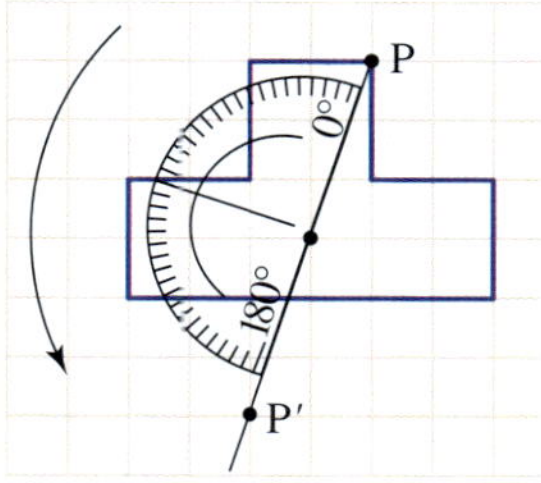

2. Repeat the first step for several other points and then complete the image of the shape by using these points as a guide. Highlight the image with a different colour.

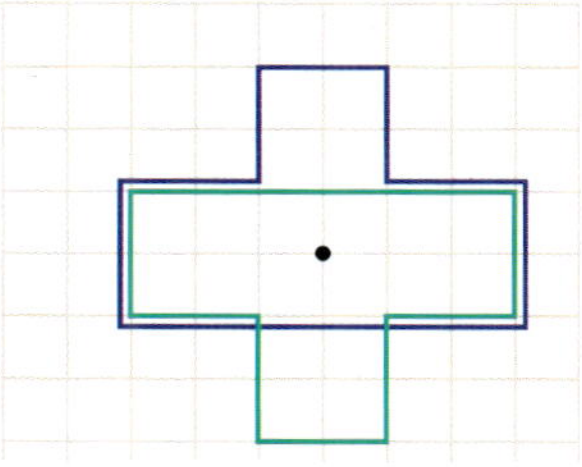

Note that for the half-turn rotation shown in the example above, it was not really necessary to specify the direction of rotation. If we were to rotate the above shape in a clockwise direction, the image would turn out exactly the same as for the anticlockwise rotation. Can you think of some other rotations which do not require specification of the direction?

If a turn of more than 180° is required and a circular protractor (a protractor that can measure angles larger than 180°) is unavailable, it is easier to perform an equivalent rotation in the opposite direction. For example, if we need to rotate a shape $\frac{3}{4}$ of a turn in a clockwise direction, it is the same as rotating it $\frac{1}{4}$ turn in an anticlockwise direction (this can be easily done using a semicircular protractor).

remember

1. There are three types of transformations: translations, reflections and rotations.
2. A translation takes place when an object is moved up, down, right or left without flipping, turning or changing size.
3. A reflection is an exact image of a shape, as seen in a mirror.
4. A reflection is as far behind the mirror as the shape is in front of it.
5. A rotation is an image created by turning an object about a set point.
6. To specify a rotation, we need to state the angle size and direction of the rotation.

EXERCISE 11H Transformations

WORKED Example 20

Cabri Geometry — Translation

1 For each of the following shapes, specify the translation that took place.

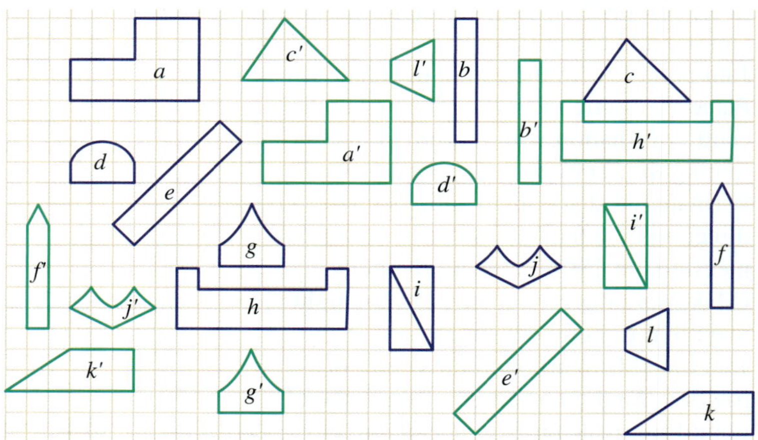

WORKED Example 21

SkillSHEET 11.5 — Translation of a point

2 Given the following translations, use graph paper to show the original shape and the position of the shape after its translation.

a

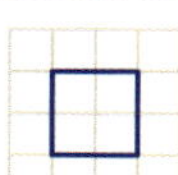

4 left

b

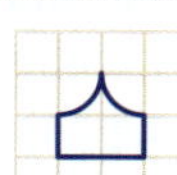

2 right, 3 up

c

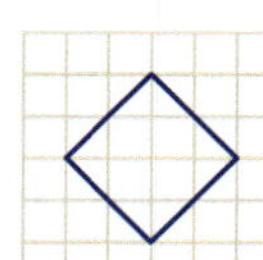

1 down, 8 left

d

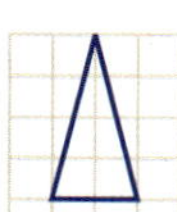

2 down, 2 left, 3 up, 4 right

e

4 right, 3 up, 6 right, 2 down

f

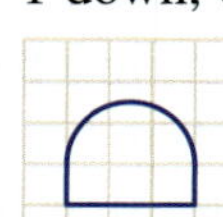

2 right, 4 up, 6 right, 5 down, 3 right, 2 up

3 For each of the following shapes, show the final position after the specified translations.

a
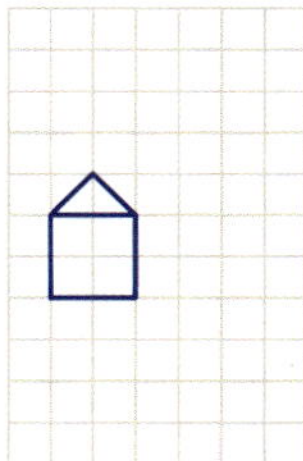
3 up, 2 left

b
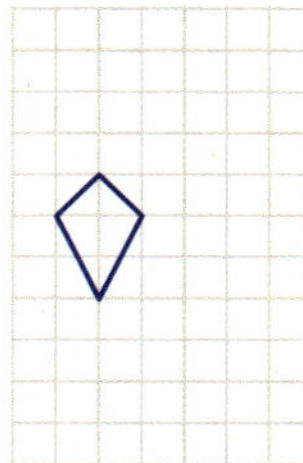
4 down, 3 left

c
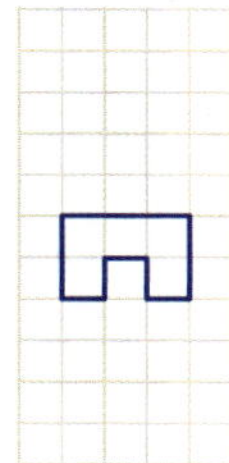
4 right, 2 down, 3 right

d
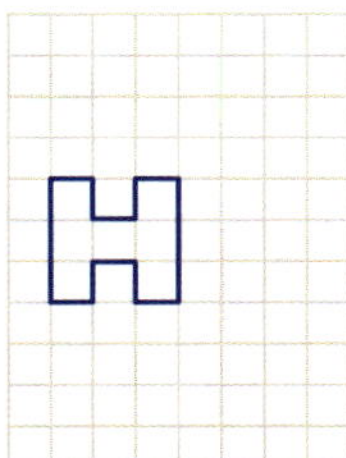
7 up, 2 left,
3 down, 2 right

e
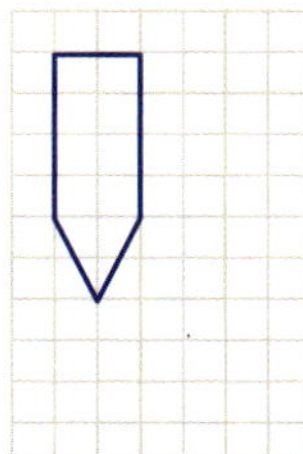
5 down, 2 right,
3 down, 3 right

f
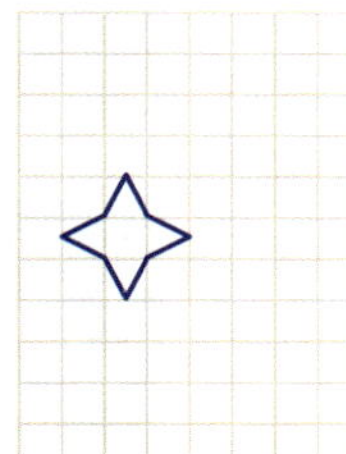
8 down, 2 up,
3 right, 5 left,
2 up, 3 left

4 For each of the following, write a single translation that would have taken the object from its starting point to its final position.

a
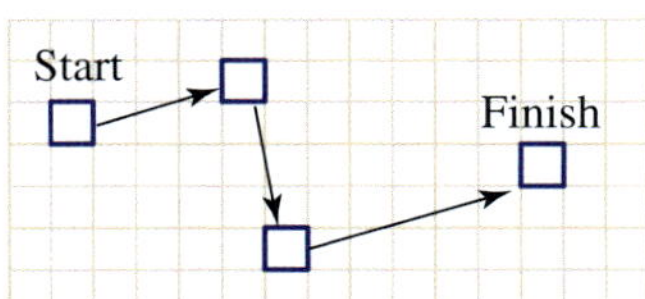

b
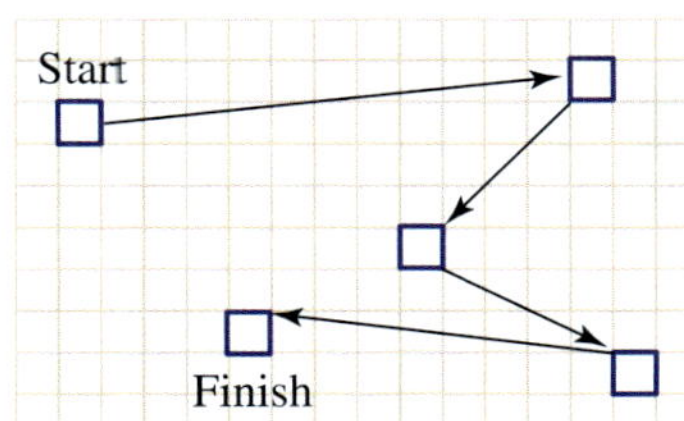

c
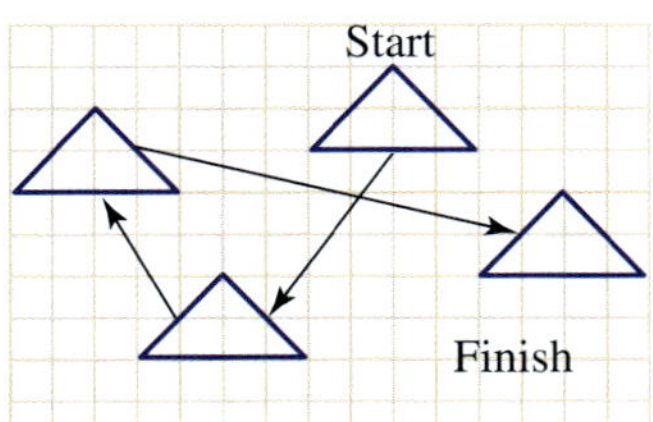

d
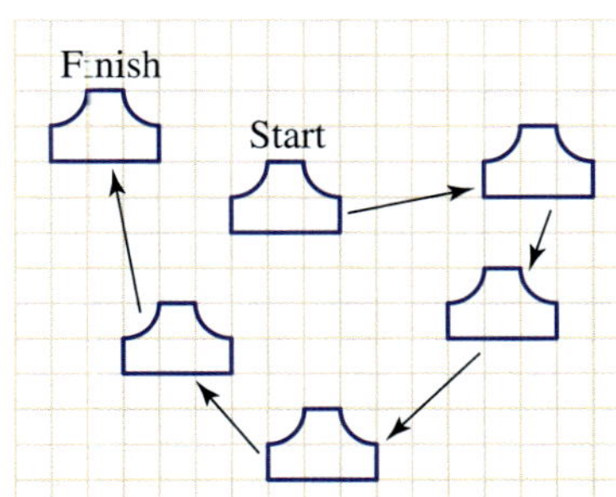

5 **multiple choice**

If an object is translated 4 units right, 3 units up, 6 units right and 7 units down, its final position (as compared to its initial position) would be:

A 7 right, 13 up
B 10 right, 10 up
C 10 right, 10 down
D 10 right, 4 down
E 13 right, 7 up

6 Explain the translation that has occurred in each of the following situations.

a

b

c

d

WORKED Example 22

7 Draw reflections of each of the following objects in the dotted line.

a
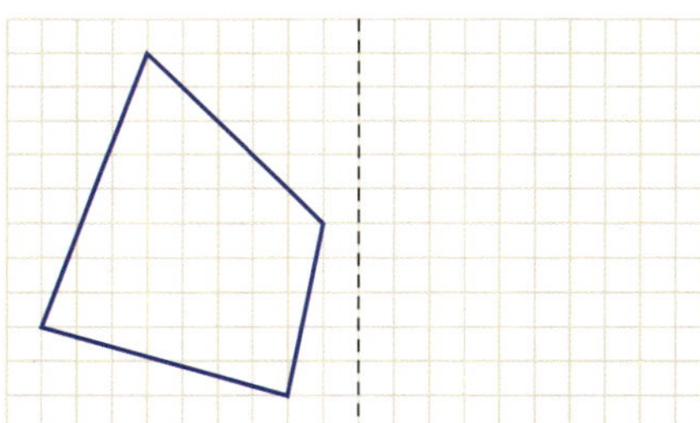

b
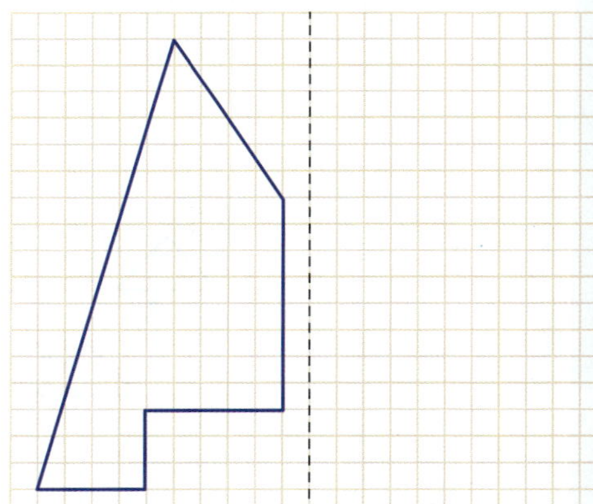

c
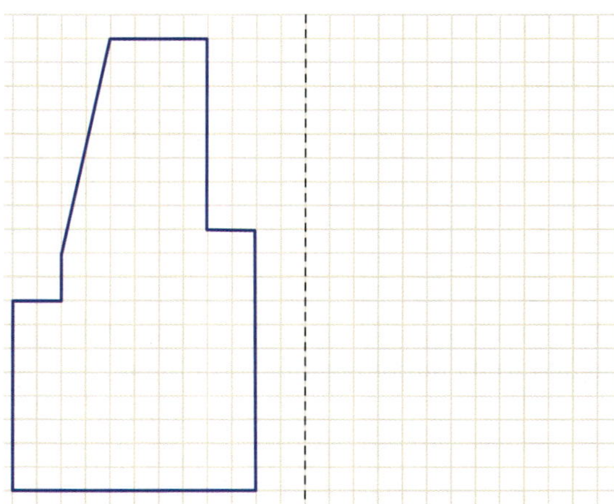

d
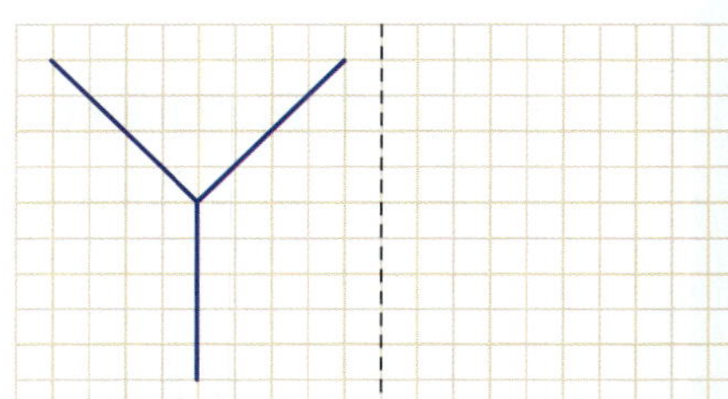

8 Draw reflections of each of the following shapes.

a

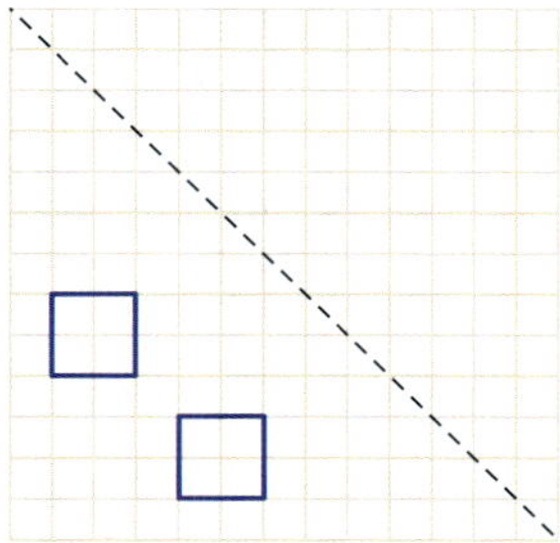

b

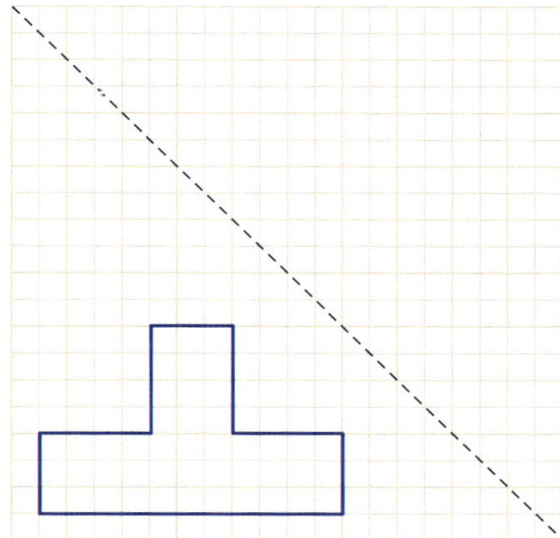

c

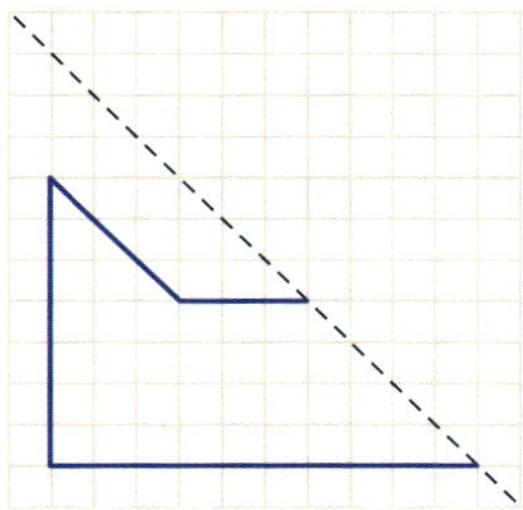

d

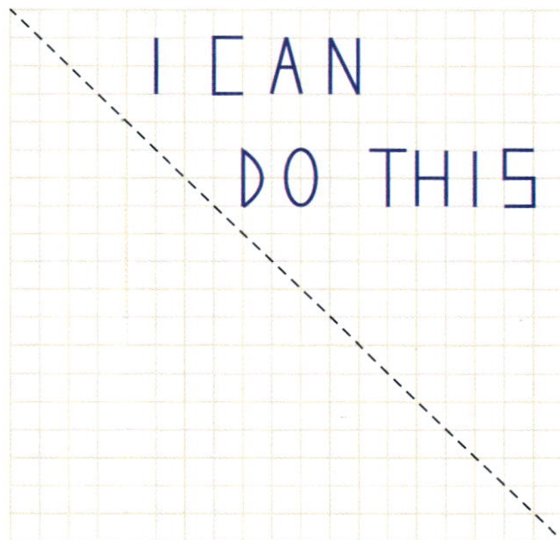

11.6 SkillSHEET Reflection of a point

Cabri Geometry Reflection

9 For each of the following, indicate where a mirror should be placed to produce the given image.

a

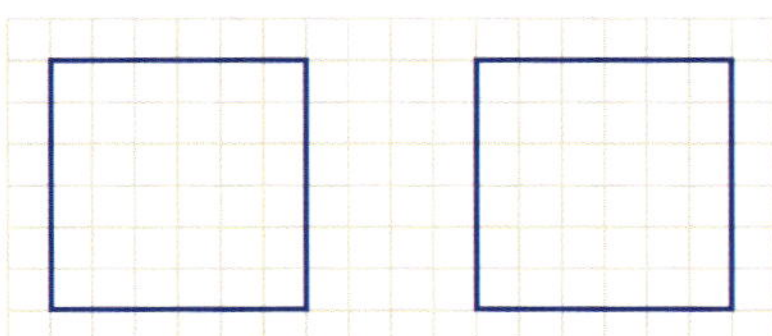

b

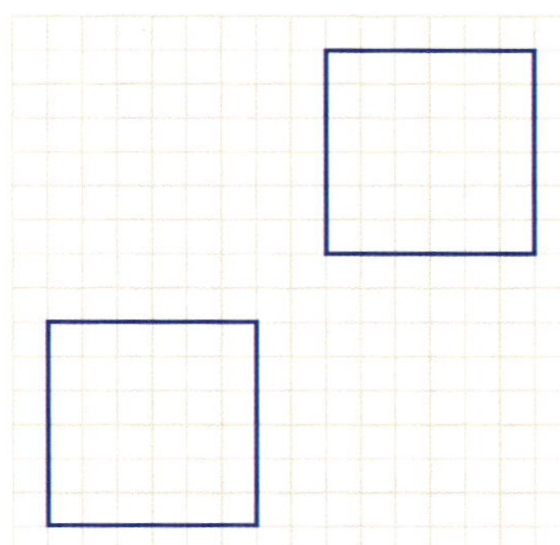

c

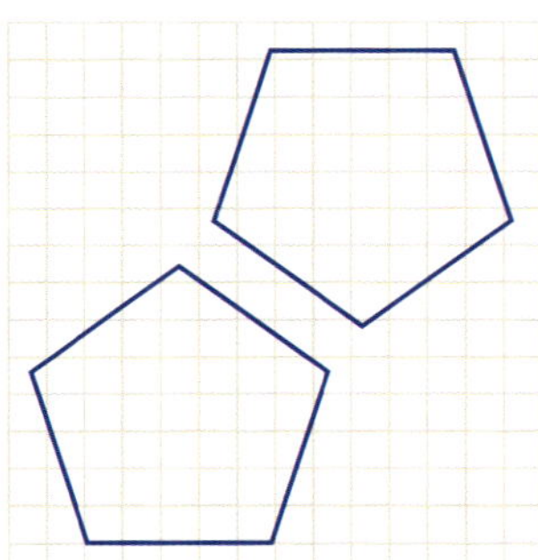

d

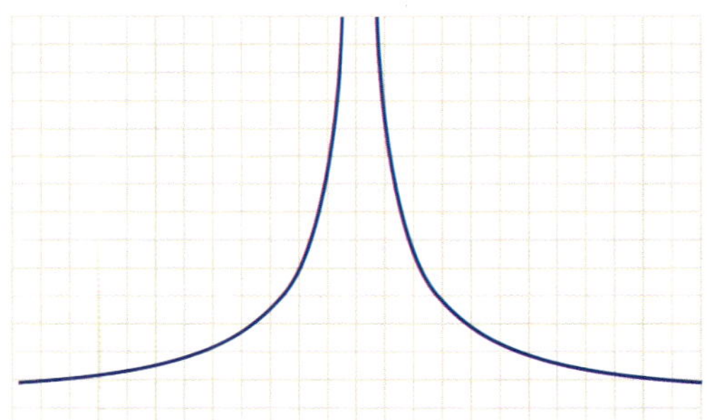

e

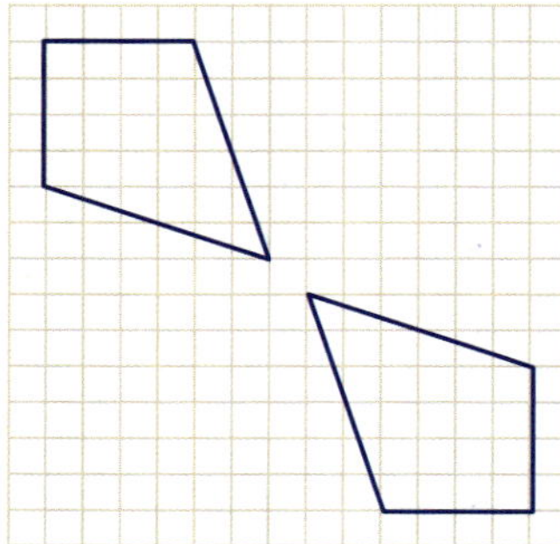

f

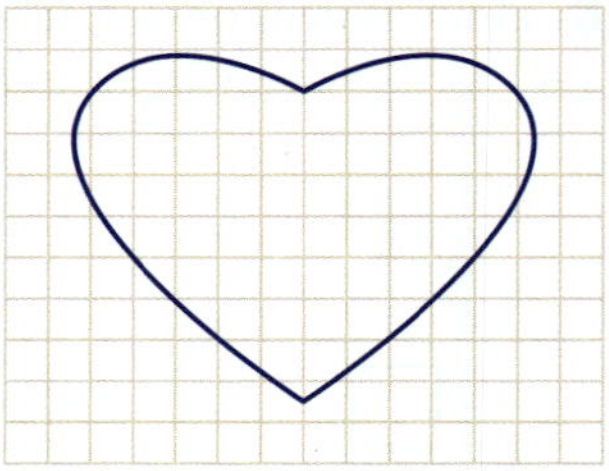

10 For each of the following, draw the reflected images. Note that some pictures can have images on both sides of the dotted line.

a

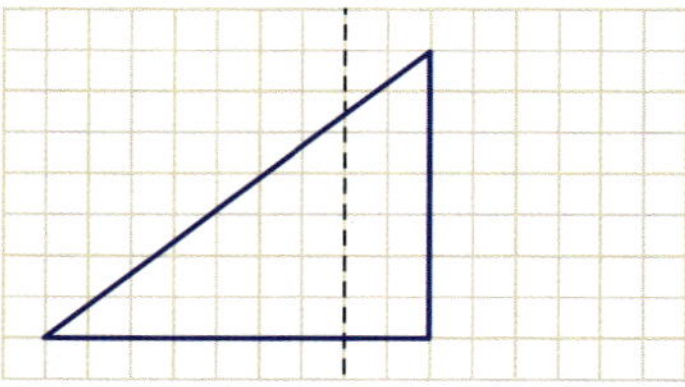

b

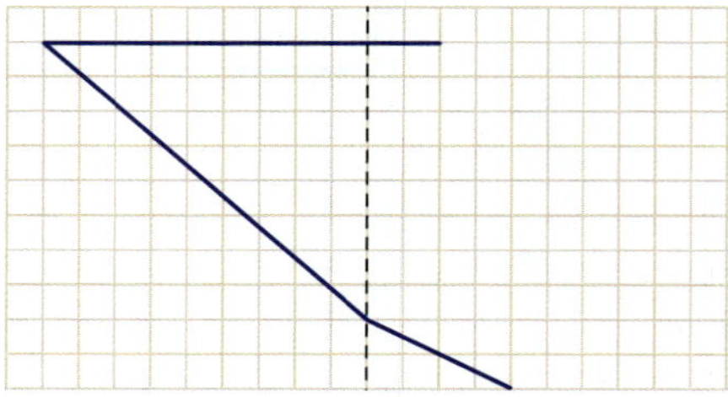

c

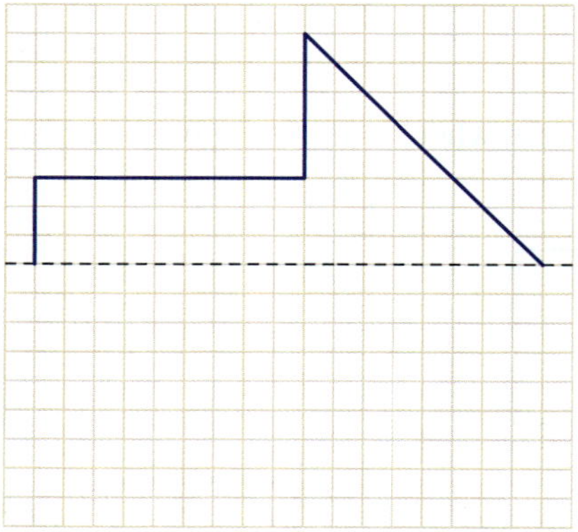

d

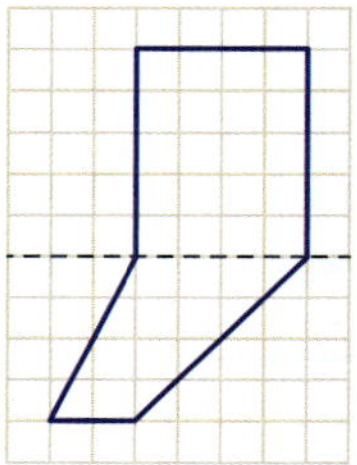

e

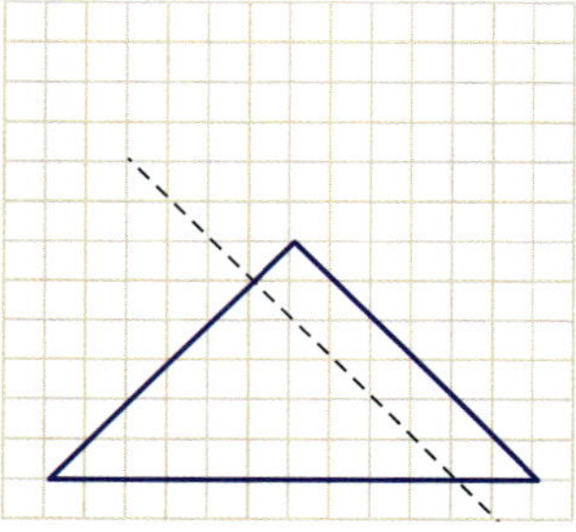

f

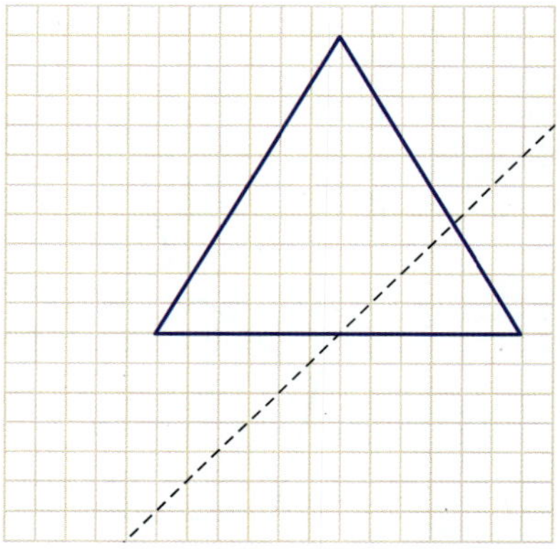

11 A kaleidoscope has beautiful patterns created by a number of mirrors. Some kaleidoscopes have 2 or 3 mirrors, while others have more. Complete the following kaleidoscope pictures by using mirrors to produce reflected images.

a
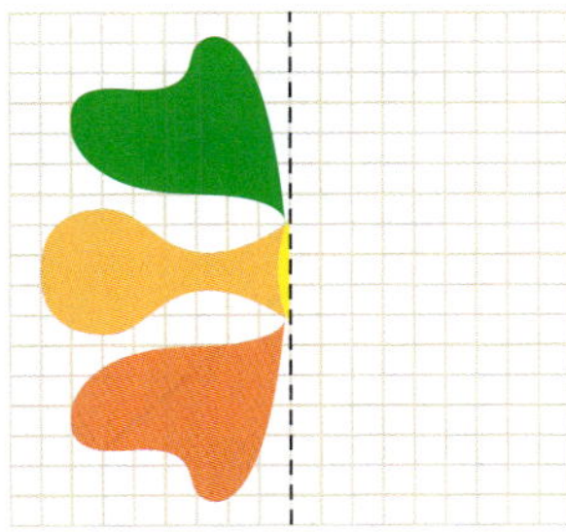

b

c

d
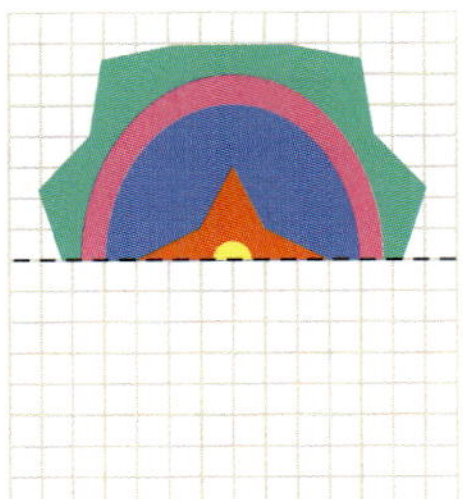

e

f
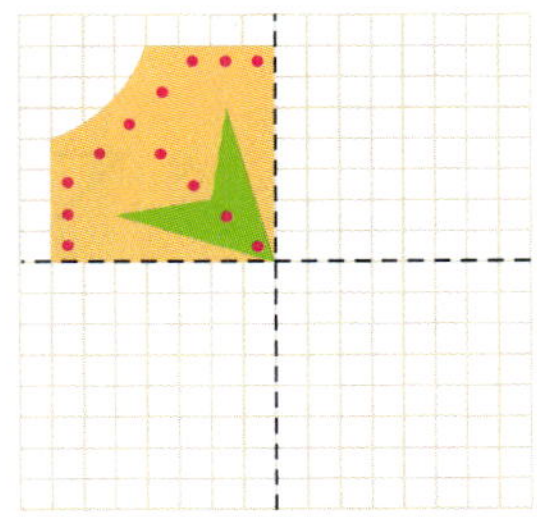

WORKED Example 23

12 Specify the rotation that has taken place in each of the following figures. (Give two possible answers in each case.)

Cabri Geometry — Rotation

a
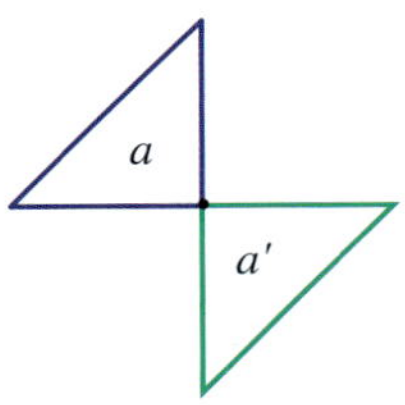

b
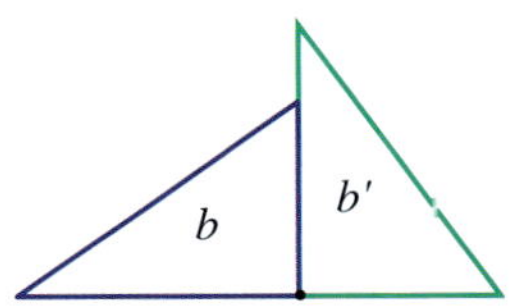

c
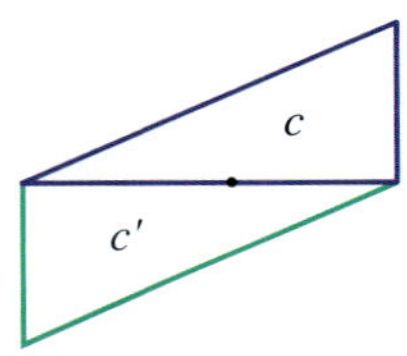

d
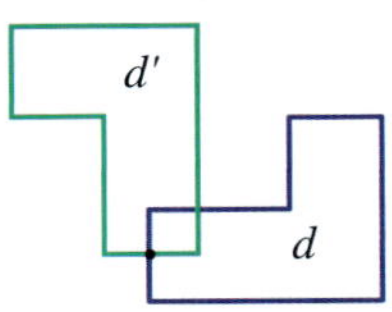

e
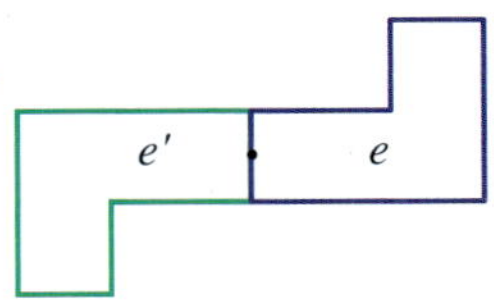

f
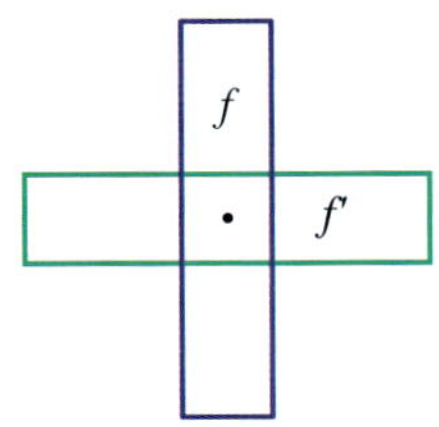

g
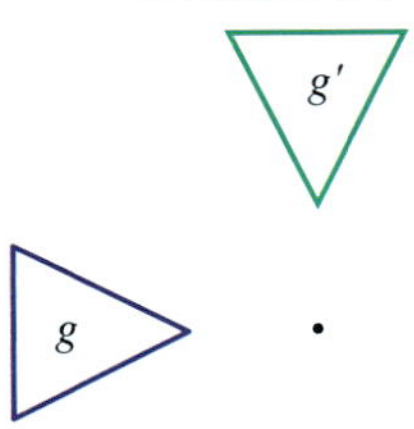

h

i
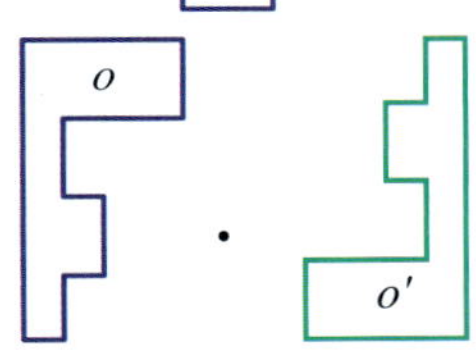

WORKED Example 24

13 Show the final position of each of the following shapes after a half-turn rotation about the dot.

11.7 SkillSHEET — Rotation of a point

a

b
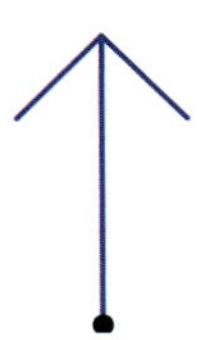

c

d
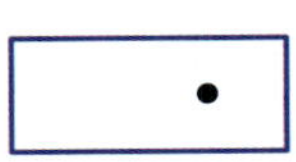

e
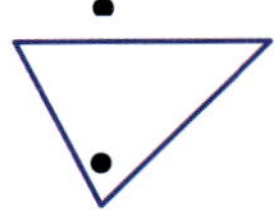

f
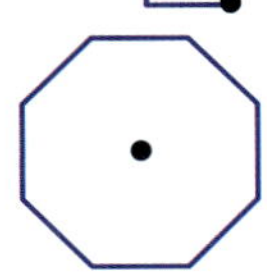

14 Show the image of the shape after each of the following rotations.

a  $\frac{1}{2}$ turn

b 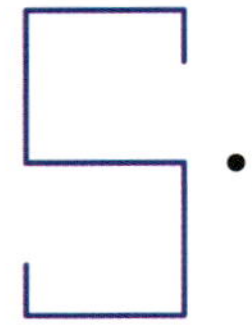 $\frac{1}{4}$ turn, anticlockwise

c  $\frac{1}{4}$ turn, clockwise

d 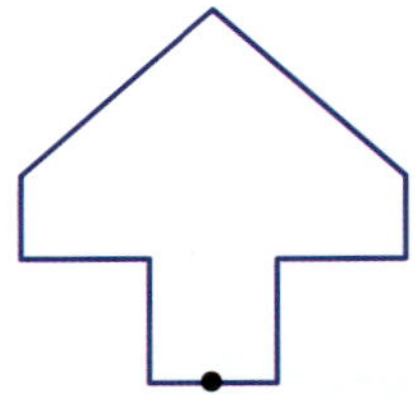 $\frac{1}{2}$ turn

e 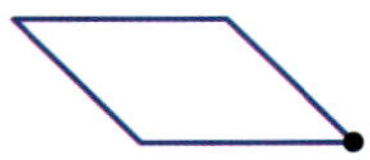 $\frac{1}{4}$ turn, clockwise

f 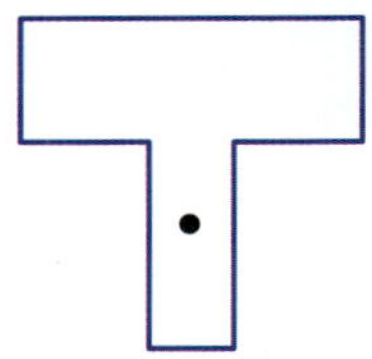 $\frac{3}{4}$ turn, anticlockwise

15 What is the least number of degrees that the figure needs to be rotated so that the image will match the original object? (Ignore minor irregularities.)

a 

b 

c 

d 

GAME time: Plane and solid shapes — 002

WorkSHEET 11.3

COMMUNICATION Transformation design

Use a square sheet of paper and follow the instructions below.

1 Fold D onto A. 2 Fold C onto B. 3 Fold C/B onto D/A.

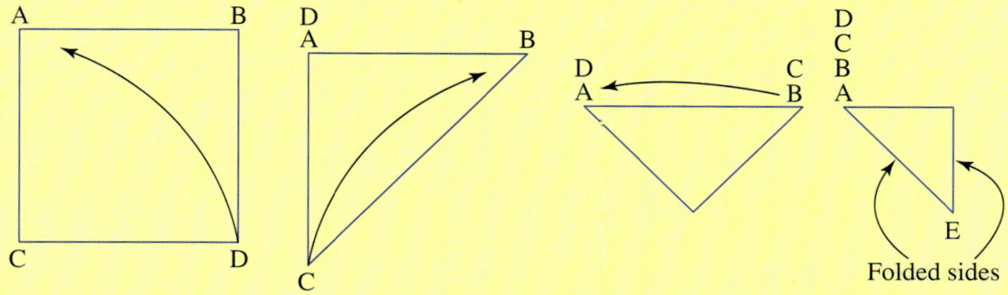

4 Cut out a different shape from each of the two folded sides.

5 Open out your sheet of paper.

6 Draw in all axes of symmetry.

7 Which part of your design would be the 'building block' if you were to create the whole design by:

a reflection?

b rotation?

8 Can you use a smaller section as the building block and use both reflection and rotation to obtain the whole design?

Tessellations

Tessellations are repeated patterns that cover a flat surface with no gaps.

Some simple shapes tessellate by themselves. Others need a combination of 2 or more shapes. Many beautiful designs can be created using tessellations.

Shapes that tessellate by themselves include equilateral triangles, rectangles, regular hexagons, certain trapeziums and squares. Three examples are shown below.

Can you think of some other shapes that would tessellate by themselves?

WORKED Example 25

Use a rectangle to produce 3 different tessellation patterns.

THINK	DRAW
1 Place rectangles next to each other.	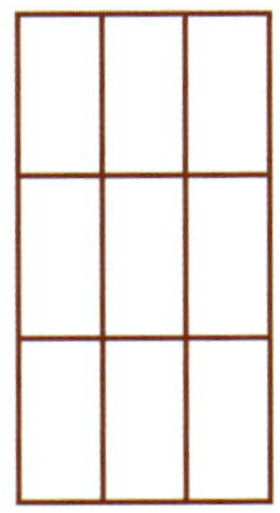
2 Place rectangles in groups of 2 with a different orientation between groups.	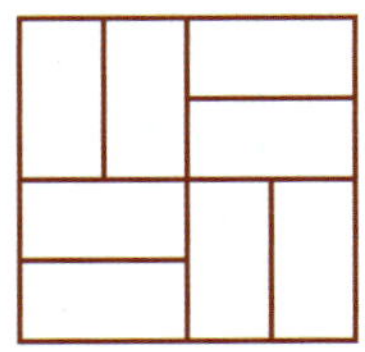
3 Arrange rectangles in a V shape.	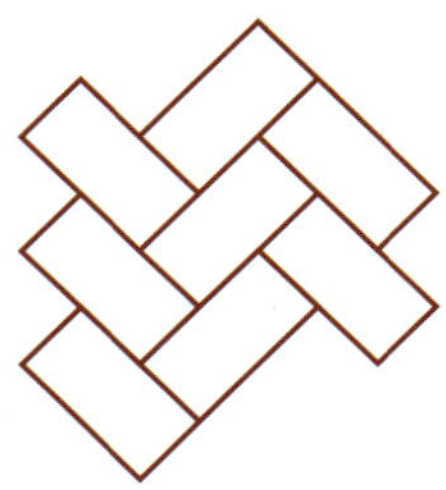

remember

A *tessellation* is a pattern, made by repeated use of a shape or group of shapes, which covers a surface so that there are no gaps.

Cabri Geometry: Tessellating triangles and squares

Cabri Geometry: Tessellating hexagons

Cabri Geometry: Will regular octagons tessellate?

Cabri Geometry: Will regular pentagons tessellate?

EXERCISE 11I Tessellations

WORKED Example 25

1 a Add 4 more rectangles to each of the designs below to show how the rectangles tessellate.

i 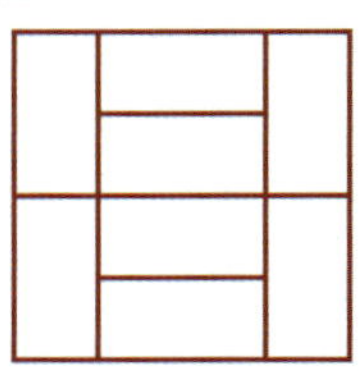

ii 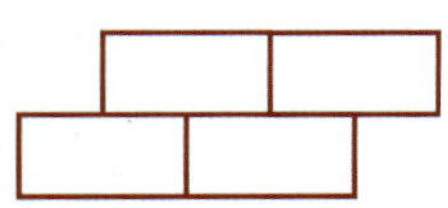

b Create 2 other tessellating patterns using rectangles.

2 Use an equilateral triangle to produce a tessellating design.

3 Use a regular hexagon to produce a tessellating design.

4 Which of the following shapes can make tessellations?
a square b rectangle c T-shape d circle e L-shape f U-shape

5 Use the following shapes to produce a tessellation.

a
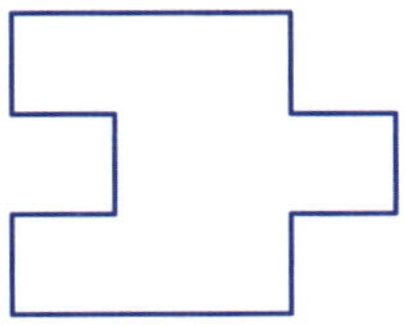
b
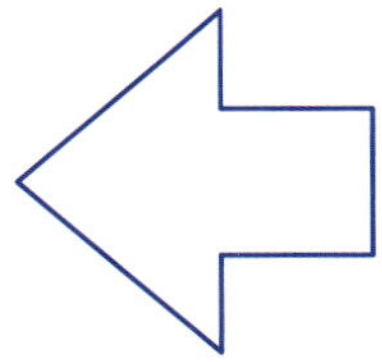
c
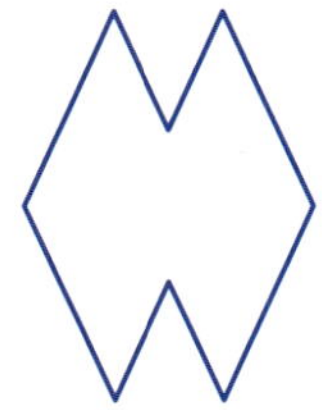

6 Use both shapes in each case to produce a tessellation.

a
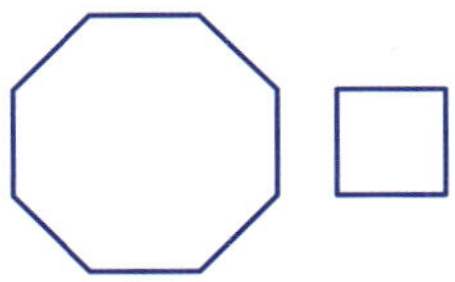
b
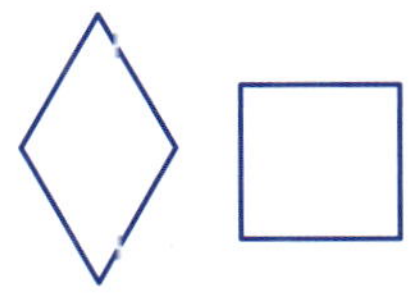

7 Modifying tessellations made of rectangles can create some interesting patterns. A rectangle can be changed by adding and removing parts. It will still tessellate, provided that the changes are balanced; that is, if a part is added on one side of a rectangle, exactly the same part is removed from its opposite side as shown at right.

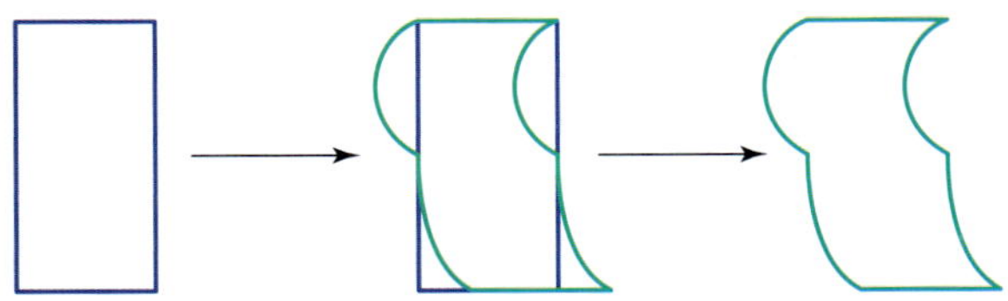

Complete the balanced changes for each of the following rectangles and draw the final shape. Show that the final shape tessellates.

a
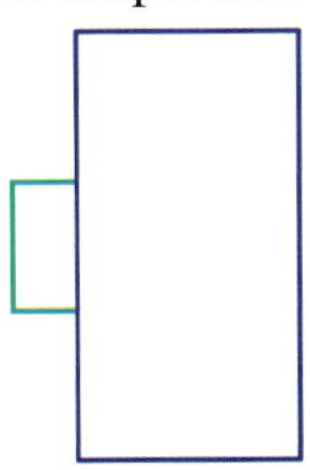
b
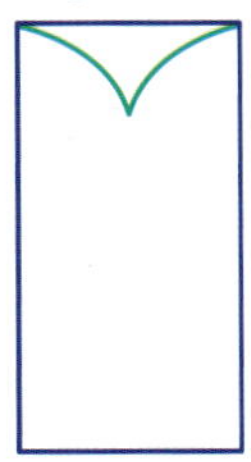
c
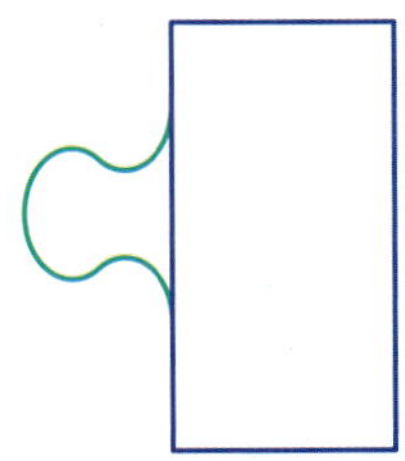
d

e
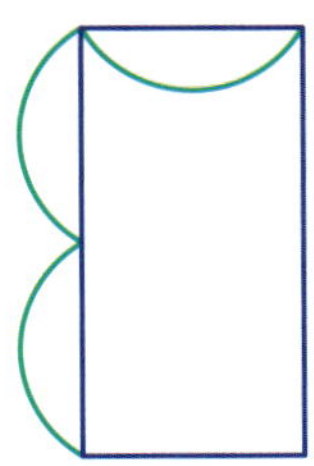
f
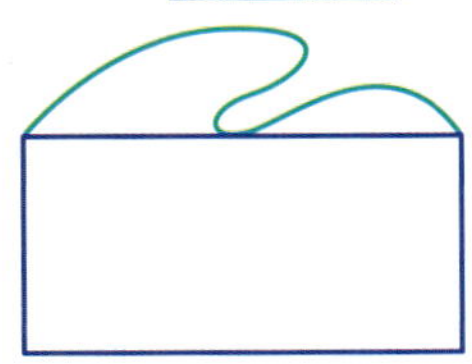
g
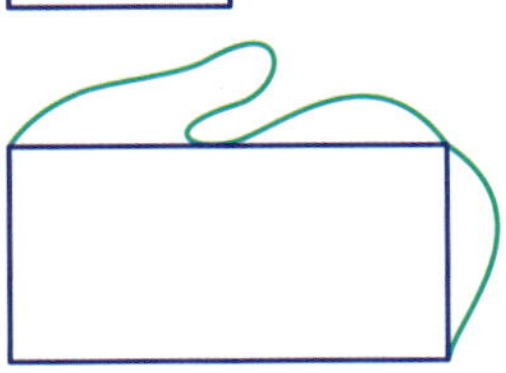
h
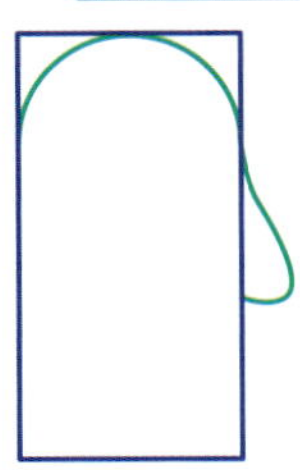
i
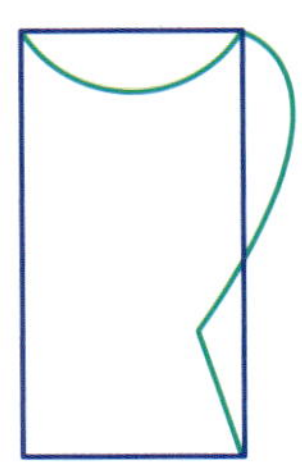

8 Use a rectangle with balanced changes to create your own tessellation designs like the example at right.

9 Other shapes that can also be modified are shown below.

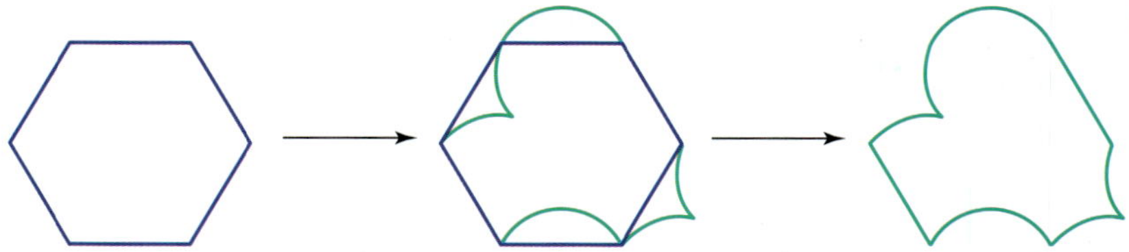

Complete the modifications for each of the following.

a
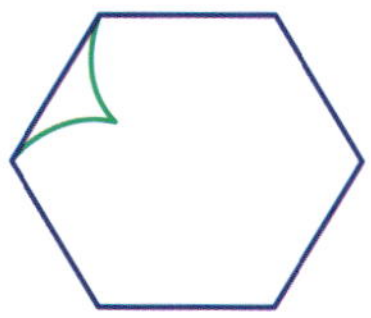

b
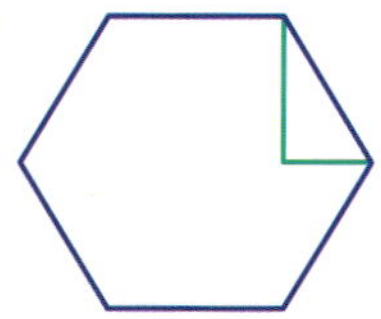

c
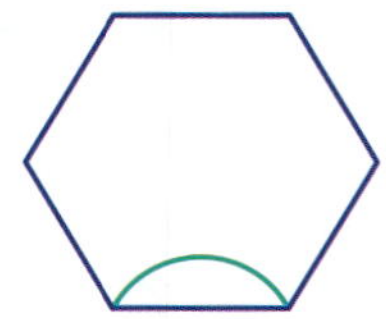

d
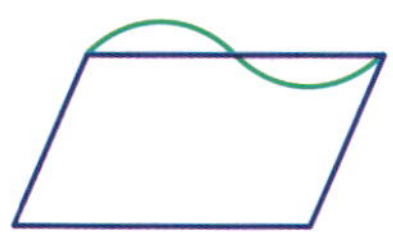

e
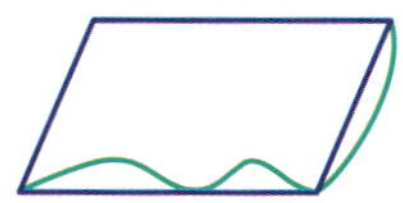

f
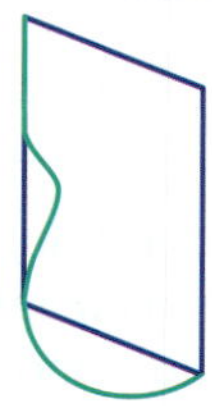

10 Find some other shapes that can be modified to form tessellations.

11 Create your own tessellation using a modified basic shape.

DESIGN

Designing a paved outdoor area

Many paths and entertaining areas are paved with bricks of the same size and shape laid in an interesting design. One such pattern is shown.

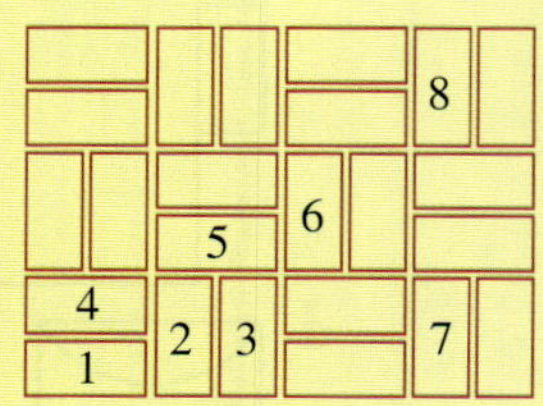

To design this pattern, the basic shape of the brick undergoes one or more transformations to different positions.

1 What transformation has occurred to brick 1 to obtain the position of brick 2?
2 What transformation has occurred to brick 2 to obtain the position of brick 3?
3 Describe how the position of brick 3 can be obtained in relation to brick 1.
4 For each of the other numbered bricks in the pattern, describe what transformations to brick 1 are needed to obtain the position of that brick.
5 Explain why this pattern is an example of a tessellation.
6 Design your own paved entertaining area. You may like to use one or more shapes which tessellate to form your pattern. Describe what transformations are involved.

summary

Copy the sentences below. Fill in the gaps by choosing the correct word or expression from the word list that follows.

1 An obtuse-angled triangle has one angle ____________ 90°.

2 An ____________ triangle has two sides equal in length and two angles equal in size.

3 An equilateral triangle has all sides ____________ in length and all angles equal in size.

4 An ____________ triangle has all angles less than 90°.

5 A ____________ triangle has no equal sides or angles.

6 A right-angled triangle has one angle equal to 90° (that is, a ______ angle).

7 A kite has two pairs of ____________ sides, equal in length. The angles between ____________ sides of a kite are equal in size.

8 A ______ has two pairs of opposite sides equal in length and four 90° angles.

9 A square has four ____________ and four 90° angles.

10 A ____________ has one pair of parallel sides.

11 A rhombus has four equal sides. Opposite angles of a rhombus are ______.

12 A parallelogram has two pairs of ____________ sides. Opposite sides of a parallelogram are equal in length and ____________ angles are equal in size.

13 A polygon is any ____________ 2-dimensional shape with ____________ edges.

14 A ________ polygon has all sides equal in length and all angles equal in size.

15 A polyhedron is a ________ object. Each face of a polyhedron is a ________.

16 The sum of the angles in any triangle is equal to ____________ degrees.

17 An ____ angle of a triangle and an interior angle adjacent to it are ____.

18 The sum of the angles in any quadrilateral is equal to ____________ degrees.

19 A plan which can be cut out and folded to make a ____________ is called a ____________ of that shape.

20 The bird's eye view of an object is its ____________ view.

21 The ________ view is drawn when standing directly in front of an object.

22 The ____________ is the left, or the right view of an object.

23 A two-dimensional representation of a three-dimensional object can be made using a single-point ____________.

24 Euler's Rule for a polyhedron states that the number of _________ plus the number of _________ minus the number of _________ equals 2.

25 Isometric _____________ result in an image that is exactly the same shape and size as the original object.

26 _____________ is moving an object up, down, right or left without flipping, turning or changing size.

27 An axis of symmetry is a line that can be drawn on a shape so that if the shape is _____________ along this line, the two sides will coincide exactly.

28 An axis of symmetry can be thought of as a line that acts like a _____________. One side of the object appears reflected in the line of symmetry.

29 A figure has _____________ symmetry if it can be turned less than 360° and still match the original figure.

30 Some shapes have _____________ axis of symmetry, while others have none.

31 Reflection of a shape produces an _____________ which is seen as in a mirror.

32 A reflected image is as far behind the 'mirror' line as the _____________ is in front of it.

33 Rotation produces an image created by turning the object about _____________.

34 To specify the rotation, we need to state the size of the angle and the _____________ (clockwise or anticlockwise) the shape is to be rotated.

35 Any rotation can be specified in two different ways by stating _____________ of rotation in a clockwise and anticlockwise direction.

36 A tessellation is a pattern made by _____________ of a shape or group of shapes so that there are no gaps.

WORD LIST

faces	greater than	front	folded
adjacent	180	regular	more than one
polygon	straight	exterior	transformations
closed	top	isosceles	object
net	edges	scalene	a set point
supplementary	side elevation	parallel	direction
acute-angled	unequal	trapezium	image
rectangle	equal sides	perspective	translation
right	equal in size	the size of the angle	mirror
opposite	equal	rotational	repeated use
vertices	360		
polyhedron	3-dimensional		

CHAPTER review

1 Name the following triangles according to the length of their sides. 11A

a

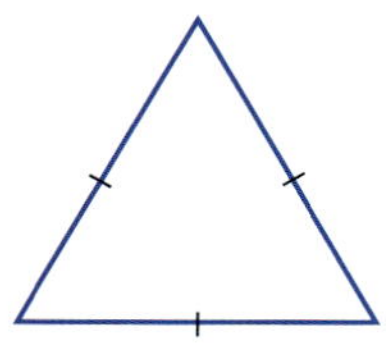

b

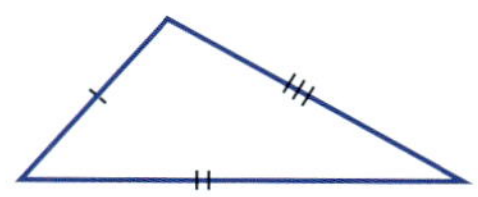

c

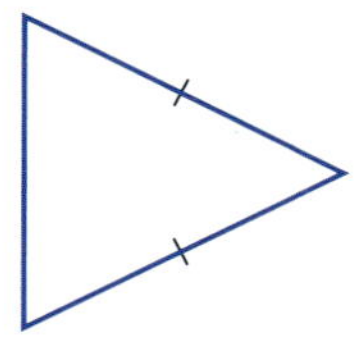

2 Name the following triangles according to the size of their angles. 11A

a

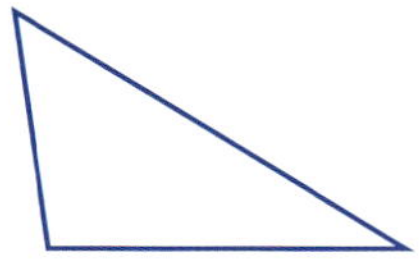

b

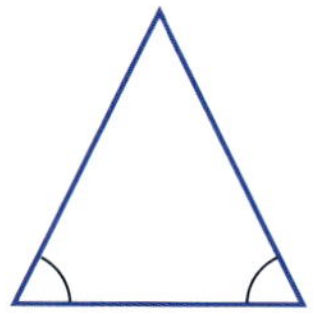

c

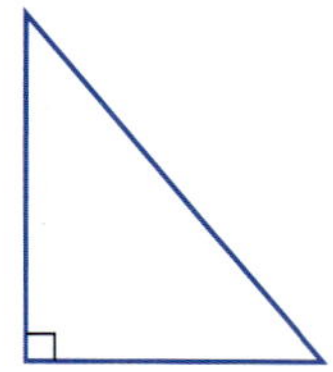

3 Find the value of the pronumeral in each of the following. 11A

a

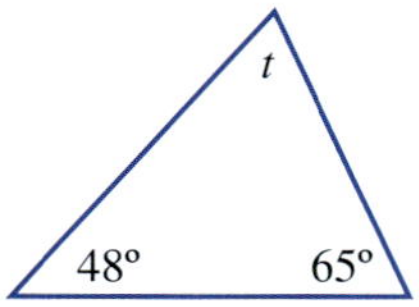

b

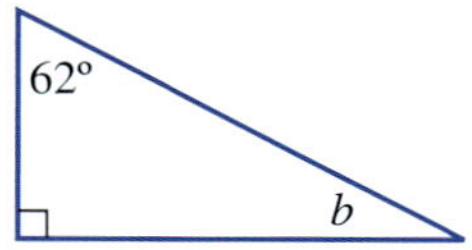

c

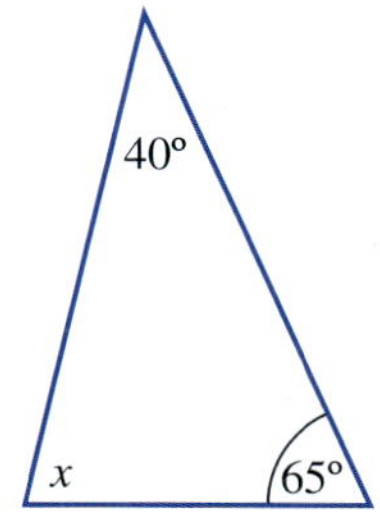

4 Find the value of the pronumeral in each of the following. 11A

a

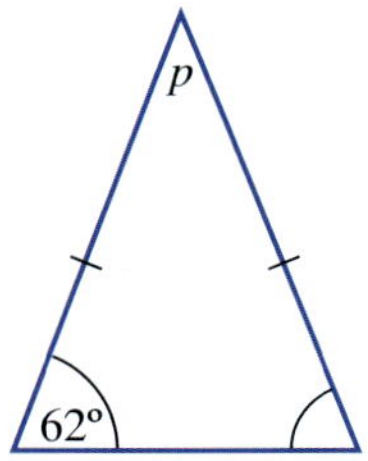

b

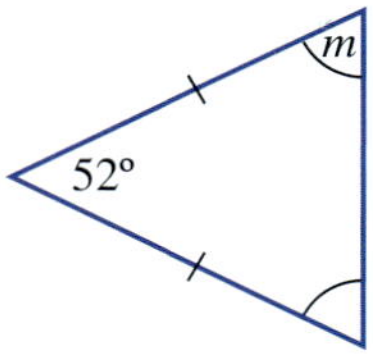

c

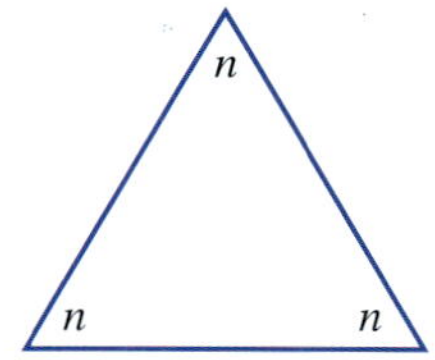

5 The Indian teepee shown has an angle of 46° at its peak. What angle, w, does the wall make with the floor? 11A

11B

6 Name the following quadrilaterals, giving reasons for your answers.

a
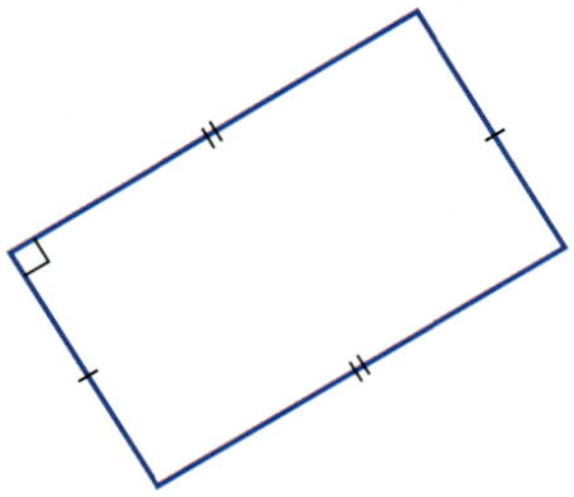

b
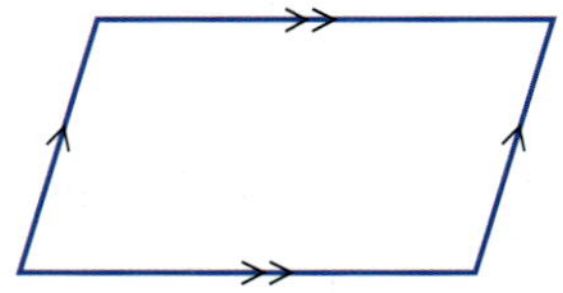

c
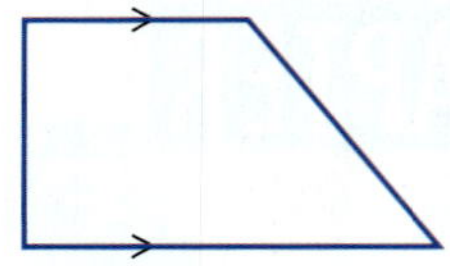

d
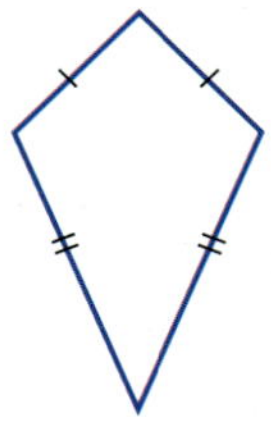

e
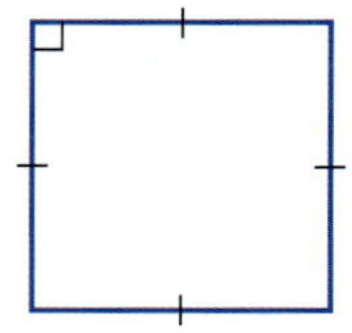

f
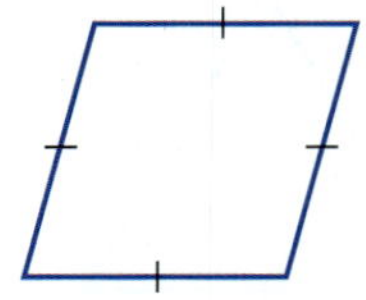

11B

7 Find the value of the pronumeral in each of the following.

a
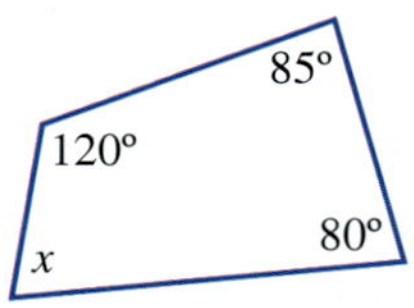

b
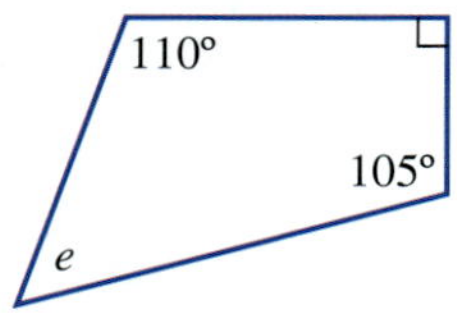

c
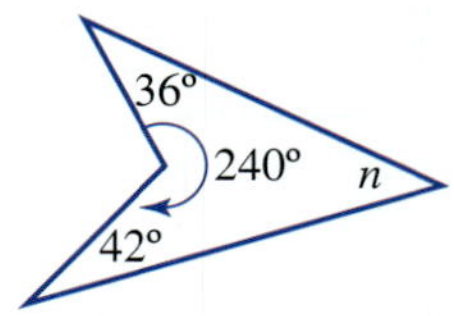

11B

8 Find the value of the pronumeral in each of the following.

a
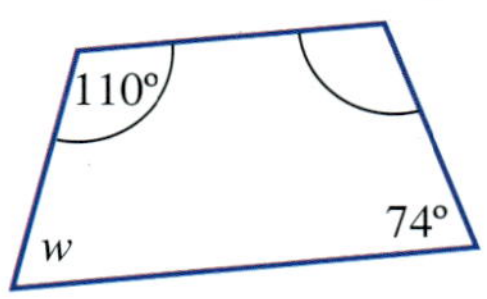

b
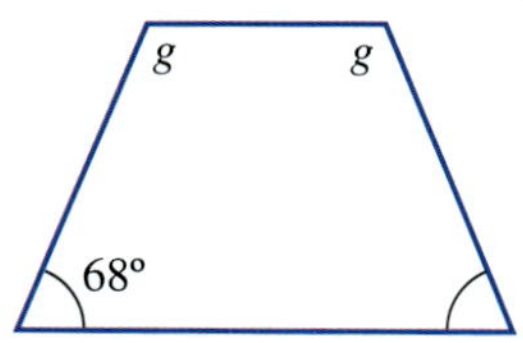

c
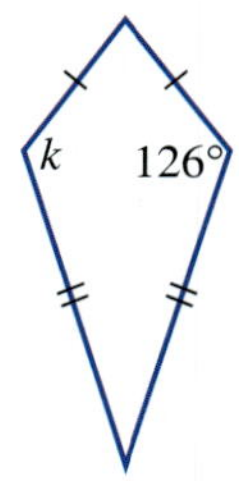

11B

9 A circus trapeze attached to a rope is shown. Find the size of angle t.

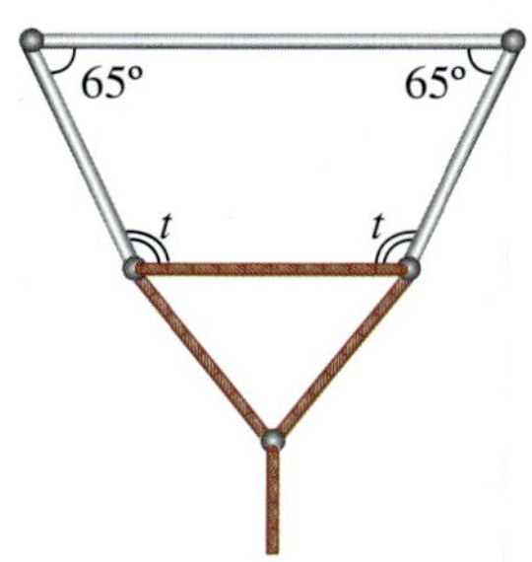

11C

10 Name the following polygons.

a
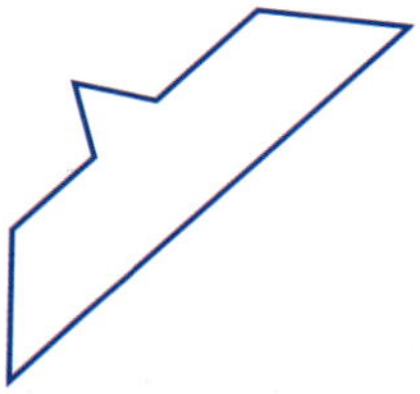

b
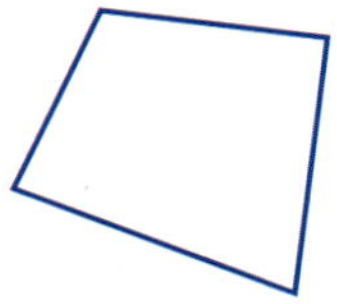

c
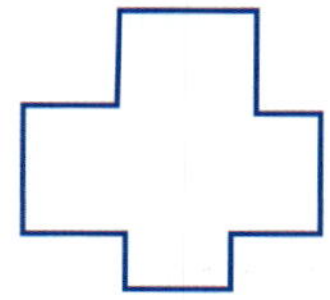

11C

11 Draw two different examples of a hexagon.

12 Draw a circle of radius 5 cm. Mark points on its circumference every 120°. Join the dots together to construct a triangle. **11D**

13 Draw a circle of radius 5 cm. **11D**

a To construct a hexagon you need to mark 6 points on the circle's circumference. Calculate the number of degrees between each point.

b Mark off 6 points on the circle and construct a hexagon.

14 Draw the front, side and top views of each of these solids. **11E**

a

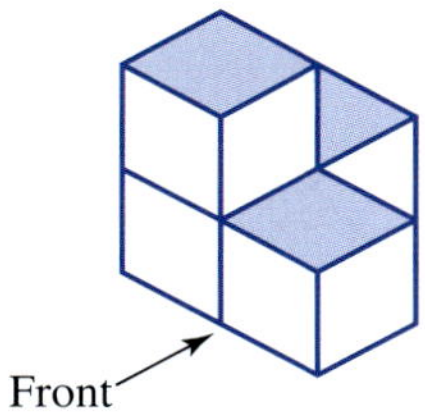

b

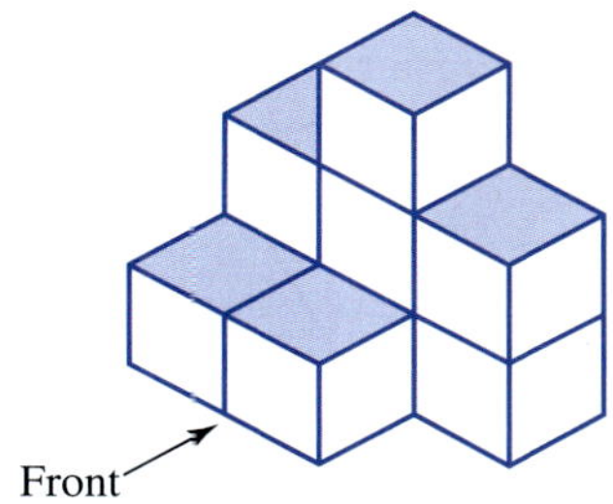

15 The front, side and top view of a solid are shown. Construct this solid, using blocks. **11E**

a

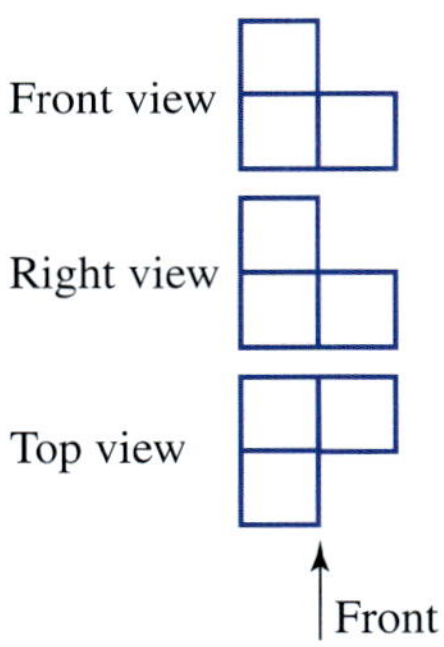

b

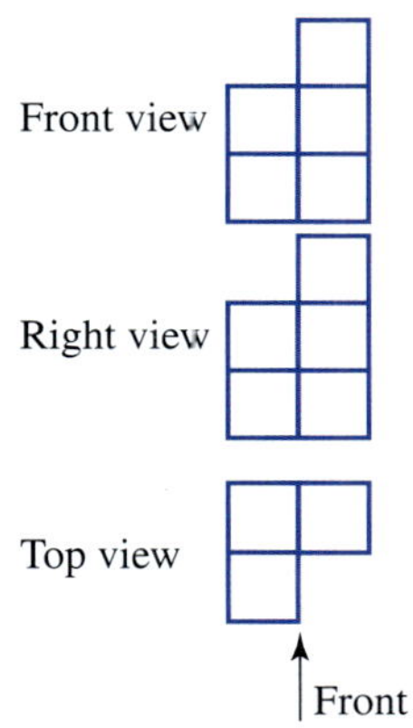

16 Construct a perspective drawing of a triangular prism. **11E**

17 For each polyhedron shown write down: **11F**

- **i** the number of faces
- **ii** the number of vertices
- **iii** the number of edges.

a

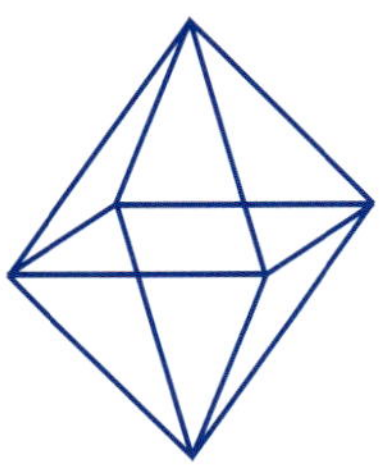

b

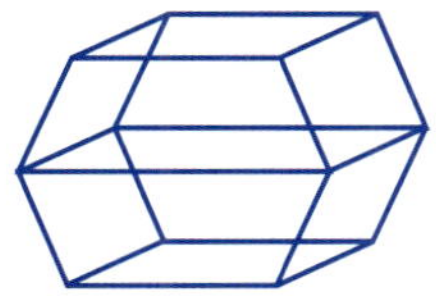

18 **a** A polyhedron has 10 faces and 8 vertices. How many edges does it have? **11F**

b A polyhedron has 6 faces and 12 edges. How many vertices does it have?

c Is it possible to have a polyhedron with 10 faces, 6 edges and 4 vertices? Give reasons for your answer.

11G **19** **multiple choice**

How many axes of symmetry does this shape have?

A 0 **B** 1 **C** 2
D 3 **E** 4

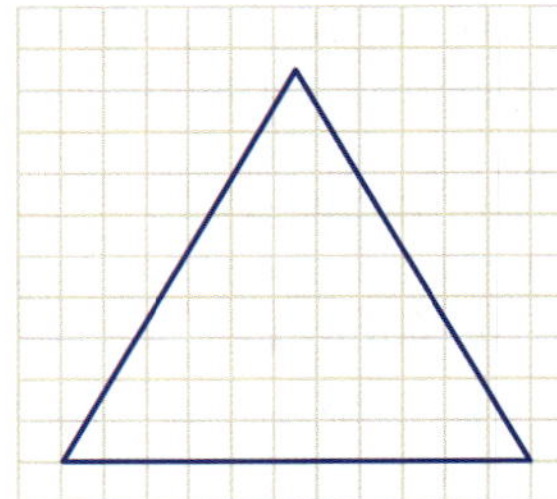

11G **20** Copy each of the shapes below and draw in its axis of symmetry.

a

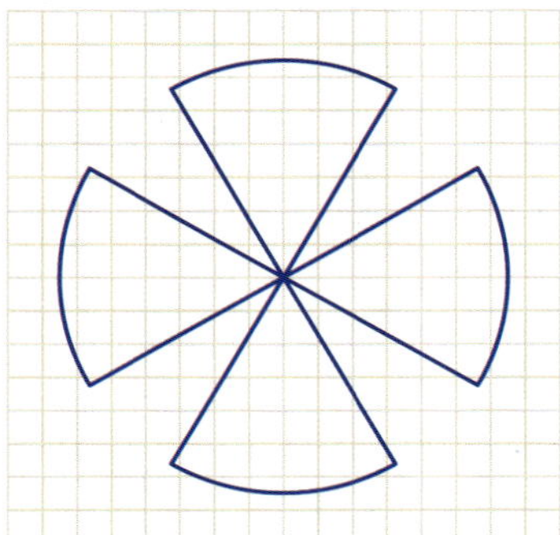

b

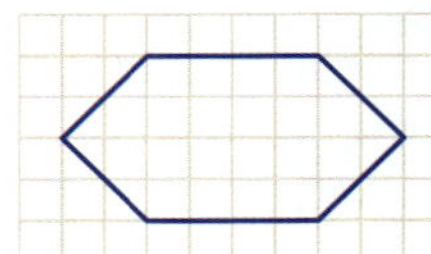

11H **21** Translate each of the following shapes as required:

a

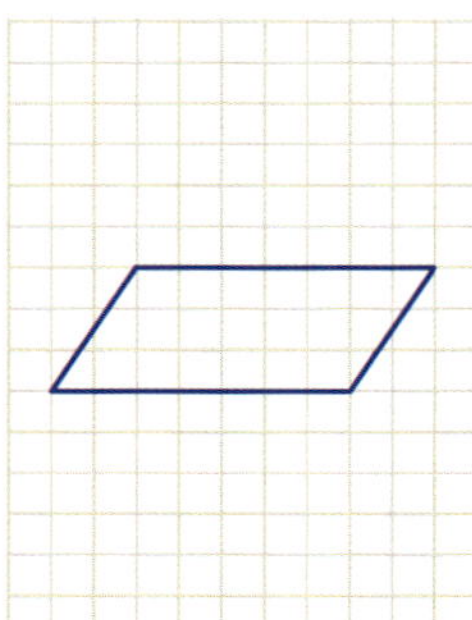

2 up, 3 left

b

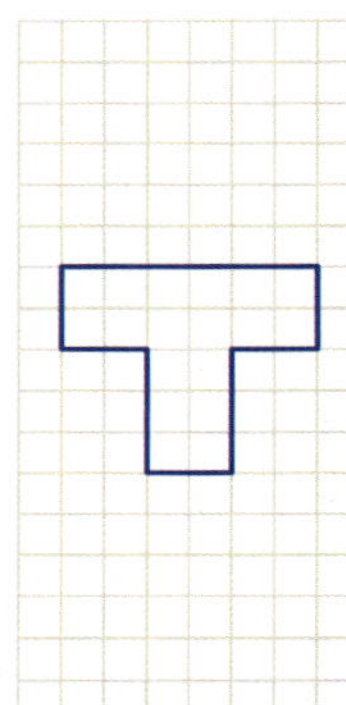

5 right, 3 down

c

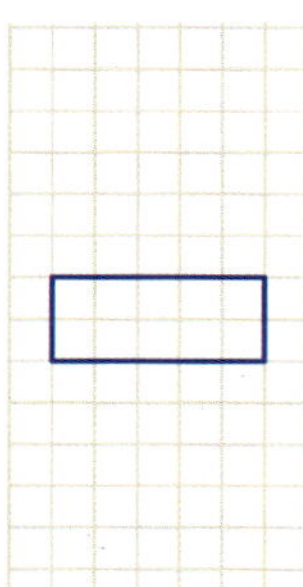

6 down, 2 left, 3 up, 4 right

d

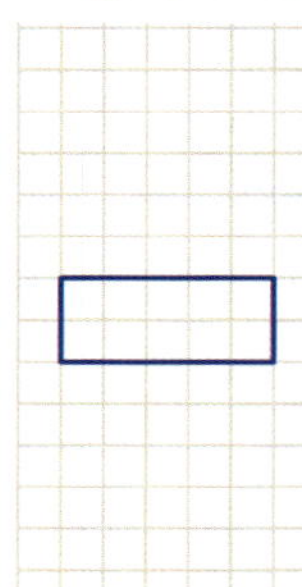

5 right, 6 up, 2 left, 3 up

11H **22** What single translation is described by each of the following sets of translations?

a 3 up, 4 left, 1 down, 2 left
b 6 down, 5 right, 2 up, 3 right
c 12 up, 3 right, 5 down, 3 left, 4 down, 8 right

23 Copy each of the following onto graph paper and draw the reflection in the line shown.

a

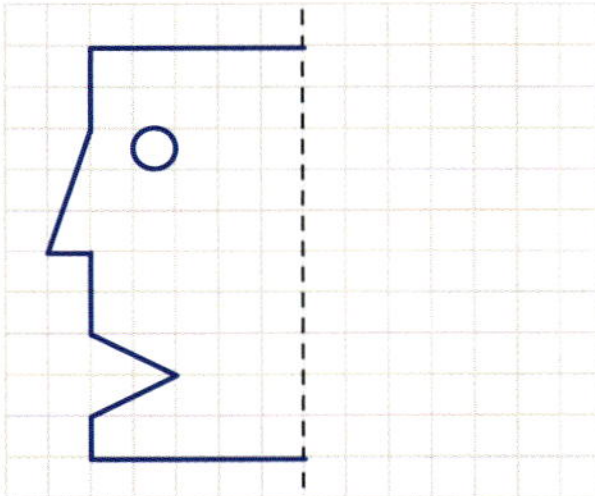

b

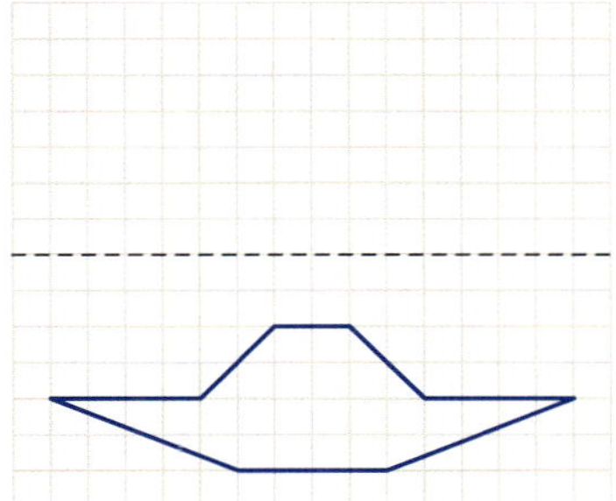

c

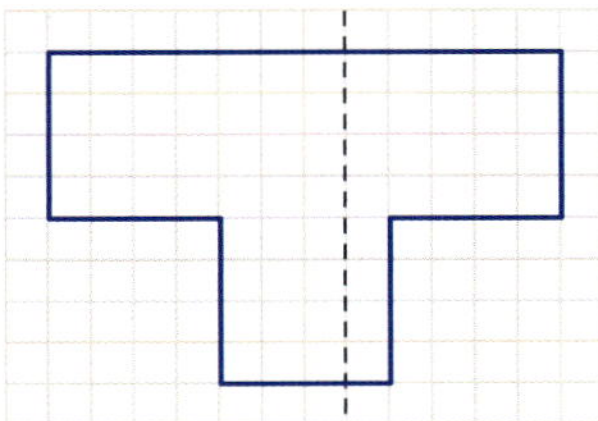

d

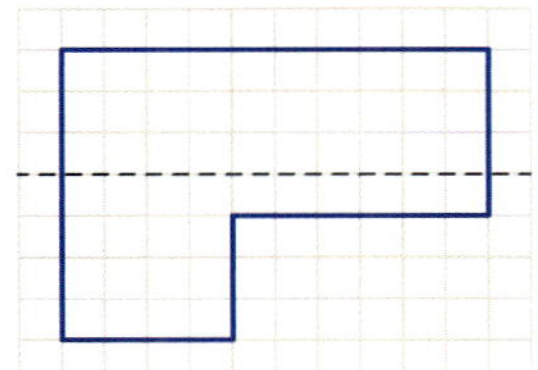

24 Copy each of the following shapes and show where the mirror should be placed to produce the given image.

a

b

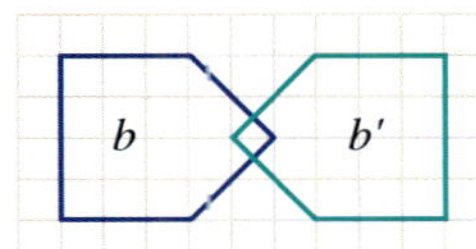

c

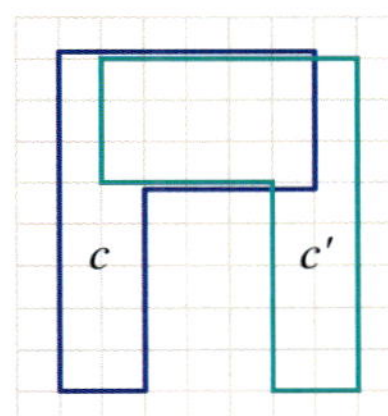

25 Draw the image of each of the following shapes after the specified rotations about the dot.

a

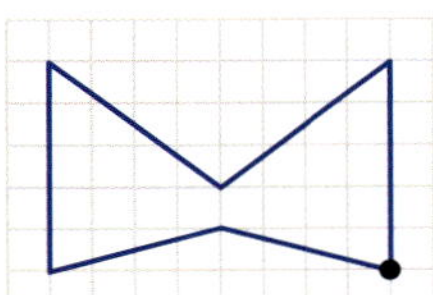

$\frac{1}{4}$ turn clockwise

b

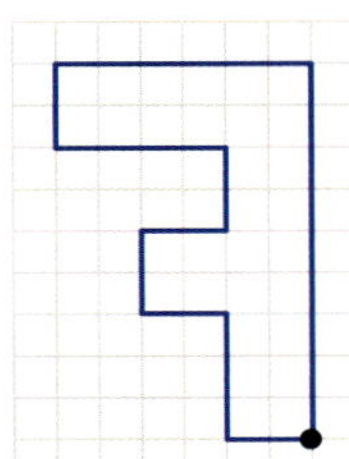

$\frac{1}{2}$ turn

c

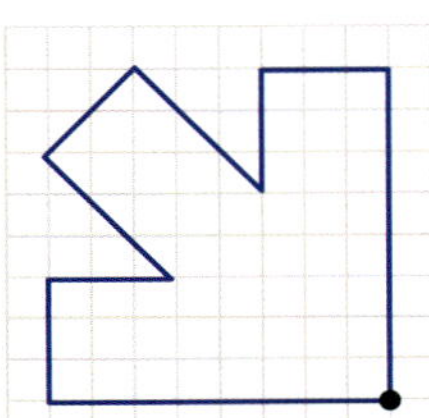

$\frac{1}{4}$ turn anticlockwise

26 Draw the image of the shape at right after a rotation in a clockwise direction of:

a 45° **b** 135° **c** 315°.

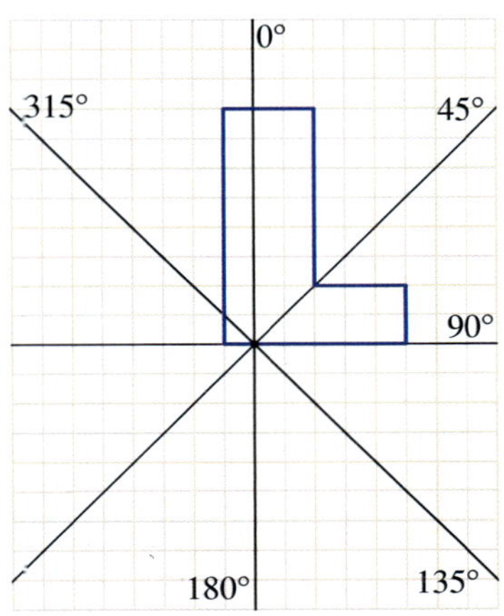

11H

27 For each of the following shapes, find the image after a translation of 5 units left, 3 units up, followed by a reflection in the dotted line.

a

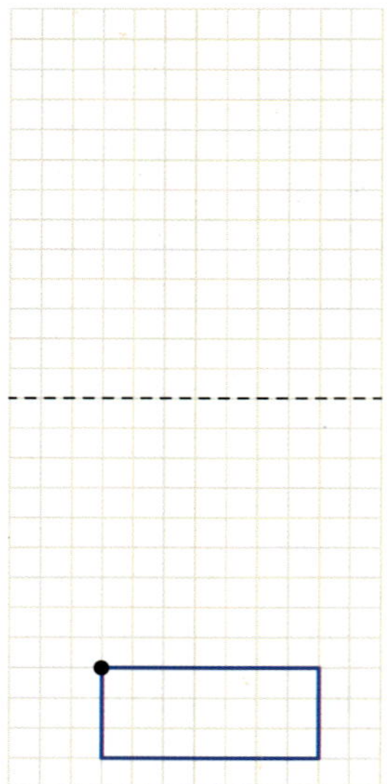

b

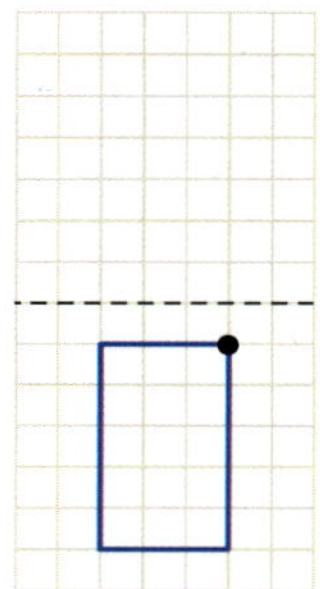

c

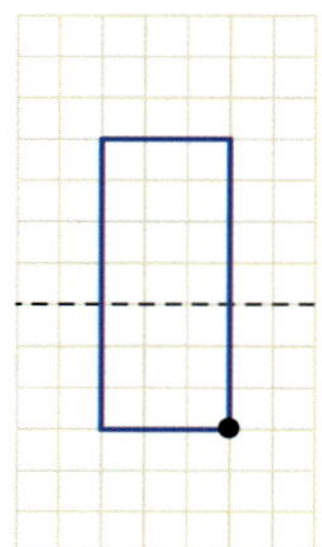

11H

28 For each of the shapes in question **9** above, find the image after reflection in the dotted line, followed by a $\frac{3}{4}$ turn in an anticlockwise direction.

11I

29 **multiple choice**

Which of the following shapes will not tessellate?

A A rectangle
B A square
C A semicircle
D An equilateral triangle
E A regular hexagon

11I

30 For each of the following, draw the final shape by completing the balanced changes, so that a tessellating design can be created.

a

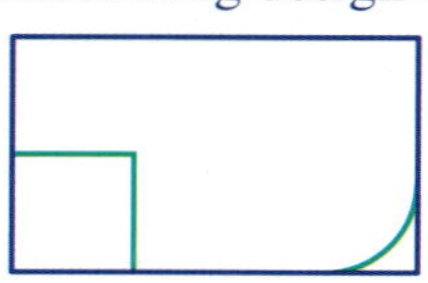

b

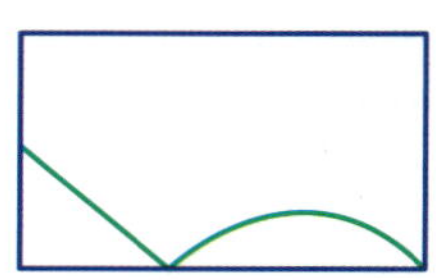

c

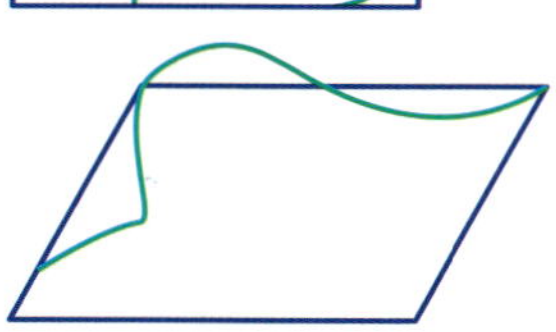

d

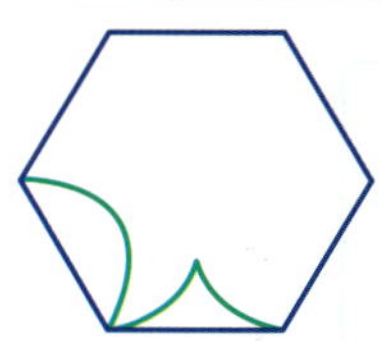

Show how each final shape tessellates by drawing part of the pattern.

test yourself CHAPTER 11

12 Maps, networks and coordinates

The tourist office in Melbourne is located next to Parliament House. A group of American tourists have requested a tour of the following places of interest: Rialto Towers Observation Deck, Shrine of Remembrance, Crown Entertainment Complex, Melbourne Aquarium, Melbourne Cricket Ground, Queen Victoria Markets and the Victorian Arts Centre. Since they have limited time, the tour group want to visit each one but not pass any of the attractions more than once.

Can you plan a travel route for them?

Are you ready?

Try the questions below. If you have difficulty with any of them, extra help can be obtained by completing the matching SkillSHEET. Either click on the SkillSHEET icon next to the question on the *Maths Quest 7* CD-ROM or ask your teacher for a copy.

SkillSHEET 12.1

Converting units of length

1 Convert each of the following to the units given in brackets.

a 50 cm (mm) **b** 400 m (km)
c 7000 cm (m) **d** 0.25 km (m)

SkillSHEET 12.2

Multiplying and dividing by powers of 10

2 Calculate the following.

a 0.025×1000 **b** $2.6 \times 10\,000$
c $3.5 \div 100$ **d** $450 \div 1000$

SkillSHEET 12.3

Angles around a circle (as in a compass)

3 State in which direction of the compass each of the following angles lies.

a 45° **b** 180° **c** 270° **d** 90°

SkillSHEET 12.4

Constructing angles with a protractor

4 Using a protractor, construct the following angles.

a 10° **b** 95° **c** 200° **d** 300°

SkillSHEET 12.5

Measuring angles with a protractor

5 Measure each of the following angles with a protractor.

a

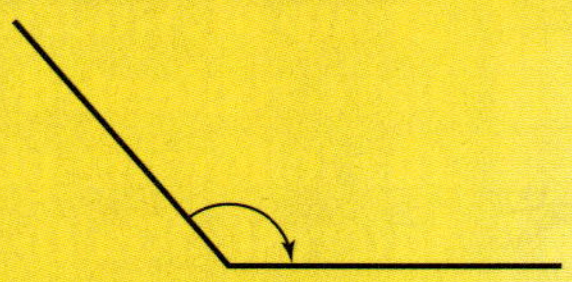

b

c

d

SkillSHEET 12.6

Plotting coordinate points

6 State the size and the direction of the moves from the origin of a Cartesian plane needed to locate each of the following points.

a A(3, 2) **b** B(5, 0) **c** C(0, 7) **d** D(2.5, 1.5)

Interpreting maps — scale

Below is a map of Noneck Island, marked with several natural features. The only remaining structures made by humans are the lighthouse, and the Temple of Doom.

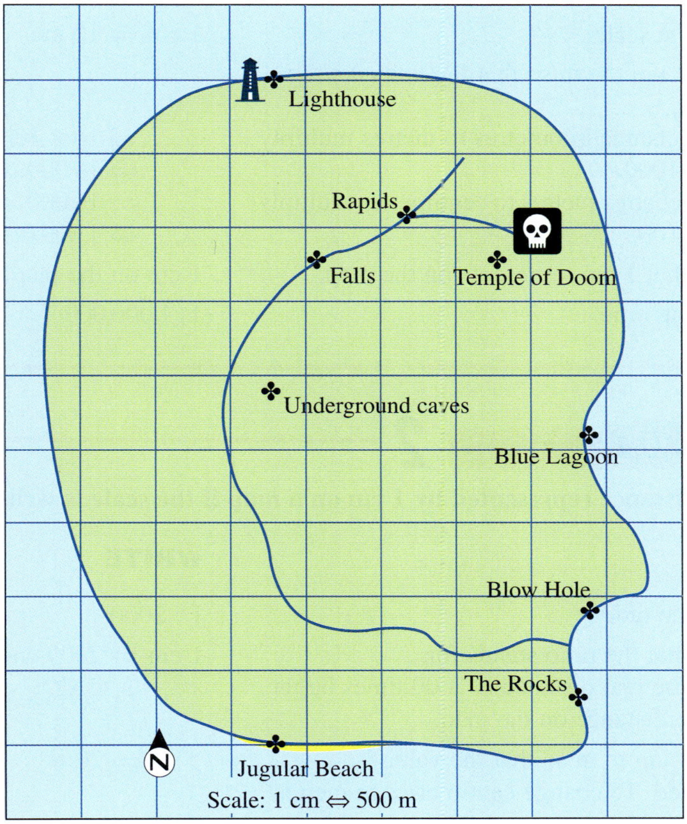

The map shows us the direction of north, so we can work out the compass bearings of south, east, and west for ourselves. The **scale**, 1 cm ⇔ 500 m, really means that 1 cm on the map represents 500 metres on the island.

On the map, the lighthouse is 9 cm due north of Jugular Beach, so the actual distance is:

$$9 \times 500 = 4500 \text{ m}$$
$$= 4.5 \text{ km}$$

The distance from the caves to the Temple of Doom is 3.5 cm, so the actual distance is:

$$3.5 \times 500 = 1750 \text{ m}$$
$$= 1.75 \text{ km}$$

Our scale tells us that 1 cm represents 500 m or, making the units the same, 1 cm represents 50 000 cm.

This is written as a ratio, 1 : 50 000 and scales written this way are found on many maps. This ratio tells us that the real distance is 50 000 times as large as the one shown in the diagram.

Ratios compare quantities of the same kind.

WORKED Example 1

Change the scale 1 cm ⇔ 10 km into a ratio.

THINK	WRITE
1 Write the scale.	1 cm ⇔ 10 km
2 Convert km to cm so that the units are the same.	
(a) To change kilometres to metres multiply by 1000.	⇔ 10 × 1000 m ⇔ 10 000 m
(b) To change metres to centimetres multiply by 100.	⇔ 10 000 × 100 cm ⇔ 1 000 000 cm
3 State what 1 cm represents on the map.	1 cm on the map stands for 1 000 000 cm.
4 Write the ratio.	1 : 1 000 000

WORKED Example 2

Find the distance represented by 1 cm on a map if the scale is written as 1 : 2000.

THINK	WRITE
1 Write the ratio.	1 : 2000
2 State what the ratio stands for. *Note:* The real distance is 2000 times larger than the distance on the map.	1 cm ⇔ 2000 cm ⇔ 2000 ÷ 100 m
3 Convert cm to m so that the value is simplified. To change centimetres to metres divide by 100.	⇔ 20 m
4 Give the answer.	The distance represented by 1 cm on the map is 20 m.

WORKED Example 3

A map is to be drawn with a scale showing that 1 cm represents 80 km. If the real distance from London to Paris is 344 km, how far apart will they be on the map?

THINK	WRITE
1 Write the scale.	1 cm ⇔ 80 km
2 Determine how many lots of 80 km there are in 344 km.	⇔ $\frac{344}{80}$ ⇔ 4.3
3 Write the ratio.	1 : 4.3
4 Give the answer.	The distance on the map is 4.3 cm.

remember

1. On a map, the scale '1 cm ⇔ 500 m' means that 1 cm on the map stands for 500 m of actual distance.
2. By making the units the same, a scale can be written as a ratio. The ratio 1 : 50 000 means that 1 cm on the diagram stands for 50 000 cm (500 m) of actual distance.
3. Ratios compare quantities of the same kind.

EXERCISE 12A Interpreting maps — scale

WORKED Example 1

SkillSHEET 12.1 Converting units of length

1 Change the given scale into a ratio in each of the following.

a 1 cm ⇔ 1 km b 1 cm ⇔ 20 km c 1 cm ⇔ 50 km
d 1 cm ⇔ 400 m e 1 cm represents 100 km f 1 cm represents 10 cm
g 1 cm ⇔ 1 m h 1 cm ⇔ 2000 km i 1 cm ⇔ 450 m

WORKED Example 2

SkillSHEET 12.2 Multiplying and dividing by powers of 10

2 Find the distance represented by 1 cm on a map if the scale is written as:

a 1 : 50 b 1 : 5000 c 1 : 200
d 1 : 50 000 e 1 : 1 000 000 f 1 : 40 000
g 1 : 30 000 h 1 : 800 000 i 1 : 10 000

Mathcad Map scales

3 **multiple choice**

A map has a scale of 1 cm ⇔ 5 km.

a 7 cm on the map will represent:

A 35 cm B 12 km C 1200 km D 35 km E 3.5 km

b 5.6 cm on the map will represent:

A 28 km B 25.6 km C 560 km D 28 cm E 2.8 km

EXCEL Spreadsheet Map scales

EXCEL Spreadsheet Map scales (DIY)

4 **multiple choice**

A map has a scale of 1 : 20 000.

a 3 cm on the map will represent:

A 60 000 km B 6 km C 6 m D 600 m E 60 km

b 5.9 cm on the map will represent:

A 5.9 km B 1.18 km C 590 m D 118 000 km E 118 km

WORKED Example 3

5 A map is to be drawn with a scale showing that 1 cm represents 50 km. If the actual distance from Adelaide to Port Augusta is 140 km, how far apart will they be on the map?

6 Complete this table for a map drawn to the scale 1 cm ⇔ 500 km.

		Actual distance	Distance on map
a	Melbourne–Perth	2700 km	
b	Sydney–Canberra	220 km	
c	Darwin–Alice Springs	1640 km	
d	Brisbane–Gold Coast	77 km	
e	Adelaide–Hobart	1150 km	

7 This is a diagram of a snooker table drawn using the scale 1 : 40. Find:

a the real length and width of the table

b the radius of the semicircle.

8 This is a diagram of a 'singles' tennis court drawn using the scale 1 : 250.

a Find the actual length and width of the tennis court.

b What is the total length of all the lines?

9 Here are 3 separate maps of well-known resort islands in the Pacific Ocean.

NEW CALEDONIA

TAHITI

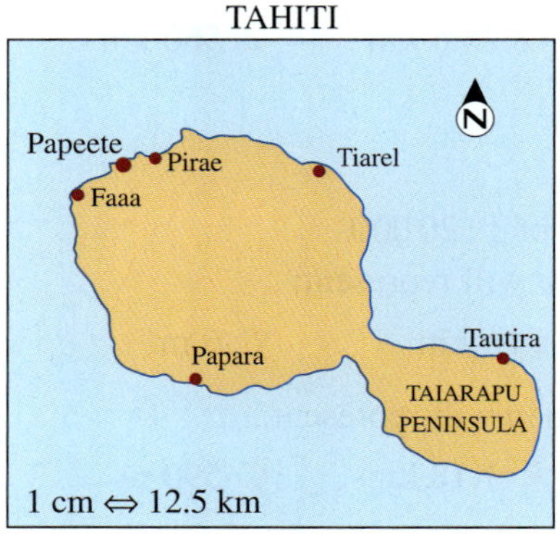

OAHU

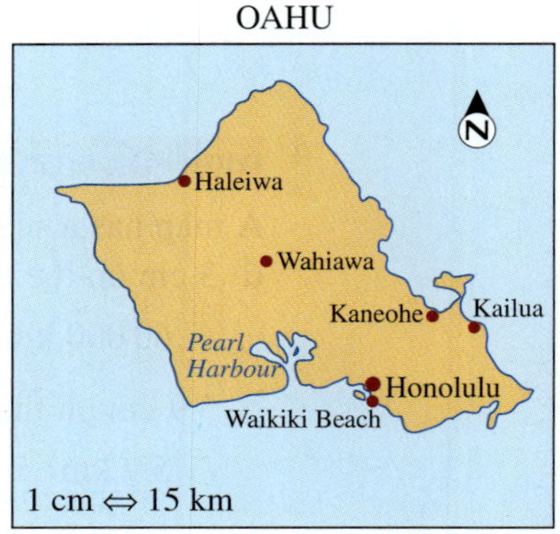

Find the distance from:

a Kone to Noumea **b** Papeete to Papara **c** Honolulu to Haleiwa.

10 This map shows a portion of the United States of America, drawn to a scale of 1 : 32 000 000.

By first measuring the distance on the map, find the approximate distance between these cities.

a Seattle and San Diego
b Houston and Atlanta
c New York and Los Angeles
d Nashville and Memphis

11 Shown below is a draftsman's rough sketch of the floor marking on a squash court.

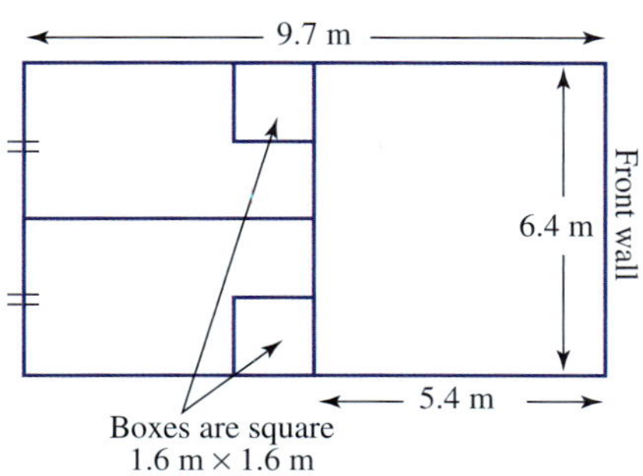

Carefully make a ruled plan using a scale of 1 cm ⇔ 1 m.

12 The dimensions of a netball court are roughly shown on the diagram at right. Using a scale of 1 cm ⇔ 3 m, draw a neat, accurately scaled diagram of the court. Carefully make a ruled plan using a scale of 1 cm ⇔ 1 m.

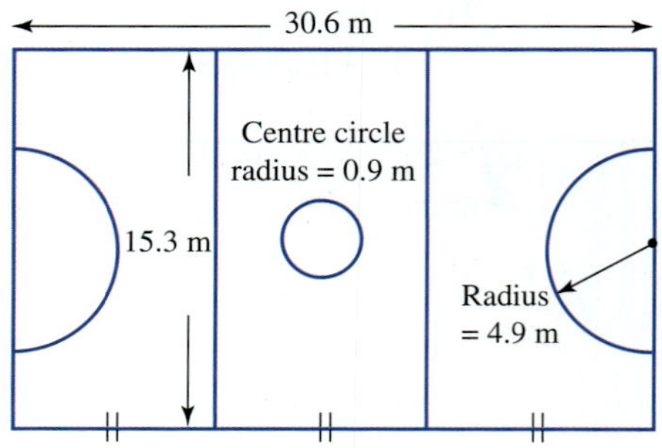

13 Choose a suitable scale, and make a scale drawing of a table tennis table, given the information on this photograph.

14 Choose a suitable scale and make an accurate scale drawing of a world class soccer ground with the dimensions shown on the plan below.

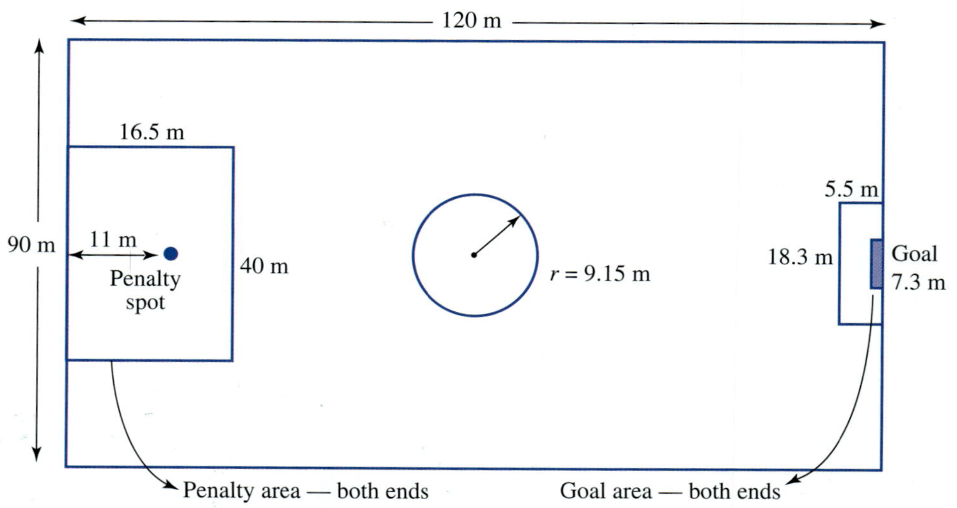

Did you hear about the boy who sat under a cow?

Find the distances represented by the map lengths given with each scale. Each distance and its letter give the puzzle code.

1 : 5000
2 cm on a map will represent:
A

1 : 20 000
5 cm on a map will represent:
D

1 : 400
15 cm on a map will represent:
E

1 : 10 000
4 cm on a map will represent:
G

1 : 100
8 cm on a map will represent:
H

1 : 2000
4 cm on a map will represent:
N

1 : 12 500
4 cm on a map will represent:
O

1 : 80
5 cm on a map will represent:
P

1 : 200
5 cm on a map will represent:
T

1 : 1000
60 cm on a map will represent:
A

1 : 500
8 cm on a map will represent:
E

1 : 1 000 000
2 cm on a map will represent:
H

1 : 40 000
5 cm on a map will represent:
O

1 : 500 000
2 cm on a map will represent:
T

1 : 150
4 cm on a map will represent:
A

1 : 8000
50 cm on a map will represent:
E

1 : 20 000
1 cm on a map will represent:
H

1 : 250 000
2 cm on a map will represent:
T

200 m	60 m	400 m	500 m	5 km	100 m	4 m	6 m	10 km
2 km	80 m	10 m	8 m	40 m	20 km	4 km	600 m	1 km

Location on Earth

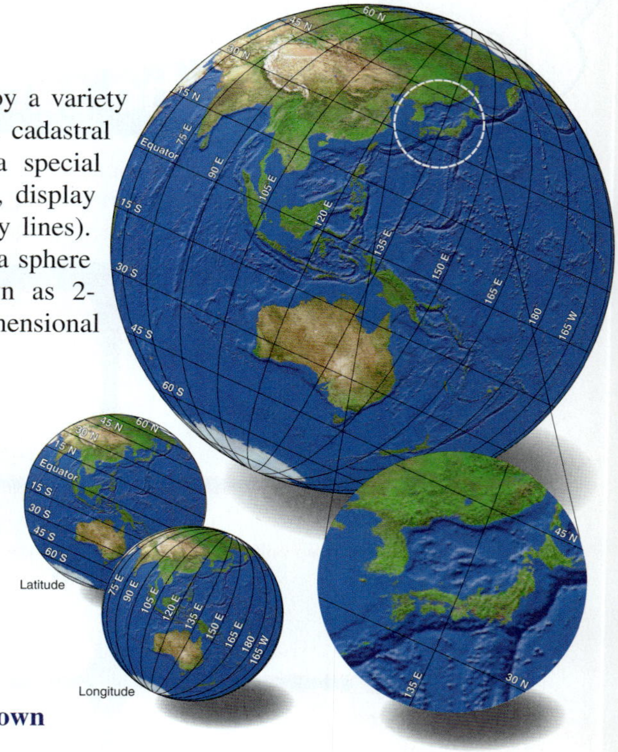

Features on the Earth are displayed by a variety of maps — topographic, contour and cadastral to name a few. Each type serves a special purpose (such as to identify position, display geographic features or show boundary lines). The shape of the Earth approximates a sphere and these maps are generally drawn as 2-dimensional representations of 3-dimensional features.

Latitude and longitude

The Earth's axis is the line through the centre of the Earth joining the North Pole and the South Pole. The Equator is an imaginary line running around the circumference of the Earth midway between the poles. It is assigned a value of 0° latitude.

All other lines running around the Earth parallel to the Equator are known as parallels of latitude.

The North Pole is 90°N and the South Pole is 90°S of the Equator.

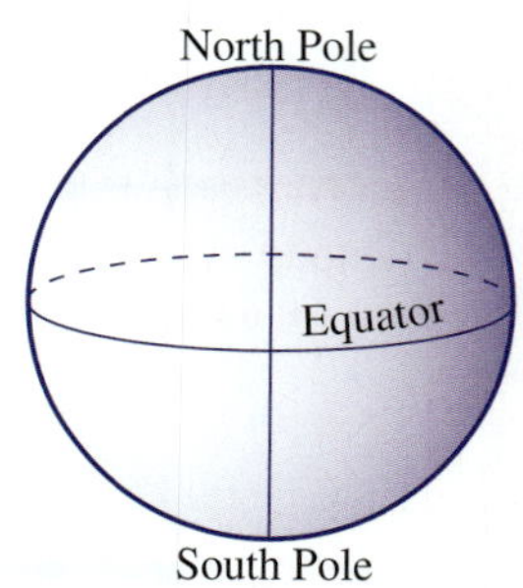

Lines running on the surface of the Earth from the North Pole directly to the South Pole are known as meridians of longitude.

The line of reference is the Greenwich Meridian (0°) which passes through Greenwich, a suburb of London. Places on the Earth can then be described as either east or west of the Greenwich Meridian. The **meridian** of longitude opposite the Greenwich Meridian is the International Date Line. It is 180° either east or west of the Greenwich Meridian.

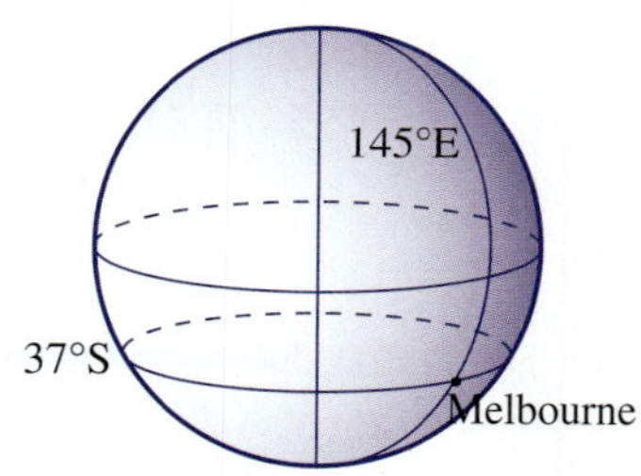

Any place on the Earth can be located by a pair of coordinates. The first coordinate represents its parallel of latitude and the second its meridian of longitude. For example, the coordinates of Melbourne are 37°S 145°E, meaning that Melbourne lies 37° south of the Equator and 145° east of the Greenwich Meridian.

The world map is generally represented in two dimensions, showing both parallels of latitude and meridians of longitude.

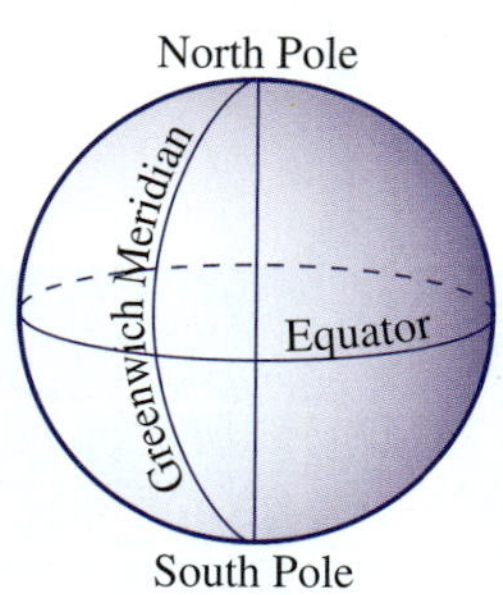

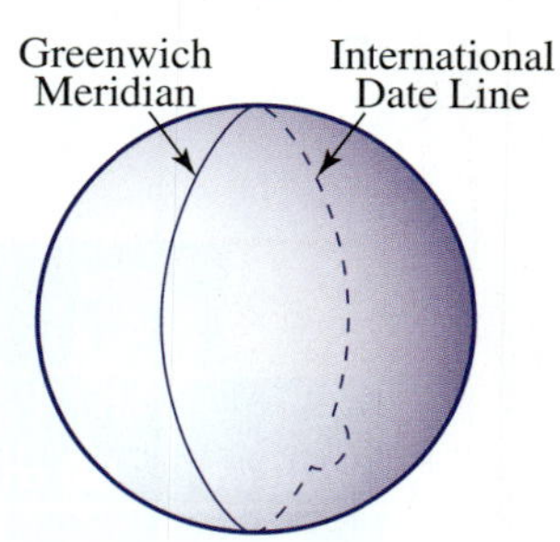

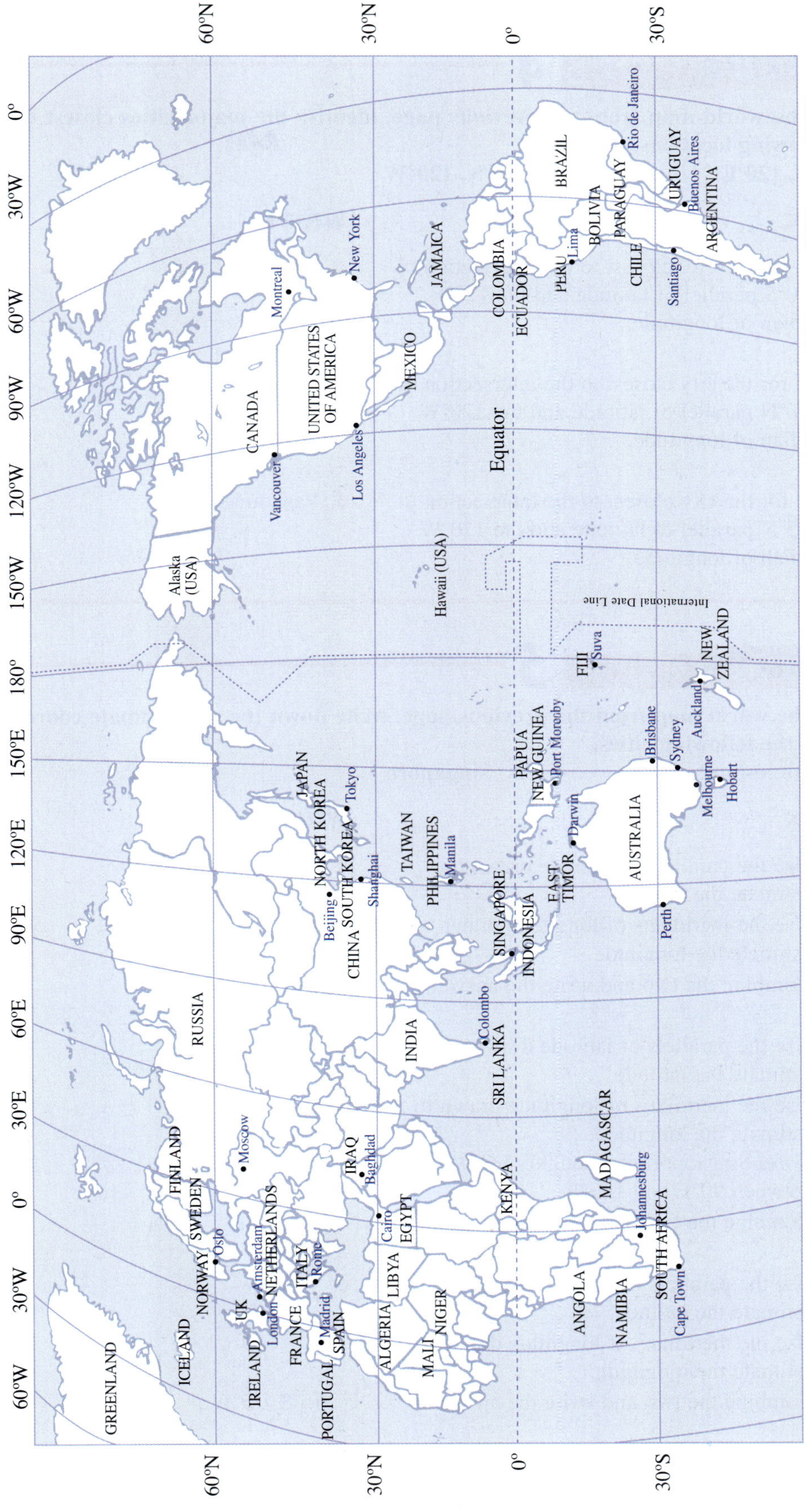
60°N
30°N
0°
30°S
60°W
30°W
0°
30°E
60°E
90°E
120°E
150°E
180°
150°W
120°W
90°W
60°W
30°W
0°
Equator
International Date Line
GREENLAND
ICELAND
IRELAND
UK
London
NORWAY
SWEDEN
FINLAND
Oslo
Amsterdam
NETHERLANDS
FRANCE
PORTUGAL
SPAIN
Madrid
ITALY
Rome
Moscow
RUSSIA
IRAQ
Baghdad
Cairo
EGYPT
LIBYA
ALGERIA
MALI
NIGER
KENYA
ANGOLA
NAMIBIA
SOUTH AFRICA
Cape Town
Johannesburg
MADAGASCAR
INDIA
SRI LANKA
Colombo
CHINA
Beijing
SOUTH KOREA
NORTH KOREA
Shanghai
JAPAN
Tokyo
TAIWAN
PHILIPPINES
Manila
SINGAPORE
INDONESIA
EAST TIMOR
Darwin
PAPUA NEW GUINEA
Port Moresby
AUSTRALIA
Perth
Brisbane
Sydney
Melbourne
Hobart
Auckland
NEW ZEALAND
FIJI
Suva
Hawaii (USA)
Alaska (USA)
CANADA
Vancouver
Montreal
UNITED STATES OF AMERICA
Los Angeles
New York
MEXICO
JAMAICA
COLOMBIA
ECUADOR
PERU
Lima
BRAZIL
BOLIVIA
PARAGUAY
CHILE
Santiago
URUGUAY
Buenos Aires
ARGENTINA
Rio de Janeiro

WORKED Example 4

Using the world map from the previous page, identify the major cities closest to each of the following locations.

a **30°S, 120°E** **b** **30°N, 120°W** **c** **45°N, 120°W**

THINK	WRITE
a Look for the city closest to the intersection of the 30°S parallel of latitude and the 120°E meridian of longitude.	**a** Perth
b Look for the city closest to the intersection of the 30°N parallel of latitude and the 120°W meridian of longitude.	**b** Los Angeles
c Look for the city closest to the intersection of the 45°N parallel of latitude and the 120°W meridian of longitude.	**c** Vancouver

WORKED Example 5

Using the world map from the previous page, write down the approximate coordinates of each of the following cities.

a **Johannesburg** **b** **Singapore** **c** **Lima**

THINK	WRITE
a 1 Use the parallels of latitude drawn to estimate the latitude.	**a**
2 Use the meridians of longitude drawn to estimate the longitude.	
3 Combine the two and write the answer.	30°S, 30°E
b 1 Use the parallels of latitude drawn to estimate the latitude.	**b**
2 Use the meridians of longitude drawn to estimate the longitude. *Note:* Singaopore lies roughly halfway between 90°E and 120°E.	
3 Combine the two and write the answer.	0°, 105°E
c 1 Use the parallels of latitude drawn to estimate the latitude.	**c**
2 Use the meridians of longitude drawn to estimate the longitude.	
3 Combine the two and write the answer.	15°S, 75°W

COMMUNICATION Parallels and meridians

Some parallels of latitude and meridians of longitude are of particular significance. Research to find answers to the following.

1 Identify the parallels of latitude for the following.

a Arctic Circle
b Antarctic Circle
c Tropic of Cancer
d Tropic of Capricorn

2 Two meridians of longitude are particularly important — the Greenwich Meridian and the International Date Line.

a How did the Greenwich Meridian come to be chosen as the reference meridian? What difference would it make if it were relocated to run through another city, say, Melbourne?

b Why is the International Date Line not a smooth continuous curve? What is the significance of the line?

Grid references

Smaller areas of land are generally displayed by using a grid reference map. Because these references refer to a square region rather than to points on a map, they are also known as *area references*.

Maps of this type are usually found in a street directory or an atlas. The reference for each square is given by two coordinates — a letter and a number.

WORKED Example 6

Joe is holidaying in Brisbane and has planned to visit various sites. Using the map of the Gold Coast on page 514, give the grid reference for each of the following sites Joe plans to visit.

a Southport **b Movie World** **c Bond University**

THINK	WRITE
a 1 Locate Southport.	**a**
2 Read the horizontal and vertical coordinates.	
3 Write the answer.	C5
b 1 Locate Movie World.	**b**
2 Read the horizontal and vertical coordinates.	
3 Write the answer.	B4
c 1 Locate Bond University.	**c**
2 Read the horizontal and vertical coordinates.	
3 Write the answer.	C6

Scale 1:258 600
0
10
Kilometres
A
B
C
D
E
1
2
3
4
5
6
7
8
North
Stradbroke
Island
Long
Island
Russell
Island
Lagoon
Island
Logan
River
Creek
Beenleigh
Beenleigh Rum
Distillery
Stapylton
Windaroo
Sandy
Albert
Yatala
Darlington Park
International
Raceway
Steiglitz
Rocky Point
Sugar Mill
Woongoolba
Cobby
Cobby
Island
Short
Island
Tabby
Tabby
Island
Eden
Island
Crusoe
Island
Jumpinpin
Kangaroo
Island
Southern
Moreton Bay Islands
National Park
Norwell
Jacobs
Well
Ormeau
Pimpama
Kingsholme
PACIFIC
Ck
26
Pimpama
River
Woogoompah
Island
South
Conservation
Stradbroke
Park
Island
Coral
Sea
Willow Vale
Coomera
Tamborine
National Park
Hotham
Sanctuary
Cove
Coomera
Island
Dreamworld
Wongawallan
Wonga wallan
Hope Island
Creek
River
Oxenford
Paradise Point
Hollywell
22
Upper Coomera
Coombabah
Movie World
Wet'nWild
Water World
HIGHWAY
Helensvale
Runaway Bay
Cable Ski World
Eagle Heights
Heritage
Centre
GOLD COAST
Biggera Waters
The Broadwater
North
Tamborine
Guanaba
Gaven
Arundel
11
Labrador
The Spit
Maudsland
Sea World
Southport
Guanaba
Mt Tamborine
Clagiraba
Mt Nathan
Ernest
NERANG RD
Main Beach
SOUTHPORT
Nerang
Surfers
Paradise
Land
NERANG
BEAUDESERT
River
RD
Carrara
Bundall
BROADBEACH
Broadbeach
Jupiters
Casino
Advancetown
Gilston
PACIFIC
Warfare
Mermaid Beach
27
Hinze
Dam
Tallai
15
Merrimac
Mermaid
Waters
Centre
Creek
Mudgeeraba
Robina
Miami
Advancetown
Lake
Bond
University
Burleigh Heads
Burleigh
Waters
War
Museum
David Fleay Wildlife Park
Bonsai World
HIGHWAY
Palm Beach
Beechmont
Nerang
Neranwood
Bonogin
Elanora
Currumbin
Currumbin Bird Sanctuary
Numinbah
Valley
Mudgeeraba
Bonogin
Tallebudgera
Currumbin
Waters
Tugun
Bilinga
Coolangatta
Coolangatta
Airport
Tallebudgera
Valley
Queensland
Olsons
Bird Gardens
Tallebudgera
Cobaki
Broadwater
Tweed
Heads
Lamington
Binna Burra
Lodge
National
Park
Springbrook
National
Park
Little
Nerang
Piggabeen
N.S.W.
Tweed Heads
West
Currumbin
Rock Pool
Currumbin
Currumbin
Valley
Cobaki
Bilambil
Heights
Terranora
Broadwater
Fingal
Head
COPYRIGHT © UNIVERSAL PRESS PTY LTD (PUBLISHER) 2000

WORKED Example 7

Name the major feature located at each of the following grid references in the Gold Coast map.

a D3 **b** E8 **c** B3

THINK	WRITE
a 1 Look for the major feature in column D and row 3.	**a**
2 Write the answer.	South Stradbroke Island
b 1 Look for the major feature in column E and row 8.	**b**
2 Write the answer.	Tweed Heads
c 1 Look for the major feature in column B and row 3.	**c**
2 Write the answer.	Dreamworld

remember

1. The location of a point on the Earth can be identified by its parallel of latitude and its meridian of longitude.
2. Parallels of latitude run parallel to the Equator.
3. Meridians of longitude run between the North Pole and the South Pole.
4. The Equator is 0° latitude. The North Pole is 90°N and the South Pole is 90°S.
5. The Greenwich Meridian is 0° longitude. The International Date Line is 180°E or W.
6. Each square in a grid reference is defined by two coordinates — a letter and a number.

EXERCISE 12B Location on Earth

Use the world map on page 511 to answer questions 1 and 2.

WORKED Example 4

1 Write down the name of the city closest to each of the following pairs of coordinates.

a 30°N, 30°E **b** 30°N, 120°E **c** 15°S, 135°E
d 45°N, 75°W **e** 50°N, 0° **f** 37°S, 175°E
g 35°N, 140°E **h** 40°N, 115°E **i** 20°S, 45°W

WORKED Example 5

2 State the approximate latitude and longitude of each of the following major cities or islands.

a Melbourne **b** New York **c** Jamaica **d** Johannesburg
e Rome **f** Buenos Aires **g** Baghdad **h** Moscow
i Singapore **j** Suva

WORKED Example 6

3 Using the grid map of Victoria and its surrounding states on the next page, give the grid reference for each of the following.

a Mt Kosciuszko **b** Port Campbell **c** Lake Tyrrell
d Wilsons Promontory **e** The Grampians **f** Phillip Island
g Mt Hotham **h** Portland Bay **i** Ninety Mile Beach

WORKED Example 7

4 Name the major city/town(s) located at each of the following grid references in the map of Victoria and its surrounding states.

a C1 b B2 c C2 d B1 e A1
f B3 g D2 h A2 i D3

10 QUICK QUESTIONS 1

1 Change the scale 1 cm ⇔ 1 m into a ratio.

2 What distance is represented by 1 cm on a map if the scale is written as 1 : 250 000?

3 A map has a scale showing that 1 cm represents 1 km. If two towns are 500 m apart, how far apart will they be on the map?

4 A map has a scale of 1 : 50 000. A measurement of 3.5 cm on the map will represent what distance?

Use the world map on page 511 to answer questions **5** to **9**.

5 What city in the world is closest to the location 15°S, 180°E?

6 In which hemisphere of the world does Australia lie?

7 Does Italy lie east or west of the UK?

8 Does New Zealand lie east or west of Australia?

9 State the approximate latitude and longitude of:

a Montreal b Auckland.

10 a Use the Gold Coast grid map on page 514 to give the coordinates of Sanctuary Cove.

b Which major town is located in grid reference B5 of the Gold Coast grid map on page 514?

Maps, bearings and plans

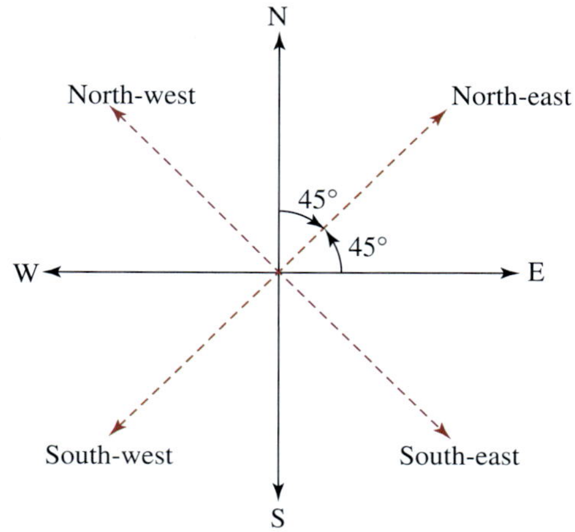

Bearings

North, south, east and west are called the cardinal points of the compass. By halving the angle between north and east we can create the direction north-east, and other directions can be created in a similar manner.

The directions or bearings marked on the diagram at right are all separated by 45° angles. Another way of naming directions is the '*true bearing*' method, that records all bearings as angles between 0° and 360°. This is how it works:

Imagine you are standing at point P, and you are facing north. By turning clockwise through 45° you will face north-east. We say that the *true bearing* of north-east is 45° and we write it as 45° T. You can see that the true bearing for due east is 90° T, due south is 180° T, and due west is 270° T.

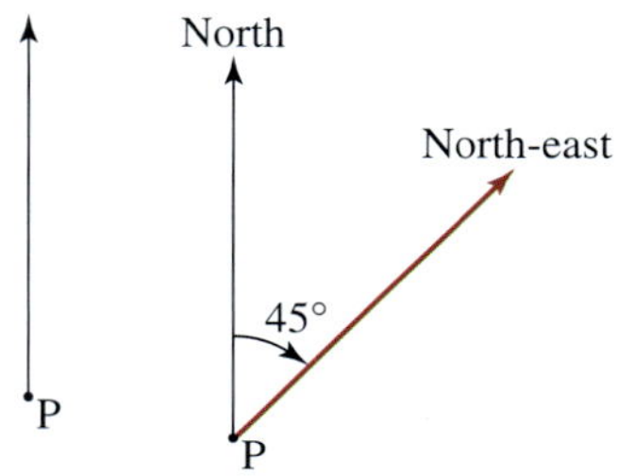

To avoid confusion, true bearings are often written with 3 digits. For example, 2° T is shown as 002° T and 91° T is shown as 091° T.

WORKED Example 8

Draw a diagram to show a true bearing of 100° T.

THINK

1. Start with a point, P, and show the direction of north.
2. Using a protractor, measure an angle of 100° clockwise from the line pointing north.
   (a) Line up the centre of the protractor's baseline with the line pointing north.
   Note: You will need to rotate the paper.
   (b) Measure an angle of 100° (using the outside scale) and mark this on the diagram.
   (c) Draw a line extending from the point P, to the marking.
   (d) Label the angle on the diagram.

DRAW

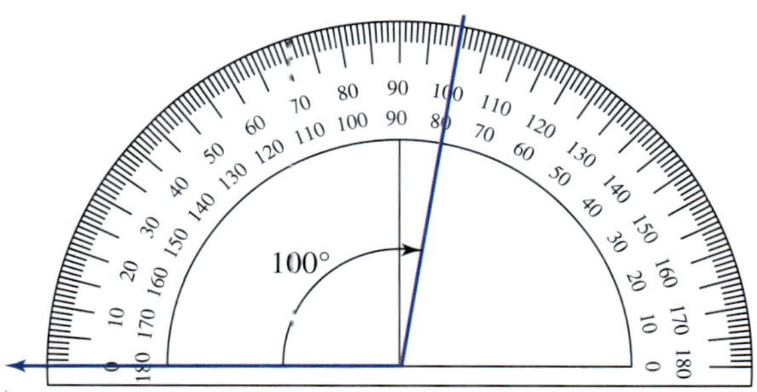

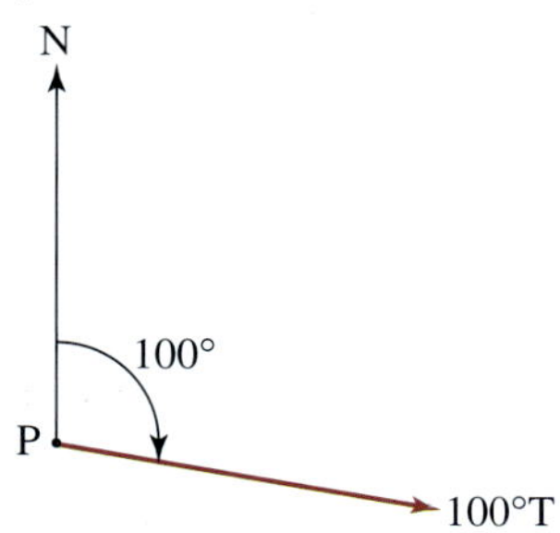

WORKED Example 9

Write south-west (SW) as a true bearing.

THINK	WRITE/DRAW
1 Show south-west on a diagram. *Note:* South-west is exactly halfway between south and west.	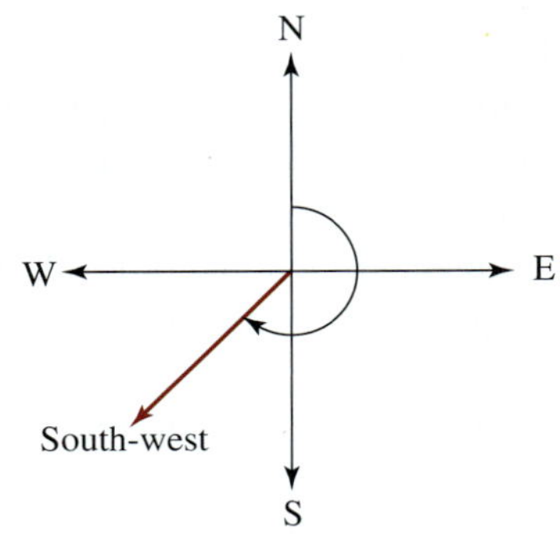
2 Work out the angle from north moving in a clockwise direction. The angle required is 90° + 90° + 45° = 225°.	SW = 225° T

WORKED Example 10

Using the map shown at right, find the bearing from Lord Howe Island to:

a **Norfolk Island**

b **Brisbane.**

THINK	WRITE/DRAW
a 1 Trace the map and mark in Lord Howe Island and Norfolk Island.	**a**
2 From Lord Howe Island rule a line north, and another line through Norfolk Island to create an angle.	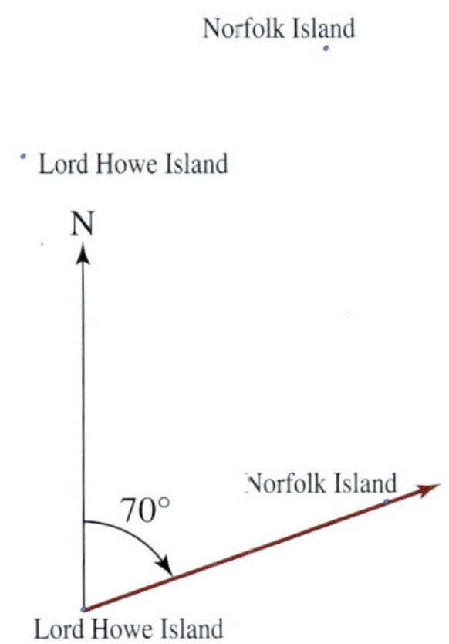
3 Measure this angle clockwise from north and give the answer.	The bearing is 070° T.
b 1 Trace the map and mark in Lord Howe Island and Brisbane.	**b** 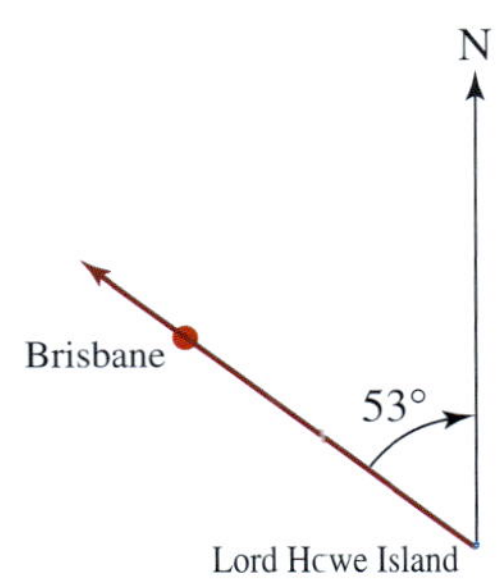
2 From Lord Howe Island rule a line north, and another line through Brisbane to create an angle.	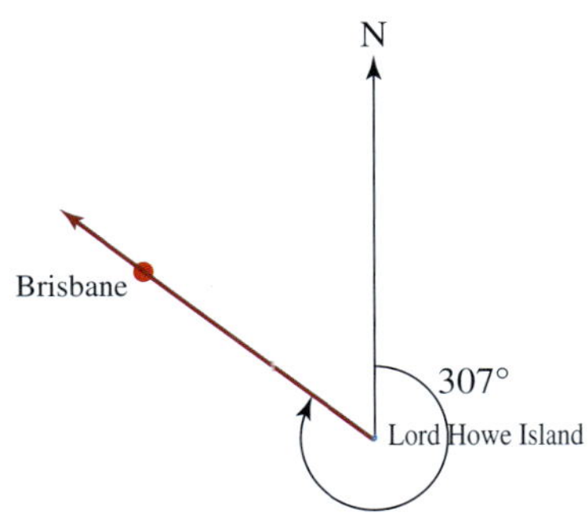
3 To measure this angle clockwise from north, first measure the acute angle, then subtract it from 360°. Give the answer.	Bearing = 360° − 53° = 307° The bearing is 307° T.

Maps

Different types of maps are designed for specific purposes. We have looked at geographic maps. These show only the important features of the area they cover, and are generally drawn to a very small scale. We will now consider topographic maps and synoptic maps.

Topographic maps

Topographic maps display the elevation of an area — mountains, plateaux, valleys, ridges and so on. These shapes are shown by **contour lines** that join points at the same height above sea level. If the contour lines are very close together, this indicates that the land is falling or rising more steeply than if the contours are spaced further apart. The elevation measurements are usually in metres.

WORKED Example 11

Use the topographic map shown here to answer the following questions.

a Give the contour interval.

b What is the height of point E?

c Compare the height of point C with that of point D.

d Name the steepest slope on the map.

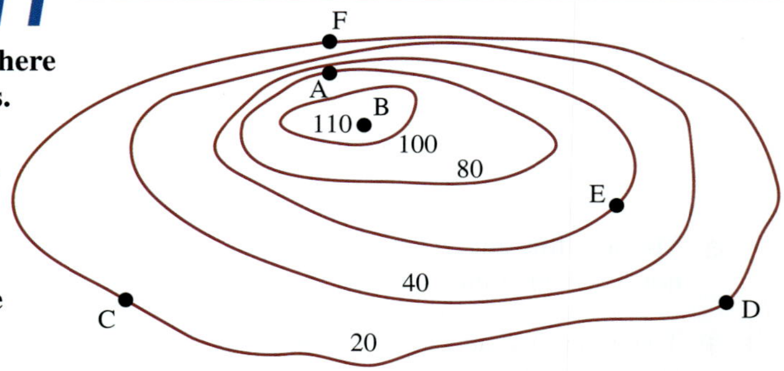

THINK	WRITE
a The contour interval is the vertical distance between adjacent contours. Calculate the difference between two adjacent contours.	**a** The contour interval is 20 m.
b Point E lies on the contour between the 40-m level and the 80-m level. Therefore add the contour interval to the smaller value (or subtract the contour interval from the larger value).	**b** The point E is 60 m above sea level.
c Consider the contours on which points C and D lie and answer the question.	**c** Points C and D lie on the same contour, therefore they are each 20 m above sea level.
d Look for a section on the map where the contours are close together. *Note:* The steepest slope occurs where the contours are closest together on the map.	**d** The steepest slope occurs from F rising to A on the map. Here, the contours are closest together.

Synoptic maps

These maps look similar to topographic maps. The contours of topographic maps connect points with the same elevation; while isobars on synoptic maps connect places of equal air pressure. These maps are used to show the weather conditions of an area. The air pressure is measured in millibars; a high air pressure indicates fine weather, while low air pressure indicates unstable weather. Extreme lows can develop into cyclones. When two air masses with similar temperature and moisture meet, they are indicated on the map as a line with triangles.

WORKED Example 12

Use the synoptic map shown to answer the following questions.

a What type of weather is likely in Perth?

b Indicate which region might expect rain.

c Why is it likely to be fine in most of Queensland?

d What do the two lines with triangles represent on the map?

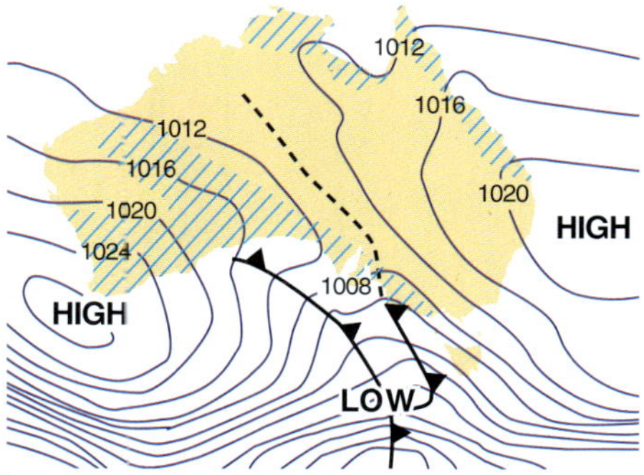

THINK	WRITE
a 1 Observe the map carefully and check whether it indicates high or low pressure around Perth.	**a**
2 Answer the question.	A high pressure around Perth indicates that the weather could be fine.
b 1 Observe the map carefully and check for a region indicating low pressure.	**b**
2 Answer the question.	A region of low pressure surrounds Tasmania, Victoria and the lower parts of South Australia. These areas could expect rain.
c 1 Observe the map carefully and consider the type of air pressure covering most of Queensland.	**c**
2 Answer the question.	A high air pressure covers most of Queensland. This is why the area could be fine.
d 1 Observe the map carefully and look for the lines with triangles.	**d**
2 Answer the question.	Two lines with triangles are in the region towards the southern part of Australia. They indicate that two sets of air masses with similar temperature and moisture are meeting.

Plans

It is essential to draw up a plan before commencing a major project. Plans involve financial commitments, time constraints and construction drawings. For Melbourne's 2006 Commonwealth Games a great deal of planning and coordination of projects was required.

Locations including the historic Melbourne Cricket Grounds (MCG), The Docklands Precinct, Melbourne Exhibition Centre, Melbourne Sports and Aquatic Centre (MSAC), the Rod Laver Arena in Melbourne Park and even the Melbourne Botanic Gardens will be used for various events. Consider the plan for the Melbourne 2006 Commonwealth Games venues on the following page.

WORKED Example 13

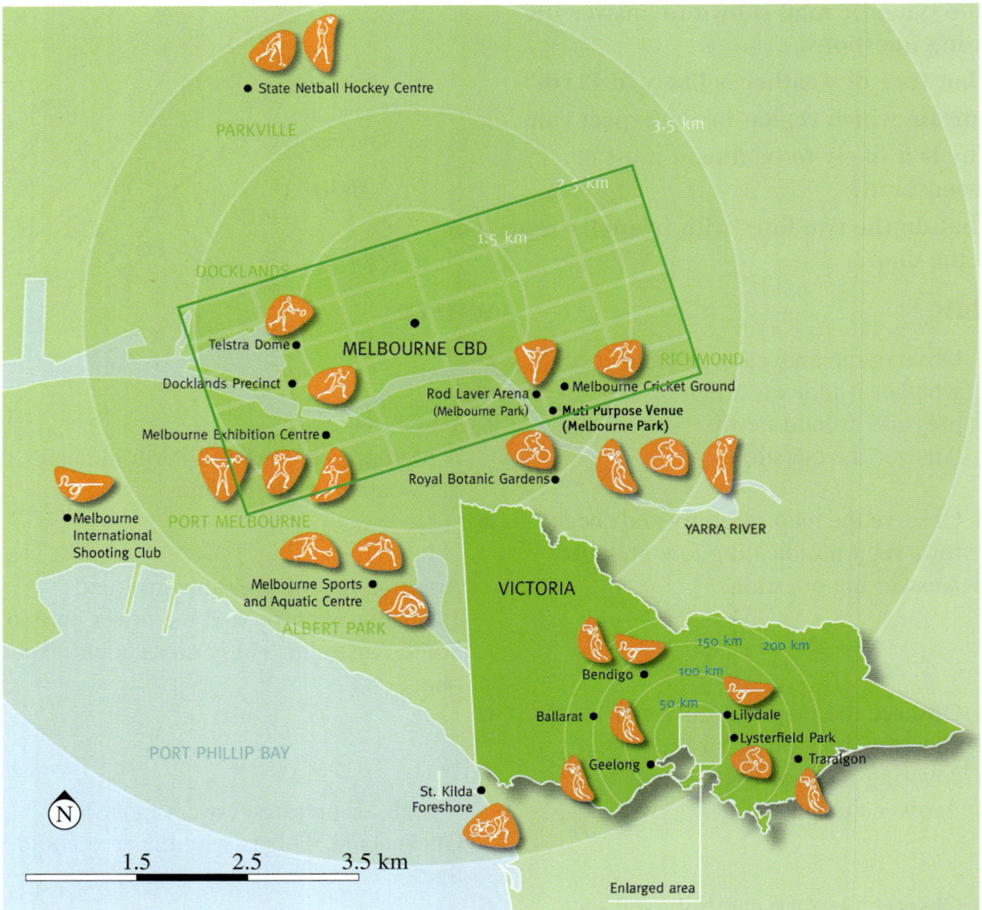

Use the enlarged area of the plan for the 2006 Commonwealth Games to answer the following questions.

a What is the scale of the plan?

b What is the approximate area of the Melbourne CBD and its surroundings represented by the green rectangle?

c In what direction is the Docklands Precinct from the Telstra Dome?

THINK	WRITE
a 1 Measure the scale legend.	**a**
2 Write the scale.	4.4 cm ⇔ 3.5 km
3 Convert kilometres to centimetres so that the units are the same. To change kilometres to centimetres multiply by 100 000.	4.4 cm ⇔ 3.5 × 100 000 cm 4.4 cm ⇔ 350 000 cm
4 Write the scale as a ratio.	4.4 : 350 000 44 : 3 500 000
5 Simplify the ratio.	11 : 875 000

THINK	WRITE
b 1 Measure the Melbourne CBD and its surroundings and specify values obtained.	**b** The Melbourne CBD and surrounding area has an approximate length of 6.3 cm and width of 3 cm.
2 Write the ratio and state what it represents.	The ratio 11 : 875 000 tells us that the real distance is approximately 79 545 (that is, $\frac{875\,000}{11}$) times larger than the distance on the map.
3 Convert each of the measurements taken to actual values.	Real length $= 6.3 \times \frac{875\,000}{11}$ cm $= 501\ 136.3636$ cm $= 5.01$ km Real width $= 3 \times \frac{875\,000}{11}$ cm $= 238\ 636.3636$ cm $= 2.39$ km
4 Calculate the area.	Area $= lw$ $= 5.01 \text{ km} \times 2.39 \text{ km}$ $= 11.9739 \text{ km}^2$
5 Answer the question in a sentence.	The approximate area of the Melbourne CBD and its surroundings is 11.9739 km^2.
c 1 Look for a compass direction on the map. 2 Locate the Docklands Precinct and answer the question. *Note:* The Docklands Precinct is below the Telstra Dome.	**c** The Docklands Precinct is towards the south of the Telstra Dome.

remember

1. The *true bearing* is given by the angle measured from north in a clockwise direction. It is written using three digits.
2. Topographic maps display the elevation of an area as shown by contour lines.
3. Synoptic maps are used to show the weather conditions of an area. Isobars on these maps connect places of equal air pressure.

12.3 SkillSHEET Angles around a circle (as in a compass)

12.4 SkillSHEET Constructing angles with a protractor

Cabri Geometry True bearings

EXERCISE 12C Maps, bearings and plans

WORKED Example 8

1 Draw a diagram to show the true bearing of each of the following.

a 10° T b 85° T c 120° T d 155° T
e 200° T f 240° T g 275° T h 350° T

WORKED Example 9

2 Write south-east (SE) as a true bearing.

3 **multiple choice**

a The direction due south could be written as:

A 180° T **B** 0° T **C** 90° T **D** 270° T **E** 360° T

b The direction SW could be written as:

A 45° T **B** 270° T **C** 135° T **D** 315° T **E** 225° T

c The direction NW could be written as:

A 270° T **B** 315° T **C** 45° T **D** 225° T **E** 135° T

d The direction NE could be written as:

A 130° T **B** 45° T **C** 145° T **D** 135° T **E** 315° T

4 For each of these diagrams, find the true bearings represented by the red arrow.

a

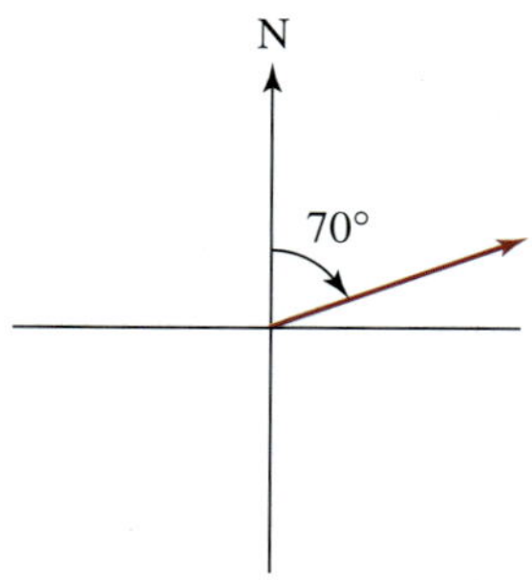

b

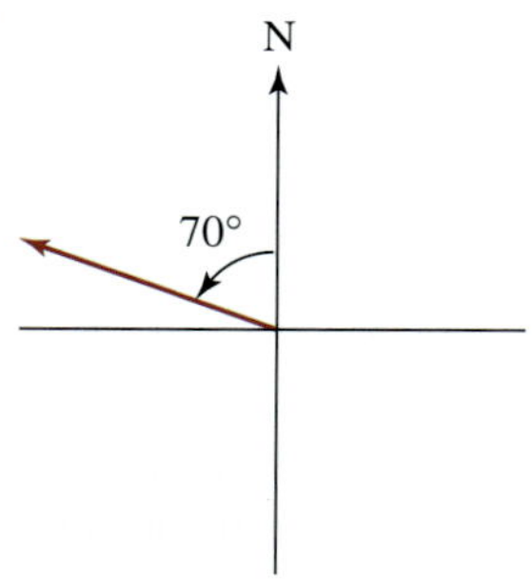

c

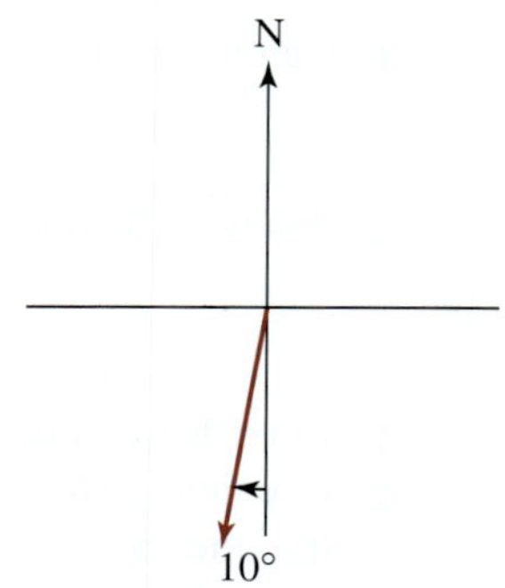

d

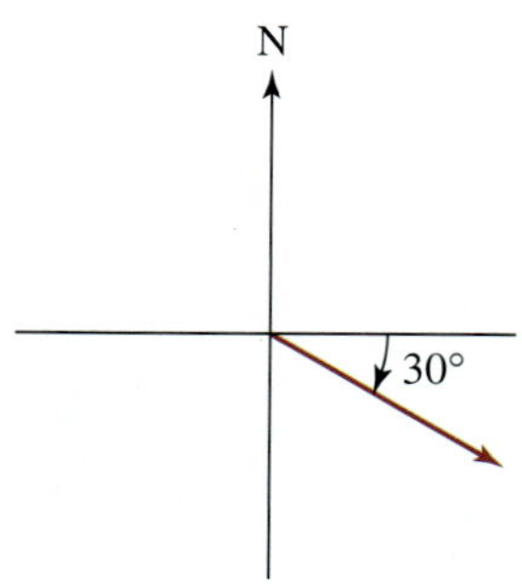

e

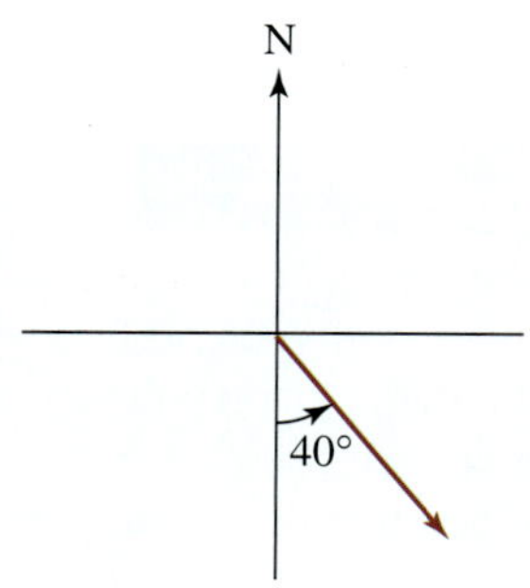

f

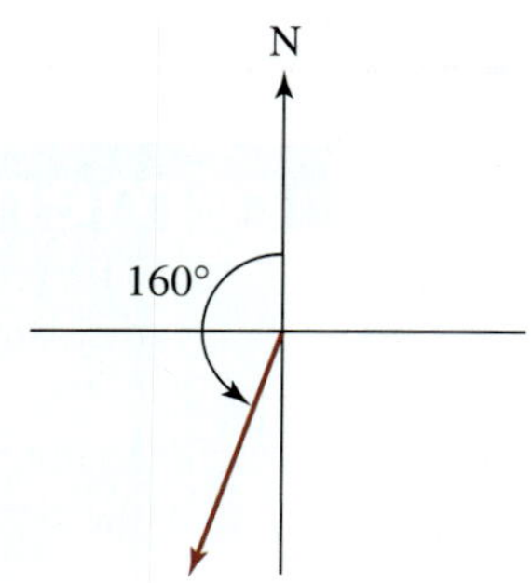

g

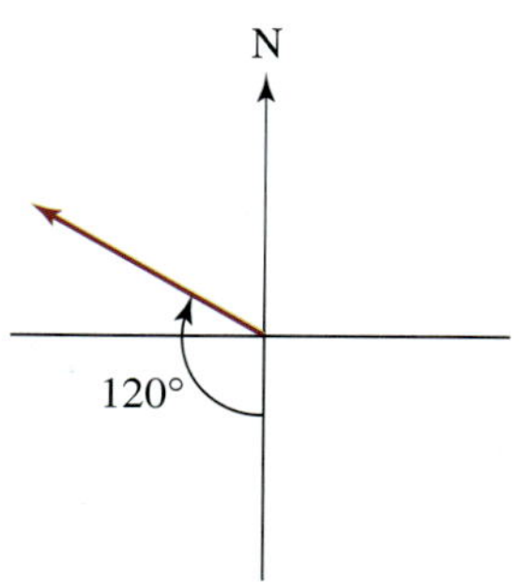

h

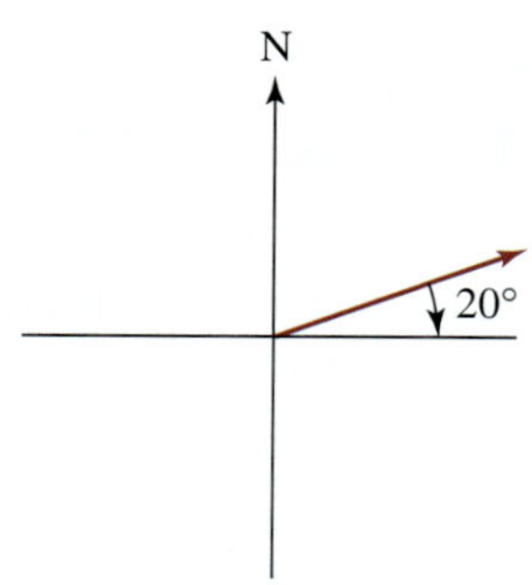

5 Captain Dinosaur, the last great pirate landed his treasure at Cutlass Cove, on Treasure Island, at the point marked X.

1 square length ⇔ 1 km

After landing he travelled 3 km south, then 2 km east, 4 km south, 5 km east and 6 km north, where he stopped to bury his treasure.

a Mark his trail on the map.

b How far was he from X when he stopped to bury his treasure?

c He returned directly to X. On which true bearing did he walk?

WORKED Example 10

6 Alice Springs is located close to the middle of Australia.

12.5 SkillSHEET Measuring angles with a protractor

a Using the map given below find the bearing from Alice Springs of Burketown.

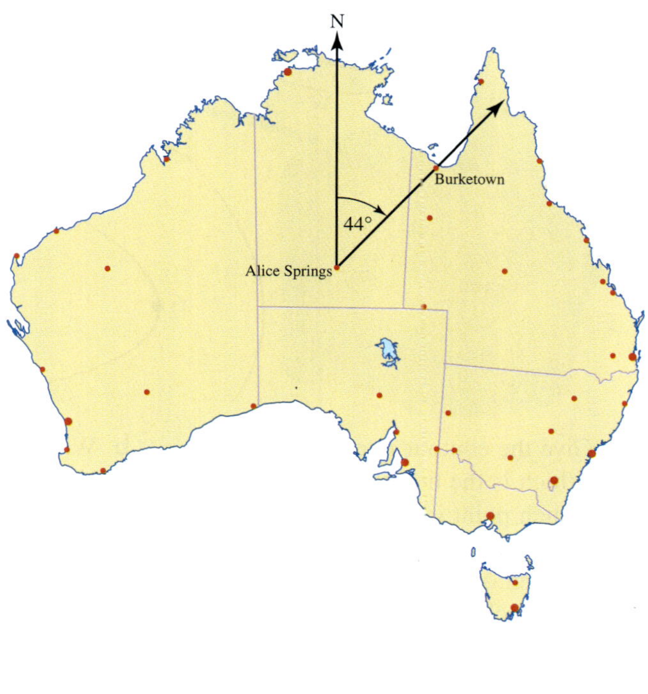

b All the towns listed in the table below lie on the coastline and their bearing from Alice Springs is given. Mark them on the map.

	Town	True bearing
i	Townsville	73° T
ii	Rockhampton	93° T
iii	Coffs Harbour	115° T
iv	Hobart (south coast of Tasmania)	152° T
v	Eucla	210° T
vi	Geraldton	250° T
vii	Dampier	278° T
viii	Derby	303° T
ix	Darwin	347° T

WORKED Example 11

7 Use this topographic map to answer the following questions.

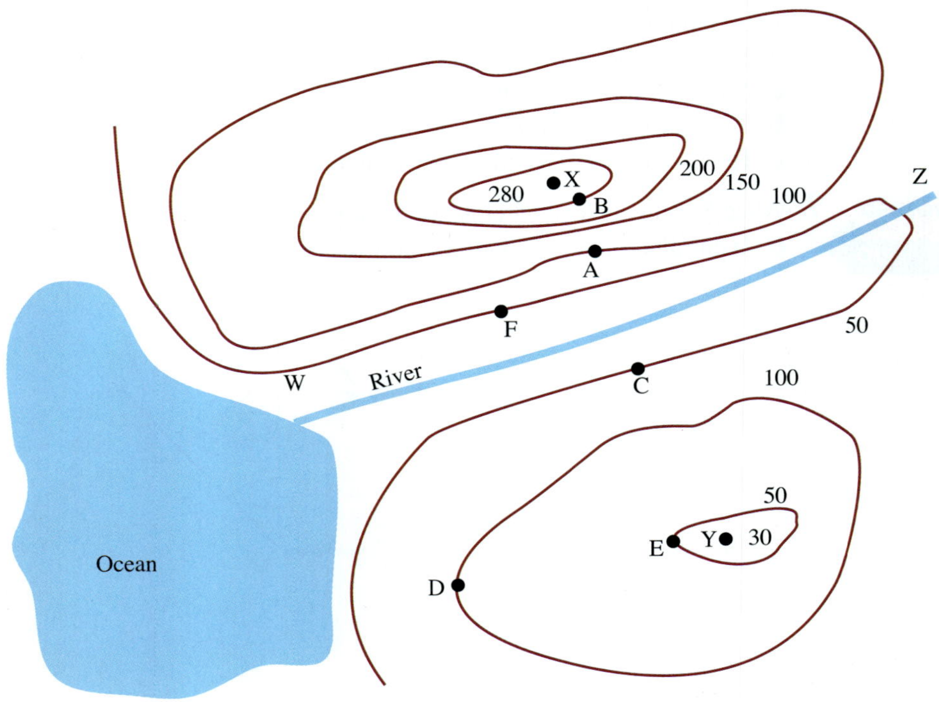

a Give the contour interval.

b What is the height of point F?

c Which is the highest point?

d Name the steepest slope.

e Which point is closest to the origin of the river?

f Describe the region DE.

g Describe the point Y.

h Give an overall description of the area, standing in a position facing the river with your back to the ocean.

8 Draw contour sketches to depict the topology of:
   a a mountain
   b a plateau
   c a cliff.

WORKED Example 12

9 Describe the weather throughout Australia as indicated by the synoptic map shown.

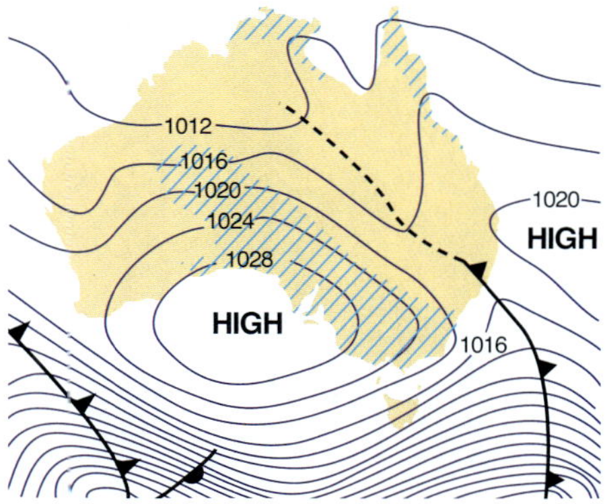

WORKED Example 13

10 Use the enlarged are of the plan of the 2006 Commonwealth Games site from page 522 to answer the following questions.
   a What is the true bearing of Rod Laver Arena from the Melbourne CBD?
   b In what direction is the Melbourne Exhibition Centre from the MCG?
   c How far is the Melbourne Sports and Aquatic Centre from the Royal Botanic Gardens?
   d How much further is the State Netball Hockey Centre than the St Kilda Foreshore from the Melbourne CBD?

WorkSHEET 12.1

COMMUNICATION

Our national capital

Study the aerial photograph of Canberra on the following page to answer these questions.

1 What is unique about the street patterns in Canberra?

2 What is the scale of the aerial photograph?

3 What direction is the War Memorial from Parliament House?

4 How far is the National Gallery from the High Court?

5 If the Prime Minister were to walk to work, what distance would he have to walk?

6 On which side of Lake Burley Griffin is the city centre?

7 What direction is Parliament House from the city centre?

8 What area does Parliament House occupy?

9 The Prime Minister wants to attract tourists to Canberra. Write a speech for him to deliver on television persuading visitors to spend time in our national capital.

Central Canberra

1. Australian National University
2. Australian War Memorial
3. Captain Cook Memorial Water Jet
4. Commonwealth Park
5. Carillon (belltower)
6. Blundell's Farmhouse
7. Lake Burley Griffin
8. National Library
9. National Science and Technology Centre
10. High Court
11. Australian National Gallery
12. Old Parliament House
13. Canberra Nature Park
14. Prime Minister's Lodge
15. Parliament House
16. Jewish Memorial Centre
17. Central Business District

Networks

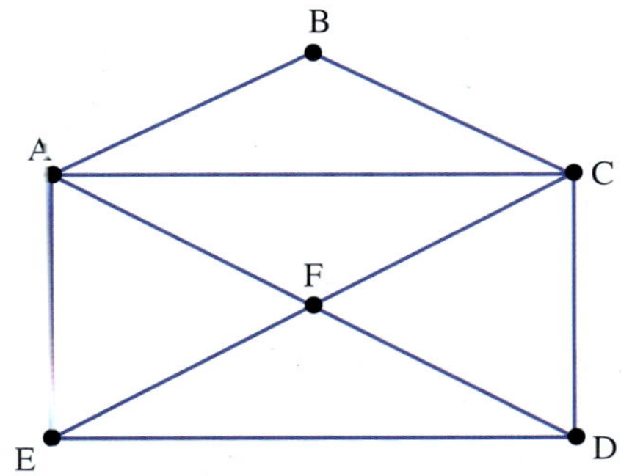

Have you ever tried to trace the following picture without lifting your pen or going over the same line twice?

The secret is to find the place to start. This picture can be traced, but only if you start at D or E. Try it!

The diagram shows an example of a **network**, which is made up of lines and dots called **edges** and **vertices**. The vertices or dots are labelled A, B, C, D, E, and F. (One dot is called a vertex.) The edges are the lines that join the vertices to one another. They can be straight lines, curves, or even loops that join a vertex to itself. Check the network above and you will see that it has 6 vertices and 10 edges. Because we can trace it without lifting the pen or going over an edge twice we say that the network is **traversable**.

A small change, like removing one edge, can mean that we are no longer able to trace over the network without lifting the pen. The network shown at right is no longer traversable.

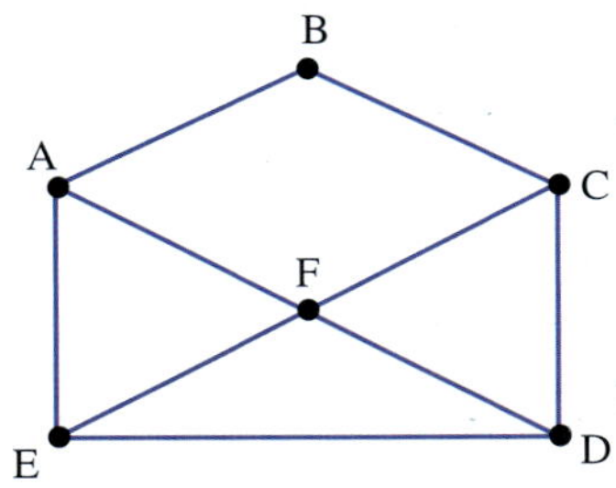

WORKED Example 14

For the networks drawn below:

i list the number of vertices

ii list the number of edges

iii state whether the network is traversable.

a

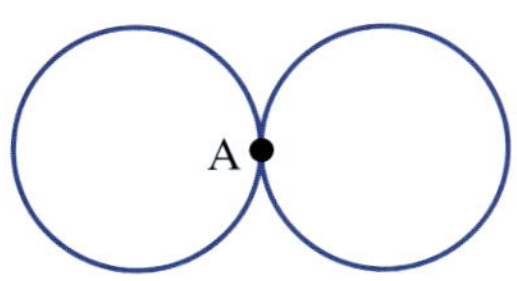

b

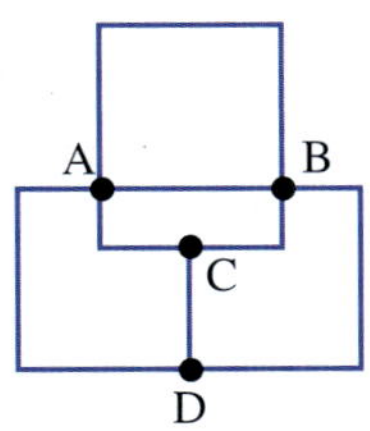

THINK	WRITE
a i Count the number of dots. Each dot corresponds to a vertex.	**a** i There is one vertex.
ii Count the number of lines that join to dots. Two curved lines join A to itself. Each line corresponds to an edge.	ii There are 2 edges.
iii Check whether the network can be traced without lifting your pen.	iii The network is traversable.

Continued over page

THINK	**WRITE**
b i Count the number of dots. There are 4 dots. Each dot corresponds to a vertex.	**b i** There are 4 vertices.
ii Count the number of lines that join to dots. Each line corresponds to an edge. Two lines join A and B. One line joins A and C. One line joins B and C. There are 3 more edges joining D.	**ii** There are 7 edges.
iii Check whether you can trace over the network without lifting your pen. This can be done if you start at C or D.	**iii** The network is traversable.

remember

1. A *network* is made up of vertices (dots) joined together by edges (lines).
2. A network is traversable if you can trace over it without lifting your pen.

EXERCISE 12D Networks

WORKED Example 14

1 For the network shown at right:

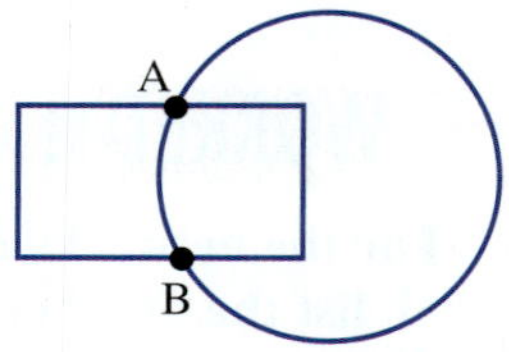

a list the number of vertices
b list the number of edges
c state whether the network is traversable.

2 Discover which of the networks below are traversable.

a
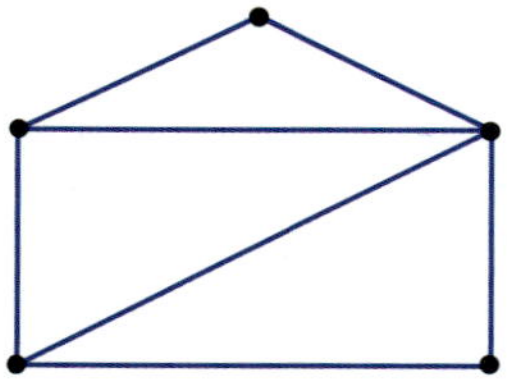

b
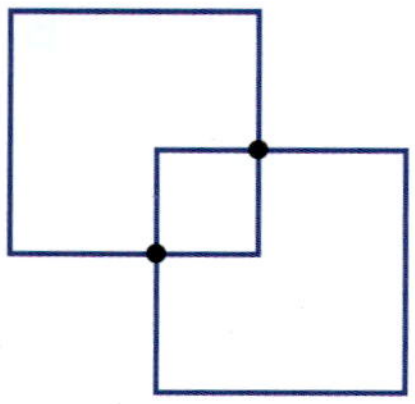

c
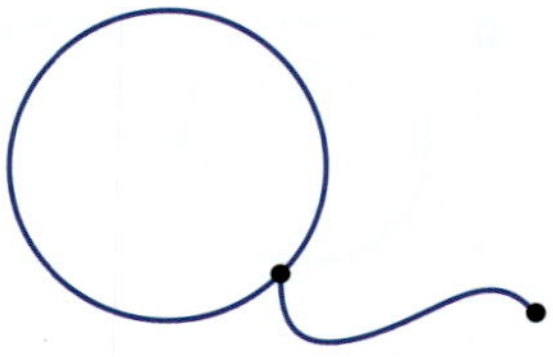

d
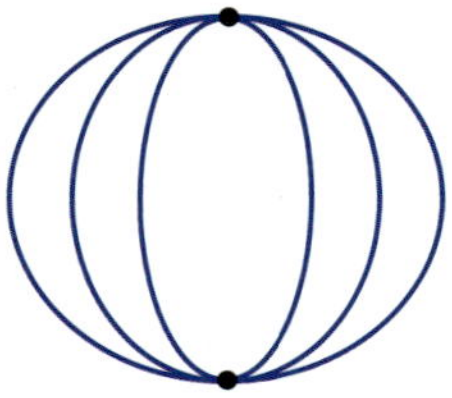

e
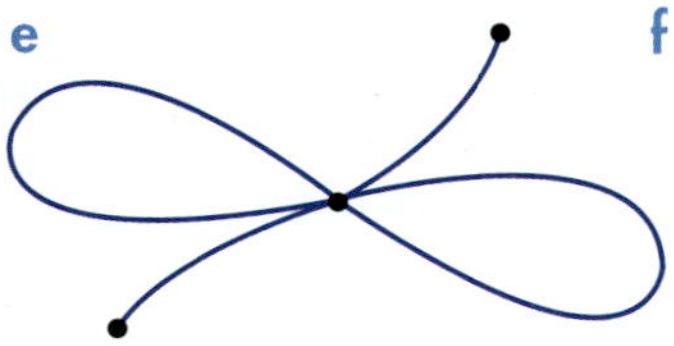

f
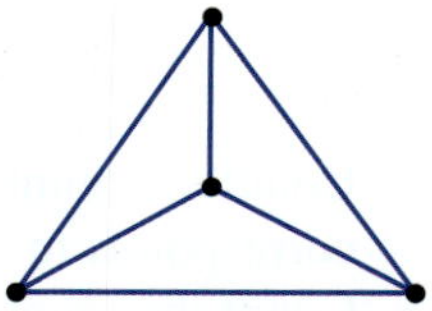

g

h
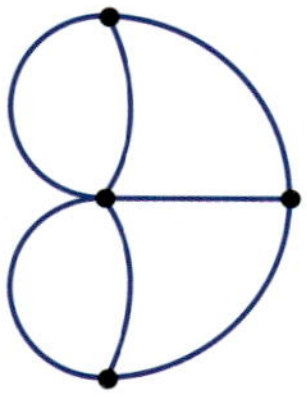

i
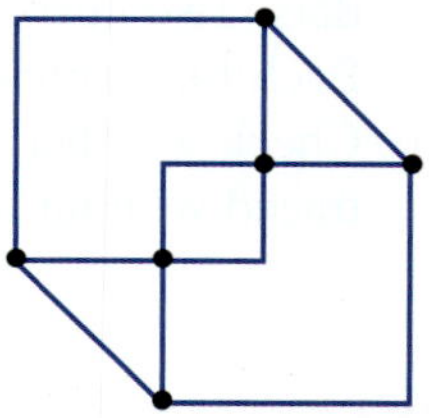

j 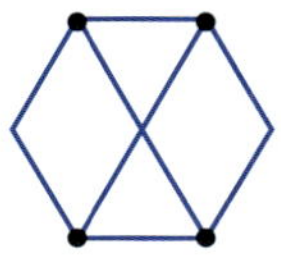

k 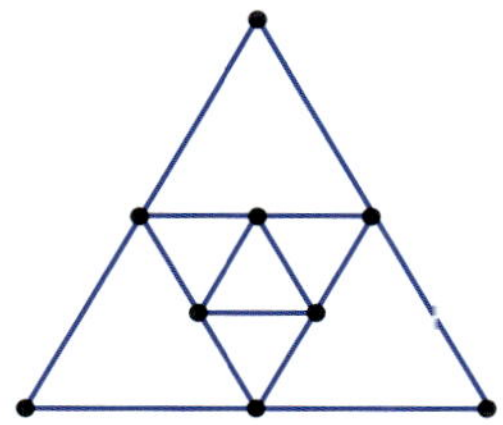

l 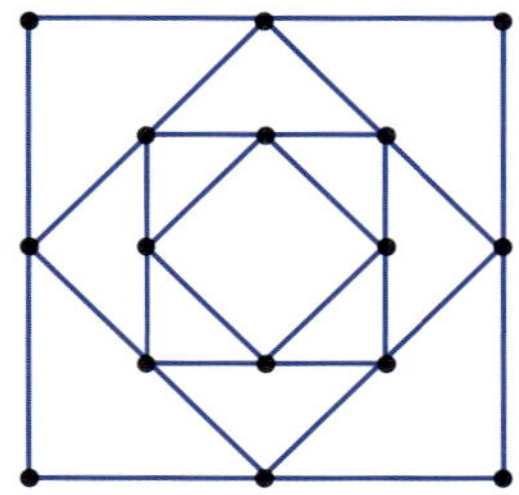

3 For each of the networks in question **2**, complete the information in the table below. The first one is done for you.

	Number of vertices	Number of edges	Traversable?
a	5	7	Yes
b			
c			
d			
e			
f			
g			
h			
i			
j			
k			
l			

4 Draw a network which fills the description given in the table.

	Number of vertices	Number of edges
a	3	3
b	6	3
c	7 (Traversable)	4
d	7 (Not traversable)	4

5 Try to draw a non-traversable network with:
a 2 vertices **b** 3 vertices **c** 4 vertices **d** 5 vertices.

THINKING

Traversable or not traversable?

If we can trace a network without lifting the pen or going over an edge twice, we say that the network is traversable. How can we decide whether a network is traversable without going through lots of trials? To do this we need to look for a pattern.

Consider the following examples.

Networks which are traversable

With these networks, you have to start and finish at the same vertex. (You can start at any vertex.)

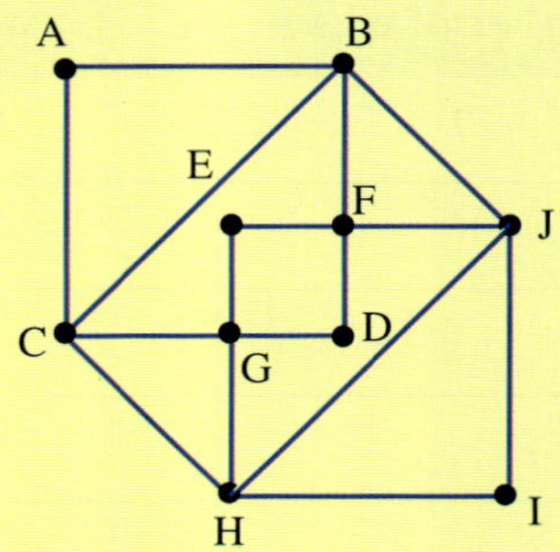

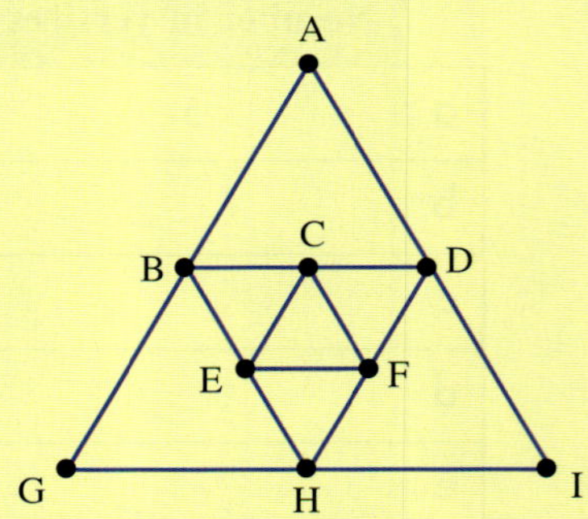

With these networks, you need to start and finish at different vertices. (Start at A and finish at B.)

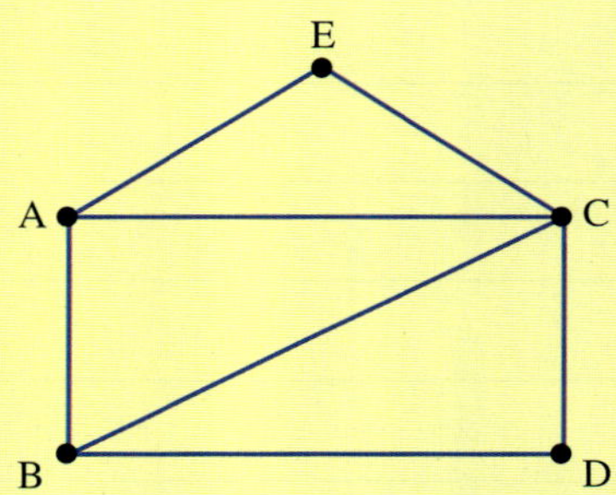

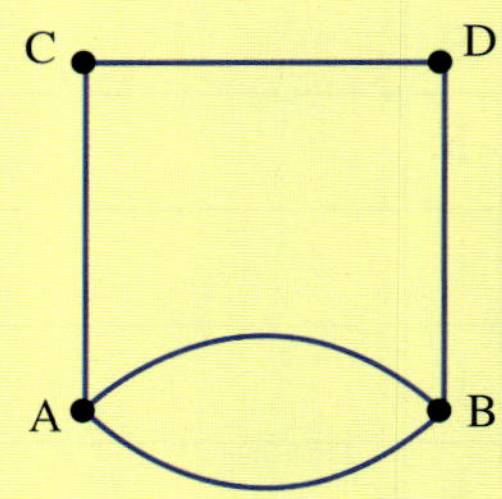

Networks that are not traversable

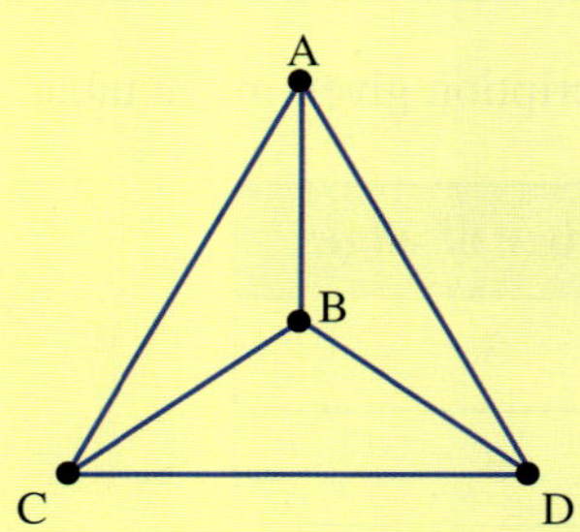

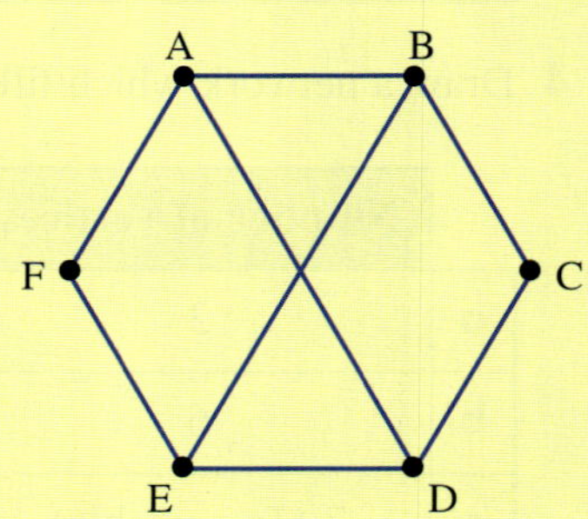

A vertex is said to be even if the number of paths leading to the vertex is even. For example,

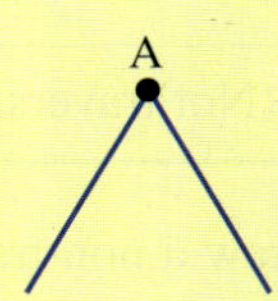

A vertex is said to be odd if the number of paths leading to the vertex is odd. For example,

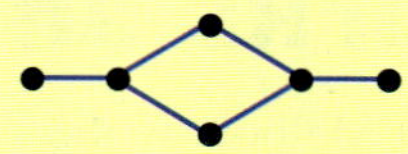

1 For each network shown on the page opposite and the one above, determine whether a vertex is even or odd.

2 For each network shown, how many vertices are even and how many are odd?

3 Copy and complete these sentences:

A network is traversable (starting and finishing at the same vertex) if there are ____________ odd vertices.

A network is traversable (starting at one vertex and finishing at another) if there are ____________ odd vertices.

A network is not traversable if there are more than____________ odd vertex/vertices.

4 Classify the following networks by writing the letter in the appropriate column of the table.

A

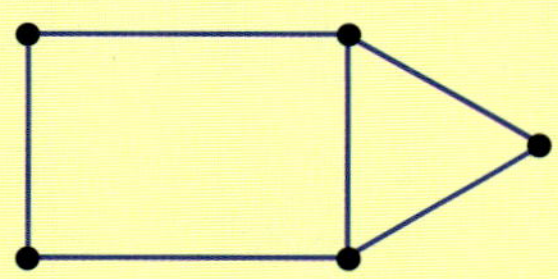

B

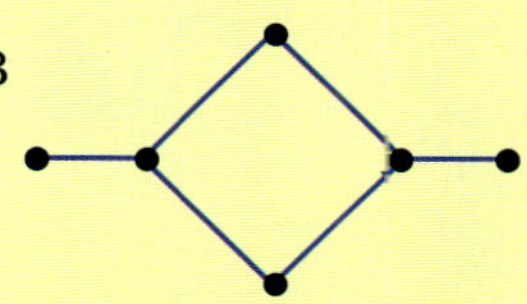

C

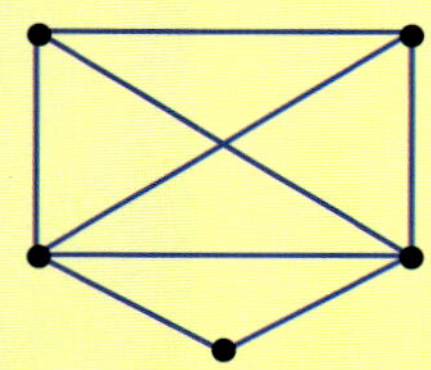

D

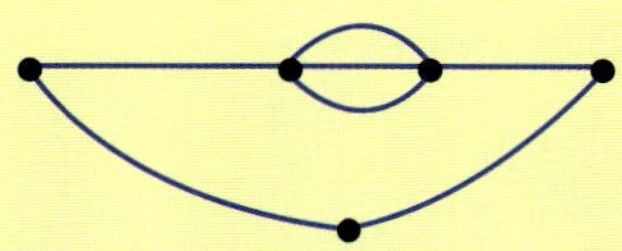

E

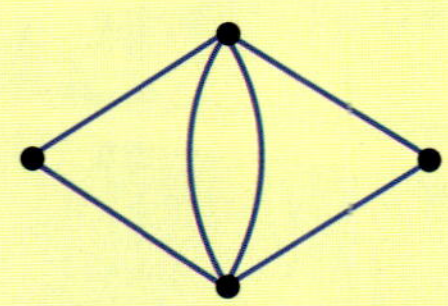

F

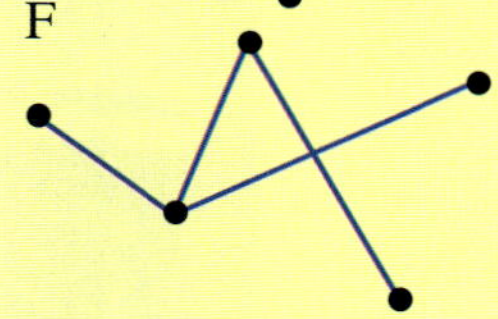

Traversable (start and finish at same vertex)	**Traversable (start and finish at a different vertex)**	**Not traversable**

a Can you add a path to the network at right to make it traversable, starting at one vertex and finishing at another?

b Can you add another path to make the network traversable from any vertex?

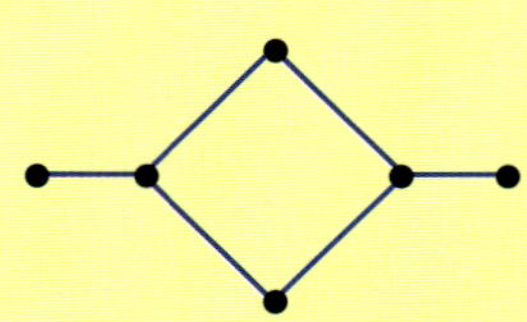

Networks and maps

In Melbourne the system of buses, trams and trains is called the Metlink. A map showing part of the Metlink train system in Melbourne is shown at right, drawn to a scale of 1 : 160 000.

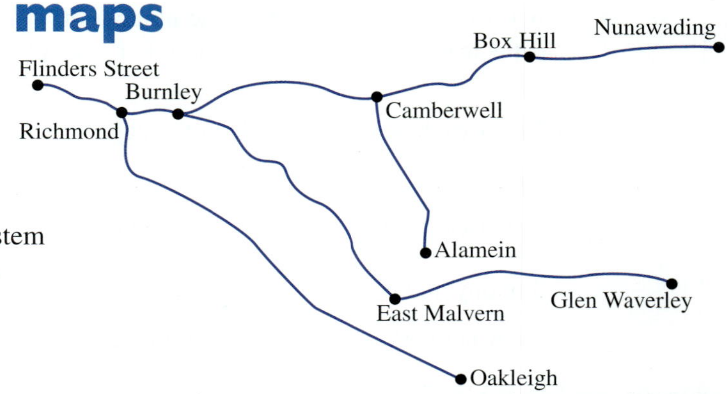

The Metlink publishes a *network* map which shows the same portion of the train system like this:

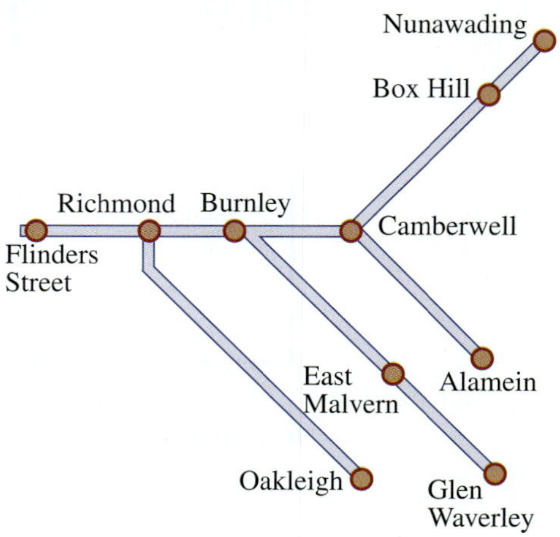

This network is not drawn to scale. Hence, it would be of no use at all for finding distances or directions. However, it does tell us that if you are travelling by train to Nunawading then you will pass through Camberwell and Box Hill. It also tells us that if you wish to travel to East Malvern then you will need to catch a train on the Glen Waverley line. In other words both diagrams tell us how vertices (stations) are linked by edges (tracks), but the second diagram is easier to read.

Network diagrams can be used in many similar situations to convey important information other than that related to finding distance or direction.

WORKED Example 15

In the house plan shown at right, the entry from one room to another is highlighted in red. Draw a network diagram that shows how the rooms are linked to each other.

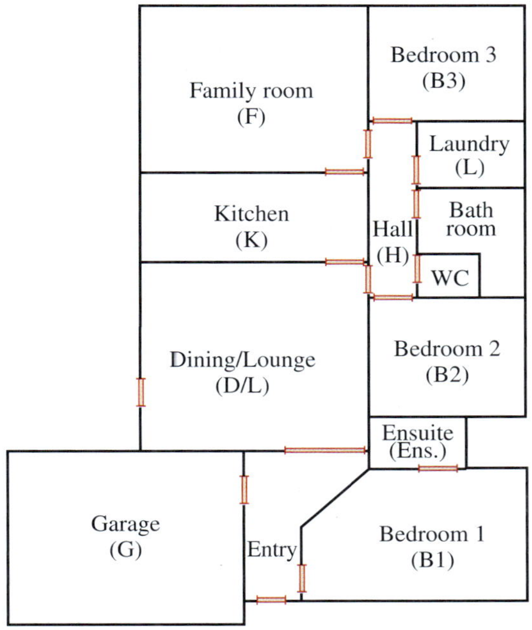

THINK

DRAW

1. Each room will be a vertex represented by a dot. Rooms are joined together by doors, so each door will be an edge, represented by a line.
2. Draw a rough sketch, starting with the family room. There are doors into the kitchen and the hallway. Mark these as 2 edges and 2 vertices.

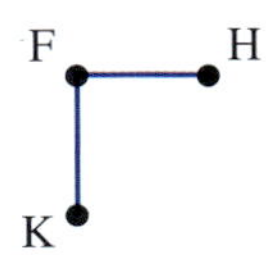

3. The hallway opens into bedroom 3, laundry, bathroom, W.C., bedroom 2 and dining/lounge area.
Mark these as 6 edges and 6 vertices.

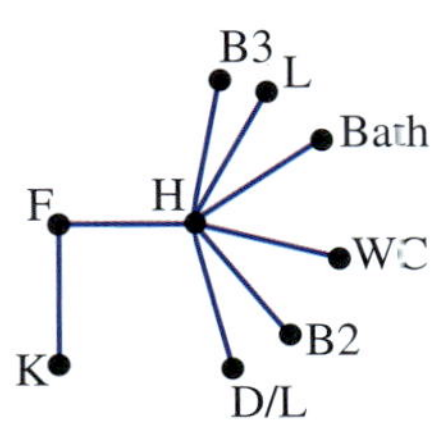

4. The dining/lounge joins back to the kitchen (making a ring) and also to the entry. The entry joins to the garage and to bedroom 1 and bedroom 1 joins to the ensuite.
Mark these as edges and vertices checking that the diagram shows all 13 rooms and 13 doorways.

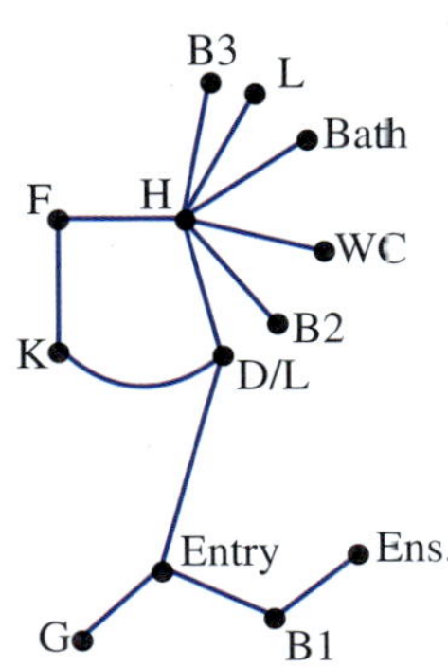

Continued over page

THINK

5. Complete the final draft, starting at the loop.

DRAW

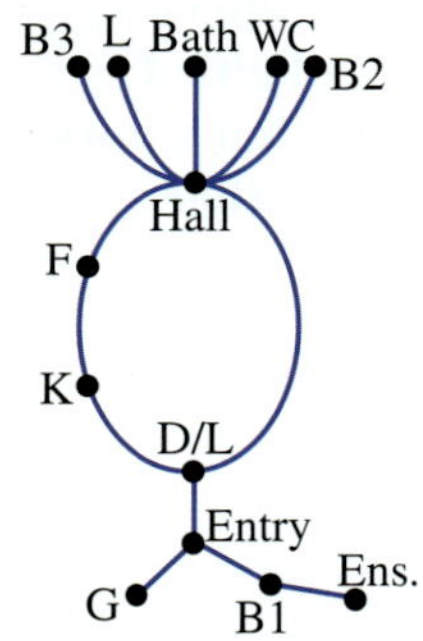

WORKED Example 16

Five friends live close together as shown on the street map at right.
Draw a network diagram to show, simply, the links between the houses marked on each street map.

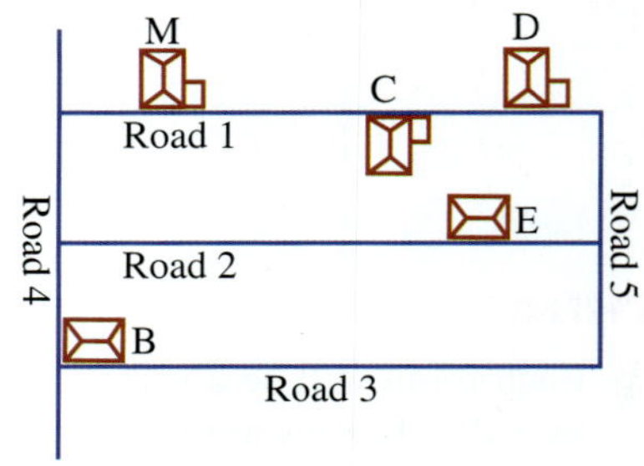

THINK

1. Each house will be a vertex. Houses are connected by roads, so each road will be an edge.
2. A person from M can travel directly to C, E and B, but cannot visit D without passing by one of the other houses.

DRAW

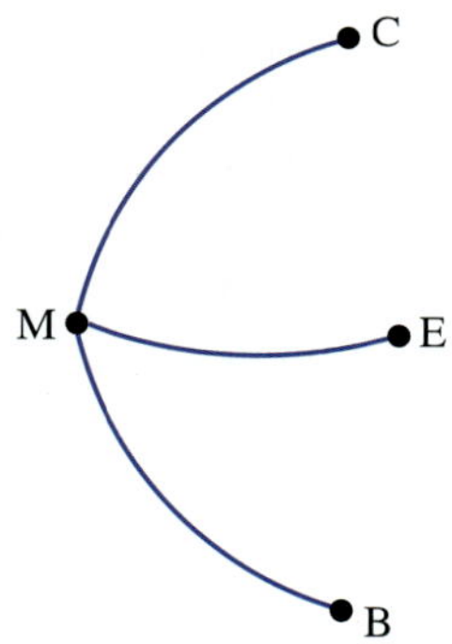

3. A person from C can only travel directly to M and D. The house at D links directly to C, E, and B. The house at E links directly to D and M and straight to B by 2 different paths.

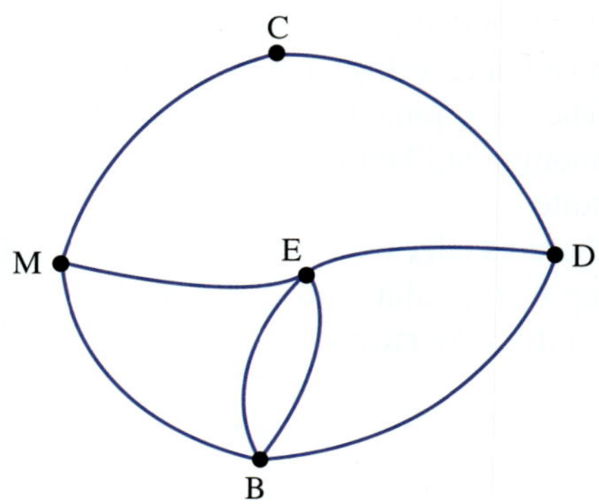

Directed networks

In some networks, the lines joining the points have direction that is indicated by arrows on the lines. Networks like these are called directed networks. They are particularly useful in planning projects and determining the order in which activities should be undertaken for the project to be completed on time. For example, you must eat a meal before you can wash the dirty dishes.

WORKED Example 17

The following directed network shows how Jack plans his morning to arrive at school on time.

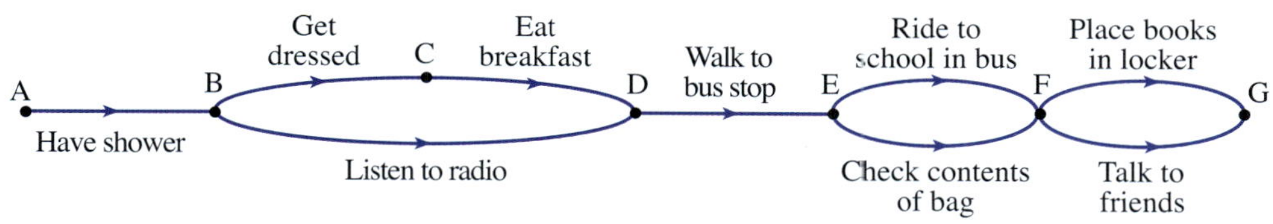

a **What must Jack do before he gets dressed?**
b **What occurs at the same time as Jack is listening to the radio?**
c **What is Jack also doing as he is riding to school in the bus?**
d **What does Jack do before he walks to the bus stop?**
e **After his bus ride to school, what does Jack do?**

THINK	WRITE
a 1 Look for the activity before 'getting dressed'.	**a**
2 Answer the question.	Jack has a shower before he gets dressed.
b 1 Look for activities that begin and end during the time Jack is listening to the radio.	**b**
2 Answer the question.	While listening to the radio, Jack first gets dressed and then eats breakfast.
c 1 Look for activities that begin and end during the 'ride to school' activity.	**c**
2 Answer the question.	Jack is checking the contents of his bag as he is riding to school.
d 1 Look for all the activities that happen before the 'walk to bus stop' activity.	**d**
2 Answer the question.	Before Jack walks to the bus stop he has a shower, gets dressed, eats breakfast and listens to the radio.
e 1 Look for activities that occur after the 'ride to school' activity.	**e**
2 Answer the question.	After Jack rides to school on the bus, he places his books in his locker while he is talking to friends.

remember

1. *Networks* can be used as maps that show how features are joined together.
2. Networks do not indicate the distances between features.
3. Directed networks are those that show an order for the completion of activities.

EXERCISE 12E Networks and maps

WORKED Example 15

1 In each of the house plans below, the entry from one room to another is highlighted in red. For each plan draw a network diagram that shows how the rooms are linked to each other.

a

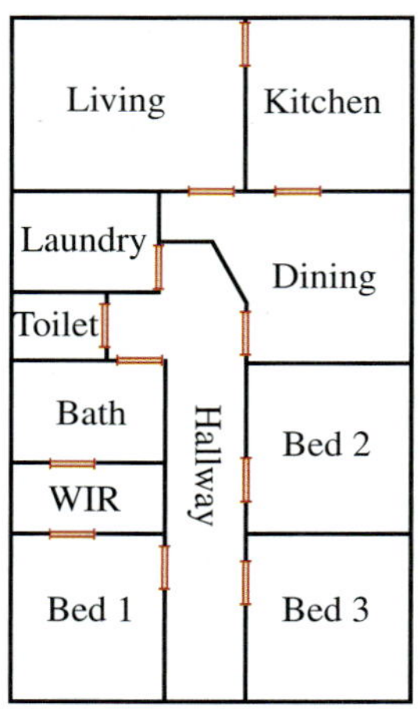

b

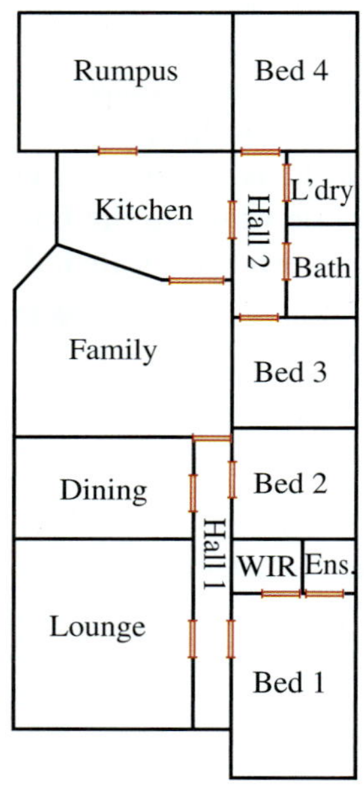

c

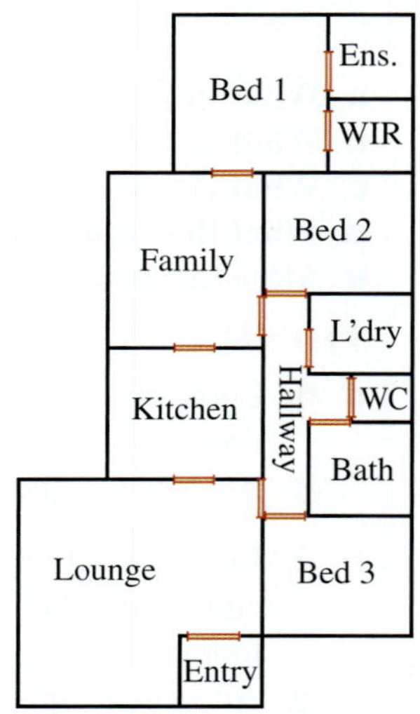

2 Here is the plan of a 2-storey house.

Show the house plan as a network diagram.
Hint: Use the top of the stairs as one room and the bottom of the stairs as another. The staircase is an entry between these two rooms.

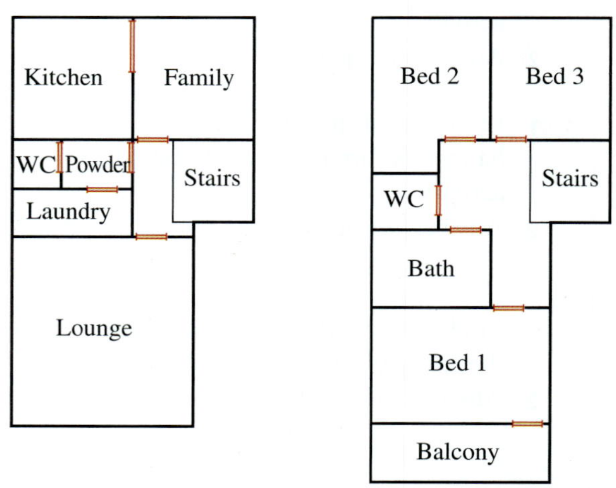

UPSTAIRS DOWNSTAIRS

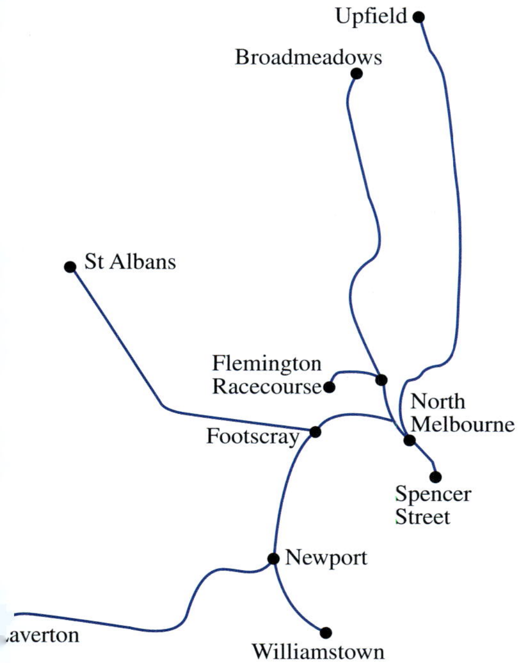

3 The map at left shows some of the northern and western rail network for the Metlink. Produce a neat simplified map of the network.

WORKED Example 16

4 Draw a network diagram to show, simply, the links between the houses marked on each street map.

a

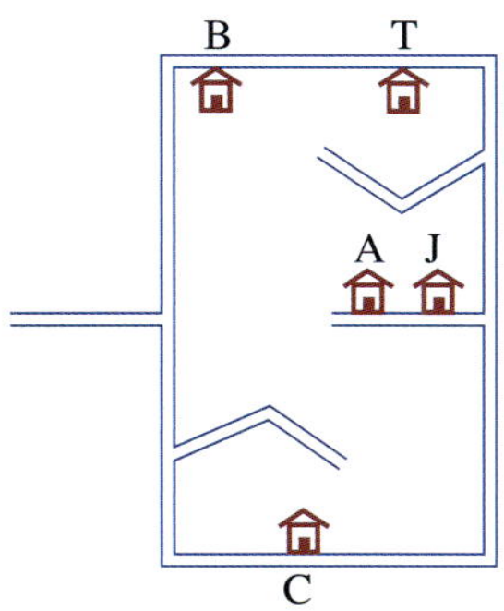

b

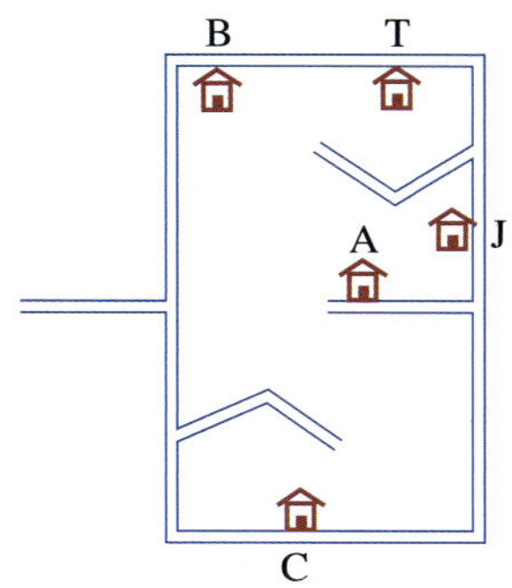

c

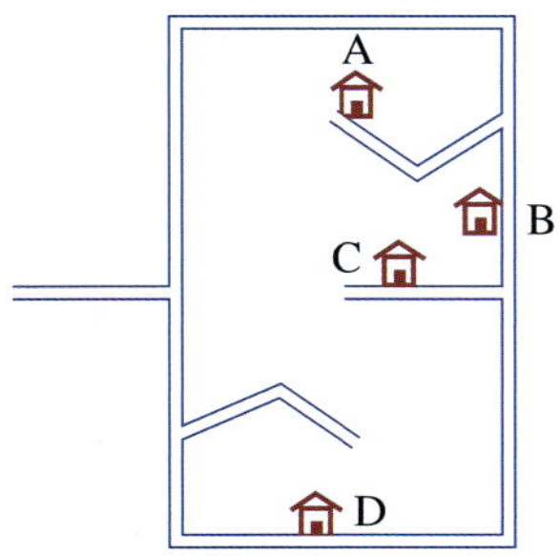

d

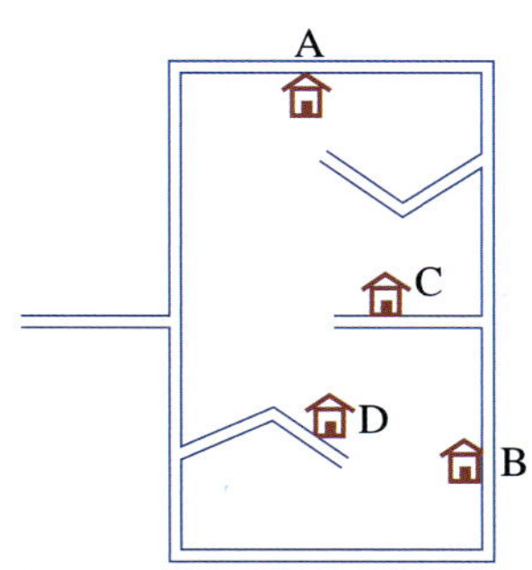

e

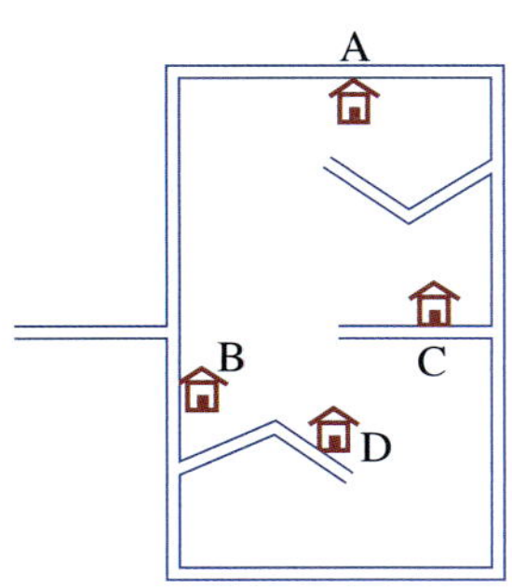

f

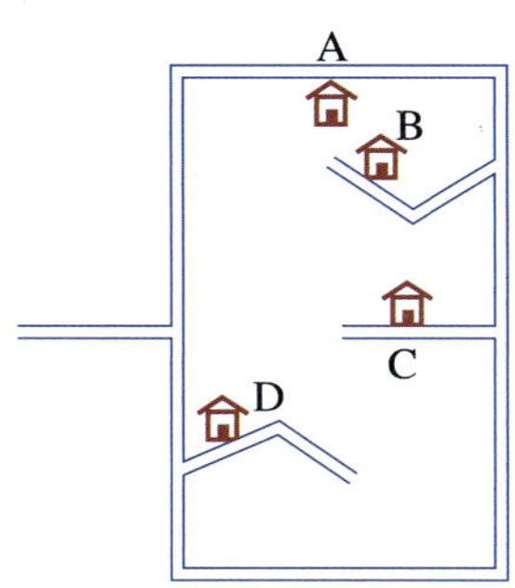

5 The map below shows Australia's major highways.

a Draw a rough draft of a simple network diagram of the highway system and highlight the largest complete loop.

b Change the highlighted loop into a large smooth curve and redraw the network neatly and carefully.

WORKED Example 17

6 Consider the map of Australia in question **5**. Monique and her family are travelling from Melbourne to Darwin by car. They do not wish to travel over the Nullarbor Plains. Draw a simple directed network showing the possible routes they could take. They will not backtrack on any section of the journey.

7 Julie is surprising her parents by making toast and coffee for their breakfast. Following is a list of the activities she undertakes.

- Boil water
- Slice bread
- Set table
- Toast bread
- Make coffee
- Eat breakfast
- Butter toast

Copy this network and insert the activities from the list in their correct order.

WorkSHEET 12.2

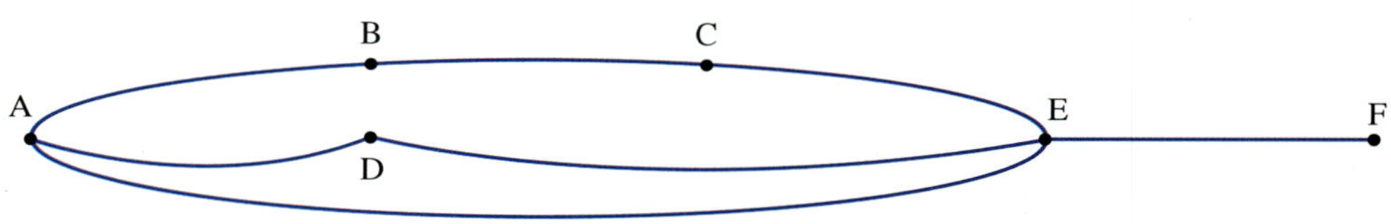

THINKING

Four colour problem

Francis Guthrie wrote to his younger brother Frederick, in 1852, asking him to answer what seemed a simple puzzle. The puzzle question was, 'Can every map drawn on the plane be coloured with four (or fewer) colours so that no two regions having a common border have the same colour?' Frederick couldn't solve it so he asked the distinguished mathematician Augustus De Morgan to solve it. He, in turn, asked William Rowan Hamilton, who also failed to solve it.

The four colour problem became known as the most easily stated problem, but most difficult to solve. It remained unsolved until 1976 when Kenneth Appel and Wolfgang Hanken solved the problem using a computer to analyse 1482 different maps.

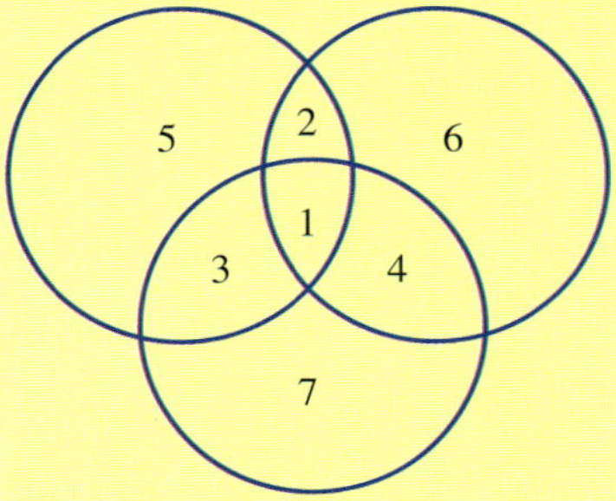

1 What is the minimum number of colours required to colour the seven regions on the right so that no two colours touch along a border? (The same colours may touch only at a single point.)
Try these steps to get you started.
   a First colour region 5. This means that regions 2 and 3 cannot be the same colour as region 5.
   b Next colour regions 2 and 3 in a different colour to region 5.
   c Region 1 can be coloured the same as region 5 as we wish to minimise colours.
   d Region 6 can be coloured exactly the same as region 5 and region 4 can be coloured exactly the same as region 3.
   e Finally, region 7 is coloured the same as region 5.
How many colours did you use?

Now try these. Trace each diagram first. What is the minimum number of colours needed in each case?

2

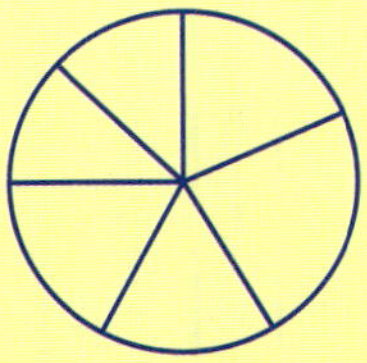

3

4

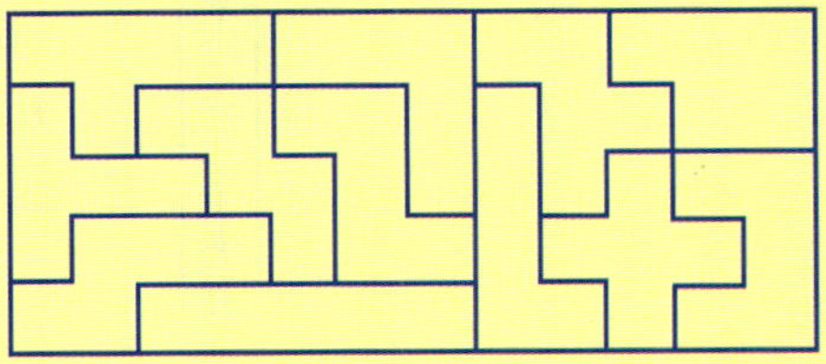

5

(continued)

6 Colour the map using the minimum number of colours. How many colours are needed?

MATHS QUEST CHALLENGE

1 List the path to be travelled in this network so that every edge is covered exactly once, starting and finishing at the same point.

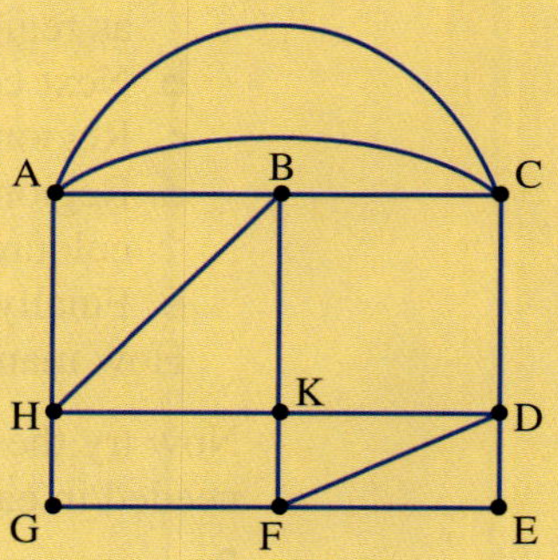

2 The diagram shows a network of one-way streets. At each intersection, the traffic divides equally between the two directions. All cars exit from B or C. If 288 cars enter the network at point A, how many of these cars exit from C?

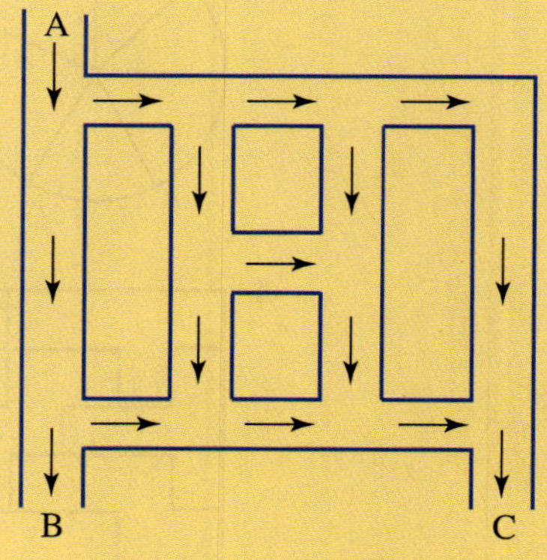

3 If you won one million dollars and decided to spend the money at the rate of one dollar per minute, how long would it take you to spend all the money?

10 QUICK QUESTIONS 2

1 What is meant when a network is described as traversable?

2 Is this network traversable?

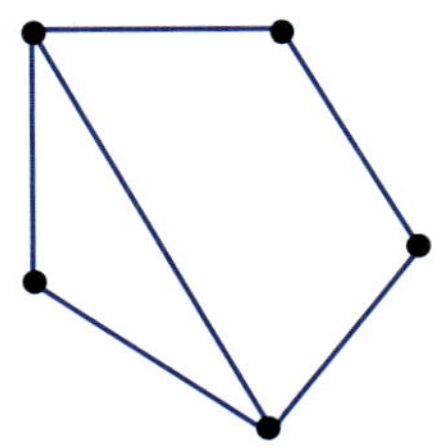

3 Is this network traversable?

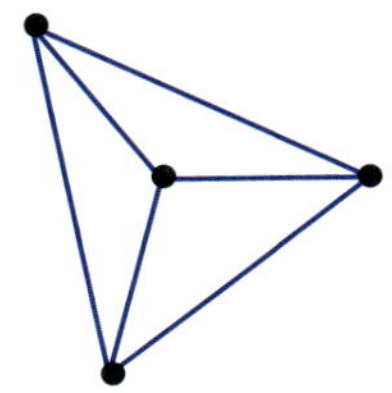

4 For the network in question 3, how many vertices and edges are there?

5 Draw a network with 4 vertices and 8 edges.

6 Draw a network with 7 vertices and 13 edges.

7 Six people live in the same neighbourhood. Show their positions on a network diagram.

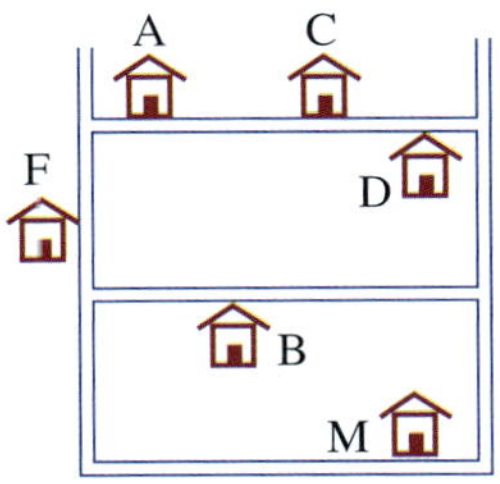

8 For the network in question 7, state the number of vertices and the number of edges.

9 a Show the house plan below as a network diagram.

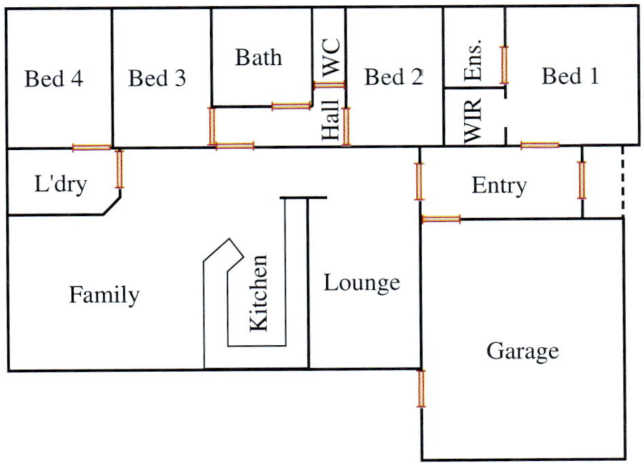

(Treat the hallway and garage as separate rooms.)

b State the number of vertices and the number of edges in the house plan above.

10 Draw a directed network diagram to show the activities involved in making a cup of hot chocolate or Milo. Make your diagram as detailed as possible.

The Cartesian plane

You have probably used a street directory, such as the one shown in the diagram below, at one time or another. To help us to find streets or other locations, each map is divided into squares and every square has it own name. The name of each square is given by 2 coordinates, a letter and a number, as with the map we looked at in worked example 6.

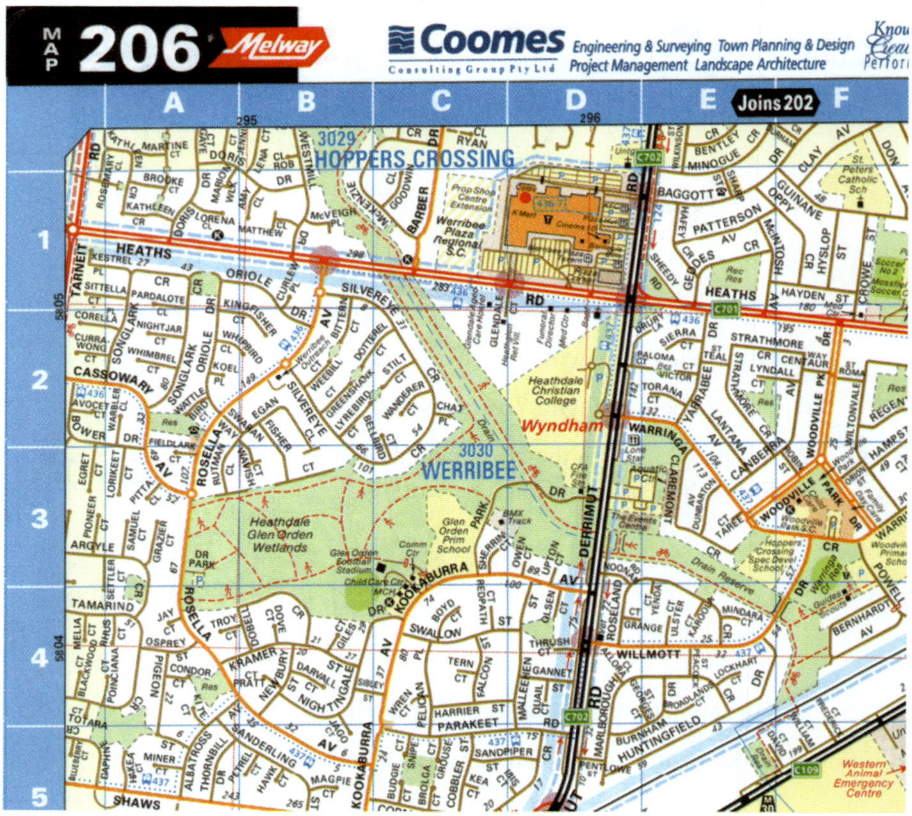

Can you find the square called A3? It contains such streets as Lorikeet Court, Rosella Avenue, Fieldlark Court and includes a part of the Heathdale Glen Orden Wetlands.

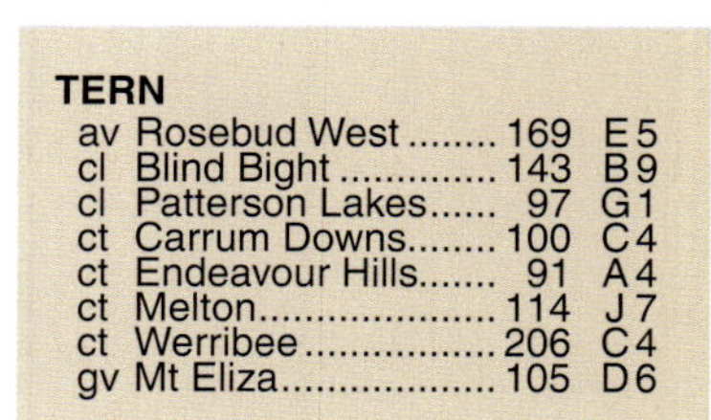

Can you find Tern Court somewhere on the map? This may be quite difficult. In this case checking the index is useful, because it tells us that Tern Court, Werribee is on Map 206 in the square with coordinates C4. Can you find Tern Court now?

In the 17th century, a mathematical genius called René **Descartes** (1596–1650) realised that any point on a plane could be found precisely, by using 2 numbers as coordinates, rather than a letter and a number. This is how Descartes' system works: we first rule two straight lines called **axes**.

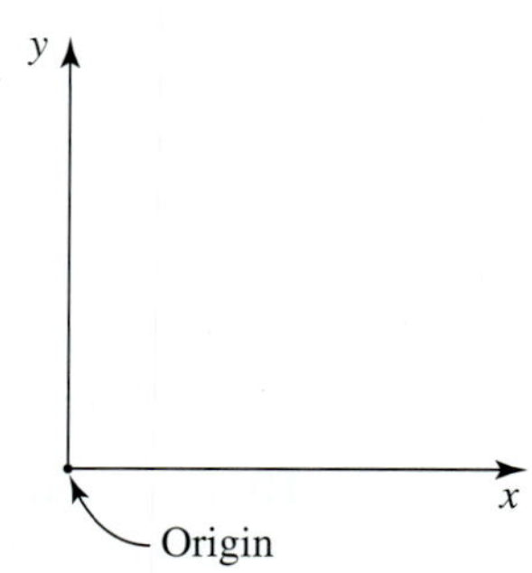

The **horizontal** line is called the ***x*-axis** and the **vertical** line is called the ***y*-axis**. The point where the two lines intersect is called the **origin**. We then carefully mark out even spaces along both axes and number them starting with zero at the origin. The distance between each mark is called one unit and the axes can extend as far as you wish. The area between the axes forms the **Cartesian plane**.

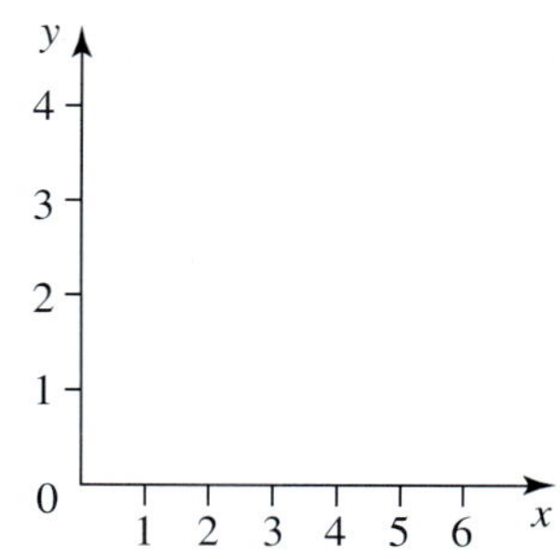

Imagine that you are an ant standing at the origin. How would you get to the point marked with * on the Cartesian plane shown at right?
You could first walk across 4 units, and then up 3 units. We say that the point with * has **coordinates** (4, 3). We always write the coordinates in brackets, and separate the two numbers with a comma.

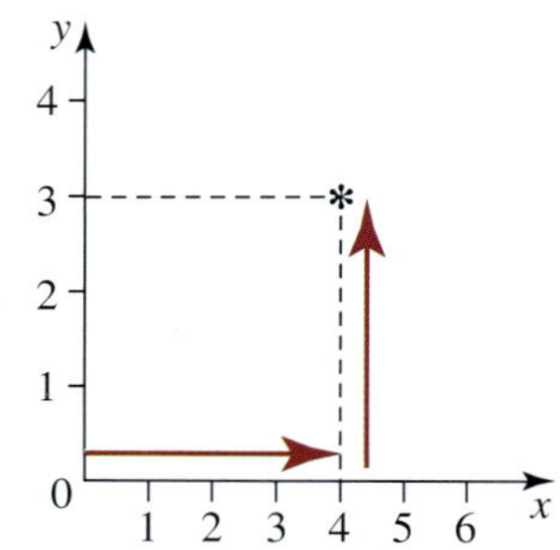

The two numbers are called *Cartesian coordinates*, after Descartes.

It's easy to find a point if we know its Cartesian coordinates.
For example, to find the point (2, 5) we simply start at the origin, walk across 2 units and then climb up 5 units.

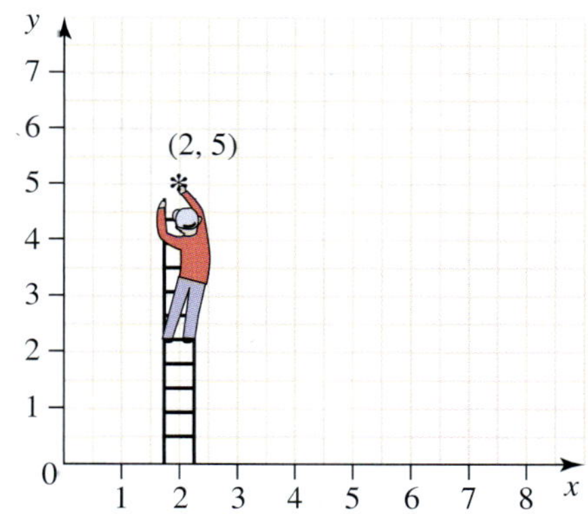

WORKED Example 18

Draw a Cartesian plane with axes extending from 0 to 6 units. Mark the following points with a dot, and label them.

a (2, 4) **b** (5, 0) **c** (0, 2) **d** $(3\frac{1}{2}, 1)$

THINK

1. First rule up and label the axes.
2. Mark each point.

a (2, 4) means starting at the origin, go across 2 units, and then up 4 units.

b (5, 0) means go across 5 units and up 0 units. It lies on the *x*-axis.

c (0, 2) means go across 0 units and up 2 units. It lies on the *y*-axis.

d $(3\frac{1}{2}, 1)$ means go across $3\frac{1}{2}$ units and up 1 unit.

Label each point.

DRAW

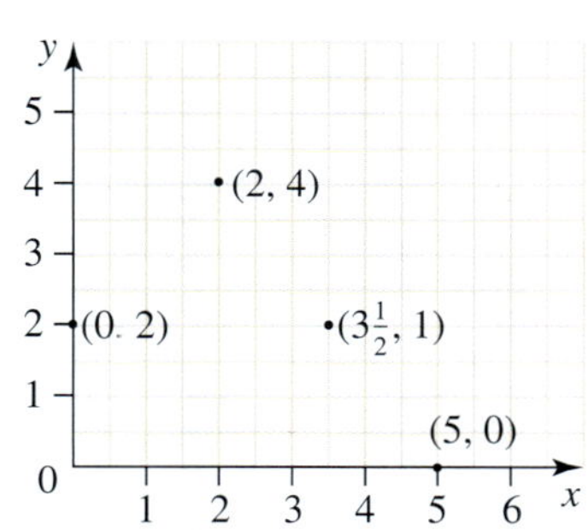

WORKED Example 19

Find the Cartesian coordinates for each of the points A, B, C, and D.

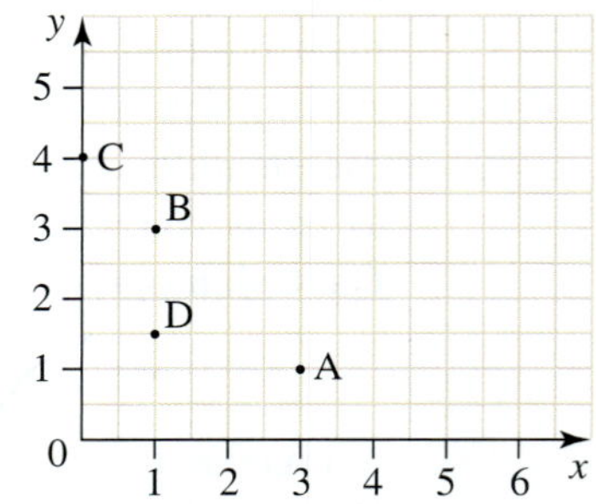

THINK	WRITE
Point A is 3 units across and 1 unit up	A is at (3, 1)
Point B is 1 unit across and 3 units up.	B is at (1, 3)
Point C is 0 units across and 4 units up.	C is at (0, 4)
Point D is 1 unit across and $1\frac{1}{2}$ units up.	D is at $\left(1, 1\frac{1}{2}\right)$

Hint: **To help remember the order in which Cartesian coordinates are measured, think about using a ladder. Remember we must always walk across with our ladder and then climb up it.**

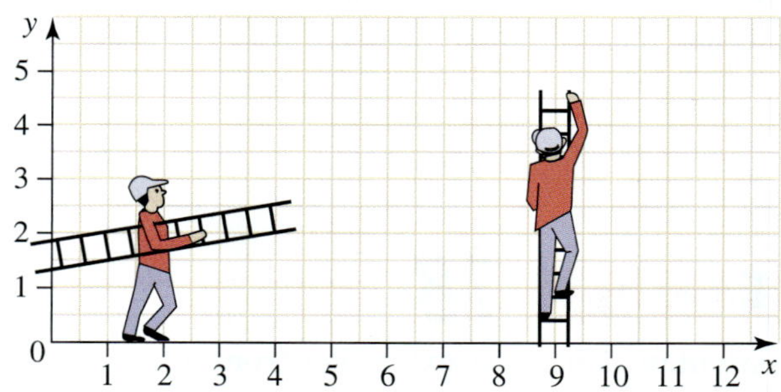

remember

1. *Cartesian coordinates* can be used to locate any point on a plane.
2. Cartesian axes need to be carefully ruled and measured.
3. Cartesian coordinates consist of a horizontal x-axis and a vertical y-axis which meet at a point called the origin.
4. The point (2, 3) is located 2 units across and 3 units up from the origin on the Cartesian plane.

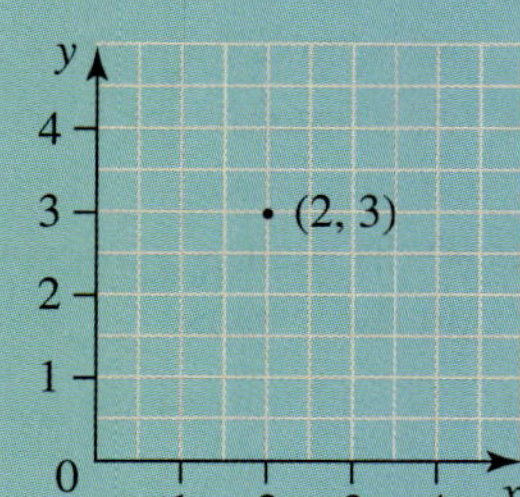

EXERCISE 12F The Cartesian plane

1 Look again at the Melway map 206 on page 544.
In which square would you find each of the following:
- **a** Hoppers Crossing station?
- **b** Glen Orden Primary School?
- **c** Kookaburra Avenue?

2 On Melway map 206 (page 544) what is the name of the longest street passing through:
- **a** G3?
- **b** C2?
- **c** B5?

3 Each of these sets of Cartesian axes (except one) has something wrong with it. From the list below, match the mistake in each diagram with one of the sentences.
- **A** The units are not marked evenly.
- **B** The y-axis is not vertical.
- **C** The axes are labelled incorrectly.
- **D** The units are not marked on the axes.
- **E** There is nothing wrong.

a
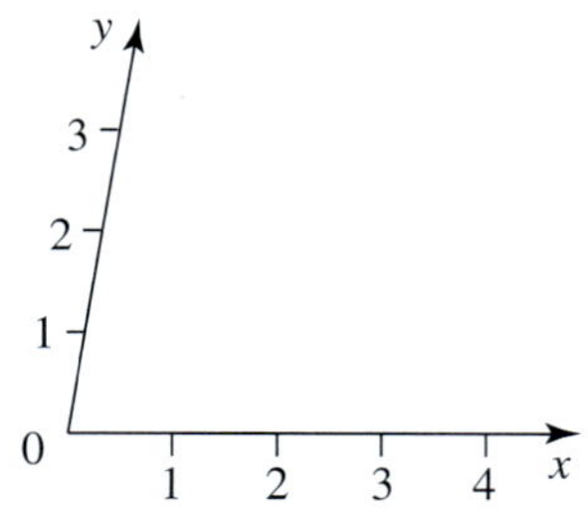

b
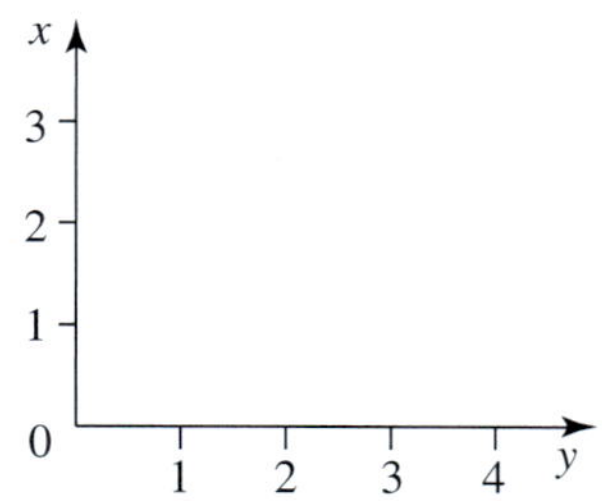

c
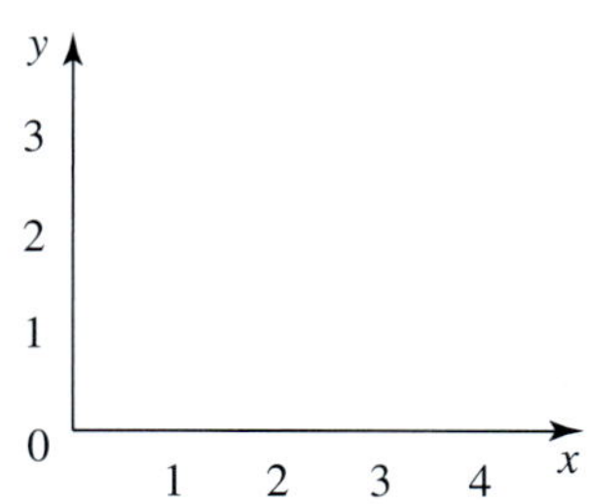

d
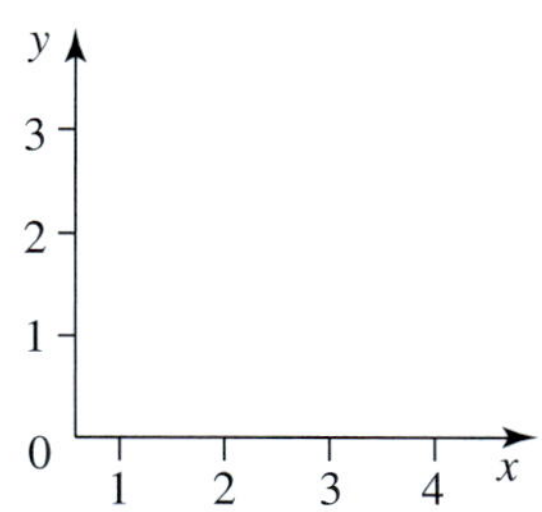

e
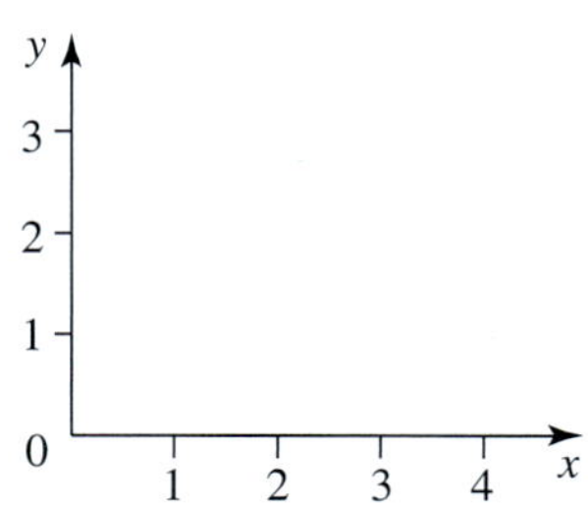

f
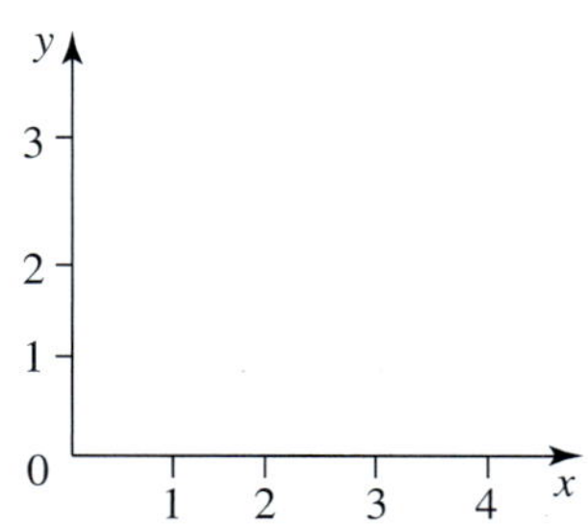

WORKED Example 18

4 Draw a Cartesian plane with axes extending from 0 to 5 units. Mark the following points with a dot, and label them.
- **a** (4, 3)
- **b** (1, 4)
- **c** (3, 3)
- **d** (2, 0)
- **e** (0, 4)
- **f** (0, 0)

12.6 SkillSHEET Plotting coordinate points

Cabri Geometry — Cartesian coordinates

WORKED Example 19

5 Find the Cartesian coordinates for each of the points A–L.

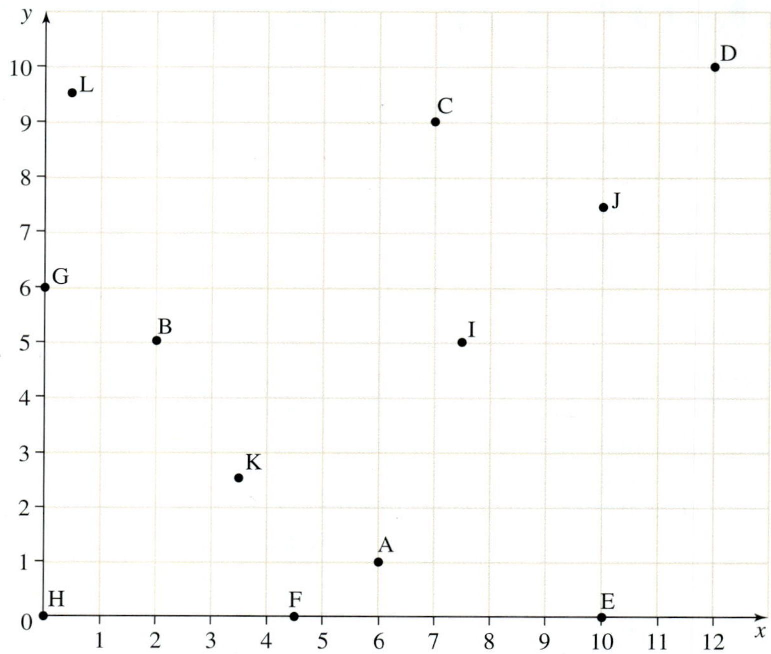

EXCEL Spreadsheet — Plotting points

Mathcad — Cartesian plane

6 From the diagram at right, write down the coordinates of 2 points which:

a have the same x-coordinate

b have the same y-coordinate.

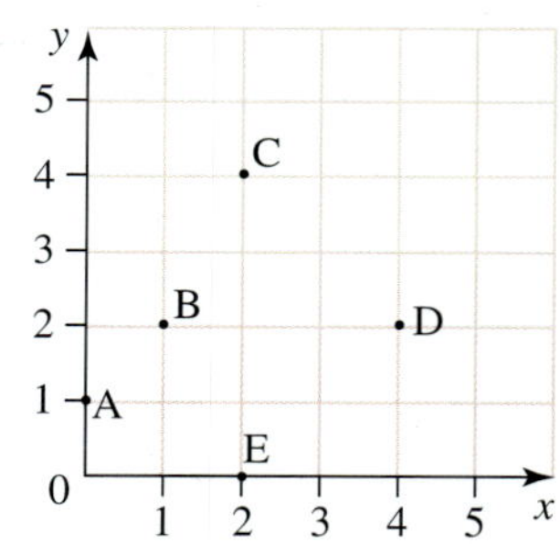

GAME time — Maps, networks and coordinates — 001

7 Messages can be sent in code using a grid like the one drawn below, where the letter B is represented by the coordinates (2, 1).

Use the diagram to decode the answer to the following riddle.

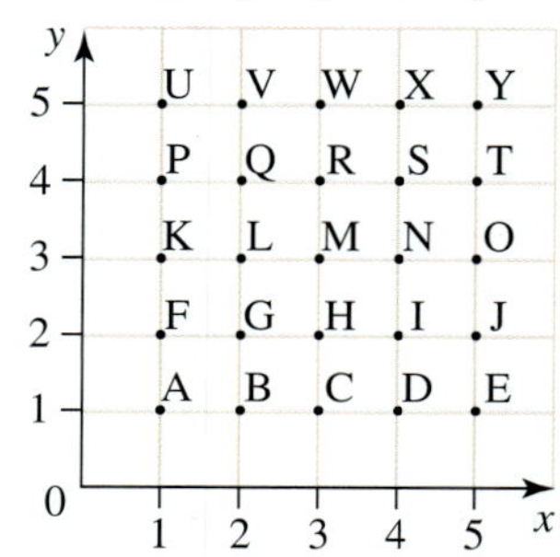

Q. Where did they put the man who was run over by a steamroller?

A. (4, 2)(4, 3) (3, 2)(5, 3)(4, 4)(1, 4)(4, 2)(5, 4)(1, 1)
(2, 3)(4, 2)(4, 3) (3, 5)(1, 1)(3, 4)(4, 1)(4, 4) (4, 4)(4, 2)(4, 5)
(4, 4)(5, 1)(2, 5)(5, 1)(4, 3) (5, 1)(4, 2)(2, 2)(3, 2)(5, 4) (1, 1)(4, 3)(4, 1)
(4, 3)(4, 2)(4, 3)(5, 1)

8 Rule up a Cartesian plane with both axes extending from 0 to 10 units. Plot the following points and join them in the order given to make a geometric figure. Name each shape.

a (2, 2)–(5, 2)–(2, 6)–(2, 2)

b (4, 4)–(8, 4)–(6, 8)–(4, 4)

c (1, 1)–(10, 1)–(8, 9)–(2, 9)–(1, 1)

d (0, 0)–(8, 0)–(10, 10)–(2, 10)–(0, 0)

9 Here is an exercise which may require care and concentration. On graph paper or in your exercise book rule up a pair of Cartesian axes. The x-axis must go from 0 to 26 and the y-axis from 0 to 24. Plot the following points and join them in the order given.
(0, 15)–(4, 17)–(9, 22)–(10, 21)–(12, 24)–(16, 22)–(15, 21)–(18, 19)–(20, 24)–(22, 18)–(26, 12)–(26, 10)–(23, 4)–(20, 3)–(18, 4)–(14, 7)–(11, 7)–(4, 6)–(2, 7)–$(2\frac{1}{2}, 8)$–(0, 15)
Complete the picture by joining (19, 2)–(21, 2)–(20, 0)–(19, 2).

GAME time
Maps, networks and coordinates — 002

WorkSHEET 12.3

MATHS QUEST CHALLENGE

1 What is the area of a rectangle formed by connecting the points (2, 1), (7, 1), (7, 4) and (2, 4) on a Cartesian plane?

2 The street map at right shows all the possible streets Martin can use to travel from home to school. He always travels due north or due east. How many different routes can Martin use to get to school?

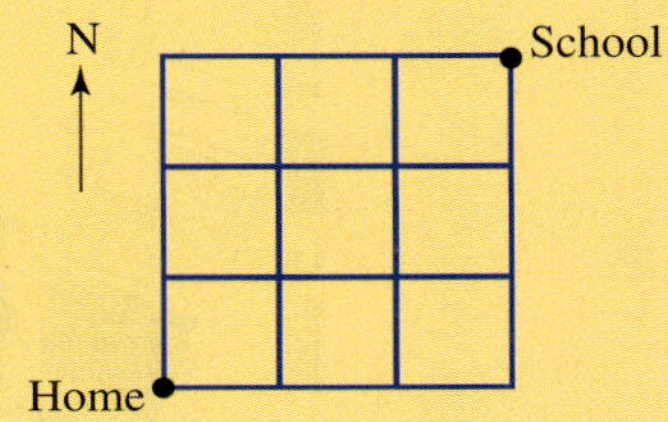

COMMUNICATION

Touring the attractions

At the start of the chapter, you looked at the problem of organising a travel route for a tour group. The group of American tourists will start from the tourist office next to Parliament House and visit the Rialto Towers Observation Deck, Shrine of Remembrance, Crown Entertainment Complex, Melbourne Aquarium, Melbourne Cricket Ground, Queen Victoria Market and the Victorian Arts Centre.

1 From the map on the following page, list the map references for each attraction. For example, the Queen Victoria market is located at F2.
2 Draw a simple network diagram linking all the attractions.

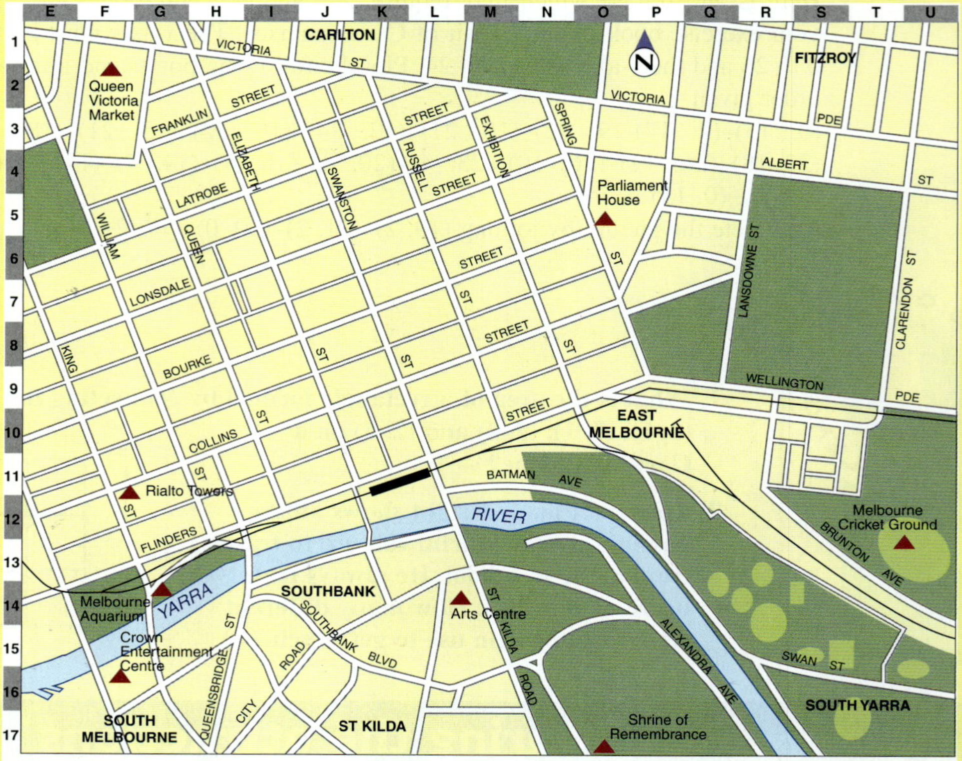

Since they have limited time, the tour group want to visit each one but not pass any of the attractions more than once. They will start and finish at the tourist office (Parliament House).

3 Suggest possible travel routes they could take.

4 Compare your answer with another student. Is there more than one way?

The network at right is actually drawn to scale (1 cm ⇔ 500 m). The lines do not follow the roads but give an indication of how far the group needs to travel between the attractions.

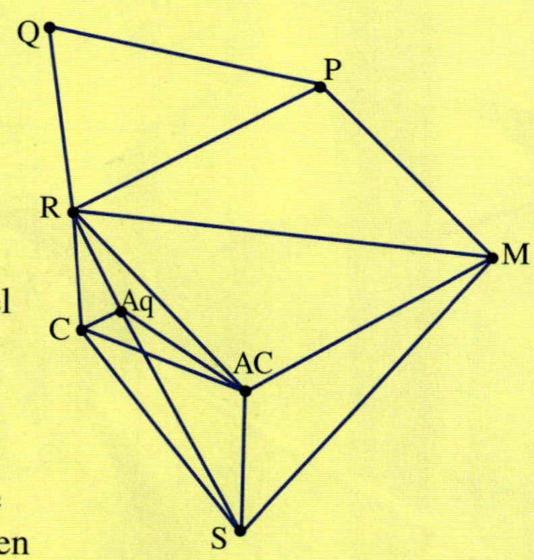

Key
P Parliament House
Aq Melbourne Aquarium
C Crown Entertainment Complex
AC Victorian Arts Centre
S Shrine of Remembrance
Q Queen Victoria Market
R Rialto Towers
M Melbourne Cricket Ground

5 Use the scale to calculate the distance to be travelled between each attraction.

6 For the tourist routes you have listed, calculate the total distance.

7 Which tourist route has the shortest travelling distance and so allows the tourists more time at each attraction?

Sick jokes!

1. Which rich Arab invented flavoured chips?
2. What do you get if you cross a frog with a chair?
3. What do you get if you cross a car with a book?
4. Why did the rooster refuse to fight?

To find the answers plot the letters below at the coordinates indicated. The answers will appear at the y values corresponding to the joke number.

ANSWERS

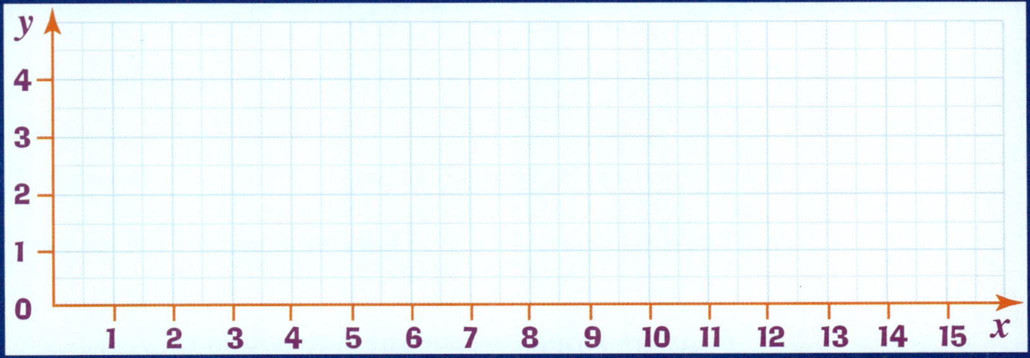

Ⓐ at (4, 4), (6, 2), (12, 3), (5, 1), (2, 2), (7, 4), (13, 1), (3, 3), (0, 3).
Ⓑ at (7, 3). Ⓒ at (12, 4), (9, 4). Ⓓ at (7, 2). Ⓔ at (11, 1), (14,4), (1, 4).
Ⓖ at (12, 1), (10, 3). Ⓗ at (14, 3), (10,4), (0, 4). Ⓘ at (9, 1), (8, 3), (11, 4). Ⓚ at (13, 4). Ⓛ at (12, 2), (3, 1). Ⓝ at (1, 3), (15, 4), (10, 1), (6, 1).
Ⓞ at (10, 2), (6, 3), (9, 3), (11, 2), (5, 2). Ⓟ at (13, 3). Ⓡ at (11, 3), (14, 1).
Ⓢ at (5, 4), (1, 1), (8, 2). Ⓣ at (4, 1), (5, 3), (4, 2), (9, 2). Ⓤ at (2, 1), (4, 3). Ⓥ at (8, 1). Ⓦ at (3, 4). Ⓨ at (15, 3).

summary

Copy the sentences below. Fill in the gaps by choosing the correct word or expression from the word list that follows.

1 The scale 1 : 20 000 tells us that the actual ____________ is 20 000 times as large as the ____________.

2 The location of a point on the Earth can be identified by its ____________ of latitude and its ____________ of longitude.

3 Parallels of latitude run parallel to the ____________.

4 Meridians of longitude run between the ____________ and the ____________.

5 The Equator is ____________ latitude. The North Pole is ____________ and the South Pole is ____________.

6 The ____________ is 0° longitude. The ____________ is 180°E or W.

7 Each square in a grid reference is defined by two ____________ — a letter and a number.

8 A true bearing is given as an ____________ between 0° and 360°.

9 The angle given in a true bearing indicates the number of degrees turned ____________ from north.

10 Maps displaying contour lines are known as ____________ maps.

11 The lines on a map connecting places of equal air pressure are known as ____________.

12 Site plans detail features such as ____________, layout and ____________.

13 Networks have points called ____________ and lines called ____________.

14 A network is ____________ if it can be traced without lifting the pen.

15 Network diagrams don't contain real information about distance and ____________, but show how features are ____________ together.

16 Networks which show an order for completion of activities are known as ____________ networks.

17 **On a Cartesian plane the two ____________ meet at a point called the ____________.**

18 **The horizontal axis is called the *x*-axis and the ____________ axis is called the *y*-axis.**

19 **Cartesian coordinates can be used to locate any ____________ on a plane.**

20 **The point (5, 1) would be 5 units ____________ from the origin and 1 unit up.**

WORD LIST

angle	North Pole	coordinates	edges
clockwise	Greenwich Meridian	South Pole	joined
diagram	parallel	90°N	origin
scale	Equator	90°S	point
0°	International Date Line	isobars	traversable
topographic	meridian	across	vertical
distance		axes	vertices
direction		direction	directed

CHAPTER review

12A

1 Write these scales as ratios.

a 1 cm represents 3 m

b 1 cm ⇔ 2 km

12A

2 What would 1 cm on a map represent if the map scale was:

a 1 : 1000?

b 1 : 80?

12A

3 This is a rough sketch of a volleyball court, not drawn to scale, but with the real dimensions marked. Using the scale 1 cm ⇔ 2 m, produce a neat ruled scale drawing of the court.

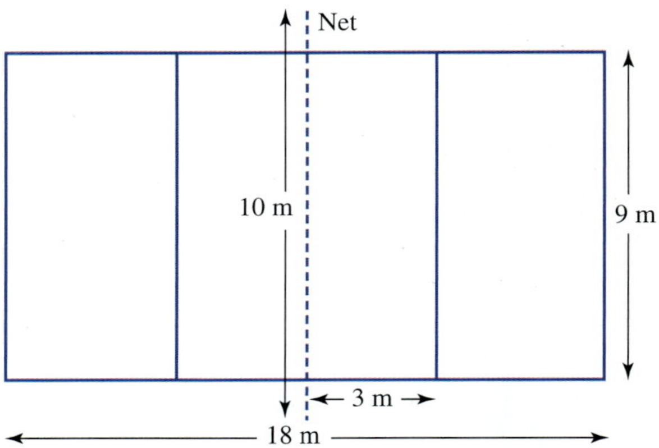

12A

4 This map of the British Isles is drawn on a scale of 1 : 9 000 000.

a What is the actual distance that is represented by 1 cm on the map?

b How far is it from London to Aberdeen?

c How far is it from Liverpool to Dublin?

Use the map of Australia to answer questions 5 to 7.

5 Give the approximate latitude and longitude of each of the following. 12B

a Rockhampton b Canberra c Broome d Hobart

6 Give the grid reference for each place in question 5. 12B

7 Which city lies closest to each of the following pairs of coordinates? 12B

a 19°S, 146°E b 11°S, 131°E c 35°S , 138°E d 34°S, 151°E

8 Draw a diagram to show the true bearing of the following. 12C

a 150° true b 300° true

9 Using the map of the British Isles in question 4, find the true bearing from Manchester of: 12C

a Edinburgh b Belfast c Grimsby.

10 Use this topographic map to answer the following questions. 12C

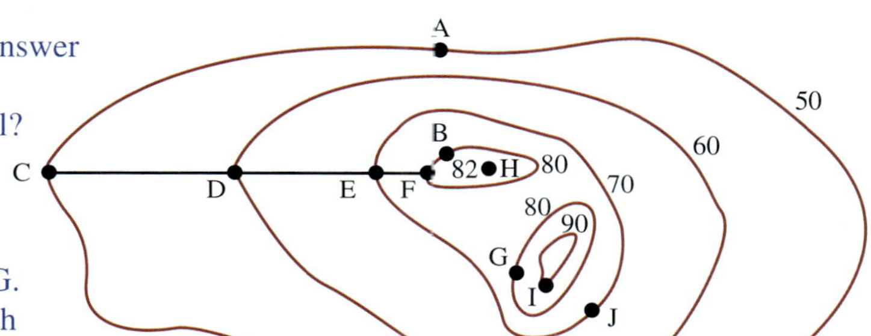

a What is the contour interval?

b Name the highest point marked on the map.

c Compare the height of point D with that of point G.

d Along the line CDEF, which section has the gentlest slope?

e If you walked in a straight line from H to J, would you walk mainly over flat or hilly country?

12C

11 Use this synoptic map to answer the following questions. This is an actual weather map for Australia for 12 February 1999.

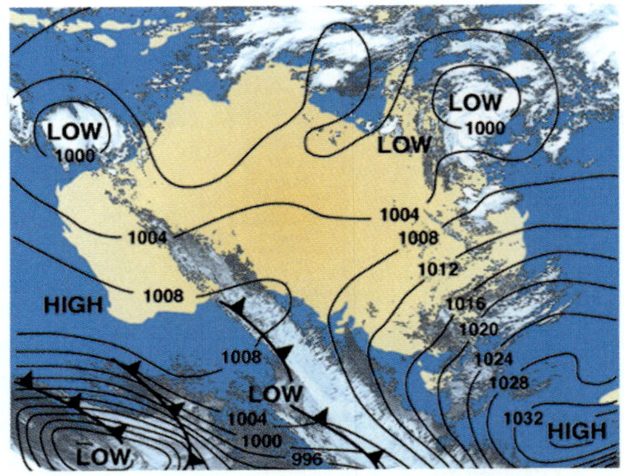

a Tropical cyclone *Rona* had just crossed the east coast of Australia. What was its air pressure?

b Where is another tropical cyclone forming?

c Describe the weather in the southern region of the continent.

d Alice Springs is experiencing unsettled weather. Why is this so?

12C

12 Using this tourist map of central London, find:

a the scale of the map

b the direction of the British Museum from Westminster Abbey

c the approximate distance between 10 Downing Street and the front of Buckingham Palace.

Questions **13** and **14** refer to the plan for the 2006 Commonwealth Games site below.

13 **a** What is the true bearing of Telstra Dome from the Melbourne CBD? **12C**
b In what direction is the Melbourne International Shooting Club from the MCG?
c How far is the Docklands Precinct from Multi Purpose Venue (Melbourne Park)?

14 A shuttle bus service has been organised to transport spectators to and from most of the events at the 2006 Commonwealth Games. The route (highlighted in red) of the shuttle bus may be considered to be a network. **12D**
a State the number of vertices of the network.
b State the number of edges of the network.
c Determine whether the network is traversable and if so specify a possible path.

15 For each of the networks below, give the number of edges and vertices, and state whether the network is traversable. **12D**

a

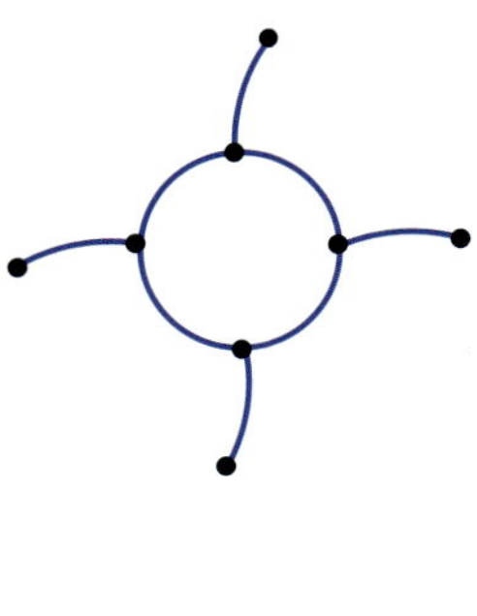

b

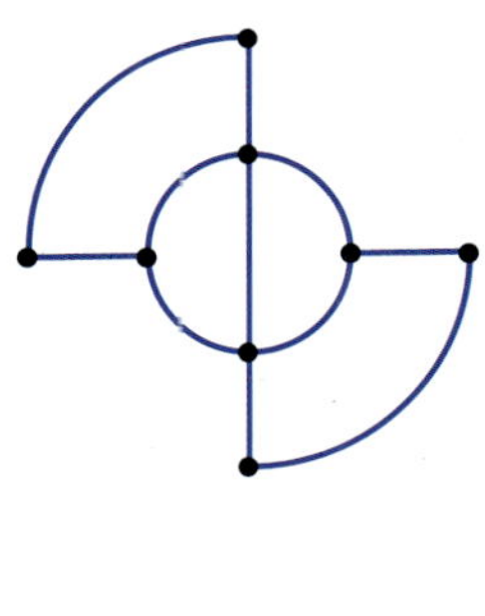

12E

16 Using each room as a vertex, draw a network diagram to show how the rooms of this house link together.

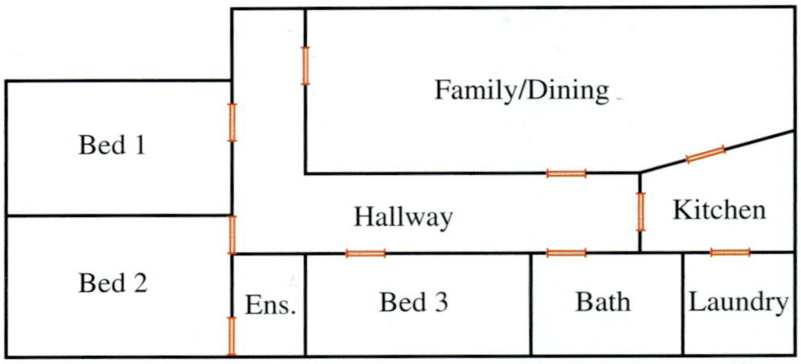

12E

17 Look at this map of Australian States and Territories. From Western Australia you can cross directly into South Australia or the Northern Territory.

Using each state or territory as a vertex, show on a network diagram how you could drive around Australia through every state and territory. (You can drive to Tasmania by putting your car on a ferry in Victoria.)

12E

18 Draw a simple directed network showing the activities you would undertake on a school morning in preparing yourself to arrive at school on time. Remember that you often perform more than one task at a time and not simply one activity after another.

12F

19 The diagram at right gives the key to a code by matching letters to coordinates. Use the key to write in code this famous statement by Descartes:
'I think therefore I am.'

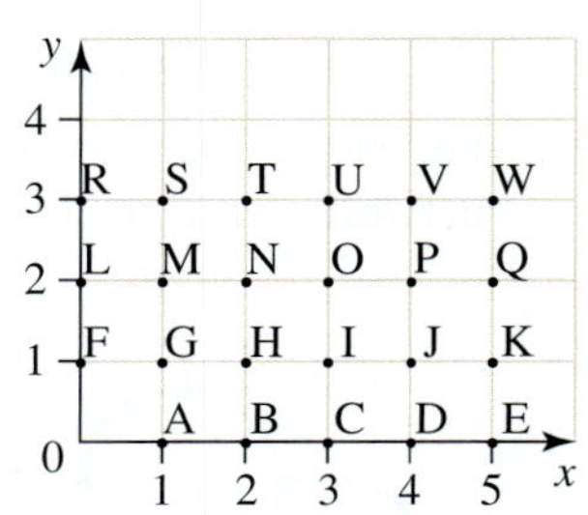

12F

20 On a carefully ruled pair of Cartesian axes join the following points in the order given. It reveals a symbol used by an ancient mathematical secret society.
(1, 0)–(6, 4)–(0, 4)–(5, 0)–(3, 6)–(1, 0)

test yourself CHAPTER 12

Data

13

Daniel conducted a survey of his class to record the number of people in each classmate's family. His results were: 3, 6, 7, 5, 4, 4, 7, 3, 2, 4, 5, 3, 4, 4, 6, 5, 5, 3, 3, 4, 6, 4, 4, 5. How many families contain 3 people? What family size is the most common?

The data as listed are difficult to work with so we need to present them in a way that makes them easier to analyse.

This chapter looks at dealing with data in different ways.

Are you ready?

Try the questions below. If you have difficulty with any of them, extra help can be obtained by completing the matching SkillSHEET. Either click on the SkillSHEET icon next to the question on the *Maths Quest 7* CD-ROM or ask your teacher for a copy.

SkillSHEET 13.1

Distinguishing qualitative from quantitative data

1 State whether the following pieces of data are numerical or non-numerical.

a The monthly temperatures
b The brands of soft drinks
c The types of dogs
d The balance in a bank account

SkillSHEET 13.2

Distinguishing discrete from continuous data

2 State whether the following pieces of data are discrete or continuous.

a The daily temperature
b Your height
c The time taken to swim 100 m
d The number of children in your family

SkillSHEET 13.4

Reading scales

3 How much is each interval worth in the following scales?

a 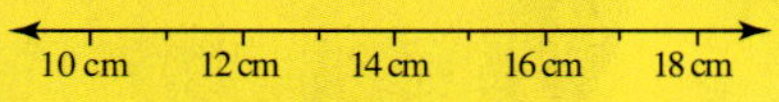

b 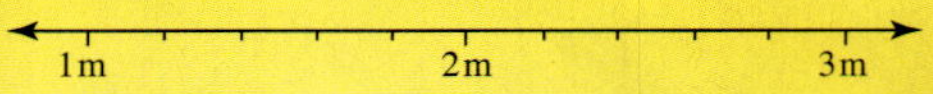

c 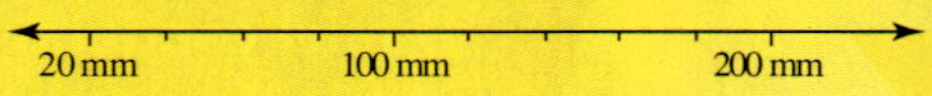

d 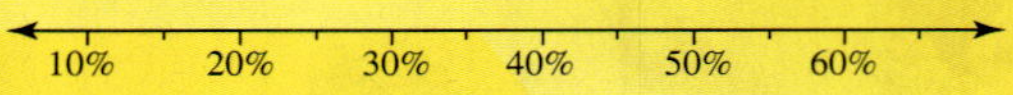

SkillSHEET 13.5

Forming fractions

4 Express the following angles as fractions of a circle (in simplest form).

a 30° **b** 45° **c** 270° **d** 300°

SkillSHEET 13.6

Multiplying fractions by a whole number

5 Simplify the following.

a $\frac{2}{3}$ of 360° **b** $\frac{5}{12}$ of 360° **c** $\frac{7}{24}$ of 360° **d** $\frac{5}{72}$ of 360°

SkillSHEET 13.10

Constructing angles with a protractor

6 Using a protractor, construct the following angles.

a 25° **b** 125° **c** 250° **d** 300°

SkillSHEET 13.13

Finding the mean of ungrouped data

7 Find the average of the following scores.

a 1, 2, 3, 4, 5
b 11, 12, 13, 14, 15
c 1, 7, 9, 4, 2
d 180, 426, 392, 874

SkillSHEET 13.15

Finding the median

8 Arrange the following scores in ascending order, then state the middle score.

a 4, 8, 3, 9, 2
b 27, 16, 2, 9, 11
c 5, 6, 4, 8, 2, 9, 4
d 9, 9, 8, 7, 8, 4, 9

Classifying data

Each day people in all types of professions are presented with various forms of information called **data**, which will assist them in answering questions such as:

How many cans of each type of drink will the school canteen sell in a week? What will be the ratio of the various flavours sold? How will this vary during the days of the week and months of the year?

Which states of Australia need extra funding?

Should a particular film be released in cinemas or go straight to video?

Which products are teenagers between the ages of 13 and 15 most likely to purchase?

Would lowering the highway speed limit to 90 km/h result in fewer road casualties?

Governments, businesses and individuals need to make good decisions based on the best available data. It is important that we are able to organise, analyse and interpret data.

Statistics is the branch of mathematics that deals with the *collection*, *organisation*, *display*, *analysis* and *interpretation* of data, which are usually presented in numerical form.

The collection and organisation of data will be covered in *Maths Quest 8 for Victoria.*

Data collected by government bodies, businesses or individuals are analysed to find out what has been happening. This information is very useful for making conclusions and decisions about what to do in the future. In this chapter we will only consider univariate data; that is, data involving one type of variable at a time, for example comparing heights.

Data may be classified in the following ways.

Data			
Qualitative Data which are placed in *categories*; that is, non-numerical form; such as hair colour, type of vehicle, and so on.		**Quantitative** Data which are in *numerical* form; such as height, number of children in the family, and so on.	
Nominal	**Ordinal**	**Discrete**	**Continuous**
Need *sub-groups* to complete the description, such as hair colour: blond, brown and so on.	Need a *ranking* to order the description, such as achievement levels: very high, high, satisfactory and so on.	*Counted* in exact values, such as goals scored in a football match, shoe size and so on. Values are often, but not always, whole numbers.	*Measured* in a continuous decimal scale, such as mass temperature, length and so on.

WORKED Example 1

Classify each of the following data using two selections from the following descriptive words: qualitative, quantitative, nominal, ordinal, discrete and continuous.

a the number of students absent from school

b the types of vehicle using a certain road

c the various pizza sizes available at a local takeaway

d the room temperature at various times during a particular day

THINK	WRITE
a 1 Determine whether the data are qualitative or quantitative.	**a** The data are quantitative as absences are represented by a number.
2 Determine whether the data are discrete or continuous.	The data are discrete as the number of absences can be counted and are exact values.
b 1 Determine whether the data are qualitative or quantitative.	**b** The data are qualitative as the types of vehicle need to be placed in non-numerical categories.
2 Determine whether the data are nominal or ordinal.	The data are nominal as there is no ranking or order involved.
c 1 Determine whether the data are qualitative or quantitative.	**c** The data are qualitative as the pizza sizes need to be ranked in order ranging from small to family.
2 Determine whether the data are nominal or ordinal.	The data are ordinal as pizzas are ranked in order of size.
d 1 Determine whether the data are qualitative or quantitative.	**d** The data are quantitative as room temperature is represented by a number.
2 Determine whether the data are discrete or continuous.	The data are continuous as temperature can assume any value and measurement is involved.

remember

1. Surveys use representative samples of the target population.
2. Samples are chosen at random.
3. Data may be classified under the following headings.
   (a) Qualitative
   Data are placed in *categories* (non-numerical form).
   • Nominal: need other *sub-groups*.
   • Ordinal: need a *ranking* system.
   (b) Quantitative
   Data are in *numerical* form.
   • Discrete: *counted* in exact values.
   • Continuous: *measured* in a continuous decimal scale.

EXERCISE 13A Classifying data

13.1 SkillSHEET Distinguishing qualitative from quantitative data

1 Match each word with its correct meaning:

a discrete	**i** placed in categories or classes
b qualitative	**ii** counted in exact values
c ordinal	**iii** data in the form of numbers
d continuous	**iv** needs further names to complete the description
e quantitative	**v** needs a ranking order
f nominal	**vi** measured in decimal numbers

WORKED Example 1

13.2 SkillSHEET Distinguishing discrete from continuous data

2 Classify each of the following data using two words selected from the following descriptive words: qualitative, quantitative, nominal, ordinal, discrete and continuous.

a the population of your town or city
b the types of motorbike in a parking lot
c the heights of people in an identification line-up
d the masses of babies in a group
e the languages spoken at home by students in your class
f the time spent watching TV
g the number of children in the families in your suburb
h the air pressure in your car's tyres
i the number of puppies in a litter
j the types of radio program listened to by teenagers
k the times for swimming 50 metres
l the quantity of fish caught in a net
m the number of CDs you own
n the types of shops in a shopping centre
o the football competition ladder at the end of each round
p the lifetime of torch batteries
q the number of people attending the Big Day Out
r final Year 12 exam grades
s the types of magazine sold at a newsagency
t hotel accommodation rating

3 Write a sentence explaining the difference between discrete and continuous data. Give an example of each.

4 List 2 examples of each of the following types of data:
a quantitative, discrete
b qualitative, ordinal
c quantitative, continuous
d qualitative, nominal.

5 **multiple choice**

Data representing shoe (or rollerblade) sizes can be classified as:
A quantitative, continuous
B qualitative, nominal
C qualitative, ordinal
D quantitative, discrete
E None of the above

6 **multiple choice**

The data, points scored in a basketball game, can be best described as being:
A discrete
B continuous
C qualitative
D ordinal
E nominal

7 **multiple choice**

An example of qualitative, ordinal data would be the:
A heights of buildings in Melbourne
B number of pets in households
C type of pets in households
D birthday month of students in Year 7
E number of hours spent playing sport

MATHS QUEST CHALLENGE

1 A fisheries and wildlife officer released 200 tagged trout into a lake. A week later, the officer took a sample of 50 trout and found that 8 of them were tagged. The officer can use this information to estimate the population of trout in the lake. How many trout are in the lake?

2 Consider the series of gears shown. In what direction will gear 5 rotate if gear 1 rotates in an anticlockwise direction?

3 Colleen had a problem. She wanted to bake a pie in the oven for exactly 10 minutes but only had 2 egg timers. One was a 4-minute timer and the other was a 7-minute timer. How can she use these timers to measure exactly 10 minutes?

COMMUNICATION

How many red M&Ms?

If we take a large pack of M&Ms and look at its contents, all the M&Ms in the pack make up what we call the *population*. A handful of M&Ms from the pack would be called a *sample* of the population. We are going to observe the results of repeated sampling of the population. From these results, we should be able to make reasonably accurate estimates of the composition of the population (our M&Ms).

1 Empty the contents of a large pack of multi-coloured M&Ms into a container. Count the number in the pack then make sure they are mixed well.

2 Without looking at the contents in the container, draw out a handful of M&Ms. Count the number of red M&Ms in the sample and the total number of M&Ms in the sample.

3 Replace the M&Ms in the container and mix well.

4 Repeat this process of drawing out a handful of M&Ms, counting the number of red M&Ms in the sample and the total number of M&Ms. Continue until you have data for ten (10) samples.

Sample number	Number of red M&Ms	Number of M&Ms
1		
2		
3		
4		
5		
6		
7		
8		
9		
10		
Total		

5 Draw up a table to collate the data collected from the ten samples.

6 If the M&Ms in the population have been mixed well, the proportion of red M&Ms in the sample should be close to the proportion of red M&Ms in the population.

$$\frac{\text{Number of red M\&Ms in sample}}{\text{Number of M\&Ms in sample}} = \frac{\text{number of red M\&Ms in population}}{\text{number of M\&Ms in population}}$$

Because the sample sizes are quite small, a better estimate is obtained by using the 'total' row for the samples. Having counted the number of M&Ms in the pack (population), you can now substitute into this equation to determine an estimate of the number of red M&Ms in the pack. What is your estimate?

7 Count the number of red M&Ms in our pack. How does this compare with your answer in part 6?

8 Repeat the experiment for different coloured M&Ms. How accurate are your results?

9 How could you adjust the experiment to estimate the total number of M&Ms in the pack without counting? Write up instructions for your experiment.

Displaying data in tables

Data can be displayed in a variety of forms. Throughout this chapter we will see different ways of presenting data using tables, graphs and pictures.

Frequency distribution tables

One way of presenting data is using a frequency distribution table. **Frequency** is the number of times a result or piece of data occurs. The frequency distribution table consists of three columns, headed score, tally and frequency.

WORKED Example 2

A particular class was surveyed to find out the number of pets per household and the data were recorded. The raw data were: 0, 3, 1, 2, 0, 1, 0, 1, 2, 4, 0, 6, 1, 1, 0, 2, 2, 0, 1, 3, 0, 1, 2, 1, 1, 2.

a Organise the data into a frequency distribution table.
b How many households were included in the survey?
c How many households have fewer than 2 pets?
d Which is the most common number of pets?
e How many households have 3 or more pets?
f What fraction of those surveyed had no pets?

THINK

a 1 Draw a frequency distribution table comprising three columns, headed score (that is, the number of pets), tally and frequency.
2 In the first column list the possible number of pets per household (that is, 0 to 6).
3 Place a stroke in the tally column each time a particular score is noted.
Note: A score of 5 is denoted as a 'gate post' (that is, four vertical strokes and one diagonal stroke 卌).
4 Write the total tally strokes for each pet in the frequency column.
5 Calculate the total of the frequency column.

WRITE

a

Score	Tally	Frequency
0	卌 II	7
1	卌 IIII	9
2	卌 I	6
3	II	2
4	I	1
5		0
6	I	1
	Total	26

THINK

b The total of the frequency column gives the number of households surveyed.

WRITE

b Twenty-six households were surveyed.

THINK

c 1 Calculate the number of households which have fewer than 2 pets.
Note: Fewer than 2 means 0 pets or 1 pet.
2 Answer the question.

WRITE

c Fewer than two pets $= 7 + 9$
$= 16$

Sixteen households have fewer than 2 pets.

THINK

d 1 Make a note of the highest value in the frequency column and check which score it corresponds to.
2 Answer the question.

WRITE

d The score with the highest frequency (that is, 9) corresponds to one pet.

The most common number of pets is one.

THINK

e 1 Calculate the number of households which have 3 or more pets.
Note: 3 or more means 3, 4, 5 or 6.
2 Answer the question.

f 1 Write the number of households with no pets.
2 Write the total number of households surveyed.
3 Define the fraction and substitute the known values into the rule.
4 Answer the question.

WRITE

e 3 or more pets $= 2 + 1 + 0 + 1$
$= 4$

Four households have 3 or more pets.

f Households with no pets = 7

Total number of households surveyed = 26

$$\frac{\text{Households with no pets}}{\text{Total number of households surveyed}} = \frac{7}{26}$$

Of the households surveyed $\frac{7}{26}$ have no pets.

Sometimes, the data we have may contain far too many individual, numerical values. Rather than construct a frequency distribution table which lists each individual value, we use a range of values, called a class interval, as our category. For example, the range 100–104 may be used to cater for all the values that lie within the range and include 100 and 104. Similarly 105–109 may be used to cater for all the values that lie within and include 105 and 109.

WORKED Example 3

The data below show the ages of a number of mobile phone owners:
12, 11, 21, 12, 30, 26, 13, 15, 29, 16, 17, 17, 17, 21, 19, 12, 14, 16, 43, 18, 51, 25, 30, 28, 33, 62, 39, 40, 30, 18, 19, 41, 22, 21, 48, 31, 33, 33, 34, 41, 18, 17, 31, 43, 42, 17, 46, 23, 24, 33, 27, 31, 53, 52, 25

a Draw a frequency table to classify the given data. Use a class interval of 10; that is, ages 11–20, 21–30 and so on, as each category.
b How many people were surveyed?
c Which age group had the largest number of people with mobile phones?
d Which age group had the least number of people with mobile phones?
e How many people in the 21–30 age group own a mobile phone?

THINK

a 1 Draw a frequency distribution table.
2 In the first column, list the possible age groups; that is, 11–20, 21–30 etc.
3 Systematically go through the results and place a stroke in the tally column each time a particular age group is noted.
4 Write the total tally of strokes for each age group in the frequency column.
5 Calculate the total of the frequency column.

WRITE

a

Age group	Tally	Frequency
11–20	卌 卌 卌 ||||	19
21–30	卌 卌 卌	15
31–40	卌 卌	10
41–50	卌 ||	7
51–60	|||	3
over 60	|	1
	Total	55

Continued over page

THINK	WRITE
b The total of the frequency column gives us the number of people surveyed.	**b** A total of 55 people were surveyed.
c 1 Make note of the highest value in the frequency column and check which age group it corresponds to.	**c** The 11–20 age group has the highest frequency; that is, a value of 19.
2 Answer the question.	The 11–20 age group has the most number of people with mobile phones.
d 1 Make note of the lowest value in the frequency column and check which age group it corresponds to. *Note*: There may be more than one answer.	**d** The over 60 age group has the lowest frequency; that is, a value of 1.
2 Answer the question.	The over 60 age group has the least number of people with mobile phones.
e 1 Check the 21–30 age group in the table to see which frequency value corresponds to this age group.	**e** The 21–30 age group has a corresponding frequency of 15.
2 Answer the question.	Fifteen people in the 21–30 age group own a mobile phone.

remember

1. *Frequency* is the number of times a result or piece of data occurs.
2. A frequency distribution table consists of 3 columns, headed category, tally and frequency.

EXERCISE 13B Displaying data in tables

SkillSHEET 13.3 Frequency distribution tables

WORKED Example 2

1 The number of children per household in a particular street is surveyed and the data recorded. The raw data are:
0, 8, 6, 4, 0, 0, 0, 2, 1, 3, 3, 3, 1, 2, 3,
2, 3, 2, 1, 2, 1, 3, 0, 2, 2, 4, 2, 3, 5, 2.

a Organise the data into a frequency distribution table.
b How many households are included in the survey?
c How many households have no children?
d How many households have at least 3 children?
e Which is the most common number of children?
f What fraction of those surveyed have 4 children?

2 Rosemary decided to survey the participants of her local gym about their preferred sport. She asked each participant to name one preferred sport and recorded her results: hockey, cricket, cricket, tennis, scuba diving, netball, tennis, netball, swimming, netball, tennis, hockey, cricket, lacrosse, lawn bowls, hockey, swimming, netball, tennis, netball, cricket, tennis, hockey, lacrosse, swimming, lawn bowls, swimming, swimming, netball, netball, tennis, golf, hockey, hockey, lacrosse, swimming, golf, hockey, netball, swimming, scuba diving, scuba diving, golf, tennis, cricket, cricket, hockey, lacrosse, netball, golf.

EXCEL Spreadsheet — Frequency tally tables

EXCEL Spreadsheet — Frequency tally tables (DIY)

- **a** Can you see any problems with the way Rosemary has displayed the data?
- **b** Organise Rosemary's results into a frequency table to show the participants' preferred sports.
- **c** From the frequency table, find:
  - **i** the most preferred sport
  - **ii** the least preferred sport.
- **d** Did any sport(s) have the same frequency?

3 Complete a frequency distribution table for each of the following.

- **a** Andrew's scores in mathematics tests this semester are: 6, 9, 7, 9, 10, 7, 6, 5, 8, 9.
- **b** The number of children in each household of a particular street are:
  2, 0, 6, 1, 1, 2, 1, 3, 0, 4, 3, 2, 4, 1, 0, 2, 1, 0, 2, 0.
- **c** The masses (in kilograms) of students in a certain Year 7 class are:
  46, 60, 48, 52, 49, 51, 60, 45, 54, 54, 52, 58, 53, 51, 54, 50, 50, 56, 53, 57, 55, 48, 56, 53, 58, 53, 59, 57.
- **d** The heights of students in a particular Year 7 class are:
  145, 147, 150, 150, 148, 145, 144, 144, 147, 149, 144, 150, 150, 152, 145, 149, 144, 145, 147, 143, 144, 145, 148, 144, 149, 146, 148, 143.

4 Use the frequency distribution table to answer the questions.

- **a** How many participated in the survey?
- **b** What was the most frequent score?
- **c** How many scored less than 3?
- **d** How many scored 3 or more?
- **e** What fraction of those surveyed scored 3?

Score	Tally	Frequency
0	II	2
1	𝍸	5
2	III	3
3	𝍸 𝍸 I	11
4	𝍸 III	8
5	IIII	4
	Total	

5 A random sample of 30 families was surveyed to find the number of high-school-aged children in each family. Below are the raw data collected:
2, 1, 1, 0, 2, 0, 1, 0, 2, 0, 3, 1, 1, 0, 0, 0, 1, 4, 1, 0, 0, 1, 2, 1, 2, 0, 3, 2, 0, 1.

- **a** Organise the data into a frequency distribution table.
- **b** How many families have no children of high school age?
- **c** How many have 2 or more children of high school age?
- **d** Which score has the highest frequency?
- **e** What is the greatest number of high-school-aged children in the 30 families surveyed?
- **f** What fraction of families had 2 children of high school age?

WORKED Example 3

6 Draw a frequency table to classify the following data on house prices. Use a class interval of 10 000; that is prices \$100 000 to \$109 000 and so on for each category. The values are: \$100 000, \$105 000, \$110 000, \$150 000, \$155 000, \$106 000, \$165 000, \$148 000, \$165 000, \$200 000, \$195 000, \$138 000, \$142 000, \$153 000, \$173 000, \$149 000, \$182 000, \$186 000.

7 Draw a frequency table to classify the following data on students' heights. Use a range of values (such as 140–144) as each category.
The values are: 168 cm, 143 cm, 145 cm, 151 cm, 153 cm, 148 cm, 166 cm, 147 cm, 160 cm, 162 cm, 175 cm, 168 cm, 143 cm, 150 cm, 160 cm, 180 cm, 146 cm, 158 cm, 149 cm, 169 cm, 167 cm, 167 cm, 163 cm, 172 cm, 148 cm, 151 cm, 170 cm, 160 cm.

8 Complete a frequency table for all vowels in the following paragraphs.
Australian Rules Football is a ball game played by two teams of eighteen players with an ellipsoid ball on a large oval field with four upright posts at each end. Each team attempts to score points by kicking the ball through the appropriate posts (goals) and prevent their opponents from scoring. The team scoring the most points in a given time is the winner. Usually this period is divided into four quarters of play.

Play begins at the beginning of a quarter or after a goal, with a tap contest between two opposing players (rucks) in the centre of the ground after the umpire either throws the ball up or bounces it down.

Questions 9, 10 and 11 refer to the following information.
A real estate agent has listed all the properties sold in the area in the last month as shown below. She wants to know what has been the most popular type of property from the following: 2 bedroom house, 4 bedroom house, 3 bedroom house, 2 bedroom unit, 4 bedroom house, 1 bedroom unit, 3 bedroom house, 2 bedroom unit, 3 bedroom house, 1 bedroom unit, 2 bedroom unit, 3 bedroom house, 3 bedroom house, 3 bedroom house, 2 bedroom unit, 1 bedroom unit.

9 Complete a frequency table for the list given and work out which type of property was most popular.

10 **multiple choice**

The least popular type of property is the:

A 1 bedroom unit
B 2 bedroom unit
C 2 bedroom house
D 3 bedroom house
E 4 bedroom house

11 **multiple choice**

The property which is half as popular as a 2 bedroom unit is the:

A 4 bedroom house
B 3 bedroom house
C 2 bedroom house
D 1 bedroom unit
E none of these

12 **multiple choice**

The frequency column of a frequency table will:

A add up to the total number of categories
B add up to the total number of results given
C add up to the total of the category values
D display the tally
E none of these

COMMUNICATION

How do they rate?

1 Determine the most popular television show among the students in your class by conducting a survey.
 a Draw a frequency table of the results.
 b Which television show was the most popular?
 c Which television show was the least popular?
 d Were any of the television shows equally popular?
 e Were there any students who did not have a favourite television show?

2 Plan and conduct a survey of each student in your class. Rate 4 different pop groups or performers (of your choice) on a scale ranging from excellent to very poor.
 a Comment on how the survey was conducted including information on:
  i the categories by which the performers were assessed
  ii the scale indicators
  iii any difficulties encountered.
 b Draw a frequency table of the results.
 c Comment on the results obtained.

Understanding graphical displays

Our world is becoming increasingly complex. Data or numerical information are used in almost every occupation and are very much a part of everyday life.

The skills needed to collect, summarise and interpret data are therefore important so we can understand the use of data in newspapers, on television and in our future education and careers.

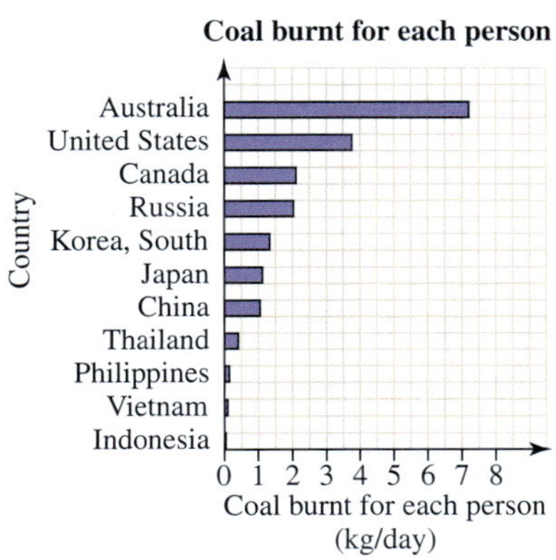

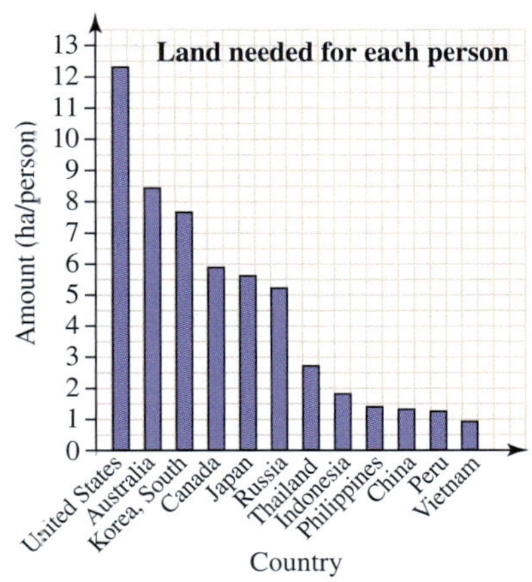

Graphs are a very useful way of displaying data, or numerical information. Newspapers regularly contain graphs to show us certain types of data. Use of graphs makes the data easier to understand than written information. The reasons for this are obvious. Pictures and images are often quicker to read and digest than paragraphs of writing.

However, without skills in reading graphs, we may receive the wrong impression from a graph or may have little understanding of what it is showing. This section of the chapter will focus on the use of graphs, reading them accurately and understanding what they can and cannot tell the reader.

All graphs should have the following features:

1. **a *title* — to tell us what the graph is about**
2. **clear *labels* for the axes — to explain what is being shown**
3. ***evenly scaled axes* — if the graph has numerical axes, they must have a scale, which must stay constant for the length of the axes and the units that are being used should be indicated**
4. **legends — these are not always necessary, but are necessary when any symbols or colours are used to show some element of the graph.**

To complete the following exercise, you will need to look very carefully at the graphs given for each question. Answering the questions will help you to understand more about the process of reading graphs.

remember

All graphs should have the following features:

1. a *title*
2. clearly *labelled* and *evenly scaled axes*
3. *legends* — when any symbols or colours are used to show some element of the graph.

EXERCISE 13C Understanding graphical displays

SkillSHEET 13.4 Reading scales

1 Telephone bills often include a graph showing how much your previous bill totals. Use the column graph at right to answer the following questions.

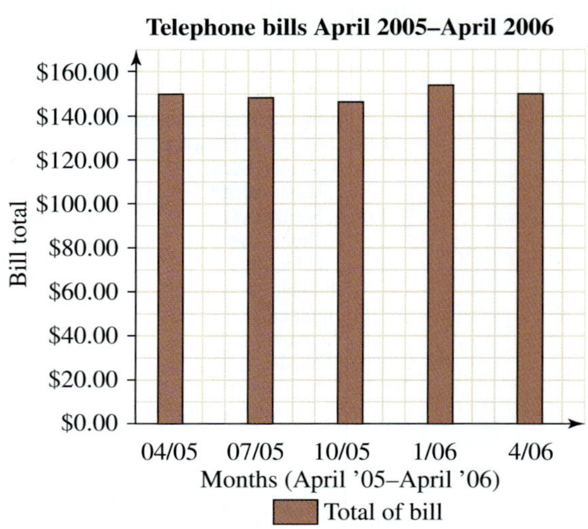

- a What is the title of this graph?
- b What is the horizontal axis label?
- c What is the vertical axis label?
- d How often does this person receive a phone bill?
- e In which month was the bill the highest?
- f Was each bill for roughly the same amount?
- g If you answered 'yes' to part f, approximately how much was the amount?
- h Why would it be useful to receive a graph like this with your phone bill?
- i If the next bill was for $240.09, would this be normal? Why?
- j How much (approximately) do phone calls from this phone cost per month?

2 Samantha and Thomas surveyed students from their school to find out how they travelled to school each day. Their results are shown in the pie graph at right.

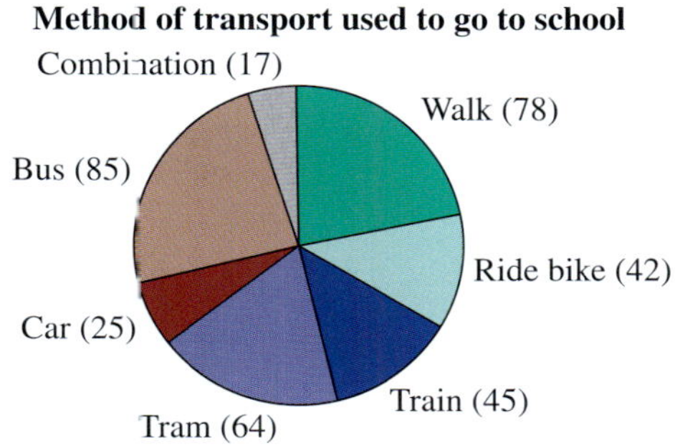

- **a** How many students were surveyed?
- **b** What is the most common method of transport?
- **c** What is the least common method of transport?
- **d** In which sector of the pie graph would a person who used both tram and train be recorded?
- **e** **i** How many students travelled by car?
  - **ii** Write this amount as a fraction of the total number of students surveyed.
  - **iii** What percentage of students travelled by car?
- **f** How many more students travelled by tram than by bike?

3 An apple producer records her sales for a 12-week period.

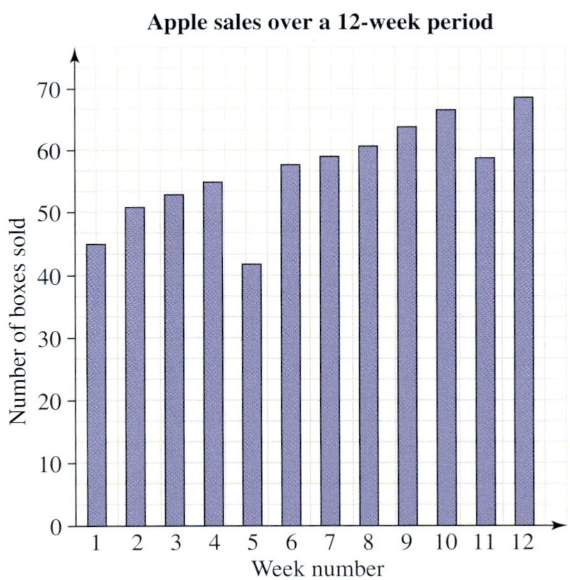

- **a** How many boxes were sold in the first week?
- **b** How many boxes were sold in the fifth week?
- **c** How many boxes were sold in the eighth week?
- **d** The values for some weeks may be unusual. Which ones might be unusual? Explain your answer.
- **e** What might cause unusual values in a graph like this?
- **f** Does the graph indicate that apple sales are improving? Explain your answer.

4 The number of cars speeding at various locations were recorded on a particular day. The results are shown in the graph below.

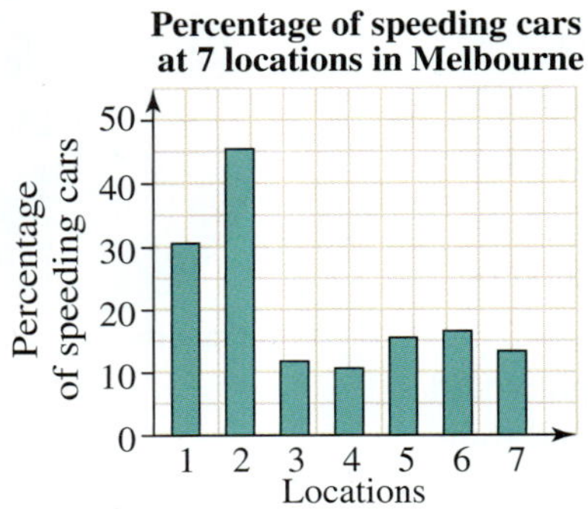

Legend

Code	Location
1	Monash Freeway, Glen Iris
2	Monash Freeway, Mulgrave
3	Maroondah Highway, Ringwood
4	Mt Alexander Rd, Essendon
5	Springvale Rd, Springvale
6	Nepean Highway, Carrum
7	Sydney Rd, Coburg

a At which location was the percentage of speeding cars greatest?
b What was the percentage of speeding cars there?
c Can we tell how many cars were speeding there?
d Why has the percentage of speeding cars, rather than number of speeding cars, been graphed?
e Why might the percentage of speeding cars be so much larger in locations 1 and 2?
f Do we know from the graph what the speed limits were in each location?
g Do we know from the graph by how much the cars exceeded the speed limit?

5 Jemima and Stephen counted the cars in the supermarket car park, dividing them into colour categories. Their results are shown in the graph at right.

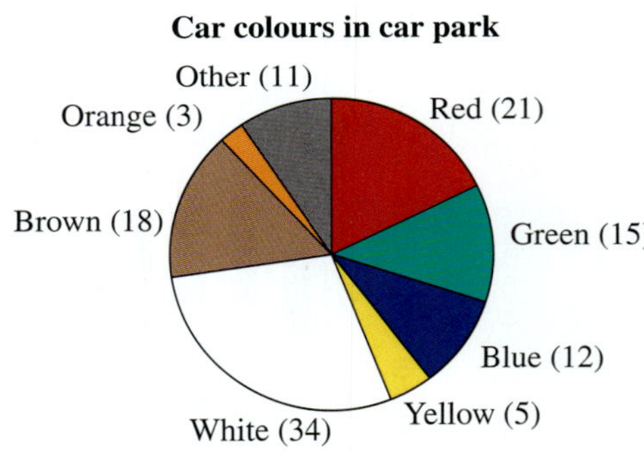

a What was the most common colour?
b What was the least common colour?
c How many cars were blue?
d How many more green cars were there than yellow cars?
e **i** How would we represent zero (or no) cars of a particular colour using this graph?
ii What type of graph could be used to represent no cars of a particular colour category?
f How many cars were in the car park altogether?
g Why would there be a category called 'other'?

6 This line graph shows the change in the value of some shares on the stock market over a day.

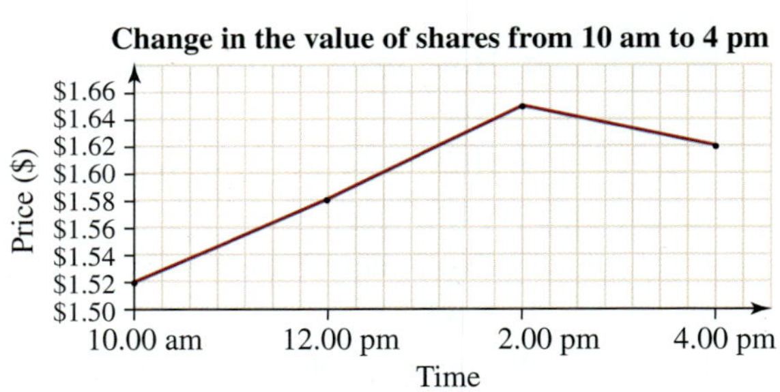

a How is this graph different from a bar or column graph? (Column graphs are used in questions 1, 3 and 4.)
b What was the share price at 10.00 am?
c What was the share price at 4.00 pm?
d Did the price rise continually throughout the day?
e Why is a line graph useful to show changes over time?

7 The chart below displays the climate of Melbourne, based on average monthly maximum temperatures and average daily rainfall for each month. The temperature is recorded in degrees Celsius and the rainfall in millimetres.

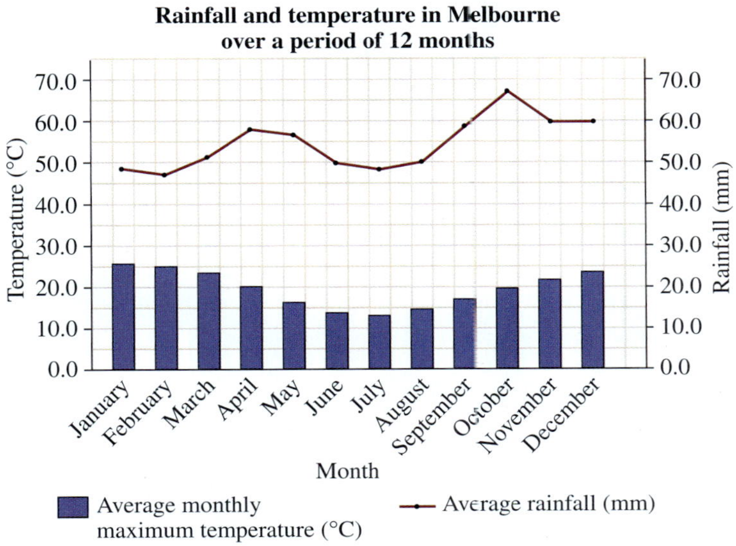

a In which month(s) is the average temperature highest?

b In which month(s) is the temperature lowest?

c In which month(s) is the rainfall highest?

d In which month is the rainfall lowest?

e With each of the 4 seasons, we expect particular characteristics. For example, in a Melbourne summer, we expect high temperatures and low rainfall. What does the graph show about the characteristics of a Melbourne:

- **i** autumn (March, April, May)?
- **ii** winter (June, July, August)?
- **iii** spring (September, October, November)?

8 In their physical education class the girls in a Year 7 class were asked to sprint for 10 seconds. The teacher recorded their results on 2 different days. They are shown in the graph at right.

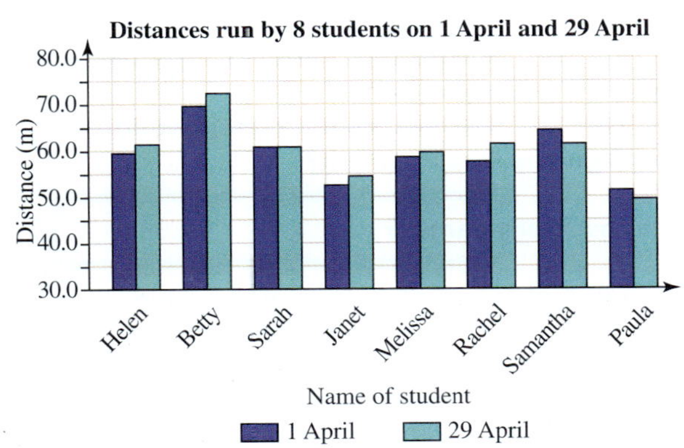

a Why are there 2 columns for each girl?

b Which girl ran the fastest on either day?

c How far did she run on each day?

d Which girl improved the most?

e Were there any students who did not improve? Who were they?

f Could this graph be misleading in any way? Explain your answer.

g Why might the graph's vertical axis start at 30 m?

WorkSHEET 13.1

What did Jacques Cousteau invent in 1942?

Answer the questions below.

The question number, and its answer letter gives the puzzle answer code.

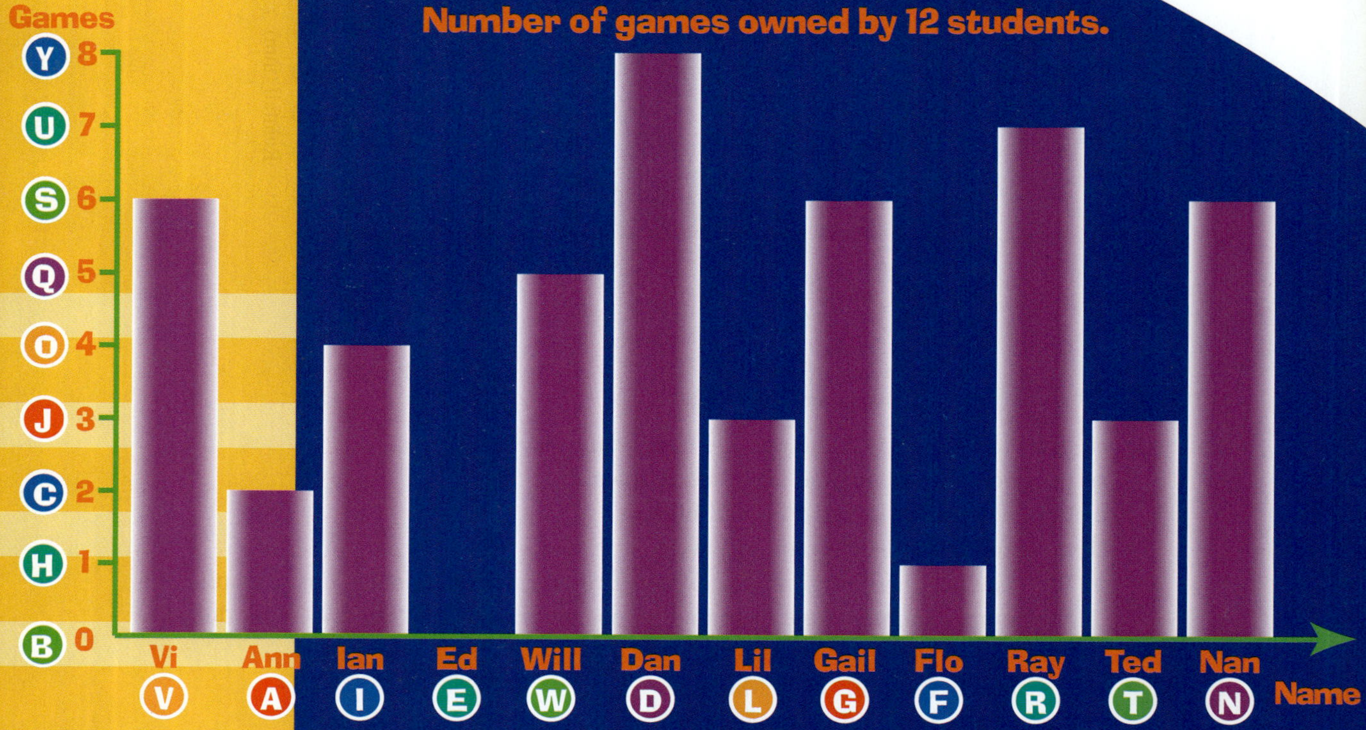

1. Who has the same number of games as Lil?
2. How many games does Flo have?
3. The person with no games.
4. Who has 3 games less than Will?
5. Will has how many games?
6. The second highest number of games owned.
7. Ted and this person have the same number of games.
8. Gail and Vi and this person have the same number of games.
9. This person has 5 more games than Flo and it's not Nan or Vi.
10. This person has only one game.
11. How many games does Ian have?
12. Who has seven games?
13. Who has the most games?
14. Who has 3 games less than Dan?
15. This person has one more game than Lil and Ted.
16. She has the third highest number of games and it's not Gail or Nan.

1	2	3	4	5	6	4	7	6	8	9	10	11	12

6	8	13	3	12	14	4	1	3	12	13	15	16	15	8	9

10 QUICK QUESTIONS 1

1 State whether the political parties represented in Parliament are an example of qualitative or quantitative data.

2 Are the salaries of the members of Parliament discrete or continuous data?

3 Is the basketball ladder at the end of each round an example of nominal or ordinal data?

4 A survey is conducted on the colours used in an artwork. Are the data nominal, ordinal, discrete or continuous?

Questions **5** to **8** relate to the column graph at right which shows the cost of grapes (per kilogram) over a number of months.

Cost of grapes (per kg) between January and May

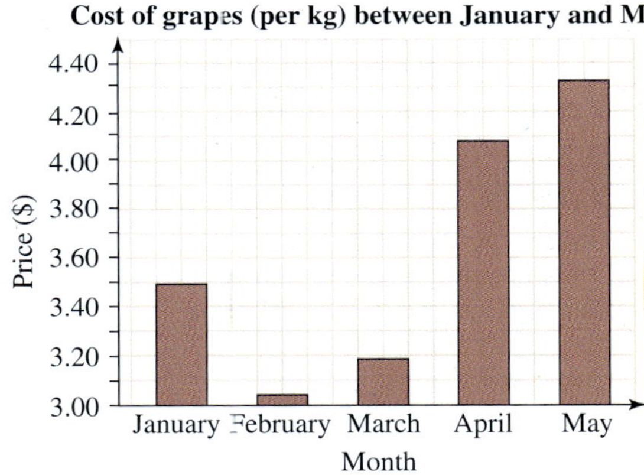

5 What was the price of grapes in March?

6 In which month did grapes cost the least?

7 During how many months were the cost of grapes recorded?

8 In which month was the cost of grapes the highest, and what was this cost?

9 Complete a frequency table for the following data from a survey about sports which are played by students. Forty students were surveyed to obtain the data:
cricket, tennis, basketball, none, football, basketball, basketball, cricket, none, tennis, tennis, cricket, football, tennis, basketball, basketball, hockey, football, cricket, none, football, tennis, tennis, hockey, none, basketball, basketball, basketball, cricket, cricket, basketball, football, football, none, cricket, hockey, tennis, hockey, none and basketball.

10 **a** Construct a frequency table to show the ages of the 577577following people who were in a picture theatre. Use a range of values (that is, 0–4 years, 5–9 years, 10–14 years and so on) to construct the table using these data:
15, 20, 48, 36, 61, 4, 17, 51, 26, 38, 40, 12, 61, 32, 18, 29, 53, 15, 22, 58, 47, 16, 31, 36, 26, 49, 6, 15, 28, 24, 36, 46, 60, 23, 29, 11, 42, 18, 25, 18, 46, 39, 56, 57, 43, 62, 24, 16, 11.

b Use the frequency table in question **9** to identify the age group with the largest number of people in the picture theatre.

Displaying data as graphs

Column and **bar graphs** are the simplest types of graphs we can use to show the results of data collection. We use categories to divide our results into groups and we count the number of results (frequency) for each category to graph them. It is easiest to graph a set of data when you have completed a frequency table. A frequency table makes the information you need easy to read.

Column graphs

Column graphs should be presented on graph paper and have:

1. **a title**
2. **labelled axes which are clearly and evenly scaled**
3. **columns of the same width**
4. **an even gap between each column**
5. **the first column beginning half a unit (that is, half the column width) from the vertical axis.**

WORKED Example 4

Beth surveyed the students in her class to find out their preferences for the school uniform. Her results are shown in the table below.

Type of uniform	Tally	Frequency
White shirt and black skirt/trousers	~~||||~~ |||	8
Blue shirt and black skirt/trousers	||||	4
Blue shirt and navy skirt/trousers	~~||||~~ ~~||||~~ ||	12
White shirt and navy skirt/trousers	~~||||~~	5
	Total	29

Construct a column graph to display the results.

THINK

1. Rule a set of axes on graph paper. Provide a title for the graph. Label the horizontal and vertical axes.
2. Scale the horizontal and vertical axes. *Note*: Leave a half interval at the beginning and end of the graph; that is, begin the first column half a unit from the vertical axis.
3. Draw the first column so that it reaches a vertical height corresponding to 8 people. Label the section of the axis below the column as 'White shirt and black skirt/trousers'.
4. Leave a gap (measuring one column width) between the first column and the second column.
5. Repeat steps 3 and 4 for each of the remaining uniform types.

DRAW

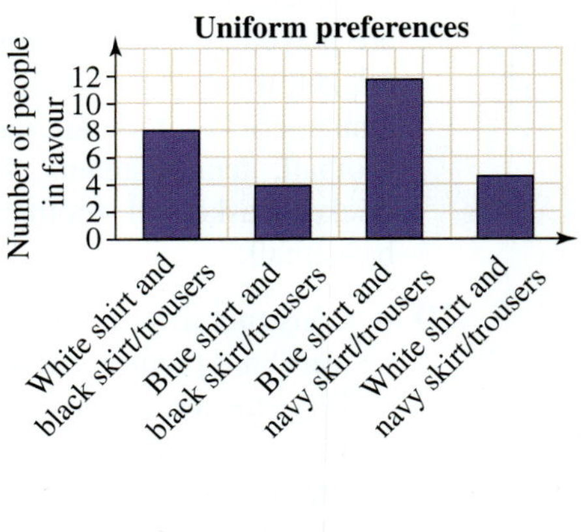

Bar graphs

Bar graphs are drawn in a similar manner to column graphs. However, there is one major difference. To draw a bar graph, numbers are placed on the horizontal axis and categories on the vertical axis. Therefore, instead of having vertical columns we have horizontal bars.

When drawing bar graphs, they should be presented on graph paper and have:

1. **a title**
2. **labelled axes which are clearly and evenly scaled**
3. **horizontal bars of the same width**
4. **an even gap between each horizontal bar**
5. **the first horizontal bar beginning half a unit (that is, half the bar width) above the horizontal axis.**

WORKED Example 5

This table shows the weekly expenses of a typical Australian family. Construct a bar graph to display the following data.

Item	Amount ($)
Food and drink	140
Rent or mortgage	200
Transport	70
Clothing	50
Recreation	70
Other	160

THINK

1. Rule a set of axes on graph paper. Provide a title for the graph and label the axes.
2. Scale the horizontal and vertical axes. *Note:* Leave a half interval at the beginning and end of the graph; that is, begin the first horizontal bar half a unit above the horizontal axis.
3. Draw the first horizontal bar so that it reaches a horizontal distance corresponding to $140. Label the section of the axis beside the horizontal bar as 'Food and drink'.
4. Leave a gap between the first horizontal bar and the second horizontal bar.
5. Repeat steps 3 and 4 for each of the remaining items.

DRAW

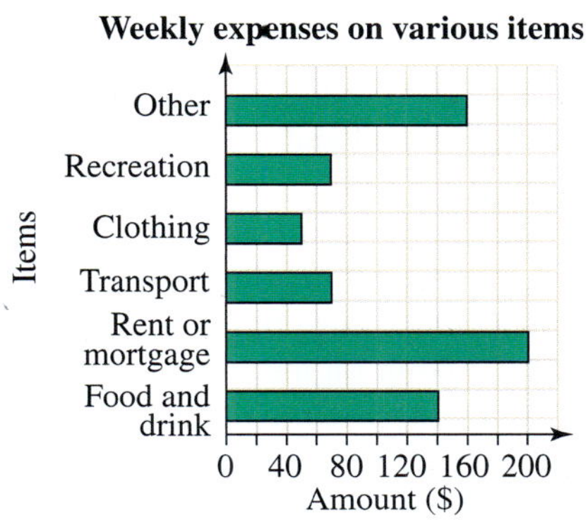

Line graphs

Line graphs are graphs which are best suited to displaying changes in values over a period of time. To draw a line graph, choose scales for both the vertical and horizontal axes, in the same way as for a column graph's vertical axis. Then simply plot points from a table of values and join them with straight lines. Line graphs also provide useful information about what happens in between particular points of data.

Line graphs should be drawn on graph paper and include:

1. **a title**
2. **a horizontal axis which is evenly scaled and labelled (usually as time)**
3. **a vertical axis which is evenly scaled and labelled**
4. **a line which joins successive plotted points.**

WORKED Example 6

The following table of values shows the number of goals scored in a ball game.

Time in minutes (horizontal axis)	0	10	20	30	40	50	60	70	80	90	100
Number of goals (vertical axis)	0	5	12	17	19	25	28	30	35	38	42

a **Draw a line graph showing the results from the above table of values.**
b **How many goals had been scored after 10 minutes?**
c **How long (in minutes) was the match?**
d **How many goals had been scored by half-time?**

THINK

a 1 Rule a set of axes on graph paper. Provide a title for the graph and label the axes.

2 Scale the horizontal and vertical axes.

3 Plot the first point on the axes. Locate 0 minutes on the time axis and 0 goals; mark this point.

4 Plot the next point on the axes. Locate 10 minutes on the time axis and follow an imaginary vertical line until a height representing 5 goals is reached; mark this point.

5 Repeat step 4 for each pair of remaining data.

6 Join the points with a straight line.

DRAW/WRITE

a

Goals scored in a match

Goals scored in a match

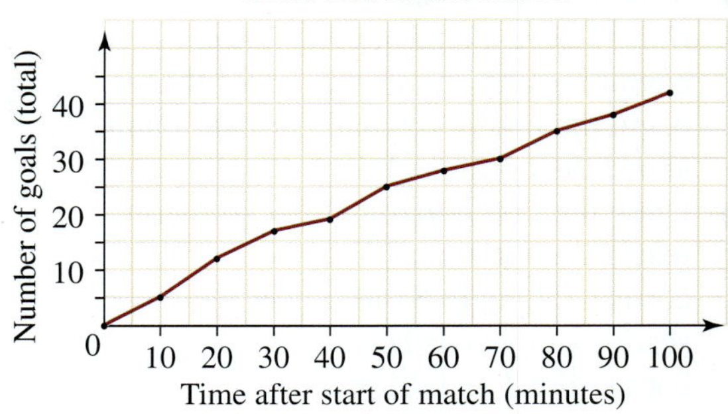

THINK	DRAW/WRITE
b Locate 10 minutes in the table of values and look directly below to see the corresponding number of goals. Alternatively, the graph may be used. Read vertically up from 10 minutes on the time axis until the point on the graph is reached and then read across until the corresponding goal value on the vertical axis is obtained.	**b** Five goals were scored after 10 minutes.
c Look at the last time value in the table of values or refer to the graph to find the length of the match.	**c** The match was 100 minutes long.
d 1 Determine when half-time occurs; that is, divide the length of the match by 2.	**d** Half-time $= \frac{\text{length of match}}{2}$
2 Evaluate.	$= \frac{100}{2}$ minutes $= 50$ minutes
3 Using either the graph or the table of values, obtain the corresponding number of goals.	Half-time occurs after 50 minutes which corresponds to 25 goals.
4 Answer the question.	Twenty-five goals were scored by half-time.

Pie graphs

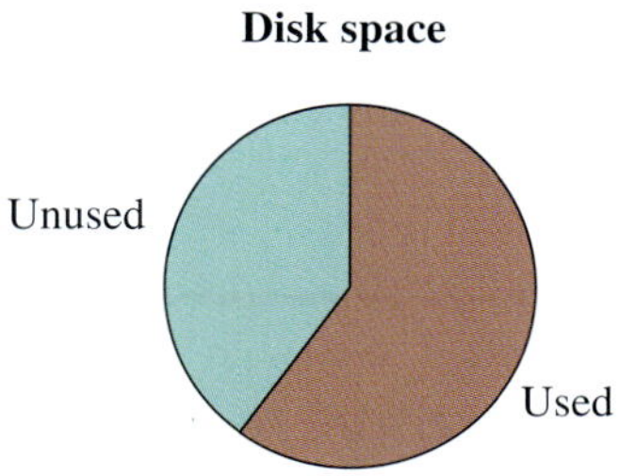

Pie graphs or pie charts are quite distinct from any of the graphs we have looked at or worked with so far. These graphs have no vertical or horizontal axes. Pie graphs are used to display the different parts or portions which make up one whole of an object or event. They are sometimes called sector graphs as they are made up of sectors of a circle. Consider the pie graph at right.

Reading pie graphs like this one can be difficult as there is a limited amount of information. We are able to determine the proportion (fraction) or **percentage** of used and unused disk space; however, we don't know the size of the disk. When drawing the pie graphs, it is important to remember that the more information you can show, the easier it will be to read. To read an unlabelled graph, you will need a protractor (remember that there are 360° in a circle).

To determine information about a sector, follow these steps.
1. **Measure the size of the angle of each sector of the graph.**
2. **Express each of the angles as a fraction of 360°.**
3. **Simplify the fraction if possible. This is called the proportion.**
4. **Multiply the fraction by 100%. This is called the percentage.**

In the given pie graph, 60% of the disk space is used, while 40% is unused. When drawing pie graphs or interpreting information from them, it is important to be able to convert angles of a circle to percentages and percentages to angles of a circle.

WORKED Example 7

a Calculate the required angle size for the following:

i $\frac{7}{15}$ ii 20%.

b Calculate the fraction and percentage of a circle that an angle of 54° represents.

THINK	WRITE
a i Convert the given fraction to an angle by multiplying by 360°.	**a i** Angle = $\frac{7}{15} \times 360°$ $= \frac{2520}{20}$ = 168°
ii 1 Convert the given percentage to a fraction; that is, express as a fraction over 100.	**ii** $20\% = \frac{20}{100}$
2 Convert the fraction to an angle by multiplying by 360°.	Angle = $\frac{20}{100} \times 360°$ $= \frac{7200}{100}$ = 72°
b 1 Convert the given angle to a fraction by dividing by 360°.	**b** 54° written as a fraction of a circle is $\frac{54}{360}$
2 Simplify the fraction.	$\frac{54}{360} = \frac{3}{20}$
3 Convert the fraction to a percentage by multiplying by 100%.	$= \frac{3}{20} \times 100\%$
4 Evaluate.	= 15%
5 Answer the question.	An angle of 54° is equivalent to the fraction $\frac{3}{20}$ or 15%.

WORKED Example 8

Look at the pie graph on the right.

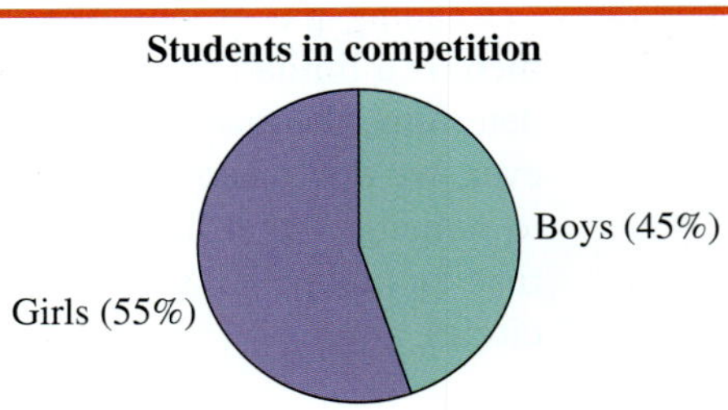

a What percentage of students in the competition are boys?

b What percentage of students in the competition are girls?

c What is the total of the percentages?

d Using a protractor, determine the size of the angle of the sector showing the boys in the competition.

e Using calculations, determine the size of the angle of the sector showing the boys in the competition. Comment on the results obtained.

THINK	WRITE
a Read the value from the pie graph and answer the question.	**a** 45% of the students in the competition are boys.
b Read the value from the pie graph and answer the question.	**b** 55% of the students in the competition are girls.
c 1 Add the two percentages obtained in parts **a** and **b**.	**c** Total = boys + girls $= 45 + 55$ $= 100$
2 Answer the question.	The total of the percentages is 100%.
d Use a protractor to measure the angle of the boys' sector.	**d** The sector showing the boys in the competition has an angle of 162°.
e 1 Convert the given percentage to a fraction; that is, express as a fraction over 100.	**e** $45\% = \frac{45}{100}$
2 Convert the fraction to an angle by multiplying by 360°.	$\text{Angle} = \frac{45}{100} \times 360°$ $= \frac{16\,200}{100}$ $= 162°$
3 Compare the answers obtained in parts **d** and **e**.	Both methods give the same result. However, the calculations would give a more accurate answer every time.

WORKED Example 9

For the following data showing the number of cars of various colours sold by the local dealer:

a copy and complete the following table of values

b construct a pie graph to display the data.

Colour of car sold	Number of cars	Fraction	Angle size
Red	6		
Blue	6		
Green	9		
Purple	3		
Orange	3		
White	9		
Total	36		

Continued over page

THINK	WRITE
a 1 Determine the fraction of the total number of cars represented by red cars, by dividing the number of red cars by the total number of cars.	**a** Fraction of red cars $= \dfrac{\text{number of red cars}}{\text{total number of cars}}$ $= \frac{6}{36}$ $= \frac{1}{6}$
2 Repeat step 1 for each of the coloured cars.	Fraction of blue cars $= \frac{6}{36}$ $= \frac{1}{6}$ Fraction of green cars $= \frac{9}{36}$ $= \frac{1}{4}$ Fraction of purple cars $= \frac{3}{36}$ $= \frac{1}{12}$ Fraction of orange cars $= \frac{3}{36}$ $= \frac{1}{12}$ Fraction of white cars $= \frac{9}{36}$ $= \frac{1}{4}$
3 Add each of the fractions to obtain the total. *Note:* The sum should equal one.	Total $= \frac{1}{6} + \frac{1}{6} + \frac{1}{4} + \frac{1}{12} + \frac{1}{12} + \frac{1}{4}$ $= 1$
4 Convert each fraction to an angle by multiplying by 360°.	Red sector $= \frac{1}{6} \times 360° = 60°$ Blue sector $= \frac{1}{6} \times 360° = 60°$ Green sector $= \frac{1}{4} \times 360° = 90°$ Purple sector $= \frac{1}{12} \times 360° = 30°$ Orange sector $= \frac{1}{12} \times 360° = 30°$ White sector $= \frac{1}{4} \times 360° = 90°$
5 Add each of the angles to obtain the total sum and enter in the table. *Note:* The sum should equal 360°.	Total $= 60° + 60° + 90° + 30° + 30° + 90°$ $= 360°$

THINK | **WRITE**

6 Enter all results into the table.

Colour of car sold	Number of cars	Fraction	Angle represented by fraction
Red	6	$\frac{1}{6}$	60°
Blue	6	$\frac{1}{6}$	60°
Green	9	$\frac{1}{4}$	90°
Purple	3	$\frac{1}{12}$	30°
Orange	3	$\frac{1}{12}$	30°
White	9	$\frac{1}{4}$	90°
Total	36	1	360°

b 1 Use a pair of compasses to draw a circle and clearly mark the centre.

2 Draw a horizontal line extending from the centre of the circle to the rim of the circle. This is called the radius.

3 Line up the centre of the protractor's baseline (that is, where the vertical, 90° line, intersects with the horizontal baseline) with the centre of the circle.

4 Measure the angle of the red sector calculated in part **a** (that is, 60°) and mark this on the diagram.

5 Label this sector as red, shade it and place the number 6 (that is, the number of red cars) in brackets next to the word 'red'.

b

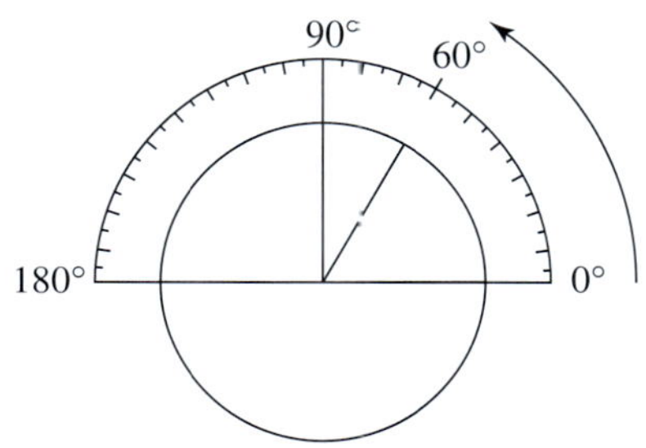

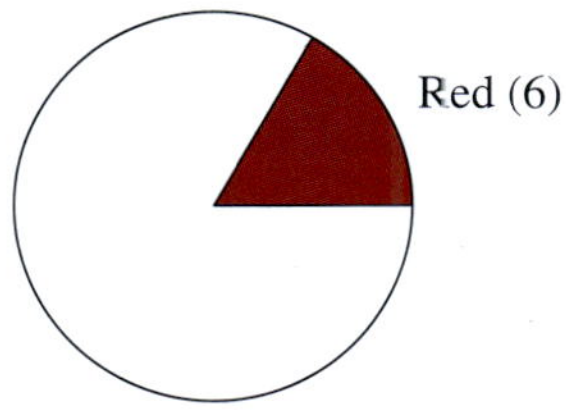

Continued over page

THINK | **WRITE**

6 Line up the centre of the protractor's base with the centre of the circle. Position the baseline of the protractor so that it is now directly above the line which denotes the end of the first sector. *Note*: You may need to rotate your page to mark the angle.

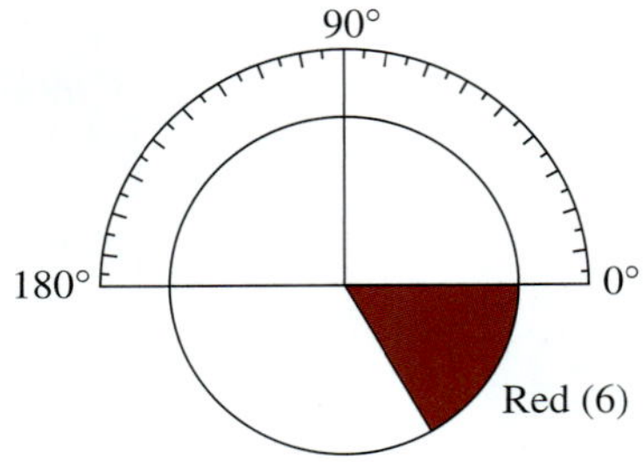

7 Measure the angle of the blue sector calculated in part **a** (that is, 60°) and mark this on the diagram.

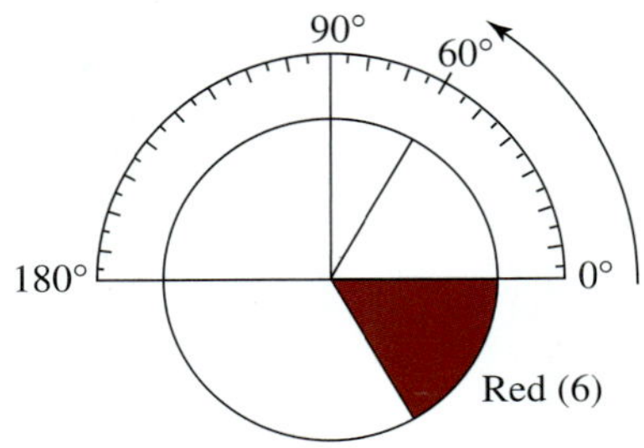

8 Label the sector as blue, shade it and place the number 6 in brackets next to the word 'blue'.

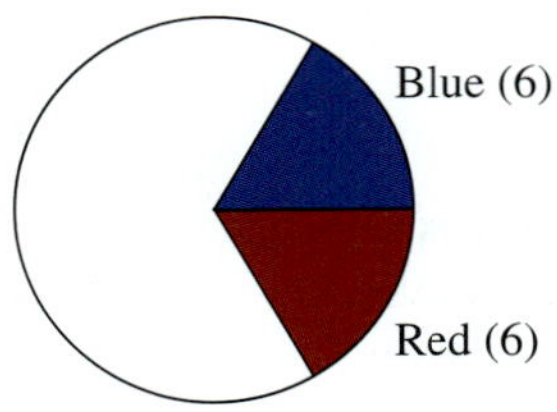

9 Repeat steps 6, 7, and 8 for the remaining sectors.

Number of cars of various colours from a total of 36 cars

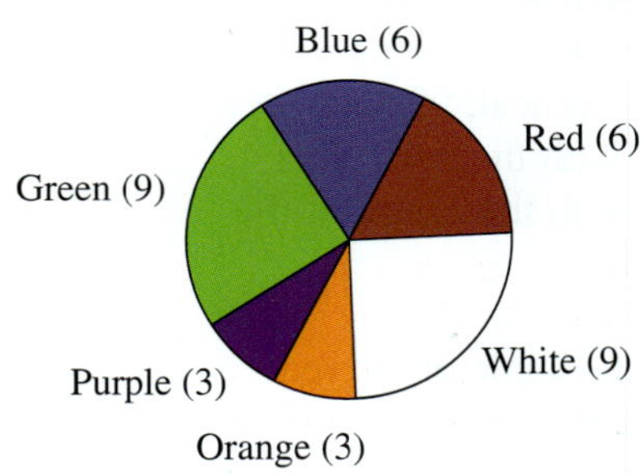

Dot plots

Dot plots (or **line plots**) can be likened to picture graphs where each piece of data or score is represented by a single dot.

Dot plots consist of a horizontal axis that is labelled and evenly scaled, and each data value is represented by a dot.
Dot plots give a quick overview of a particular distribution. They show clustering, extreme values, and help to determine whether data should be grouped.

If a score is repeated in a dot plot, a second dot is placed directly above the previous one. Once all values have been recorded, the data points, if neatly drawn and evenly spaced, resemble columns placed over a number line.

WORKED Example 10

Over a 2-week period, the number of packets of potato chips sold from a vending machine each day was recorded:
10, 8, 12, 11, 12, 18, 13, 11, 12, 11, 12, 12, 13, 14.
a Draw a dot plot of the data.
b Comment on the distribution.

THINK

a 1. Use a scaled number line to include the full range of data recorded.
2. Place a dot above the appropriate scale number for each value recorded.

b Comment on interesting features of the dot plot, such as the range, clustering, extreme values and any practical conclusions that fit the situation.

WRITE

a

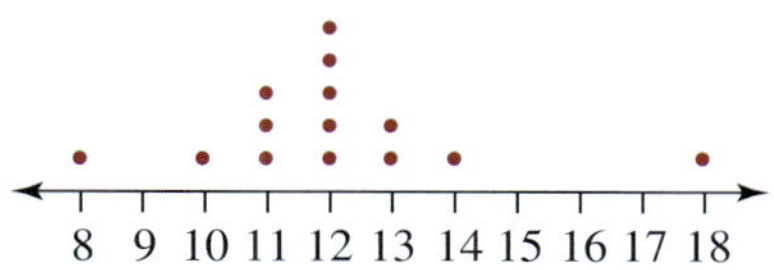

b For the given dot plot:
The scores extend from 8 to 18; that is, a range of ten.
Mostly between 11 to 13 packets were sold.
Sales of 8 and 18 packets of chips were extremely low.
A provision of 20 packets of chips each day should cover the most extreme demand.

remember

1. Graphs should be drawn on graph paper for greater accuracy.
2. *Column graphs* must have:
   (a) a clear, bold title
   (b) labelled axes which are clearly and evenly scaled
   (c) columns of the same width
   (d) an even gap between each column
   (e) the first column beginning half a unit from the vertical axis.
3. *Bar graphs* must have:
   (a) a clear, bold title
   (b) labelled axes which are clearly and evenly scaled
   (c) horizontal bars of the same width
   (d) an even gap between each horizontal bar
   (e) the first horizontal bar beginning half a unit above the horizontal axis.

(*continued*)

4. *Line graphs* must be drawn on graph paper and include:
   (a) a title
   (b) horizontal and vertical axes which are evenly scaled and labelled
   (c) a line which joins successive plotted points.
5. The horizontal axis of a line graph usually displays time.
6. To find the angle in a circle for a given fraction, multiply by 360°.
7. When converting percentages to angles in a circle, first change the percentage to a fraction then multiply by 360°.
8. An angle can be written as a fraction of 360°.
9. To draw a *pie graph* or pie chart, follow these steps:
   (a) draw a circle with a pair of compasses
   (b) use a ruler to mark the radius
   (c) measure the angle which corresponds to the first category and label it
   (d) measure the other angles around the circle
   (e) shade each sector of the pie graph
   (f) include a legend explaining the shading or, alternatively, label each sector.
10. Dot plots (or line plots) give a quick overview of a distribution: they show clustering, extreme values, and help us to see if data should be grouped.

EXERCISE 13D Displaying data as graphs

EXCEL Spreadsheet Column graphs

Mathcad Column graphs

EXCEL Spreadsheet Column graphs (DIY)

WORKED Example 4

1 Beth surveyed the students in her class to find out their method of travelling to school. Her results are shown in the table at right.
Construct a column graph to display the data.

Transport	Tally	Frequency
Car	卌 卌 卌	15
Tram	卌 ||||	9
Train	卌 卌 卌 |||	18
Bus	卌 |||	8
Bicycle	|||	3
	Total	53

2 Construct a column graph to display the data at right, showing the types of pets owned by the people in a certain street.

Type of pet	Number owned
Dog	16
Cat	12
Budgerigar	5
Rabbit	6
Snake	1
Guinea pig	3
Canary	3
Pony	1
Mouse	5

3 Construct a column graph to display the data at right, showing the number of runs scored by Australian batsmen in a recent cricket test innings.

Batsman	Number of runs
M. Hayden	112
D. Martyn	18
R. Ponting	7
J. Langer	3
S. Katich	2
A. Gilchrist	26
S. Warne	62
G. McGrath	23

4 Construct a column graph to display the data at right, showing the mean daily maximum temperatures for each month in Cairns, Queensland.

Month	Mean daily maximum temperature (°C)
January	31.8
February	31.5
March	30.7
April	29.3
May	27.6
June	25.9
July	25.6
August	26.4
September	27.9
October	29.7
November	30.8
December	31.8

WORKED Example 5

5 The table at right shows the number of students absent from school each day in a fortnight. Construct a bar graph to display the data.

EXCEL Spreadsheet Bar graphs

EXCEL Spreadsheet Bar graphs (DIY)

Day	Number of students absent
Monday	15
Tuesday	17
Wednesday	20
Thursday	10
Friday	14
Monday	16
Tuesday	14
Wednesday	12
Thursday	5
Friday	14

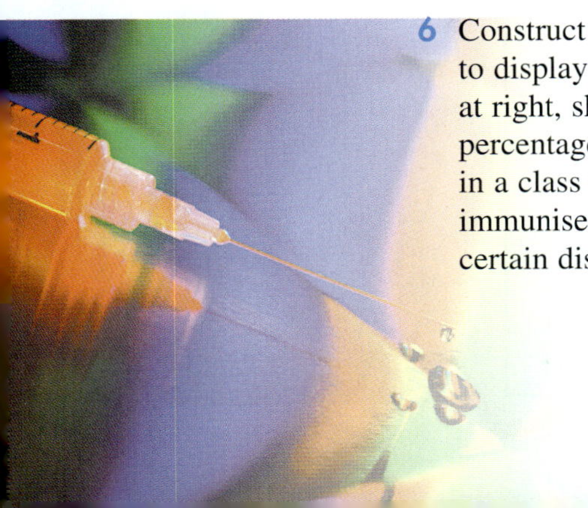

6 Construct a bar graph to display the data at right, showing the percentage of children in a class who are immunised against certain diseases.

Disease	Percentage immunised
Measles	95
Mumps	68
Chickenpox	52
Rubella	65
Hepatitis A	12
Hepatitis B	15
Tetanus	85

EXCEL Spreadsheet: Line graphs

Mathcad: Line graphs

EXCEL Spreadsheet: Line graphs (DIY)

WORKED Example 6

7 The table of values at right shows the number of bacteria after an infection started in an animal.
  a Construct a line graph to display the data.
  b How many bacteria were there when the animal first became infected?
  c How long did it take for this number to double?
  d How long would it be before the number of bacteria was 128 000?

Time after infection (min)	Number of bacteria
0	1 000
1	2 000
2	4 000
3	8 000
4	16 000
5	32 000
6	64 000

8 The table of values at right shows the number of people in a stadium over a period of time when a cricket match is played.
  a Construct a line graph to display the data.
  b What time would you think the match started?
  c Which time would have been busiest for gate attendants?
  d What was the maximum number of people in the stadium throughout the day?
  e If entry costs were $27.50 per person, how much money in total was paid for entry tickets?

Time	Number of people in stadium
10.00 am	150
11.00 am	2000
12.00 pm	3500
1.00 pm	10 500
2.00 pm	24 000
3.00 pm	24 100
4.00 pm	24 200
5.00 pm	24 100
6.00 pm	2000
7.00 pm	0

9 The table of values below shows the height of a tree after planting.
  a Construct a line graph to display the data.
  b How tall was this tree when it was planted?
  c Has this tree grown at the same rate every year? (Did it increase by the same amount each time?)
  d Why might it have grown faster in some years than other

Number of years since tree planted	Height of tree (m)
0	0.25
1	0.35
2	0.45
3	0.58
4	0.78
5	1.00
6	1.20
7	1.56
8	1.89
9	2.05
10	2.48
11	2.95

10 The following table of values shows the first 10 overs of a limited overs (50 overs) cricket match and the number of runs scored after each over.

13.5 SkillSHEET Forming fractions

Over	Total runs scored after each over
0	0
1	4
2	6
3	9
4	14
5	19
6	25
7	27
8	29
9	35
10	42

EXCEL Spreadsheet Pie graphs

13.6 SkillSHEET Multiplying fractions by a whole number

a Draw a line graph to display these data.
b What was the score after 1 over?
c What was the score after 10 overs?
d Would it be reasonable to predict the team's total score would be 5 times the score after 10 overs? Explain your answer.

13.7 SkillSHEET Simplifying fractions

WORKED Example 7a

11 Calculate the required angle size for the following.

a $\frac{1}{2}$ **b** $\frac{1}{3}$ **c** $\frac{1}{10}$ **d** $\frac{1}{12}$ **e** $\frac{1}{4}$
f $\frac{1}{6}$ **g** $\frac{1}{9}$ **h** $\frac{1}{8}$ **i** $\frac{1}{24}$ **j** $\frac{1}{60}$
k 10% **l** 17% **m** 12% **n** 92% **o** 60%

13.8 SkillSHEET Converting a fraction to a percentage

WORKED Example 7b

12 Calculate what fraction and percentage of a circle is represented by the following angles.

a 90° **b** 45° **c** 60° **d** 30°
e 20° **f** 10° **g** 15° **h** 36°
i 72° **j** 24° **k** 12° **l** 9°
m 18° **n** 180° **o** 120°

13.9 SkillSHEET Measuring angles with a protractor

Mathcad Pie graphs

WORKED Example 8

13 Look at the pie graph on the right.

a What percentage of votes went to Danielle?
b What percentage of votes went to Andrew?
c What percentage of votes went to Christiana?
d What is the total of the percentages?
e Using a protractor, determine the size of the angle of the sector showing the votes received by:
i Danielle **ii** Andrew **iii** Christiana.

Election results

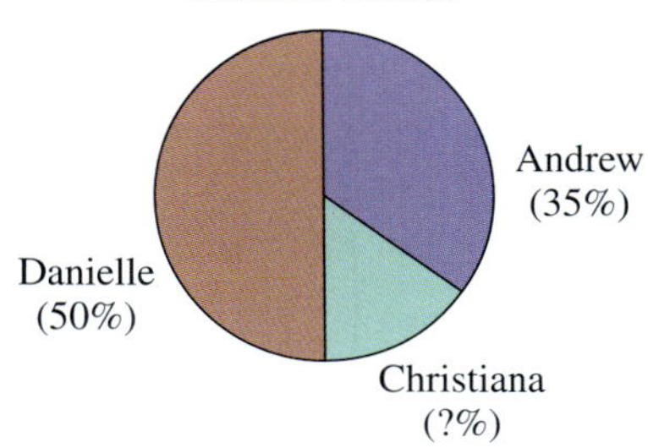

EXCEL Spreadsheet Pie graphs (DIY)

f Using calculations, determine the size of the angle of the sectors showing the votes for the 3 candidates.
g Comment on the results obtained.

14 Look at the pie graph at right.

a From the graph, state the total number of dogs in the pound.
b How many labradors were in the pound?
c How many Rottweilers were in the pound?
d What fraction of the total number of dogs were terriers?
e What angle should the terrier sector be?
f Which type of dog was the fewest in number?

Type of dogs in the pound

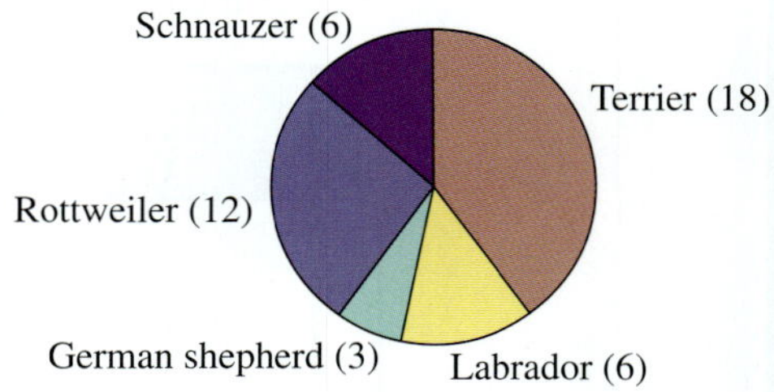

15 Look at the pie graph at right showing the candidates and the percentage of votes received in a club election recently.

a Who won the election?
b What percentage of votes did Paula receive?
c Can this pie graph tell us how many people voted in the election? Explain your answer.
d What would be the size of the angle, to 2 decimal places, for the sector representing Brigid's votes?

Votes received

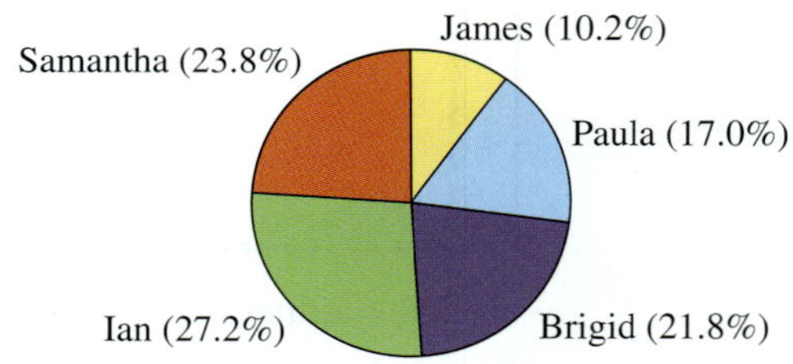

SkillSHEET 13.10 Constructing angles with a protractor

SkillSHEET 13.11 Drawing pie graphs

WORKED Example 9

16 For the following table of values, showing the number of kilograms of meat of various types which a butcher sold in a day:

a copy and complete the table of values
b construct a pie graph to display the data.

Type of meat	Amount sold (kg)	Fraction	Angle size (°)
Lamb	10		
Beef	45		
Pork	5		
Chicken	15		
Turkey	10		
Rabbit	5		
Total	90		

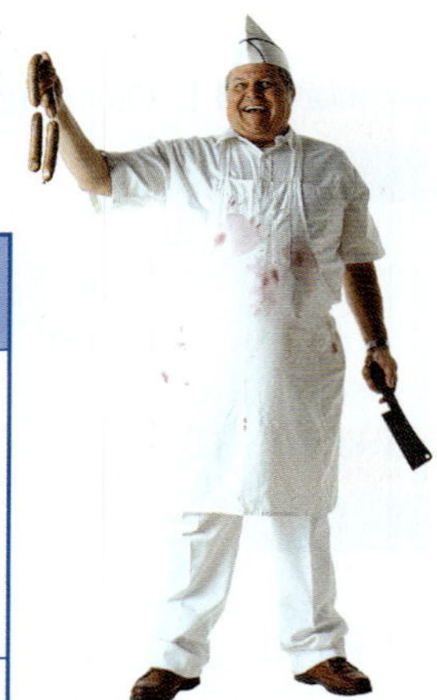

17 For the following table of values, showing the popularity of certain breakfast cereals, complete the fraction and angle size columns before drawing a pie graph to display the data.

Type of cereal	Number who like it	Fraction	Angle size (°)
Corn Flakes	18		
Oats	9		
Rice Crunches	9		
Muesli	6		
Super Sugary Cereal	24		
Frosted Circle Crisps	6		
Total	72		

18 For the following table of values, showing the hobbies of a group of students, complete the angle size column before drawing a pie graph to display the data. The first percentage has been converted for you.

Hobby	Percentage of total	Angle size (°)
Stamp collecting	20	$\frac{20}{100} \times 360° = 72°$
Photography	15	
Rollerblading	20	
Skateboarding	30	
Model making	15	
Total	100	

19 For the following table of values, showing the sports played by a group of students, complete the angle size column before drawing a pie graph to display the data.

Sport	Percentage of total	Angle size (°)
Cricket	20	
Hockey	25	
Netball	40	
Soccer	10	
Tennis	5	
Total	100	

WORKED Example 10

20 Over a 2-week period, the number of packets of potato chips sold from a vending machine each day was recorded as follows:
15, 17, 18, 18, 14, 16, 17, 6, 16, 18, 16, 16, 20, 18.
a Draw a dot plot of the data.
b Comment on the distribution.

13.12 SkillSHEET Choosing an appropriate scale for dot plots

EXCEL Spreadsheet Dot plots (DIY)

21 Draw a dot plot for each of the following sets of data:
a 2, 0, 5, 1, 3, 3, 2, 1, 2, 3
b 18, 22, 20, 19, 20, 21, 19, 20, 21
c 5.2, 5.5, 5.0, 5.8, 5.3, 5.2, 5.6, 5.3, 6.0, 5.5, 5.6
d 49, 52, 60, 55, 57, 60, 52, 66, 49, 53, 61, 57, 65, 62, 64, 48, 51, 60.

GAME time Data — 001

22 Compare and comment on the range, clustering and extreme values (if any) for the dot plots in question **21**.

WorkSHEET 13.2

COMMUNICATION

Walking billboard

Top driver's a walking billboard

By PAUL GOVER, motoring editor

The world's most expensive billboard isn't in London or Manhattan.

It's the space on the head, chest and arms of Michael Schumacher.

It costs mega-money to buy advertising space on the body of Mr Formula One, the Ferrari star expected to win the Australian Grand Prix on Sunday afternoon.

His personal salary is in the $50 million-plus range. The Formula One budget for the Ferrari team is more than $150 million a year.

Schumacher's hat alone cost a German bank more than $10 million for a three-year lease. It outbid McDonald's for the space.

A lot of the advertising on Schumacher's racing suit is reserved for the major backers of the Ferrari team, led by Marlboro and Shell.

A cigarette company also ensures the race suits are a suitable red.

Even when local laws ban its name, the distinctive fag packet logo takes its place.

There are arm patches for the team's minor sponsors, including Tic-Tac lollies, while FedEx tipped in an estimated $10 million last year to claim the shoulders.

Source: *Herald Sun*, 10 March 2000

Read the article and then answer the following questions.

1 Where can the world's most expensive billboard be found?

2 How much did it cost to lease Michael Schumacher's hat? For what period of time will it be leased?

3 Who are the major backers of the Ferrari team? How much do they each contribute?

4 Draw up a table displaying each of the listed sponsors of the Ferrari team and the amount they contribute.

5 Present the data displayed in the table as a column graph.

6 Present the data displayed in the table as a pie graph.

7 Investigate how much Michael Schumacher is worth today.

8 Investigate sponsorship deals of your favourite sportsteam or sportsperson.

10 QUICK QUESTIONS 2

1 a Complete a frequency table for the following data from a survey on a group of people's preferred takeaway food. The raw data obtained were: pizza, chicken, hamburgers, fish and chips, hamburgers, pizza, hamburgers, hamburgers, pizza, chicken, pizza, hamburgers, chicken, fish and chips, hamburgers, hamburgers, chicken, pasta, pizza, fish and chips, chicken, pasta, hamburgers, pasta, pizza, fish and chips, chicken, pizza, pizza, hamburgers, pasta, pizza, pasta, hamburgers, chicken, pizza, pasta, pasta, hamburgers, pasta, hamburgers, pizza, pizza, hamburgers, chicken, pasta, pizza, fish and chips, hamburgers and pasta.

b How many people were surveyed?

c What is the most preferred takeaway food?

d What is the least preferred takeaway food?

e Calculate the fraction of each preference.

Questions **2** to **6** relate to the table of values at right which shows the accumulation of frequent flyer points for a traveller.

Date	Number of points
1.8.05	5 000
1.9.05	5 800
1.10.05	7 800
1.11.05	7 800
1.12.05	10 800
1.1.06	12 500

2 After choosing an appropriate scale, draw a line graph to display the data.

3 How many frequent flyer points did the traveller have on 1.8.05?

4 How many frequent flyer points did the traveller earn in the next month?

5 Are there any parts of the graph which are horizontal?

6 When do they occur? What does this show?

Questions **7** to **10** relate to the pie graph shown. The pie graph shows the seasons in which a number of different plants flower.

Flowering seasons

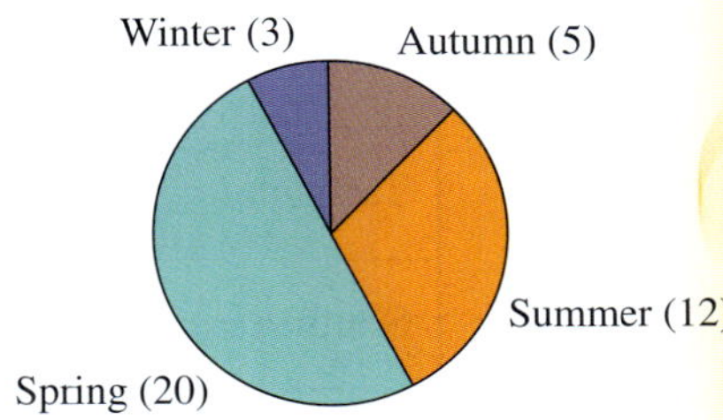

7 How many flowers are included in the group?

8 What percentage of the flowers flower in spring?

9 Calculate the size of the angle needed to represent the number of plants which flower in summer.

10 Use a protractor to check your answer to question **9**.

THINKING

Pictographs

Pictographs (or picture graphs) represent data by using simple, eye-catching pictures. There should always be a key or legend to show the number each picture represents.

1 The pictograph at right represents the most popular videos hired from Mathsville video stores.

Most Popular Videos

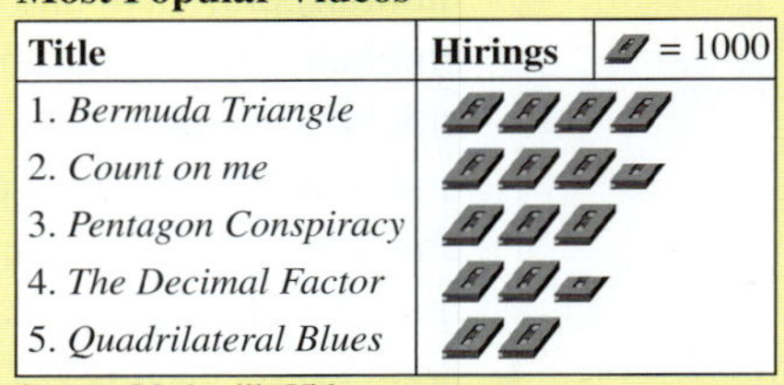

Source: Mathsville Video stores.

- **a** How many times was the number one video hired out?
- **b** Which video had hirings of 3000 or less?
- **c** What were the total hirings of the top 5 videos?
- **d** How many more hirings of the *Pentagon Conspiracy* video were there than the *Quadrilateral Blues* video?
- **e** Express the hiring of the *Pentagon Conspiracy* compared with the *Quadrilateral Blues* as a fraction.
- **f** What percentage of the total hiring was the hiring of *The Decimal Factor* video?
- **g** How many symbols would be used to show 1900 hirings of the video, *The Pythagoras Theory*?

2 **a** Display the data in the table below as a picture graph.

Feral cat survey	
State	**Number of cats (to the nearest 5000)**
Queensland	70 000
New South Wales	115 000
Victoria	90 000
Tasmania	25 000
South Australia	45 000

From the picture graph, answer the following:

- **b** List the states in descending order of feral cat numbers.
- **c** Express the number of feral cats in South Australia compared with the number of feral cats in Victoria as a fraction.
- **d** What percentage of Australia's feral cats are in Queensland?

Summary statistics

Summary statistics are used to summarise data so we can make sense of them, without necessarily graphing them. Given below are some key words which will be used throughout the exercise.

The **range** of a set of values is the difference between the highest and lowest value in the data. For example, in a set of test results, the lowest might be 34 while the highest might be 99. The range in that case would be 99 − 34 = 65.

The **mean** of a set of data is what many of us think of as the average. It is the exact middle of a set of numbers and is found by adding the data values and dividing by the number of values in the set of data.

The **median** is the middle value of the data, when the values are arranged in numerical order.

The **mode** of a set of data is the most common value.

Mean

The mean or average of a set of values is the sum of all the values divided by the number of values.

WORKED Example 11

For each of the following sets of data, calculate the mean:

a 5, 5, 6, 4, 8, 3, 4

b 0, 0, 0, 1, 1, 1, 1, 1, 4, 4, 4, 4, 5, 5, 5, 7, 7.

THINK	WRITE
a 1 Calculate the total of the given values.	**a** Total of values $= 5 + 5 + 6 + 4 + 8 + 3 + 4$ $= 35$
2 Count the number of values.	Number of values = 7
3 Define the rule for the mean.	$\text{Mean} = \dfrac{\text{total of values}}{\text{number of values}}$
4 Substitute the known values into the rule and evaluate.	$= \frac{35}{7}$ $= 5$
b 1 Calculate the total of the given values. Take note of the number of times each value occurs. That is, 0 occurs 3 times (3×0), 1 occurs 5 times (5×1), 4 occurs 4 times (4×4), 5 occurs 3 times (3×5), 7 occurs 2 times (2×7).	**b** Total of values $= 3 \times 0 + 5 \times 1 + 4 \times 4 + 3 \times 5 + 2 \times 7$ $= 0 + 5 + 16 + 15 + 14$ $= 50$
2 Count the number of values. *Note*: Although zero has no numerical value, it is still counted as a piece of data and must be included in the number of values tally.	Number of values = 17

Continued over page

THINK	WRITE
3 Define the rule for the mean.	$\text{Mean} = \dfrac{\text{total of values}}{\text{number of values}}$
4 Substitute the known values into the rule and evaluate.	$= \frac{50}{17}$ $= 2.941\,176\,471$
5 Round the answer to 1 decimal place. *Note*: The mean doesn't necessarily have to be a whole number or included in the original data.	$= 2.9$

Median

The median is the middle value for an odd number of data and the average of the two middle values for an even number of data.

When determining the median:

1. The values must be arranged in numerical order.
2. There are as many values above the median as there are below it.
3. For an even number of values, the median may not be one of the listed scores.

WORKED Example 12

Find the median for the following sets of data:

a 5, 4, 2, 6, 3, 4, 5, 7, 4, 8, 5, 5, 6, 7, 5

b 8, 2, 5, 4, 9, 9, 7, 3, 2, 9, 3, 7, 6, 8.

THINK	WRITE
a 1 Arrange the values in ascending order.	**a**
2 Select the middle value. *Note*: There are an odd number of values; that is, 15. Hence, the eighth value is the middle number or median.	2, 3, 4, 4, 4, 5, 5, ⑤ 5, 5, 6, 6, 7, 7, 8
3 Answer the question.	The median of the scores is 5.
b 1 Arrange the values in ascending order.	**b**
2 Select the middle values. *Note*: There are an even number of values; that is, 14. Hence, the sixth and seventh values are the middle numbers.	2, 2, 3, 3, 4, 5, ⑥ ⑦ 7, 8, 8, 9, 9, 9

THINK	WRITE
3 Obtain the average of the two middle values. *Note:* Add the two middle values and divide the result by 2.	Median = $\frac{6+7}{2}$ $= \frac{13}{2}$ (or $6\frac{1}{2}$)
4 Answer the question.	The median of the scores is $6\frac{1}{2}$ or 6.5.

Mode

The mode is the most common value in a set of data.

Some sets of data have more than one mode, or no mode at all; that is, there is no value which corresponds to the highest frequency, as all values occur once only.

WORKED Example 13

Find the mode of the following scores:

a 2, 3, 4, 5, 5, 6, 6, 7, 8, 8, 8, 9
b 12, 18, 5, 17, 3, 5, 2, 10, 12
c 42, 29, 11, 28, 21.

THINK	WRITE
a 1 Look at the set of data and highlight any values that have been repeated.	**a** 2, 3, 4, (5), (5), (6), (6), 7, (8), (8), (8), 9
2 Choose the value which has been repeated the most.	The numbers 5 and 6 occur twice. However, the number 8 occurs three times.
3 Answer the question.	The mode for the given set of values is 8.
b 1 Look at the set of data and highlight any values that have been repeated.	**b** (12), 18, (5), 17, 3, (5), 2, 10, (12)
2 Choose the value(s) which have been repeated the most.	The number 5 occurs twice. The number 12 occurs twice
3 Answer the question. *Note:* Some sets have more than one mode. The data set is called bimodal as two values were most common.	The mode for the given set of values is 5 and 12.
c 1 Look at the set of data and highlight any values that have been repeated.	**c** 42, 29, 11, 28, 21
2 Answer the question. *Note:* No mode is not the same as having a mode which equals 0.	No values have been repeated. The set of data has no mode since none of the scores correspond to a highest frequency. Each of the numbers occurs once only.

Range

The range of a set of values is the difference between the highest and lowest values.

WORKED Example 14

Find the range of the following data.
12, 76, 35, 29, 16, 45, 56

THINK	WRITE
1 Obtain the highest and lowest values.	Highest value = 76 Lowest value = 12
2 Define the range.	Range = highest value − lowest value
3 Substitute the known values into the rule.	= 76 − 12
4 Evaluate.	= 64
5 Answer the question.	The set of values has a range of 64.

Graphics Calculator tip! Finding the mean, median, mode and range

Finding the mean, median, mode and range

To enter data into the TI–83 graphics calculator, press STAT and select **1:Edit**. Type your data in the L1 column. Press ENTER after each entry. (You can type over existing numbers or press DEL to delete.) To find the mean and median of your data, press STAT then arrow across to select **CALC** and **1:1 – Var Stats** and press ENTER. The mean is the value given for $\bar{x}$. Scroll down to find the median. This is shown as **Med=**.

You will also be able to calculate the range by viewing the values for **minX** (the lowest value) and **maxX** (the highest value).

To assist you in finding the mode you can sort the data list into ascending order. Press STAT select **2:Sort A(**, then press 2nd **L1** (to sort List 1) and press ENTER. (Note that selecting **3:Sort D(** sorts the data in descending order.) You can view the sorted data by pressing STAT **1:Edit**. (Highlighting **L1** shows the full list of the data across the bottom of the screen for easier viewing.)

The screens below show the summary statistics for the data: 3, 4, 8, 4, 5, 6.

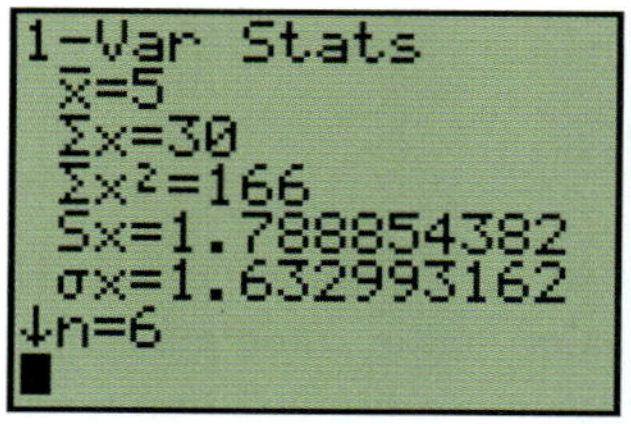

Mean = 5

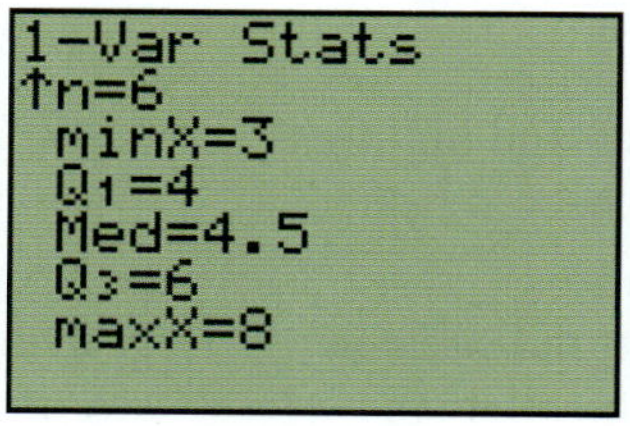

Median = 4.5
Range = 8 − 3 = 5

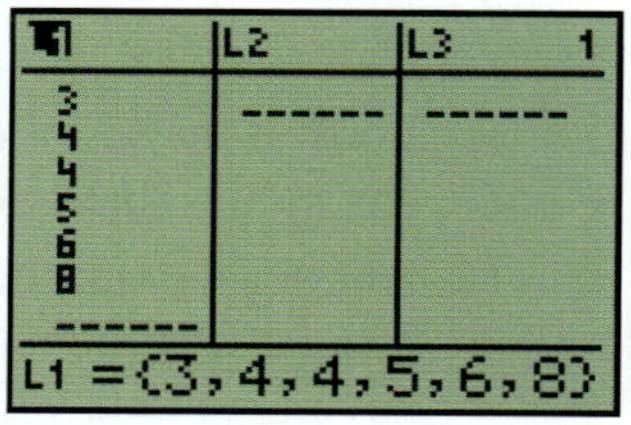

Mode = 4

remember

1. To determine the *mean* of data in a list simply obtain the total of all the values and divide by the number of values.
2. The *median* is the middle value for an odd number of data and the average of the two middle values for an even number of data which are arranged in numerical order. There are as many values above the median as there are below it.
3. The *mode* is the most common value. If two values are most common, they are called bimodal.
4. The *range* is the difference between the highest and lowest value in the data.

EXERCISE 13E Summary statistics

WORKED Example 11a

SkillSHEET 13.13 Finding the mean of ungrouped data

Mathcad: Summary statistics

EXCEL Spreadsheet: Mean

EXCEL Spreadsheet: Mean (DIY)

1 For each of the following sets of data, calculate the mean.

a 3, 4, 5, 5, 6, 7 **b** 5, 6, 7, 5, 5, 8 **c** 4, 6, 5, 4, 2, 3
d 3, 5, 6, 8, 7, 7 **e** 5, 4, 4, 6, 2, 3 **f** 2, 2, 2, 4, 3, 5
g 12, 10, 13, 12, 11, 14 **h** 11, 12, 15, 17, 18, 11 **i** 12, 15, 16, 17, 15, 15
j 10, 14, 12, 12, 16, 14

2 For each of the following sets of data, calculate the mean.

a 2, 4, 5, 6, 8 **b** 9, 8, 7, 4 **c** 4, 1, 5, 4, 2, 3, 2
d 4, 1, 5, 2 **e** 7, 5, 4, 1, 6, 7 **f** 5, 7, 7, 9, 2
g 2, 4, 3, 5, 9, 1 **h** 7, 1, 4 **i** 7, 6, 4, 8, 2, 9
j 4, 5, 8, 2, 4, 7, 7, 3

3 **multiple choice**

The mean for the data 5, 5, 6, 7, 2 would be found by:

A adding all the results and multiplying by the number of results
B adding all the results and dividing by the number of results
C adding all the results
D choosing the middle result
E ordering the results, then choosing the middle result

WORKED Example 11b

4 For each of the following sets of data, calculate the mean.
Hint: Use the grouping of values to help you.

a 9, 9, 7, 7 **b** 2, 2, 2, 4, 4, 4 **c** 4, 4, 3, 3, 5, 5
d 1, 1, 2, 3, 3 **e** 1, 2, 2, 4, 4, 5 **f** 1, 2, 2, 5, 5, 6, 7
g 9, 9, 8, 8, 7, 1, 1, 1, 1 **h** 3, 3, 3, 1, 1, 1, 2, 2, 2 **i** 4, 4, 5, 5, 8, 8, 1, 1, 9
j 2, 2, 2, 3, 3, 3, 3, 6

5 **multiple choice**

When finding the mean of a set of data:

A zeroes do not matter at all
B zeroes must be counted in the number of results
C zeroes must be added to the total as they will change it
D zeroes will make the mean zero
E none of these is true

6 For each of the following sets of data, calculate the mean.

a 5, 0, 4, 6, 0, 5, 1 **b** 3, 4, 7, 1, 5, 0, 8 **c** 8, 4, 17, 2, 0, 0, 4
d 5, 1, 2, 1, 0, 3, 2 **e** 0, 5, 0, 1, 4, 8 **f** 5, 3, 0, 4
g 4, 1, 0, 9, 6 **h** 4, 1, 5, 0, 14, 12 **i** 11, 11, 0, 15, 0, 17
j 15, 1, 7, 4, 1, 0, 0

7 For each of the following sets of data, calculate the mean. Express your answers as decimals, rounded off to 2 decimal places where appropriate.

a 5, 1, 4, 0, 3, 8 **b** 4, 4, 1, 0, 2 **c** 5, 5, 1, 3, 0, 3
d 2, 5, 0, 1, 8 **e** 3, 3, 2, 4, 0, 0 **f** 3, 8, 5, 1, 4, 2, 0, 3, 5
g 4, 5, 1, 4, 12, 15 **h** 6, 5, 1, 6, 5, 2 **i** 9, 4, 1, 1, 3, 8, 3, 0, 1
j 4, 0, 0, 5, 2, 4

SkillSHEET 13.14 Finding the location of the median

8 **multiple choice**

For the following set of data, 2.6, 2.8, 3.1, 3.7, 4.0, 4.2:

A the mean value for the data will be above 4.2
B the mean value for the data will be below 2.6
C the mean value for the data will be between 2.6 and 3.0
D the mean value for the data will be between 3.0 and 4.0
E the mean value for the data will be between 4.0 and 4.2

SkillSHEET 13.15 Finding the median

WORKED Example 12a

9 Find the middle value (median) for the following sets of data, by carefully ordering the values first.

EXCEL Spreadsheet Median

a 3, 3, 4, 5, 5, 6, 7 **b** 1, 2, 2, 3, 4, 8, 9
c 1, 2, 5, 6, 8, 8, 9 **d** 2, 2, 2, 3, 3, 4, 5
e 5, 5, 6, 6, 7, 7, 8, 9, 9 **f** 7, 7, 7, 10, 11, 12, 15, 15, 16
g 4, 3, 5, 3, 4, 4, 3, 5, 4 **h** 1, 2, 5, 4, 1, 1, 1, 2, 5
i 1, 2.5, 5, 3.4, 1, 2.4, 5 **j** 1.2, 1.5, 1.4, 1.8, 1.9

EXCEL Spreadsheet Median (DIY)

WORKED Example 12b

10 Find the middle value (median) for the following sets of data, by carefully ordering the values first. Note there is an even number of values.

a 1, 1, 2, 2, 4, 4 **b** 1, 2, 2, 2, 4, 5 **c** 4, 5, 5, 5, 6, 7
d 4, 5, 7, 7, 8, 9 **e** 1, 2, 2, 3, 3, 4 **f** 2, 4, 4, 6, 8, 9
g 1, 5, 7, 8 **h** 2, 4, 5, 7, 8, 8, 9, 9 **i** 1, 4, 7, 8
j 1, 5, 7, 8, 10, 15

EXCEL Spreadsheet Mode

WORKED Example 13

11 Find the mode for each of the following sets of data.

a 3, 3, 4, 4, 4, 5, 6 **b** 2, 9, 8, 8, 4, 5 **c** 1, 1, 2, 2, 2, 3
d 4, 6, 4, 2, 7, 3 **e** 2, 4, 3, 6, 2, 4, 2 **f** 4, 8, 8, 3, 3, 4, 3, 3
g 6, 2, 12, 10, 14, 14, 10, 12, 10, 12, 10, 12, 10
h 7, 9, 4, 6, 26, 71, 3, 3, 3, 2, 4, 6, 4, 25, 4
i 2, 2, 3, 4, 4, 9, 9, 9, 6 **j** 3, 7, 4, 3, 4, 3, 6, 3

EXCEL Spreadsheet Mode (DIY)

12 Find the mode for each of the following sets of data.

a 2, 0, 4, 0, 6, 2, 0 **b** 4, 2, 5, 6, 8, 4
c 35, 6, 6.2, 7.0, 6.2, 3.4 **d** 15, 14, 11, 14, 17, 14
e 1.2, 1.5, 1.3, 1.02, 1.5, 1.4 **f** 3.7, 3.5, 3.8, 3.08, 3.5
g 100, 101, 105, 102, 100, 100, 102, 105, 102, 103
h 29, 53, 47, 48, 35, 45, 37, 45, 42, 46, 27, 45
i 11.2, 11.4, 11.5, 11.6, 11.2, 11.7, 11.8
j 11, 15, 14.9, 15, 16.2, 11.32, 15, 14.2, 17

13 **multiple choice**

Which of the following is a correct statement?

A The mean, median and mode for any set of data will always be the same value.
B The mean, median and mode for any set of data will never be the same value.
C The mean, median and mode for any set of data must always be close in value.
D The mean, median and mode for any set of data are usually close in value.
E None of these statements is true.

WORKED Example 14

14 **a** Find the range of the following: 15, 26, 6, 38, 10, 17.
b Find the range of the following: 12.8, 21.5, 1.9, 12.0, 25.4, 2.8, 1.3.

15 **multiple choice**

The range of the following set of numbers: 16, 33, 24, 48, 11, 30, 15, is:

A 48 **B** 59 **C** 37 **D** 20 **E** 11

16 Eleanor wanted to know what her mathematics test average was. Her teacher said that she used the mean of her test results to calculate the end-of-year mark. Eleanor's test results (percentages) were: 89, 87, 78, 75, 89, 94, 82, 93, 78.
What was her mathematics test mean?

17 The number of shoes inspected by a factory worker in an hour was counted over a number of days' work. The results are as follows: 105, 102, 105, 106, 103, 105, 105, 102, 108, 110, 102, 103, 106, 107, 108, 102, 105, 106, 105, 104, 102, 99, 98, 105, 102, 101, 97, 100.

What is the mean number of shoes checked by this worker in one hour? Round your answer to the nearest whole number.

18 The number of students in the cafeteria each lunchtime was surveyed for 2 weeks. The results were as follows: 52, 45, 41, 42, 53, 45, 47, 32, 52, 56.

What was the mean number of students in the cafeteria at lunchtime in that fortnight? Round your answer to the nearest whole number.

19 A cricketer had scores of 14, 52, 35, 42 and 47 in her last 5 innings. What is her mean score?

20 Tom thinks that the petrol station where he buys his petrol is cheaper than the one where his friend Sarah buys her petrol. They begin to keep a daily watch on the prices for 4 weeks and record the following prices (in cents per litre).

Tom: 75.2, 72.5, 75.2, 75.3, 75.4, 75.6, 72.8, 73.1, 73.1, 73.2, 73.4, 75.8, 75.6, 73.4, 73.4, 75.6, 75.4, 75.2, 75.3, 75.4, 76.2, 76.2, 76.2, 76.3, 76.4, 76.4, 76.2, 76.0

Sarah: 72.6, 77.5, 75.6, 78.2, 67.4, 62.5, 75.0, 75.3, 72.3, 82.3, 75.6, 72.3, 79.1, 70.0, 67.8, 67.5, 70.1, 67.8, 75.9, 80.1, 81.0, 58.5, 68.5, 75.2, 68.3, 75.2, 75.1, 72.0

a Calculate the mean petrol prices for Tom and Sarah.
b Which station sells cheaper petrol on average?
c Why might Tom have been misled?

21 Peter has calculated his mean score for history to be 89%, based on five tests. If he scores 92% in the sixth test, what will his new mean score be?

22 Kim has an average (mean) score of 72 in Scrabble. He has played six games. What must he score in the next game to keep the same average?

23 A clothing company wanted to know the size of jeans that should be manufactured in the largest quantities. A number of shoppers were surveyed and asked their jeans size. The results were: 13, 12, 14, 12, 15, 16, 14, 12, 15, 14, 12, 14, 13, 14, 11, 10, 12, 13, 14, 14, 10, 12, 14, 12, 12, 10, 8, 16, 17, 12, 11, 13, 12, 15, 14, 12, 17, 8, 16, 11, 12, 13, 12, 12.

a What is the mode of these data?

b Why would the company be more interested in the mode than the mean or median values?

24 Jennifer wants to ensure that the mean height of her jump in the high jump for 10 jumps is over 1.80 metres.

a If her jumps so far have been (in metres) 1.53, 1.78, 1.89, 1.82, 1.53, 1.81, 1.75, 1.86, 1.82, what is her current mean?

b What height must she jump on the tenth jump to achieve a mean of 1.80?

c Is this likely, given her past results?

GAME time Data — 002

WorkSHEET 13.3

25 The local football team has been doing very well. They want to advertise their average score (to attract new club members). You suggest that they use the mean of their past season's game scores. They ask you to find that out for them. Here are the results.
Game scores for season (totals):
110, 112, 141, 114, 112, 114, 95, 75, 58, 115, 116, 115, 75, 114, 78, 96, 78, 115, 112, 115, 102, 75, 79, 154, 117, 62.

a What was their mean score?

b Would the mode or median have been a 'better' average to use for the advertisement?

COMMUNICATION Families

Let us look at the data presented at the start of this chapter. Daniel obtained the following data from his survey on the size of each class member's family:
3, 6, 7, 5, 4, 4, 7, 3, 2, 4, 5, 3, 4, 4, 6, 5, 5, 3, 3, 4, 6, 4, 4, 5.

1 Complete a frequency table for these data.

2 How many families contain 3 people?

3 Use your frequency table to draw a column graph.

4 What family size is the most common?

5 Draw a pie graph to display the data.

6 Calculate the: a mean b mode c median.

7 What is the range of the data?

8 Conduct a survey of your own to collect data from members of your class on the size of their families.

9 Display your data by drawing a column graph and a pie graph.

10 Calculate the mean, mode and median of your data.

11 Write a short report comparing your data with that collected by Daniel.

summary

Copy the sentences below. Fill in the gaps by choosing the correct word or expression from the word list that follows.

1 Qualitative data are arranged in categories which may be ___________ (names complete the description) or ___________ (a ranking system).

2 Quantitative data are in numerical form, either ___________ (counted) or ___________ (measured).

3 When the whole target population is contacted, it's called a ___________; otherwise just a representative ___________ is surveyed. ___________ selection prevents bias.

4 A frequency ______________________ has three columns headed: ______________________, ______________________ and frequency.

5 _____________ is the number of times a piece of data or _____________ occurs.

6 All graphs should have a _____________, be well _____________ and have _____________ axes.

7 Column graphs consist of _____________ spaced, vertical _____________ of _____________ width.

8 Bar graphs are similar to column graphs; however, they consist of ______________________.

9 _____________ graphs are drawn by joining _____________ with _____________ lines.

10 Line graphs display a _____________ in values over a _____________ of time.

11 _____________ graphs or _____________ graphs are graphs with no _____________ or horizontal _____________.

12 When converting fractions to an angle in a circle, _____________ by 360°.

13 Angles in a circle can be written as a fraction out of _____________.

14 The ____________ equals the highest value minus the lowest value.

15 The ____________ (or average) is obtained by adding the total of the values and dividing by the number of values.

16 The ____________ is the middle value of the data arranged in numerical ____________.

17 The most ____________ value is called the ____________.

WORD LIST

distribution table	continuous	sample	multiply
median	straight	columns	discrete
vertical	change	labelled	sector
range	line	pie	equal
evenly	mean	period	axes
tally	horizontal bars	scaled	ordinal
common	order	result	census
title	points	mode	random
360°	frequency	score	nominal

CHAPTER review

1 **multiple choice** 13A

The classification of data describing the number of iPods sold during the year is:

A quantitative, discrete **B** qualitative, nominal **C** qualitative, ordinal
D quantitative, continuous **E** none of these

2 **multiple choice** 13A

The classification of data which describe a person's mass is:

A quantitative, discrete **B** qualitative, nominal **C** qualitative, ordinal
D quantitative, continuous **E** none of these

3 **multiple choice** 13A

The data which describe sandwich types at a takeaway outlet are:

A qualitative, nominal
B quantitative, discrete
C qualitative, ordinal
D quantitative, continuous
E none of these

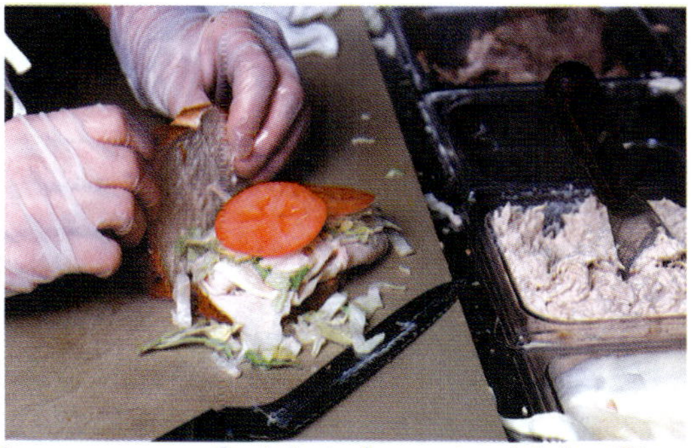

4 Eighty students in a school of 800 participated in a survey to find their favourite drinks. The results were: Coke: 20, Pepsi: 14, other soft drink: 10, milk/milk-based: 16, coffee: 9, tea: 4 and fruit juice: 7. 13B

a What was the most popular beverage?
b What was the least popular beverage?

5 Use this frequency distribution table to answer the following questions. 13B

a How many were in the class?
b What was the most frequent score?
c What was the least frequent score?
d How many scored, at most, 15?
e How many scored at least 16?

Score	Frequency
13	2
14	9
15	3
16	5
17	6
18	1

6 A random sample of 24 families was surveyed to determine the number of vehicles in each household. Here are the raw data collected: 13B

2, 0, 3, 2, 1, 0, 2, 3, 4, 2, 2, 1, 0, 1, 3, 2, 1, 0, 0, 0, 2, 2, 3, 3.

a Organise the data into a frequency distribution table.
b How many families have no vehicles in their household?
c How many families have 2 or more vehicles in their household?
d Which score has the highest frequency?
e Which is the highest score?
f What fraction of families had 2 vehicles in their household?

13B

7 This table shows the maximum and minimum daily temperatures in a city over a one-week period.

Day	1	2	3	4	5	6	7
Maximum (°C)	12	13	10	11	9	10	8
Minimum (°C)	3	3	2	1	0	4	2

Use the table to answer the following questions.

a What was the maximum temperature on day 3?

b Which day had the lowest minimum temperature?

c Which day was the coldest?

d Which day had the warmest overnight temperature?

e What was the temperature range (variation) on day 2?

f Which day had the smallest range of temperatures?

13C

8 Look at the graph at right, showing the test marks of a group of students.

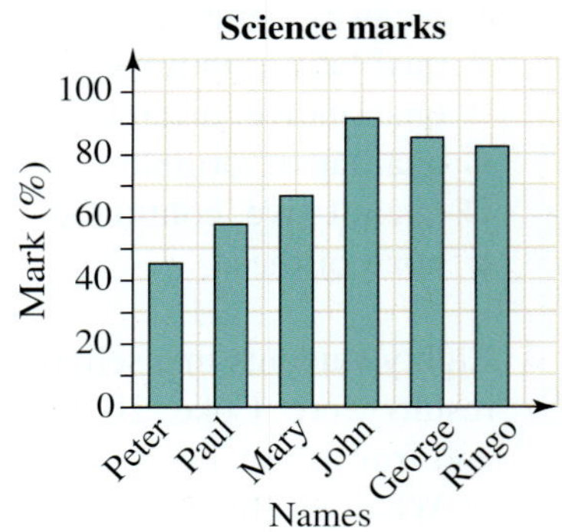

a What is the title for this graph?

b What is the label on the vertical axis?

c What is the label on the horizontal axis?

d What scale is used on the vertical axis?

e Which student has the highest mark?

f Which student(s) would have failed if the pass mark was 50%?

g What was Paul's mark?

h What was George's mark?

13C

9 Look at the graph at right, showing air quality.

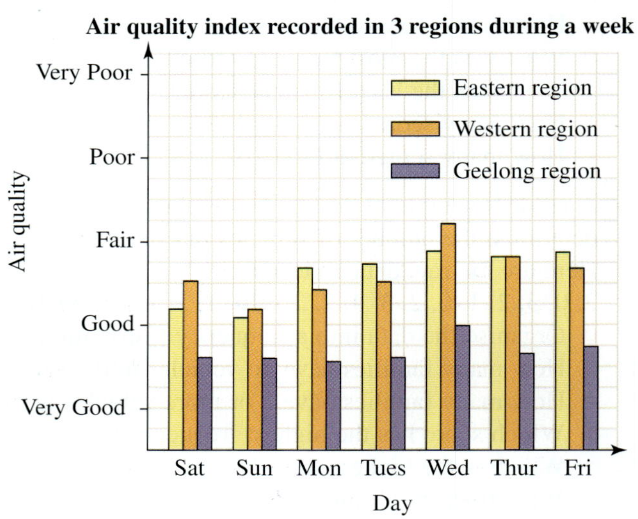

a What does the horizontal axis represent?

b What does the vertical axis represent?

c Which region had a rating that ranged from very good to good every day?

d Which day did the Western region have a fair air quality rating?

e Which region had a good rating every day?

13D

10 Use the following table of values to answer the questions below.

Month	Number of flies (thousands)
January	24
February	28
March	26
April	20
May	18
June	15
July	12
August	10
September	11
October	12
November	15
December	19

a In drawing a column graph using these data, what would be on the horizontal axis?
b In drawing a column graph, what would be on the vertical axis?
c What is the minimum value required on the vertical axis?
d What is the maximum value required on the vertical axis?
e What scale could be used for the vertical axis?
f Draw a column graph to display these data.

13D

11 Use the following table of values of students enrolled in elective classes to answer the questions below.

Subject	Number enrolled
Modern history	15
China in focus	12
Beginner's photography	16
First aid and fitness	18
Business enterprises	15
Consumer studies	12
Outdoor education	16

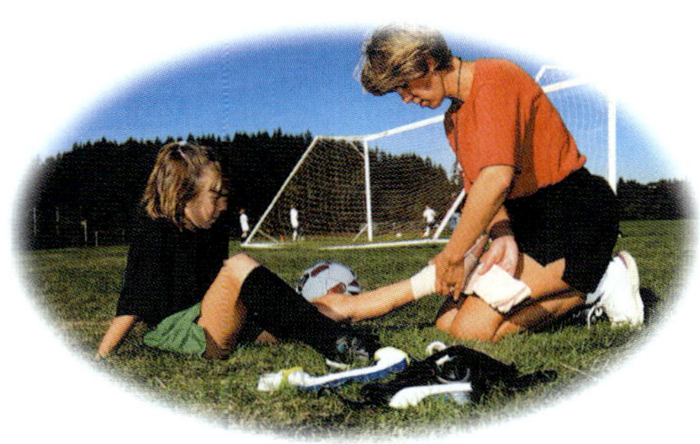

a What is the minimum value which would be required on the horizontal axis if a bar graph was to be drawn using these data?
b What is the maximum value which would be required on the horizontal axis if a bar graph was to be drawn?
c Which class has the largest number of students?
d If the maximum number of students for each of these electives is 15 students, which ones can take more students?
e Which electives have too many students?
f Draw a bar graph of the data.

13D

12 Look at the following graph, showing the change in size of an iceblock over time.

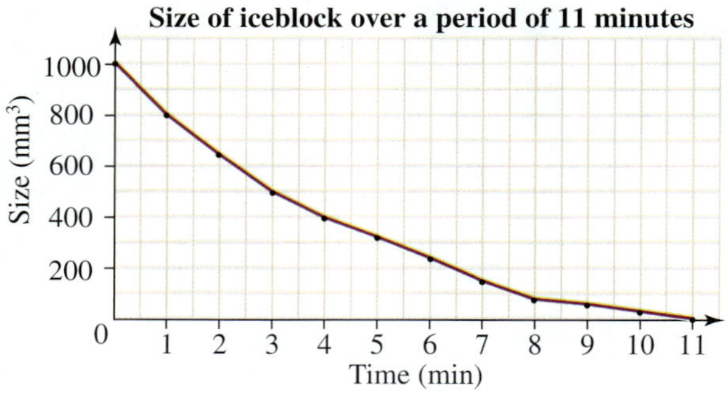

a What are the units on the vertical axis?

b What are the units on the horizontal axis?

c How large was the iceblock at the beginning?

d How long did it take for the iceblock to melt?

e How long was it before half of the iceblock melted?

f How much of the iceblock had melted by 5.5 minutes?

13D

13 a Use the table of values at right to construct a line graph of the data.

b Which axis should represent age?

c Which axis should represent height?

Age (years)	Height (cm)
1	80
2	85
3	88
4	93
5	99
6	104

13D

14 Look at the following graph showing the main reasons for people settling in Melbourne in 2005.

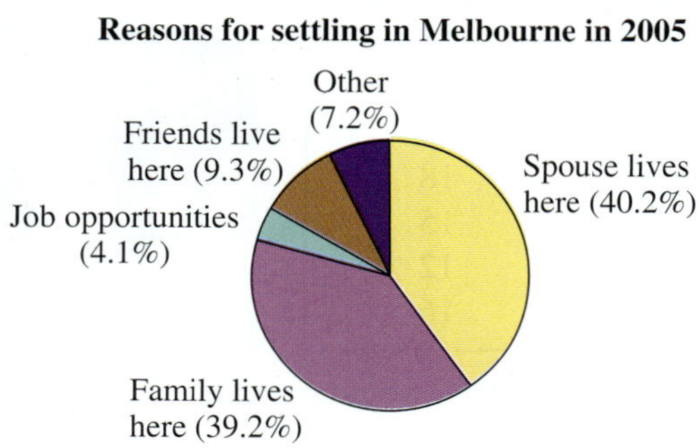

a What sort of graph is this?

b Which was the most common reason for people to migrate to Melbourne?

c The 'Other' category includes the reasons 'preferred climate', 'lifestyle' and 'employer located here'. Why do you think these reasons have been included under 'Other'?

d What should be the size of the 'job opportunities' sector?

e Why could a value of zero not be shown on a pie chart?

15 For the following table of values showing the types and numbers of animals on a particular farm: 13D

a copy and complete the following table of values

b construct a pie graph to display the following data.

Type of animal	Number on farm	Fraction	Angle size (°)
Horse	5		
Cow	25		
Duck	15		
Dog	5		
Cat	10		
Pig	40		
Sheep	20		
Total	120		

16 For the table of values at right showing the occupations of a group of people, complete the fraction and angle size column before drawing a pie graph to display the data. 13D

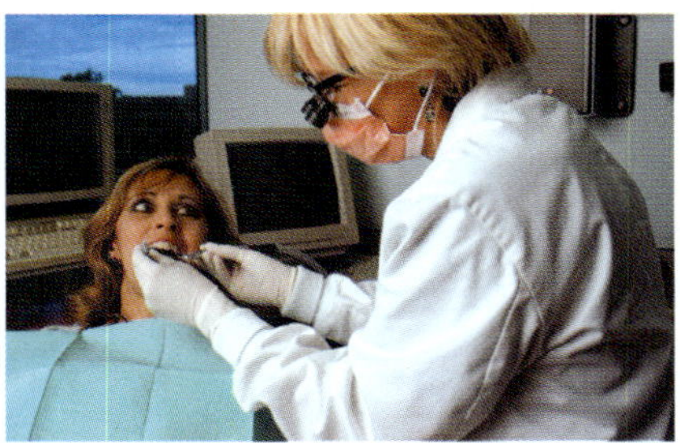

Occupation	Number of people	Fraction	Angle size (°)
Bank clerk	10		
Dentist	5		
Doctor	15		
Engineer	5		
Manager	20		
Real estate agent	10		
Secretary	20		
Solicitor	5		
Teacher	10		
Total	100		

17 The number of pets cared for by each of 20 families was surveyed, giving these data: 13B, D,E
1, 2, 3, 2, 0, 1, 0, 2, 5, 3, 2, 1, 2, 0, 2, 0, 1, 3, 2, 1.

a Organise the data into a frequency distribution table.

b Present the information as a pie graph.

c Draw a dot plot of the data and comment on the distribution.

d Find the mode, median and mean.

18 For the following sets of data, find: 13E

i the mean

ii the mode

iii the median.

a 4, 4, 4, 3, 2, 2, 3

b 2, 4, 2, 4, 3, 7, 4

c 4, 2, 6, 6, 5, 4, 6

d 4, 5, 6, 6, 5, 5, 3

e 1, 2, 1, 4, 1, 5, 1

f 5, 8, 1, 7, 7, 5, 2, 7, 5, 5

g 1, 5, 4, 5, 7

h 2, 4, 2, 5, 7, 6, 3

i 1, 7, 5, 8, 9, 5, 3

j 1, 5, 8, 7, 4, 8, 5, 6, 8

13E

19 Look at the following part of a teachers' mark book. Use it to answer the questions below.

Name	Test 1	Test 2
John	85	94
Peter	85	63
Mark	95	58
James	82	67
David	76	95
Rachel	62	85
Mary	87	75
Eve	94	63
Esther	68	68

a What was the class mean for test 1?

b What was the class mean for test 2?

c In which test did the class do better?

d How many students scored above the mean in test 1?

e How many students scored above the mean in test 2?

f What was the mode score for test 1?

g What was the mode score for test 2?

h What was the median score for test 1?

i What was the median score for test 2?

j Calculate the mean score for the two tests for each student and list them.

k What is the range of values for these means?

13E

20 Look at the following results of height measurements (in cm) for the students in class 9A:
145, 152, 148, 152, 163, 148, 165, 158, 159, 162, 145, 153, 156, 158, 157,
159, 169, 156, 156, 156, 152, 154, 128, 141, 154, 153, 156, 156, 165, 168.

a What is the range of heights?

b What is the mean height of the class?

c What is the mode for the heights?

d What is the median height of the class?

e Are these three values (mean, mode, median) similar?

f Why might this be so?

test yourself CHAPTER 13

Chance

14

Thomas is playing a game of 'Laughing clowns' at the Mathsville spring carnival. Each clown has six channels numbered 1 to 6 under its head. A red and a blue ball are given to each player for a game. A game consists of putting the balls into the clown's open mouth as its head moves from side to side. Each ball will go down into one of the six numbered channels. The score for a game is the sum of the numbers of the channels in which the two balls land. For example, if both balls land in channel 4, the score will be 4 + 4 = 8. Each total has a different prize.

What is the chance of winning the big prize? What score is Thomas most likely to get?

Are you ready?

Try the questions below. If you have difficulty with any of them, extra help can be obtained by completing the matching SkillSHEET. Either click on the SkillSHEET icon next to the question on the *Maths Quest 7* CD-ROM or ask your teacher for a copy.

SkillSHEET 14.1

Understanding 'chance' words

1 a On a number line, order the following commonly used 'chance words' from zero (impossible) to one (certain): even chance, unlikely, likely, impossible, certain.

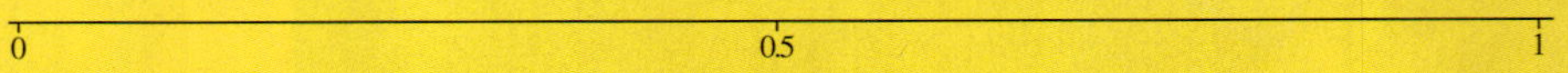

b Give an example of an event which describes each.

SkillSHEET 14.2

Comparing the likelihood of different events occurring

2 Compare the given events **a**, **b** and **c** and order them from least likely to most likely.

a The captain of the Australian cricket team will win the coin toss at the start of the match.
b There will be snow falling on Christmas day in Darwin.
c You will have at least one mathematics test this term.

SkillSHEET 14.3

Describing the chance of an event occurring as more or less than half

3 Describe the chance of each of the following events occurring as being more or less than a half.

a It will be sunny in Melbourne for most of the summer months.
b You will win a prize in your school raffle after buying one ticket.
c You will use a textbook for one of your classes today.

SkillSHEET 14.4

Listing all possible outcomes for an event

4 List all the possible outcomes for each of the following.

a Tossing a coin
b Rolling a six-sided die
c Selecting a day of the week

SkillSHEET 14.5

Understanding a standard deck of playing cards

5 Consider the cards in a standard deck of playing cards.

a How many cards are there in the deck?
b Name the four suits in the deck.
c How many aces are there in the deck?

SkillSHEET 14.6

Writing a simple probability as a fraction

6 a If there are three red pegs in a bag containing five pegs, write a fraction which describes the chance of picking a red peg from the bag.

b For a six-sided die, write a fraction which describes the chance of rolling a five.

The language of chance

The idea of *chance* is a very common one. We often discuss the chance that a certain thing will or will not happen. For example, we may say, 'There's a good chance of rain tomorrow' or 'Peter has a great chance of winning the race'. What we are really predicting is how possible or how likely it is for that thing to happen. When we speak about chance or the probability of something occurring, we are often not being exact. By using mathematics, precise values can be given to the **probability** of a particular **event** (a result that may be obtained) occurring. The chance of an event occurring is given a value ranging from and including 0 (for impossible events) to 1 (for events that are certain). A value of 0.5 would be given to something that was equally likely or unlikely to happen.

For example, the probability of Christmas Day being on June 29 this year is equal to zero because the chance of this occurring is impossible. The probability of a cricket team playing at the MCG this summer is equal to 1 because the chance of this occurring is certain. The probability that you will win a race against an equally matched competitor could be 0.5.

Particular words are associated with probability. When rating the chance of an event occurring, words including *certain*, *likely*, *even chance*, *unlikely* and *impossible* are used.

WORKED Example 1

For each of the given statements, specify whether the chance of the following events occurring is certain, likely, even chance, unlikely or impossible.

a You will compete in the next Olympics.
b Every student in Year 7 will obtain 100% in their next mathematics test.
c Each person in your class has been to the zoo.
d You flip a coin and Tails comes up.
e March is followed by April.

THINK	WRITE
a 1 Read the given statement and associate the chance of the event occurring with one of the given words from the list. Provide reasons.	a The chance of a person competing in the next Olympics is very small; however, it could happen.
2 Answer the question.	It is *unlikely* that this event will occur.

Continued over page

THINK	WRITE
b Repeat steps 1 and 2 of part **a**.	**b** Due to each student having different capabilities and the number of students involved, this situation could never happen. It is *impossible* for this event to occur.
c Repeat steps 1 and 2 of part **a**.	**c** The chance that each student in your class has been to the zoo, either with their family or primary school, is very probable. However, there may be a few students who missed out. It is *likely* this event will occur.
d Repeat steps 1 and 2 of part **a**.	**d** When you flip a coin there are only two possibilities, a Head or a Tail. So there is a 50% chance of Tails coming up. There is an *even chance* this event will occur.
e Repeat steps 1 and 2 of part **a**.	**e** This is a true statement. April always follows the month of March. It is *certain* this event will occur.

The probability scale

When assigning numerical values to probabilities, recall that zero (0) represents impossible events; that is, events which have no chance of happening. One (1) represents certain events; that is, events which must happen. A half or 0.5 represents an even chance; that is, the event is equally likely to happen as not to happen. Anything that *might* happen would be slightly below 0.5, and anything that will *probably* happen would be above 0.5.

It is very helpful to have some sort of scale at hand when assigning words or numerical values to the probability of a particular event occurring. The *probability scale* associates important words used in describing probabilities with their approximate corresponding numerical values.

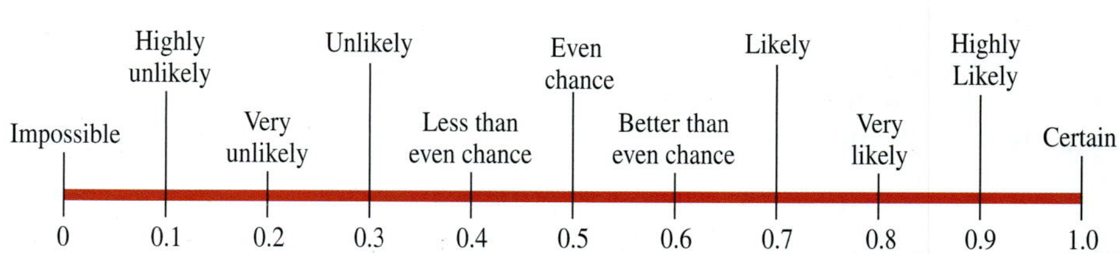

WORKED Example 2

Assign a number between and including 0 and 1 to represent the estimated probability of the following events, depending on how likely they are.

a One of two evenly matched tennis players will win the next game.

b You will guess the correct answer on a multiple choice question with four options.

c Rolling a fair die and obtaining a number less than 6.

THINK	WRITE
a 1 Determine the likelihood of an event occurring, with reasoning.	**a** Since the two players are evenly matched, one does not have an advantage over the other. Therefore, they each have an equal chance of winning the next game.
2 Express the answer as a decimal.	The probability that one player wins the game is $\frac{1}{2}$ or 0.5.
b 1 Determine the likelihood of an event occurring, with reasoning.	**b** When guessing an answer on a multiple choice question with 4 options, 1 out of the 4 possibilities will be correct. One out of 4 may be expressed as a fraction.
2 Express the answer as a decimal.	The probability of guessing the correct answer is $\frac{1}{4}$ or 0.25.
c 1 Determine the likelihood of an event occurring, with reasoning.	**c** When rolling a die there are six possibilities. They are 1, 2, 3, 4, 5, 6. A number less than 6 includes 1, 2, 3, 4, 5. Therefore, five out of the six possibilities may be rolled. Five out of six may be expressed as a fraction.
2 Express the answer as a decimal, correct to 2 decimal places.	The probability of obtaining a number less than six is $\frac{5}{6}$ or approximately 0.83.

The probabilities obtained in worked example 2 were derived through logic and most people would agree with the given answers. However, sometimes there may be a possible range of values, depending on the event, or the person answering the question. For example, the event 'Richmond will win the AFL Premiership next year' depends on a number of factors such as the team's position on the ladder this year, the players and coach they have next year, injuries to key players, or even personal factors (such as whether the person answering the question supports Richmond). Therefore, probability values ranging from 0.05 to 0.7 may be reasonable answers to expect.

remember

1. An *event* is a result that may occur.
2. *Probability* is the chance of a particular event (result) occurring.
3. The probability of an event occurring ranges from and includes 0 (impossible) to 1 (certain).
4. A probability scale associates important words used in describing probabilities with their approximate corresponding numerical values.

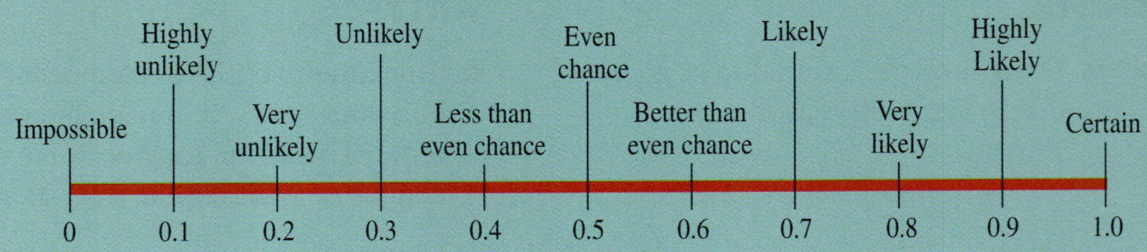

EXERCISE 14A The language of chance

SkillSHEET 14.1 Understanding 'chance' words

WORKED Example 1

1 For each of the given statements, specify whether the chance of the following events occurring is certain, likely, even chance, unlikely or impossible.

a New Year's Day will be on January 1 next year.
b Australia will experience at least one earth tremor this year.
c Water will boil in the fridge.
d There will be at least one day with a maximum temperature under 5°C in Melbourne in January.
e A horse will win the Melbourne Cup.
f There will be snow at Mt Buller this winter.
g You will grow 18 cm taller this year.
h You will win first prize in Tattslotto.
i You choose a blue ball from a bag which contains only white balls.
j You roll a fair die and obtain an odd number.
k The year 2020 will be a leap year.
l You choose a white ball from a bag which contains only white balls.
m You roll a fair die and obtain a number greater than 6.
n You choose a yellow ball from a bag containing 4 red balls and 4 yellow balls.
o You roll a fair die and obtain a number less than 5.

SkillSHEET 14.2 Comparing the likelihood of different events occurring

2 Compare the given events A, B, C, D and order them from least to most likely.

A It will be sunny in Queensland most of the time when you visit.
B Melbourne Cup Day will be on the first Tuesday in November next year.
C You will be the next Australian Idol.
D Saturn will be populated next year.

3 List five events that are:

a impossible **b** unlikely to happen **c** likely to happen **d** sure to happen.

4 **multiple choice**

The word which has the same meaning as *improbable* is the word:

A unlikely **B** impossible **C** uncertain **D** certain **E** likely

5 **multiple choice**

The word which has the same meaning as *certain* is the word:

A definite **B** possible **C** sure **D** probable **E** unlikely

WORKED Example 2

6 Assign a number between and including 0 and 1 to represent the estimated probability of the following events, depending on how likely they are.

a You flip a coin and obtain a Tail.
b You choose a red ball from a bag containing only 8 white balls.
c You guess the correct answer in a multiple choice question with 5 options.
d You roll a die and obtain an even number on a fair die.
e You flip a coin and obtain a Head.
f Your class elects one class captain from four candidates.
g You guess the incorrect answer in a multiple-choice question with five options.
h You choose a green ball from a bag containing only four green balls.
i You have science classes this year.
j St Kilda will win the AFL Premiership next year.

14.3 SkillSHEET Describing the chance of an event occurring as more or less than half

7 Match the words below with one of the numbers between 0 and 1 that are given. Choose the number depending on what sort of chance the word means, between *impossible* and *certain*. You may use a number more than once. If you are unsure, discuss your choice with another class member.

Numbers to choose from: 1, 0.75, 0.25, 0, 0.5.

a certain	**b** likely	**c** unlikely
d probable	**e** improbable	**f** definite
g impossible	**h** slim chance	**i** sure
j doubtful	**k** not able to occur	**l** more than likely
m fifty-fifty	**n** fair chance	

The sample space

The term **sample space** refers to the list of all possible **outcomes** to a probability experiment. Before we are able to calculate the probability of a particular event occurring, we need to be able to calculate the number of possible outcomes for that particular event.

Consider the case of a coin being tossed. We know that there are two possible outcomes. The coin can land either 'Heads' or 'Tails'. We write this as $S = \{\text{Heads, Tails}\}$ where S is the notation used to represent the sample space.

WORKED Example 3

A card is drawn from a standard deck. The suit of the card is then noted. List the sample space for this experiment.

THINK	WRITE
Although there are 52 cards in the deck we are concerned only with the suit. List each of the four suits as the sample space for this experiment.	$S = \{\text{clubs, spades, diamonds, hearts}\}$

On some occasions we do not need to actually list the sample space; rather we simply need to know how many elements are in the sample space. In the case where we tossed the coin, there were two possible outcomes to the experiment. We can say $n(S) = 2$; that is, the number of elements in the sample space is 2.

WORKED Example 4

A die is rolled and the number on the uppermost face is noted. How many elements are in the sample space?

THINK	WRITE
The die can show the numbers 1, 2, 3, 4, 5 or 6.	$n(S) = 6$

remember

1. The sample space represents a list of all possible outcomes to a probability experiment.
2. The sample space can be denoted by the symbol S and all outcomes are then listed within braces { }.
3. $n(S)$ represents the number of elements to the sample space.

EXERCISE 14B The sample space

WORKED Example 3

1 A spinner with 10 equal sectors labelled 1 to 10 is spun. List the sample space.

14.4 SkillSHEET Listing all possible outcomes for an event

2 For each of the following probability experiments list the sample space.
- a A coin is tossed.
- b A multiple-choice question has five alternative answers, A, B, C, D and E.
- c A soccer team plays a match and the result is noted.
- d A card is selected from the four aces in a deck.
- e An exam paper is given the grade A to F.

14.5 SkillSHEET Understanding a standard deck of playing cards

3 A card is selected from a standard deck. List the sample space if we are interested in:
- a the suit of the card chosen
- b the face of the card chosen
- c the colour of the card chosen.

4 A bag contains 8 red marbles, 9 green marbles and 2 orange marbles. A marble is selected from the bag. List the sample space.

WORKED Example 4

5 A coin is tossed. How many elements are in the sample space?

6 In each of the following, state the number of elements in the sample space.
- a A card is selected from a standard deck.
- b The first ball drawn in the Tattslotto draw. (Balls are numbered from 1 to 45.)
- c The winner of the AFL premiership. (There are 16 teams.)
- d A day of the year is selected.
- e A letter from the alphabet is selected at random.
- f The first prize in the lottery is chosen from tickets numbered 1 to 180 000.
- g A term is selected from a school year.
- h You win a medal at your chosen event at the world swimming championships.

7 **multiple choice**

From the list below select the event that has the most elements in the sample space.

- A Selecting a card from a standard deck.
- B Selecting a page at random from this book.
- C Selecting an exercise book from your school bag.
- D Selecting a student at random from your class.
- E Selecting a page at random from the phone directory.

WorkSHEET 14.1

Simple probability

Mathematically, we define *probability* as the chance of a favourable event happening, compared with that event not happening. For example, we can compare the chance of getting a Head when we toss a coin with getting a Tail. The result we want is 'getting a Head', hence we call this outcome 'favourable'; the outcome we don't want, 'getting a Tail', is called 'unfavourable'. The total number of outcomes or results we could have is 2; that is, a Head or a Tail. The number of these outcomes which is favourable is 1; that is 'getting a Head'. Therefore, the chance or probability of 'getting a Head', is 1 out of 2 possible results, or $\frac{1}{2}$.

Mathematically, we write this as Pr(getting a Head) = $\frac{1}{2}$. Obviously, we could have been just as interested in 'getting a Tail' and Pr(getting a Tail) = $\frac{1}{2}$ also.

An *outcome* is a particular result of an experiment.
A favourable outcome is one that we are looking for.
The probability of a particular result or event is defined as 'the number of favourable outcomes divided by the number of possible outcomes'.

$$\text{Pr(event)} = \frac{\text{number of favourable outcomes}}{\text{number of possible outcomes}}$$

WORKED Example 5

State how many possible outcomes there are for each of the following experiments and specify what they are:

a tossing a coin

b spinning a circular spinner with 9 equal sectors labelled from a to i as shown at right

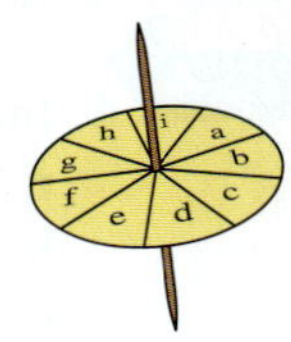

c drawing a picture card (jack, king, queen) from a standard pack of cards.

THINK	WRITE
a 1 Make a note of how many sides the coin has and what each side represents.	**a** The coin has 2 sides, a Head and a Tail.
2 Answer the question.	When tossing a coin there are two possible outcomes; they are Head or Tail.
b 1 Make a note of how many sectors the circular spinner has and what each one represents.	**b** The circular spinner has 9 sectors labelled a to i.
2 Answer the question.	When spinning the circular spinner, there are 9 possible outcomes; they are a, b, c, d, e, f, g, h or i.
c 1 Make a note of how many picture cards there are and what they are.	**c** There are 3 picture cards in each of the four suits.
2 Answer the question.	When drawing a picture card there are 12 possible results; they are: the jack, king and queen of clubs, the jack, king and queen of diamonds, the jack, king and queen of hearts and the jack, king and queen of spades.

WORKED Example 6

Christopher rolls a fair 6-sided die.

a What are all the possible results that could be obtained?

b What is the probability of obtaining:

i a 4?

ii a number greater than 2?

iii an odd number?

THINK	WRITE
a Write all the possible outcomes and answer the question.	**a** There are six possible outcomes; they are 1, 2, 3, 4, 5, 6.
b i 1 Write the number of favourable outcomes. A 4 occurs once.	**b i** Number of favourable outcomes = 1
Write the number of possible outcomes.	Number of possible outcomes = 6
2 Write the rule for probability.	$\text{Pr(event)} = \dfrac{\text{number of favourable outcomes}}{\text{number of possible outcomes}}$
3 Substitute the known values into the rule and evaluate.	$\text{Pr}(4) = \frac{1}{6}$
4 Answer the question.	The probability of obtaining a 4 is $\frac{1}{6}$.
ii 1 Write the number of favourable outcomes and the number of possible outcomes. *Note:* 'Greater than 2' implies 3, 4, 5, 6.	**ii** Number of favourable outcomes = 4 Number of possible outcomes = 6
2 Substitute the known values into the rule for probability and evaluate.	$\text{Pr(greater than 2)} = \frac{4}{6}$
3 Simplify the fraction.	$= \frac{2}{3}$
4 Answer the question.	The probability of obtaining a number greater than two is $\frac{2}{3}$.
iii Repeat steps 1 to 4 of part **b ii**. *Note:* 'An odd number' implies 1, 3, 5.	**iii** Number of favourable outcomes = 3 Number of possible outcomes = 6 $\text{Pr(an odd number)} = \frac{3}{6}$ $= \frac{1}{2}$ The probability of obtaining an odd number is $\frac{1}{2}$.

remember

The probability of a particular result or event is defined as the number of favourable outcomes divided by the number of possible outcomes. This is written as:

$$\Pr(\text{event}) = \frac{\text{number of favourable outcomes}}{\text{number of possible outcomes}}.$$

EXERCISE 14C Simple probability

WORKED Example 5

1 State how many possible outcomes there are for each of the following experiments and specify what they are.
  a Rolling a 12-sided die, numbered 1 to 12 inclusive
  b Spinning a spinner for a game that has 5 equal-sized sections, numbered 1 to 5 inclusive
  c Choosing a consonant from the word 'cool'
  d Choosing a sock out of a drawer containing 3 different socks coloured red, blue and black
  e Picking a marble out of a bag containing 5 different marbles coloured black, blue, green, red and yellow
  f Rolling an even number on a fair 6-sided die
  g Rolling an even number greater than 2 on a fair 6-sided die
  h Choosing an odd number from the first 20 counting numbers

2 List all the possible results in the following experiments. Comment on whether all results in each case are equally likely. Explain your answer.
  a Rolling a fair 6-sided die
  b Tossing a normal coin
  c Spinning a spinner where half is white and half is black
  d Spinning a spinner where half is white, a quarter is blue and a quarter is red
  e Rolling a 6-sided die that has the numbers 1, 2, 3, 4, 5, 5 on it
  f Shooting at a target where $\frac{1}{3}$ of the area is blue, $\frac{1}{3}$ green and $\frac{1}{3}$ red
  g Obtaining an even number when spinning a circular spinner with 11 equal sectors labelled 1 to 11
  h Choosing a vowel in the word 'mathematics'

SkillSHEET 14.6 Writing a simple probability as a fraction

WORKED Example 6

3 Christina rolls a fair 10-sided die with faces numbered from 1 to 10.
  a What are all the possible results that could be obtained?
  b What is the probability of obtaining:
    i a 9?
    ii a number less than 7?
    iii a prime number?
    iv a number greater than 3?
    v a multiple of 3?
    vi a number greater than 10?
    vii an even number greater than 4?
    viii an odd number divisible by 3?

4 Leo has been given a bag of marbles to play with. Inside the bag there are 3 blue, 6 red, 4 green and 7 black marbles.
  a How many marbles are in the bag?

b If Leo takes out one marble from the bag what is the:
  i Pr(getting a red marble)?
  ii Pr(getting a green marble)?
  iii Pr(getting a black marble)?
  iv Pr(getting a blue marble)?

c How many marbles in the bag are either blue or black?

d Find Pr(getting a blue or a black marble).

e Find Pr(getting a green or red marble).

f Find Pr(getting a green, red or blue marble).

g Find Pr(getting a green, red, blue or black marble).

h Explain your answer to part g.

5 There is a valuable prize behind 2 of the 5 doors in the game show 'Pick Your Door'. What is the probability that a player choosing any door will win a valuable prize?

6 In a raffle where there is only 1 prize (a car), 100 000 tickets have been sold, at a cost of \$5.00 each. What is the chance of winning the prize for a buyer who:

a purchases only 1 ticket?

b purchases 20 tickets?

c purchases 50 tickets?

d purchases all the tickets?

e Would the buyer who buys all the tickets have made a wise purchase? Explain.

7 **multiple choice**

A circular spinner is shown at right. When it is spun, the probability of obtaining an orange sector is:

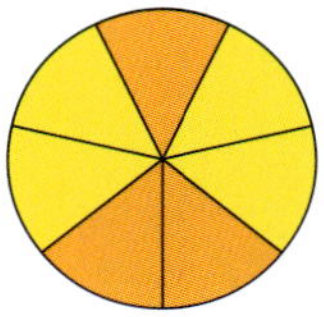

A $\frac{4}{7}$ B $\frac{1}{7}$ C $\frac{3}{4}$

D $\frac{1}{2}$ E $\frac{3}{7}$

8 **multiple choice**

For an octagonal spinner with equal sectors numbered 1 to 8, the chance of getting a number *between* 2 and 7 is:

A $\frac{5}{8}$ B $\frac{3}{8}$ C $\frac{3}{4}$ D $\frac{1}{2}$ E $\frac{1}{4}$

9 A pack of playing cards is shuffled and a card is chosen at random (in no particular order or pattern).
Find the probability that the card chosen is:

a a black card (that is, spades or clubs)

b an ace

c a diamond

d a picture card (that is, a jack, king, queen)

e the queen of hearts

f a diamond or a black card

g not a king

h a club, diamond, heart or spade

i not a spade

j red and a ten.

COMMUNICATION

Parachute landing

Jim operates a parachute school. Being a man who is interested in statistics, he keeps a record of the landing position of each jump from first-time parachutists. With experience, parachutists are able to land on a particular spot with great accuracy. However, first-time parachutists do not possess this ability. He has marked out his landing field with squares, as shown below.

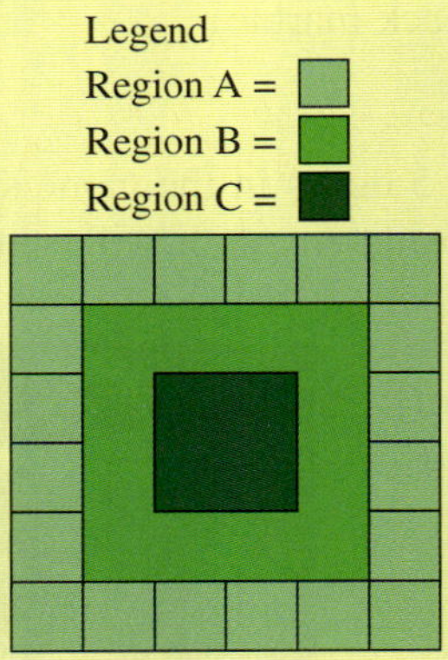

We are going to look at the areas of each of the regions A, B and C. To do this, we will determine each of the areas in terms of one of the small squares in region C. We will say that each small square has an area of 1 square unit.

1 What is the area of Jim's whole paddock (in square units)?

2 Determine the areas of regions A, B and C (in square units).

3 Assuming that the parachutist lands in the field, calculate the probability that the landing will occur in:
   a region A   b region B   c region C.
   These represent theoretical probabilities.

4 Jim's records indicate that, from 5616 jumps of first-time parachutists, the landing positions were:
   a 592 in region C   b 1893 in region B   c 3131 in region A.
   Comment on these results in comparison with the probabilities you calculated in question 3.

5 On paper or cardboard, draw up a grid of the same pattern as Jim's landing field. Surround the perimeter with books to act as a fence.

6 Take a small piece of paper (smaller than one of the small squares in region A) and drop it from a small height above your grid. If the paper overlaps into two regions on its fall, decide in which region it lies most.

7 Draw a table to record the landing region for 50 trials.

8 Calculate the probability of the paper landing in each of the regions.

9 Compare these experimental results with the theoretical probabilities you calculated earlier. Comment on any similarities/differences.

What was special about the "USS Enterprise"?

Calculate the probability of randomly selecting each of the lettered items from the groups below. Each letter and its probability gives the puzzle code.

Pr (A) =

Pr (C) =

Pr (D) =

Pr (E) =

Pr (F) =

Pr (H) =

Pr (I) =

Pr (L) =

Pr (N) =

Pr (O) =

Pr (P) =

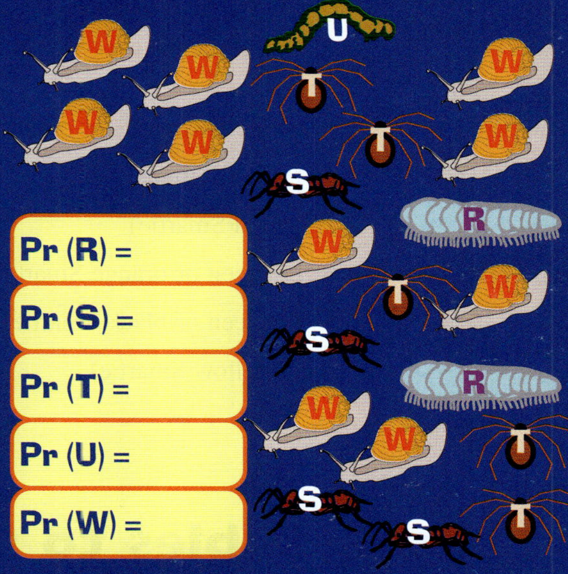

Pr (R) =

Pr (S) =

Pr (T) =

Pr (U) =

Pr (W) =

$\frac{1}{5}$	$\frac{5}{22}$	$\frac{5}{11}$	$\frac{1}{4}$	$\frac{2}{11}$	$\frac{5}{22}$	$\frac{3}{10}$	$\frac{3}{8}$	$\frac{1}{10}$	$\frac{1}{5}$	$\frac{1}{11}$	$\frac{2}{11}$	$\frac{5}{22}$		
$\frac{1}{2}$	$\frac{1}{22}$	$\frac{5}{16}$	$\frac{2}{5}$	$\frac{3}{8}$	$\frac{1}{4}$	$\frac{1}{11}$	$\frac{1}{3}$	$\frac{1}{6}$	$\frac{5}{11}$	$\frac{3}{8}$	$\frac{1}{11}$	$\frac{3}{8}$	$\frac{1}{16}$	
$\frac{1}{4}$	$\frac{1}{5}$	$\frac{1}{11}$	$\frac{5}{16}$	$\frac{1}{11}$	$\frac{1}{4}$	$\frac{1}{10}$	$\frac{5}{22}$	$\frac{5}{16}$	$\frac{1}{4}$	$\frac{1}{11}$	$\frac{1}{11}$	$\frac{1}{5}$	$\frac{3}{8}$	$\frac{1}{11}$

10 QUICK QUESTIONS 1

1 Match the term on the left with the appropriate numerical value on the right.

Unlikely	0
Highly likely	0.13
Impossible	1
Certain	0.5
Fair chance	0.89

For questions 2 to 4, list all of the possible outcomes in the following experiments and then state whether all outcomes in each question are equally likely.

2 Choosing a marble out of a bag containing 1 white and 1 red marble

3 Selecting the gender of a student from a class containing 12 boys and 14 girls

4 Rolling a die with the numbers 2, 3, 3, 4, 4, 4 on it

Questions 5 to 10 refer to the following information.

In a box there are 14 Smarties. Four of them are green, 2 red, 3 black, 1 pink and 4 orange.

5 List all of the possible outcomes if I select 1 Smartie.

If I select 1 Smartie at random, find the following.

6 Pr(pink Smartie)

7 Pr(red Smartie)

8 Pr(green or orange Smartie)

9 Pr(green or black Smartie)

10 Pr(yellow Smartie)

Using tables to show sample spaces

Sometimes it is not easy for us to memorise or remember all of the possible results that can be obtained from an experiment — it is best to write them down in an ordered manner. As mentioned previously, the *sample space* is the name given to the complete set of outcomes or results obtained from an experiment. Sample space is usually shown as an ordered list of all possible results enclosed in a pair of curled brackets, { }.

Some experiments have two steps or stages that give a pair of results, such as when you toss 2 coins, or toss a coin and roll a die, or roll 2 dice. When the result is a combined one (that is, from two things, such as a coin and a die) we usually write the combined result as an *ordered pair*, in brackets, separated by a comma. For example, the ordered pair, (H, 6) would correspond to obtaining a Head on the coin and a 6 on the die.

If an experiment was conducted involving 2 coins or 2 dice, where the two results could look the same, for example (6, 6), it is a good idea to use a key to show which comes first; that is (red die, blue die). Obviously, if they are called 'die 1' and 'die 2', then die 1 is listed first within the brackets.

Two-way tables

The sample space of two-step experiments may be displayed in a table called a **two-way table**. This table displays results which are obtained for each step of a particular experiment and then allows us to determine the chances of various combinations.

WORKED Example 7

a **Draw a two-way table and list the sample space for the experiment 'tossing a coin and rolling a die'.**
b **State how many different outcomes or results are possible.**
c **Determine the probability of obtaining:**
  i **a Head**
  ii **a Tail and an even number**
  iii **a 5**
  iv **a Tail and a number greater than 2.**

THINK

a 1 Rule a table consisting of 7 rows and 3 columns. Leave the first cell blank.
2 Label the second and third cells of the first row as H and T respectively.
3 Label cells 2 to 7 of the first column as 1, 2, 3, 4, 5, 6 respectively.
4 Combine the outcome pairs in the order in which they occur in each of the remaining cells; that is, the first event result followed by the second event result.
5 Answer the question.

WRITE

a

	H	T
1	H 1	T 1
2	H 2	T 2
3	H 3	T 3
4	H 4	T 4
5	H 5	T 5
6	H 6	T 6

The sample space for the experiment 'tossing a coin and rolling a die' is {(H, 1), (H, 2), (H, 3), (H, 4), (H, 5), (H, 6), (T, 1), (T, 2), (T, 3), (T, 4), (T, 5), (T, 6)}.

THINK

b Count the number of different outcomes and answer the question.

WRITE

b There are 12 different outcomes.

Continued over page

THINK	WRITE
c i 1 Write the number of favourable outcomes. Write the number of possible outcomes. *Note:* The favourable outcomes featuring a Head are (H, 1), (H, 2), (H, 3), (H, 4), (H, 5) and (H, 6).	**c i** Number of favourable outcomes = 6 Number of possible outcomes = 12
2 Write the rule for probability.	$\text{Pr(event)} = \dfrac{\text{number of favourable outcomes}}{\text{number of possible outcomes}}$
3 Substitute the known values into the rule and evaluate.	$\text{Pr(Head)} = \frac{6}{12}$
4 Simplify the fraction.	$= \frac{1}{2}$
5 Answer the question.	The probability of obtaining a Head is $\frac{1}{2}$.
ii Repeat steps 1 to 5 of part **c i**. *Note:* The favourable outcomes featuring a Tail and an even number are (T, 2), (T, 4) and (T, 6).	**ii** Number of favourable outcomes = 3 Number of possible outcomes = 12 $\text{Pr(event)} = \dfrac{\text{number of favourable outcomes}}{\text{number of possible outcomes}}$ $\text{Pr(Tail and even number)} = \frac{3}{12}$ $= \frac{1}{4}$ The probability of obtaining a Tail and an even number is $\frac{1}{4}$.
iii Repeat steps 1 to 5 of part **c i**. *Note:* The favourable outcomes featuring a five are (H, 5) and (T, 5).	**iii** Number of favourable outcomes = 2 Number of possible outcomes = 12 $\text{Pr(event)} = \dfrac{\text{number of favourable outcomes}}{\text{number of possible outcomes}}$ $\text{Pr(five)} = \frac{2}{12}$ $= \frac{1}{6}$ The probability of obtaining a five is $\frac{1}{6}$.
iv Repeat steps 1 to 5 of part **c i**. *Note:* The favourable outcomes featuring a Tail and a number greater than 2 are (T, 3), (T, 4), (T, 5) and (T, 6).	**iv** Number of favourable outcomes = 4 Number of possible outcomes = 12 $\text{Pr(event)} = \dfrac{\text{number of favourable outcomes}}{\text{number of possible outcomes}}$ $\text{Pr(Tail and number greater than 2)} = \frac{4}{12}$ $= \frac{1}{3}$ The probability of obtaining a Tail and a number greater than 2 is $\frac{1}{3}$.

remember

1. A *sample space* is a list of all the possible outcomes obtained from an experiment and it is shown as a list enclosed in a pair of curled brackets { }.
2. An *ordered pair* (a, b) displays the result of a two-step experiment.
3. A *two-way table* allows us to display and obtain the sample space in a logical manner.

EXERCISE 14D Using tables to show sample spaces

1 Write down the sample space as an ordered list for each of the following simple or one-step experiments:

a rolling a 6-sided die
b spinning a spinner which can land on any of the numbers from 1 to 10
c choosing an item from a menu that contains fruit salad, cheesecake, mudcake and cheese platter
d choosing a number which is a multiple of 5 in the first 50 counting numbers
e choosing an Australian state or territory for a holiday destination
f picking the correct answer in a true/false question
g choosing a king from a pack of standard cards
h choosing an instrument from the following list: guitar, drum, saxophone, piano and trumpet.

WORKED Example 7

2 **a** Draw a two-way table and list the sample space for the experiment 'spinning a circular spinner divided into 3 equal sectors labelled A, B, C and rolling a die'.
b State the number of different outcomes or results.
c Determine the probability of obtaining:
- **i** the letter A
- **ii** the number 4
- **iii** a number greater than 2
- **iv** a number which is a multiple of 3
- **v** an odd number
- **vi** the letter C and a prime number
- **vii** the letter A, B or C
- **viii** any number except the number 6
- **ix** the letter B and a number less than 3
- **x** a number greater than 6.

3 **a** Draw a table to show the sample space for the experiment 'tossing 2 coins at once'. (*Hint:* Call the first coin 'Coin 1' and the other 'Coin 2'.)
b How many possible results are there for this experiment?
c How many times does the result (H, H) appear?
d How many times does the result (T, T) appear?
e How many times does a result with a Tail and a Head in any order appear?
f What is Pr(H, H)?
g What is Pr(T, T)?
h What is Pr(getting a Tail and a Head in any order)?

4 a Draw a table to show the sample space for the experiment 'tossing a 5c coin and tossing a 10c coin'.
Note: In this case order will matter.
b How many possible results are there?
c What is Pr(getting a H on the 5c coin and getting a T on the 10c coin)?
d What is Pr(getting a H on the 5c coin and a H on the 10c coin)?
e What is Pr(getting a T on the 5c coin and a H on the 10c coin)?
f What is Pr(getting a T on the 5c coin and a T on the 10c coin)?

5 a Draw a table to show the sample space for the experiment 'tossing a coin and rolling a 10-sided die'.
b How many possible results are there?
c What is Pr(getting a H on the coin and a 6 on the die)?
d What is Pr(getting a H on the coin and an even number on the die)?
e What is Pr(getting either a H or a T on the coin and an even number on the die)?
f What is Pr(getting a T on the coin and a number divisible by 3 on the die)?
g What is Pr(getting a number less than 3)?
h What is Pr(getting a number greater than 5)?
i What is Pr(getting a H on the coin and a number less than 6 on the die)?
j What is Pr(getting either a H or a T on the coin)?

6 a Draw a table to show the sample space for the experiment 'rolling a red die and a blue die'.
b How many possible results are there?
(*Note:* There is a difference between a 5 on the red and a 5 on the blue die.)
c What is Pr(getting 1 on both dice)?
d What is Pr(getting 1 on the red die and 6 on the blue die)?
e What is Pr(getting 1 on any die)?
f What is Pr(getting an even number on the red die and an odd number on the blue die)?
g What is Pr(getting an even number on both dice)?
h What is Pr(getting an odd number on both dice)?
i What is Pr(getting 2 numbers whose sum is 6)?
j What is Pr(getting 2 numbers whose sum is 1)?
k What is Pr(getting 2 numbers whose sum is 13)?

7 a Draw a table to show the sample space for the following experiment 'picking a marble out of a bag containing a red, a blue and a green marble and tossing a coin'.
b How many possible results are there?
c What is Pr(picking out a green marble and getting a Head)?
d What is Pr(picking out a red marble and getting a Head)?
e What is Pr(picking out a blue marble and getting a Head)?
f What is Pr(picking out a blue or green marble and getting a Head)?
g What is Pr(picking out a blue or red marble and getting a Head)?
h What is Pr(picking out a blue or red or green marble and getting a Head)?
i How is your answer to part **h** related to the probability of getting a Head if you were tossing only a coin? Can you explain why this is the case?

8 **multiple choice**

Two dice are rolled simultaneously. The probability of obtaining the sum of 7 (by adding the results of the two dice being rolled simultaneously) is:

A $\frac{1}{12}$ **B** $\frac{7}{36}$ **C** $\frac{1}{6}$ **D** $\frac{1}{4}$ **E** 0

9 **multiple choice**

A circular spinner divided into seven equal sectors is spun and a die is rolled. The total number of outcomes is:

A 7 **B** 13 **C** 42 **D** 21 **E** 6

10 For the events given below, determine the following, without listing the sample space.

i State how many rows and columns would be needed to draw up a table representing the sample space.

ii State the number of possible outcomes in the sample space.

a Picking a day in January from a calendar and tossing a coin

b Tossing a coin and shooting a dart at a board with 3 zones

c Choosing a pencil from a set of 72 and rolling a 6-sided die

d Rolling a 10-sided die and rolling a 6-sided die

e Choosing a member from a class of 30 students and rolling a 6-sided die

f Choosing a politician from a list of 100 and tossing a coin

WorkSHEET 14.2

Experimenting with chance

The probability of events that we have examined so far has been based on the number of favourable results that can be obtained out of the total possible number of results of an experiment. This is called *theoretical* or *pure probability*; the answers given for various probabilities are those that the 'rules' teach us. When we actually try some of the experiments we may find that we get surprising results. For example, we know that the chance of getting a Head when we toss a coin is $\frac{1}{2}$. However, if we toss a coin 10 times, because the result is one of pure chance, we may not get the expected 5 Heads and 5 Tails. This *experimental probability* result is likely to be different from the theoretical result, because any experiment needs to be performed a large number of times before the experimental results will come close to or equal the theoretical results. This can be achieved only if the coin is perfectly fair or unbiased and the tossing style does not change throughout the experiment.

However, experimental probability is most useful in real life, because usually only numerous experiments can give us a clue to the chances of something actually happening. In many areas of science, medicine, computing and business, repeated tests are necessary to determine the chances accurately.

The chances of many things occurring are determined not only by the number of outcomes, but by other factors. For example, the chance of you beating your friend in a game of tennis could theoretically be $\frac{1}{2}$. There are 2 possible winners and you are one of them.

However, it is obvious that other factors are involved here. Your friend having a weekly tennis lesson and being a rising star on the court would be a big factor. You may have played 50 matches before and your friend won 40 of them. The chance of you winning at $\frac{10}{50}$ (simplified to $\frac{1}{5}$) is a much more reasonable estimate, based on the **long-term trend** (also referred to as the long run) observed over the 50 matches.

The *long-term trend* (that is, the trend observed for results from a very large number of trials) shows that results obtained through experimental probability will match those of theoretical probability.

In this section you will be given a number of experiments to perform. You will find that the more times you do the experiment, the closer your result should be to the theoretical answer for it. Many times it will be useful for you to work in a group and even to obtain answers from the whole class (that is, to pool results). This will ensure that the number of repetitions of the experiment will be large.

WORKED Example 8

a **Copy the table below. Toss a coin 10 times and record the results in row 1 of the table.**

Experiment number	Heads		Tails	
	Tally	Count	Tally	Count
1 2 3 4 5 6				
	Total		Total	

b **What is the probability of obtaining a Head from your experiment?**
c **What is the probability of obtaining a Tail from your experiment?**
d **How do these values compare with the theoretical results?**
e **Repeat step a another 5 times and combine all of your results.**
f **How does the combined result compare with the theoretical results?**

THINK

a Toss a coin 10 times and record the results in the first row of the table.
Notes: (a) Place a stroke in the appropriate tally column each time an outcome is obtained. Five is denoted by a 'gatepost'; that is, 4 vertical strokes and 1 diagonal stroke (卌).
(b) The same coin must be used throughout the experiment. The style of the toss and the surface the coin lands on must be the same.

WRITE

a

Exp. No.	Heads		Tails	
	Tally	Count	Tally	Count
1	||||	4	卌 |	6
2	||||	4	卌 |	6
3	卌 ||	7	|||	3
4	|||	3	卌 ||	7
5	||||	4	卌 |	6
6	卌 |||	8	||	2
	Total	30	Total	30

THINK

b 1 Calculate the probability of obtaining a Head for this experiment using the rule.

WRITE

b $\Pr(\text{event}) = \dfrac{\text{number of favourable outcomes}}{\text{number of possible outcomes}}$

$\Pr(\text{Heads}) = \dfrac{\text{number of Heads obtained}}{\text{total number of tosses}}$

THINK	WRITE
2 Substitute the given values into the rule.	$\text{Pr(Heads)} = \frac{4}{10}$
3 Evaluate and simplify.	$= \frac{2}{5}$
4 Convert the fraction to a percentage by multiplying by 100%.	As a percentage $\frac{2}{5} = \frac{2}{5} \times 100\%$ $= \frac{200}{5}\%$ $= 40\%$
5 Answer the question.	The probability of obtaining a Head in this experiment is $\frac{2}{5}$ or 40%.
c **1** Calculate the probability of obtaining a Tail for this experiment.	**c** $\text{Pr(Tails)} = \frac{\text{number of Tails obtained}}{\text{total number of tosses}}$
2 Substitute the given values into the rule and simplify.	$\text{Pr(Tails)} = \frac{6}{10}$ $= \frac{3}{5}$
3 Convert the fraction to a percentage by multiplying by 100%.	As a percentage $\frac{3}{5} = \frac{3}{5} \times 100\%$ $= \frac{300}{5}\%$ $= 60\%$ The probability of obtaining a Tail in this experiment is $\frac{3}{5}$ or 60%.
d Compare the results obtained in parts **b** and **c** with the theoretical results.	**d** The experimental value obtained for the Pr(Head) is $\frac{2}{5}$ (or 40%) and the Pr(Tail) is $\frac{3}{5}$ (or 60%). The theoretical value of these probabilities is $\frac{1}{2}$ (or 50%). Therefore, the experimental probabilities differ from the theoretical probabilities by 10%.
e **1** Repeat the procedure of part **a** 5 times.	**e** Refer to results in table in part **a**.
2 Calculate the total number of Heads and Tails obtained and enter the results in the table.	
f **1** Calculate the probability of obtaining a Head for this experiment.	**f** $\text{Pr(Heads)} = \frac{\text{number of Heads obtained}}{\text{total number of tosses}}$
2 Substitute the given values into the rule and simplify.	$\text{Pr(Heads)} = \frac{30}{60}$ $= \frac{1}{2}$

Continued over page

THINK	**WRITE**
3 Convert the fraction to a percentage by multiplying by 100%.	As a percentage $\frac{1}{2} = \frac{1}{2} \times 100\%$ $= \frac{100}{2}\%$ $= 50\%$
4 Answer the question.	The probability of obtaining a Head in this experiment is $\frac{1}{2}$ or 50%. $\text{Pr(Tails)} = \frac{\text{number of Tails obtained}}{\text{total number of tosses}}$ $\text{Pr(Tails)} = \frac{30}{60}$ $= \frac{1}{2}$ As a percentage $\frac{1}{2} = \frac{1}{2} \times 100\%$ $= \frac{100}{2}\%$ $= 50\%$ The probability of obtaining a Tail in this experiment is $\frac{1}{2}$ or 50%.
5 Compare the combined result obtained with the theoretical results.	The combined results in this experiment produced probability values that were equal to the theoretical probability values. Therefore, the long-term trend of obtaining a Head or Tail when tossing a coin is equal to $\frac{1}{2}$.

EXCEL Spreadsheet: Simulating coin tosses

Instead of physically tossing a coin, we can use technology to simulate or copy this process. The Excel file provided on the *Maths Quest 7* CD can be used to simulate the coin tosses. The spreadsheet screen below shows the number of Heads and Tails obtained in one simulation of tossing a coin 100 times. New sets of results can be obtained for the 100 tosses by pressing the F9 key. The number of times a coin is tossed can also be changed.

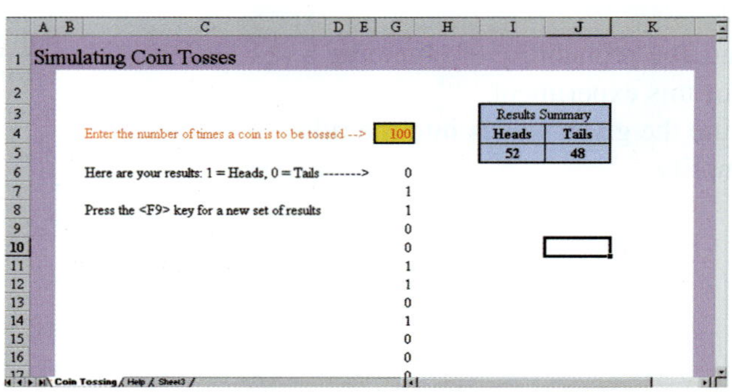

Simulations

Experimental probability involving simple devices such as coins, dice, a pack of cards and spinners, to name a few, may be used to model or mimic real life situations. When we run an experiment which tries to predict results for another, more complex situation we are performing a **simulation**.

Simulations give us the opportunity of studying various outcomes by performing experiments. These outcomes may otherwise be impractical because of time constraints, lack of resources, financial restrictions or personal risk factors. Techniques involving experimental probability are quite useful and, as mentioned previously, are employed in many everyday practical situations.

WORKED Example 9

Alex wants to know how many packets of cereal she must purchase in order to collect 4 different types of plastic toys during a promotion.

a Design an experiment which will simulate the given situation, providing details of the equipment required and procedure involved.

b Discuss the fairness of the experiment and findings.

THINK	WRITE
a 1 Determine the sample space. *Note:* There are 4 toys, each of which is equally likely to be found in a packet of cereal.	**a** Sample space = {toy 1, toy 2, toy 3, toy 4} The 4 events are equally likely.
2 Decide on the equipment required for this experiment. *Note:* Choose an item which produces 4 possibilities; that is, a spinner or 4 different coloured balls in a box, etc.	A circular spinner divided into 4 equal sectors and labelled 1, 2, 3, 4 will be used to simulate the outcomes of obtaining a toy from the cereal packet. Sector 1 represents toy 1 (T1). Sector 2 represents toy 2 (T2). Sector 3 represents toy 3 (T3). Sector 4 represents toy 4 (T4).
3 Give details of how the experiment will be conducted.	Spin the pointer on the circular spinner until all 4 toys are represented; that is, until T1, T2, T3 and T4 are obtained. Repeat this procedure another 19 times.

Continued over page

THINK

4 Record the results in a table comprising 3 columns, headed: 'Experiment number', 'Results' and 'Number of packets'.

WRITE

Experiment number	Results	Number of packets
1	T4 T3 T4 T3 T2 T4 T4 T1	8
2	T3 T4 T4 T3 T4 T1 T2	7
3	T1 T4 T4 T2 T1 T4 T4 T3	8
4	T1 T4 T4 T4 T4 T4 T2 T3	8
5	T1 T4 T3 T4 T1 T2	6
6	T1 T4 T1 T2 T1 T2 T4 T2 T4 T4 T1 T1 T4 T3	14
7	T1 T4 T4 T2 T2 T4 T4 T3	8
8	T4 T2 T2 T3 T2 T4 T3 T2 T4 T4 T1	11
9	T3 T4 T2 T4 T2 T2 T4 T2 T1	9
10	T4 T3 T3 T1 T1 T2	6
11	T3 T1 T4 T3 T1 T1 T1 T3 T1 T4 T1 T3 T1 T1 T3 T2	16
12	T3 T2 T1 T1 T4	5
13	T3 T4 T1 T2	4
14	T4 T2 T3 T3 T4 T1	6
15	T4 T1 T4 T3 T3 T4 T4 T3 T1 T2	10
16	T1 T1 T2 T1 T1 T3 T4	7
17	T4 T2 T3 T2 T4 T2 T4 T4 T4 T4 T3 T4 T4 T3 T4 T3 T4 T4 T3 T4 T2 T1	22
18	T1 T2 T1 T1 T4 T3	6
19	T2 T4 T2 T4 T1 T4 T2 T4 T1 T1 T2 T2 T1 T3	14
20	T1 T1 T1 T1 T4 T3 T2	7
	Total	182

b 1 Comment on the fairness of the experiment.

b In order to ensure fairness throughout the experiment, the pointer of the spinner was spun from the wider end by the same person each time. Twenty experiments were performed in total; however, this amount could be increased.

2 Answer the original question. Write the total number of packets and total number of experiments.

Total number of packets
$= 8 + 7 + \ldots + 14 + 7$
$= 182$
Total number of experiments = 20

THINK	**WRITE**
3 Write the rule for the average number of packets.	Average = $\dfrac{\text{total number of packets}}{\text{total number of experiments}}$
4 Substitute the known values into the rule.	Average number of packets = $\frac{182}{20}$
5 Evaluate.	$= 9.1$
6 Round the value to the nearest whole number. *Note:* Since we are dealing with packets we must work with whole numbers.	≈ 9
7 Summarise your findings. *Note:* Comment on points of interest; that is, the maximum and minimum number of packets that need to be purchased.	The average number of packets of cereal Alex must purchase to obtain each of the 4 plastic toys is 9. From this experiment, the minimum number of packets needed to obtain the 4 toys is 4 while the largest number of packets needed is 22.

It is important to note that this is only one possible solution. Each time the experiment is conducted, different results will be obtained.

The previous worked example is time consuming and tedious; however, we can repeat the experiment using our calculator or spreadsheets.

Generating random numbers using the graphics calculator

To generate random whole numbers (also called integers) on the TI-83 follow these steps. First determine the value of the smallest and largest numbers to be used and the number of values to be generated. In worked example 9, the smallest value was 1, the largest value 5 and about 20 values were required.

Generating random numbers

Press MATH, scroll across with the arrow to select the **PRB** menu, then scroll down with the arrow to choose option **5: randInt(** and press ENTER. Then type in **1, 5, 20)** and press ENTER again. The calculator will generate a row of 20 random numbers between 1 and 5. You will need to use the arrow to scroll across to the right to find more than the numbers shown on this screen. Simply press ENTER for another row of 20 random numbers.

Generating random numbers on an Excel spreadsheet

EXCEL Spreadsheet: Generating random numbers

The random number generator [=rand()] on the spreadsheet generates a decimal number between 0 and 1. We need to convert this decimal number to a whole number between 1 and 5.

Enter the formula **=INT(RAND()*5)+1** into cell **A1**.

Multiplying the random decimal number by 5 produces a random decimal number between 0 and 5. The **INT** function will change the decimal to a whole number. We finally add 1 to make it a whole number between 1 and 5.

Use the **Fill Down** function to fill this formula down to cell A20. You should now have 20 random numbers as shown in this screen dump.

= =INT(RAND()*5)+1

	A	B	C	D	E
1	2				
2	3				
3	4				
4	2				
5	2				
6	2				
7	1				
8	3				
9	2				
10	1				
11	4				
12	5				
13	5				
14	2				
15	5				
16	4				
17	4				
18	3				
19	5				
20	3				

Note that every time you perform an action on this spreadsheet, the random numbers will change. We can stop this by following these instructions.

1. Choose **Options** from the **Tools** menu.
2. Select the calculation tab and click the manual calculation radio button.
3. You can obtain a recalculation of your random numbers by pressing **F9**.

remember

1. The *long-term trend* (that is, the trend observed for results from a very large number of trials) shows that results obtained through experimental probability will match those of theoretical probability.
2. A *simulation* is the use of simple devices such as coins, dice, a pack of cards and spinners to model or mimic a real life situation.

EXERCISE 14E Experimenting with chance

EXCEL Spreadsheet: Simulating coin tosses

EXCEL Spreadsheet: Simulating coin tosses (DIY)

WORKED Example 8

1 a Copy the table below. Toss a coin 10 times and record the results in the first row of the table.

b What is the probability of obtaining a Head from your experiment?

c What is the probability of obtaining a Tail from your experiment?

Experiment number	Heads Tally	Heads Count	Tails Tally	Tails Count
1				
2				
3				
4				
5				
6				
	Total		Total	

d How do these values compare with the theoretical results?
e Repeat step a another 5 times and combine all your results.
f How does the combined result compare with the theoretical results?

GC program—TI Coin flip

2 Use your results from question 1 to answer the following.
a The long-term trend of the probability of obtaining a Head on the toss of a coin is the Pr(Heads) from your experiment.
What is the long-term trend of the probability after:
i 10 tosses of the coin? ii 20 tosses of the coin?
iii 30 tosses of the coin? iv 60 tosses of the coin?

GC program—Casio Coin flip

b Obtain a classmate's 60 results. Combine these with yours. State the long-term trend of Pr(Heads) obtained.
c Combine your pair's 120 results with another pair's. State the long-term trend of Pr(Heads) obtained.
d Finally, obtain the results obtained by whole class for this experiment. (Make sure nobody's results are counted twice). You should have 60 tosses per person. State the long term trend of Pr(Heads) obtained.
e Copy and complete the table below.

14.7 SkillSHEET Converting a fraction to a percentage

	Heads		**Tails**	
Number of tosses	**Pr(heads)**	**Pr(heads) as percentage**	**Pr(tails)**	**Pr(tails) as percentage**
10				
20				
30				
60				
120				
240				
Whole class (specify number of tosses)				

f Comment on the changes of the long-term trend value of Pr(Heads) as you toss the coin more times.

3 Simulate a 5-colour spinner using a graphics calculator's **RandInt(1,5)** function or the **(= RAND()*5+1)** formula with an Excel spreadsheet on your computer.
a What is the chance of getting any one of the 5 colours when you spin the spinner (theoretically)?
b Spin the spinner 10 times and, using a table such as the one below, record your results.

Colour	**1**	**2**	**3**	**4**	**5**
Number of times it occurs					

c From your results, list the probabilities of obtaining each colour. For example, divide the number of times a particular colour was obtained by the total number of spins, (that is, 10).

d Why might these probabilities not be the same as the theoretical probability would suggest?

e Spin your spinner and record the results for another 10 spins.

f Spin your spinner so that you have 100 results. Is the experimental probability closer to the pure probability? Why might this be? Discuss.

4 If you wanted to create a device that would give a theoretical probability of achieving a particular result as $\frac{1}{4}$, how many sections would a spinner such as this need to be divided into?

5 How would you divide or colour a spinner if you wanted to achieve the probability of a success equal to $\frac{3}{10}$?

6 For the spinner at right, what would be the probability of getting the red section?

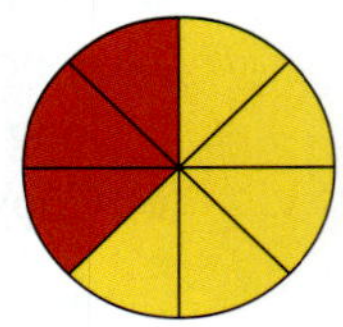

WORKED Example 9

7 Repeat the experiment described in worked example 9.

GC program—TI: Rolling a die

GC program—Casio: Rolling a die

8 You have calculated previously that the chance of getting any particular number on a 6-sided die is $\frac{1}{6}$. You have 2 different coloured dice. Is there any difference in your dice apart from colour? Could one be biased (more likely to give a particular result than theory says it should)?

a Design a test to determine whether the dice you have are fair. Write down what you are going to do.

b Perform your test, and record your results.

c Calculate the probability of getting each of the numbers on each of your dice, based on your tests and on the long-term trend you have observed.

d What does your test say about your 2 dice? Are there any things that need to be considered before giving your answer? (Perhaps your dice have slightly uneven shapes or something that might cause them to lean towards one result more than others.)

9 Use a box of Smarties or marbles, or technology such as a graphics calculator or an Excel computer program, for this question. Count the number of items of each colour in the box before you start.

a Calculate the theoretical probability of getting a particular colour if you pick 1 out of the box without looking.

b Design an experiment to determine the probability of getting a particular colour out of the box, using the long-term trend.

c Why is it important that if you take Smarties (marbles or a colour in the simulation) out of the box for this experiment, you must put them back each time?

d Could you use something other than your box of Smarties, marbles or technology to determine this probability? What other things could you use to simulate this experiment?

GAME time: Chance — 002

10 QUICK QUESTIONS 2

For questions 1 to 3, state whether each of the events is considered certain, highly likely, even chance, very unlikely or impossible.

1 One of your parents will let you have the day off school tomorrow.

2 You will have something to drink during the day.

3 If you choose one of two pairs of socks from your drawer without looking, you will select the black pair, not the white.

4 Draw a table to show the sample space for rolling a 6-sided die and selecting a marble from a box containing one blue, one green, one black and one white marble.

For questions 5 to 10, use the sample space table from question 4.

5 How many possible results are there?

6 How many results include selecting a black marble?

7 Find Pr(3, white).

8 Find Pr(5, green).

9 Find Pr(even number, black).

10 Find Pr(odd number, green or blue).

MATHS QUEST CHALLENGE

1 In how many different ways can change be given for a 50 cent coin using only 20 cent, 10 cent and 5 cent coins?

2 In your sock drawer there are only 4 identical blue socks and 5 identical black socks. What is the smallest number of socks you can remove from the drawer in the dark so that you will be absolutely sure of having:

a a pair of black socks?

b a pair of blue socks?

c a black pair and a blue pair of socks?

3 Sabita remembered that her mother's car registration plate had 2 letters followed by 3 digits. She recalled that the letters were S and C and that the digits were 8, 3 and 7 but couldn't remember the order. What combination of letters and numbers could her mother's car registration plate have? Make a list of the possibilities.

COMMUNICATION

Petals

Sally and Robin are playing a game they call Petals. It is played on a hexagonal grid in the shape of a flower with petals.

Starting with a marker at the centre of the flower, C, a die is rolled in turn by two players. The marker is moved around the flower from one hexagonal shape to another. The direction of movement of the marker is determined by the number showing on the uppermost face of the rolled die.

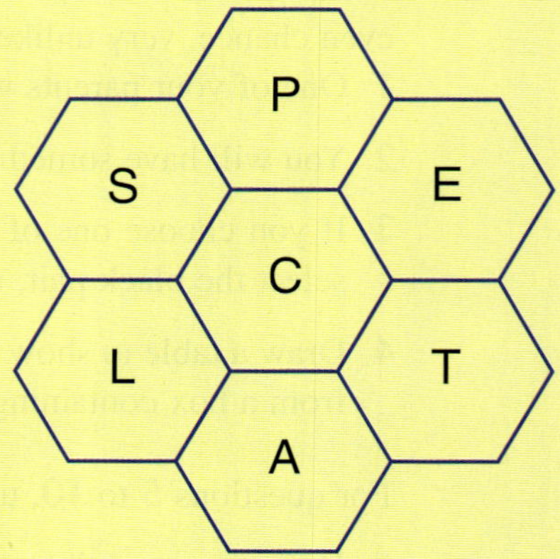

- 1 means move in a northerly direction
- 2 means move in a north-easterly direction
- 3 means move in a south-easterly direction
- 4 means move in a southerly direction
- 5 means move in a south-westerly direction
- 6 means move in a north-westerly direction

There comes a point when the marker moves off the grid. The first player to do this is the winner of the game. For example, rolls of 1, 5, 4, 2, 3, 5, 5 in succession, starting from C would result in the path C, P, S, L, C, T, A, off grid. You can exit the grid in any direction. There are 18 exit points from the hexagonal petals.

1 Draw the grid and practise playing with a partner, taking turns with a die.

2 After mastering the game, each player should draw up the following tables to record the progress of games.

Die roll	1	2	3	4	5	6
Tally						
Frequency						

Exit petal	P	E	T	A	L	S
Tally						
Frequency						

Game	**Number of moves to exit**		**Win/Lose**
	Tally	**Frequency**	
1			
2			
3			
…			
10			

Play 10 games to gather sufficient data for analysis.

3 Each player should make individual calculations.

- Which die face turned up most frequently for you?
- Which exit petal was the most common?
- What is the average number of moves it took before you exited the grid?
- How many of the ten games did you win?

4 Collate the calculations of all the pairs in your class. Can you see any trends? Write the results of your experiment.

Fair games

Your ability to win a particular game depends on the type of game; that is, whether it is based on skill or luck. Games may be classified into distinct groups: some are largely dependent on skill, others involve a combination of skill and luck, while others are totally dependent on luck. Games of chance are purely dependent on luck. It is therefore important to have an understanding of whether the game being played allows each person an equal chance of winning.

A fair game is one which allows each player an equal chance of winning.

WORKED Example 10

List the possible outcomes (that is, the sample space) of the following games and then make a comment regarding the fairness of the game.

a **Samira and Liana play a game which requires them to roll a die. Samira wins the game if she rolls an even number while Liana wins if she rolls an odd number.**

b **Samira and Liana play another game which requires them to roll a die. Samira wins the game if she rolls a number which is a multiple of 3 while Liana wins if she rolls a prime number.**

THINK	WRITE
a 1 List the sample space of each of the players. 2 Count the number of possible ways each player can win. 3 Answer the question.	**a** Samira: {2, 4, 6} Liana: {1, 3, 5} Samira has 3 chances of winning the game. Liana has 3 chances of winning the game. Yes, the game is fair since each player has the same number of chances of winning.
b 1 List the sample space of each of the players. 2 Count the number of possible ways each player can win. 3 Answer the question.	**b** Samira: {3, 6} Liana: {2, 3, 5} Samira has 2 chances of winning the game. Liana has 3 chances of winning the game. No, the game is not fair since each player does not have the same number of chances of winning. Liana has a greater chance of winning.

remember

A fair game is one which allows each player an equal chance of winning.

EXERCISE 14F Fair games

WORKED Example 10

1 List the possible outcomes (that is, the sample space) of the following games and then make a comment regarding the fairness of the game.

a Samira and Liana play a game which requires them to roll a die. Samira wins the game if she rolls an even number while Liana wins if she rolls a number greater than 3.

b Samira and Liana play a game which requires them to roll a die. Samira wins the game if she rolls a number which is not a multiple of 2 while Liana wins if she rolls a number less than 3.

2 Consider the possible outcomes of each the following games and then state whether the games are fair.

a Thu and Anh roll a 10-sided die. Thu wins if she obtains a number which is a multiple of 4, and Anh wins if she obtains a multiple of 6.

b Naomi and Wei roll a 10-sided die. Naomi wins if she obtains a number which is a multiple of 5, and Wei wins if she obtains a number divisible by 4.

c Rohan and Kristyn roll a 10-sided die. Rohan wins if he obtains a number which is less than 6, and Kristyn wins if she obtains a number greater than 5.

d Lucy and James roll a 10-sided die. Lucy wins if she obtains a number which is less than 9, and James wins if he obtains a number greater than 4.

e Laura and Kirrilee roll a 10-sided die. Laura wins if she obtains a number which is a divisor of 6, and Kirrilee wins if she obtains a square number.

f Jason and Casey roll a 10-sided die. Jason wins if he obtains a prime number, and Casey wins if he obtains an even number less than 10.

g Christine and Kim roll a 10-sided die. Christine wins if she obtains an odd number which is less than 6, and Kim wins if she obtains a number which is a factor of 9.

h Caroline and Mark roll a 10-sided die. Caroline wins if she obtains a number which is less than 7, and Mark wins if he obtains a number greater than 5.

i Virginia and Marolin roll a 10-sided die. Virginia wins if she obtains a number which is both a factor of 4 and 10, and Marolin wins if she obtains a number which is both a factor of 6 and 10.

j Kim and Eddy roll a 10-sided die. Kim wins if he obtains a prime number, and Eddy wins if he obtains a odd number.

3 **multiple choice**

The statement which does not result in a fair game is:

A obtaining the outcome (H, H) or (T, T) compared with obtaining one Head when a coin is tossed twice

B obtaining the outcome {1, 2, 3} compared with {4, 5, 6} when a die is rolled

C obtaining the outcome {1, 3} compared with {4, 6} when a die is rolled

D obtaining the outcome (H, H) compared with obtaining one Head when a coin is tossed twice.

E obtaining the outcome 'multiples of 6' compared with obtaining 'multiples of 7' from the set of the first 35 counting numbers.

4 Using the spinner illustrated at right, design a game, with appropriate rules, which is:

a equally fair to each player
b unfair to one or more of the players (depending on the number of players)
c Repeat parts a and b another 3 times using different requirements.

5 Design your own game which illustrates both a fair and unfair outcome. Be sure to clearly define the rules of the game and the outcome for each player. Also list any equipment required.

WorkSHEET 14.3

THINKING

I win!

Natalie and Sasha are playing a number of games with a standard deck of cards. The cards are shuffled. Natalie chooses a card, makes a note of it and returns it to the deck. The cards are then reshuffled and Sasha chooses a card, makes a note of it and returns it to the deck.

For each of the following games:

a Consider the possible outcomes and state whether the game is fair.

b Calculate the probability of each person winning the game.

1 Natalie wins if she obtains an ace, and Sasha wins if he obtains a king.

2 Natalie wins if she obtains a spade and Sasha wins if he obtains a red card (diamonds or hearts).

3 Natalie wins if she obtains a black card (spades or clubs) and Sasha wins if he obtains a picture card.

4 Natalie wins if she obtains a picture card (jack, queen or king) and Sasha wins if he obtains a club.

5 Natalie wins if she obtains a diamond and Sasha wins if he obtains an ace or picture card (jack, queen or king).

6 Natalie wins if she obtains a diamond or heart and Sasha wins if he obtains a club or spade.

COMMUNICATION

Clowning around

Thomas is playing a game of 'Laughing clowns' at the Mathsville carnival. Each clown has six channels numbered 1 to 6 under its head. A red and a blue ball are given to each player for a game. Each game consists of putting the balls into the clown's open mouth as its head moves from side to side. Each ball will go down into one of the six numbered channels. The score for a game is the sum of the numbers of the channels in which the two balls land. For example, if both balls land in channel 4, the score will be 4 + 4 = 8. Each total has a different prize.

1 Copy and complete this two-way table.

		Red					
		1	2	3	4	5	6
Blue	1	(1, 1)	(1, 2)	(1, 3)	(1, 4)	(1, 5)	(1, 6)
	2						
	3						
	4						
	5						
	6						

2 What is the number of events in the sample space $n(S)$?

3 **a** What is the smallest total possible?
b What is the largest total possible?

4 Complete another two-way table that displays the totals and check your answers to part **c**. The first two cells have been done for you.

		Red					
		1	2	3	4	5	6
Blue	1	1 + 1 = 2	1 + 2 = 3				
	2						
	3						
	4						
	5						
	6						

5 What is the probability of obtaining a score totalling:
a 7? b 11? c 2?

6 a Are all the scores equally likely?
b Is this a fair game?

7 If you were the 'Laughing Clowns' operator and wanted to make a profit what score would you choose to win the big prize? Explain your answer.

8 What score is Thomas most likely to get?

summary

Copy the sentences below. Fill in the gaps by choosing the correct word or expression from the word list that follows.

1 ____________ tells us how ____________ it is for something to ____________.

2 ____________ is defined as the chance of a particular ____________ or result occurring.

3 The probability of an event occurring ranges from and ____________ ____________ (impossible) to 1 (____________).

4 The probability ____________ associates important words used in describing probability with their approximate corresponding ____________ values.

5 An outcome is a particular ____________ of an experiment.

6 ________________________ are the ones we are looking for.

7 $\text{Pr(event)} = \dfrac{\_\_\_\_\_\_\text{ of favourable outcomes}}{\text{number of }\_\_\_\_\_\_\text{ outcomes}}$

8 A ________________________ is the list of all possible outcomes obtained from an ____________ and enclosed in a pair of curled ____________ { }.

9 A ________________________ table allows us to obtain the sample space in a ____________ manner.

10 An ________________________, such as (H, 6) shows a result from a two-step experiment in a logical manner.

11 The ____________________________________ shows that results obtained through ____________ probability will match those of ____________ probability.

12 Experimental probability uses simple devices such as ____________, ____________, a pack of ____________ and spinners to ____________ real life situations.

13 We perform a ____________ when we run an experiment and try to ____________ results for another more ____________ situation.

WORD LIST

two-way	complex	predict	theoretical
chance	experimental	0	event
cards	long-term trend	imitate	sample space
probability	result	coins	ordered pair
brackets	possible	certain	simulation
dice	experiment	likely	happen
favourable outcomes	scale	number	numerical
	logical	includes	

CHAPTER review

1 For each of the given statements, specify whether the chance of the following events occurring is certain, likely, even chance, unlikely or impossible. **14A**

- **a** Australia is in the southern hemisphere.
- **b** You will still be alive in the next century.
- **c** Obtaining an even number on a circular spinner numbered from 1 to 16.
- **d** England is in the southern hemisphere.
- **e** You roll a fair die and obtain a number less than or equal to 4.
- **f** Humans can survive without water indefinitely.

2 List three events where the chance of each occurring would be: **14A**

a impossible **b** certain **c** improbable.

3 Copy the number line below and place the list of words on it to indicate what sort of chance each number would represent. **14A**

Words to place: certain, very unlikely, unlikely, likely, highly likely, highly unlikely, even chance, impossible, very likely.

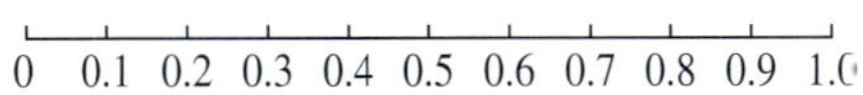

4 List the sample space for the following experiments: **14B**

- **a** tossing a coin
- **b** drawing a marble out of a bag containing red, green and blue marbles
- **c** rolling a 6-sided die
- **d** spinning a circular spinner numbered from 1 to 10
- **e** drawing a 4 card from a pack of playing cards
- **f** answering a true/false question in a test
- **g** winning a race in which there are 9 runners
- **h** choosing a year level to represent the school at a public event.

5 For each of the following, state the number of elements in the sample space. **14B**

- **a** The first prize in a raffle is chosen from tickets numbered from 1 to 5000.
- **b** A card selected from the red cards in a standard deck.
- **c** A circular spinner numbered 1 to 5 is spun.
- **d** A day is selected in the month of July.

6 A six-sided die, labelled 1, 2, 3, 3, 4, 5 is rolled. Find: **14C**

- **a** Pr(getting a 5)
- **b** Pr(getting a 3)
- **c** Pr(getting an even number)
- **d** Pr(getting an odd number)
- **e** Pr(getting a number greater than or equal to 3)
- **f** Pr(getting a number less than 3)
- **g** Pr(getting a 0)
- **h** Pr(getting a 6).

14C

7 A person has a normal pack of cards and draws one out. Find:

a Pr(picking a heart)

b Pr(picking a spade)

c Pr(picking a picture card)

d Pr(picking a card with a number less than 5 on it). Do not count aces as a number.

14D

8 Use a table to show the sample space for the experiment 'tossing a coin and rolling a die'. Find:

a Pr(a tail and a number less than 3)

b Pr(a head and a number greater than 3)

c Pr(a tail and an even number)

d Pr(a head and a prime number).

14D

9 Use a table to show the sample space for the experiment 'spinning a spinner with 10 sectors and rolling a die'. Find:

a Pr(the sum totals to an even number)

b Pr(odd number on the spinner and an even number on the die)

c Pr(both digits having a value less than 5)

d Pr(the sum totals to a value between 7 and 10)

e Pr(the sum totals to a value greater than or equal to 13 but less than 16)

f Pr(the sum totals to a prime number).

14E

10 Write a sentence to explain what the difference is between theoretical probability and experimental probability.

14E

11 Jane and her father have decided to play a game of chess against each other. Neither has played before but each has been given the same explanation of the rules of the game.

a What would be the probability of either of them winning a game?

b Explain your answer.

Since her first game with her father, Jane has joined a chess club and competes regularly against various players. Her father, on the other hand, plays occasionally. Over a number of years Jane and her father have played 340 games. Jane has won 255 of these games.

c Given this long-term trend, what would be the probability that the father wins the next game?

d Does the probability remain the same over a period of time? Explain your answer.

14E

12 Danielle wants to know how many packets of cereal she must purchase in order to collect 6 different limited edition Disney cards. Design an experiment that will simulate the given situation. Provide clear instructions and details of 2 different devices that may be used for the simulation.

14F

13 Consider the possible outcomes of each of the following games which involve rolling a 6-sided die and then state whether each game is fair.

a Nadia wins if 1 or 2 is rolled and Alex wins if 3, 4, 5 or 6 is rolled.

b Cassandra wins if 1, 2 or 3 is rolled and Jessica wins if 4, 5 or 6 is rolled.

c John wins if a multiple of 3 is rolled and Jennifer wins if a multiple of 3 is not rolled.

d Paola wins if an even number is rolled and Serge wins if an odd number is rolled.

test yourself CHAPTER 14

Strategies for problem solving

Contents

CONTENTS

Introduction to problem solving

In mathematics classes and in our everyday lives, we are presented with many problems and situations that can be solved using mathematical processes. When we 'solve problems', there are five main processes that we can use:

1. questioning
2. applying strategies
3. communicating
4. reasoning
5. reflecting.

To solve a simple problem, we may need to apply only one or two of the processes. For example, imagine you are asked to keep the score in a basketball game. You might draw up a table with two columns — one for each team's score — then enter the points scored as the game progresses and add up the two columns at the end. In this way, you would have focused on two processes:

1. applying a strategy — you chose to use a table to collect and summarise the data
2. communicating — the information was recorded in a way that could be checked and understood by the teams.

However, what if the players want to know how many points each of them scored? What if the coach needs to know how many fouls were awarded against each player or the team? If you had asked yourself these questions before applying a strategy, the result might have been a more detailed and useful record of the game.

The five processes in problem solving are interrelated. By practising the skills involved in using all five processes, you will learn to tackle new mathematics problems with confidence and arrive at the correct and complete solution using the most appropriate methods.

Questioning

Why do you think that teachers often say: 'Read through the question a number of times before you begin'? What does this achieve? The aim is to make you think about the information that is needed and what will be the best approach to the problem. You are questioning and looking ahead to what may be produced by certain strategies. Be clear about the problem and the solution that is required. Consider whether you have come across a similar problem before that could be useful. Is there more than one method that you could use? Which method would be best to solve this problem?

Applying strategies

To apply a strategy when problem solving is to follow a plan or a method. The goal is to complete an investigation, answer a question or find a solution to a problem. Examples of strategies that can be used to achieve this include:

1. drawing up tables and diagrams
2. finding patterns of numbers

3. making use of technology, such as a computer spreadsheet
4. working backwards from the answer
5. using a process of elimination
6. looking at similar but simpler problems
7. using trial and error (guess and check).

Communicating

Another person who reads your work should be able to follow your method or strategy. It is important to learn to present data, explanations and solutions in a clear and concise form, using correct mathematical terms and appropriate diagrams.

Reasoning

When you reach a solution to a particular problem, make sure that you check it and can prove with mathematical reasoning that it is correct and that the method was appropriate.

Reflecting

Whenever you use the processes of problem solving, there is an opportunity to reflect on what you did and why, and to improve your understanding of the strategies and concepts. Consider whether the solution could have been obtained in a different or better way. Learn from the experience and look forward with confidence to using the knowledge gained when you are faced with the next problem or situation.

Throughout this textbook there are many questions and problems that can be answered using the five processes. On the following pages, some examples and practice questions are supplied. These will help you to focus on improving your skills relating to each process. The examples will also help you to become aware of the need to use all the processes to arrive at the correct solution in the most appropriate and efficient way.

Strategies for investigation and problem solving

Questioning skills

Before beginning work on a problem, devise some questions that ensure you set off on an effective path to solving it. Questions can help us to organise the information, test an idea, add another key piece of information and/or decide what to do next. There may be more than one question that could be asked.

The first important question is, 'What is this problem asking me to do or find?'. In order to answer this question, it is good practice to read the question at least twice.

Create a table, then look for a pattern or a result

A table is a way of organising or grouping numbers. You should consider the number of rows and columns that will be needed and label them appropriately. A well-designed table helps you to see any patterns or results in the numbers you have organised, and also demonstrates to others how you were able to arrive at your solution. There are many different ways of presenting information in a table.

WORKED Example 1

Amad operates a petrol station on a busy highway. He is investigating the purchasing habits of his customers. From the sample of purchases below, calculate the most frequently purchased amount of petrol. First, round each amount to the nearest 5 litres.

14.7 L 22.0 L 32.2 L 8.1 L 23.5 L 15.5 L 27.9 L 44.0 L
27.2 L 32.9 L 25.5 L 42.4 L 23.7 L 31.7 L 39.7 L

THINK	WRITE
1 Read the question at least twice and take note of all the important facts.	There are 15 pieces of data showing the individual quantities of petrol purchased.
2 Identify the solution required.	The question asks for the most frequently purchased amount of petrol.
3 Decide how the 'amounts' of petrol can be arranged in a table.	Each 'amount' of petrol can be organised into multiples of 5 L.
4 State what the smallest and largest amounts of petrol are. *Note:* The smallest and largest amounts give us an indication of the range of values required for the table.	The smallest amount is 8.1 L, which rounds to 10 L; the largest amount is 44.0 L, which rounds to 45 L.
5 Create a table showing each amount as it appears in the given list and place a tick in the column representing the amount rounded to the nearest 5 litres.	
6 Total the ticks in each column and look for the multiple of 5 L with the most ticks.	

Rounded amount

Actual amount	10 L	15 L	20 L	25 L	30 L	35 L	40 L	45 L
14.7 L		✔						
22.0 L			✔					
32.2 L					✔			
8.1 L	✔							
23.5 L				✔				
15.5 L		✔						
27.9 L					✔			
44.0 L								✔
27.2 L					✔			
32.9 L					✔			
25.5 L				✔				
42.4 L							✔	
23.7 L				✔				
31.7 L					✔			
39.7 L							✔	
TOTAL	1	2	1	3	5	0	2	1

THINK	WRITE
7 Answer the question.	To the closest multiple of 5 L, the amount of petrol that is purchased the most often is 30 L.

Another way of showing the information required in worked example 1 could be in a table like this:

Amount	Tally	Frequency
10 L	|	1
15 L	||	2
20 L	|	1
25 L	|||	3
30 L	~~||||~~	5
35 L		0
40 L	||	2
45 L	|	1

Can you think of other ways?

Try these

Construct a table to solve each of the following problems.

1 A new mobile phone company is investigating the habits of telephone users. From the sample of telephone calls below, find out the most popular length of phone call. First, round each call time to the nearest half minute.

132 s	10 s	43 s	71 s	243 s	52 s
142 s	332 s	6 s	38 s	63 s	32 s
132 s	32 s	43 s	52 s	352 s	101 s
124 s	28 s	153 s	10 s	243 s	34 s

2 In a survey, 24 people were asked: 'What is the maximum distance you would walk to a train station?'. A person who says '800 m' will walk up to 800 m, but no more. What is the longest distance (as a multiple of 100 m) that at least $\frac{3}{4}$ of the people would walk to the train station?
The survey responses are:

100 m	200 m	250 m	1 km	$\frac{1}{2}$ km	$\frac{3}{4}$ km
600 m	500 m	450 m	100 m	1.2 km	800 m
1.5 km	1.4 km	200 m	300 m	1.2 km	350 m
900 m	750 m	300 m	650 m	320 m	100 m.

3 The club coach at a local cycling track was overheard saying that he felt at least half the cyclists were cycling at a speed of 30 km/h or more. The speeds (in km/h) of the club cyclists were recorded as follows.

31, 22, 40, 12, 26, 39, 49, 23, 24, 38, 27, 16, 25, 37, 19, 25, 45, 23, 17, 20, 34, 19, 24, 15, 40, 39, 11, 29, 33, 44, 29, 50, 18, 22, 51, 24, 19, 20, 30, 49, 29, 17, 25, 37, 25, 18, 34, 21, 20, 18

Is the coach correct in making this statement? First round each of these speeds to the nearest 5 km/h.

4 A clothing store records the dress sizes sold during a day in order to cater for the popular sizes. The results for a particular day are:

12, 14, 10, 12, 8, 12, 16, 10, 8, 12, 10, 12, 18, 10, 12, 14, 16, 10, 12, 12, 12, 14, 18, 10, 14, 12, 12, 14, 14, 10

Rebecca is in charge of marketing and sales. She uses these figures in ordering future stock. From these figures she decides on the following ordering strategy.
Order:

- the same number of size 8, 16 and 18 dresses
- three times this number of size 10 and size 14 dresses
- five times as many size 12 dresses as size 8, 16 and 18.

Comment on Rebecca's strategy.

5 A new vending machine stocking water and sports drinks was installed at the local gymnasium. The number of bottles sold each day over the first 5 weeks after its installation is shown.

4, 39, 31, 31, 50, 43, 60, 45, 57, 61, 18, 26, 3, 52, 51, 59, 33, 51, 27, 62, 30, 40, 3, 30, 37, 9, 33, 44, 53, 16, 22, 6, 42, 33, 19

The gymnasium realises that the number of daily sales will vary, depending particularly on the weather. The machine is stacked with 60 bottles at the start of each day. The supplier advises that, because of electricity costs and other running expenses, the machine is not worthwhile having unless at least three-quarters of the bottles are sold on average 2 days per week. Round these sales to the nearest 5 and analyse the figures to determine whether the gymnasium should keep the vending machine.

Draw a diagram, then look for a pattern or a result

When information is represented in the form of a diagram, it can be easier to study all the information at once. There are many types of diagrams, so no single diagram is necessarily the best.

WORKED Example 2

The Davies family have a hobby-farm, which they use on weekends and school holidays. Their home is on one side of a large freshwater lake and their farm is on the other. It takes about 3 hours to drive around the lake to the farm. Alternatively, there is a train that goes about halfway around the lake; the rest of the trip requires a bus. There are still other options: the Davies could drive to the lake, where there is a ferry service that will take them across; from the other side they could catch a bus, a taxi or just walk to the farm. How many different ways can the Davies travel from their home to their hobby-farm?

THINK / **WRITE**

1. Read the question at least twice and take note of all the important facts. *Note:* The information regarding the time it takes to drive to the farm is not relevant.

   The Davies' home and farm are on opposite sides of the lake. They are able to travel from their home to the farm in a number of ways: drive by car; take the train, then the bus; drive to the lake, take the ferry across the lake, then choose to take either the bus or taxi or to walk.

2. Identify the solution required.

   The question asks to find the number of ways to travel from the home to the farm.

3. Start a diagram by showing the position of their home H, the farm F, and the lake L.

   H L F

4. Draw an arrow from H to F around L to represent the car trip.

5. Draw a different arrow to represent the train and bus combination.

   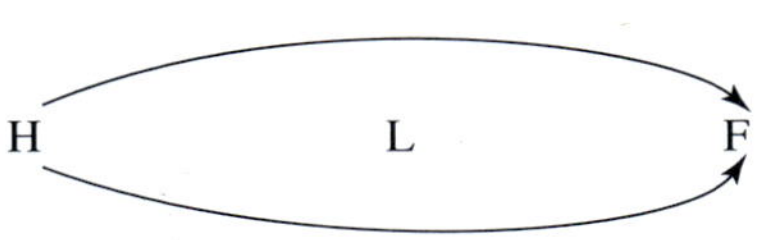

6. An arrow from H through L could represent the car trip to the lake followed by the ferry ride. Three arrows from L could represent travelling by bus, travelling by taxi or just walking.

   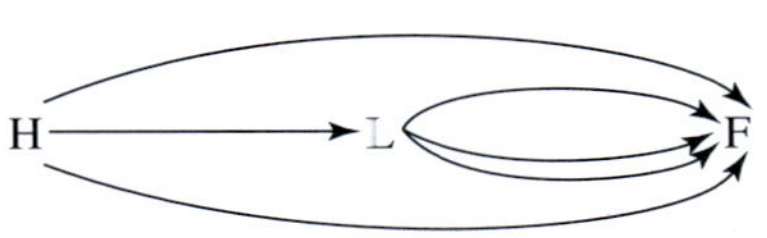

7. Each different path from H to F needs to be counted.

8. Answer the question.

   There are 5 different ways for the Davies family to travel from their home to their hobby-farm.

Try these

1 Nadia's family live on a property in outback Victoria. Nadia's Aunt Emily lives in the closest town, which is 45 km from the property. Nadia goes to a university in Melbourne, which is 570 km away. From university, Nadia communicates with her parents in many ways: she can write a letter; send an email; send a fax; telephone her family at home; or telephone her Aunt, who may decide either to drive over to the property with the message or telephone Nadia's parents later that night. How many ways are there for Nadia to communicate with her parents from university?

2 Michael studies elective music. His assignment this term is to compose a piece of music using as many instruments as he chooses, but only those that he can play. Michael plays the acoustic guitar, the piano, the double bass and the electric bass. How many choices does Michael have?

3 Alex has one brother, one sister, a mother and a father. The family owns a 5-seater car. When the family goes out, the parents always sit in the front two seats. There are three seats behind that. How many different seating arrangements are there?

4 Sam has a set of wooden cube blocks he uses to build different shapes. His sister Chris challenged him to see how many different ways he could stack 4 cubes. He can only stack them on top of each other or side by side – not one behind the other. The arrangements must be different and not simply a mirror image or rotation of another shape. Draw the different arrangements possible. Ignore the numbers on the blocks in your answer.

5 Kelly has a piece of string 32 cm long. He uses it to form rectangular shapes each with the length and width a whole number of centimetres. How many different rectangular shapes can he form?

Use a pattern of numbers, making use of technology such as a computer spreadsheet

Repetitive tasks are well suited to spreadsheets because once a spreadsheet is set up, the repetitive tasks are achieved in an instant.

A spreadsheet can list patterns of numbers from which a result can be found.

WORKED Example 3

Carla enjoys sport. She is going to invite those people in her class who play both tennis and netball to her home. In her class of 30 students, 11 play tennis and 22 play netball. How many play both sports? How many will be invited if Carla is one of the people who play both sports?

THINK	WRITE
1 Read the question at least twice and take note of all the important facts.	Students who play both tennis and netball will be invited to Carla's house. In Carla's class there are 30 students of whom 11 play tennis and 22 play netball. Carls plays both sports.
2 Identify the solution required.	The question asks to find how many students play both sports and how many students will be invited to Carla's house (keeping in mind she plays both tennis and netball).

THINK

3. Set up a spreadsheet and label the first cell of each column. In cell A1 type 'Tennis and netball'; in cell B1 type 'Tennis only'; in cell C1 type 'Netball only'; and in cell D1 type 'Total'. (Do not include the quotation marks.)
4. The first column is the list of possible numbers of people who play both sports; it cannot be more than 11, since that is the total of all who play tennis. We can call this the *common amount*. In cells A2, A3, A4 …, enter 1, 2, 3 …, up to 11 in cell A12.
5. In the second column, we will calculate the numbers who play tennis only, by subtracting the common amount from 11. In cell B2, enter **=11-A2**. Highlight cells B2 to B12 and from the **Edit** menu select **Fill Down** to perform the same calculation with the remaining cells in this column.
6. Similarly, enter the calculation to obtain the C column for those who play netball only. In cell C2, enter **=22-A2**. Then highlight cells C2 to C12 and fill down the column.
7. Each cell in the D column is the sum of the preceding three cells in the same row. In cell D2, enter **=sum(A2:C2)**. Then highlight cells D2 to D12 and fill down the column.
8. Now look for the row with the set of values that fit the given information. We know that there should be 30 people altogether, so look for the answer 30 in the D column.
9. Answer the questions.

WRITE

	A	B	C	D
1	Tennis and netball	Tennis only	Netball only	Total
2	1	10	21	32
3	2	9	20	31
4	3	8	19	30
5	4	7	18	29
6	5	6	17	28
7	6	5	16	27
8	7	4	15	26
9	8	3	14	25
10	9	2	13	24
11	10	1	12	23
12	11	0	11	22

In Carla's class of 30 students, 3 play both tennis and netball (she is one of these), 8 play tennis only and 19 play netball only. Carla will invite two people from her class.

Try these

1 In a class of 22 students, each attends swimming practice, or lifesaving, or both. If 16 attend swimming practice and 9 attend lifesaving, find the number of students who attend both.

2 In a class of 30 students, each chooses to study French or German or both languages. If 24 study French and 18 study German, find the number of students who study both languages.

3 Vernon works in a restaurant where people choose to work Monday to Friday, or weekends, or Monday to Sunday. There are 15 employees altogether. If 13 work Monday to Friday and 14 work weekends, find the number of employees who *do not* work Monday to Sunday.

4 There are 20 chocolate bars on the table. If 12 contain nuts and 14 contain fruit, find the numbers of chocolate bars that contain both fruit and nut.

5 The students in a Year 7A were voting for a class captain. There were only two candidates. When the votes were counted, Rachel received 84% of the votes and Susi received 28% of the votes. It was discovered that some students had voted twice. If the class consisted of 25 students, how many of them voted twice?

Work backwards from the answer

If there is a sequence of steps for which we know the final result, then a useful strategy may be to work backwards from this final result or answer. We start with the last step of the sequence.

WORKED Example 4

Luke was paid for mowing the lawn at his neighbour's home. However, for weeding and trimming the garden, his older brother Tom received $4 more than Luke did. Luke's mother spent $45 on groceries. If this is five times the amount that Tom is paid, how much did Luke receive?

THINK	WRITE
1 Read the question at least twice and take note of all the important facts.	Luke was paid for mowing the lawn. Tom was paid $4 more than Luke for weeding and trimming. Their mother spent $45 on groceries: this amount is 5 times Tom's pay.
2 Identify the solution required.	The question asks to find how much Luke was paid.
3 From the given information, list the amount of money that is known. Use this to work out how much Tom received.	Luke's mother has spent $45. This is 5 times the amount Tom received. So Tom received $\frac{1}{5}$ of $45, that is, $\frac{1}{5} \times \$45 = \$9$.
4 Work backwards from Tom's payment to calculate Luke's payment.	Luke received $4 less than Tom.
5 Answer the question.	Luke received $\$9 - \$4 = \$5$.

This problem could also have been solved using a flow chart and then backtracking. For example, Tom is paid $4 more than Luke. The groceries are 5 times this amount and total to $45.

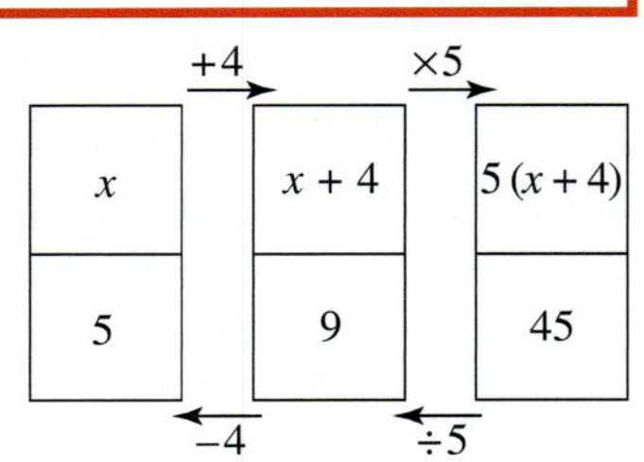

Try these

1 Jo and Anton made a certain number of telephone calls over a 2-month period. There is a charge of 25 cents for each call, and the monthly rental of the phone line is $13.60. The bill for the 2 months comes to $39.45. How many calls did they make in the 2-month period?

2 Christine and Con made a certain number of telephone calls over a 3-month period. There is a charge of 22 cents for each call, and the monthly rental of the phone line is $15.20. The bill for the 3 months comes to $138. How many calls did they make in the 3-month period?

3 Stan bought 2 bottles of juice and 6 packets of chips. The total bill was $14. If the juice cost $2.80 per bottle, what did he pay for a packet of chips?

4 Alvin bought 3 bottles of juice and 8 packets of chips. The total bill was $18.50. If the chips cost $1.75 per packet, what did he pay for a bottle of juice?

5 A jug is $\frac{1}{5}$ full of water. Jack added more water to the jug until it was $\frac{3}{4}$ full. At that stage there was 1.5 L of water in the jug. How much water was originally in the container?

Use a process of elimination

When using a process of elimination we remove or eliminate possible solutions that do not match the given information. We first write down all the possible combinations or solutions in a grid or table. From the information supplied, we cross out (eliminate) those combinations that do not match.

WORKED Example 5

Harriett is trying to find the house where she grew up 35 years ago. The house number is either 27a or 27 or 72a and the street name is either North Avenue or North Road or North Lane. Assuming the house is still there, help Harriett by matching the house numbers with the street names. Use each number only once.

North Avenue has no odd-numbered house numbers because there is a golf course on one side of the avenue. Number 27a shares a block with number 27 and number 72a shares a block with number 72. There are no shared blocks in North Lane.

THINK	WRITE
1 Read the question at least twice and take note of all the important facts.	The possible house numbers are 27a, 27 or 72a. The possible streets are North Avenue, North Road or North Lane. North Avenue has no odd numbers. Numbers 27a and 72a are on shared blocks. No shared blocks are in North Lane.
2 Identify the solution required.	The question asks to find options for the location of Harriett's house.

Continued over page

THINK

3 Rule up a grid of possibilities.

4 Go through each clue and begin the process of elimination.
(a) Eliminate (cross out) 27a and 27 against North Avenue as there are no odd numbered houses in North Avenue.
(b) Eliminate 27a and 72a against North Lane, as there are no shared blocks in North Lane.
(c) Tick 72a against North Avenue and 27 against North Lane since there are no other options.
(d) Eliminate 27 and 72a from North Road as these numbers have been allocated.
(e) Tick 27A against North Road as there are no other options.

5 List the resulting combinations from the grid.

6 Answer the question.

WRITE

	27a	72a	27
North Avenue	✖		✖
North Road			
North Lane	✖	✖	

	27a	72a	27
North Avenue	✖	✔	✖
North Road	✔	✖	✖
North Lane	✖	✖	✔

The house numbers and matching street names are 72a North Avenue, 27a North Road and 27 North Lane.

Try these

1 Mrs Flushcombe needs to find Fernando before lunch. Fernando is in either a French, a German or a Visual Arts lesson. These lessons take place in rooms K1, L1 or L2. We know that the Kitchen, K1, is never used by a Visual Arts class. The French class has too many students to fit in K1 (Kitchen) or L1 (the smaller language room). Match each lesson with its room.

2 Brandon works after school delivering milk. In Belleview Road he delivers two 2-litre containers of milk to one house, 1 litre of skim milk to another house, and three 1-litre containers of milk to another. The houses on his delivery run are numbers 5, 7 and 18. Brandon remembers that the houses next to each other both have multiple orders (more than one item), and the largest containers of milk are delivered to the house with the smallest house number. Match each house number with its milk order.

3 Four girls — Eugenie, Florence, Kim and Anthea — each prefer a different means of communication. Match the girls to one of the following: home phone, mobile wih SMS messaging, email, mail. Use the following information: both Eugenie and Florence have no computer, though Florence is the only one with a mobile phone with SMS; also, Australia Post can collect mail from Kim and Florence.

4 Alan, Bob and Cam each have one of the following coins – a 50c piece, a $1 coin or a $2 coin. Alan and Cam's coins are circular and Alan's coin is smaller in size than Cam's. What coin does each of the boys have?

5 Penny, Teresa and Raelene are friends in high school and each has one brother. Their three brothers are Steve, Matt and Fred. Penny's brother is at primary school. Fred has finished school and is working as an apprentice carpenter. Steve attends the same school as his sister. Fred's sister is not Raelene. Match the girls with their brothers.

Look at similar but simpler problems

If you are overwhelmed by the size of the numbers involved in a question, try to solve a similar but simpler question. This can be achieved by changing the numbers in the original question to smaller numbers. After finding the answer to the simpler question, the same method can be used to solve the original problem.

WORKED Example 6

The World Cup is a world championship soccer tournament that is held every 4 years and is played over several rounds. In the first round of finals, 24 teams compete against one another. The teams are divided into six groups of four teams. Within each group, each team plays the other three teams. Find the total number of games played in the first round of the World Cup finals.

THINK	WRITE
1 Read the question at least twice and take note of all the important facts.	There are 24 soccer teams competing. The teams are divided into 6 groups; each group consisting of 4 teams. In each group, each team plays 3 other teams.
2 Identify the solution required.	The question asks to find the total number of games played.
3 Consider a simpler competition with only one group of four teams.	Consider a simpler problem with only one group of 4 teams.
4 Use a diagram to work out how many games are played within this group if each team plays every other team. Team 1 plays Teams 2, 3 and 4. Team 2 plays Teams 3 and 4. (Don't count playing Team 1 again.) Team 3 plays Team 4. (Don't count playing Teams 1 and 2 again.)	Each team is to play every other team in the group. 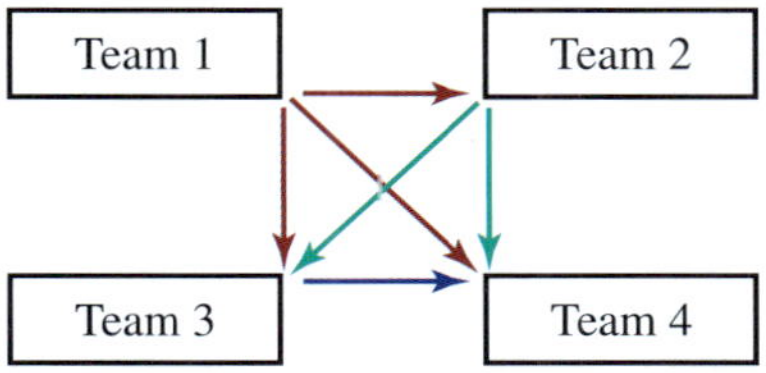 With 4 teams playing each other, there is a total of 6 games.
5 Consider how many games would be played if there were two groups of four teams.	For two groups of four teams, there would be a total of 2×6 (that is, 12) games.
6 Relate this result to the original problem where there are six of these groups.	For six groups of four teams, there would be a total of 6×6 (that is, 36) games.
7 Answer the question.	

Try these

1 By joining all the vertices of a decagon, how many diagonals are formed?

2 What is the angle sum (sum of all the angles) of a decagon?

3 How many different three-way telephone calls are possible among a group of five friends?

4 Imagine a simplified form of car numberplate where the plate consists of a letter followed by a number; that is, A0 up to Z9. How many numberplates could be issued under this system? Remember that all numberplates issued have to be different.

5 Ten people were sitting around a round table. How many handshakes were required so that each person had shaken hands with everyone else exactly once?

Use trial and error (guess and check), making use of technology such as a computer spreadsheet

Sometimes it may not be easy to solve a problem directly; in this case we can use a strategy by which we guess at the solution. We test this value (guess), using the available information supplied in the problem, to check whether it is the solution. Even if it is not the solution to the problem, this process provides us with further information that we can use to try another, better-informed guess.

We can continue to guess and check until we reach the solution. Since this can be a lengthy process, we can use technology such as a spreadsheet to provide instant feedback on our checking.

WORKED Example 7

Jono is 12 years older than Jade. The sum of their ages is 48. How old is Jade?

THINK

1 Read the question at least twice and note all of the important facts.

2 Identify the solution required.

3 Since Jade's age is the unknown value, guess any number (say 20). On a spreadsheet, type the heading 'Jade's age' in cell A1. Enter 20 in cell A2.

4 Jono's age is 12 more than Jade's age, so type the heading 'Jono's age' in cell B1 and enter **=A2+12** in cell B2.

5 As we are interested in finding the sum of their ages, type the heading 'Sum of ages' in cell C1 and enter **=A2+B2** in cell C2.

WRITE

Jono is 12 years older than Jade.
The sum of Jono and Jade's ages is 48.

The question asks to find the age of Jade.

	A	B	C	D	E
1	Jade's age	Jono's age	Sum of ages		
2	20	32	52		
3	19	31	50		
4	18	30	48		

THINK	WRITE
6 Change the number in cell A2 to make a new guess and observe what happens to cell C2. This is our check. Alternatively, to keep a record of previous guesses, enter a new value in the next row of column A, and highlight and **Fill Down** in columns B and C.	
7 When 48 appears in column C, we know that our guess must be correct! (You may need to use decimal numbers as well as whole numbers as your guess in some cases.)	
8 Answer the question.	Jade is 18 years old.

Try these

1 Rachel and Jackson are 7 years apart in age. Jackson is older than Rachel. The sum of their ages is 51. Find Rachel's age.

2 Eric and Colin are partners in owning a taxi. They share the driving so that the taxi is always on the road. Yesterday, Colin drove for $2\frac{1}{2}$ hours longer than Eric. For how long did Eric drive yesterday?

3 Marcus and Melanie pooled their funds and purchased shares on the Stock Exchange. Melanie invested \$350 more than Marcus. Together they invested \$2780. Find how much Marcus invested.

4 The sum of two numbers is 32 and their product is 247. What are the two numbers?

5 At a take-away food stall Martin purchased 3 hamburgers and 2 drinks while Ken purchased 2 hamburgers and 3 drinks. If Martin paid \$16.15 and Ken paid \$14.85, what are the individual costs of a hamburger and a drink?

Communicating, reasoning and reflecting

Once you have applied a strategy, there remains the task of communicating your findings. (You are usually expected to support your findings with some reasoning.) As an example of communication, consider the following solution to a question.

$$10 - 2 \times 3 = 4$$

What was the question?

When looking at a solution, you should be able to work backwards and gain a fair idea of the question. This equation could be the solution for any one of the following questions.

- Calculate $10 - 2 \times 3$.
- What is the remainder when 3 is doubled and the result is subtracted from 10?
- Jack bought 2 pens for \$3 each. If he handed the cashier \$10, how much change did he receive?

It is important to understand that solving problems involves much more than just writing numbers on a page. Words should accompany the mathematics and these words should be in the form of appropriate English and use correct mathematical terms. Another danger spot occurs in the use of the equals (=) sign. Care should be taken that when this sign is used, the mathematics following is indeed equal. In fact, it is important to communicate correctly in all cases – not just in solving problems. After providing a solution to a question, it is good practice to review your solution to see whether another person could understand your work without first reading the question.

Finally, if you take the time to reflect on your work, you may increase your understanding of the problem and the strategies you used to solve it. You may be able to connect this to previous experiences as well as to future problems you will have to tackle.

In summary, to solve problems successfully, the following steps should be followed.

1. Read the question at least twice and take note of all the important facts.
2. Identify the solution required.
3. Decide on, and apply an appropriate strategy (such as create a table, draw a diagram, use technology, work backwards, use a process of elimination, look at similar but simpler problems or use trial and error).
4. Communicate the whole solution using appropriate language and mathematical terms.
5. Support the solution with mathematical reasoning.
6. Reflect on the solution. Does it answer the question and does it make sense?

WORKED Example 8

Tom makes house calls to fix electrical appliances. His charges involve a call-out fee of $35 plus a cost of $12.50 per quarter of an hour for service. Any replacement parts are added to the cost. The bill to fix the Adams family's washing machine amounted to $145. If the replacement part cost $72.50, how long did Tom spend fixing the machine?

THINK	WRITE
1 Read the question at least twice and take note of all the important facts.	The call-out fee is \$35 and each $\frac{1}{4}$ hour of service is \$12.50. The total cost can be expressed as an equation: Total cost = Call-out fee + service fee + replacement parts The bill totals \$145 and a replacement part was \$72.50.
2 Identify the solution required.	The question asks to find the service time.
3 Consider an appropriate strategy. In this case, working backwards from the total bill cost seems the best one. The service cost can be obtained by subtracting both the call-out fee and replacement parts from the total cost.	Service cost = Total cost – call-out fee – replacement parts = \$145 – \$35 – \$72.50 = \$37.50 $\frac{1}{4}$ hour rate = \$12.50 Number of $\frac{1}{4}$ hours = \$37.50 ÷ \$12.50 = 3 Therefore, service time = $\frac{3}{4}$ hours

THINK	WRITE
4 Communicate the solution.	Tom spent $\frac{3}{4}$ hours or 45 minutes repairing the washing machine.
5 Provide mathematical reasoning.	Check the answer using the original equation: Total cost = Call-out fee + service fee + replacement parts = \$35 + 3 × \$12.50 + \$72.50 = \$145
6 Reflect on the solution.	The solution seems reasonable and answers the question posed.

Try these

1 Minh was asked to buy enough ice-creams for everyone in his class. There are 31 people in the class. At the supermarket, identical ice-creams come in packets of 3 or packets of 5. Minh is told that the cost of one ice-cream is the same from either packet. What packets or combination of packets should Minh buy?

2 Geoff wants to establish a rectangular vegetable garden bed on his farm. His 20-m long work shed will act as one of the boundaries. (*Note:* The garden will not necessarily be this long.) He has 24 metres of fencing which he plans to use to fence the other three sides. How could Geoff use this fencing to enclose the largest possible area?

3 Bronwyn was born on a Wednesday in March. It was not a leap year, so it had 365 days. Investigate the days of the week on which Bronwyn's first birthday could have fallen the next year.

4 In a class of 30 students, 22 play basketball and 18 play volleyball. Some of the students play both sports. How many students play volleyball only?

5 Alexander was born on a Monday. The next year was a leap year (366 days). Investigate the days of the week on which Alexander's 1st birthday could have fallen. Was Alexander born in the month of February?

6 The product of two consecutive numbers is 1190. What are the two numbers and what is their sum?

7 During summer, there is 1 hour of daylight saving in New South Wales but no daylight saving in Queensland. Peta lives very close to the New South Wales–Queensland border. One day in summer, Peta walked across the border into Queensland. Investigate how she would need to change the time on her watch.

8 The land-line telephone numbers in Victoria start with 03 (the area code) followed by 8 digits:
03 _ _ _ _ _ _ _ _.

a How many telephone numbers are possible using this system?

b How many telephone numbers are possible if the first digit following the area code can only be an 8 or a 9?

Glossary

2-dimensional representation (net) — plan or net that can be folded to form a polyhedron

3-dimensional object — shape that has depth, as well as length and width; also called a polyhedron

Acute angle — angle greater than 0° but less than 90°

Acute-angled triangle — triangle where each angle is acute; that is, less than 90°

Angle — rotation between a pair of lines that meet at a point

Anticlockwise — movement in the opposite direction to that of the hands of a clock

Area — amount of surface enclosed by the shape; measured in square units, such as square metres, m^2, and square kilometres, km^2

Arithmagon — triangular figure in which the two numbers at the ends of each line add to the number along the line

Ascending — increasing from smallest to largest

Axes — horizontal and vertical lines with scales that form the Cartesian plane

Axis of symmetry — line that divides a shape into two identical parts

Backtracking — process of working backwards through a flow chart. Each step is the inverse operation of the corresponding step in the flow chart. It can be used to solve equations.

Bar graph — graph drawn in a similar way to column graphs with horizontal bars instead of vertical columns. Categories are graphed on the vertical axis and the frequencies (numbers) on the horizontal axis.

Bearings — directions from one place to another

Bird's eye view — view observed from the top of the object looking down on it

Bisect (angles) — to divide an angle into two equal, smaller angles

C (Roman numeral) — represents 100

Capacity — maximum amount of fluid that can be contained in an object. The units may be the same as for volume or litres and millilitres may be used.

Cartesian plane — area formed by a horizontal line with a scale (x-axis) joined to a vertical line with a scale (y-axis). The point of intersection of the lines is called the origin.

Categorical (data) — data that involve grouping or classifying things, not numbers; for example, grouping hair colour

Classify — to group objects based on similarities

Clockwise — movement in the same direction to that of the hands of a clock

Column graph — graph in which equal width columns are used to represent the frequencies (numbers) of different categories

Compass — instrument with magnetic needle pointing north — used to measure direction

Complementary angles — sum (add) to 90°

Composite numbers — numbers with more than two factors

Composite shape — figure made up of more than one basic shape

Continuous (data) — data that take on any value and are able to be measured; for example, the heights of people in a class

Contour lines — lines that join points of equal elevation above sea level

Conventional compass bearing — directions measured in degrees from the north-south line in either a clockwise or anticlockwise direction. To write the bearing we need to state whether the angle is measured from the north or south, how many degrees there are in the angle and whether the angle is measured in the direction of east or west.

Convert — to change from one unit to another

Coordinates (x, y) — two numbers that give the position of a point on the Cartesian plane. The first number is the x-coordinate and the second number is the y-coordinate.

Cross-sectional area (shape) — area of the face created by slicing a 3-dimensional solid

Cube — polyhedron with 6 faces; all faces are squares of the same size

Cube root ($\sqrt[3]{\ }$) — number that multiplies by itself three times to equal the original number. Finding a cube root is the opposite of cubing a number.

Cubic numbers — numbers with three equal factors; for example, $8 = 2 \times 2 \times 2$ and $27 = 3 \times 3 \times 3$

D (Roman numeral) — represents 500

Data — numerical information

Decagon — 10-sided polygon. It also has 10 angles.

Decomposition method — method of subtraction. The larger number is decomposed by taking 10 from the tens column and adding it to the units column. It may also involve taking 100 from the hundreds column and adding it to the tens column.

Degrees (°) — units used to measure the size of an angle

Degrees Celsius (°C) — units used to measure temperature

Denominator — bottom term of a fraction. It shows the total number of parts into which the whole has been divided.

Descartes — René Descartes was a mathematician in the 17th century (1596–1650) who first proposed the Cartesian plane and the use of Cartesian coordinates to locate points on a plane.

Descending — decreasing from largest to smallest

Digits — first nine counting numbers and zero

Discrete (data) — data that take on exact values that may be counted; for example, the number of televisions sold each week

Dodecagon — 12-sided polygon. It also has 12 angles.

Dodecahedron — polyhedron with 12 faces

Dot plot — consists of a horizontal axis labelled and evenly scaled, with each data value represented by a dot

Edges (networks) — lines in networks that join the vertices to one another

Edges (shapes) — lines where two faces of a polyhedron meet

Equation — mathematical statement containing a left-hand side, a right-hand side and an equals sign between them

Equilateral triangle — triangle with all sides equal in length. All angles will be equal too.

Equivalent fractions — fractions that look different, but represent the same amount. Equivalent fractions can be formed by multiplying (or dividing) both the numerator and denominator by the same (non-zero) value.

Estimation — approximate answer to a question based on information given

Euler's Rule — rule that links the number of faces, F, the number of vertices, V, and the number of edges, E, of a polyhedron: Euler's Rule is $F + V - E = 2$

Evaluate — to obtain an answer

Even numbers — numbers divisible by 2. Numbers that end in 0, 2, 4, 6 or 8

Event (probability) — result that may be obtained

Expression — mathematical statement made up of pronumerals (letters) and numbers. Expressions do not contain an equals sign.

Exterior angle — angle formed when any side of a closed shape is extended outwards. The exterior angle and the interior angle adjacent to it are supplementary (add up to 180°).

Faces (shapes) — flat surfaces that make a polyhedron

Factor pairs — pairs of numbers which multiply together to equal another number; they are factors of the same number.

Factor tree — diagram that displays the prime factors of a composite number. Each branch shows a factor of all numbers above it. The last numbers at the bottom of the tree are the prime factors of the original number.

Factors — whole numbers that divide exactly into another whole number, without any remainder

Fair game (probability) — game that allows each player an equal chance of winning

Favourable outcome (probability) — outcome that we are looking for

Flow charts — diagrams that illustrate the process by which an input number becomes an output number in an equation or formula

Formula — rule that is written using pronumerals

Frequency — number of times a result or piece of data occurs

Front elevation — view observed from directly in front of an object

Front view — *see* front elevation

Greater than (>) — symbol used to represent one number being larger than another number

Hectare (ha) — area equal to 10 000 m^2, area of a square of side length of 100 m, unit commonly used to measure land area

Heptagon — 7-sided polygon. It also has 7 angles.

Hexagon — 6-sided polygon. It also has 6 angles.

Highest Common Factor (HCF) — largest factor which is common to all the numbers given

Horizontal — lying flat or level (with the horizon). A horizontal line is drawn straight across the page.

I (Roman numeral) — represents 1

Icosagon — 20-sided polygon. It also has 20 angles.

Icosahedron — polyhedron with 20 faces

Illustration — example or picture which makes something clearer

Improper fraction — fraction whose numerator is larger than its denominator, for example $\frac{5}{4}$

Index (or power) — indicates the number of times a base is to be multiplied when written in index form

Index notation (index form) — form of writing the product of prime factors. It has a base and an index or power. The index indicates how many times the base is being multiplied by itself.

Interior angle — angle inside a shape

Interpret — to understand or give meaning to; to interpret a graph means to understand what the graph is telling us

Inverse operation — operation that 'undoes' a previous operation. Addition and subtraction are inverse operations; and multiplication and division are inverse operations of each other.

Irregular quadrilateral — quadrilateral that does not have any special properties

Isobars — lines on synoptic maps that connect places of equal air pressure

Isometric — type of transformation in which the size and shape of the original object does not change

Isosceles triangle — triangle that has two sides of equal length. The two angles opposite the equal sides will also be equal.

Kite — quadrilateral in which two pairs of adjacent sides are equal in length and one pair of opposite angles (those between the sides of unequal length) are equal

L (Roman numeral) — represents 50

Legends — definitions of different symbols or colours used when drawing graphs

Less than (<) — symbol used to represent one number being smaller than another number; for example, $\frac{1}{5} < \frac{3}{5}$

Like terms — terms that contain exactly the same pronumeral (letter) part; for example $5a$ and $33a$ are like terms but $4ab$ is not.

Line graph — graph containing points joined with line segments

Line plot — *see* dot plot

Long run (probability) — *see* long-term trend

Long-term trend (probability) — trend observed for results from a very large number of trials. This tends to match the results of theoretical probability.

Lowest Common Multiple (LCM) — lowest multiple that two or more numbers have in common

M (Roman numeral) — represents 1000

Mass — describes how much matter makes up an object. The base unit of mass is the gram, g.

Mean (summary statistics) — sum of all the values (scores) divided by the number of values — also called the average

Median (summary statistics) — middle value if the number of data is odd; or the average of the two middle values if the number of data is even. Data must first be in numerical order.

Meridian — imaginary line on the Earth's surface passing through the North and South Poles and running perpendicular to the equator

Meridians of longitude — imaginary lines running on the Earth's surface from the North Pole to the South Pole: for example, the Greenwich Meridian has a longitude of 0° and the International Date Line has 180°E or 180°W longitude

Metric — system of measurement based on the number 10. The base unit of length in this system is the metre, m.

Mixed number — number that is made up of a whole number and a fraction, for example $2\frac{3}{4}$

Mode (summary statistics) — most common value in a set of data

Multiples — answers obtained when a number is multiplied by other whole numbers; for example, 4, 8, 12, 16 are multiples of 4

Nets — 2-dimensional plans of polyhedra

Network — series of dots (vertices) joined together in some way by lines (edges). The lines can be straight lines, curves or loops and the dots are labelled. Networks can be used as maps that show how features are joined together.

Nonagon — 9-sided polygon. It also has 9 angles.

Numerator — top term of a fraction. It shows how many parts there are.

Obtuse angle — angle greater than 90° but less than 180°

Obtuse-angled triangle — triangle that has one angle greater than 90°

Octagon — 8-sided polygon. It also has 8 angles.

Octahedron — polyhedron with 8 faces

Odd numbers — numbers not divisible by 2; numbers that end in 1, 3, 5, 7 or 9

Ordered pair (probability) — pair of items written in brackets separated by a comma; for example, when tossing a coin and rolling a die, an outcome might be a 'Head' and a '6'. This would be written as (H, 6) in ordered pair notation.

Ordered pair (x, y) — *see* coordinates

Ordinal data — data that involve a rating or ranking; for example, the ratings or grades on a report card where the scale may range from A^+ to UG

Origin — point where the x-axis and the y-axis intersect on the Cartesian plane; that is, the point with coordinates (0, 0)

Outcomes (probability) — particular result of an experiment

Pair of compasses — instrument that can be used to draw angles, circles and distances

Palindromes — numbers or words that read the same forwards and backwards; for example, 898 or mum

Parallel (lines) — lines running in exactly the same direction and therefore never meeting. They are usually drawn with arrowheads.

Parallelogram — quadrilateral with both pairs of opposite sides parallel to each other. Rectangles, squares and rhombuses are parallelograms.

Parallels of latitude — imaginary lines running around the Earth parallel to the equator: for example the equator is of latitude 0°, the North Pole is of latitude 90°N and the South Pole is of latitude 90°S

Pentagon — 5-sided polygon. It also has 5 angles.

Percentage — amount out of 100 or per hundred; for example, 50% means 50 out of 100 or $\frac{50}{100}$. They are used to represent part of a quantity.

Perigon (revolution) — angle that equals 360° (a full circle)

Perimeter — distance around the outside (border) of a shape

Perpendicular lines — lines that intersect at right angles (90°)

Pie graph (pie charts or sector graphs) — data are graphed on a circle with each category being represented by a sector of the circle

Polygon — 2-dimensional, closed shape with 3 or more sides, each of which is a straight line

Polyhedron — 3-dimensional shape in which each flat surface (face) is a polygon. The plural term is polyhedra.

Power (or index) — *see* index

Prime number — counting number that has exactly two factors, itself and 1

Prism — solid shapes with identical opposite ends joined by straight edges. Rectangular prisms have rectangles as identical ends and triangular prisms have triangles.

Probability — likelihood or chance of a particular event (result) occurring

$$\Pr(event) = \frac{\text{number of favourable outcomes}}{\text{number of possible outcomes}}$$

The probability of an event occurring ranges from 0 (impossible—will not occur) to 1 (certainty—will definitely occur).

Product — result of a multiplication

Pronumeral — letter used in place of a number

Proper fraction — fraction whose numerator is smaller than its denominator, for example $\frac{3}{4}$

Protractor — instrument used to measure angles in degrees

Quadrilateral — 2-dimensional, closed shape formed by four straight lines

Quotient — result of a division

Range (summary statistics) — difference between the highest and lowest values (scores)

Rates — measurements used to describe what can occur in a certain amount of time or how quickly a quantity changes: rate = $\frac{\text{quantity}}{\text{time}}$

Reciprocal — the reciprocal of a number is obtained by first expressing it as a fraction then tipping the fraction upside down; for example, $\frac{3}{4}$ is the reciprocal of $\frac{4}{3}$. A mixed number must be converted to an improper fraction first and then tipped.

Rectangle — parallelogram in which the opposite sides are equal in length and all angles are equal to 90°

Recurring decimals — decimals that have one or more digits repeated continuously; for example 0.999... They can be expressed exactly by placing a dot or horizontal line over the repeating digits as in this example. 8.343 434 = $8.\dot{3}\dot{4}$ or $8.\overline{34}$

Reflection — exact image of an object, as seen in a mirror; as far behind the mirror as the object is in front of it. Reflections always have reversed orientations; right appears left and left appears right.

Reflex angle — angle greater than 180° but less than 360°

Regular polygon — polygon with all sides equal in length and all angles equal in size

Revolution — *see* perigon

Rhombus — parallelogram in which all sides are equal and opposite angles are equal

Right angle — angle that equals 90°

Right-angled triangle — triangle that has one of its angles equal to 90° (a right angle). A small square is placed in that angle to mark the right angle.

Rotation — object that is turned about a certain point. This point (centre of rotation) needs to be identified, along with the size of the angle of rotation and whether the object is to be turned clockwise or anticlockwise.

Rounding — technique used to give an approximate answer to a question. If the second digit is 0, 1, 2, 3 or 4, the first digit stays the same. If the second is 5, 6, 7, 8 or 9, the first digit is rounded up to the next value. This also applies to decimals. For example, 2.456 rounded to 2 decimal places is 2.46.

Rule — pattern that links a set of input numbers to a set of output numbers in algebra; for example, 'adding 5' is a rule

Sample space (probability) — the complete set of outcomes or results obtained from an experiment. It is shown as a list enclosed in a pair of braces (curled brackets) { }.

Scale (in relation to maps) — ratio that states the actual distance represented by a distance on a map; for example, 1 cm ⇔ 500 m and 1: 50 000 mean the same thing, that is, 1 cm on a map represents an actual distance of 500 m (or 50 000 cm).

Scalene triangle — triangle that has no equal sides. The angles will all be unequal too.

Side elevation — view observed when standing directly to one side of an object. The left view or the right view of the object can be drawn.

Side view — *see* side elevation

Simulation (probability) — experiment that uses simple devices such as coins, dice, a pack of cards and spinners to model or mimic a real life situation

Solution — the answer obtained when solving an equation

Solving equations — finding a value for a pronumeral that makes one side of an equation equal to another

Speed — special rate that tells us how fast something is moving; that is, how quickly distance changes over a period of time: speed = $\frac{\text{distance}}{\text{time}}$. Common units are kilometres per hour (km/h) and metres per second (m/s).

Square — parallelogram in which all sides are equal and all angles are equal to 90°

Square numbers — numbers that have two equal factors; for example, $9 = 3 \times 3$ and $36 = 6 \times 6$

Square root ($\sqrt{\ }$) — number that multiplies by itself to equal the original number. Finding a square root is the opposite of squaring a number; for example $\sqrt{49} = 7$.

Straight angle — angle that equals 180°

Substitution — process by which a number replaces a pronumeral in a formula

Summary statistics — calculations made with data in order to summarise the information or get meaning from them

Supplementary angles — sum (add) to 180°

Synoptic maps — maps that show the weather conditions of an area

Term — part of an equation or expression. Terms can be expressed as one or more than one pronumeral, or a number only. Examples include $5x$, $3ab$, xyz, 8 and so on.

Terminating decimals — decimals that have a fixed number of decimal places; for example 0.6 and 2.54

Tessellations — repeated patterns that cover a flat surface without any gaps

Tetrahedron — polyhedron with 4 faces. It is also called a triangular-based pyramid because it has a triangle at the bottom and a point at the top.

Time — measurement used to work out how long we have been doing things or how long something has been happening. Common time units are: 1 minute = 60 seconds, 1 hour = 60 minutes, 1 day = 24 hours and 1 year = 365 days (366 in a leap year).

Timelines — lists that show when events occur

Time zones — regions on the Earth's surface that run from the North Pole to South Pole. The time everywhere in one zone is the same. The time of day in zones next to each other differs by 1 hour. There are 24 time zones around the Earth, giving 24 hours of time or 1 day. Generally, travelling east means you would set your watch ahead and travelling west means setting your watch back.

Timetable — lists that show when events are scheduled to happen; for example, the train timetable shows when and where trains should be each day.

Top view — *see* bird's eye view

Topographic maps — maps that show the elevations of features of an area such as mountains and valleys by the use of contour lines

Transformation — ordered set of directions that enables a shape to be moved from one place to another. At the start the shape is called the original object and after transformation it is called the image. Examples of transformations are translations, reflections and rotations.

Translation — horizontal (left/right) or vertical (up/down) movement

Trapezium — quadrilateral in which one pair of opposite sides is parallel

Traversable (networks) — networks that can be traced over without lifting the pen

Triangle — 2-dimensional, closed shape formed by three straight lines

True bearings — directions that are written as the number of degrees (3 digits) from north in a clockwise direction, followed by the word 'true' or 'T'; for example, due east would be 090° true or 090° T.

Two-way table (probability) — displays the sample space of 2-step experiments in a logical manner

Undecagon — 11-sided polygon. It also has 11 angles.

V (Roman numeral) — represents five

Vertex (angles) vertices (plural) — point where two lines meet

Vertex (shapes) vertices (plural) — point where 3 or more edges of a polyhedron meet

Vertical — upright. A vertical line is drawn straight up and down the page; that is, it is drawn at right angles (perpendicular) to the horizontal.

Vertically opposite angles — special angles formed when two straight lines intersect. The two non-adjacent angles are called vertically opposite angles. These angles are equal in size.

Vinculum — horizontal bar separating the numerator from the denominator

Volume — amount of space a 3-dimensional object occupies. The units used are cubic units, such as cubic metres, m^3, and cubic centimetres, cm^3. A volume of 1 m^3 is the space occupied by a cube of side length 1 m.

X (Roman numeral) — represents 10

***x*-axis** — horizontal line with a scale that is part of the Cartesian plane

***y*-axis** — vertical line with a scale that is part of the Cartesian plane

Answers

Chapter 1 Number skills and the history of number

Are you ready?

1 a Sixty (60) b Fifty thousand (50 000)
c Three hundred (300) d Eight thousand (8000)
2 a 19, 91, 99, 199, 919, 991
b 201, 121, 120, 112, 102, 21, 12
3 a 15 b 27 c 26
d 4 e 7 f 9
4 a 32 b 30 c 63
d 36 e 16 f 66
5 a 315 b 322 c 5652
6 a 14 b 163 remainder 1 c 93
7 a 19 b 26 c 22
8 a 500 b 2000 c 70

Exercise 1A — Place value

1 a 900 + 20 + 5 b 1000 + 60 + 2
c 20 000 + 8000 + 400 + 60 + 9
d 40 + 3 e 500 000 + 2000 + 30 + 9
f 800 000 + 2 g 1 000 000 + 80 000 + 100
h 20 000 + 2000 + 200 + 20 + 2
2 a Seven hundred and sixty-five
b Nine thousand, one hundred and five
c Ninety thousand, four hundred and fifty
d One hundred thousand, two hundred and thirty-six
3 a 495 b 2670 c 24 000 d 109 605
4 a i Two hundred ii Forty
iii Eight
b i Two ii One hundred
iii Thirty
c i Ten ii Forty
iii Nine
d i Four ii Twenty
iii One hundred
5 A 6 B
7 a 48 987, 8569, 742, 647, 28
b 72 167, 58 625, 47 890, 32 320, 12 947
c 7647, 7476, 6747, 6477, 4776
d 8880, 8808, 8800, 8088, 8080, 8008
8 a 9, 58, 743, 68 247, 1 258 647
b 34 108, 58 610, 60 000, 78 645, 84 364
c 1290, 1902, 2019, 2910, 9021, 9201
d 111, 211, 212, 221, 222, 1112
9 a i \$6000 ii \$340 000 iii \$58 000
b i \$430 K ii \$7 K iii \$800 K
c Check with your teacher.

History of mathematics: The abacus — c 500 BC to now!

1 Either from the Greek word 'abax' or from the Phoenician word 'abak'
2 Sorabon 3 Nepohualtzitzin
4 It was won by Kiyoshi Matsuzaki on the abacus.

Exercise 1B — Adding and subtracting whole numbers

1 a 15 b 24 c 37
d 121 e 210 f 240
g 472 h 1730 i 7458
j 18 220 k 175 000 l 20
m 40 n 50
2 a 99 b 999 c 4183
d 72 648 e 3 485 551 f 13 629
3 a 978 b 1155 c 766
d 257 e 1335 f 11 747
g 14 462 h 332 941 i 1 701 553
j 1 053 944
4 \$247 5 3032 pages 6 \$71
7 a 180 b 200 c 280
d 460 e 790 f 827
8 a 8 days b 2003 km c 3492 km d 2754 km
9 6670 km
10 a 7, 6, 14 b 15, 10, 15 c 8, 20, 36
11 a 6 b 8 c 23
d 5 e 60 f 125
g 814 h 900 i 700
j 10 500 k 70 l 35
m 500 n 63 o 5
p 26 q 39
12 a 4 b 20 c 26
d 55 e 40 f 0
g 73 h 1150
13 a 44 b 35 c 35 123
d 86 e 584 599 f 1831
g 267 h 4413 i 65 439
14 a 40 672 b 2634 c 69 397
d 56 176 e 1553 f 517 526
g 349 818 g 19 088 744
15 15 16 32 922 17 31
18 a 215 million b 220 million
19 a 26 m b 26 m c 52 m
d Check with your teacher.
20 a 172 km b 353 km
21 a 17 b 25 c 23 d 27 e 6
f 4 g 5 h 0 i 17 j 17
k 6 l 8 m 35

Maths Quest challenge (page 17)

1 a 638 321 − 488 417 = 149 904
b 3492 − 1685 = 1807 (Other answers possible)
2 5 849 318 3 7 days
4 735 + 1689 = 2424 or 641 + 8752 = 9393

Exercise 1C — Multiplying whole numbers

1 a 12 b 45 c 22
d 56 e 96 f 110
g 54 h 132 i 72
2 a 26 b 45 c 50
d 39 e 100 f 90
g 32 h 70 i 42

j 63 k 108 l 75
m 72 n 90 o 27
p 90 q 100 r 80

3 a 128 b 116 c 1233
d 2571 e 24 600 f 111 888
g 63 104 h 1888 i 16 120
j 63 582 k 374 055 l 492 720

4 a 2288 b 3007 c 1652
d 912 e 7093 f 64 600
g 21 571 h 97 478 i 2 161 485
j 4 870 460 k 27 529 460 l 1 725 805

5 a 80 b 240 c 1900
d 4500 e 900 f 67 000

6 a 900 b 1830 c 3100
d 16 800 e 9500 f 172 000

7 a 8000 b 21 000 c 480 000
d 7200 e 48 000 f 5 500 000
g 6 300 000 000 h 144 000 000 i 5 600 000
j 54 000 000 k 440 000 l 13 200 000

8 a 306 b 747 c 5652
d 7425 e 456 f 208

9 a 5600 b 560 c 5040

10 a 4320 b 6660 c 11 250
d 28 800 e 3600 f 57 600

11 a 150 b 300 c 300
d Multiply first by 6 then double the answer.

12 a 180 b 840 c 216 d 1260
e 350 f 640 g 198 h 680

13 a 90 b 27 c 117
d Multiply by 10 and by 3 then add the answers.

14 a 325 b 390 c 312 d 1326

15 a $72 b $1728

16 $132 000

17 a $675 b $1200 c $2700

18 490 km

19 a 4420 litres b 22 100 litres

20 524 seconds

21 a 34 hours b $1105 c $1885

22 a 10 km b $175 c 120 km
d $9100 e 400 papers f 20 800 papers

10 Quick Questions 1

1 MMMCC 2 515 3 9584
4 130 5 2224 6 2996
7 7176 8 711 9 1840
10 560

Maths Quest challenge (page 23)

1 3 goals, 9 behinds; 4 goals, 8 behinds; 7 goals, 7 behinds

2 a 1089, 2178, 3267, 4356, 5445, 6534, 7623, 8712, 9801
b 10 989, 21 978, 32 967, 43 956, 54 945, 65 934, 76 923, 87 912, 98 901
c 109 989, 219 978, 329 967, 439 956, 549 945, 659 934, 769 923, 879 912, 989 901, 1 099 989, 2 199 978, 3 299 967, 4 399 956, 5 499 945, 6 599 934, 7 699 923, 8 799 912, 9 899 901

Exercise 1D — Dividing whole numbers

1 a 4 b 3 c 4
d 9 e 7 f 8
g 12 h 8 i 4
j 4 k 20 l 13
m 3 n 2 o 3
p 2 q 2 r 4
s 2 t 5 u 15

2 a 10 b 6 c 56
d 35 e 84 f 60
g 3 h 44 i 54

3 a 485 b 6942 c 6208
d 57 319 e 2761 f 80 045
g 85 064 remainder 2
h 86 047 remainder 3
i 8643 remainder 1
j 904 075 remainder 5
k 93 155 remainder 2
l 60 074 remainder 2

4 a 259 b 968 c 83 015
d 6703 e 621 f 2153
g 642 remainder 1
h 30 246 remainder 12
i 5384 remainder 7

5 a 700 b 10 c 7
d 900 e 9 f 500
g 5000 h 7 i 800

6 a $27 = 5 \times 5 + 2$ b $68 = 13 \times 5 + 3$
c $82 = 8 \times 10 + 2$ d $156 = 26 \times 6$
e $784 = 71 \times 11 + 3$ f $230 = 15 \times 15 + 5$
g $458 = 22 \times 20 + 18$ h $3625 = 97 \times 37 + 36$

7 a 58 b 29 c 29
d Divide by 10 and halve the answer.

8 a 14 b 92 c $39\frac{1}{2}$ d 32

9 a 28 km b 14 km

10 a $208 b $48

11 a 5000 cm b 20 kites c Yes

12 186 cows

13 Number of tables required: 19; total number of courses to be served: 608; number of waiters required: 8: total charge for catering: $8360

14 a 950 blocks b $794 750

15 $16 per night

16 a 23 750 kg b 95 bales c $95 000

Exercise 1E — Order of operations

1 Tran

2 a 5 b 20 c 3 d 30
e 59 f 25 g 7 h 96
i 15 j 24 k 32 l 18
m 76 n 27 o 26 p 7
q 12 r 96 s 27 t 0
u 18 v 3

3 a $(12 - 8) \div 4 = 1$
b $(4 + 8) \times 5 - 4 \times 5 = 40$
c $3 + 4 \times (9 - 3) = 27$
d $3 \times (10 - 2) \div 4 + 4 = 10$
e $12 \times (4 + 2) - 12 = 60$
f $17 - 8 \times 2 + 6 \times (11 - 5) = 37$
g $10 \div (5 + 5) \times 9 \times 9 = 81$
h $(18 - 3) \times 3 \div 5 = 9$

4 C 5 D

6 a $6 + 2 \times (4 - 3) \times 2 = 10$
b $(6 + 2) \times 4 - 3 \times 2 = 26$
c $6 + (2 \times 4 - 3) \times 2 = 16$
d $6 + 2 \times 4 - 3 \times 2 = 8$, no brackets required

Maths Quest challenge (page 30)

1 43 pebbles
2 Same as the digit you started with. This is because you are dividing by 111 111 each time.

Exercise 1F — Estimation

1 28 000 000
2 Estimation table:

	Estimate	Estimated answer	Actual answer
a	500 + 1000	1500	1449
b	30 000 + 80 000	110 000	115 670
c	200 000 + 700 000	900 000	907 448
d	1000 + 6000	7000	7429
e	30 + 80	110	112
f	60 000 + 50 000	110 000	104 266
g	90 000 − 40 000	50 000	55 880
h	300 − 200	100	127
i	500 000 − 200 000	300 000	296 049
j	900 000 − 400 000	500 000	553 120
k	60 000 − 20 000	40 000	44 682
l	90 000 − 40 000	50 000	50 888
m	40 × 200	8000	7128
n	9000 × 9	81 000	77 679
o	90 × 400	36 000	37 584
p	600 × 10 000	6 000 000	7 894 033
q	30 000 × 40	1 200 000	1 149 954
r	200 × 60	12 000	13 320
s	30 000 ÷ 1000	30	32.907 58
t	60 000 ÷ 3000	20	24.325 483
u	900 000 ÷ 3000	300	263.003 03
v	8000 ÷ 4000	2	2.295 287 4
w	70 000 ÷ 1000	70	53.385 505
x	40 000 ÷ 200	200	205.605 56

3 a D b B c B

4 a 8000 b 200 c 100
d 400 000 e 700 000 f 11 000
g 6000 h 40 i 8 000 000
j 100 k 40 000 000 l 500
5 a ii b iv c i d iv
6 a 1500 tickets b $11 250
7 a Bread rolls $75
Hamburgers $50
Tomato sauce $4
Margarine $4
Onions $4
Tomatoes $8
Lettuce $5
b $150 c $600 d $450

10 Quick Questions 2

1 2790 2 4554 3 69
4 4637 5 6 pieces with 6 left over
6 24 7 11
8 7 + 2 × (11 − 4) 9 30 000 10 80

Exercise 1G — Subsets (special groups) of numbers

1 81
2 a 25 b 49 c 196
d 400 e 2500 f 10 000
3 16, 25, 64, 81, 100, 400
4 a 64 b 121 c 225 d 900
5 a Third b Seventh
c Twelfth d Fiftieth
6 a 5 b 9 c 13 d 15
7 36
8 1, 3, 6, 10, 15, 21, 28, 36, 45, 55
9 $F_1 = 1$ $F_2 = 1$ $F_3 = 2$ $F_4 = 3$
$F_5 = 5$ $F_6 = 8$ $F_7 = 13$ $F_8 = 21$
$F_9 = 34$ $F_{10} = 55$ $F_{11} = 89$ $F_{12} = 144$
$F_{13} = 233$ $F_{14} = 377$ $F_{15} = 610$ $F_{16} = 987$
$F_{17} = 1597$ $F_{18} = 2584$ $F_{19} = 4181$ $F_{20} = 6765$
10 a 65, 64 b 24, 25 c 3026, 3025
11 a 1 b 1 c 1
12 a 2 b 4 c 7
13 See table at foot of this page.
14 a 1, 2, 4, 8, 16, 32, 64, 128, 256, 512, 1024
b The total of each row is double the previous.
c 2048

13

Row 0	1
Row 1	1 1
Row 2	1 2 1
Row 3	1 3 3 1
Row 4	1 4 6 4 1
Row 5	1 5 10 10 5 1
Row 6	1 6 15 20 15 6 1
Row 7	1 7 21 35 35 21 7 1
Row 8	1 8 28 56 70 56 28 8 1
Row 9	1 9 36 84 126 126 84 36 9 1
Row 10	1 10 45 120 210 252 210 120 45 10 1

15 a and b

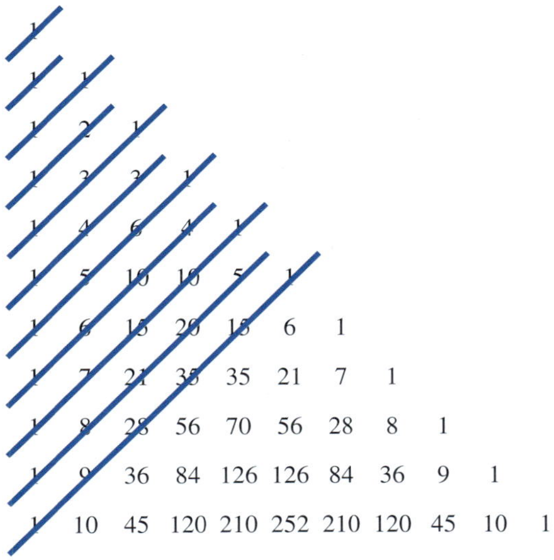

c Diagonal 1 — 1
Diagonal 2 — 1
Diagonal 3 — 2
Diagonal 4 — 3
Diagonal 5 — 5
Diagonal 6 — 8
Diagonal 7 — 13
Diagonal 8 — 21
Diagonal 9 — 34
Diagonal 10 — 55
Diagonal 11 — 89

d The sum of each diagonal row produces a Fibonacci number.

16 a 5 petals b 21 petals
These are Fibonacci numbers.

17 a Answers will vary. b Answers will vary.
c 101, 111, 121, 131, 141, 151, 161, 171, 181, 191, 202, 212, 222, 232, 242

18 a i 77 ii 99 iii 646
b 132, not a palindrome; 363 is a palindrome; 2 steps needed.
c i 363, 2 steps ii 909, 2 steps
iii 44 044, 6 steps iv 13 431, 4 steps
v 9339, 3 steps vi 475 574, 4 steps

Maths Quest challenge (page 43)

1 9983 2 10 237
3 26
4 One solution is 98 765 + 1234 = 99 999.
5 Some examples are $2 \times 2 + 2 + 2 \div 2 = 7$;
$(66 - 6 \times 6) \div 6 + \frac{6+6}{6} = 7$.

Summary

1 fives, subtract, add 2 10, Hindu–Arabic
3 place value 4 number pairs
5 12 6 zeros
7 brackets, subtraction
8 estimating 9 Square
10 Triangular 11 two
12 Pascal's

Chapter review

1 a 7 b 40 000 c 1 000 000 d 90
2 a 300 + 90 + 2 b 4000 + 100 + 9
c 40 000 + 2000 + 1 d 100 000 + 20 000
3 349, 394, 934, 943, 3499, 3994
4 1101, 1100, 1011, 110, 101
5 a 127 b 3187 c 7641
d 4 217 656
6 a 306 km b 322 km c 482 km
7 a 2 b 35 c 150 d 14
8 a 415 b 217 c 60 322
d 37 758 e 92 957 f 95 545
9 a 150 b 8400 c 1240
d 16 800 e 603 f 589
10 a 596 b 8420 c 5021
11 a 8 b 3 c 84
d 45 e 1 f 22
12 a 405 b 476 c 2015
13 a 700 b 7 c 410
14 a 448 b 3136 c 98
15 Brackets
Multiplication and division
Addition and subtraction
16 a 5 b 38 c 6
d 5 e 70 f 34
g 54 h 29
17 a 14 000 b 5000 c 50
d 420 000 e 120 000 f 10 000
g 3 h 210 000
18 1, 4, 9, 16, 25, 36, 49, 64, 81, 100
19 a 121 b 256 c 4 d 9
20 28
21 1, 1, 2, 3, 5, 8, 13, 21, 34, 55
22 Answers will vary.
23 2882

Chapter 2 Number theory

Are you ready?

1 a 16 b 144 c 49
2 a 6 b 9 c 5 d 8
3 a 125 b 1331 c 27
4 a 6 b 9 c 2 d 4
5 a 107 b 85 c 26
6 a 4, 8, 12, 16, 20 b 6, 12, 18, 24, 30
7 a 1 b 8 c 4
8 a 5, 6, 7 b 14, 16, 18, 20
c 37, 39, 41

Exercise 2A — Squares, square roots, cubes and cube roots

1 a 64 b 121 c 225 d 625
2 9801
3 a 64, 81 b 169, 196
4 a 16, 36, 64 b 81
5 a 36 and 49 b 49 and 64
c 81 and 100 d 144 and 169
6 a 5 b 9 c 12 d 20
7 a 70 b 120 c 600 d 400
8 7 and 8

9 a 3 and 4 b 9 and 10
c 14 and 15 d 1 and 2
10 B 11 A
12 64
13 a 8 b 27 c 216 d 1000
14 1, 8, 27, 64, 125, 216, 343, 512, 729, 1000
15 2
16 a 4 b 6 c 7 d 9
17 a 9 b 75 c 700 d 25
e 14 f 45

Maths Quest challenge (page 53)

1 169, 196, 225 2 7 and 9

Exercise 2B — Index notation

1 7^4
2 a 2^4 b 8^6 c 10^5 d 3^9
e 6^7 f 13^3 g 12^6 h 9^8
3 $4^5 \times 6^3$
4 a $2^2 \times 3$ b $2^2 \times 3^4$
c $2^4 \times 5^2$ d $2^3 \times 7$
e $3^3 \times 5 \times 11^2$ f $5^3 \times 7^2 \times 13$
g $2^3 \times 3^2 \times 5$ h $2^2 \times 3^2 \times 5^3$
5 $6 \times 6 \times 6 \times 6 \times 6$
6 a $11 \times 11 \times 11$
b $4 \times 4 \times 4 \times 4 \times 4 \times 4 \times 4 \times 4 \times 4$
c $5 \times 5 \times 5 \times 5 \times 5 \times 5$
7 243
8 a 1024 b 2401 c 729 d 1024
9 A 10 D 11 72
12 a 5184 b 972 c 16
d 567 e 2717 f 112
13 a 5^{11} b 7^8 c 4^{13}
d 2^{21} e 11^{24} f 9^{15}
g $3^6 \times 13^6$ h $2^4 \times 5^4$ i $5^5 \times 7^5$

Exercise 2C — Multiples

1 a 3, 6, 9, 12, 15
b 6, 12, 18, 24, 30
c 100, 200, 300, 400, 500
d 11, 22, 33, 44, 55
e 15, 30, 45, 60, 75
f 4, 8, 12, 16, 20
g 21, 42, 63, 84, 105
h 25, 50, 75, 100, 125
i 14, 28, 42, 56, 70
j 12, 24, 36, 48, 60
k 50, 100, 150, 200, 250
l 30, 60, 90, 120, 150
m 33, 66, 99, 132, 165
n 120, 240, 360, 480, 600
o 45, 90, 135, 180, 225
p 72, 144, 216, 288, 360
2 10, 20, 100, 70 3 21, 7, 70, 35
4 16, 64, 160, 32, 80 5 70, 35, 140, 105, 700
6 Three of: 8, 24, 32, 48, 56, 72, 80, 88, 96
7 9, 18, 27, 36, 45, 54, 63, 72, 81, 90, 99
8 102, 108, 114, 120, 126, 132, 138, 144, 150, 156
9 a C b A
10 a 4, 8, 12, 16, 20, 24, 28, 32, 36, 40
b 6, 12, 18, 24, 30, 36, 42, 48, 54, 60
c 12, 24, 36
d 12
11 a 3, 6, 9, 12, 15, 18 b 9, 18, 27, 36, 45, 54
c 9, 18 d 9
12 a 3 b 30 c 70 d 48 e 60
13 a F b T c T d F
14 a 6 days
b 6 is the smallest multiple of both 2 and 3.
15 a 35 minutes
b 35 is the smallest multiple of both 5 and 7.
16 a 12 hours
b 12 is the smallest multiple of both 4 and 6.
17 20
18 a 6, 12, 18 (or any other multiple of 6)
b 6 c 30
19 a 12 b 6 c 7
20

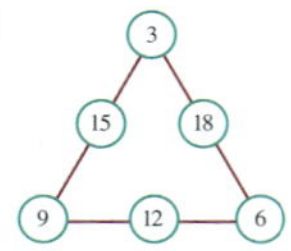

Maths Quest challenge (page 60)

1 84 2 60 3 12

10 Quick Questions 1

1 169 2 121, 144, 169, 196
3 6 4 4 and 5
5 216 6 4
7 4^{10} 8 $7^3 \times 8^4$
9 15, 30, 45, 60, 75 10 12

Exercise 2D — Factors

1 a 1, 2, 3, 4, 6, 12
b 1, 2, 4, 8
c 1, 2, 4, 5, 8, 10, 20, 40
d 1, 5, 7, 35
e 1, 2, 4, 7, 14, 28
f 1, 2, 3, 4, 5, 6, 10, 12, 15, 20, 30, 60
g 1, 2, 4, 5, 10, 20, 25, 50, 100
h 1, 2, 3, 4, 6, 8, 9, 12, 18, 24, 36, 72
i 1, 3, 13, 39
j 1, 5, 17, 85
k 1, 2, 4, 19, 38, 76
l 1, 3, 23, 69
m 1, 3, 9, 11, 33, 99
n 1, 2, 5, 10, 25, 50, 125, 250
o 1, 3, 17, 51
p 1, 3, 5, 7, 15, 21, 35, 105
2 a 20: 1, 20; 2, 10; 4, 5
b 18: 1, 18; 2, 9; 3, 6
c 36: 1, 36; 2, 18; 3, 12; 4, 9; 6, 6
d 132: 1, 132; 2, 66; 3, 44; 4, 33; 6, 22; 11, 12
3 15
4 a B b A
5 3, 4
6 a i 1, 3, 5, 15 ii 1, 5, 7, 35 iii 1, 5
b HCF = 5
7 a 1, 3, 7, 21 b 1, 2, 4, 7, 8, 14, 28, 56
c HCF = 7
8 a 1, 3, 9, 27 b 1, 3, 5, 15 c HCF = 3
9 a 1, 7 b 1, 2, 4, 7, 14, 28 c HCF = 7
10 a 1, 2, 3, 4, 6, 8, 12, 16, 24, 48
b 1, 2, 3, 5, 6, 10, 15, 30
c HCF = 6

11 9 12 7 13 4 14 9

15 a 300 b 420 c 660 d 176
e 154 f 165 g 300 h 360
i 630

16 a 900 b 1100 c 600 d 1800
e 3100 f 3300 g 2800 h 6000
i 13 000

17 4 rows of 5 students or 5 rows of 4 students; 2 rows of 10 students or 10 rows of 2 students; 1 row of 20 students or 20 rows of 1 student

18 D

19 a 3 metres b 10

20 a Daniella b 10 minutes

Maths Quest challenge (page 68)

1 60 2 12 or 30 3 42

History of mathematics: Eratosthenes of Cyrene (c. 270 BC to 194 BC)

1 Main librarian at the Alexandrian museum
2 The Earth's circumference
3 It was based on longitude and latitude.
4 Ptolemy III, ruler of Egypt
5 $2^{1\,398\,269} - 1$
6 He became blind when he was older.

Exercise 2E — Prime numbers

1 a
1: 1
2: 1, 2
3: 1, 3
4: 1, 2, 4
5: 1, 5
6: 1, 2, 3, 6
7: 1, 7
8: 1, 2, 4, 8
9: 1, 3, 9
10: 1, 2, 5, 10
11: 1, 11
12: 1, 2, 3, 4, 6, 12
13: 1, 13
14: 1, 2, 7, 14
15: 1, 3, 5, 15
16: 1, 2, 4, 8, 16
17: 1, 17
18: 1, 2, 3, 6, 9, 18
19: 1, 19
20: 1, 2, 4, 5, 10, 20

b 2, 3, 5, 7, 11, 13, 17, 19

2 23, 29, 31, 37

3 No. The number 2 would always be a factor if the number is even.

4 a Composite b Prime
c Composite d Composite
e Prime f Composite
g Composite h Prime
i Composite j Prime
k Composite l Prime

5 a F b F c F d T
e T f F g F h F

6 a A b C c C d C

7 11 and 13; 17 and 19

8 a 3, 5, 7. Prime numbers greater than 2 are always odd numbers and the difference between 2 odd numbers is an even number. So 2, 4 and 6 can be differences between consecutive primes.
b 2: 17 and 19; 4: 19 and 23; 6: 23 and 29; other answers possible

9 a $365 = 5 \times 73$ b $187 = 11 \times 17$

Maths Quest challenge (page 73)

1 773 2 47 3 61 4 432

5 Angus will be 27 and his Dad will be 81.

6 98 641

7 A = 4, B = 8, C = 16, D = 13

Exercise 2F — Prime factors and factor trees

1 a i 3, 5 ii $15 = 3 \times 5$

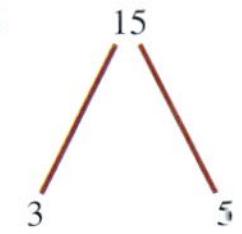

b i 2, 3, 5 ii $30 = 2 \times 3 \times 5$

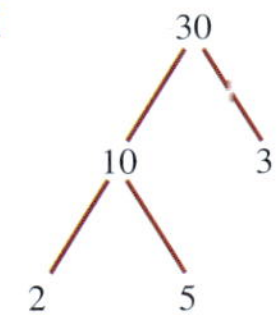

c i 2, 3 ii $24 = 2 \times 2 \times 2 \times 3$

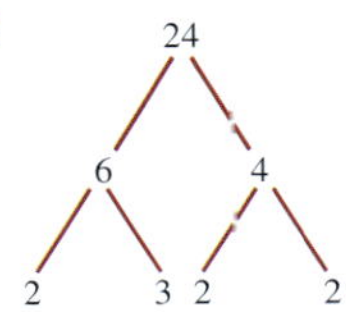

d i 2, 5 ii $100 = 2 \times 2 \times 5 \times 5$

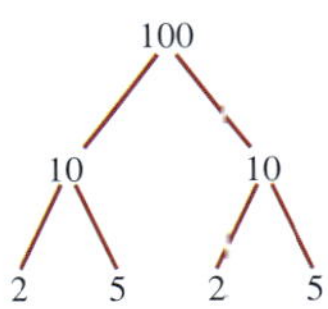

e i 7 ii $49 = 7 \times 7$

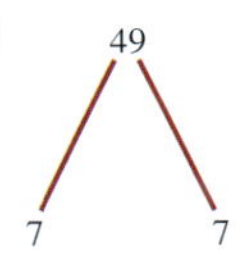

f i 2, 3 ii $72 = 2 \times 2 \times 2 \times 3 \times 3$

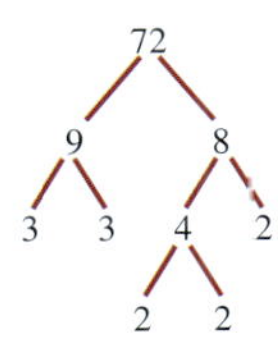

g i 2, 3 ii $18 = 2 \times 3 \times 3$

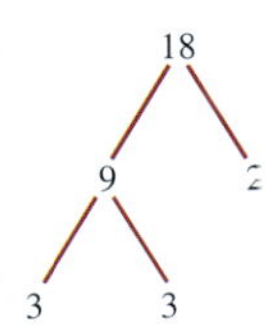

h i 2, 7 ii $56 = 2 \times 2 \times 2 \times 7$

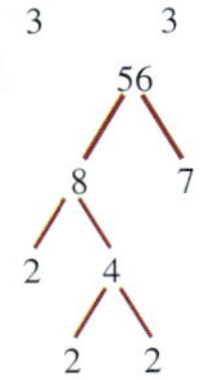

i i 3, 5 ii $45 = 3 \times 3 \times 5$

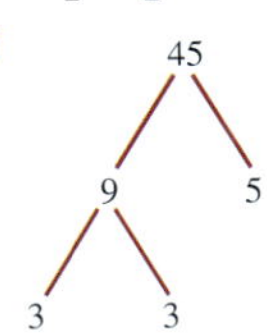

j i 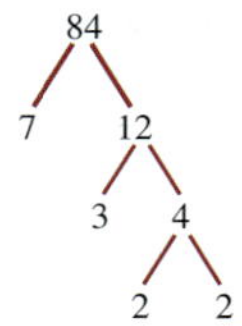

2, 3, 7 **ii** $84 = 2 \times 2 \times 3 \times 7$

k i 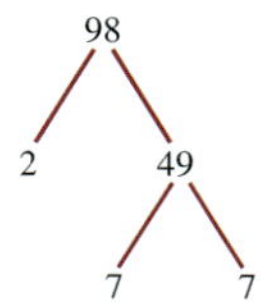

2, 7 **ii** $98 = 2 \times 7 \times 7$

l i 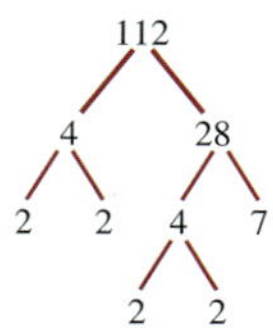

2, 7 **ii** $112 = 2 \times 2 \times 2 \times 2 \times 7$

2 a i 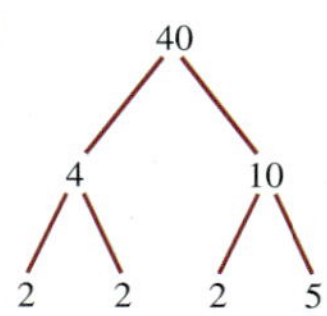

2, 5 **ii** $40 = 2 \times 2 \times 2 \times 5$

b i 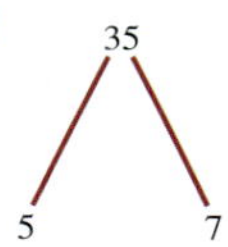

5, 7 **ii** $35 = 7 \times 5$

c i 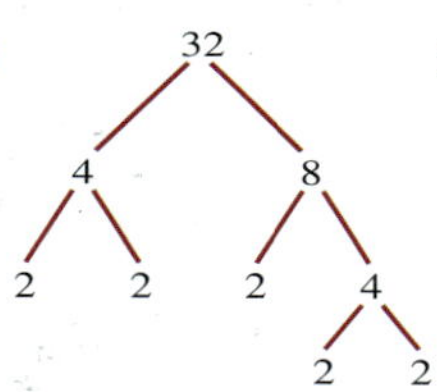

2 **ii** $32 = 2 \times 2 \times 2 \times 2 \times 2$

d i 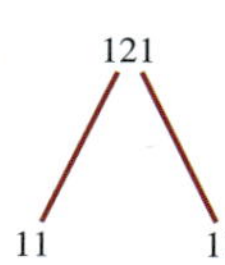

11 **ii** $121 = 11 \times 11$

e i 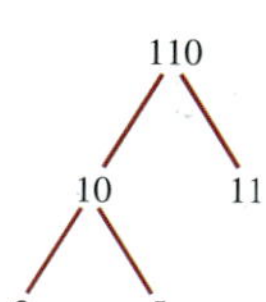

2, 5, 11 **ii** $110 = 2 \times 5 \times 11$

f i 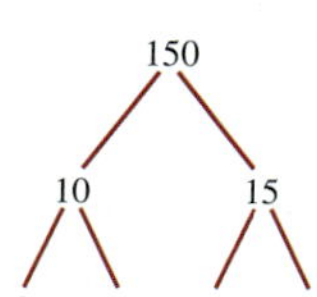

2, 3, 5
ii $150 = 2 \times 3 \times 5 \times 5$

g i 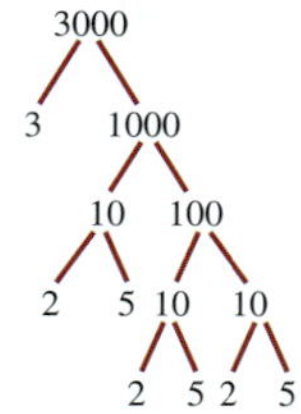

2, 3, 5
ii $3000 = 2 \times 2 \times 2 \times 3 \times 5 \times 5 \times 5$

h i 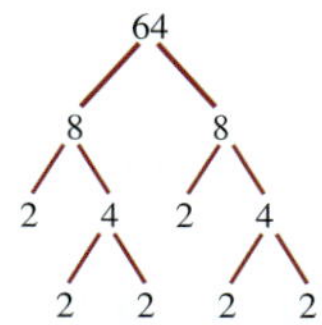

2
ii $64 = 2 \times 2 \times 2 \times 2 \times 2 \times 2$

i i 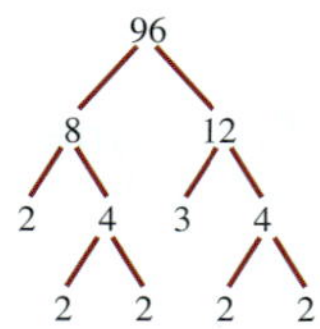

2, 3
ii $96 = 2 \times 2 \times 2 \times 2 \times 2 \times 3$

j i 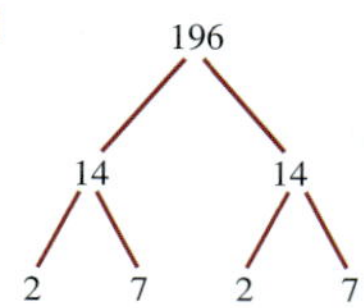

2, 7
ii $196 = 2 \times 2 \times 7 \times 7$

k i 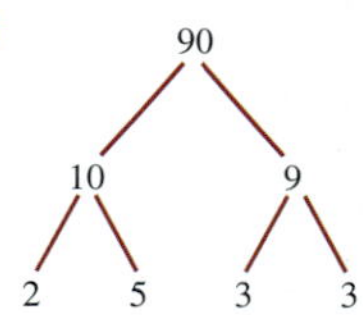

2, 3, 5
ii $90 = 2 \times 3 \times 3 \times 5$

l i 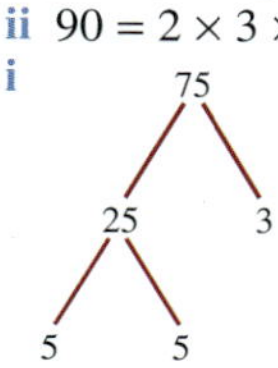

3, 5 **ii** $75 = 3 \times 5 \times 5$

3 a A **b** D **c** C **d** B

4 a 2, 3 **b** 2, 5 **c** 2, 3, 7 **d** 3
e 2, 3 **f** 3, 13 **g** 3 **h** 2, 3, 5
i 2, 3, 5 **j** 2, 3, 5 **k** 2, 5 **l** 2, 5

5 a T **b** F **c** T **d** F
e T **f** F

6 a $2^2 \times 3 \times 5$ **b** 2×5^2 **c** 3×5^2
d $2^2 \times 5 \times 11$ **e** $2^6 \times 3$ **f** $2^3 \times 3^2$
g $2^2 \times 31$ **h** $2^3 \times 5^2$

10 Quick Questions 2

1 1, 3, 5, 9, 15, 45
2 1, 2, 4, 5, 8, 10, 16, 20, 40, 80
3 1 and 100, 2 and 50, 4 and 25, 5 and 20, 10 and 10
4 2, 3, 5, 7, 11, 13, 17, 23, 29, 31, 37
5 Composite
6 101
7 **a** $40 = 2^3 \times 5$ **b** $70 = 2 \times 5 \times 7$

8 $96 = 2^5 \times 3$

9 $500 = 2^2 \times 5^3$

10 a

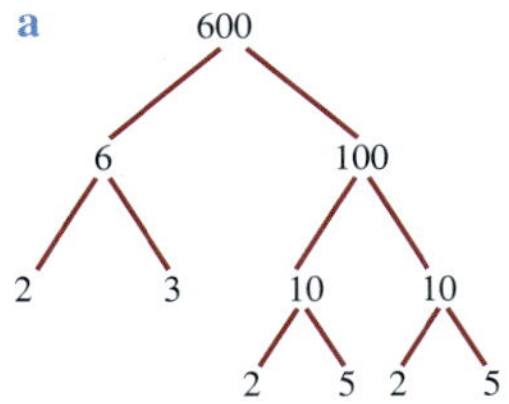

b

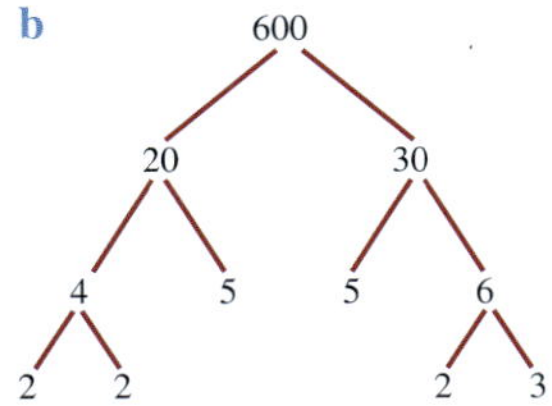

$600 = 2 \times 2 \times 2 \times 3 \times 5 \times 5$

c

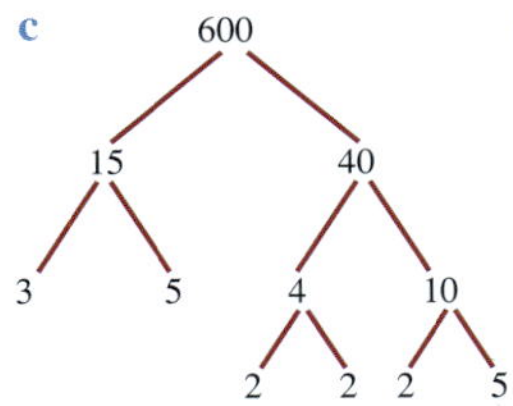

Maths Quest challenge (page 77)

1 a $28 = 2 \times 2 \times 7$. Proper factors are 1, 2, 4, 7, 14. $1 + 2 + 4 + 7 + 14 = 28$ so 28 is perfect.

b $496 = 2 \times 2 \times 2 \times 2 \times 31$. Proper factors are 1, 2, 4, 8, 16, 31, 62, 124, 248. $1 + 2 + 4 + 8 + 16 + 31 + 62 + 124 + 248 = 496$ so 496 is perfect.

2 $999 = 3 \times 3 \times 3 \times 37$. Proper factors are 1, 3, 9, 27, 37, 111, 333. $1 + 3 + 9 + 27 + 37 + 111 + 333 = 484$ so 999 is not a perfect number.

3 $8128 = 2 \times 2 \times 2 \times 2 \times 2 \times 2 \times 127$. Proper factors are 1, 2, 4, 8, 16, 32, 64, 127, 254, 508, 1016, 2032, 4064. $1 + 2 + 4 + 8 + 16 + 32 + 64 + 127 + 254 + 508 + 1016 + 2032 + 4064 = 8128$ so 8128 is a perfect number.

4 a 12, 18, 20, 24, 30, 36, 40, 42, 48, 54
b None

Exercise 2G — Divisibility

1 a Yes b No c Yes d No
e Yes f Yes g Yes h No
i Yes j No k No l Yes

2 a Yes b Yes c Yes d Yes
e No f No g Yes h Yes
i No j Yes k No l Yes

3 a Yes b No c Yes d Yes
e No f No g Yes h Yes
i Yes j No

4 a Yes b Yes c No d Yes
e Yes f No g No h Yes
i Yes j Yes k No l Yes

5 a Yes b No c Yes d No
e Yes f Yes g No h No
i Yes j Yes

6 a Yes b Yes c No d Yes
e Yes f Yes g No h Yes
i No j No k No l No

7 a Yes b Yes c No d Yes
e No f Yes g No h No
i Yes j Yes

8 a Yes b No c Yes d Yes
e No f Yes g No h Yes
i No j Yes k No l Yes

9 D

10 a Yes b No c Yes

11 326 is not divisible by 4. The tennis club would not be able to make teams of 4 exactly; there would be some players left over.

12 No, 142 is divisible by 2 but not by 3 ($1 + 4 + 2 = 7$ which is not divisible by 3) so 142 is not divisible by 6.

13 a $3^2 \times 5^2 \times 7$ b $2^3 \times 7 \times 11$
c $2^3 \times 3^3 \times 5$ d $2 \times 3^3 \times 7^2$

14 a $11 \times 697 = 7667$ b $7 \times 1317 = 9219$

15 c 103

Maths Quest challenge (page 82)

1 36 2 18 3 96

Summary

1 2, cubic
2 square root, cube root
3 Index notation
4 multiple
5 factor
6 prime number
7 composite number
8 one
9 prime, factor tree
10 even, 3, two, 5, divisible, 8, sum, 0

Chapter review

1 a 36 b 196 c 361 d 6400

2 a 9 and 16 b 25 and 36
c 100 and 121 d 225 and 256

3 a 7 b 16 c 30 d 37

4 8 and 9

5 10 is not a square number.

6 a 64 b 343 c 3375 d 27 000

7 a 3 b 5 c 10 d 20

8 a 3^4 b 5^6 c 7^7

9 a 416 b 8 c 73

10 a 2×3^3 b $5^2 \times 6^4$ c $2 \times 5^3 \times 9^3$

11 a 243 b 343 c 4096 d 161 051

12 a 7^{12} b 3^{13} c 13^{16}
d 9^{20} e $4^8 \times 6^8$ f $12^3 \times 15^3$

13 a 11, 22, 33, 44, 55 b 100, 200, 300, 400, 500
c 5, 10, 15, 20, 25 d 20, 40, 60, 80, 100
e 13, 26, 39, 52, 65 f 35, 70, 105, 140, 175

14 a 12 b 30 c 28 d 40

15 120 seconds or 2 minutes

16 a 12, 24, 36 b 12 c 60

17 a 1, 2, 4, 8, 16
b 1, 3, 9, 27
c 1, 2, 5, 10, 25, 50
d 1, 2, 3, 6, 7, 14, 21, 42
e 1, 2, 3, 4, 6, 9, 12, 18, 36
f 1, 2, 3, 4, 6, 8, 9, 12, 18, 24, 36, 72

18 a 1, 24; 2, 12; 3, 8; 4, 6
b 1, 40; 2, 20; 4, 10; 5, 8
c 1, 48; 2, 24; 3, 16; 4, 12; 6, 8
d 1, 21; 3, 7
e 1, 99; 3, 33; 9, 11
f 1, 100; 2, 50; 4, 25; 5, 20; 10, 10

19 5 metres

20 2, 3, 5, 7, 11, 13, 17, 19, 23, 29

21 There are 4 single-digit prime numbers: 2, 3, 5, 7.

22 53

23 a 3 and 11 b 3 and 7 c 5 d 2 and 31

24 $280 = 2 \times 2 \times 2 \times 5 \times 7$ or $280 = 2^3 \times 5 \times 7$

25 a $44 = 2 \times 2 \times 11$ **b** $132 = 2 \times 2 \times 3 \times 11$
c $150 = 2 \times 3 \times 5 \times 5$
d $360 = 2 \times 2 \times 2 \times 3 \times 3 \times 5$

26 6, 12

27 a Yes **b** No **c** No **d** Yes

28 a i A number is divisible by 2 if it ends in an even number.
ii A number is divisible by 5 if it ends in 0 or 5.
iii A number is divisible by 10 if it ends in 0.
b The test for each of these numbers involves looking at the last digit.

29 c 90 **d** 294

30 a 162 **c** 459 **d** 49 725

31 a T **b** T **c** F **d** T
e F **f** T **g** T **h** F
i T **j** F **k** T

Chapter 3 Fractions

Are you ready?

1 a Proper fraction **b** Improper fraction
c Mixed number **d** Improper fraction
e Mixed number **f** Proper fraction

2 a $\frac{10}{15}$ **b** $\frac{6}{7}$

3 $\frac{1}{3} = \frac{2}{6} = \frac{3}{9} = \frac{4}{12}$

4 a 3, 6, 9, 12, 15, 18, 21, 24
b 4, 8, 12, 16, 20, 24, 28, 32
c 12, 24 **d** 12

5 a $\frac{1}{5}, \frac{3}{5}, \frac{4}{5}, 1, \frac{7}{5}$ **b** $0, \frac{5}{12}, \frac{7}{12}, \frac{11}{12}, \frac{13}{12}$

6 a 1, 2, 3, 6, 9, 18 **b** 1, 2, 3, 4, 6, 12
c 1, 2, 3, 6 **d** 6

7 $2\frac{1}{2}$ **8** $\frac{4}{3}$

9 a $\frac{8}{6} = \frac{4}{3}$ or $1\frac{1}{3}$ **b** $\frac{2}{12} = \frac{1}{6}$
c $\frac{2}{3}$ **d** $3\frac{3}{4}$

Exercise 3A — Understanding fractions

1 a $\frac{3}{4}$ **b** $\frac{15}{16}$ **c** $\frac{7}{8}$ **d** $\frac{5}{9}$

2 a $\frac{1}{3}$ **b** $\frac{3}{4}$
c The flags belong to Italy and Taiwan.

3 a i $\frac{3}{11}$ **ii** $\frac{2}{11}$ **b** $\frac{3}{11}$

4 a 3 **b** 3 **c** 4 **d** 5
e 6 **f** 9 **g** 11 **h** 12
i 7

5 a $\frac{20}{30}$ $\frac{14}{21}$ $\frac{40}{60}$ **b** $\frac{12}{15}$ $\frac{36}{45}$ $\frac{16}{20}$ $\frac{28}{35}$ $\frac{80}{100}$
c $\frac{56}{64}$ $\frac{21}{24}$ **d** $\frac{35}{50}$ $\frac{21}{30}$ $\frac{14}{20}$ $\frac{140}{200}$

6 a $\frac{9}{15}$ $\frac{12}{20}$ $\frac{21}{35}$ **b** $\frac{30}{40}$ $\frac{18}{24}$ $\frac{27}{36}$
c $\frac{7}{28}$ $\frac{6}{24}$ $\frac{3}{12}$ **d** $\frac{15}{18}$ $\frac{35}{42}$ $\frac{40}{48}$

7 a $\frac{5}{6} = \frac{10}{12} = \frac{15}{18} = \frac{20}{24}$ **b** $\frac{5}{8} = \frac{10}{16} = \frac{15}{24} = \frac{20}{32}$
c $\frac{3}{10} = \frac{6}{20} = \frac{9}{30} = \frac{12}{40}$ **d** $\frac{2}{3} = \frac{4}{6} = \frac{6}{9} = \frac{8}{12}$
e $\frac{1}{8} = \frac{2}{16} = \frac{3}{24} = \frac{4}{32}$ **f** $\frac{7}{8} = \frac{14}{16} = \frac{21}{24} = \frac{28}{32}$

8 a $\frac{3}{5}$ **b** $\frac{7}{8}$ **c** $\frac{1}{5}$ **d** $\frac{1}{8}$ **e** $\frac{1}{2}$
f $\frac{3}{4}$ **g** $\frac{5}{8}$ **h** $\frac{5}{8}$ **i** $\frac{4}{5}$

9 a $>$ **b** $<$ **c** $>$ **d** $>$
e $<$ **f** $<$

10 a D **b** B

11 a $\frac{3}{10}, \frac{5}{10}, \frac{2}{10}; \frac{1}{5}, \frac{3}{10}, \frac{1}{2}$ **b** $\frac{9}{24}, \frac{12}{24}, \frac{8}{24}; \frac{1}{3}, \frac{3}{8}, \frac{1}{2}$
c $\frac{10}{15}, \frac{12}{15}, \frac{6}{15}; \frac{6}{15}, \frac{2}{3}, \frac{4}{5}$ **d** $\frac{45}{60}, \frac{40}{60}, \frac{28}{60}; \frac{7}{15}, \frac{2}{3}, \frac{3}{4}$

12 a

Sally Hanna

Sonya Vanessa

b Vanessa Vegetarian
c They ate the same amount.

Exercise 3B — Simplifying fractions

1 a $\frac{1}{2}$ **b** $\frac{2}{3}$ **c** $\frac{7}{8}$ **d** $\frac{4}{5}$ **e** $\frac{4}{5}$
f $\frac{3}{4}$ **g** $\frac{9}{10}$ **h** $\frac{7}{8}$ **i** $\frac{5}{6}$ **j** $\frac{24}{25}$
k $\frac{7}{8}$ **l** $\frac{7}{10}$ **m** $\frac{11}{12}$ **n** $\frac{11}{25}$ **o** $\frac{3}{8}$

2 a $4\frac{3}{5}$ **b** $7\frac{5}{6}$ **c** $10\frac{1}{2}$ **d** $5\frac{1}{6}$ **e** $3\frac{7}{8}$
f $1\frac{11}{12}$ **g** $6\frac{3}{4}$ **h** $5\frac{1}{3}$ **i** $3\frac{1}{5}$ **j** $2\frac{1}{4}$

3 A **4** $\frac{21}{28} = \frac{3}{4}$

5 a $\frac{5}{9}$ **b** $\frac{4}{9}$

6 7A: $\frac{1}{5}$, 7B: $\frac{3}{20}$, 7C: $\frac{8}{25}$, 7D: $\frac{2}{25}$, 7E: $\frac{1}{4}$

7 a 40
b Mark: $\frac{5}{40} = \frac{1}{8}$, David: $\frac{10}{40} = \frac{1}{4}$, Samantha: $\frac{8}{40} = \frac{1}{5}$, Jules: $\frac{6}{40} = \frac{3}{20}$, Ahmed: $\frac{4}{40} = \frac{1}{10}$, Darren: $\frac{7}{40}$

Maths Quest challenge (page 99)

1 $\frac{7293}{14\,586}, \frac{6792}{13\,584}$ **2** $\frac{5823}{17\,469}$

Exercise 3C — Improper fractions and mixed numbers

1 a

$2\frac{1}{2}$

b

$1\frac{1}{3}$

c

$1\frac{5}{6}$

d  $1\frac{5}{8}$

e  3

f  $1\frac{1}{6}$

2 a Improper fraction **b** Mixed number
c Proper fraction

3 a 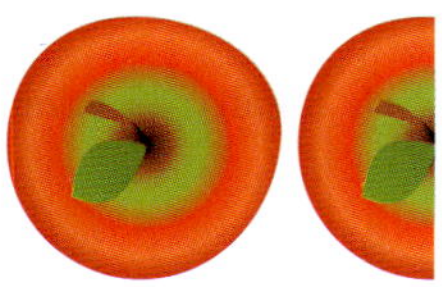 $\frac{3}{2} = 1\frac{1}{2}$

b 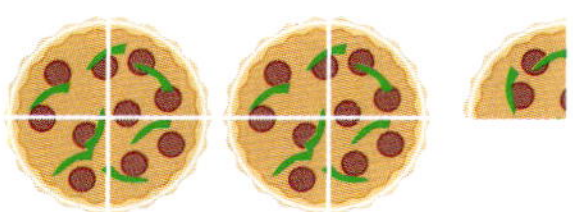 $\frac{9}{4} = 2\frac{1}{4}$

4 a $1\frac{2}{5}$ **b** $2\frac{3}{4}$ **c** $10\frac{1}{2}$ **d** $3\frac{9}{10}$ **e** $4\frac{1}{4}$
f $8\frac{4}{11}$ **g** $2\frac{3}{13}$ **h** $2\frac{7}{8}$ **i** $33\frac{1}{3}$ **j** $12\frac{1}{2}$
k $6\frac{2}{3}$ **l** $3\frac{1}{3}$

5 a $1\frac{3}{5}$ **b** $3\frac{3}{8}$ **c** $5\frac{1}{2}$ **d** $10\frac{7}{9}$ **e** $8\frac{2}{7}$
f $11\frac{7}{10}$ **g** $8\frac{3}{8}$ **h** $8\frac{2}{9}$ **i** $2\frac{7}{12}$ **j** $14\frac{5}{6}$

6 a C **b** B **c** D

7 Kim: $2\frac{1}{4}$, Carly: $2\frac{3}{4}$ **8** $6\frac{3}{8}$ **9** $4\frac{3}{4}$

10 a 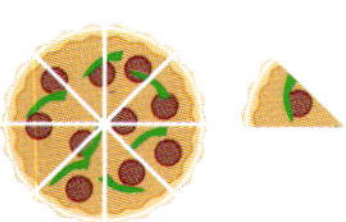 $\frac{9}{8}$

b  $\frac{15}{4}$

c  $\frac{13}{6}$

d  $\frac{7}{6}$

e  $\frac{21}{4}$

f  $\frac{8}{5}$

11 a $\frac{8}{7}$ **b** $\frac{9}{5}$ **c** $\frac{8}{3}$ **d** $\frac{10}{3}$ **e** $\frac{25}{4}$
f $\frac{37}{9}$ **g** $\frac{23}{2}$ **h** $\frac{31}{8}$ **i** $\frac{39}{5}$ **j** $\frac{47}{5}$

12 a $\frac{7}{2}$ **b** $\frac{2}{5}$ **c** $\frac{29}{5}$ **d** $\frac{47}{7}$ **e** $\frac{29}{10}$
f $\frac{43}{12}$ **g** $\frac{28}{5}$ **h** $\frac{67}{7}$ **i** $\frac{21}{11}$ **j** $\frac{53}{6}$

13 a D **b** D **c** A

14 a 25 **b** 77

15 a > **b** > **c** < **d** > **e** <
f < **g** < **h** < **i** <

10 Quick Questions 1

1 $\frac{1}{4}, \frac{3}{12}, \frac{8}{32}, \frac{16}{64}, \frac{19}{76}, \frac{31}{124}$ **2** $\frac{4}{10}, \frac{6}{15}, \frac{8}{20}$

3 $\frac{3}{5}$ **4** $\frac{44}{50} = \frac{22}{25}$ **5** $8\frac{2}{3}$ **6** $\frac{5}{4}$ **7** $1\frac{1}{3}$ **8** $\frac{27}{7}$

9 $\frac{2}{5}, 1\frac{3}{5}, \frac{14}{5}, \frac{15}{5}, 3\frac{3}{5}, 4\frac{2}{5}, \frac{28}{5}, 6\frac{1}{5}$ **10** 10

Exercise 3D — Adding and subtracting fractions

1 a $\frac{4}{5}$ **b** $\frac{6}{11}$ **c** $\frac{5}{6}$ **d** $\frac{31}{50}$ **e** $\frac{9}{25}$
f $\frac{11}{12}$ **g** $\frac{7}{8}$ **h** $\frac{6}{7}$ **i** $\frac{29}{100}$

2 a $\frac{1}{2}$ **b** $\frac{1}{4}$ **c** $\frac{1}{3}$ **d** $\frac{1}{2}$ **e** $\frac{1}{2}$
f 1 **g** $\frac{3}{4}$ **h** $\frac{3}{4}$ **i** $\frac{8}{9}$

3 a $\frac{3}{4}$ **b** $\frac{1}{4}$ **c** $\frac{3}{7}$ **d** $\frac{3}{4}$ **e** $\frac{2}{3}$
f $\frac{37}{40}$ **g** $\frac{13}{27}$ **h** $\frac{16}{35}$ **i** $\frac{2}{5}$

4 a 6 **b** 15 **c** 10 **d** 12 **e** 30
f 18 **g** 8 **h** 24 **i** 35

5 a $\frac{1}{6}$ **b** $\frac{4}{15}$ **c** $\frac{9}{10}$ **d** $\frac{7}{12}$ **e** $\frac{19}{30}$
f $\frac{17}{18}$ **g** $\frac{5}{8}$ **h** $\frac{11}{24}$ **i** $\frac{11}{35}$

6 a $\frac{11}{12}$ **b** $\frac{19}{56}$ **c** $\frac{1}{20}$ **d** $\frac{5}{6}$ **e** $\frac{5}{24}$
f $\frac{5}{18}$ **g** $\frac{23}{36}$ **h** $\frac{39}{44}$ **i** $\frac{34}{55}$ **j** $1\frac{8}{15}$
k $\frac{1}{4}$ **l** $\frac{11}{40}$ **m** $1\frac{7}{15}$ **n** $1\frac{13}{28}$ **o** $\frac{17}{63}$
p $\frac{23}{39}$ **q** $\frac{13}{60}$ **r** $1\frac{17}{42}$ **s** $\frac{2}{15}$ **t** $\frac{7}{33}$

7 a D **b** C

8 a $\frac{7}{12}$ **b** $\frac{2}{5}$ **c** $\frac{19}{24}$ **d** $\frac{7}{30}$
e $\frac{7}{16}$ **f** $\frac{1}{3}$ **g** $1\frac{11}{12}$ **h** $1\frac{11}{12}$

9 a $\frac{4}{15}$ b $\frac{11}{15}$

10 $1\frac{1}{4}$ hours

11 a $\frac{3}{4}$ b $\frac{7}{12}$ c 1

12 a $\frac{2}{3}$ b $\frac{23}{30}$

13 a $4\frac{7}{10}$ b $1\frac{1}{6}$ c $3\frac{5}{12}$ d $13\frac{3}{8}$
e $1\frac{1}{10}$ f $6\frac{1}{4}$

14 a $1\frac{1}{3}$ b $1\frac{1}{2}$ c $1\frac{5}{8}$ d $1\frac{7}{10}$
e $3\frac{2}{3}$ f $1\frac{11}{12}$

15 a $2\frac{17}{35}$ b $3\frac{7}{8}$ c $3\frac{13}{18}$ d $3\frac{1}{24}$

16 a $7\frac{7}{12}$; >7 b $3\frac{2}{3}$; >3 c $4\frac{1}{2}$; >4 d $4\frac{1}{4}$; >4
e $3\frac{3}{10}$; >3 f $4\frac{1}{2}$; >4

17 $1\frac{5}{6}$ litres **18** $6\frac{1}{4}$ hours **19** $1\frac{5}{16}$ inches

20 a $3\frac{15}{16}$ cm b $2\frac{15}{16}$ cm

Exercise 3E — Multiplying fractions

1 a $\frac{1}{8}$ b $\frac{2}{9}$ c $\frac{6}{25}$ d $\frac{20}{27}$
e $\frac{11}{26}$ f $\frac{54}{77}$ g $\frac{121}{144}$ h $\frac{40}{81}$
i $\frac{5}{24}$ j $\frac{4}{9}$ k $\frac{55}{72}$ l $\frac{25}{132}$
m $\frac{49}{64}$ n $\frac{8}{39}$ o $\frac{9}{50}$ p $\frac{8}{21}$

2 a $\frac{1}{6}$ b $\frac{1}{5}$ c $\frac{5}{14}$ d $\frac{1}{10}$ e $\frac{4}{5}$
f 1 g $\frac{1}{2}$ h $\frac{12}{35}$ i $\frac{4}{9}$ j $\frac{1}{7}$
k $\frac{1}{3}$ l $\frac{2}{5}$ m $\frac{4}{21}$ n $\frac{1}{20}$ o $\frac{1}{4}$

3 a
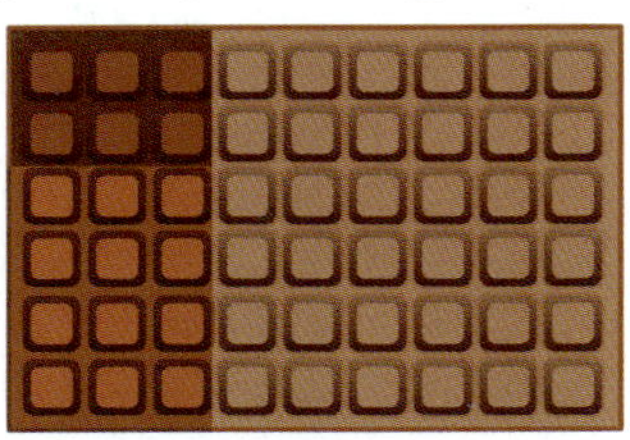
Answers may vary.
b $\frac{1}{9}$

4 a 2 b $2\frac{1}{2}$ c $7\frac{1}{2}$ d $1\frac{1}{4}$

5 a $\frac{5}{6}$ b $1\frac{7}{32}$ c $2\frac{1}{4}$ d $4\frac{19}{32}$ e $7\frac{7}{10}$
f $7\frac{4}{15}$ g $5\frac{5}{8}$ h $1\frac{17}{35}$ i $3\frac{3}{10}$ j $2\frac{3}{5}$
k 18 l $16\frac{1}{2}$ m $17\frac{1}{2}$ n 26 o $9\frac{33}{50}$
p $37\frac{5}{21}$ q $34\frac{2}{3}$ r 10 s $2\frac{11}{12}$ t $26\frac{18}{35}$

6 $\frac{3}{8}$ cup sugar
$\frac{3}{8}$ cup margarine
$\frac{3}{4}$ cup peanut butter
3 eggs
$2\frac{1}{4}$ cups milk
$3\frac{3}{4}$ cups self-raising flour
$\frac{3}{8}$ teaspoon baking soda

7 a $\frac{1}{8}$ b $\frac{1}{2}$ c $\frac{25}{36}$ d $\frac{9}{14}$ e $\frac{2}{9}$
f $\frac{1}{5}$ g 16 h 28 i 45 j 18
k 15 l 72 m $4\frac{1}{2}$ n $13\frac{1}{2}$ o $\frac{5}{6}$
p $5\frac{1}{7}$ q $\frac{3}{4}$ r $6\frac{1}{2}$ s 1 t $\frac{3}{5}$

8 a $\frac{3}{4}$ b 36 litres

9 Emily drank 2 litres. Tracy drank $\frac{1}{2}$ litre.
Jonathan drank $\frac{1}{3}$ litre.

10 a Zoe 32 minutes b Sarah 35 minutes

11 $\frac{1}{24} \times 2400 = \$100$

12 a 90 minutes b $\frac{90}{180} = \frac{1}{2}$
c $140 + 26\frac{2}{3} = 166\frac{2}{3}$ minutes

13 a

b

c $\frac{1}{8}$ d $\frac{1}{2} \times \frac{1}{4} = \frac{1}{8}$

14 a 2 b 4 c 10

Exercise 3F — Dividing fractions

1 a $\frac{4}{3}$ b $\frac{7}{2}$ c $\frac{3}{5}$ d $\frac{12}{5}$ e $\frac{5}{6}$
f $\frac{2}{9}$ g $\frac{3}{5}$ h $\frac{3}{10}$ i 5 j 12
k 5 l 11 m $\frac{1}{5}$ n $\frac{1}{20}$ o $\frac{1}{13}$
p 1

2 a $\frac{3}{5}$ b $\frac{7}{31}$ c $\frac{2}{15}$ d $\frac{5}{22}$ e $\frac{7}{65}$
f $\frac{7}{27}$ g $\frac{8}{27}$ h $\frac{3}{16}$ i $\frac{9}{92}$ j $\frac{10}{69}$

3 a 1 b 1 c 1 d 1
e 1 f 1

4 Reciprocal, one

5 a $\frac{3}{8}$ b $\frac{16}{21}$ c $1\frac{4}{5}$ d $\frac{1}{8}$ e $\frac{24}{55}$
f $\frac{77}{108}$ g $4\frac{2}{3}$ h $\frac{3}{10}$ i $\frac{1}{2}$ j 1
k $\frac{2}{35}$ l $\frac{1}{32}$ m $\frac{1}{9}$ n $\frac{3}{11}$ o $\frac{7}{270}$

6 a $1\frac{2}{3}$ b 2 c $\frac{9}{14}$ d 1 e $1\frac{9}{11}$
f $22\frac{1}{3}$ g $\frac{1}{20}$ h $\frac{18}{35}$ i $2\frac{10}{77}$ j 3
k $2\frac{29}{40}$ l $1\frac{1}{6}$ m $1\frac{8}{15}$ n 4 o $2\frac{1}{14}$

7 $\frac{3}{16}$ kg

8 a 22 b 96

9 a $6\frac{2}{3}$ laps b 12 minutes

10 Quick Questions 2

1 $\frac{8}{15}$ 2 $1\frac{2}{15}$ 3 $5\frac{13}{20}$ 4 $8\frac{1}{8}$ 5 $\frac{2}{15}$
6 $2\frac{2}{3}$ 7 $\frac{3}{20}$ 8 $\frac{5}{6}$ 9 $4\frac{4}{7}$ 10 $24\frac{1}{5}$

Exercise 3G — Mixed operations with fractions

1 a $1\frac{11}{24}$ b $1\frac{1}{3}$ c $\frac{27}{40}$ d $\frac{17}{40}$ e $11\frac{6}{35}$
f $1\frac{11}{27}$ g $2\frac{11}{35}$ h $\frac{5}{8}$ i $3\frac{1}{20}$

2 a $2\frac{4}{5}$ b $2\frac{1}{2}$ c $6\frac{11}{20}$ d $4\frac{5}{24}$ e $2\frac{1}{6}$
f $\frac{19}{30}$ g $10\frac{3}{4}$ h $\frac{1}{2}$ i $\frac{1}{4}$

3 a $\frac{23}{24}$ b $4\frac{20}{21}$ c $2\frac{13}{16}$ d $2\frac{1}{12}$ e $\frac{9}{32}$
f $\frac{5}{18}$ g $1\frac{35}{36}$ h $2\frac{2}{7}$ i $\frac{24}{25}$ j $1\frac{5}{7}$

Maths Quest challenge (page 126)

1 26 2 3 3 $\frac{3}{4}$ 4 $32 000

5 The result is a number which is smaller than the original number.

6 $\frac{15}{25}$

Exercise 3H — Powers and square roots of fractions

1 a $\frac{1}{4}$ b $\frac{1}{9}$ c $\frac{4}{25}$ d $\frac{25}{36}$ e $\frac{9}{16}$
f $\frac{49}{100}$ g $\frac{25}{144}$ h $\frac{9}{121}$ i $\frac{9}{64}$

2 a $\frac{1}{64}$ b $\frac{1}{8}$ c $\frac{8}{125}$ d $\frac{27}{1000}$ e $\frac{8}{27}$
f $\frac{1}{216}$

3 a $1\frac{7}{9}$ b $5\frac{1}{16}$ c $2\frac{1}{4}$ d $1\frac{21}{100}$ e $4\frac{21}{25}$
f $13\frac{4}{9}$

4 a $\frac{4}{5}$ b $\frac{7}{10}$ c $\frac{3}{8}$ d $\frac{5}{6}$ e $1\frac{1}{2}$
f $1\frac{1}{3}$ g $1\frac{1}{4}$ h $1\frac{3}{5}$ i $1\frac{1}{7}$

5 a $\frac{13}{16}$ b $\frac{7}{36}$ c $\frac{1}{9}$ d $1\frac{7}{9}$ e $\frac{11}{18}$
f $\frac{1}{10}$

Maths Quest challenge (page 130)

1 7 2 30 m 3 $\frac{2}{25}$ 4 $\frac{37}{83}$, $\frac{44}{120}$

5 $\frac{23}{24}$, $\frac{13}{24}$ 6 11 seconds

7 Jeff's answer is incorrect. The answer should be $\frac{19}{20}$.
Instead of adding equivalent fractions with the same denominator, Jeff has added the numerators together and added the denominators together to obtain his answer.

Summary

1 numerator 2 denominator
3 same 4 Ascending
5 Descending 6 Equivalent fractions
7 simplify 8 Improper, remainder
9 numbers 10 common
11 multiply, cancel 12 reciprocal
13 tip, mixed

Chapter review

1 $\frac{3}{20}$

2 a $\frac{6}{16}$, $\frac{9}{24}$, $\frac{12}{32}$ b $\frac{8}{18}$, $\frac{12}{27}$, $\frac{16}{36}$
c $\frac{10}{22}$, $\frac{15}{33}$, $\frac{20}{44}$ d $\frac{12}{14}$, $\frac{18}{21}$, $\frac{24}{28}$

3 a $\frac{2}{4}=\frac{3}{6}$ b $\frac{3}{9}=\frac{5}{15}$ c $\frac{9}{12}=\frac{21}{28}$
d $\frac{4}{10}=\frac{16}{40}$ e $\frac{5}{20}=\frac{25}{100}$ f $\frac{30}{100}=\frac{300}{1000}$

4 a Leo b Ricky

5 a $\frac{6}{7}$ b $\frac{3}{4}$ c $\frac{9}{10}$ d $\frac{3}{4}$
e $\frac{1}{2}$ f $\frac{1}{3}$ g $\frac{3}{5}$ h $\frac{1}{40}$

6 a $6\frac{4}{5}$ b $3\frac{2}{3}$ c $4\frac{3}{4}$ d $1\frac{1}{5}$

7 $\frac{3}{12}=\frac{1}{4}$

8 7.1 = $\frac{1}{8}$ 7.2 = $\frac{1}{4}$ 7.3 = $\frac{5}{32}$ 7.4 = $\frac{5}{32}$
7.5 = $\frac{7}{40}$ 7.6 = $\frac{11}{80}$

9 a $5\frac{1}{3}$ b $4\frac{1}{5}$ c $1\frac{5}{7}$ d $5\frac{1}{2}$
e $1\frac{1}{4}$ f $11\frac{3}{7}$ g $6\frac{1}{9}$ h $7\frac{1}{5}$

10 a $\frac{11}{4}$ b $\frac{79}{8}$ c $\frac{26}{7}$ d $\frac{35}{6}$
e $\frac{11}{3}$ f $\frac{99}{10}$ g $\frac{47}{12}$ h $\frac{17}{2}$

11 4 trays

12 a $\frac{23}{24}$ b $\frac{17}{45}$ c $1\frac{1}{15}$ d $6\frac{3}{4}$ e $4\frac{17}{24}$
f $1\frac{13}{14}$ g $3\frac{7}{9}$ h $2\frac{2}{3}$ i $6\frac{13}{40}$ j $3\frac{43}{72}$
k $1\frac{47}{70}$ l $\frac{76}{99}$

13 a $3\frac{2}{3}$ b $6\frac{5}{18}$

14 a $2\frac{4}{5}$ kilograms b 8 kilograms
c $\frac{3}{10}$ kilogram

15 a $\frac{20}{63}$ b $\frac{6}{25}$ c $\frac{5}{8}$ d $\frac{7}{36}$ e 14
f 64 g 44 h 22 i $\frac{1}{6}$ j $16\frac{2}{3}$

16 a $\frac{1}{7}$ b $\frac{1}{5}$ c $\frac{3}{20}$ d $\frac{3}{5}$
e $13\frac{2}{3}$ f $1\frac{1}{2}$

17 $150

18 a $\frac{2}{5}$ b $\frac{7}{2}$ c 9 d $\frac{1}{12}$ e $\frac{1}{124}$
f 4 g $\frac{4}{23}$ h $\frac{2}{17}$ i $\frac{5}{37}$ j $\frac{9}{22}$
k $\frac{8}{29}$ l $\frac{5}{21}$

19 **a** $1\frac{1}{2}$ **b** $1\frac{1}{3}$ **c** $\frac{7}{36}$ **d** $\frac{3}{70}$ **e** $\frac{5}{16}$
f $1\frac{3}{5}$ **g** $1\frac{53}{115}$ **h** $\frac{217}{267}$ **i** $1\frac{73}{147}$

20 **a** 9 laps **b** 10 min

21 **a** $\frac{23}{54}$ **b** $\frac{1}{4}$ **c** $\frac{11}{16}$ **d** $1\frac{1}{2}$
e 4 **f** $5\frac{1}{2}$

22 **a** $\frac{16}{25}$ **b** $\frac{81}{100}$ **c** $\frac{9}{16}$ **d** $\frac{49}{64}$ **e** $\frac{1}{27}$
f $\frac{125}{216}$ **g** $2\frac{1}{4}$ **h** $5\frac{1}{16}$ **i** $2\frac{14}{25}$

23 **a** $\frac{9}{10}$ **b** $\frac{5}{7}$ **c** $1\frac{2}{3}$ **d** $1\frac{1}{2}$

Chapter 4 Decimals

Are you ready?

1 **a** 0.3 **b** 0.05 **c** 0.007
2 **a** 1.35 **b** 0.484
3 **a** 5 **b** 18 **c** 57
4 **a** 6.7 **b** 1.37 **c** 14.02
5 **a** 3.96 **b** 5.576 **c** 26.77
6 **a** 5.3 **b** 0.32 **c** 5.17
7 **a** 5.55 **b** 0.45 **c** 7.496
8 **a** 5.0 **b** 11.04 **c** 63.55
9 **a** 14.56 **b** 35.1 **c** 16 039
10 **a** 2.3 **b** 4.16 **c** 3.147
11 **a** 1.435 **b** 0.271 **c** 0.038 664

Exercise 4A — Place value

1 **a** $\frac{2}{10}$ **b** $\frac{2}{100}$ **c** $\frac{2}{100}$ **d** $\frac{2}{10}$
e $\frac{2}{1000}$ **f** $\frac{2}{1000}$ **g** $\frac{2}{10\,000}$ **h** $\frac{2}{10\,000}$
i 20 **j** 2 **k** $\frac{2}{10\,000}$ **l** $\frac{2}{100\,000}$

2 **a** $\frac{9}{10}$ **b** $\frac{9}{10}$ **c** $\frac{9}{100}$ **d** $\frac{9}{100}$
e $\frac{9}{100}$ **f** $\frac{9}{1000}$ **g** $\frac{9}{10\,000}$ **h** $\frac{9}{100\,000}$
i 9 **j** 9 **k** 900 **l** 90

3 **a** Four units, 4; one tenth, $\frac{1}{10}$; $4 + \frac{1}{10}$
b One unit, 1; eight tenths, $\frac{8}{10}$; five hundredths, $\frac{5}{100}$; $1 + \frac{8}{10} + \frac{5}{100}$
c Two tenths, $\frac{2}{10}$; seven hundredths, $\frac{7}{100}$; one thousandth, $\frac{1}{1000}$; $\frac{2}{10} + \frac{7}{100} + \frac{1}{1000}$
d Nine units, 9; two hundredths, $\frac{2}{100}$; $9 + \frac{2}{100}$
e One ten, 10; six units, 6; one thousandth, $\frac{1}{1000}$; $16 + \frac{1}{1000}$
f Three units, 3; four tenths, $\frac{4}{10}$; two thousandths, $\frac{2}{1000}$; seven hundred thousandths, $\frac{7}{100\,000}$; $3 + \frac{4}{10} + \frac{2}{1000} + \frac{7}{100\,000}$

4 **a** $\frac{4}{10}$ **b** $2 + \frac{7}{10}$
c $6 + \frac{8}{10}$ **d** $5 + \frac{2}{10} + \frac{3}{100}$
e $\frac{7}{10} + \frac{6}{100} + \frac{3}{1000}$ **f** $2 + \frac{1}{10} + \frac{8}{1000}$
g $10 + 9 + \frac{9}{10} + \frac{1}{100}$ **h** $\frac{1}{10} + \frac{1}{100} + \frac{1}{10\,000}$
i $7 + \frac{2}{10} + \frac{9}{100} + \frac{6}{1000} + \frac{4}{10\,000}$
j $\frac{3}{10} + \frac{3}{100} + \frac{2}{10\,000} + \frac{4}{100\,000}$
k $300 + \frac{3}{100}$
l $10 + 2 + \frac{2}{10} + \frac{7}{100} + \frac{6}{1000} + \frac{4}{10\,000} + \frac{3}{1\,000\,000}$

5 **a** $(2 \times 1) + (4 \times \frac{1}{10}) + (7 \times \frac{1}{100})$
b $(3 \times 1) + (6 \times \frac{1}{10}) + (9 \times \frac{1}{100})$
c $(1 \times 1) + (2 \times \frac{1}{10}) + (5 \times \frac{1}{100})$
d $(5 \times 10) + (6 \times 1) + (1 \times \frac{1}{100})$
e $(3 \times 10) + (9 \times 1) + (1 \times \frac{1}{100})$
f $(1 \times 10) + (6 \times 1) + (7 \times \frac{1}{100})$
g $(7 \times 1) + (1 \times \frac{1}{10}) + (2 \times \frac{1}{100}) + (3 \times \frac{1}{1000})$
h $(5 \times 1) + (9 \times \frac{1}{10}) + (8 \times \frac{1}{100}) + (7 \times \frac{1}{1000})$
i $(1 \times 10) + (3 \times 1) + (4 \times \frac{1}{10}) + (8 \times \frac{1}{100}) + (2 \times \frac{1}{1000})$
j $(3 \times \frac{1}{10}) + (6 \times \frac{1}{1000}) + (2 \times \frac{1}{10\,000})$
k $(1 \times \frac{1}{10}) + (3 \times \frac{1}{100}) + (5 \times \frac{1}{10\,000})$
l $(5 \times \frac{1}{10}) + (2 \times \frac{1}{10\,000})$

6 **a** 5.8 **b** 3.7 **c** 0.34 **d** 0.25
e 10.1 **f** 103.668 **g** 28.12 **h** 0.348
i 65.872 **j** 1.829 021 **k** 0.999 99 **l** 2.0

7 **a** 2.15 **b** 7.28 **c** 512.39
d 0.147 **e** 41.14 **f** 100.042
g 13.29 143 **h** 0.375 12 **i** 0.94
j 11.44 **k** 31.73 **l** 0.12

8 **a** 9.415 **b** 11.307 **c** 84.2134
d 17.7771 **e** 0.6122 **f** 2.002
g 0.006 3678 **h** 49.272 **i** 3.212
j 0.202 **k** 0.231 **l** 4.170

9 **a** 2 **b** 1 **c** 3 **d** 2
e 3 **f** 4 **g** 4 **h** 1
i 8 **j** 0 **k** 0 **l** 0

10 **a** C **b** D

11 **a** **i** Tenths **ii** Yes
b **i** Units **ii** Yes
c **i** Thousandths **ii** No
d **i** Hundredths **ii** No
e **i** Tenths **ii** No
f **i** Tenths **ii** Yes
g **i** Units **ii** No
h **i** Hundredths **ii** Yes
i **i** Units **ii** Yes
j **i** Tens **ii** Yes

12 **a** F **b** T **c** T **d** F
e T **f** F **g** T **h** F

13

		Tens	Units		Tenths	Hundredths	Thousandths
a	0.205	0	0	.	2	0	5
b	1.06	0	1	.	0	6	0
c	74.108	7	4	.	1	0	8
d	0.108	0	0	.	1	0	8
e	50.080	5	0	.	0	8	0

Exercise 4B — Comparing decimals

1 **a** 0.32 **b** 0.91 **c** 0.59
d 0.99 **e** 0.901 **f** 0.666
g 0.1440 **h** 0.039 90 **i** 2.559
j 10.0233 **k** 0.110 49 **l** 0.102 36

2 **a** > **b** < **c** < **d** <
e < **f** < **g** < **h** >
i > **j** < **k** < **l** >

3 **a** 0.17, 0.21, 0.33, 0.39, 0.45
b 0.12, 0.16, 0.19, 0.22, 0.27
c 0.314, 0.391, 0.413, 0.420, 0.502
d 0.613, 0.624, 0.677, 0.690, 0.710
e 0.803, 0.807, 0.811, 0.821, 0.902
f 0.1002, 0.1033, 0.1164, 0.1196, 0.1245
g 0.9087, 0.9189, 0.9812, 0.9864, 0.9943
h 0.4004, 0.4076, 0.4100, 0.4139, 0.4826
i 4.0292, 4.5097, 4.6031, 4.6249, 4.802
j 13.0229, 13.0291, 13.0294, 13.0299, 13.0929
k 0.003, 0.004 65, 0.005 02, 0.0056, 0.009
l 0.507, 0.5079, 0.595, 0.61, 0.617

4 **a** 0.48, 0.39, 0.36, 0.31, 0.19
b 0.38, 0.35, 0.27, 0.16, 0.02
c 0.99, 0.97, 0.95, 0.91, 0.90
d 0.29, 0.13, 0.09, 0.07, 0.02
e 1.291, 1.288, 1.279, 1.273, 1.264
f 0.491, 0.442, 0.437, 0.433, 0.406
g 0.390, 0.372, 0.318, 0.317, 0.309
h 0.602, 0.591, 0.573, 0.556, 0.502
i 0.8889, 0.8823, 0.8448, 0.8217, 0.8207
j 0.7695, 0.7657, 0.6024, 0.1079, 0.0307
k 1.843 21, 1.702 96, 1.486 59, 1.486 13, 1.349 54
l 12.9092, 12.392, 12.289 50, 12.208 64, 12.002 36

5 E **6** D **7** E

8 **a** Shelley, 0.21 seconds **b** Carolyn, 0.06 seconds
c Mara, 0.11 seconds **d** Shelley, Carolyn, Robyn
e Carolyn, Shelley, Kyah **f** Mara, Jenika, Shelley
g Yes, Shelley

Exercise 4C — Converting decimals to fractions

1 **a** $\frac{3}{10}$ **b** $\frac{1}{2}$ **c** $\frac{9}{10}$ **d** $\frac{21}{10}$ **e** $\frac{2}{5}$
f $\frac{4}{5}$ **g** $\frac{6}{25}$ **h** $\frac{11}{25}$ **i** $\frac{49}{100}$ **j** $\frac{63}{100}$
k $\frac{251}{500}$ **l** $\frac{617}{1000}$ **m** $\frac{3}{25}$ **n** $\frac{3}{10}$ **o** $\frac{16}{25}$
p $\frac{7}{25}$ **q** $\frac{441}{500}$ **r** $\frac{591}{625}$ **s** $\frac{9209}{10\,000}$ **t** $\frac{4621}{10\,000}$
u $\frac{3}{4}$ **v** $\frac{3}{25}$ **w** $\frac{143}{500}$ **x** $\frac{117}{250}$

2 **a** $1\frac{3}{10}$ **b** $1\frac{3}{5}$ **c** $2\frac{7}{10}$ **d** $9\frac{2}{5}$
e $1\frac{1}{5}$ **f** $2\frac{4}{5}$ **g** $4\frac{1}{5}$ **h** $8\frac{1}{2}$
i $2\frac{13}{100}$ **j** $6\frac{12}{25}$ **k** $5\frac{27}{100}$ **l** $19\frac{91}{500}$
m $12\frac{21}{50}$ **n** $3\frac{3}{20}$ **o** $6\frac{1}{4}$ **p** $9\frac{7}{50}$
q $12\frac{843}{1000}$ **r** $16\frac{341}{500}$ **s** $2\frac{4917}{10\,000}$ **t** $4\frac{1693}{10\,000}$
u $37\frac{41}{200}$ **v** $18\frac{129}{200}$ **w** $24\frac{69}{200}$ **x** $100\frac{3}{625}$

3 **a** B **b** A **c** C **d** D **e** E

Exercise 4D — Rounding

1 **a** 0.32 **b** 0.86 **c** 1.25 **d** 13.05
e 7.13 **f** 100.81 **g** 71.26 **h** 0.01
i 0.19 **j** 19.70 **k** 0.40 **l** 1.00

2 **a** 0.4 **b** 0.9 **c** 9.3 **d** 25.3
e 300.1 **f** 12.8 **g** 99.9 **h** 8.9
i 17.6 **j** 0.9 **k** 94.0 **l** 1.0

3 **a** 2.39 **b** 14.0 **c** 0.03 **d** 0.8765
e 64.2952 **f** 0.382 **g** 96.3 **h** 3.04
i 8.90 **j** 47.880 **k** 0.10 **l** 0.49

4 **a** C **b** D **c** D **d** A

5 **a** 11 **b** 8 **c** 4 **d** 93
e 112 **f** 22 **g** 42 **h** 2138
i 0 **j** 1 **k** 1 **l** 41

6 **a** 10 **b** 80 **c** 50 **d** 140
e 260 **f** 180 **g** 310 **h** 1480
i 10 020 **j** 209 720 **k** 20 **l** 10

7 **a** 300 **b** 200 **c** 900 **d** 100
e 100 **f** 6800 **g** 900 **h** 9800
i 1700 **j** 1 462 800 **k** 100 **l** 0

8 **a** 3000 **b** 6000 **c** 12 000 **d** 19 000
e 28 000 **f** 9000 **g** 1000 **h** 497 000
i 13 000 **j** 1000 **k** 130 000 **l** 50 000

9 $27.70 **10** 25°C

11 3.568 metres. We could not measure a length of 3.567 82 m accurately with the measuring tools we have available.

10 Quick Questions 1

1 $\frac{5}{100}$ or 5 hundredths

2 One and nine tenths and six hundredths and three thousandths and two ten thousandths.

3 4.328 **4** 4.098 < 4.908

5 0.0709, 0.8, 1.61, 4.15, 6.08, 7.03, 24.5, 324.8

6 $\frac{67}{100}$ **7** $\frac{2}{5}$ **8** $7\frac{41}{500}$

9 13.63 **10** 13.6

Exercise 4E — Adding and subtracting decimals

1 **a** 3.5 **b** 2.69 **c** 13.239 **d** 102.1858

2 **a** 6.77 **b** 21.906 **c** 2012.333 **d** 261.449 06

3 **a** 4.6 **b** 6.9 **c** 1.59 **d** 3.471
e 17.280 **f** 8.9685 **g** 17.3271 **h** 20.4329
i 1056.7863

4 a 11.69 b 105.22 c 42.031 d 18.951
e 105.958 f 192.8949 g 20.9916 h 2618.9564
5 a B b A c A d D
6 $167.00 7 $4.70
8 20.05 kilometres 9 8.1 kilometres 10 $6.60
11 a 0.60 b 4.205 c 0.0806 d 1.3368
e 5.02 f 89.19 g 87.09 h 2.733
12 a 3.34 b 5.93 c 3.849 d 0.327
e 4.146 f 27.83 g 1.934 h 3.289
i 1.2509 j 14.5326 k 16.766 l 38.238
m 4.1136 n 0.1951 o 0.772 16
13 a C b B c B d B
14 $10.35 15 50.48 kilograms
16 11.73 metres 17 48.38 seconds
18 31.28 metres
19 a

Team	1	2	3	4	5	6
Time (seconds)	153	159.3	148.9	166.6	155.9	155.8

b Team 3 c 4.1 seconds

Exercise 4F — Multiplying decimals by a whole number

1 a 14.0 b 125.6 c 146.7
d 61.2 e 111.70 f 428.67
g 190.26 h 518.96 i 26.068
j 6.384 k 0.794 43 l 113.541 464
2 a 312.9 b 327.0 c 10.54
d 54.74 e 152.83 f 232.74
g 268.018 h 321.522 i 30.2586
j 3366.636 k 1139.1432 l 1577.736
m 7287.336 n 69 223.98 o 924 682.0608
3 a 142.3 b 86.52 c 31.49 d 9.9
e 274.815 f 0.023
4 One, right
5 a 827.8 b 496.2 c 27.84 d 1302.6
e 529.43 f 29 684.5
6 Two, right
7 a 249.8 b 602.49 c 1486.1
d 83 020.91
8 Three, right
9 a 4 b 5 c 6
10 a 132.7 b 2.64 c 80.49 d 4.86
e 74.57 f 1277.29 g 823 h 1970
i 30
11 a 64.8 b 1389.6
c 58 906.43 d 270.8
e 217 148.96 f 842 619
g 82 049.6783 h 3 268 904.3267
i 984 326.641 j 59 027 683.017
k 278.498 32 l 460
m 5 290 000 n 39 486 000
12 a 3500c b 12 700c c 1100c d 2535c
e 5820c f 11 015c
13 a 1092 b 4548 c 13 450 d 38 340
e 1016 f 5973 g 147 200 h 437 400
14 $750 15 108 litres

Maths Quest challenge (page 160)

1 5 cents 2 1.25 3 3.7, 6.8, 12.5

Exercise 4G — Multiplying, squaring and obtaining square roots of decimals

1 a 0.84 b 2.94 c 0.32
d 7.76 e 0.140 f 0.192
g 0.385 h 0.441 i 0.326
j 4.304 k 0.2082 l 0.3486
m 2.8992 n 4.905 64 o 10.016 395
2 a 0.0001 b 0.000 012 c 0.063
d 0.000 222 e 0.000 002 44 f 0.000 36
g 0.392 h 0.000 6062 i 0.000 0042
3 a 0.3 b 0.851 c 2.538
d 6.557 e 16.848 f 60.300
g 55.872 h 28.652 i 12.3095
j 18.9644 k 79.7720 l 35.6896
m 4.2036 n 1.699 24 o 14.892
4 a E b B c A d B
5 $5.50
6 a 3205 cents b $32.05
7 0.375 litres 8 260.7569 square metres
9 a 0.004 b 1.69 c 4.2025
d 0.3 e 0.4 f 0.07

Maths Quest challenge (page 164)

1 191.75 cm 2 0.6 3 13

Exercise 4H — Dividing decimals by whole numbers

1 a 0.6 b 3.1 c 2.9 d 0.54
e 1.61 f 4.71 g 2.42 h 0.788
i 1.983 j 1.147 k 4.0701 l 8.3729
m 6.8594 n 31.6176 o 3.13 p 3.12
q 2.231 r 1.222 s 6.421 t 11.0311
2 a 1.85 b 4.75 c 1.46 d 2.45
e 1.25 f 6.95
3 a 6.72 b 3.59 c 8.14 d 10.25
e 9.99 f 0.0546 g 0.0989 h 10.204
i 3.5492
4 One, left
5 a 1.569 b 0.249 c 0.106
d 6.077 e 23.401 f 0.348
g 0.2480 h 0.00312 i 0.003 581
6 Two, left
7 a $3.65 b $1.70 c $56.85
d $0.75 e $0.90 f $63.50
8 a 0.013 98 b 0.029 876 c 0.504 87
d 0.767 9062 e 0.140 09 f 6.001 65
9 a 3 b 4 c 5 d 6
10 a 1.407 b 9.6813
c 9.8506 d 0.6208
e 3.592 87 f 0.023 4978
g 0.053 320 667 h 0.000 090 769
i 0.103 454 97 j 0.802 4056
k 0.000 1527 l 0.000 583
m 0.000 072 05 n 0.000 000 0032
o 0.000 000 048 7
11 a 0.265 b 0.79 c 0.235
d 0.595 e 0.001 465 f 0.001 74
12 a C b B c A d B
13 $0.60 14 $4.50 15 0.675 metre

10 Quick Questions 2

1 6.01 2 21.28 3 0.94
4 4062.39 5 133.815 6 1.351 968
7 1.5625 8 1.3568 9 0.06
10 36.65 kilometres

Exercise 4I — Converting fractions to decimals and recurring decimals

1 a 0.75 b 0.5 c 0.8 d 0.05
e 0.2 f 0.25 g 0.375 h 0.02
i 0.32
2 a $2.\dot{5}$ b $0.\dot{6}$ c $12.\dot{8}$ d $49.\dot{1}$
e $0.\dot{2}\dot{6}$ f $0.\dot{4}\dot{1}$ g $0.\dot{9}1\dot{3}$ h $8.\dot{6}41\dot{8}$
i $0.040\overline{12}$ j $133.9\overline{462}$ k $1.8\dot{3}$ l $0.12\dot{7}$
3 a 0.17 b 0.3 c 0.1 d 0.13
e 0.18 f 0.4 g 0.417
h 0.142 857 i 0.47
4 a B b A c D d A
5 a 0.5 b $0.\dot{3}$ c 0.25 d 0.2 e $0.\dot{1}$
6 a 3.5 b $4.\dot{6}$ c 0.75 d 3.8 e $0.\dot{7}$

Exercise 4J — Dividing decimals

1 a 5 b 7 c 8 d 9
e 4 f 11.1 g 0.31 h 5.1
i 2.2 j 66 k 72.1 l 12.31
m 0.852 n 0.336 77 o 1.367
2 a 1564.08 b 4.254 c 20.01
d 12.13 e 143.134 f 1786.1
g 0.4782 h 15 771.987 i 1630.95
3 a D b D c B d D
4 20 bottles 5 71.5 or 71 people
6 26 7 201

Summary

1 decimal 2 place
3 less 4 greater
5 first 6 right, zero
7 dot, line 8 numerator, denominator
9 answer, question 10 lined up
11 divide 12 ten, left

Chapter review

1 a 7 tenths b 7 hundredths
c 7 tenths d 7 thousandths
e 7 ten thousandths
f 7 hundred thousandths
g 7 units h 7 thousandths
2 a $(2 \times 1) + (6 \times \frac{1}{10}) + (4 \times \frac{1}{100})$
b $(3 \times \frac{1}{10}) + (6 \times \frac{1}{100}) + (9 \times \frac{1}{1000})$
c $(1 \times 10) + (8 \times 1) + (4 \times \frac{1}{10}) + (6 \times \frac{1}{1000})$
d $(9 \times 10) + (6 \times 1) + (3 \times \frac{1}{10}) +$
$(4 \times \frac{1}{100}) + (2 \times \frac{1}{1000}) + (8 \times \frac{1}{10\,000})$
3 a 6.4 b 0.943 c 12.26 d 4.11
4 a 0.459 b 12.806 c 1.603 d 2.793
5 a > b < c < d <
e > f < g < h >
6 a 0.13, 0.22, 0.34, 0.71, 0.86
b 0.124, 0.247, 0.258, 0.274, 0.285
c 0.826, 0.834, 0.859, 0.888, 0.891
d 0.217, 0.356, 0.358, 0.365, 0.385
7 a $\frac{4}{5}$ b $\frac{17}{100}$ c $\frac{9}{25}$ d $\frac{187}{1000}$
e $\frac{1}{8}$ f $\frac{71}{125}$ g $\frac{41}{200}$ h $\frac{19}{20}$
8 a $1\frac{1}{2}$ b $4\frac{3}{5}$ c $3\frac{12}{25}$ d $5\frac{1}{4}$
e $2\frac{3}{4}$ f $2\frac{5}{8}$ g $1\frac{14}{25}$ h $8\frac{8}{25}$
9 a 1.3 b 2.05 c 13.86 d 0.048
e 1.9260 f 17.90
10 a 14 b 29 c 87 d 100
11 a 9.1 b 10.09 c 6.993 d 15.647
e 215.72 f 41.737 45
12 $53.30
13 a 5.3 b 9.07 c 2.04 d 90.659
e 0.949 f 1.575
14 1.125 litres
15 a 18.6 b 42.03 c 66.0180 d 5.3494
16 a 2.53 b 667.48 c 5122.32
d 0.926 08
17 a 2.6 b 13.45 c 6.45 d 182.94
e 14 662.81 f 48.0643 g 839.204 h 368
18 a 1.312 b 0.516 c 0.783
d 0.000 18 e 0.87 f 2.7168
g 14.9084 h 1.023 36
19 $3.68 or $3.70
20 a 6.25 b 0.0009 c 0.8 d 0.05
21 a 0.3 b 0.41 c 2.08 d 73.62
e 0.16 f 33.11
22 a 1.4623 b 10.236
c 9.612 347 d 0.200 32
e 264.983 0025 f 34.6294
23 a 0.05 b 0.625 c 0.3125 d 0.125
24 a $4.\dot{5}$ b $0.\overline{82}$ c $19.\overline{2781}$
d $83.01\dot{6}\dot{2}$
25 a $0.\overline{076\,923}$ b $0.\overline{384\,615}$
26 a 8 b 51 c 11
d 34.3 e 894.73 f 6200.762
g 2.413 h 21 426.78

Chapter 5 Patterns and algebra

Are you ready?

1 a Add 2 to obtain the next number in the sequence. The next three numbers are 15, 17, 19.
b Subtract 4 to obtain the next number in the sequence. The next three numbers are 12, 8, 4.
c Multiply by 2 to obtain the next number in the sequence. The next three numbers are 48, 96, 192.
d Divide by 10 to obtain the next number in the sequence. The next three numbers are 100, 10, 1.
2 a 16 b 23 c 2 d 2
3 a Each first number is one more than the previous first number and each second number is one more than the previous second number. Each second number is always 6 more than the matching first number.
b Each first number is two more than the previous first number and each second number is six more than the previous second number. Each second number is three times the matching first number.
4 a 3 + 5 b 7 × 8 c 6 − 2 d 5 + 9

5 a

Number of squares	1	2	3	4	5	6
Number of sides	4	8	12	16	20	24

b 40

Exercise 5A — Using rules

1 a

Input	4	5	6	10	14
Output	0	1	2	6	10

b

Input	1	2	6	13	107
Output	13	14	18	25	119

c

Input	1	10	51	60	144
Output	2	20	102	120	288

d

Input	3	12	21	66	141
Output	1	4	7	22	47

2 a

Input	2	3	4	10	17
Output	9	11	13	25	39

b

Input	2	4	5	20	100
Output	12	28	36	156	796

c

Input	2	3	5	12	43
Output	25	30	40	75	230

d

Input	3	4	5	8	25
Output	0	11	22	55	242

e

Input	0	1	5	8	10
Output	0	1	25	64	100

f

Input	0	1	4	7	12
Output	4	5	20	53	148

3 a

Input	1	3	27	2	53	191
Output	3	5	29	4	55	193

b

Input	3	25	56	0	74	1753
Output	23	45	76	20	94	1773

c

Input	7	15	96	9	17	109
Output	2	10	91	4	12	104

d

Input	1	6	321	2	7	103
Output	4	24	1284	8	28	412

e

Input	3	13	21	1	10	50
Output	8	28	44	4	22	102

f

Input	2	5	16	3	5	20
Output	3	18	73	8	18	93

g

Input	3	15	273	6	162	63
Output	1	5	91	2	54	21

h

Input	1	5	17	2	7	12
Output	1	25	289	4	49	144

i

Input	2	7	15	3	9	11
Output	34	134	294	54	174	214

j

Input	7	19	31	23	51	63
Output	1	4	7	5	12	15

Maths Quest challenge (page 186)

1

8	1	6
3	5	7
4	9	2

2 7, 3, 4, 6

History of mathematics: The development of algebra

1 The word algebra comes from the title of a book called *Al-jabr wa'l muqabala* written by the mathematician Al-Khwarizmi.
2 Omar Khayyam
3 Istanbul
4 Iraq

Exercise 5B — Writing and finding formulas

1 a $b = 4h$ b $m = 4f$
c $r = 5a$ d $m = 4t$
e $x = 5(k + 4)$ f $k = 6w - 2$
g $t = 4(20 - g)$ h $b = 10a - 5$
i $d = 6f + 7$ j $h = 9(x + 5)$
k $y = 8(3b + 6)$ l $y = 8p - 6$
m $g = 2(3r + 17)$ n $j = 18(5h - 4)$

2 a $g = f + 9$ b $b = a - 3$
c $t = 2k + 1$ d $y = 7x - 5$
e $g = 4a + 2$ f $t = 9m + 11$
g $w = 3 \times p + 2$ ($w = 3p + 2$)
h $x = 5 \times t - 8$ ($x = 5t - 8$)
i $p = 3m - 1$
j $b = s \times 10 - 7$ ($b = 10s - 7$)

3 a $b = a + 3$ b $w = t + 5$
c $a = m - 4$ d $t = 2s + 3$
e $g = 2f - 1$ f $c = 3s + 1$
g $a = 4d - 3$ h $g = 5s - 4$
i $e = 11f - 20$ j $q = 20p + 4$
k $t = 4b$ l $p = 3u - 1$

4 C 5 C 6 D 7 E 8 C
9 D 10 A 11 D 12 C

10 Quick Questions 1

1

Input	2	5	14	20	23	36
Output	6	9	18	24	27	40

2

Input	6	8	11	16	24	32
Output	12	18	27	42	66	90

3

Input	15	25	30	40	65	70
Output	3	5	6	8	13	14

4 Add 6 to each input number.
5 Multiply each input number by 4.

Input	1	4	7	12	13	18	24	31
Output	4	16	28	48	52	72	96	124

6 Add 1 to the input number then multiply the result by 5 or multiply the input number by 5 and then add 5.
7 $k = 8w + 12$ 8 $g = 7m - 4(t + 11)$
9 $b = a - 2$ 10 $m = 4t + 4$

Exercise 5C — Substitution

1 a

x	2	3	11	18	40
y	7	8	16	23	45

b

n	17	20	22	43	700
m	0	3	5	26	683

c

x	1	3	40	70	123
y	2	4	41	71	124

d

h	1	5	11	22	62
p	6	14	26	48	128

e

b	5	15	20	60	100
t	8	38	53	173	293

f

t	1	3	7	8	40
s	8	16	32	36	164

g

x	1	3	40	70	123
y	0	2	39	69	122

h

g	1	7	8	19	154
s	2	62	72	182	1532

i

m	3	4	7	11	12
t	9	8	5	1	0

j

f	1	8	11	12	111
p	6	48	66	72	666

k

y	0	1	3	4	6
m	12	10	6	4	0

l

r	4	11	12	16	804
k	18	102	114	162	9618

2 a i $m = 2$ ii $m = 3$ iii $m = 0$ iv $m = 100$
b i $m = 11$ ii $m = 1$ iii $m = 197$ iv $m = 13$
c i $m = 19$ ii $m = 7$ iii $m = 67$ iv $m = 247$
d i $m = 21$ ii $m = 13$ iii $m = 1$ iv $m = 25$
3 a i $m = 4$ ii $m = 2$ iii $m = 26$ iv $m = 152$
b i $m = 20$ ii $m = 40$ iii $m = 45$ iv $m = 0$
c i $m = 40$ ii $m = 36$ iii $m = 0$ iv $m = 4$
d i $m = 5$ ii $m = 125$ iii $m = 35$ iv $m = 75$
e i $m = 7$ ii $m = 1$ iii $m = 15$ iv $m = 63$
f i $m = 17$ ii $m = 23$ iii $m = 32$ iv $m = 77$
g i $m = 9$ ii $m = 3$ iii $m = 303$ iv $m = 216$
h i $m = 16$ ii $m = 28$ iii $m = 13$ iv $m = 49$
i i $m = 26$ ii $m = 8$ iii $m = 44$ iv $m = 2$

10 Quick Questions 2

1 $c = 6k + 3t$ 2 $r = 17 \times y + 8 \times h - 2 \times m$
3 $w = 2y - 1$ 4 $c = 3f$

5

n	1	5	9	16	18	23
h	6	10	14	21	23	28

6

r	2	8	15	16	32	41
c	20	44	72	76	140	176

7 $g = 30$ 8 $m = 21$
9 $d = 75$ 10 21

Maths Quest challenge (page 199)

1 5 years 2 16 boys
3 9 trips: (1) 2 boys cross to opposite side, (2) 1 boy rows back, (3) 1 man crosses to opposite side, (4) 1 boy rows back, (5) 2 boys cross to opposite side, (6) 1 boy rows back, (7) remaining man crosses to opposite side, (8) 1 boy rows back, (9) both boys cross to opposite side.

Exercise 5D — Terms, expressions and equations

1 a $c + 3$ b $2c$ c $3c + 2$
d $2c + 5$ e $4c + 3$
2 a $4c + 34$ b 242 c \$48.40
3 a $2m$ b $2m + n$ c $4m + 3n$
d $m + 2n + 5$ e $3m + 2n + 8$
4 a 34 b 46 c 104 d 46 e 83
5 a $B + 2$ b $T - 3$ c $D + 6$
d $K - 5$ e $G + N + W$ f $D + H$
g $2N$ h $C + H$ i $12 - G$
j $4D$ k $H + 6$ l $Z - G$
m BF n $3M + Y$
6 a True b True c False d True
e True f True g True h False
i False j True k True l False
7 a B b C c D d D
e B f A g A
8 a $A + B + C$ b $A - C$ c AB
d ABC e $\frac{B}{C}$ f $\frac{A + C}{B}$
g $A + 3$
9 a $T - X$ b $\$4B$ c $45G$
d $\frac{\$T}{5}$ e $R + M$ f $\frac{Y}{6}$
g $\$11M$ h $A + H$ i $52 - J$
j $12B$ k $\frac{B + G}{2}$ l $A + H + K$
m $4R$
10 a $x + 8 = 1$ b $x - 11 = 63$ c $8x = 128$
d $\frac{x}{7} = 5$ e $A + H = 6.49$ f $52 - Q = 19$
g $99M = 792$ h $4R + 4D = 164$

Exercise 5E — Like terms

1 a True b True c False d False
e True f False g True h False
i True j True k True l False
m True n True o True p True
q False r False
2 a E b B c C d D
e C f A g D
3 a True b False c False d True
e False f True g False h True
i True j True
4 a $5a$ b $14y$ c $15c$ d $2u$
e $20e$ f $5t$ g $14ab$ h $6fg$
i $3e$ j $7t$ k $5f$ l 0
m $18x$ n $12m$ o $8a$
5 a $25ab - 7$ b $3y$
c Cannot be simplified d $30i - 2$
e $14t - 3$ f $13r$

g Cannot be simplified h Cannot be simplified
i $16f + 5$ j $4 + 2g$
k t l 0
m $2mno$ n $14pq$
o $11pr + 2$ p Cannot be simplified
q $11t - 5$ r t
s $8aw$ t Cannot be simplified
u $11t - 8$ v $4t + 3$
w $10g - 2$ x $22r + 6$

6 a 0 b 0 c 0 d 8 e 7
f 2 g 0 h 7 i $6y$

7 a $5a$, 35 b $9a$, 63 c $4a$, 28
d $10a$, 70 e $10a$, 70 f $10a$, 70
g $17 + 8a$, 73 h $5a + 2$, 37 i $2a$, 14
j a, 7 k 0, 0 l $17a - 16$, 103

Maths Quest challenge (page 210)

1 A = 2, B = 9, C = 1

2 D = 5, E = 2 or D = 6, E = 3

3

734	765	836	867	928	938
+ 734	+ 765	+ 836	+ 867	+ 928	+ 938
1468	1530	1672	1734	1856	1876

4 9 567 + 1 085 = 10 652

5 940 + 739 + 8 925 = 10 604

Exercise 5F — Problem solving using algebra

1 a

T	1	2	3	4	5
D	3	4	5	6	7

b 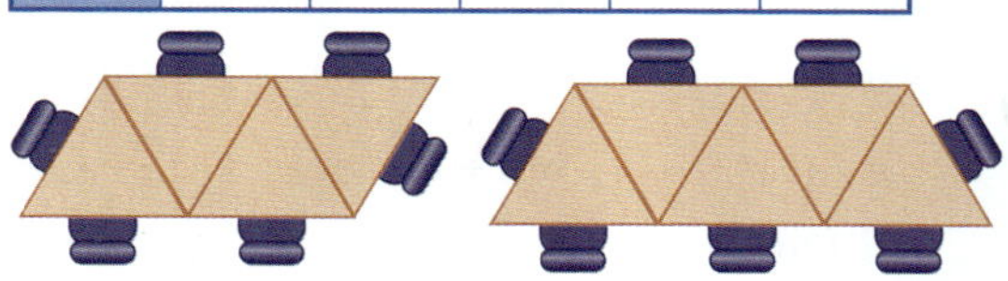

c $D = T + 2$ d $D = 26$

2 a

B	1	2	3	4	5
W	4	6	8	10	12

b 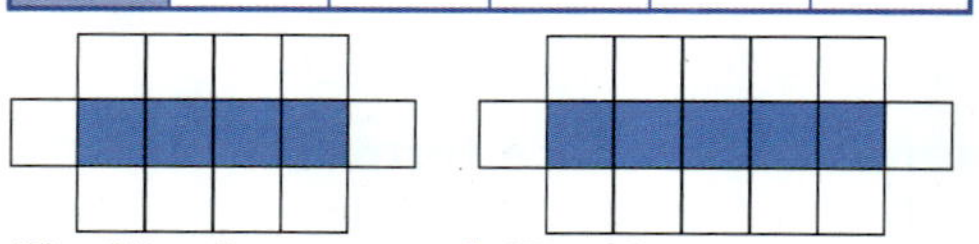

c $W = 2B + 2$ d $W = 22$

3 a

S	1	2	3	4	5
P	4	7	10	13	16

b 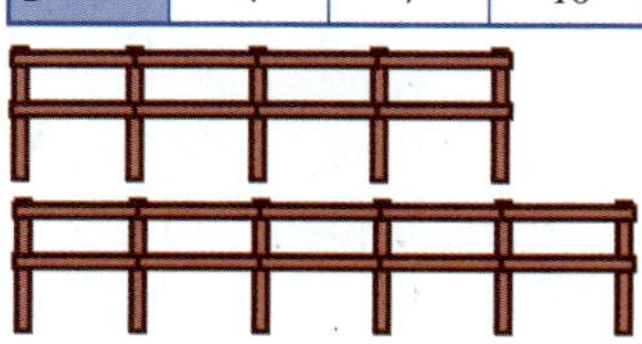

c Formula is $P = 3S + 1$
d $P = 661$ pieces of timber
e $2544.85

4 a, b

A	1	2	3	4	5
L	4	7	10	13	16

c 

d Formula is $L = 3A + 1$
e $L = 226$ light panels

5 a

B	1	2	3	4	5
S	6	11	16	21	26

b 

c Formula is $S = 5B + 1$
d $S = 61$ sleepers

Summary

1 output 2 pronumeral
3 formula, equation 4 multiplying
5 front 6 substitution
7 terms 8 equals
9 pronumerals 10 simplifying
11 expression, like terms 12 evaluate

Chapter review

1 a

Input	1	3	4	7	13
Output	4	12	16	28	52

b

Input	1	4	6	7	15
Output	8	11	13	14	22

c

Input	3	4	7	10	13
Output	3	5	11	17	23

d

Input	0	4	8	16	20
Output	0	1	2	4	5

2 a

Input	15	16	33	10	14
Output	8	9	26	3	7

b

Input	1	4	15	5	33
Output	10	19	8	22	106

c

Input	1	3	6	4	43
Output	12	16	22	18	96

d

Input	8	11	15	13	38
Output	11	17	25	21	7

3 a C b B

4 a $y = 9x - 4$ b $q = 3p + 6$

5 a $y = x + 6$ b $d = c - 5$
c $h = 5g + 2$ d $n = 4m - 1$

6 a

x	1	3	4	7	105
y	0	2	3	6	104

b

x	1	2	6	7	12
d	3	6	18	21	36

c

x	0	2	4	5	6
h	6	4	2	1	0

d

x	0	1	3	4	7
n	3	14	36	47	80

7 a $35x + 18$ b 1068

8 a $M - C$ b \$3B c XY d $G + 15$
e $D + 1$ f $12H$ cents g $5T$

9 a $7g$ b $6y$ c $9h$
d $5ag$ e $3gy$ f 0
g $5y$ h $7t + 6$ i $18gh$
j $11t - 2m$ k $4m$ l $15g + 12$
m $4h + 4t$ n $10b + 7c$ o $8axy$

10 a $10x$, 50 b $5x - 4$, 21 c $23x$, 115
d $3x$, 15 e $3x$, 15 f $x + 16$, 21
g $8x + 7$, 47 h $11 + 7x$, 46 i $6x$, 30

11 a

T	1	2	3	4
S	3	5	7	9

b

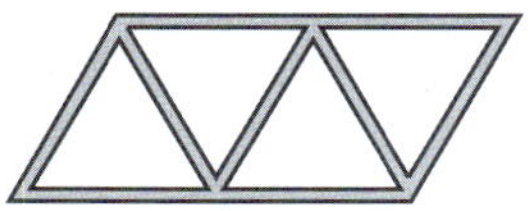

c $S = 2T + 1$ d $S = 51$

Chapter 6 Equations

Are you ready?

1 a 3 b 11 c 15 d 27

2 a $\square + 12 = 30$ b $\square - 9 = 6$
c $\square \times 4 = 48$ d $\square \div 8 = 7$

3 a 22 b 9 c 1
d 7 e 5 f 12

4 a -2 b $+5$ c $\div 7$ d $\times 3$

5 a $5x + 6y$ b $7x + 9y$ c $10x + 4$ d $4x + 6$

6 a True b False c False
d True e False f True

Exercise 6A — Using inverse operations

1 a $x + 7 = 11$ b $x + 3 = 5$ c $x + 12 = 12$
d $x + 5 = 56$ e $x - 7 = 1$ f $x - 11 = 11$
g $x - 4 = 7$ h $x - 8 = 0$ i $2x = 12$
j $6x = 30$ k $5x = 30$ l $6x = 12$
m $\frac{x}{7} = 1$ n $\frac{x}{3} = 100$ o $\frac{x}{5} = 2$
p $\frac{x}{7} = 0$ q $15 - x = 2$ r $52 - x = 8$
s $\frac{21}{x} = 7$ t $x^2 = 100$

2 a $x = 11$ b $y = 9$ c $m = 5$
d $m = 30$ e $w = 1$ f $m = 274$
g $w = 7$ h $k = 0$ i $b = 7$
j $w = 18$ k $k = 4$ l $h = 25$
m $b = 8.8$ n $c = 4.2$ o $x = 2.8$

3 a i 5 $\xrightarrow{\times 3}$ 15 $\xrightarrow{+1}$ 16
ii 5 $\xrightarrow{+1}$ 6 $\xrightarrow{\times 3}$ 18

b i 15 $\xrightarrow{\div 5}$ 3 $\xrightarrow{+10}$ 13
ii 15 $\xrightarrow{+10}$ 25 $\xrightarrow{\div 5}$ 5

c i 7 $\xrightarrow{-5}$ 2 $\xrightarrow{\times 2}$ 4
ii 7 $\xrightarrow{\times 2}$ 14 $\xrightarrow{-5}$ 9

d i 4 $\xrightarrow{\div 2}$ 2 $\xrightarrow{+6}$ 8
ii 4 $\xrightarrow{+6}$ 10 $\xrightarrow{\div 2}$ 5

e i 30 $\xrightarrow{\div 3}$ 10 $\xrightarrow{-9}$ 1
ii 30 $\xrightarrow{-9}$ 21 $\xrightarrow{\div 3}$ 7

f i 0 $\xrightarrow{+6}$ 6 $\xrightarrow{\div 3}$ 2
ii 0 $\xrightarrow{\div 3}$ 0 $\xrightarrow{+6}$ 6

g i 7 $\xrightarrow{+3}$ 10 $\xrightarrow{\times 5}$ 50
ii 7 $\xrightarrow{\times 5}$ 35 $\xrightarrow{+3}$ 38

h i 15 $\xrightarrow{-5}$ 10 $\xrightarrow{\times 10}$ 100
ii 15 $\xrightarrow{\times 10}$ 150 $\xrightarrow{-5}$ 145

4 a Each pair of flow charts has the same operations, but in the reverse order.
b The results are quite different.

5 a Adding and subtracting are inverse operations.
b Multiplying and dividing are inverse operations.

6 **a** 1 **b** 7 **c** 15 **d** 14
e 5 **f** 10 **g** 4 **h** 9
i 5 **j** 0 **k** 14.4522 **l** 4
m 3 **n** 17 **o** 2.59 **p** 532

Exercise 6B — Building up expressions

1 **a** $x \xrightarrow{\times 5} 5x \xrightarrow{-2} 5x - 2$

b $x \xrightarrow{\div 3} \frac{x}{3} \xrightarrow{+1} \frac{x}{3} + 1$

c $x \xrightarrow{\times 2} 2x \xrightarrow{+7} 2x + 7$

d $x \xrightarrow{\div 5} \frac{x}{5} \xrightarrow{+11} \frac{x}{5} + 11$

e $x \xrightarrow{\times \frac{1}{2}} \frac{x}{2} \xrightarrow{-3} \frac{x}{2} - 3$

f $x \xrightarrow{\div 8} \frac{x}{8} \xrightarrow{-2} \frac{x}{8} - 2$

g $x \xrightarrow{\div 7} \frac{x}{7} \xrightarrow{+9} \frac{x}{7} + 9$

h $x \xrightarrow{\times 3.1} 3.1x \xrightarrow{+1.8} 3.1x + 1.8$

2 **a** $x \xrightarrow{+5} x + 5 \xrightarrow{\times 4} 4(x + 5)$

b $x \xrightarrow{+10} x + 10 \xrightarrow{\div 3} \frac{x + 10}{3}$

c $x \xrightarrow{-2} x - 2 \xrightarrow{\div 7} \frac{x - 2}{7}$

d $x \xrightarrow{+3} x + 3 \xrightarrow{\times 9} 9(x + 3)$

e $x \xrightarrow{-2.1} x - 2.1 \xrightarrow{\times 3} 3(x - 2.1)$

f $x \xrightarrow{+4.9} x + 4.9 \xrightarrow{\div 5} \frac{x + 4.9}{5}$

g $x \xrightarrow{-\frac{1}{2}} x - \frac{1}{2} \xrightarrow{\times 7} 7(x - \frac{1}{2})$

h $x \xrightarrow{-3.1} x - 3.1 \xrightarrow{\div 1.8} \frac{x - 3.1}{1.8}$

3 **a** $x \xrightarrow{+2} x + 2 \xrightarrow{\times 6} 6(x + 2)$

b $x \xrightarrow{\times 3} 3x \xrightarrow{-2} 3x - 2$

c $x \xrightarrow{-8} x - 8 \xrightarrow{\div 12} \frac{x - 8}{12}$

d $x \xrightarrow{\times 4} 4x \xrightarrow{+1} 4x + 1$

e $x \xrightarrow{\div 7} \frac{x}{7} \xrightarrow{+3} \frac{x}{7} + 3$

f $x \xrightarrow{-3} x - 3 \xrightarrow{\div 5} \frac{x - 3}{5}$

g $x \xrightarrow{+5} x + 5 \xrightarrow{\times 6} 6(x + 5) \xrightarrow{-2} 6(x + 5) - 2$

h $x \xrightarrow{\times 3} 3x \xrightarrow{+4} 3x + 4 \xrightarrow{\div 7} \frac{3x + 4}{7}$

i $x \xrightarrow{-8} x - 8 \xrightarrow{\div 5} \frac{x - 8}{5} \xrightarrow{+9} \frac{x - 8}{5} + 9$

j $x \xrightarrow{\div 3} \frac{x}{3} \xrightarrow{+4} \frac{x}{3} + 4 \xrightarrow{\times 6} 6(\frac{x}{3} + 4)$

k $x \xrightarrow{\times 11} 11x \xrightarrow{-2} 11x - 2 \xrightarrow{\times 4} 4(11x - 2)$

l $x \xrightarrow{\times 3} 3x \xrightarrow{\div 4} \frac{3x}{4} \xrightarrow{+2} \frac{3x}{4} + 2$

4 **a** $x \xrightarrow{\times 5} 5x \xrightarrow{+9} 5x + 9$

b $x \xrightarrow{+1} x+1 \xrightarrow{\times 2} 2(x+1)$

c $x \xrightarrow{\div 6} \frac{x}{6} \xrightarrow{+4} \frac{x}{6}+4$

d $x \xrightarrow{-8} x-8 \xrightarrow{\div 7} \frac{x-8}{7}$

e $x \xrightarrow{-7} x-7 \xrightarrow{\times 12} 12(x-7)$

f $x \xrightarrow{\div 5} \frac{x}{5} \xrightarrow{-2} \frac{x}{5}-2$

g $x \xrightarrow{\times 7} 7x \xrightarrow{-12} 7x-12$

h $x \xrightarrow{+6} x+6 \xrightarrow{\div 3} \frac{x+6}{3}$

i $x \xrightarrow{+7} x+7 \xrightarrow{\times 3} 3(x+7) \xrightarrow{-5} 3(x+7)-5$

j $x \xrightarrow{\times 3} 3x \xrightarrow{+7} 3x+7 \xrightarrow{\div 2} \frac{3x+7}{2}$

k $x \xrightarrow{\times 3} 3x \xrightarrow{+1} 3x+1 \xrightarrow{\times 4} 4(3x+1)$

l $x \xrightarrow{\div 5} \frac{x}{5} \xrightarrow{+6} \frac{x}{5}+6 \xrightarrow{\times 3} 3(\frac{x}{5}+6)$

5 a $x \xrightarrow{\times 7} 7x \xrightarrow{+3} 7x+3$
$x \xleftarrow{\div 7} 7x \xleftarrow{-3} 7x+3$

b $x \xrightarrow{-2} x-2 \xrightarrow{\times 5} 5(x-2)$
$x \xleftarrow{+2} x-2 \xleftarrow{\div 5} 5(x-2)$

c $x \xrightarrow{\div 2} \frac{x}{2} \xrightarrow{+1} \frac{x}{2}+1$
$x \xleftarrow{\times 2} \frac{x}{2} \xleftarrow{-1} \frac{x}{2}+1$

d $x \xrightarrow{-5} x-5 \xrightarrow{\div 4} \frac{x-5}{4}$
$x \xleftarrow{+5} x-5 \xleftarrow{\times 4} \frac{x-5}{4}$

e $x \xrightarrow{+7} x+7 \xrightarrow{\times 2} 2(x+7)$
$x \xleftarrow{-7} x+7 \xleftarrow{\div 2} 2(x+7)$

f $x \xrightarrow{\div 6} \frac{x}{6} \xrightarrow{-3} \frac{x}{6}-3$
$x \xleftarrow{\times 6} \frac{x}{6} \xleftarrow{+3} \frac{x}{6}-3$

g $x \xrightarrow{+8} x+8 \xrightarrow{\div 3} \frac{x+8}{3}$
$x \xleftarrow{-8} x+8 \xleftarrow{\times 3} \frac{x+8}{3}$

h $x \xrightarrow{+8} x+8 \xrightarrow{\times 4} 4(x+8)$
$x \xleftarrow{-8} x+8 \xleftarrow{\div 4} 4(x+8)$

i $x \xrightarrow{\times 8} 8x \xrightarrow{-3} 8x-3$
$x \xleftarrow{\div 8} 8x \xleftarrow{+3} 8x-3$

j $x \xrightarrow{-7} x-7 \xrightarrow{\div 5} \frac{x-7}{5}$
$x \xleftarrow{+7} x-7 \xleftarrow{\times 5} \frac{x-7}{5}$

k $x \xrightarrow{+55} x+55 \xrightarrow{\times 3} 3(x+55)$
$x \xleftarrow{-55} x+55 \xleftarrow{\div 3} 3(x+55)$

l $x \xrightarrow{\div 9} \frac{x}{9} \xrightarrow{+2.1} \frac{x}{9}+2.1$
$x \xleftarrow{\times 9} \frac{x}{9} \xleftarrow{-2.1} \frac{x}{9}+2.1$

m $x \xrightarrow{\times 7} 7x \xrightarrow{-5} 7x-5$
$x \xleftarrow{\div 7} 7x \xleftarrow{+5} 7x-5$

n $x \xrightarrow{+8} x+8 \xrightarrow{\times 4} 4(x+8) \xrightarrow{+5} 4(x+8)+5$
$x \xleftarrow{-8} x+8 \xleftarrow{\div 4} 4(x+8) \xleftarrow{-5} 4(x+8)+5$

History of mathematics
Robert Recorde (1510–1558)

1 The 'equals' symbol '='
2 The letters 'ae' come from the word 'aequalis' meaning 'equal'.
3 48 years old

Exercise 6C — Solving equations using backtracking

1 a 2 b 4 c 8 d 13
e 10 f 7 g 67 h 88
i 1 j 5 k 8 l 5.3
m 10 n 21.053 o $4\frac{2}{5}$ p $\frac{31}{36}$

2 a $x = 3$ b $y = 1$ c $m = 6$ d $x = 5$
e $w = 3$ f $m = 25$ g $w = 5.7$ h $b = \frac{11}{54}$
i $a = \frac{2}{15}$

3 a $x = 1$ b $x = 18$ c $x = 23$ d $x = 4$
e $x = 3$ f $x = 12$ g $w = 1.3$ h $b = \frac{2}{3}$
i $x = \frac{7}{20}$

4 a $x = 3$ b $x = 45$ c $x = 12$ d $x = 34$
e $x = 77$ f $x = 0$ g $x = 13$ h $x = \frac{28}{11}$
i $x = \frac{2}{3}$

5 a $x = 14$ b $x = 29$ c $x = 78$ d $x = 1$
e $x = 5$ f $x = 291$ g $x = 4.65$ h $x = \frac{27}{7}$
i $x = \frac{1}{2}$

6 a $x = 10$ b $x = 3$ c $x = 48$ d $x = 50$
e $x = 6$ f $x = 21$ g $x = 7$ h $x = 6$
i $x = 11.13$ j $x = \frac{1}{10}$ k $x = 25$ l $x = 1$
m $x = 6$ n $x = 8$ o $x = 2$

7 a $x = 3$ b $x = 5$ c $x = 3$ d $x = 4$
e $x = 7$ f $x = 4$ g $x = 7$ h $x = \frac{2}{9}$
i $x = 1$

8 A

10 Quick Questions 1

1 $n + 11 = 15$ (Pronumeral may vary)
2 $k = 32$ 3 17 4 5 5 $4n - 11$ 6 $\frac{n+3}{7}$

7
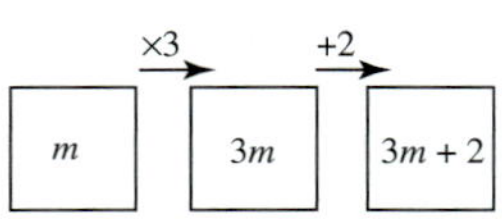

8
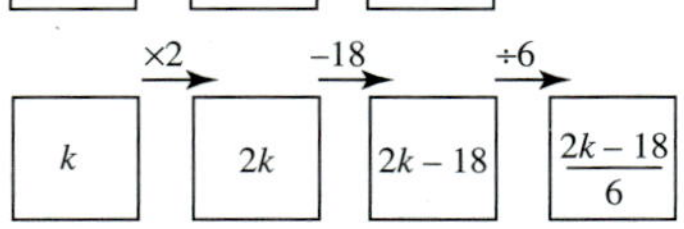

9
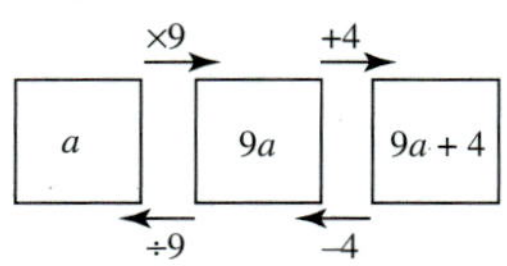

10
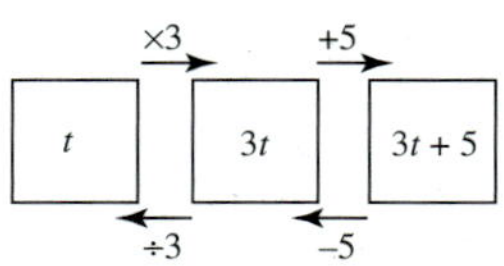

Exercise 6D — Checking solutions

1 a No b Yes c No d No
e No f Yes g No h No
i Yes j No k Yes l No

2

x	0	1	2	3	4
$2x + 3$	3	5	7	9	11

a $x = 4$ b $x = 4$ c $x = 1$

3

x	2	3	4	5	6
$5(x - 2)$	0	5	10	15	20

a $x = 4$ b $x = 6$ c $x = 8$

4 a

x	$2x + 1$	$3x - 5$
3	7	4
4	9	7
5	11	10
6	13	13
7	15	16

b $x = 6$

5 a

x	$\frac{x+3}{2}$	$2x - 6$
3	3	0
5	4	4
7	5	8
9	6	12
11	7	16

b $x = 5$

6 a Yes b No c No d Yes
e Yes f No g No h Yes
i Yes j No

7 $x = 6$

8 a $x = 3$ b $x = 14$ c $x = 10$ d $x = 9$
e $x = 3$ f $x = 20$ or $x = 0$
g $x = 5$ or $x = 0$ h $x = 6$ i $x = 4$

9 7 and 14

10 a 11 and 15 b 9 and 18 c 7 and 47
d 8 and 37 e 42 and 136 f 51 and 53
g 42 and 111 h 97 and 145 i 2.2 and 3.9
j 2.9 and 5.1 k 347 and 631 l 11.9 and 23.1

11

x	$x^2 + 4$	$4x + 1$
0	4	1
1	5	5
2	8	9
3	13	13
4	20	17

$x = 1$ and $x = 3$ are both solutions.

Maths Quest challenge (page 247)

1 10 2 24 years old
3 5 metres, 15 metres 4 7

10 Quick Questions 2

1 −8 2 $n = 17$ 3 $k = 11$ 4 $a = 19$
5 $e = 64$ 6 $m = 6$ 7 True

8

a	0	1	2	3	4	5
$7(2a + 5)$	35	49	63	77	91	105

$a = 2$

9

a	$3a + 4$	$5a - 10$
5	19	15
6	22	20
7	25	25
8	28	30
9	31	35

$a = 7$

10 $m = 19$

Exercise 6E — Solving word problems

1 a $x + 5$ b $x - 8$ c $2x$ d $\frac{x}{4}$
e $x + 1$ f $100x$ g $\frac{x}{100}$ h $7x$
i $\frac{x + 20}{2}$ j $10x$ dollars k $\frac{k + 23}{3}$

2 a D b C c C d E
e B f C

3 a $\$(2x + 7)$ b $3x$ c $4x + 3$ d $3x + 3$
e $3x + 6$ f $3x + 6$

4 Gareth is 13 years old.

5 a 12 b $\frac{x + 28}{3} = 12$ c $x = 8$
d Karina scored 8 points in the third game.

6 Callie scored 96 pins.

7 16, 17 and 18 8 7, 9 and 11 9 25, 30 and 35

10 David is 14 years old.

11 Chris jumped 156 centimetres.

12 The twins are 11 years old.

13 a Cost of a hamburger + cost of a hamburger + cost of a drink = total cost
or cost of the drink + twice the cost of a hamburger = total cost
b $\$3.75 + 2h = \$7.25$
c $d = 1.75$. The cost of a hamburger is \$1.75.

14 a i \$62 ii \$74 iii \$86
b Cost $= 50 + 12n$ c \$146
d $100 = 50 + 12n$
e $n \approx 4.2$. You could hire the windsurfer for 4 hours.
f \$2 left over
g Yes, as you would receive your deposit back. Total amount of money you would have is \$2 + \$50 = \$52.

Summary

1 Flow charts 2 backtracking
3 algebraic 4 inverse
5 dividing 6 multiplying
7 subtracting 8 adding
9 solve 10 substituting
11 word, Read, equation, Simplify, Check, answer

Chapter review

1 a $m = 5$ b $h = 6$ c $s = 19$ d $d = 20$

2 a 11 b 2 c 16 d 35

3 a 16 b 42 c 11 d 8

4 a $7x + 8$ b $\frac{x}{3} - 5$ c $6(x + 2)$
d $\frac{x - 7}{5}$ e $5(x + 3) - 9$ f $\frac{4x + 11}{8}$

5 a
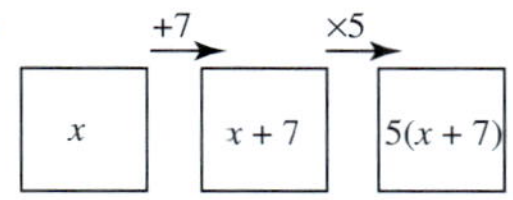

b
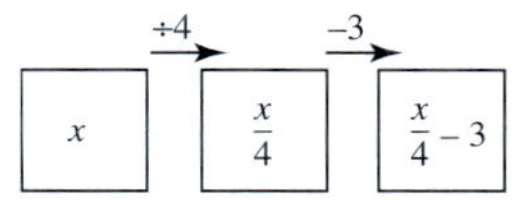

c
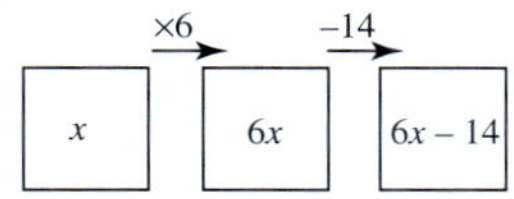

d
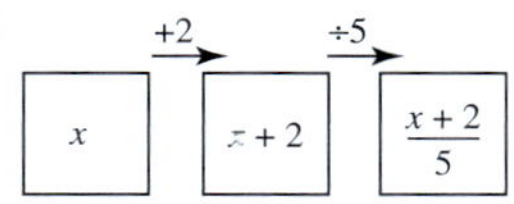

6 a
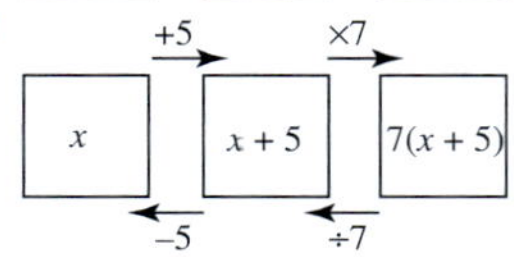

b
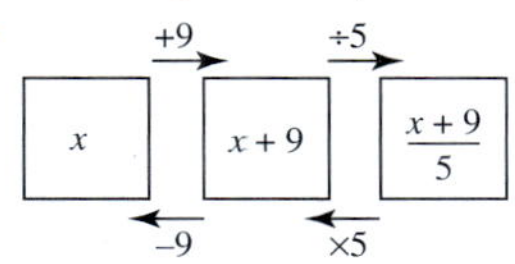

c
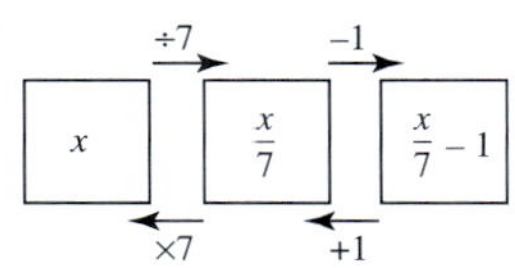

d
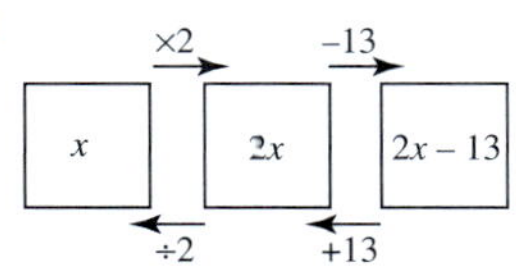

7 a $x = 2$ b $y = 10$ c $m = 60$ d $s = 18$

8 a $d = 4$ b $t = 100$ c $d = 3$ d $a = 6$

9 a $v = 3$ b $c = 12$

10 a Yes b No

11 a 32, 51 b 27, 59

12 Sophie owns 31 cards.

13 7 pieces of homework

Chapter 7 Length and perimeter

Are you ready?

1 a i cm ii km iii m iv mm
b Check with your teacher.

2 a 1 unit b 0.1 of a unit

3 a 43 mm b 3.7 cm

4 Metres

5 a 18 000 m b 6 km

6 a 2.3 cm, 25 mm, 0.2 m
b 1700 mm, 1.8 m, 185 cm

7 a 32 cm b 17 mm

Exercise 7A — Metric units of length

1 a m, cm b m, km c mm, cm d m
e m f m, cm g cm h m, cm
i cm, m j km
2 a m b cm c cm d mm
e cm, mm f km g m, cm h cm, mm
i cm, mm j mm k km, m l cm, mm
3 A
4 C
5 Various answers
6 Various answers

History of mathematics: Joseph-Louis Lagrange (1736–1813)

1 1736
2 Astronomy, sound, mechanics (calculus and probability could also be listed)
3 Supervise the development of the metric system
4 Answers may vary. Multiples of 10 are very easy to calculate.
5 Answers may vary. Measurements such as 12 inches = 1 foot and time of day are counted up to 12 hours.
6 Check with your teacher.

Exercise 7B — Reading scales and measuring length

1 a i 4 cm ii N/A
b i 7 cm ii N/A
c i 2 cm ii N/A
d i 9.5 cm ii $9\frac{1}{2}$ cm
e i 6.4 cm ii $6\frac{2}{5}$ cm
f i 10.1 cm ii $10\frac{1}{10}$ cm
g i 5.8 cm ii $5\frac{4}{5}$ cm
h i 11.6 cm ii $11\frac{3}{5}$ cm
i i 2.3 units ii $2\frac{3}{10}$ units
j i 1.6 units ii $1\frac{3}{5}$ units
2 a i 30° ii 165° iii 288°
b i 45°F ii 34°C iii 10°C iv 26°C
c i 45 km/h ii 80 km/h iii 100 km/h
d i 6.5 cm ii 19.8 cm iii 27.5 cm
e i 1250 mL (5 cups) ii 1800 mL (7 cups)
iii 500 mL (2 cups) iv 1000 mL (4 cups)
f 11
g i 130°C ii 220°C iii 360°F
3 B
4 C
5 a 5 cm b 7 cm c 3 cm d 8 cm
e 9 cm f 2.5 cm g 4.5 cm h 6.5 cm
6 a 45 mm b 86 mm c 90 mm d 35 mm
e 25 mm f 80 mm g 42 mm h 65 mm
7 a $4\frac{1}{2}$ cm b $8\frac{3}{5}$ cm c 9 cm d $3\frac{1}{2}$ cm
e $2\frac{1}{2}$ cm f 8 cm g $4\frac{1}{5}$ cm h $6\frac{1}{2}$ cm
8 a 150 million km b 60 m
c 4.6 m d 1.74 m
e i 187.5 km ii 412.5 km iii 540 km
f 136 m g 13.32 m

Maths Quest challenge (page 270)

1
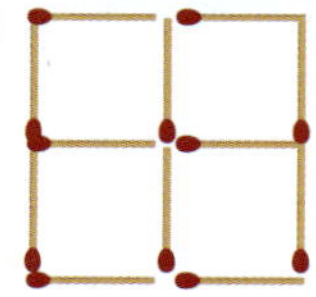
2
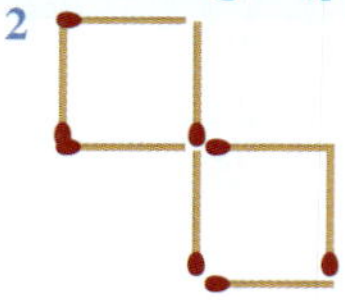
3
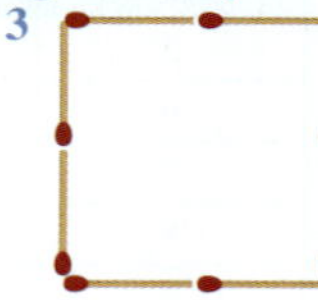
4
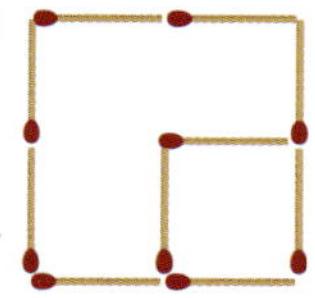
5
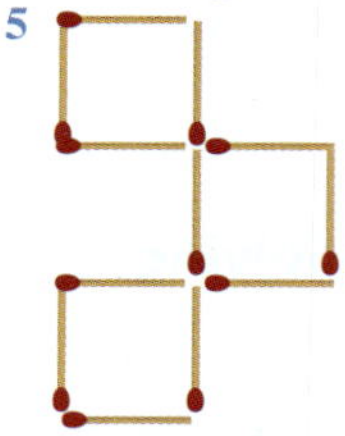

10 Quick Questions 1

1 Millimetres
2 Metres
3 Approximately 5 cm
4 60 units
5 4.2 cm or 42 mm
6 9.15 m
7 False
8 Millimetres, centimetres, metres, kilometres
9 Check with your teacher.
10 Check with your teacher.

Exercise 7C — Converting units of length

1 a 2000 m b 7000 m c 65 000 m
d 5300 m e 660 m f 900 cm
g 25 cm h 280 mm i 2000 mm
j 70 000 cm
2 a 8 km b 6.5 km c 0.7 km
d 0.05 km e 0.0105 km f 60 m
g 3 m h 0.57 m i 0.09 m
j 4.5 cm k 83.5 cm l 2560 cm
3 a 800 000 cm b 5.4 m
c 101 000 mm d 0.078 km
e 60 250 000 mm f 1.128 m
g 72 330 mm h 4.05 cm
i 5050 m j 0.030 35 mm
4 D
5 A
6 a 67 700 m b 0.0452 km c 0.56 m
d 4.5 km e 87.5 mm f 60 cm
g 7210 mm h 309 000 cm i 4.8 cm
j 11 655 mm
7 868 cm, 10.8 cm. About 80 times longer
8 Everest 8.863 km, K2 8.607 km
9 1630 cm
10 448 m
11 a 150 cm, 12.5 m, 0.02 km
b 0.445 m, 3000 mm, 350 cm
c 50 000 mm, 500 m, 50 km
d 1.7 m, 1700 cm, 0.17 km
e 0.0052 mm, 0.000 052 m, 0.052 cm
f 0.909 m, 9000 mm, 990 cm
12 a 3.75 m (375 cm) b 10.2 km (10 200 m)
c 432 cm (4320 mm) d 2.242 m (224.2 cm)
e 3.91 m (391 cm) f 25.6 m (25 600 mm)
g 47.8 km (47 800 m) h 0.001 553 km (155.3 cm)
13 a 66.4 km (66 400 m)
b 410.4 m (410 400 mm)
c 0.104 m (10.4 cm)
d 1.7 km (170 000 cm)

14 2.76 cm 15 44 16 216 m
17 1.593 m 18 Yes, 36 cm
19 215 cm, 105 cm, 155 cm, 170 cm
20 2.4 m 21 198 cm
22 40 cm 23 38

10 Quick Questions 2

1 Centimetres 2 Check with your teacher.
3 4300 m 4 87 m
5 210 000 cm
6 1345 mm, 9 m, 2000 cm, 0.04 km, 67 m
7 583 mm 8 28 cm
9 30 240 mm or 3024 cm or 30.24 m
10 1.91 m

Exercise 7D — Perimeter

1 a 8 units b 16 units c 18 units d 10 units
e 18 units f 22 units g 16 units h 12 units
i 14 units j 14 units
2 a 40 cm b 92 mm c 108 mm d 30 m
e 44 cm f 29.5 m g 74 cm h 173 mm
i 68 cm j 28 cm
3 a 16 cm b 16 cm c 18 cm d 8 cm
e 22 cm f 22 cm g 15 cm h 25 cm
i 22 cm j 18 cm
4 a 108 mm b 173 mm c 18 units
d 2995 cm e 408 mm f 180 mm
g 14 433 m h 17 579 m
5 B 6 D 7 A
8 a 156 cm b 664.2 cm
c $12\frac{9}{10}$ m or 12.9 m
9 a 14.8 cm b 8.56 m
c $32\frac{4}{5}$ mm or 32.8 mm
10 48.5 m 11 252 cm
12 a 193 m b $3667
13 a

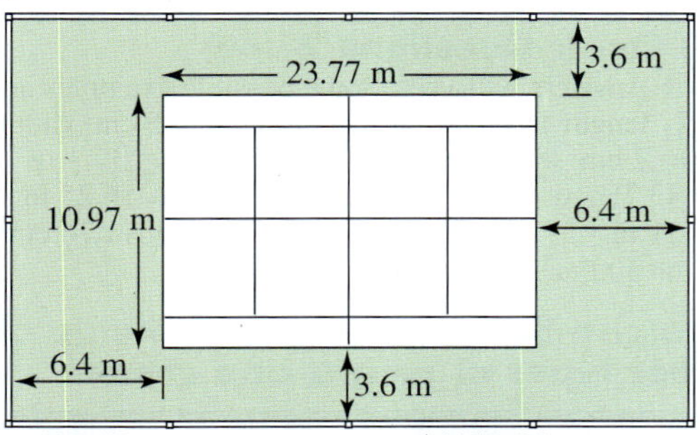

b 109.48 m c $4050.76
14 325.8 m 15 139.8 m

Maths Quest challenge (page 284)

1 5083 2 36 3 30 days 4 C

Summary

1 metres 2 zero
3 kilometre 4 metre
5 cm 6 mm
7 (top line) × 10, (bottom line) ÷ 100, ÷ 10
8 multiply 9 divide
10 1 000 000 11 100
12 distance, outside 13 same

Chapter review

1 a Kilometres or metres b Centimetres
c Millimetres
2 a 114 kg b 42 kg
c 80 kg d 95 kg
3 a i 4.5 cm ii $4\frac{1}{2}$ cm
b i 3.2 cm ii $3\frac{1}{5}$ cm
c i 2.9 cm ii $2\frac{9}{10}$ cm
4 5.525 m
5 a 0.56 m b 0.023 km c 1700 cm
d 750 000 mm e 2090 mm f 0.068 m
g 2.25 cm h 6.3 m i 82 000 km
j 0.59 cm
6 0.004 45 km, 455 cm, 44.5 m, 455 000 mm
7 a 51 200 m (51.2 km) b 3700 cm (37 m)
8 25 cm
9 8.7 m (870 cm)
10 a 46 cm b 20 cm
c 191 mm (19.1 cm) d 21 cm (210 mm)
11 243 m
12 780 cm
13 a 41 m b $346.45
14 870 cm
15 540 m

Chapter 8 Area, volume and capacity

Are you ready?

1 a i m^2 ii cm^2 iii km^2 iv ha
b Check with your teacher.
2 a 4 cm^2 b 12 cm^2 c 13 cm^2
3 a The area of a rectangle is found by multiplying the length by the width.
b 15 cm^2
4 a The area of a triangle is half the base multiplied by the height.
b 20 cm^2
5 a i cm^3 ii L iii mL iv m^3
b Check with your teacher.
6 a 6 cm^3 b 18 cm^3

Exercise 8A — Area

1 a cm^2 b cm^2 c m^2 d mm^2
e km^2 f cm^2 g m^2 h m^2
i km^2 j m^2 k ha l mm^2
2 a 14 cm^2 b 12 cm^2 c 10 cm^2 d 10 cm^2
e 10 cm^2 f 8 cm^2
3 a 16 cm^2 b 8 cm^2 c 7 cm^2 d 9 cm^2
e 6 cm^2 f $9\frac{1}{2}$ cm^2 g 12 cm^2 h 10 cm^2
4 Approximate answers — accept answers within 2 cm^2 of those listed here.
a 19 cm^2 b 24 cm^2 c 19 cm^2
5 a 23 m^2 b 27 cm^2
c 7 952 400 km^2 d 22 cm^2

Exercise 8B — Converting units of area

1 a 30 000 cm^2 b 350 mm^2 c 9 200 000 m^2
d 123 400 cm^2 e 660 mm^2 f 240 mm^2

2 a 2.3 m^2 b 2.35 cm^2 c 74 km^2
d 89 000 m^2 e 116 700 cm^2 f 200 000 m^2
3 a 5.3 ha b 0.0786 ha c 100 m^2
d 1.0646 ha e 35 km^2 f 600 m^2
g 560 000 m^2 h 14 800 m^2
4 A 5 C
6 a 12 500 m^2, 1.3 ha, 0.3 km^2
b 0.2 ha, 250 000 m^2, 1 km^2
c 0.2 m^2, 2 m^2, 200 000 cm^2
d 500 mm^2, 0.002 m^2, 100 cm^2
7 a 540 mm^2 or 5.4 cm^2
b 10.3 m^2 or 103 000 cm^2
c 22.06 cm^2 or 2206 mm^2
d 362 000 m^2 or 36.2 ha
e 8400 m^2 or 0.0084 km^2
f 6 600 000 m^2 or 6.6 km^2
8 200 000 m^2
9 a 16 000 000 000 m^2 b 1 600 000 hectares

Exercise 8C — Finding the area of a rectangle

1 a i length 4 cm, width 2 cm
ii 8 cm^2 iii 8 cm^2 iv same
b i length 6 cm, width 1 cm
ii 6 cm^2 iii 6 cm^2 iv same
c i length 6 cm, width 4 cm
ii 24 cm^2 iii 24 cm^2 iv same
d i length 8 cm, width 5 cm
ii 40 cm^2 iii 40 cm^2 iv same
2 a 15 cm^2 b 4 cm^2 c 15 m^2
d 14 km^2 e 24 cm^2 f 10 m^2
3 a l = 1.8 cm w = 1.2 cm A = 2.16 cm^2
b l = 3.6 cm w = 0.7 cm A = 2.52 cm^2
c l = 31 mm w = 12 mm A = 372 mm^2
d l = 25 mm w = 0.4 mm A = 10 mm^2
e l = 2.5 cm w = 1.3 cm A = 3.25 cm^2
f l = 25 mm w = 25 mm A = 625 mm^2
4 a 20 cm^2 b 48 m^2 c 261 mm^2
d 56 m^2 e 400 mm^2 f 72 cm^2
g 76 m^2 h 18 cm^2 i 135 mm^2
j 686 mm^2 k 48 m^2 l 325 m^2
5 17.5 m^2 6 800 cm^2 7 9.66 m^2
8 a 7.5 m^2 b 15 boxes c $528
9 104 m^2, $1300
10 75 m^2 11 $141.64 12 4920 km^2 13 3 sheets
14 135 m^2 15 39 m^2 16 864 cm^2

10 Quick Questions 1

1 cm^2 2 m^2 3 19 cm^2
4 32 cm 5 14 cm^2 6 11 m^2
7 130 cm^2 8 4 200 000 m^2 9 2436 km^2
10 6826 m^2

Exercise 8D — Finding the area of a triangle

1 a 4 cm^2 b 3 cm^2 c 40.5 cm^2
d 8 cm^2 e 50 cm^2 f 1 m^2
2 a C b A c D d A e B
3 a 416 cm^2 b 4.4 m^2 c 3.125 m^2
d 2.4 m^2 e 493 mm^2 f 583 cm^2
g 600 km^2 h 337.5 cm^2 i 152 cm^2
4 88.8 m^2 5 900 cm^2
6 6000 cm^2 (0.6 m^2) 7 12.6 m^2 8 126 m^2
9 a 212 500 km^2
b Check with your teacher.

Maths Quest challenge (page 311)

1 $2560 2 Length 9 cm, width 4 cm
3 48 mm 4 26 mm

Exercise 8E — Area of composite shapes

1 a 81 cm^2 b 5 m^2 c 16 cm^2
d 56 cm^2 e 588 cm^2 f 13.5 m^2
2 a 23.625 m^2 b $283.50
3 a 27 m^2 b $657
4 13.05 m^2 5 39.28 m^2
6 315 cm^2 7 14.25 m^2
8 $2202.40 (rounded to the nearest 5 cents)

Exercise 8F — Volume

1 a 8 cm^3 b 8 cm^3 c 6 cm^3
d 12 cm^3 e 12 cm^3 f 48 cm^3
2 B
3 a 4 cm^3 b 24 cm^3 c 15 cm^3
d 8 cm^3 e 13 cm^3 f 7 cm^3
g 35 cm^3 h 14 cm^3 i 17 cm^3
4 A

Exercise 8G — Finding the volume of a rectangular prism

1 a 12 cm^3 b 12 m^3 c 243 cm^3
d 6144 cm^3 e 4 cm^3 f 192 m^3
g 270 cm^3 h 101.25 cm^3
2 9690 cm^3 3 477 750 cm^3
4 20 cm 5 9 cm 6 12 cm
7 a 2700 cm^3 b 2 700 000 mm^3
8 4.875 m^3
9 a 28.875 cm^3 b 28 875 mm^3 c 176 mm^3
d 20 075 mm^3
10 768 m^3 11 40.8 m^3
12 Each guest receives 267 cm^3 of cake. 13 2277 cm^3

10 Quick Questions 2

1–3 Answers will vary. Some possible examples are: length 10 cm, width 3 cm; length 15 cm, width 2 cm; length 6 cm, width 5 cm.
4 43.75 cm^2 5 288 m^2 6 16.25 m^2
7 21 m^2 8 60 mm^3 9 11.76 cm^3
10 64 500 m^3

Exercise 8H — Finding the volume of other types of prisms and shapes

1

Prism	Area of base (cm^2)	Height (cm)	Volume (cm^3)
a	6	2	12
b	5	2	10
c	3	3	9
d	5	2	10
e	5	3	15
f	4	4	16
g	3	2	6
h	7	3	21
i	8	3	24
j	6	2	12

2 D 3 B
4 a 126.9 cm^3 b 45 m^3
c 108 cm^3 d 720 mm[illegible]
e 81.9 cm^3 f 15.68 c[illegible]
5 a 186 cm^3 b 136 cm^3 c 189 cm^3 d [illegible] cm^3
6 2500 m^3
7 a 1920 cm^3 b 12

Exercise 8I — Capacity

1 a 2000 mL b 3 L c 13 000 L
d 7 L e 5.5 L f 0.26 kL
g 2500 mL h 32 L i 0.055 L
j 35 mL k 420 000 mL l 990 L
m 1870 mL n 22.5 L o 0.35 L
p 100 mL q 0.025 kL r 7 mL
2 a 750 mL b 2500 kL
c 800 mL = 0.8 L d 40 000 mL = 40 L
e 6000 mL = 6000 cm^3 f 5.2 kL = 5.2 m^3
g 2450 cm^3 h 78 L
3 D 4 D
5 a 0.25 L, 2.45 L, 2.5 L, 25 000 mL
b 7.65 mL, 760 mL, 0.765 L, 7.60 L
c 0.011 L, 0.1 L, 110 mL, 1.1 L
6 16 bottles
7 1750 mL (1.75 L) 8 1620 mL (1.62 L)
9 6 L 10 50 doses
11 225 bottles
12 The 185 mL container is the better buy.
13 9.9 L 14 4.8 L 15 616 mL 16 No
17 Yes 18 50.4 L
19 Approximately 27 L (27.09 L)
20 Answers will vary. 16.4 kL = 16.4 m^3. An example could be length 2 m, width 2 m, height 4.1 m.
21 300 mL

Maths Quest challenge (page 333)

1 48 mm 2 25
3 a 8 b 12 c 6 d 1

Summary

1 Area
2 square metres, square centimetres
3 rectangle 4 multiply, 1 000 000
5 divide, 10 000 6 hectare
7 half 8 volume
9 cubic metres, cubic centimetres
10 rectangular 11 base, height
12 Capacity 13 litres, millilitres, kilolitres

Chapter review

1 a m^2 b km^2 c mm^2
d ha e cm^2 f cm^2 or mm^2
g m^2 h km^2
2 Answers will vary.
3 a 12 cm^2 b 8 cm^2 c 13 cm^2 d 2 cm^2
e 5 cm^2 f 8 cm^2 g $6\frac{1}{2}$ cm^2 h 15 cm^2
4 a 15 cm^2 b 23 cm^2
5 a 5.6 m^2 b 500 cm^2 c 3000 m^2
d 2.53 ha e 84 km^2 f 514 mm^2
g 34 cm^2 h 12 500 m^2
6 a 150 m^2 b 6 cm^2 c 2914 mm^2
7 a 5.04 cm^2 b 1.98 cm^2 c 4.8 cm^2
8 a 9 cm^2 b 9 cm^2 c 19 cm^2
d 52 cm^2 e 380 cm^2 f 11 cm^2
9 1230 cm^2 10 \$582.40
11 3.6 m^2 12 0.72 m^2
13 a 627 cm^2 b 96 m^2 c 1344 cm^2
14 a 16 m^2 b 432 mm^2 c 87.5 cm^2
15 a 18 cm^3 b 12 cm^3 c 30 cm^3
d 16 cm^3 e 20 cm^3 f 28 cm^3
16 60 m^3
17 144 m^3 18 28 cm 19 8.5 cm
20 a 36 cm^3 b 24 cm^3 c 15 cm^3
d 24 cm^3 e 24 cm^3 f 8 cm^3
21 a 112 cm^3 b 384 cm^3 c 40 cm^3 d 72 cm^3
22 15 960 cm^3
23 a 8000 mL b 420 mL c 3.3 L
d 1012 mL e 4300 L f 3400 mL
g 4.755 kL h 0.432 L
24 4.9 L
25 a 1725 mL (1.725 L) b 8 glasses
26 Approximately 3 L

Chapter 9 Time, mass and temperature

Are you ready?

1 a Four o'clock or 4.00
b A quarter past ten or 10.15
2 a Ten past eight b A quarter to nine
c Half past eight d Twenty to nine
3 a 45 minutes b 9 hours
4 a 2100 b 1000 c 1400
5 Check with your teacher.
6 a i kg ii g iii t iv mg
b Check with your teacher.

Exercise 9A — Time calculations

1 A cricket test match — 5 days, writing your name — 10 seconds, eating breakfast — 15 minutes, building a house — 6 months, flying time from Melbourne to Hong Kong — 7 hours, being in Year 7 — 1 year.
2 a 5:00 b 5:04 c 7:23
d 1:05 e 9:14 f 6:47
3 a

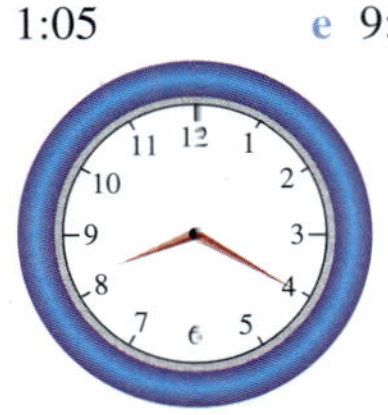

b

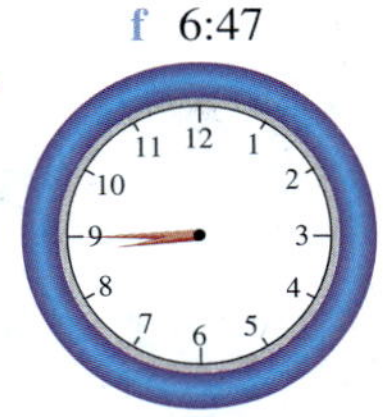

c

d

e

f

g

h

i

4 Sundial and hourglass.

5 2.25 hours = $2\frac{1}{4}$ hours which is 2 hours and 15 minutes

6 a 120 b 150 c 195 d 30
e 45 f 465 g 1440 h 12 960

7 a 3 h 20 min b 3 h 5 min c 2 h 40 min
d 3 h 50 min e 6 h 45 min f 1 h 35 min
g 10 h 10 min h 1 h 12 min i 5 h 5 min

8 a 75 min b 130 min c 110 min
d 265 min e 455 min f 192 min
g 1005 min h 330 min i 375 min

9 a 7 b 52 c 60
d 3600 e 86 400 f 24
g 8760 h 1440 i 525 600

10 a 5 pm b 3 pm c 6.30 am
d 11.50 am e 8.40 am f 5.18 pm
g 2.00 am h 2.20 am i 5.30 pm
j 6.35 am k 7.41 pm l 10.20 am

11 a 15 min b 1 h 5 min c 1 h 13 min
d 1 h 15 min e 53 min f 1 h 53 min
g 32 min h 3 h 45 min i 5 h 15 min
j 12 h k 8 h 45 min

12 a 35 h 10 min b 32 h 47 min c 13 h 31 min
d 43 h 49 min

13 C 14 D 15 A

16 48 min 17 $7\frac{1}{2}$ h 18 7.25 am

19 6.30 am 20 11.45 pm 21 10.35 pm

22 52 min 23 4 h 2 min

Maths Quest challenge (page 348)

1 25 minutes past 10 2 12 o'clock and 6 o'clock

3 10

Exercise 9B — 24-hour clock

1 10.12 am = 1012, 1212 pm = 1212,
12 midnight = 0000, 12 midday = 1200,
11.20 pm = 2320, 11.20 am = 1120

2 a 1020 b 1130 c 0510 d 0415
e 1715 f 1830 g 0830 h 2040
i 1200 j 2330 k 0435 l 1430

3 a 11.15 pm b 1.10 pm c 8.15 am
d 1.15 am e 6.18 pm f 12.20 pm
g 12.05 am h 10.05 am i 8.05 pm
j 11.35 am k 3.20 pm l 2.14 pm

4 a 12 h 49 min b 2 h 15 min c 1 h 45 min
d 45 min e 13 h 10 min f 3 h 5 min
g 7 h 30 min h 12 h 56 min i 9 h 20 min

5 6 h 31 min

6 a 3 h 15 min b 240-min tape

7 3 h 45 min

8 B 9 E 10 A 11 E 12 3 hours

13 a Basic light: 2 h 50 min, basic medium: 3 h, basic dark: 3 h 10 min
b 1.40 pm, timer set to start process at 1340.

10 Quick Questions 1

1 180 minutes 2 253 minutes

3 13 hours 7 minutes 4

5 9.22 am 6 1 hour 9 minutes 12 seconds

7 1547 8 10.22 pm

9 6 hours 48 minutes 10 1615

History of mathematics: The calendar

1 Gregorian calendar 2 Julian calendar

3 These days were removed from existence because they were skipped when the calendar was redone.

4 The two calendar systems had not been brought into line and thus one was behind the other.

Exercise 9C — The calendar

1 a 4 — April, June, September, November
b 7 — January, March, May, July, August, October, December
c 1 — February
d 365 e 14 f 12
g A year in which there are 29 days in February
h December, January, February
i June, July, August
j March, April, May
k September, October, November

2 a Saturday b 5 September c Tuesday
d 18 January, 15 February, 15 March, 19 April, 17 May, 21 June, 19 July, 16 August, 20 September, 18 October, 15 November, 20 December
e Sunday
f No, 2010 is not exactly divisible by 4 and there are 28 days in February.
g 8 July h Thursday

3 a 15 years b 7 months c 7 months d 12 days
e 2 months and 4 days or 9 weeks and 2 days or 65 days
f 6 years 2 days
g 1 month 9 days or 5 weeks 5 days or 40 days
h 1 month 3 days or 4 weeks 6 days or 34 days
i 50 years 1 day
j 110 years 2 months 8 days

4 a 35 years b 32 years c 12 years
d 197 years e 298 years f 48 years

5 a 164 years b 326 years c 1080 years
d 1200 years e 400 years f 1650 years
g 1197 years h 2440 years i 2070 years

6 15 7 19 years 8 12 years

9 a, b, e, h, n

10 a 0 b 1 c 2 d 3 e 4 f 9

11 a False b True c False d True e False
f False g True h True i False j False
k False l False m True
12 A 13 C 14 C
15 8 months 6 days 16 18 days

Exercise 9D — Timelines

1

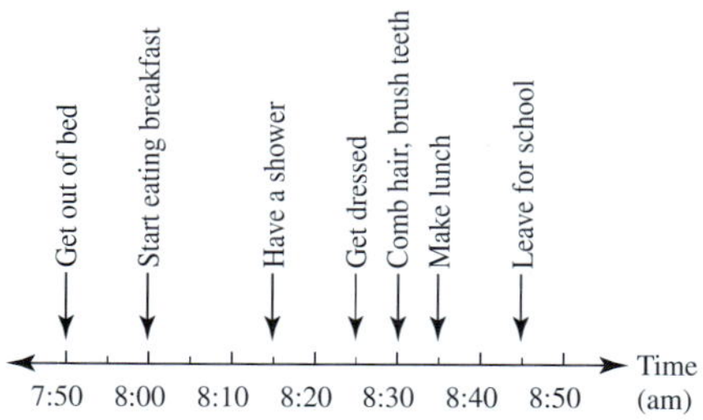

2

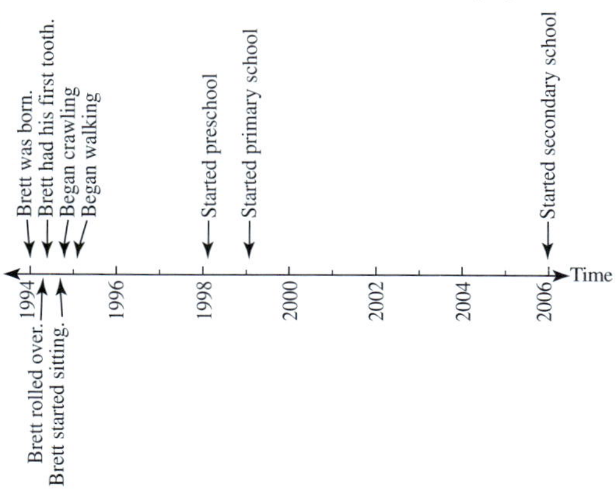

3

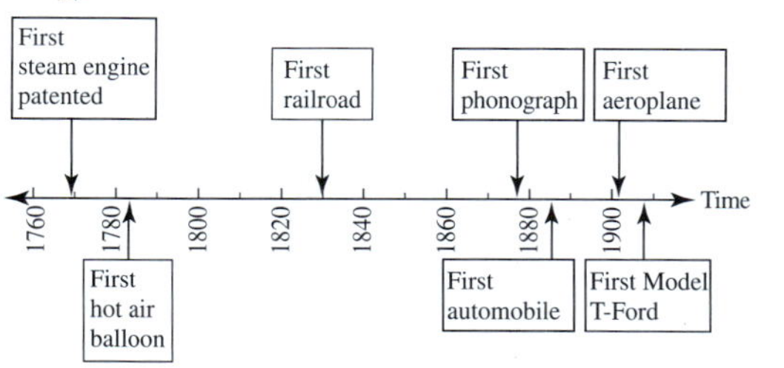

4

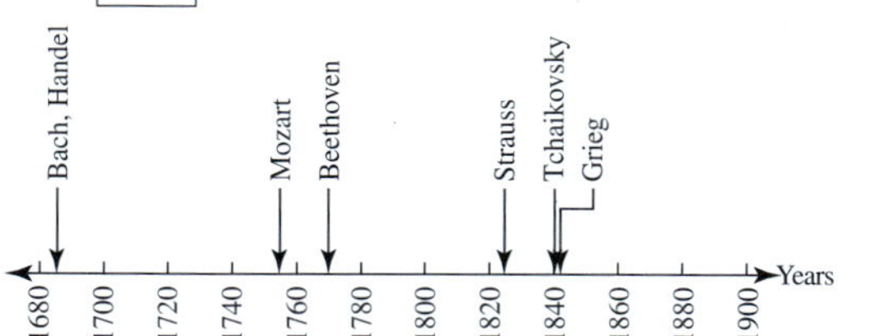

5

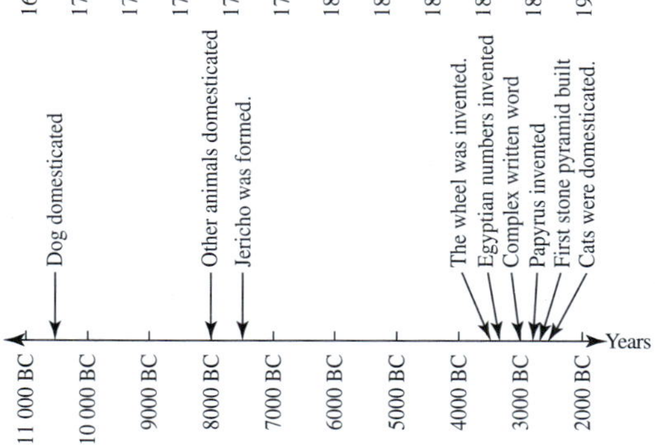

6

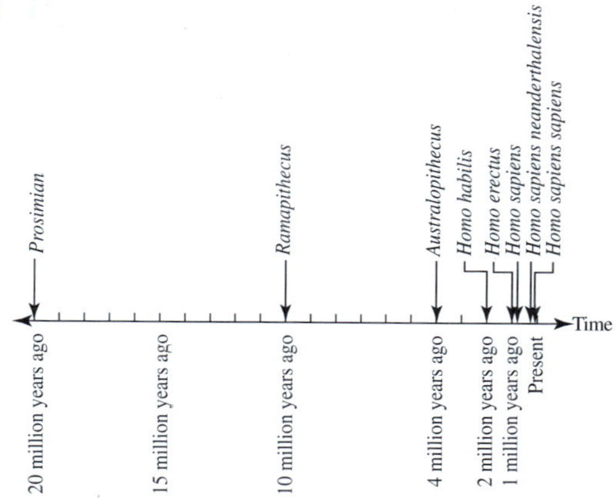

7

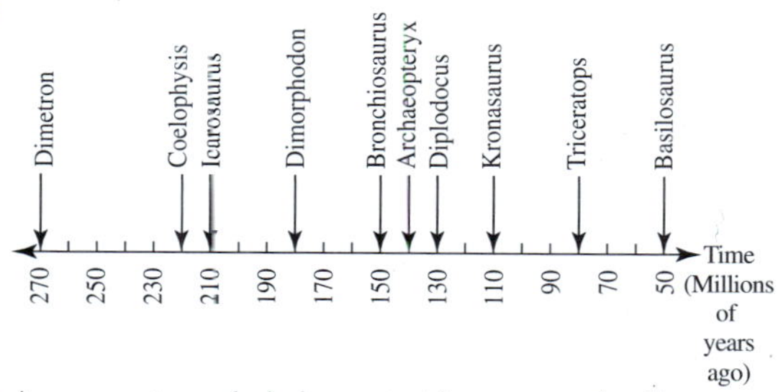

8 An accurate scale is important because a timeline shows when events occur in relation to other events. We can then compare how close (or far apart) different sets of events are.

9 An advantage of using a timeline is that you can see when things occurred in time, in relation to each other.

10 The order of the information on a timeline is very important because there should be an accurate scale to show when things occurred.

11 Various answers.

Exercise 9E — Timetables

1 One

2 12.40 pm

3 Spencer St

4 Yes, all trains finish at Warrnambool and there is a coach service after that.

5 Change to coach service.

6 The 08.48 am finishes at Mt Gambier, the 12.40 pm finishes at Port Fairy, the 17.47 finishes at Heywood Monday to Thursday and Mt Gambier on Friday.

7 Westcoaster

8 a i 1 h 17 min (Mon.–Fri.), 53 min (Sun.)
ii 3 h 42 min (Mon.–Thur.), 3 h 45 min (Fri.), 3 h 11 min (Sun.)
iii 45 min or 47 min on Friday or 44 min (Sun.)
iv 1 h 20 min (Mon.–Thurs.), 1 h 22 min (Fri.), 1 h 19 min (Sun.)
v 1 h 15 min (Mon.–Thur.), 1 h 10 min (Fri.), 1 h 21 min (Sun.)
vi 1 h 49 min (Fri.), 1 h 55 min (Sun.)

b i 3 min
ii 3 min
iii 11 min Mon.–Thurs. or 13 min on Fri. or 9 min on Sun.

iv Camperdown station, Port Fairy and Portland
v Friday and Sunday

9 1 service Monday to Thursday, 2 services on Friday, 1 service on Sunday

10 Friday

11 a Westcoaster on Friday
b 6 h 12 min

12 More people use the train.

13 3 per day 14 2 15 No

16 a 2.68 m b 1.35 am

17 11 h 33 min 18 No

19 11 h 55 min 20 No

21 a 0.14 m b 6 and 7 November

22 a 1.19 m b Tue. 1 November at 5.24 am

Exercise 9F — Time zones and flight schedules

1 a 9.00 pm b 10.30 pm
c 8.20 pm d 10 am (next day)
e 10 pm f 12.15 pm
g 1.30 am (next day) h 3.45 am (next day)
i 6.10 pm j 7.35 am (next day)
k 11.12 pm l 11.12 am

2 a 5.30 pm (previous day) b 2.00 am
c 2.00 pm d 7.50 am
e 8.40 pm (previous day) f 9.20 am
g 1.10 am h 0325 h
i 0550 h j 2015 h
k 0855 h l 0935 h

3 B

4 a 1 pm b 3 pm c 5 pm

5 a 12 midday b 10 pm c 12 midnight

6 a 3.30 am b 11.30 am c 1.30 pm

7 Saturday 8 1805 h, 1905 h

9 a 5 b 3 c 2

10 3

11 0605 h, 1105 h, 1505 h

12 a Yes b i 1 stop ii Brisbane

13 Flight 626

14 Same flight travels from Melbourne to Brisbane and then to Cairns.

15 a 5 b 7 c 6

16 Tuesday 17 4 h 45 min

18 a 3 h 15 min
b Flight 608 has a stop in Brisbane, whereas flight 644 has no stops.

19 3

20 a 7 b 5 c 6

21 55 min

22 No

23 a C b C

Maths Quest challenge (page 370)

1 2062 2 14 August

3 Connor misses the train by 20 minutes.

10 Quick Questions 2

1 495 minutes

2 28 days, 29 in a leap year — February
30 days — April, June, September, November
31 days — January, March, May, July, August, October, December

3 1218 years 4 Yes 5 55 days

6

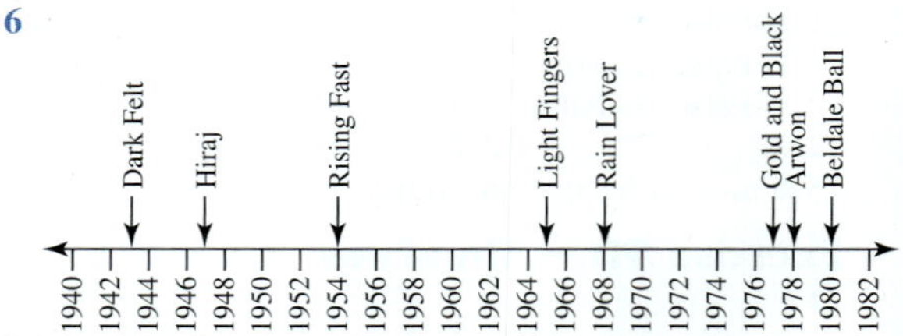

7 Monday, Tuesday, Wednesday, Thursday, Friday

8 4 hours 55 minutes

9 0900, 6 August 10 2300, 5 August

Exercise 9G — Mass

1 Feather, egg, lunch box, brick, bag of cement, car

2 a 1000 g b 2000 g c 2500 g
d 3700 g e 500 g f 720 g
g 6700 g h 109 020 g i 14 080 g
j 30 g k 25 g l 3070 g
m 5120 g n 1004 g o 10 040 g

3 a 500 g b 250 g c 750 g
d 600 g e 3500 g f 22 250 g
g 100 850 g h 10 g i 68 g

4 a 2000 kg b 2700 kg c 3800 kg
d 300 kg e 40 kg f 15 100 kg
g 7030 kg h 18 005 kg i 103 040 kg

5 a 500 kg b $333\frac{1}{3}$ kg c 250 kg
d 750 kg e 800 kg f 5250 kg
g 90 kg h 122 kg i 203 750 kg

6 a 2000 mg b 0.25 t c 1 000 000 mg
d 0.82 g e 8.4 t f 0.75 t
g 3200 g h 2500 kg i 7030 g
j 75 tonne k 5000 g l 4 500 000 kg
m 18 kg n 0.02 t

7 2935 g 8 5.1 kg 9 73.7 kg

10 900 kg 11 1.3 kg

12 a 20.33 kg b Yes, by 0.33 kg

13 a 417 kg b 117 kg
c Yes, under by 3 kg

14 a 1 kg b 3 kg c 10 g
d 500 g e 1–2 tonnes

15 The weight of the product without the container or packaging.

16 45 g 17 250-g jar

Exercise 9H — Temperature

1 a 20°C b 300°C c 1500°C
d 32.5°C e 25.8°C f 180°C
g 39.3°C h 25°C

2 a A winter's morning in Melbourne 5°C
b Microwaving some nachos 250°C
c Melting iron 1535°C
d An air-conditioned office 23°C
e A winter's maximum temperature in Darwin 30°C
f The temperature of the Sun 5500°C

3 a 113°F b 302°F c 54°F

4 a 21.1°C b 5°C c 204.4°C

5 D

6 a 53.6°F b 98.6°F c 2256.08°F

7 a 4.4°C b 1093.3°C c 32.19°C

8 a 14°C b 15.67°C

9 a 6°C b 6 hours

10 62°C

Summary

1 vertical
2 final
3 noon
4 4
5 add
6 before
7 after
8 total
9 equal
10 24
11 ahead
12 multiply
13 divide
14 milligrams, kilograms, tonnes
15 temperature
16 Celsius, $F = \frac{9}{5}C + 32$
17 Fahrenheit, $C = \frac{5}{9}(F - 32)$

Chapter review

1 a 180 min b 315 min c 450 min d 105 min e 1440 min f 4320 min
2 a 2 h 45 min b 2 h 20 min c 3 h 30 min d 3 h 40 min e 13 h 20 min f 1 h 15 min
3 a 80 min b 160 min c 190 min d 258 min e 635 min f 222 min
4 8 h 15 min
5 a 2235 h b 0715 h c 0320 h d 1642 h e 2150 h f 0105 h
6 a 12.40 pm b 7.25 am c 3.50 pm d 9.09 am e 9.21 pm f 11.20 am
7 a 9 h 20 min b 9 h 57 min c 15 h 38 min d 6 h 35 min
8 a 107 years b 26 years c 237 years d 415 years
9 a 9 years b 7 years 5 days c 2 years 2 days d 64 years 7 months 13 days
10 a, c, d, f
11

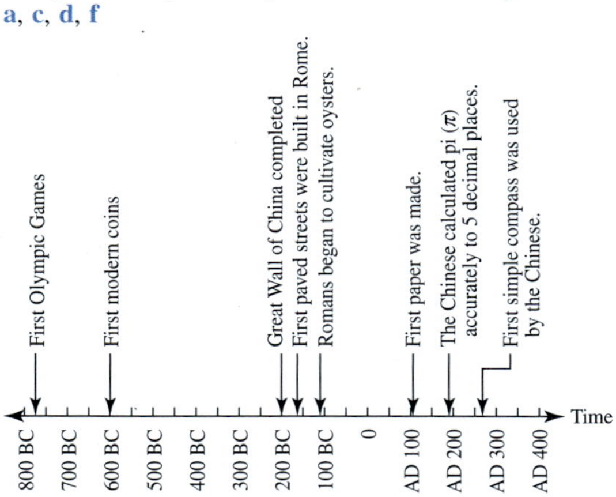

12 a 8.48 am b 3 h 12 min c 8 or 9 min
13 a 9.20 pm b 1.30 am (next day) c 7.15 am (next day) d 4.23 pm
14 a 1.30 pm b 1.30 am c 0450 h d 2235 h (previous day)
15 7.00 am GMT
16 a 3 tonnes b 25 000 g c 0.5 g d 20 kg
17 1139.4 kg
18 2075 g
19 2.75 kg
20 a 92°C b 250°C c 26.7°C
21 a 131°F b 482°F c 64.4°F
22 a 32.2°C b 82°C c 426.7°C
23 a 6°C b 1°C
24 28°C

Chapter 10 Angles

Are you ready?

1 27°
2 a C b B
3 a 45° and 55° b 30° and 40° c 90° and 100°
4 a 52° b 38° c 97°
5 Check with your teacher.
6 a B b C c F d E e A f D
7 a PQR or RQP b KDL or LDK

Exercise 10A — Measuring angles

1 a 40° b 81° c 75° d 13° e 142° f 107° g 162° h 103° i 240° j 201° k 316° l 356°
2 Discuss with your teacher as answers will vary.
3 a B b B
4 a 45° b 155° c 35° d 180° e 77° f 63°
5 a Difficult, as angle is approximately 7°
b Same as bottom section

Exercise 10B — Constructing angles with a protractor

1 a

b
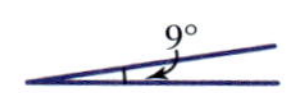

c
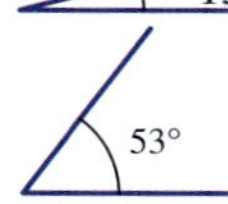

d
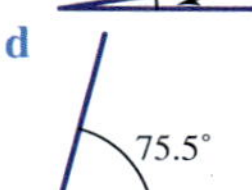

e
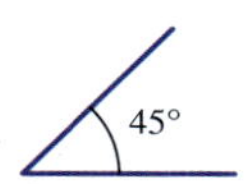

f
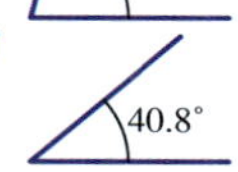

g
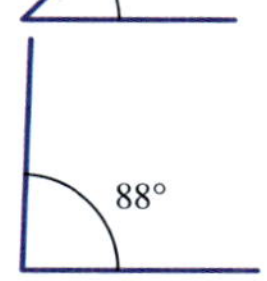

h
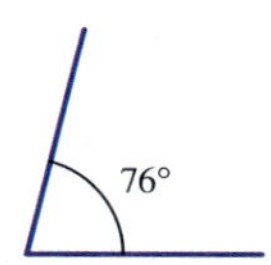

2 a
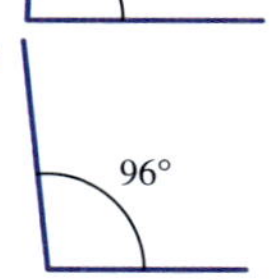

b
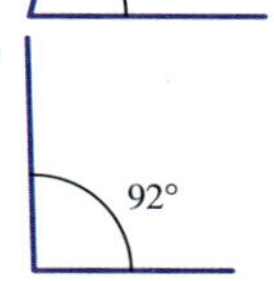

c
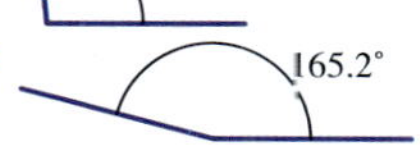

d
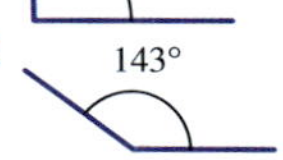

e
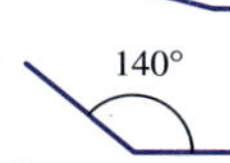

f

g

h
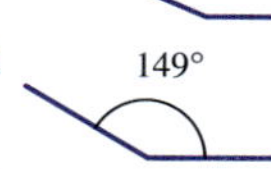

3 a
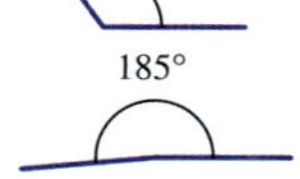

b
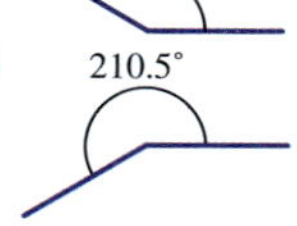

c
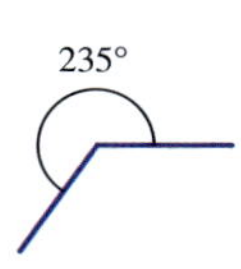

d
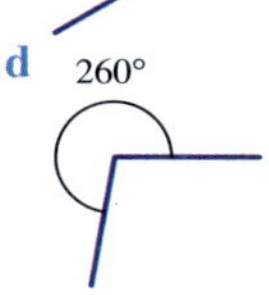

e
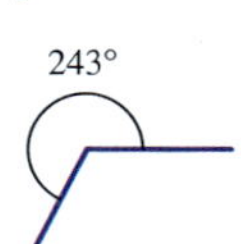

f
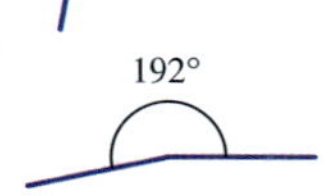

g 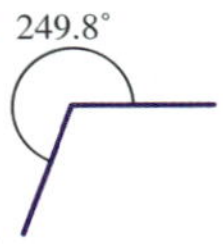

h 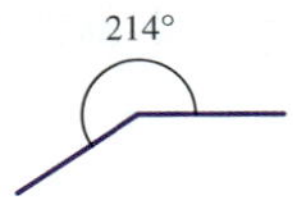

4 C

5 a 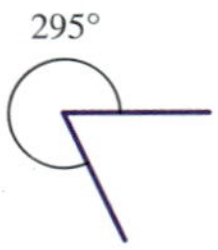

b 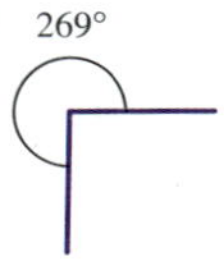

c 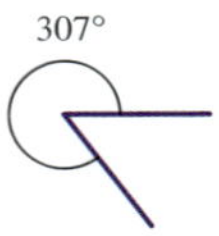

d 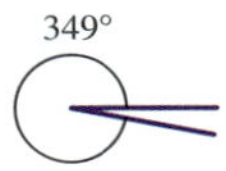

e 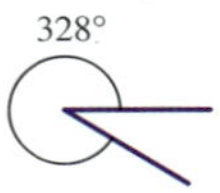

f 

g 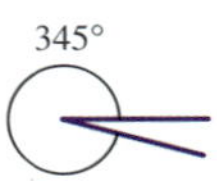

h 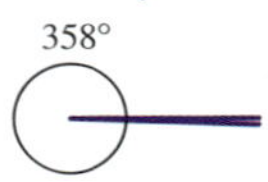

Exercise 10C — Types of angles

1 a Acute b Reflex c Obtuse
d Obtuse e Straight f Acute
g Reflex h Right i Acute
j Revolution k Obtuse l Acute
m Acute n Reflex o Acute
p Reflex

2 a 4; 3°, 45°, 65°, 69° b 2; 123°, 165°
c Yes, 90° d Yes, 180°
e 4; 234°, 300°, 270°, 210° f Yes, 360°

3 a Right b Straight line c Obtuse
d Acute e Reflex f Revolution

4 a Right b Acute c Revolution
d Obtuse

5 a Acute b Right c Obtuse
d Straight e Reflex f Revolution

6 a i Acute ii Obtuse iii Straight iv Right
v Obtuse vi Obtuse
b Various answers

7 a Angle 1: obtuse; angle 2: acute; angle 3: straight
b Angle 1: 140°; angle 2: 48°; angle 3: 180°
c Various answers

10 Quick Questions 1

1 150° 2 296°

3 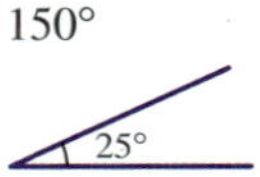

4 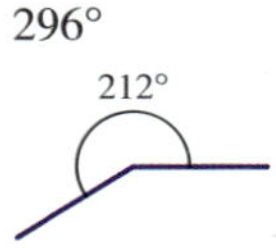

5 Obtuse 6 Reflex
7 2 acute angles 8 Many possible
9 3 obtuse angles 10 1 reflex angle

Exercise 10D — Naming angles

1 a ∠XUV or ∠VUX b ∠PQR or ∠RQP
c ∠SDX or ∠XDS d ∠GPZ or ∠ZPG
e ∠PFH or ∠HFP f ∠ZRO or ∠ORZ

2 Answers will vary; however, in each case, the middle letter should be at the vertex of the angle.

a 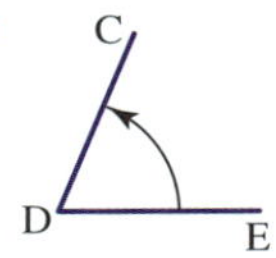

b 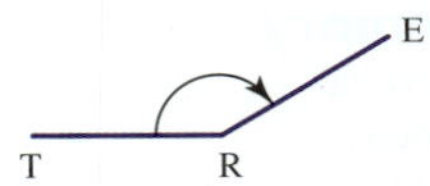

c 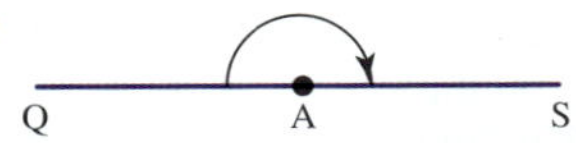

d 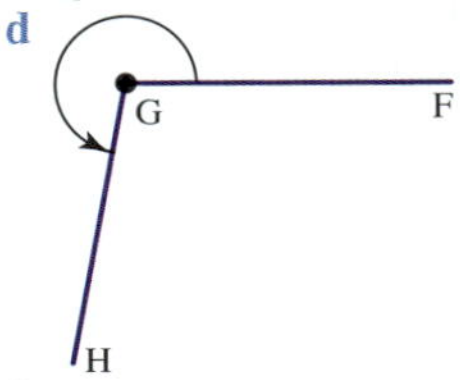

e 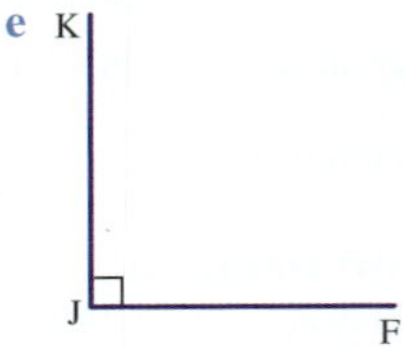

f 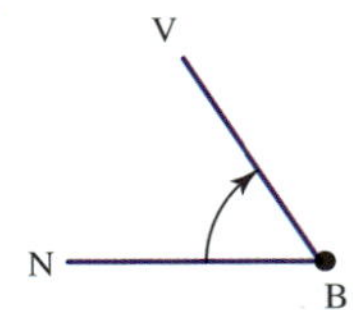

3 a Acute: ∠NOM, ∠MOY; obtuse: ∠MOX
b Acute: ∠PQS, ∠RQO; obtuse: ∠SQR, ∠SQO, ∠RQP
c Acute: ∠MLN, ∠NLO, ∠OLP, ∠PLK, ∠NLP, ∠OLK; obtuse: ∠MLO, ∠MLP, ∠NLK
d Acute: ∠POR, ∠ROT, ∠TOU, ∠UOV; obtuse: ∠POU, ∠ROU, ∠ROV
e Acute: ∠YOX, ∠WOQ, ∠POQ obtuse: ∠XOW, ∠YOP , ∠YOW, ∠XOP, ∠YOQ, ∠POW
f Acute: ∠VOT, ∠TOR, ∠ROP, ∠POZ, ∠ZOW, ∠WOV; obtuse: ∠VOR, ∠TOP, ∠ROZ, ∠POW, ∠ZOV, ∠WOT
g Acute: ∠FOG, ∠GOK ; obtuse: ∠DOG
h Acute: ∠JOH, ∠KOH, ∠POK; obtuse: ∠JOK, ∠HOP

4 In each case the middle letter should be at the vertex of the angle.
a i Acute: ∠HUT
ii Right: ∠ABC, ∠GHI
iii Obtuse: ∠DEF
iv Reflex: ∠MNO
b i ∠DEF
ii ∠ABC, ∠XWY
iii ∠PQR
iv Various answers possible.

Exercise 10E — Calculating the size of angles

1 a 78° b 58° c 2° d 45°
e 17° f 65° g 21° h 84°

2 a 105° b 174° c 68° d 132°
e 13° f 118° g 147° h 55°

3 b, c, f 4 a, c, f

5 a 55° b 18° c 45° d 78°
e 24° f 32° g 28° h 72°
i 53°

6 30°

7 a 130° b 10° c 25° d 52°
e 75° f 52° g 58° h 78°
i 79°

8 a 320° b 288° c 225° d 152°
e 270° f 205° g 130° h 120°
i 72°

9 a 32° b 130° c 6° d 100°
e 40° f 160°

10 a $x = 18^\circ$ b $b = 90^\circ$ c $c = 110^\circ$ d $d = 55^\circ$

10 Quick Questions 2

1 133° 2 Obtuse
3 ∠DVM or ∠MVD 4 ∠PQF or ∠FQP
5 56° 6 159°
7 $m = 11^\circ$ 8 $k = 55^\circ$
9 $t = 126^\circ$ 10 $w = 153^\circ$

Maths Quest challenge (page 412)

1 45° 2 150°

3 a 180° b 90° c 0°
d 30° e 150° f 15°

4 a 270° b 120° c 330°

Exercise 10F — Constructing angles with a pair of compasses

Check your constructions with your teacher.

Exercise 10G — Bearings

1 a 025° true b 207° true c 322° true
d 060° true e 192° true f 077° true
g 132° true h 261° true i 170° true
j 168° true k 291° true l 135° true

2 a N 25° E b S 27° W c N 38° W
d N 60° E e S 12° W f N 77° E
g S 48° E h S 81° W i S 10° E
j S 12° E k N 69° W l S 45° E or SE

3 a N 23° E b S 36° E c S 18° W
d S 45° W e Due west f N 45° W
g N 2° W h N 70° E i Due east
j N 42° E k Due south l S 25° W

4 a
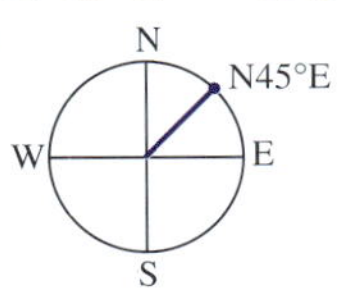

b
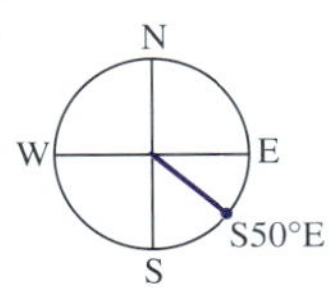

c
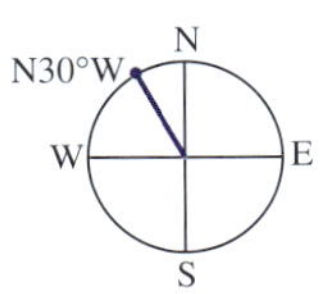

d
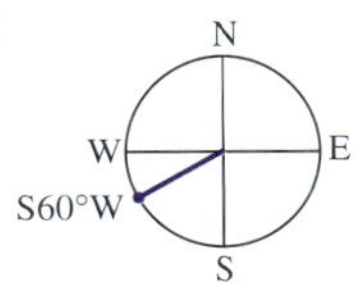

e
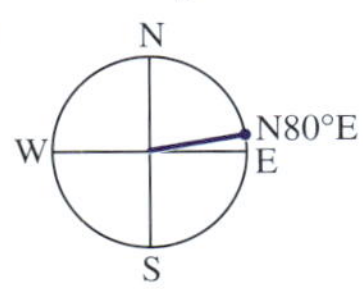

f
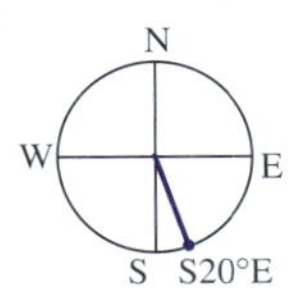

g
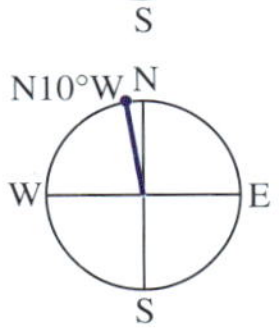

h
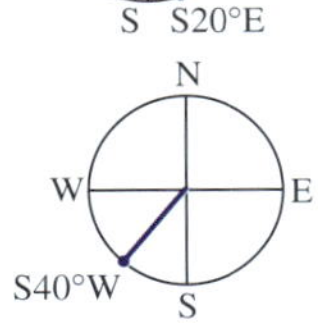

5 a 023° true b 345° true c 355° true d 088° true
e 125° true f 191° true g 106° true h 228° true

6

Point	True bearing	Compass bearing
1	090° true	due east
2	050° true	N 50° E
3	340° true	N 20° W
4	140° true	S 40° E
5	180° true	due south
6	220° true	S 40° W
7	250° true	S 70° W
8	290° true	N 70° W
9	310° true	N 50° W
10	160° true	S 20° E
11	200° true	S 20° W
12	110° true	S 70° E

7 a 138° true b 095° true
c 129° true d 162° true

8 a 275° T b 208° T c 250° T

Maths Quest challenge (page 424)

1 $16\frac{2}{3}$ m 2 250 minutes

3 Mowing lawns, $519.35

Summary

1 lines 2 a vertex, the arms
3 degrees, estimate 4 subtract it from 360°
5 'outside' 6 acute
7 right 8 obtuse
9 180° 10 reflex
11 one revolution 12 vertex of the angle
13 symbol ∠ 14 Complementary
15 Supplementary 16 circle
17 equal 18 halve
19 from north, three
20 north–south line, east or west

Chapter review

1 a 42° b 105° c 298°

2 Constructions

3 a Revolution b Right angle c Straight line
d Obtuse e Reflex f Acute

4 C 5 A 6 B 7 D

8 ∠EDF or ∠FDE

9 a 135° b 131° c 100° d 115° e 155°

10 a 78° b 50° c 34° d 52° e 20°

11 a 35° b 165° c 122° d 90° e 20°
f 120° g 14° h 45° i 30°

12 Check your constructions with your teacher.

13 Check your constructions with your teacher.

14 Check your constructions with your teacher.

15 a N 78° E b S 55° E c S 9° W d S 70° W
e N 65° W

16 a 315° true b 072° true c 125° true
d 192° true e 270° true

Chapter 11 Plane and solid shapes

Are you ready?

1 a isosceles b equilateral
c scalene

2 i triangle a is obtuse-angled
ii triangle b is acute-angled
iii triangle c is right-angled

3 a quadrilateral b hexagon
c triangle d octagon
4 F6 5 G4 6 E2

Exercise 11A — Triangles

1 a △ABC Scalene b △PQR Isosceles
c △XYZ Isosceles d △LMN Equilateral
e △RST Scalene f △EFG Isosceles

2 a Right-angled triangle b Acute-angled triangle
c Acute-angled triangle d Acute-angled triangle
e Obtuse-angled triangle f Right-angled triangle

3 a
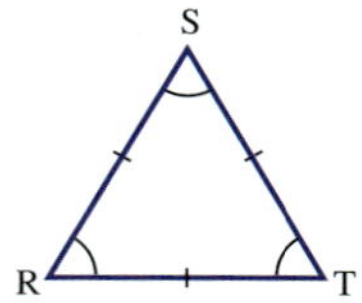

b
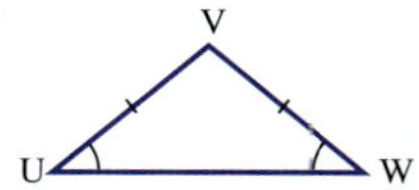

c
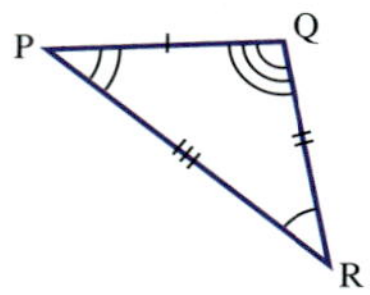

d
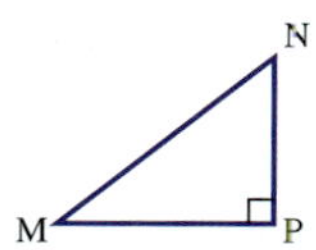

e
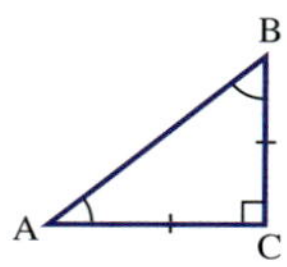

f
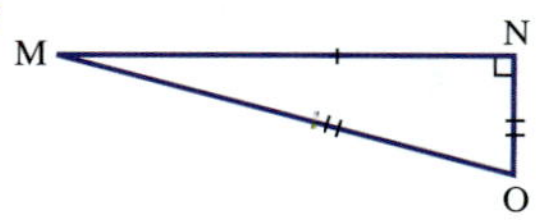

4 a A b B
5 a D b A
6 Isosceles/acute-angled triangle
7 Check with your teacher.
8 Check with your teacher.
9 a 5 b 6 c 2
10 c 10 cm (a, b, d; check with your teacher)
11 a $x = 57°$ b $g = 125°$ c $t = 44°$
d $k = 60°$ e $f = 93°$ f $z = 90°$
12 a $d = 45°$ b $b = 65°$ c $a = 50°$
13 a $c = 70°$ b $e = 76°$ c $n = 52°$
14 a $u = 66°$ b $k = 45°$ c $d = 76°$
d $t = 74°$ e $f = 57°$ f $p = 70°$
15 a $b = 20°$ b $p = 60°$ c $k = 30°$ d $p = 56°$
16 a 70° b 156° c 50° d 64°
17 a $p = 50°$; $n = 70°$ b $a = 22°$; $b = 28°$
c $x = 80°$, $y = 50°$ d $t = 35°$; $s = 55°$
e $b = 46°$ f $n = 21°$
g $m = 60°$ h $t = 28°$
18 Check with your teacher.
19 a Yes, angles add to 180°.
b No, angles add to 175°.
c No, angles add to 185°.
d Yes, angles add to 180°.
20 The sum of the angles will be greater than 180°.

Maths Quest challenge (page 444)

1 a 27 b 32
2 a 16 b 44

Exercise 11B — Quadrilaterals

1 a Square b Trapezium
c Kite d Rhombus
e Rectangle f Irregular quadrilateral
2 a E b A c E
3 a True b True c False
d True e True
4 D
5 a
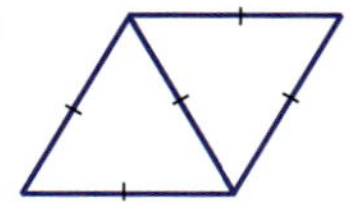
b
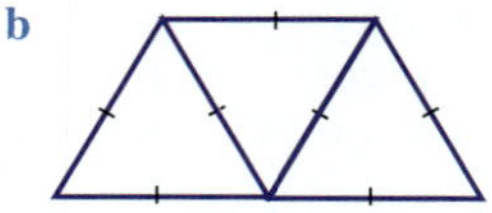
c
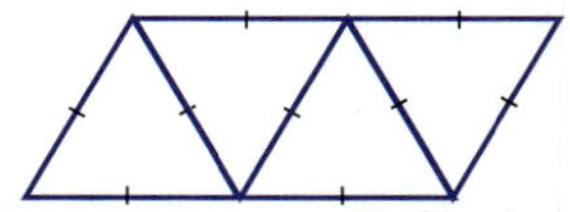
6

7 Check with your teacher.
8 a $t = 138°$ b $b = 85°$ c $t = 90°$
d $t = 202°$ e $p = 92°$ f $m = 90°$
g $s = 53°$ h $c = 66°$ i $k = 67°$
9 a $m = 78°$ (opposite angles of parallelogram equal)
b $u = 75°$ (opposite angles of parallelogram equal)
c $t = 132°$ (opposite angles of a kite equal)
d $f = 108°$ (opposite angles of a kite equal)
e $p = 63°$ (opposite angles of parallelogram equal)
f $z = 73°$ (opposite angles of parallelogram equal)
10 a $m = 113°$ b $c = 98°$; $d = 82°$
c $t = 116°$ d $p = 52°$; $m = 106°$
e $t = 103°$; $x = 91°$
f $m = 65°$; $p = 115°$; $n = 65°$
11 D 12 E 13 $p = 61°$
14 115° 15 45° and 135°
16
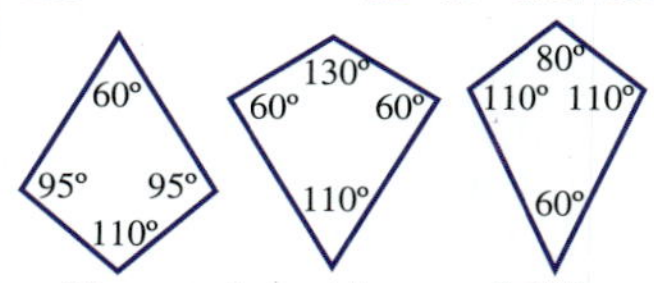

17 a Yes, angles add up to 360°.
b No, angles add up to 270°.
18 a 130° b Infinite number
c Various answers

Exercise 11C — Polygons

1 a Octagon, regular b Nonagon, irregular
c Decagon, irregular d Triangle, regular
e Heptagon, irregular f Decagon, irregular
2 a, b, c, e and f are convex.
3 Answers will vary.
4 12 sides; dodecagon
5 a B b B
6 a Triangles, hexagons, quadrilaterals
b Quadrilaterals, triangles, octagons
c Hexagons
d Quadrilaterals, hexagons, octagons
7 Answers will vary.

10 Quick Questions 1

1 Scalene 2 Obtuse angled 3
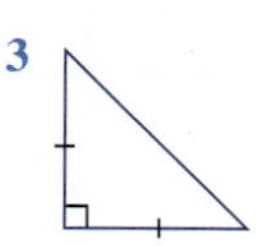
4 Trapezium 5
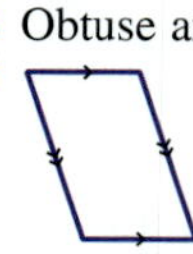
6 Rhombus 7 Parallelogram

8 Octagon, regular
9 Answers may vary — figures must have five sides and an angle greater than 180°.
10 Hexagon

Exercise 11D — Constructing polygons

1–7 Check your constructions with your teacher.
8 D

Exercise 11E — Plans and views

1 a
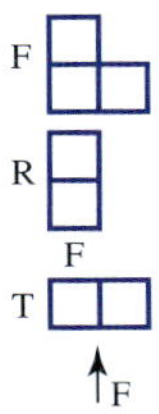

b
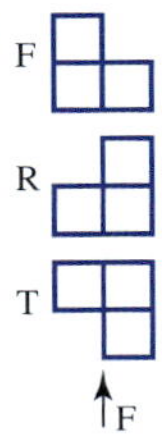

c
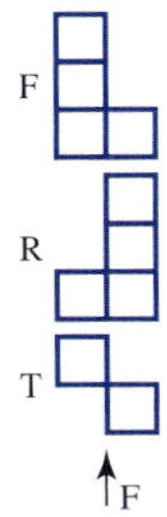

d
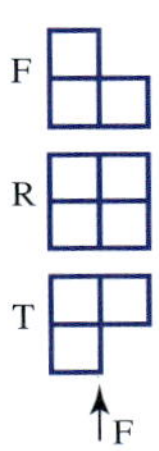

e
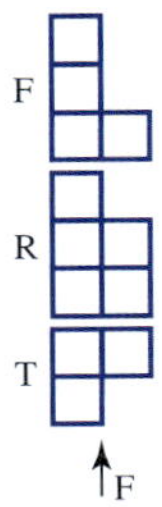

f
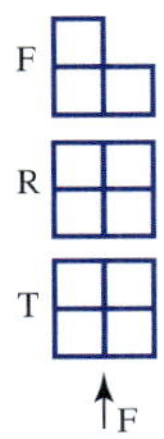

g
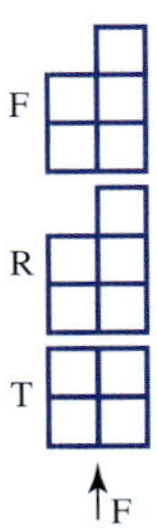

h
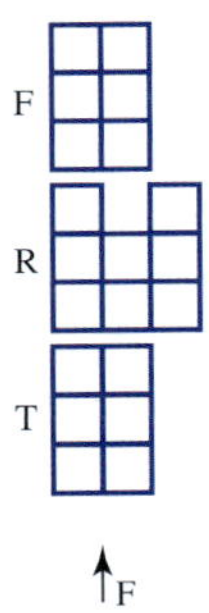

i
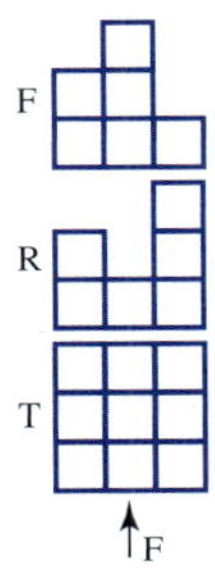

2 a
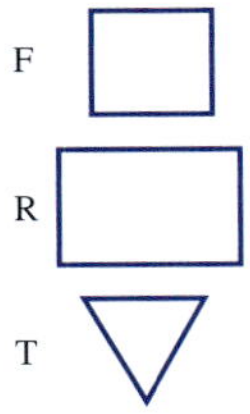

b
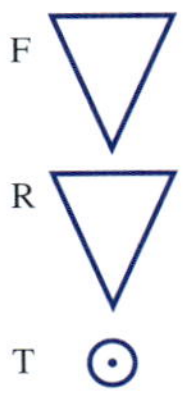

c
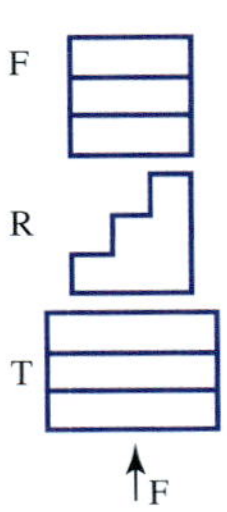

d
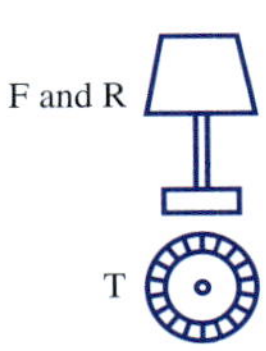

3 a
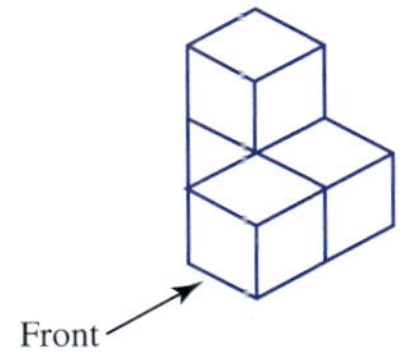

b
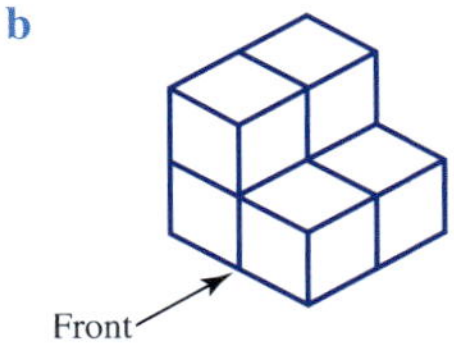

c
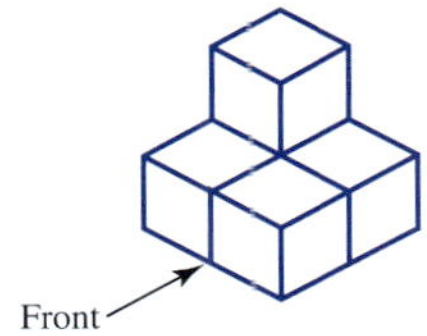

d
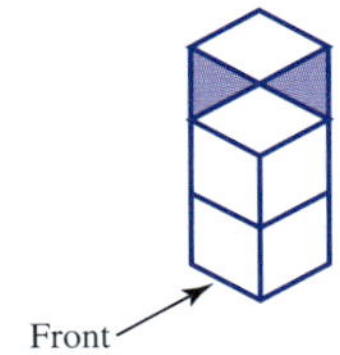

4 D
5 a Circle b Oval c Trapezium d Rectangle
6 a
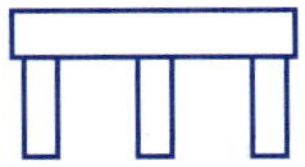

7 a It would be difficult to construct this solid.
b
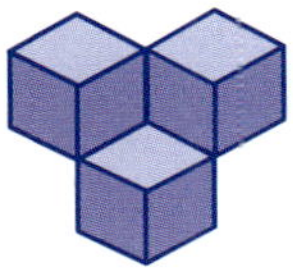
The solid would look like the one shown. The 4th cube is behind the front bottom cube, resting on the surface. The two top cubes have no cubes supporting them.

8
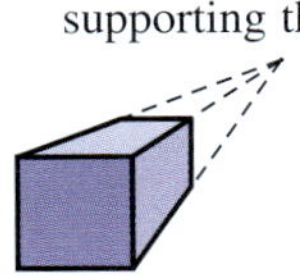

Exercise 11F — Polyhedra, nets and Euler's Rule

1 a i 6 faces, all square
ii 8 vertices
iii 12 edges
b i 8 faces (all triangles)
ii 6 vertices
iii 12 edges
c i 12 faces (all pentagons)
ii 20 vertices
iii 30 edges
2 i Cube ii Octahedron iii Dodecahedron
3 a 4 b 4 c 6
4 a 6 b 8 c 12
5 a 8 b 12 c 18
6 a 20 b 12 c 30
7

Question	Polyhedron	Number of faces (F)	Number of vertices (V)	Number of edges (E)	F + V − E
1 i	cube	6	8	12	2
1 ii	octahedron	8	6	12	2
1 iii	dodecahedron	12	20	30	2
3	tetrahedron	4	4	6	2
4	parallelepiped	6	8	12	2
5	truncated tetrahedron	8	12	18	2
6	icosahedron	20	12	30	2

8 **a** No **b** Yes

9 **a** 24 **b** 28 **c** 6 **d** 18

10 **b** **i** F:5, E:8, V:5 **ii** F:9, E:15, V:8

Exercise 11G — Axes of symmetry

1 Check with your teacher.

2 **a** 3 **b** 0 **c** 1 **d** 1
e 1 **f** 0 **g** 1 **h** 4

3 **a** 2 **b** 2 **c** 6 **d** 8
e 5 **f** 0

4 **a** A B C D E H I K M O T U V W X Y
b c i o v w x **c** H I O X o x

5 C **6** A

7 **a**

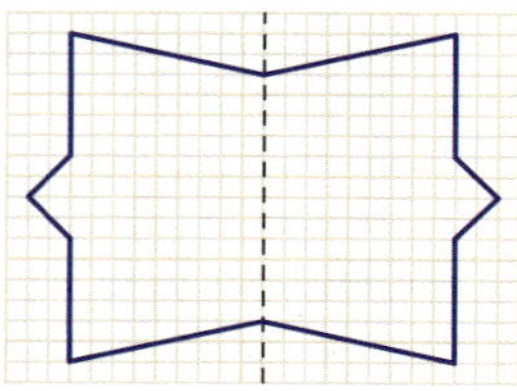

b

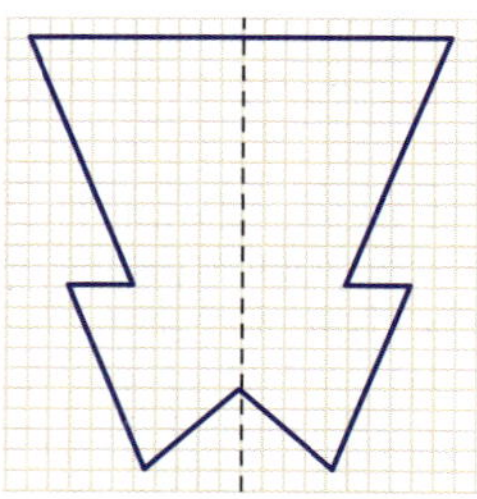

8 Check with your teacher.

Maths Quest challenge (page 475)

1 3 ducks

2

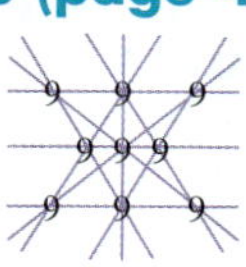

3 Thanh's father

10 Quick Questions 2

1

2 Rectangle

3 Faces = 8, vertices = 12, edges = 18

4 $F + V - E = 8 + 12 - 18 = 2$

5 Square-based pyramid

6

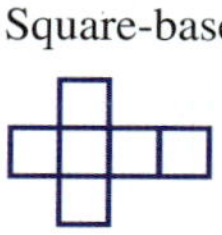

7

8 No, because Euler's rule will not work

9 4

10 **a**

b

c

d

Other solutions are possible.

Exercise 11H — Transformations

1 **a** 9 right, 4 down **b** 3 right, 2 down
c 1 up, 16 left **d** 1 down, 16 right
e 16 right, 9 down **f** 1 down, 32 left
g 7 down **h** 18 right, 8 up
i 10 right, 3 up **j** 2 down, 19 left
k 2 up, 29 left **l** 13 up, 11 left

2 **a**

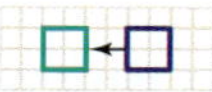

b

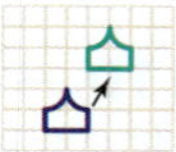

c

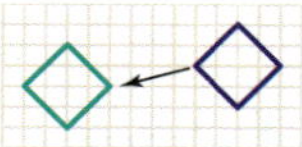

d

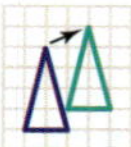

e

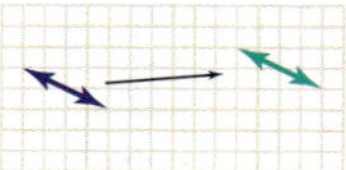

f

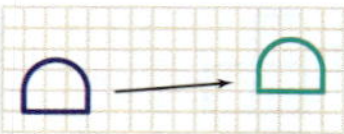

3 **a**

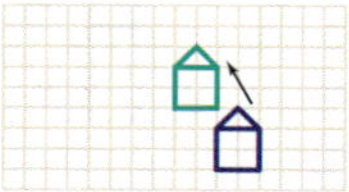

b

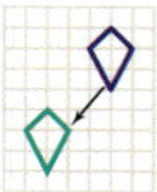

c

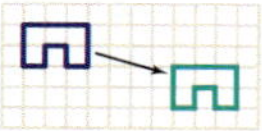

d

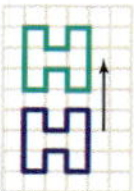

e

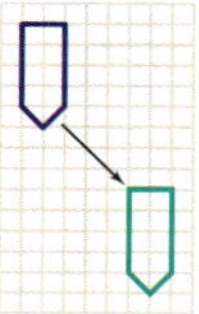

f

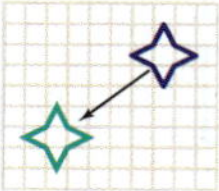

4 **a** 11 right, 1 down **b** 4 right, 5 down
c 4 right, 3 down **d** 5 left, 2 up

5 D

6 **a** Each car body is translated one car body width sideways.
b The third footprint is a translation of the first footprint. It has been translated 2 steps forward.
c The left highlighted tile has been translated 2 tiles to the right and 1 tile up to obtain the other highlighted tile.
d The position of the right starfish is a translation of the position of the left starfish. It has been translated 5 cm down and 8 cm to the right.

7 **a**

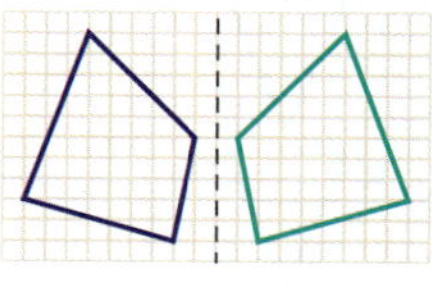

b

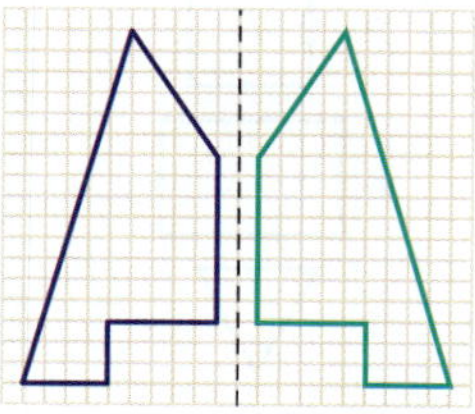

c

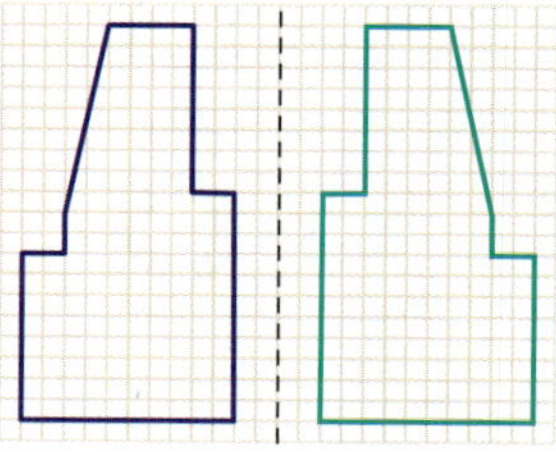

d

8 a

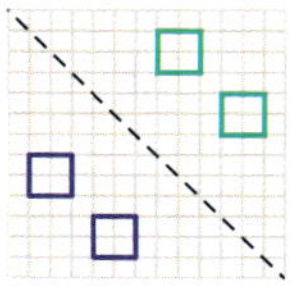

b

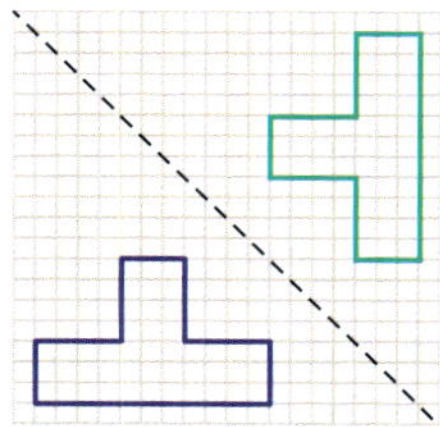

c

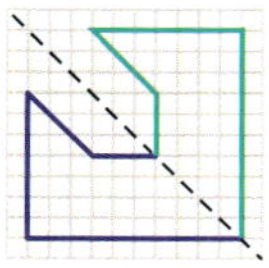

d

9 a

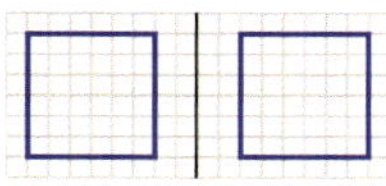

b

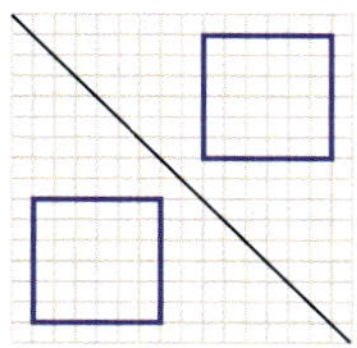

c

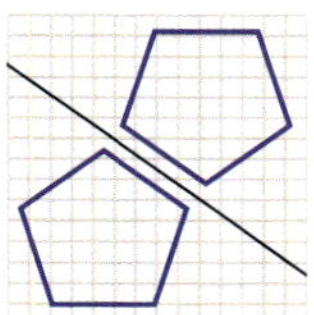

d

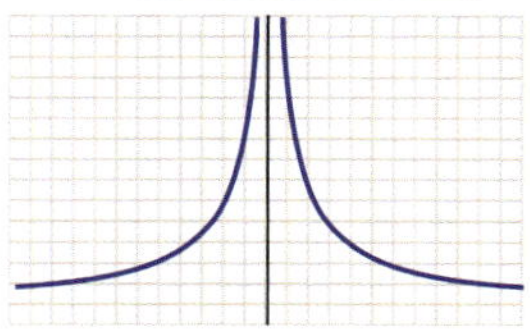

e

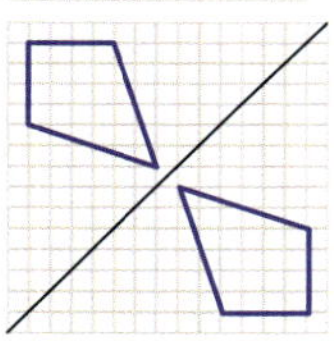

f

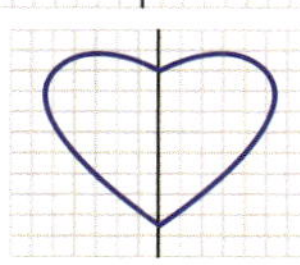

10 a

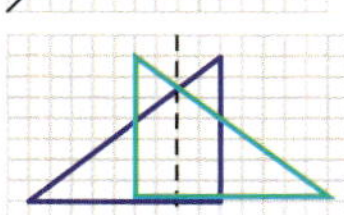

b

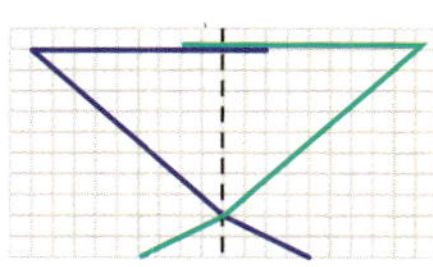

c

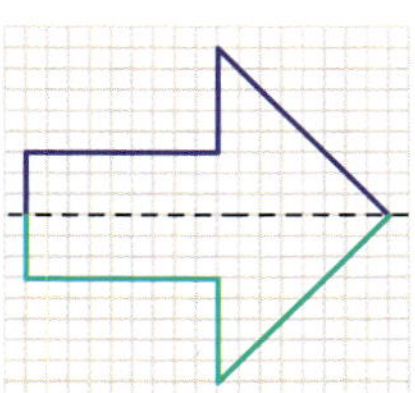

d

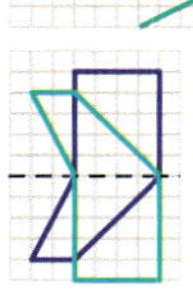

e

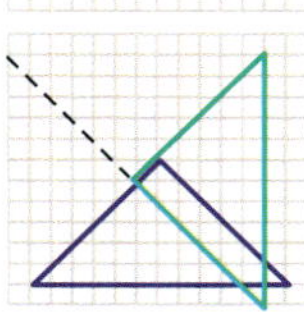

f

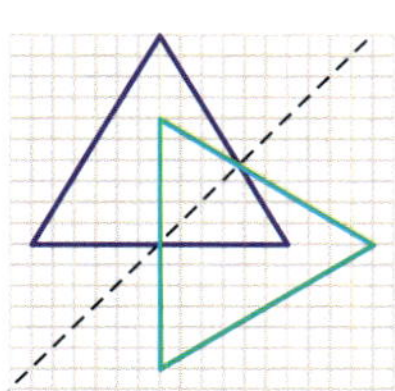

11 Check with your teacher.

12 a $\frac{1}{2}$ turn clockwise, $\frac{1}{2}$ turn anticlockwise

b $\frac{1}{4}$ turn clockwise, $\frac{3}{4}$ turn anticlockwise

c $\frac{1}{2}$ turn clockwise, $\frac{1}{2}$ turn anticlockwise

d $\frac{3}{4}$ turn clockwise, $\frac{1}{4}$ turn anticlockwise

e $\frac{1}{2}$ turn clockwise, $\frac{1}{2}$ turn anticlockwise

f $\frac{1}{4}$ turn clockwise, $\frac{3}{4}$ turn anticlockwise

g $\frac{1}{4}$ turn clockwise, $\frac{3}{4}$ turn anticlockwise

h $\frac{1}{2}$ turn clockwise, $\frac{1}{2}$ turn anticlockwise

i $\frac{1}{2}$ turn clockwise, $\frac{1}{2}$ turn anticlockwise

13 a

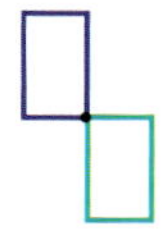

b

c

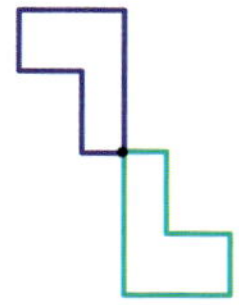

d

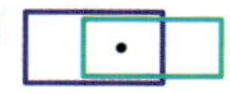

e

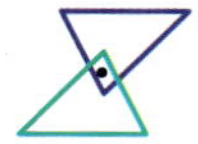

f

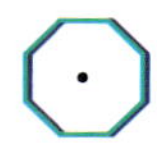

14 a

b

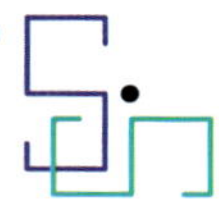

c

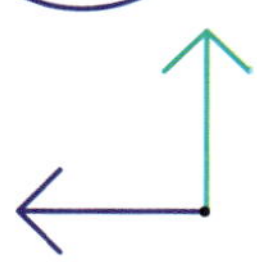

d

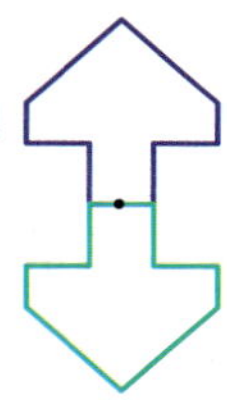

e

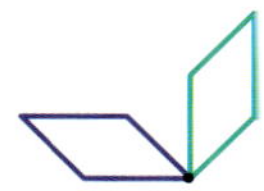

f

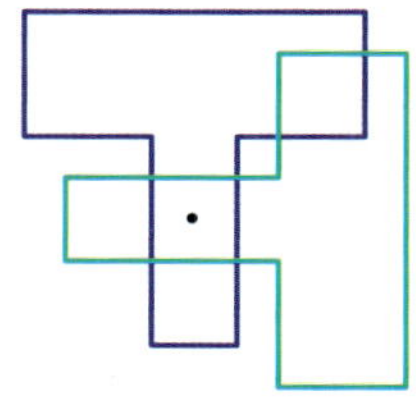

15 a $\frac{1}{5}$ turn (or 72° turn) **b** $\frac{1}{3}$ turn (or 120° turn)

c $\frac{1}{5}$ turn (or 72° turn) **d** $\frac{1}{16}$ turn (or $22\frac{1}{2}$° turn)

Exercise 11I — Tessellations

1–3 Check with your teacher.

4 a, **b**, **c**, **e**

5 a

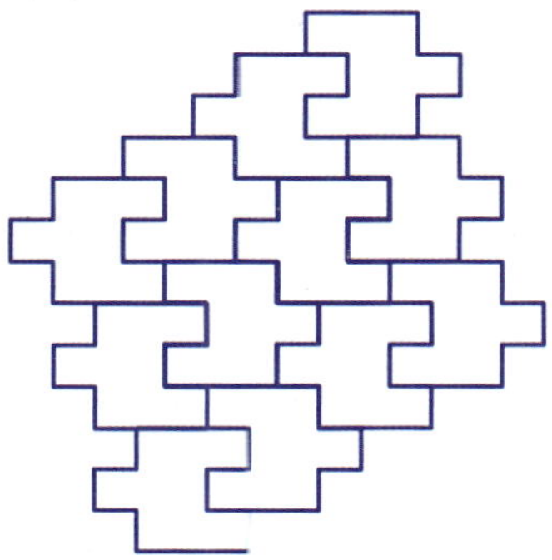

b 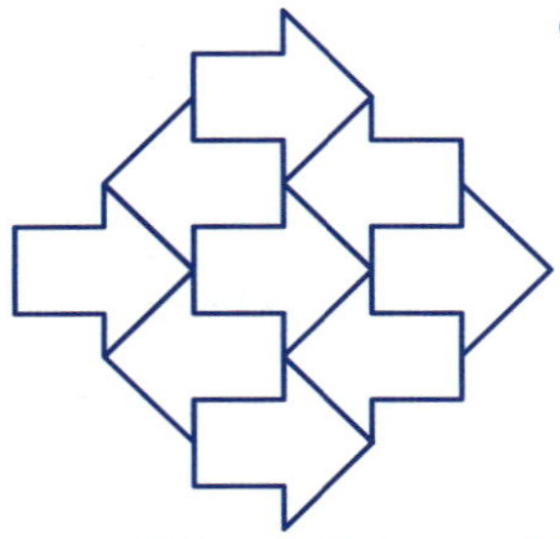

c 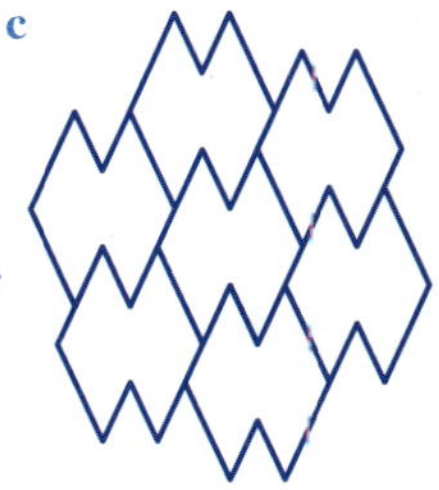

6 a 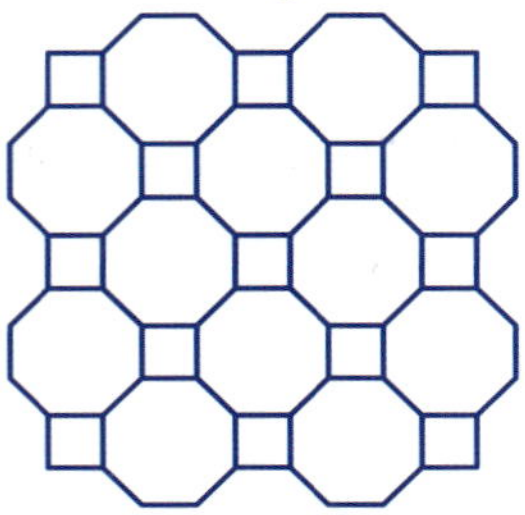

b 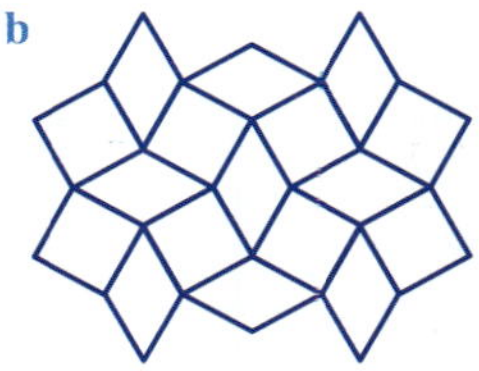

7 a 

b 

c 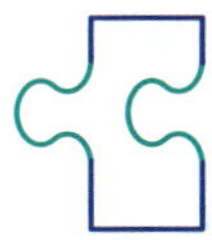

d 

e 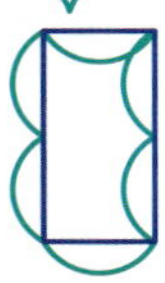

f 

g 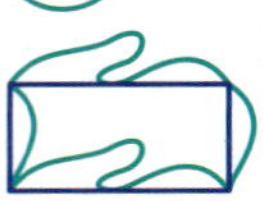

h 

i 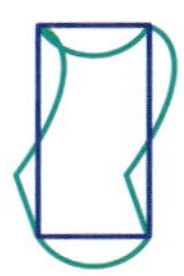

8 Check with your teacher.

9 a 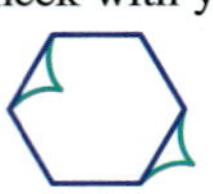

b 

c 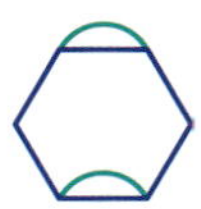

d 

e 

f 

10–11 Check with our teacher.

Summary

1 greater than
2 isosceles
3 equal
4 acute-angled
5 scalene
6 right
7 adjacent, unequal
8 rectangle
9 equal sides
10 trapezium
11 equal in size
12 parallel, opposite
13 closed, straight
14 regular
15 3-dimensional, polygon
16 180
17 exterior, supplementary
18 360
19 polyhedron, net
20 top
21 front
22 side elevation
23 perspective
24 faces, vertices, edges
25 transformations
26 Translation
27 folded
28 mirror
29 rotational
30 more than one
31 image
32 object
33 a set point
34 direction
35 the size of the angle
36 repeated use

Chapter review

1 a Equilateral triangle b Scalene triangle
 c Isosceles triangle
2 a Obtuse-angled triangle b Acute-angled triangle
 c Right-angled triangle
3 a $t = 67°$ b $b = 28°$ c $x = 75°$
4 a $p = 56°$ b $m = 64°$ c $n = 60°$
5 67°
6 a Rectangle b Parallelogram c Trapezium
 d Kite e Square f Rhombus
7 a 75° b 55° c 42°
8 a 66° b 112° c 126°
9 115°
10 a Heptagon b Quadrilateral c Dodecagon
11 Answers will vary but should have 6 sides.
12 Construction
13 a 60°

14 a 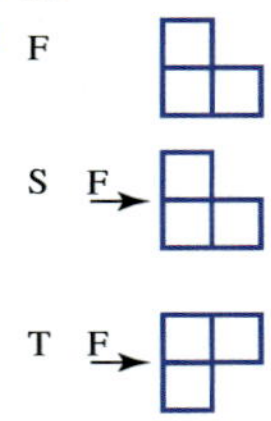

b 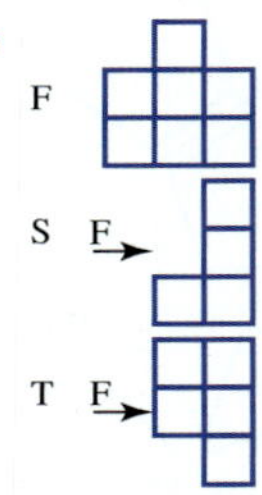

15 a 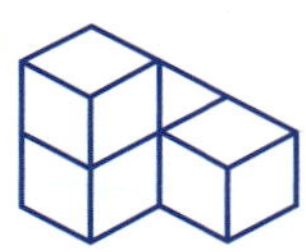

b 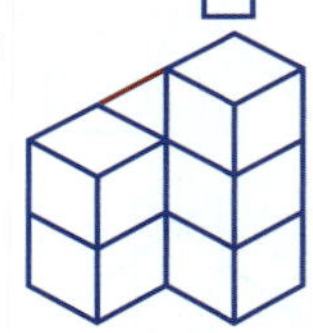

16 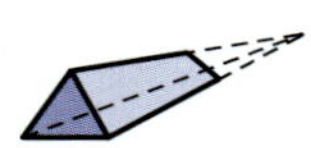

17 a i 8 ii 6 iii 12
 b i 10 ii 12 iii 20
18 a 16 b 8 c No
19 D

20 a 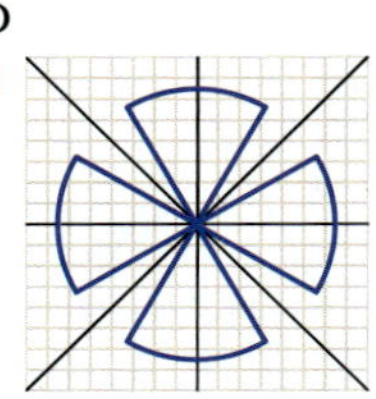

b 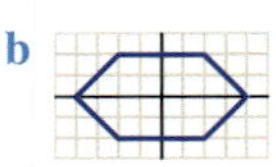

21 a 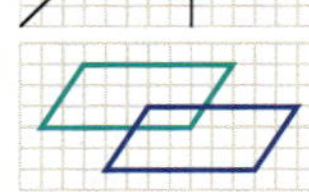

b 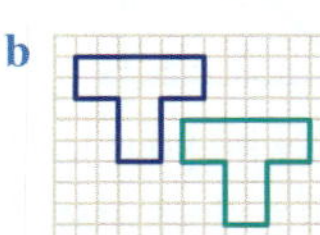

c 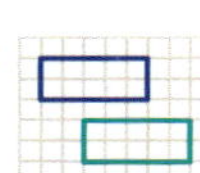

d 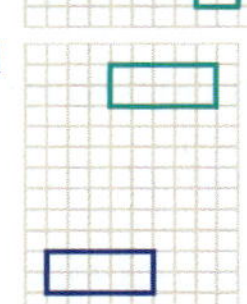

22 a 2 up, 6 left b 4 down, 8 right
c 3 up, 8 right

23 a 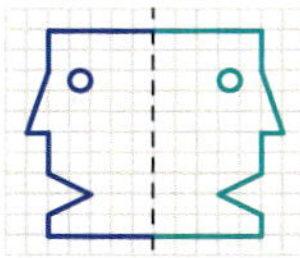 b 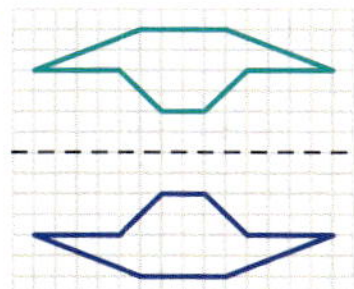
c 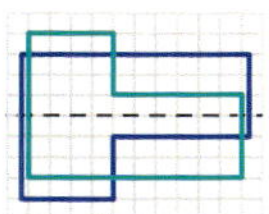 d 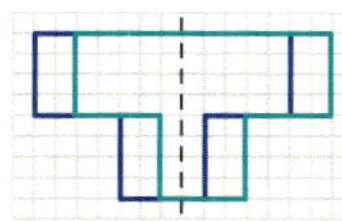

24 a 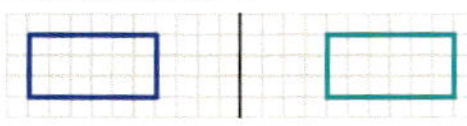 b 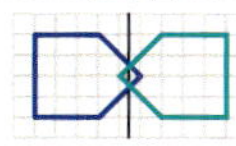
c 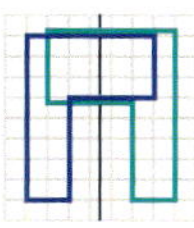

25 a 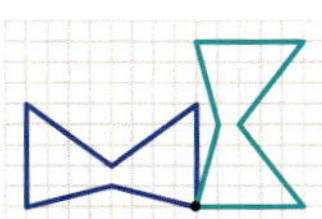 b 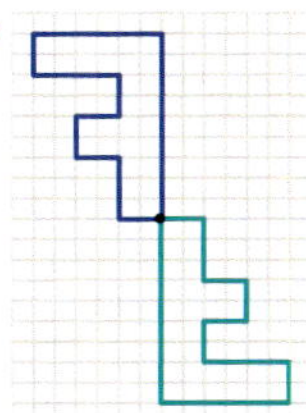 c 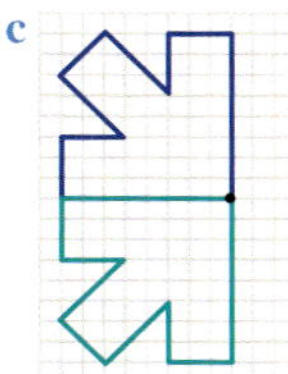

26 a 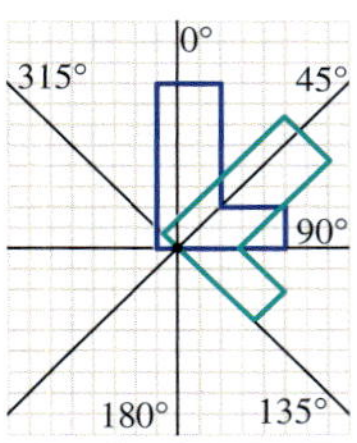

b 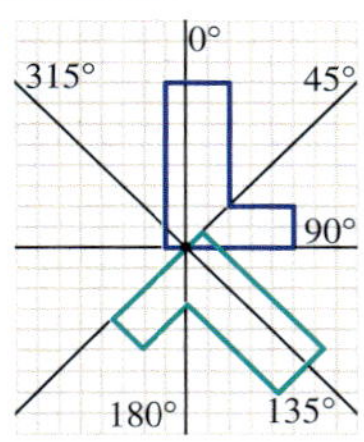

c 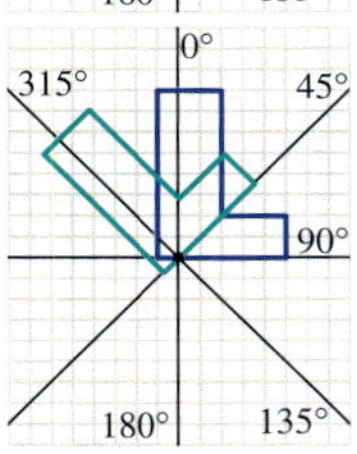

27 a 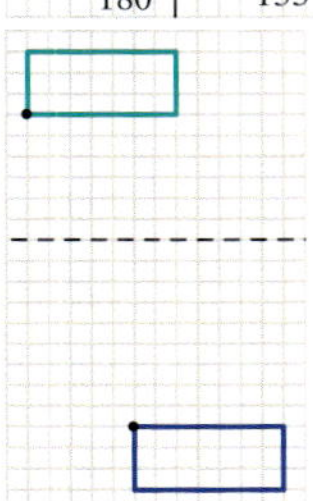 b 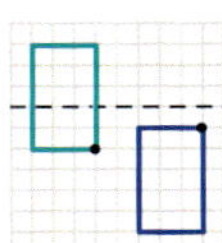 c 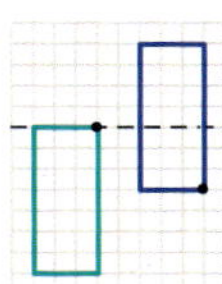

28 a 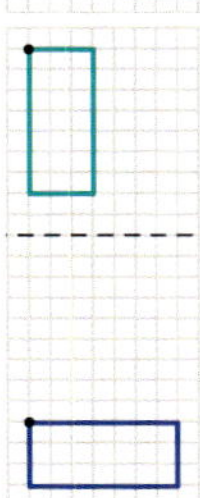 b 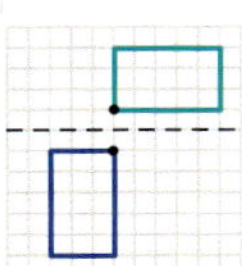 c 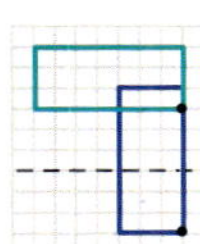

29 C

30 a 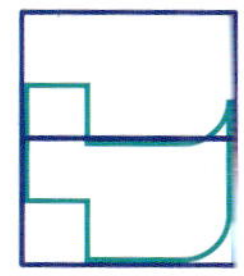 b 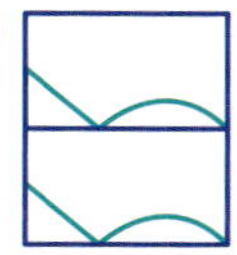
c 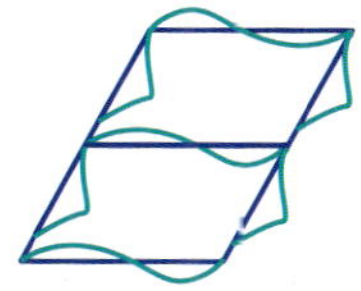 d 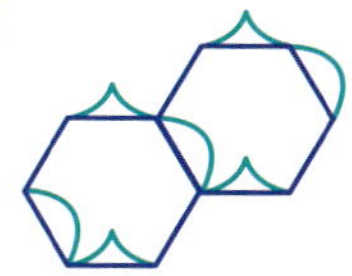

Chapter 12 Maps, networks and coordinates

Are you ready?

1 a 500 mm b 0.4 km c 70 m d 250 m
2 a 25 b 26 000 c 0.035 d 0.45
3 a NE b S c W d E
4 a 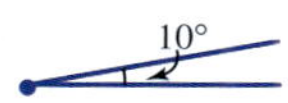

b 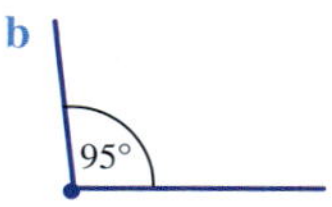

c 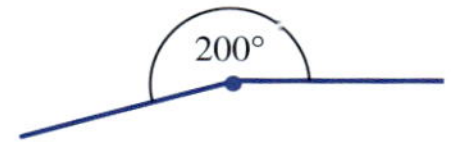

d 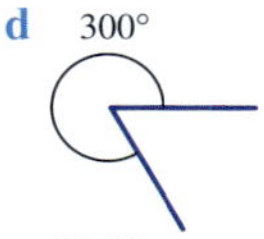

5 a 113° b 50° c 225° d 295°
6 a 3 right, 2 up b 5 right
c 7 up d 2.5 right, 1.5 up

Exercise 12A — Interpreting maps — scale

1 a 1 : 100 000 b 1 : 2 000 000
c 1 : 5 000 000 d 1 : 40 000
e 1 : 10 000 000 f 1 : 10
g 1 : 100 h 1 : 200 000 000
i 1 : 45 000
2 a 50 cm b 50 m c 2 m
d 500 m e 10 km f 400 m
g 300 m h 8 km i 100 m
3 a D b A
4 a D b B
5 2.8 cm
6 a 5.4 cm b 0.44 cm c 3.28 cm
d 0.154 cm e 2.3 cm
7 a Length = 3.6 m, width = 1.84 m b 28 cm
8 a Length = 23.75 m, width = 8.25 m b 93.5 m
9 a 240 km b 26 km c 39 km
10 a 1664 km b 1120 km c 3872 km d 256 km
11–14 Check with your teacher.

Exercise 12B — Location on Earth

1 a Cairo b Shanghai c Darwin
d Montreal e London f Auckland
g Tokyo h Beijing i Rio de Janeiro
2 a 38°S, 145°E b 40°N, 75°W
c 18°N, 76°W d 26°S, 28°E
e 42°N, 12°E f 35°S, 57°W
g 33°N, 44°E h 55°N, 40°E
i 2°N, 104°E j 18°S, 178°E
3 a E2 b B1 c B3 d D1 e B2
f C1 g D2 h A1 i D1

4 a Geelong b Ballarat
c Melbourne d Warnambool/Colac
e Portland f Mildura
g Albury/Wodonga/Wangaratta/Bairnsdale
h Mount Gambier i Griffith/Wagga Wagga

10 Quick Questions 1

1 1 : 100
2 2.5 km
3 0.5 cm
4 1.75 km
5 Suva
6 Southern
7 East
8 East
9 a 45°N, 80°W b 45°S, 170°E
10 a C3 b Nerang

Exercise 12C — Maps, bearings and plans

1 a
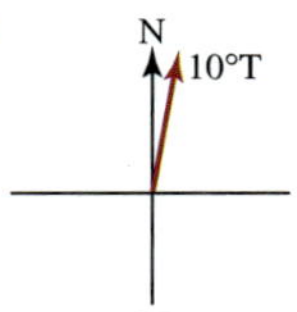

b
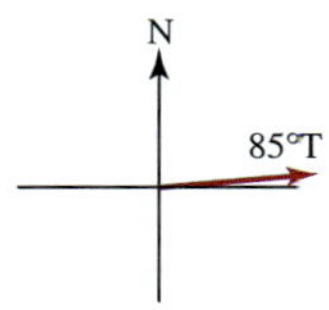

c
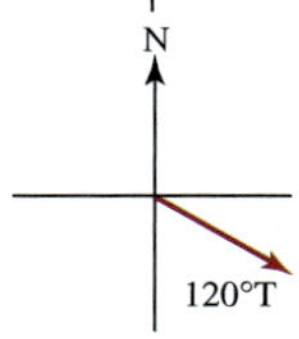

d
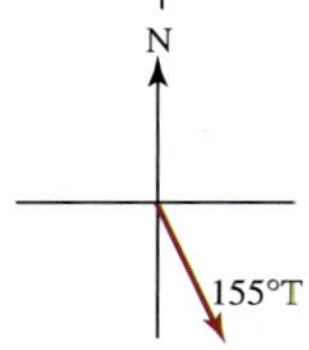

e
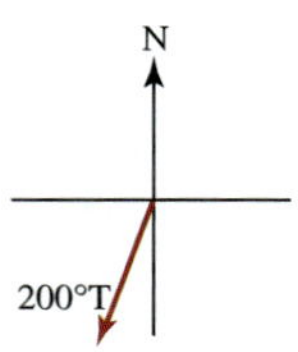

f
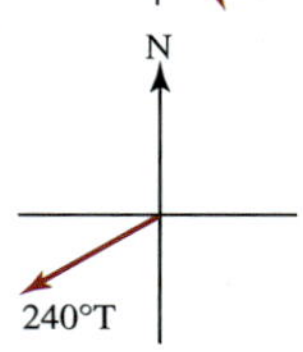

g
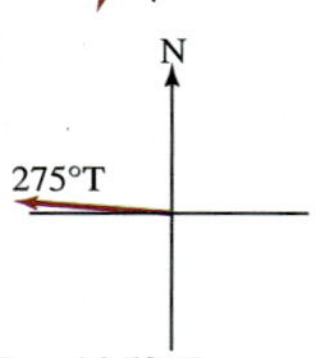

h
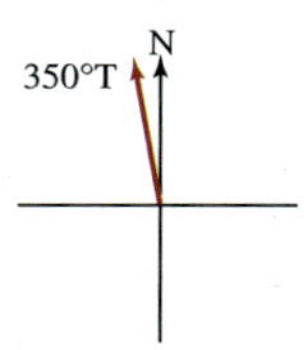

2 SE = 135° T
3 a A b E c B d B
4 a 70° T b 290° T c 190° T d 120° T
e 140° T f 200° T g 300° T h 70° T
5 a Check with your teacher.
b 7.1 km c 278° T
6 a 044° T
b
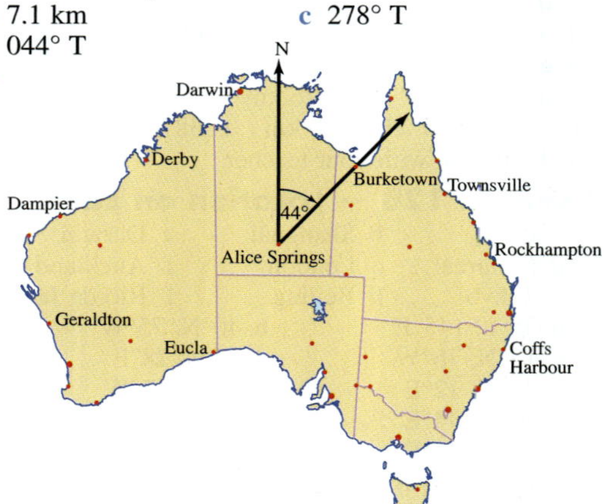

7 a 50 m b 50 m c X (280 m)
d AB e Z
f The land slopes downwards from 100 m to 50 m in moving from D to E.
g Y is in a depression 20 m below the 50 m contour.
h The land to the left of the river rises to a height of 280 m while the land to the right of the river rises to a height of 100 m then declines to a low point 30 m above sea level.

8 a
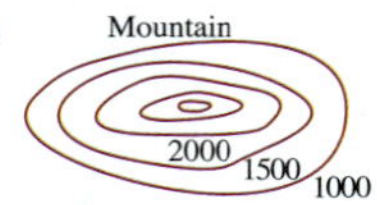

b
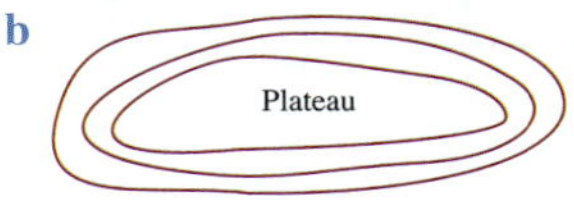

c
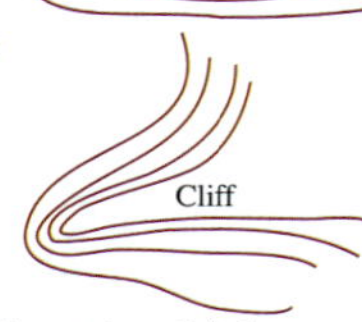

9 There is a high pressure system in the Great Australian Bight and this is causing fine weather. Another high off the Qld–NSW coast is bringing fine weather to that region. The fine weather extends to a lesser extent to the northern regions of Australia. On the whole, there is little chance of any rain throughout Australia.

10 a 123° T b S30°W c 2.86 km d 2.07 km

Exercise 12D — Networks

1 a 2 b 3 c Yes
2 a Yes b Yes c Yes
d Yes e Yes f No
g Yes h No i No
j No k Yes l Yes

3

	No. of vertices	No. of edges	Traversable?
a	5	7	Yes
b	2	4	Yes
c	2	2	Yes
d	2	6	Yes
e	3	4	Yes
f	4	6	No
g	2	5	Yes
h	4	7	No
i	6	10	No
j	4	6	No
k	9	15	Yes
l	16	28	Yes

4 Check with your teacher.
5 a–d Cannot be done.

Exercise 12E — Networks and maps

1 a
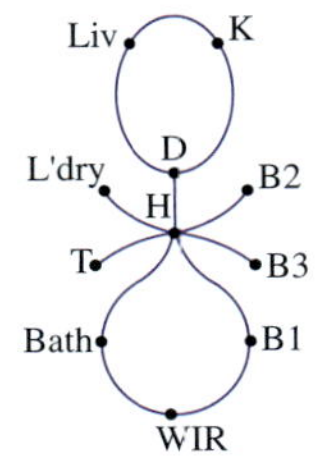

b
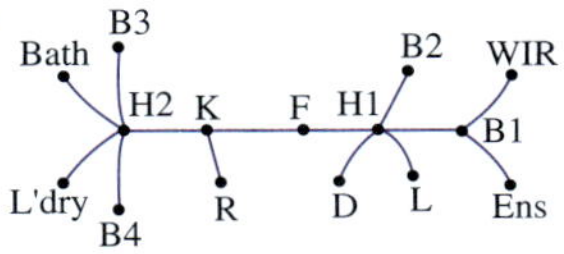

c
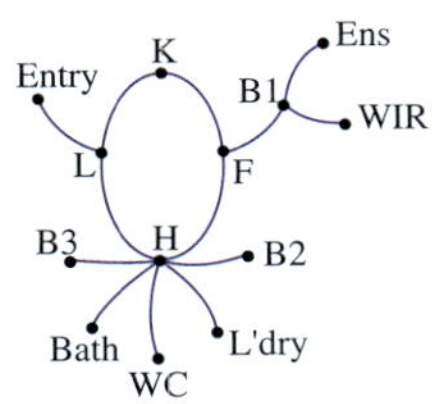

2
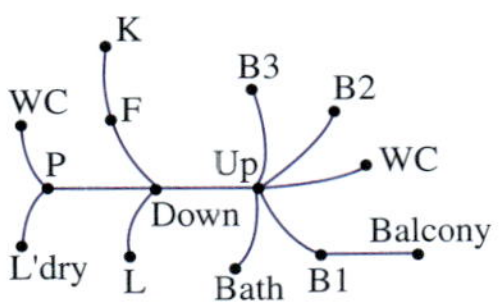

3
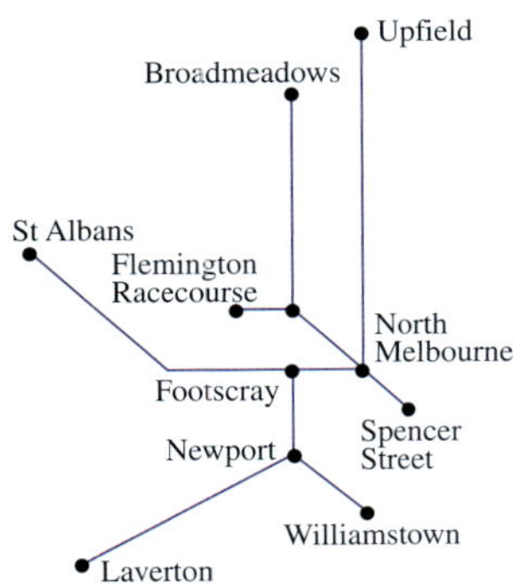

4 a
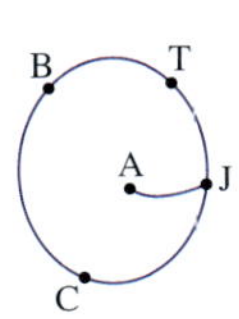

b
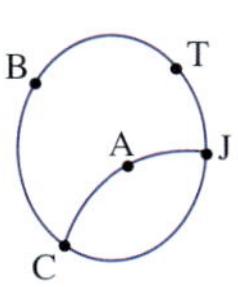

c
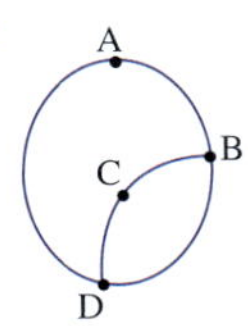

d
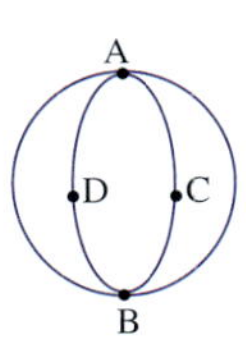

e
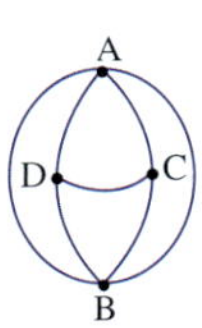

f
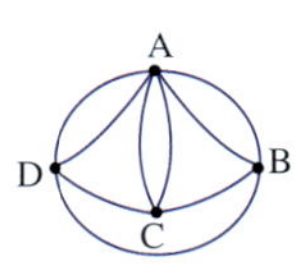

5 a, b
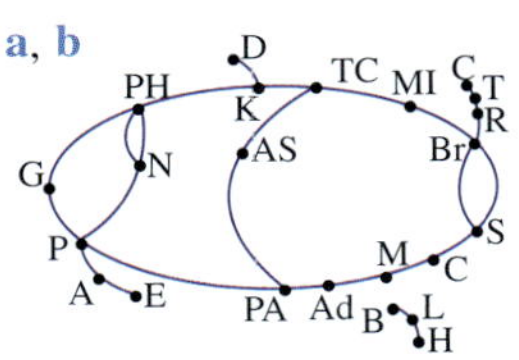

PH	Pt Hedland	M	Melbourne
N	Newman	C	Canberra
G	Geraldton	S	Sydney
P	Perth	Br	Brisbane
A	Albany	R	Rockhampton
E	Esperance	T	Townsville
PA	Pt Augusta	C	Cairns
Ad	Adelaide	MI	Mt Isa
B	Burnie	TC	Tennant Ck
L	Launceston	D	Darwin
H	Hobart	K	Katherine

6
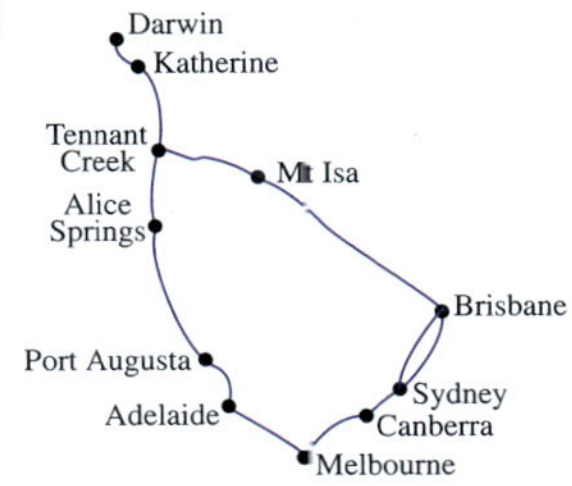

7
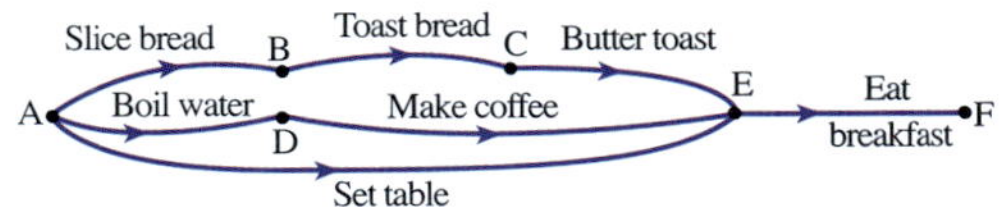

Maths Quest challenge (page 542)

1 One possible path is: A, C, A, B, C, D, E, F, D, K, F, G, H, K, B, H, A.
2 216
3 694 days, 10 hours and 40 minutes

10 Quick Questions 2

1 You can trace a network without lifting your pen or going over an edge twice.
2 Yes
3 No
4 4 vertices, 6 edges
5 Check with your teacher.
6 Check with your teacher.
7
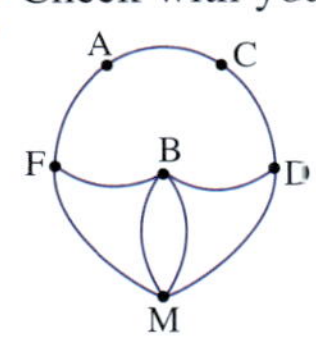

8 6 vertices, 8 edges

9 a
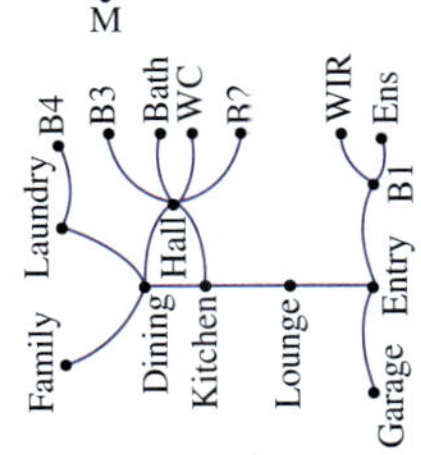

b 16 vertices, 16 edges

10 Answers will vary. Check with your teacher.

Exercise 12F — The Cartesian plane

1 a J3 b C3
c C5, C4, C3, D3, D4
2 a Powell Drive
b Silvereye Crescent
c Rosella Avenue
3 a B b C c D
d A e E f A
4
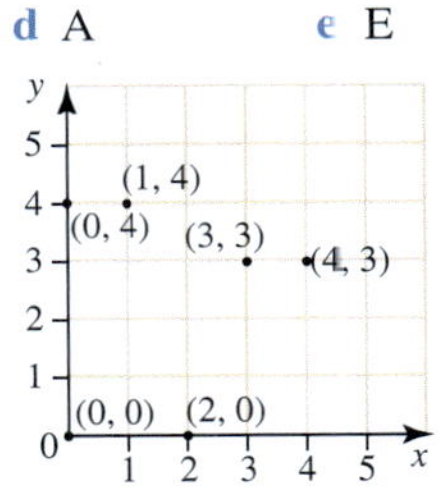

5 A(6, 1) B(2, 5) C(7, 9) D(12, 10) E(10, 0) F($4\frac{1}{2}$, 0) G(0, 6) H(0, 0) I($7\frac{1}{2}$, 5) J(10, $7\frac{1}{2}$) K($3\frac{1}{2}$, $2\frac{1}{2}$) L($\frac{1}{2}$, $9\frac{1}{2}$)

6 a E, C b B, D

7 In hospital; in wards six, seven, eight and nine

8 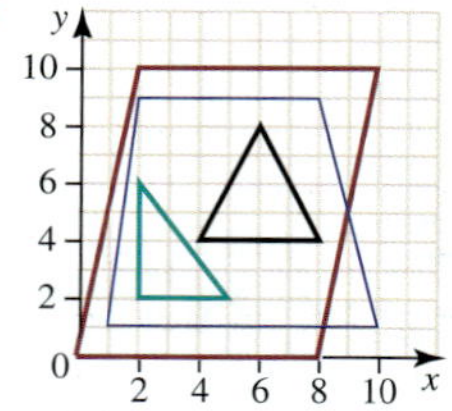

a Right-angled triangle b Isosceles triangle
c Trapezium d Parallelogram

9 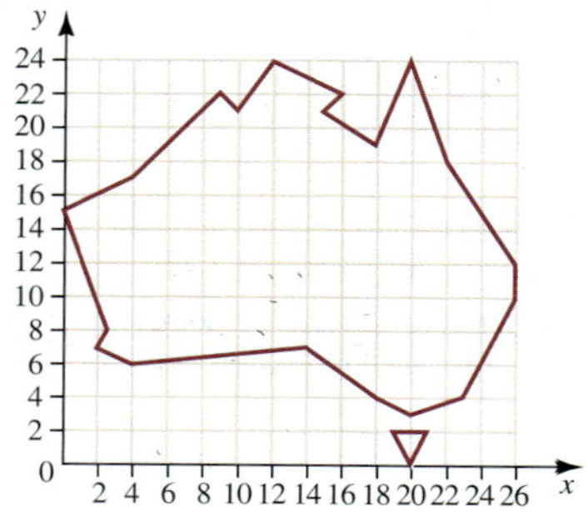

Maths Quest challenge (page 549)

1 15 square units 2 20

Summary

1 distance, diagram 2 parallel, meridian
3 Equator 4 North Pole, South Pole
5 0°, 90°N, 90°S
6 Greenwich Meridian, International Date Line
7 coordinates 8 angle
9 clockwise 10 topographic
11 isobars 12 scale, direction
13 vertices, edges 14 traversable
15 direction, joined 16 directed
17 axes, origin 18 vertical
19 point 20 across

Chapter review

1 a 1 : 300 b 1 : 200 000
2 a 10 m b 80 cm
3 In the scale drawing, the length of the court should be 9 cm and the width 4.5 cm.
4 a 90 km b 648 km c 216 km
5 a 23°S, 150°E b 35°S, 150°E
c 18°S, 122°W d 43°S, 147°E
6 a F3 b E2 c C4 d E1
7 a Townsville b Darwin
c Adelaide d Sydney

8 a 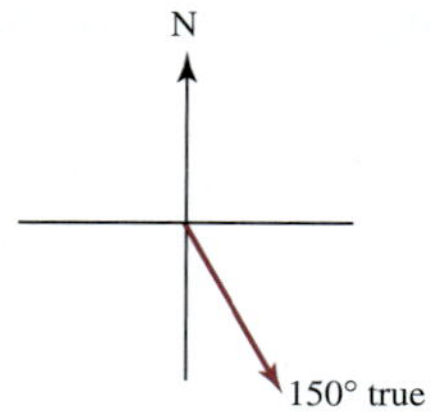

b 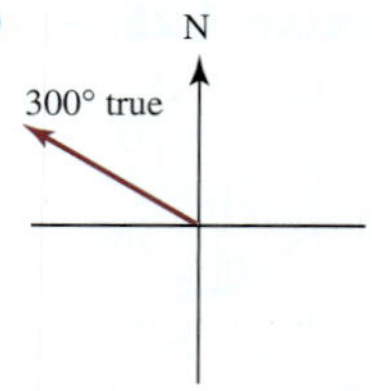

9 a 0° true b 320° true c 97° true
10 a 10 m b I
c G is 80 m above sea level and D is 60 m above, so G is 20 m higher than D.
d From C to D
e The journey would be hilly – first going down, then across flat country before then going over the rise and down the other side.
11 a 1000 millibars
b Off the central coast of WA
c A high pressure system around Perth indicates fine weather in that region. A high off New Zealand brings fine weather to the south-east. A low forming off the Great Australian Bight would bring unsettled weather to that region.
d Alice Springs lies within the 1004 millibar region, so the low air pressure would indicate unsettled conditions.
12 a 1 : 40 000 b North c About 1 km
13 a 260° T b S 75° W c 2.78 km
14 a 10
b

Venue	Number of Edges
Telstra Dome (TD)	4
Melbourne CBD (MCBD)	2
Rod Laver Arena (RLA)	3
MCG	4
Melbourne Park (MP)	4
Royal Botanical Gardens (R.B.G.)	2
Melbourne Sports and Aquatic Centre (MSAC)	4
Melbourne Exhibition Centre (MEC)	2
Docklands Precinct (DP)	3
Melbourne International Shooting Club (MISC)	2

c Yes the network is traversable.
DP – TD – MISC – MSAC – RBG – MP – MSAC – MEC – DP – RLA – MP – MCG – TD – MCBD – MCG – RLA
Note: Answers may vary this is only one possible answer.
15 a 8 edges, 8 vertices, not traversable
b 11 edges, 8 vertices, traversable

16 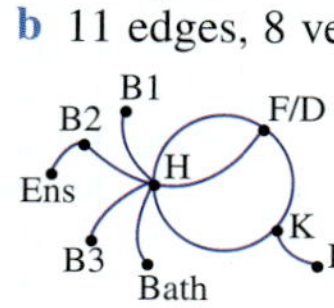

17 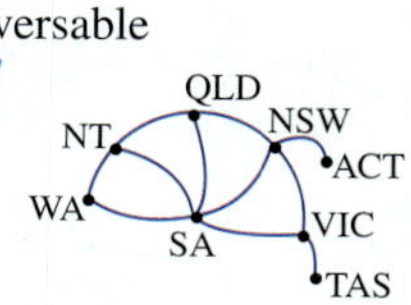

18 Check with your teacher.
19 (3, 1) (2, 3)(2, 1)(3, 1)(2, 2)(5, 1) (2, 3)(2, 1)(5, 0) (0, 3)(5, 0)(0, 1)(3, 2)(0, 3)(5, 0) (3, 1) (1,0)(1,2)

20
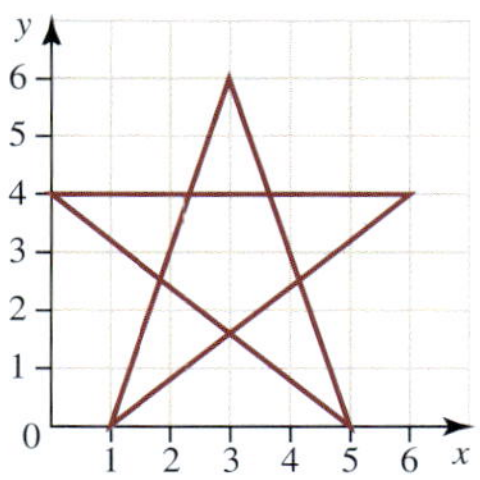

Chapter 13 Data

Are you ready?

1 a Numerical b Non-numerical
c Non-numerical d Numerical

2 a Continuous b Continuous
c Continuous d Discrete

3 a 1 cm b 0.2 m c 20 mm d 5%

4 a $\frac{1}{12}$ b $\frac{1}{8}$ c $\frac{3}{4}$ d $\frac{5}{6}$

5 a 240° b 150° c 105° d 25°

6 a
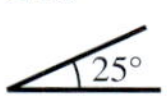

b

c
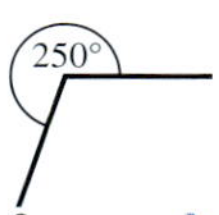

d
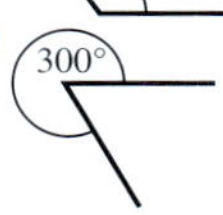

7 a 3 b 13 c 4.6 d 468

8 a 4 b 11 c 5 d 8

Exercise 13A — Classifying data

1 a ii b i c v d vi e iii f iv

2 a Quantitative, discrete b Qualitative, nominal
c Quantitative, continuous d Quantitative, continuous
e Qualitative, nominal f Quantitative, continuous
g Quantitative, discrete h Quantitative, continuous
i Quantitative, discrete j Qualitative, nominal
k Quantitative, continuous l Quantitative, discrete
m Quantitative, discrete n Qualitative, nominal
o Qualitative, ordinal p Quantitative, continuous
q Quantitative, discrete r Qualitative, ordinal
s Qualitative, nominal t Qualitative, ordinal

3 Discrete data deal with values which are exact and must be counted, such as the number of people attending a football match.
Continuous data deal with values which are measured and may assume decimal form such as the length of each football quarter.

4 Discuss with your teacher. 5 D 6 A 7 D

Maths Quest challenge (page 564)

1 1250 2 Anticlockwise

3 One possible solution: Start both timers together. When the 4-minute timer has run out, place the pie in the oven. Allow the 7-minute timer to keep running (it runs for a further 3 minutes). Then turn the 7-minute timer over and let it run through to make a total of 10 minutes.

Exercise 13B — Displaying data in tables

1 a

Score	Tally	Frequency
0	卌	5
1	IIII	4
2	卌 IIII	9
3	卌 II	7
4	II	2
5	I	1
6	I	1
7		0
8	I	1
	Total	30

b 30 c 5 d 12

e 2 f $\frac{2}{30} = \frac{1}{15}$

2 a The list is messy; it is difficult to see how many different sports there are and to gauge how many people prefer a particular sport.

b

Sport	Tally	Frequency
Hockey	卌 III	8
Cricket	卌 I	6
Tennis	卌 II	7
Netball	卌 IIII	9
Swimming	卌 II	7
Golf	IIII	4
Scuba diving	III	3
Lacrosse	IIII	4
Lawn bowls	II	2
	Total	50

c i Netball ii Lawn bowls

d Yes, tennis and swimming had a frequency of 7 and golf and lacrosse had a frequency of 4.

3 a

Score	Tally	Frequency
5	I	1
6	II	2
7	II	2
8	I	1
9	III	3
10	I	1
	Total	10

b

Score	Tally	Frequency
0	卌	5
1	卌	5
2	卌	5
3	II	2
4	II	2
5		0
6	I	1
	Total	20

c

Score	Tally	Frequency
45	𝍷	1
46	𝍷	1
47		0
48	𝍷𝍷	2
49	𝍷	1
50	𝍷𝍷	2
51	𝍷𝍷	2
52	𝍷𝍷	2
53	𝍷𝍷𝍷𝍷	4
54	𝍷𝍷𝍷	3
55	𝍷	1
56	𝍷𝍷	2
57	𝍷𝍷	2
58	𝍷𝍷	2
59	𝍷	1
60	𝍷𝍷	2
	Total	28

d

Score	Tally	Frequency
143	𝍷𝍷	2
144	𝍸 𝍷	6
145	𝍸	5
146	𝍷	1
147	𝍷𝍷𝍷	3
148	𝍷𝍷𝍷	3
149	𝍷𝍷𝍷	3
150	𝍷𝍷𝍷𝍷	4
151		0
152	𝍷	1
	Total	28

4 a 33 b 3 c 10 d 23 e $\frac{1}{3}$

5 a

Score	Tally	Frequency
0	𝍸 𝍸 𝍷	11
1	𝍸 𝍸	10
2	𝍸 𝍷	6
3	𝍷𝍷	2
4	𝍷	1
	Total	30

b 11 c 9 d 0 e 4 f $\frac{1}{5}$

6

Value (thousand dollars)	Tally	Frequency
100–109	𝍷𝍷𝍷	3
110–119	𝍷	1
120–129		0
130–139	𝍷	1
140–149	𝍷𝍷𝍷	3
150–159	𝍷𝍷𝍷	3
160–169	𝍷𝍷	2
170–179	𝍷	1
180–189	𝍷𝍷	2
190–199	𝍷	1
200–209	𝍷	1
	Total	18

7

Height	Tally	Frequency
140–144	𝍷𝍷	2
145–149	𝍸 𝍷	6
150–154	𝍷𝍷𝍷𝍷	4
155–159	𝍷	1
160–164	𝍸	5
165–169	𝍸 𝍷	6
170–174	𝍷𝍷	2
175–179	𝍷	1
180–184	𝍷	1
	Total	28

8

Vowel	Tally	Frequency
A	𝍸𝍸𝍸𝍸 𝍸𝍸𝍸𝍸𝍷	41
E	𝍸𝍸𝍸𝍸 𝍸𝍸𝍸𝍸 𝍸𝍸	50
I	𝍸𝍸𝍸𝍸 𝍸𝍸𝍸𝍷	36
O	𝍸𝍸𝍸𝍸 𝍸𝍸𝍸𝍷𝍷𝍷𝍷	39
U	𝍸𝍸𝍸	15
	Total	181

9

Type	Tally	Frequency
1 bedroom unit	𝍷𝍷𝍷	3
2 bedroom unit	𝍷𝍷𝍷𝍷	4
2 bedroom house	𝍷	1
3 bedroom house	𝍸 𝍷	6
4 bedroom house	𝍷𝍷	2
	Total	16

A 3 bedroom house was the most popular.

10 C 11 A 12 B

Exercise 13C — Understanding graphical displays

1 a Telephone bills 2005–06
b Months
c Bill total
d Every three months (quarterly)
e January 2006
f Yes

g $150

h A graph like this is useful to monitor the speeding pattern over time and to decide whether there have been any unusual increases.

i It would not be normal, but there might be reasons to explain why it was so much higher, such as overseas phone calls.

j This phone costs about $50 per month.

2 a 356 b Bus

c Combination d Combination

e i 25 ii $\frac{25}{356}$ iii 7%

f 22

3 a 45 b 42 c 61 d 5 and 11

For these weeks the sales decreased in comparison to other weeks.

e Sometimes an unusual result may be caused by seasonal effects, or unusual situations in the particular survey or area.

f It does indicate an improvement, despite these two low weeks, because there is an overall upward trend throughout the time period.

4 a Location 2 — Monash Freeway: Mulgrave

b 45%

c No

d The percentage has been graphed to give a better comparison. Some roads may have less traffic and therefore actual numbers of speeding cars do not give a good comparison.

e Locations 1 and 2 are on a major freeway and the cars are travelling faster there, without any interruptions; so they may exceed the limit more easily.

f No g No

5 a White b Orange c 12 d 10

e i A pie graph can't show 0 as a result therefore the colour would not be mentioned.

ii Column

f 119

g The other cars may not have fitted into a category previously chosen, or the inclusion of them under their separate headings may have made too many divisions in the graph.

6 a A line graph is different from a bar or column graph because it has dots which are joined by lines at the values instead of bars or columns. The dots are directly above the time position on the horizontal axis.

b $1.52 c $1.62

d The price did not rise continually, but rose and fell over time.

e A line graph is good to show changes over time, because you can see by the connecting line whether something is increasing or decreasing.

7 a January and February b July

c October d February

e i Autumn — high rainfall, low temperatures

ii Winter — low temperatures, low rainfall

iii Spring — high rainfall, warm temperatures.

8 a There are two columns for each girl because each girl ran on two occasions.

b Betty

c 1 April — 70 m, 29 April — 73 m

d Rachel

e Yes. Sarah — remained the same. Samantha and Paula sprinted at a slower pace.

f Yes, the graph could be misleading as the vertical scale begins at 30 m and not at 0 m. If the vertical scale was not read properly, it might appear that the girls covered a greater distance in a shorter period of time.

g The graph's vertical axis starts at 30 m because no-one ran less than that distance and the teacher wanted to use all the space to show small differences in the distances.

10 Quick Questions 1

1 Qualitative 2 Discrete

3 Ordinal 4 Nominal

5 $3.20 in March

6 In February grapes cost $3.05.

7 The cost was recorded over 5 months.

8 Grapes cost the most in May when they were $4.35 per kilogram.

9

Sport	Tally	Frequency
Cricket	𝍸 II	7
Football	𝍸 I	6
Tennis	𝍸 II	7
Basketball	𝍸 𝍸	10
Hockey	IIII	4
None	𝍸 I	6
	Total	40

10 a

Age	Tally	Frequency
0–4	I	1
5–9	I	1
10–14	III	3
15–19	𝍸 IIII	9
20–24	𝍸	5
25–29	𝍸 I	6
30–34	II	2
35–39	𝍸	5
40–44	III	3
45–49	𝍸	5
50–54	II	2
55–59	III	3
60–64	IIII	4
	Total	49

b The 15–19 age group

Exercise 13D — Displaying data as graphs

1

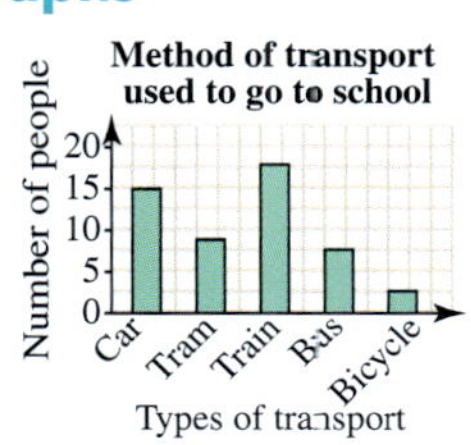

2 Types of pets in our street

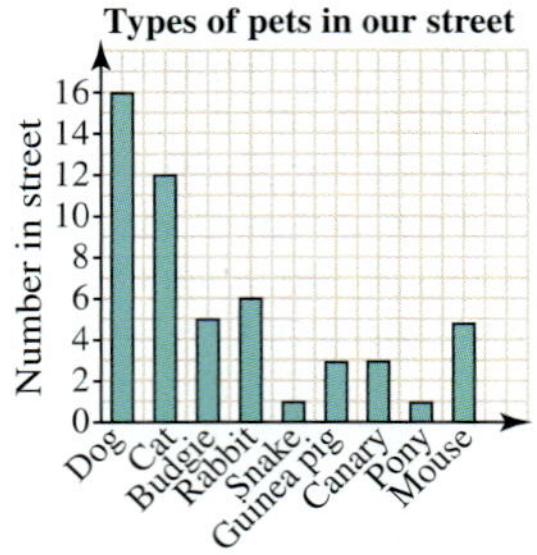

3 Runs scored by Australian batsmen

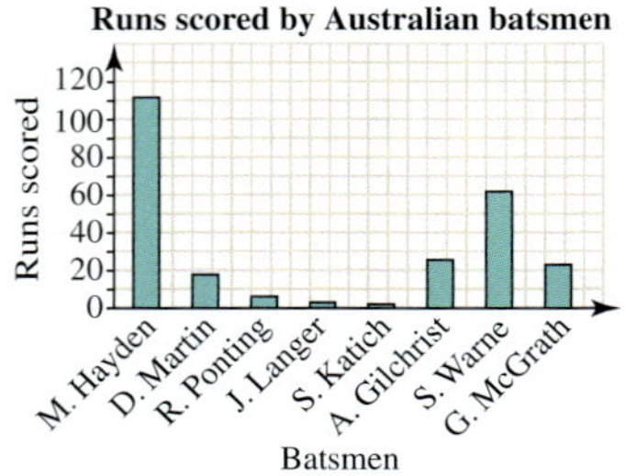

4 Mean daily maximum temperatures, Cairns

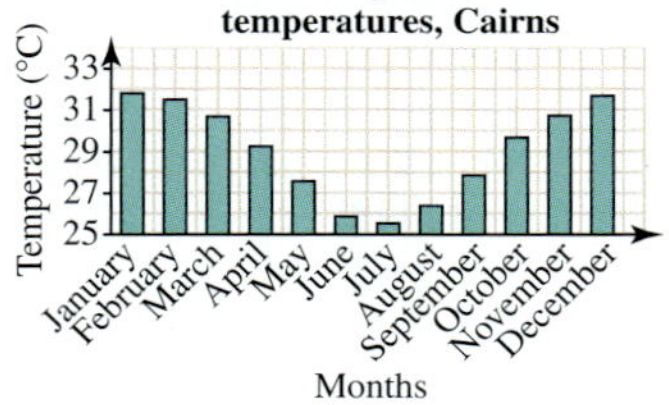

5 Students absent

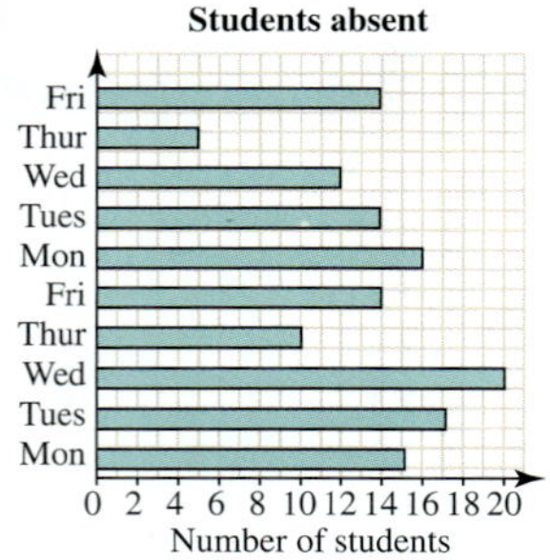

6 Percentage immunised for diseases

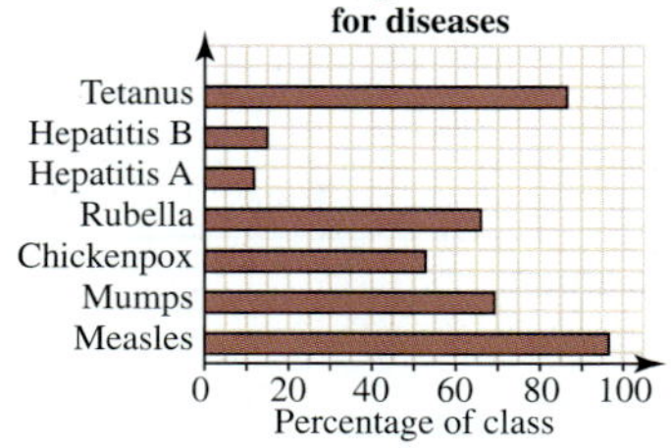

7 a Growth of bacteria

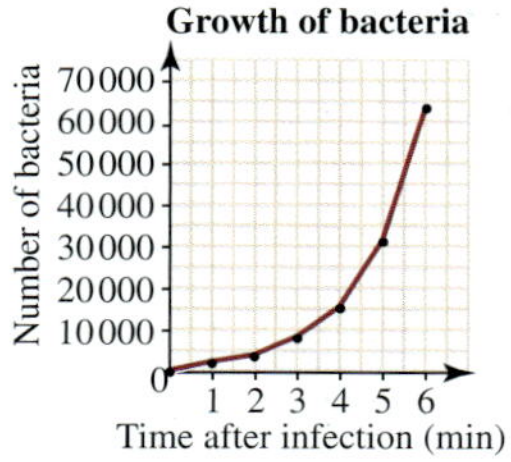

b 1000
c 1 minute
d 7 minutes

8 a Attendance at venue

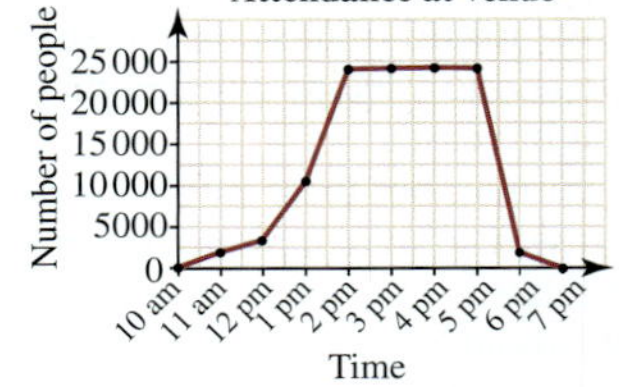

b 2 pm **c** Between 1 pm and 2 pm
d 24 200 people **e** $665 500

9 a Height of tree

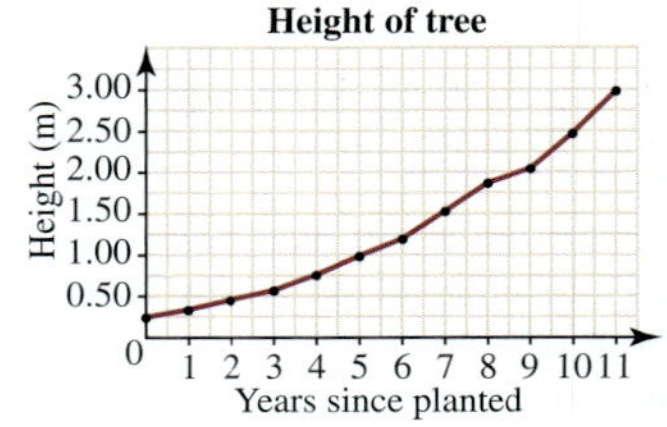

b 0.25 m **c** No
d The tree's growth depends on many environmental factors, such as water, sunlight, fertiliser.

10 a Runs scored in first 10 overs

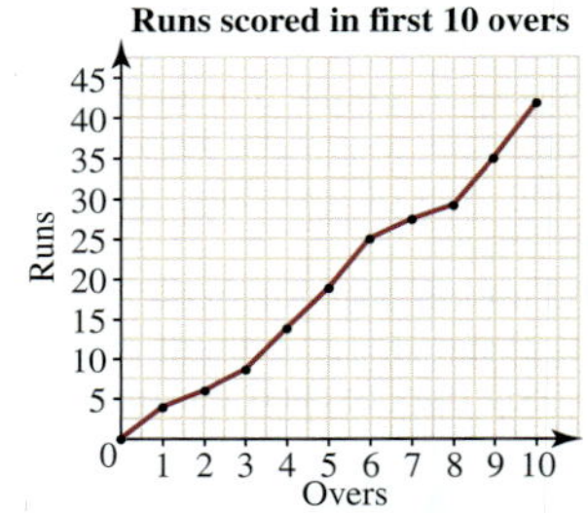

b 4 **c** 42

d No, it would not be reasonable, as there can be many variations between overs, as shown in the first 10 overs. The score after 10 overs is not exactly 10 times the score after 1 over.

11 a 180° **b** 120° **c** 36° **d** 30°
e 90° **f** 60° **g** 40° **h** 45°
i 15° **j** 6° **k** 36° **l** 61.2°
m 43.2° **n** 331.2° **o** 216°

12 a $\frac{1}{4}$, 25% **b** $\frac{1}{8}$, $12\frac{1}{2}$% **c** $\frac{1}{6}$, $16\frac{2}{3}$%
d $\frac{1}{12}$, $8\frac{1}{3}$% **e** $\frac{1}{18}$, $5\frac{5}{9}$% **f** $\frac{1}{36}$, $2\frac{7}{9}$%
g $\frac{1}{24}$, $4\frac{1}{6}$% **h** $\frac{1}{10}$, 10% **i** $\frac{1}{5}$, 20%
j $\frac{1}{15}$, $6\frac{2}{3}$% **k** $\frac{1}{30}$, $3\frac{1}{3}$% **l** $\frac{1}{40}$, $2\frac{1}{2}$%
m $\frac{1}{20}$, 5% **n** $\frac{1}{2}$, 50% **o** $\frac{1}{3}$, $33\frac{1}{3}$%

13 a 50% **b** 35% **c** 15% **d** 100%
e i 180° **ii** 126° **iii** 54°
f i 180° **ii** 126° **iii** 54°
g Both results compare well; however, the calculations are quicker and more accurate.

14 a 45 **b** 6 **c** 12 **d** $\frac{2}{5}$
e 144° **f** German shepherd

15 a Ian **b** 17%
c No, this pie graph gives percentages only, not actual values or numbers of people who voted.
d 78.48°

16 a

Type of meat	Amount sold (kg)	Fraction	Angle size (°)
Lamb	10	$\frac{1}{9}$	40
Beef	45	$\frac{1}{2}$	180
Pork	5	$\frac{1}{18}$	20
Chicken	15	$\frac{1}{6}$	60
Turkey	10	$\frac{1}{9}$	40
Rabbit	5	$\frac{1}{18}$	20
Total	90	1	360

b Type of meat sold

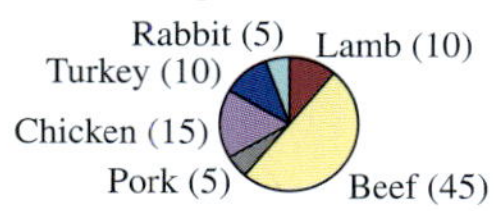

17

Type of cereal	Number who like it	Fraction	Angle size (°)
Corn Flakes	18	$\frac{1}{4}$	90
Oats	9	$\frac{1}{8}$	45
Rice Crunches	9	$\frac{1}{8}$	45
Muesli	6	$\frac{1}{12}$	30
Super Sugary Cereal	24	$\frac{1}{3}$	120
Frosted Circle Crisps	6	$\frac{1}{12}$	30
Total	72	1	360

Breakfast cereals

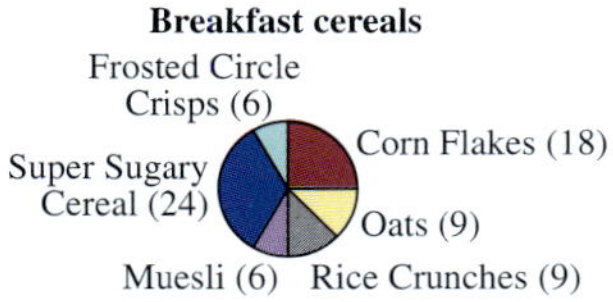

18

Hobby	Percentage of total	Angle size (°)
Stamp collecting	20	72
Photography	15	54
Rollerblading	20	72
Skateboarding	30	108
Model making	15	54
Total	100	360

Hobbies

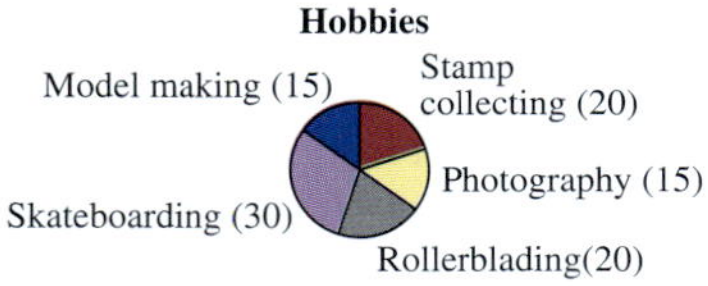

19

Sport	Percentage of total	Angle size (°)
Cricket	20	72
Hockey	25	90
Netball	40	144
Soccer	10	36
Tennis	5	18
Total	100	360

Sports

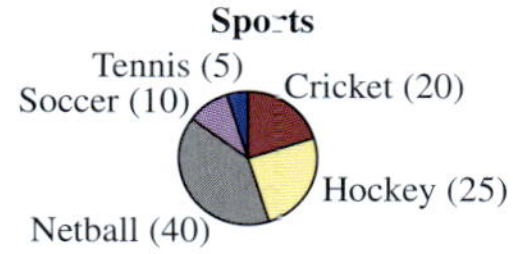

20 a

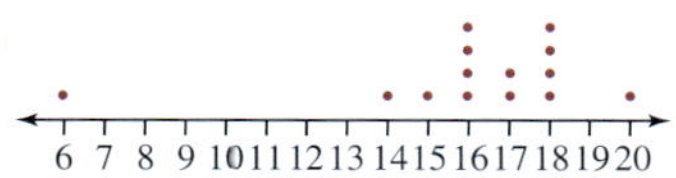

b The scores lie between and include 6 to 20; that is, a range of fourteen.
Mostly between 16 to 18 packets were sold. Sales of 6 and 20 packets of chips were extremely low. A provision of 20 packets of chips each day should cover the most extreme demands.

21 a

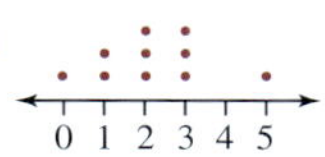

b

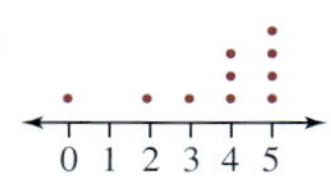

c

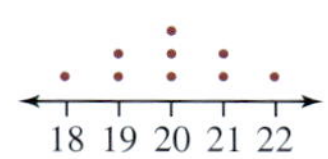

d

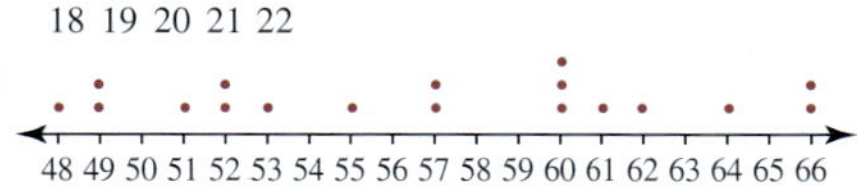

22 Check with your teacher.

10 Quick Questions 2

1 a

Food	Tally	Frequency
Pasta	~~||||~~ ~~||||~~	10
Pizza	~~||||~~ ~~||||~~ |||	13
Chicken	~~||||~~ |||	8
Hamburgers	~~||||~~ ~~||||~~ ||||	14
Fish and chips	~~||||~~	5
	Total	50

a 50
b Hamburgers
c Fish and chips
d Pasta = $\frac{1}{5}$, pizza = $\frac{13}{50}$, chicken = $\frac{4}{25}$,
hamburgers = $\frac{7}{25}$, fish and chips = $\frac{1}{10}$

answers 13D → 13D

2

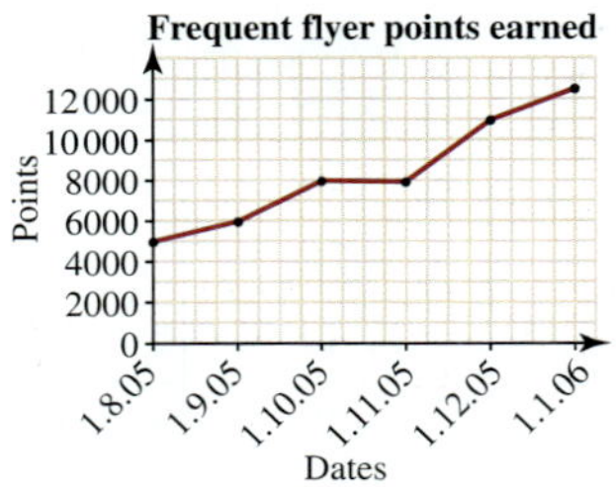

3 5000 4 800 5 Yes

6 Between 1.10.05 and 1.11.05
The traveller earned no points in that month.

7 There are 40 flowers included in the group.

8 $\frac{20}{40} \times \frac{100}{1} = 50\%$. 50% of the group flower in spring.

9 $\frac{12}{40} \times 360° = 108°$. Angle of 108° is needed to represent summer flowering plants.

10 Angle measured with protractor should be approximately 108°.

Exercise 13E — Summary statistics

1 a 5 b 6 c 4 d 6
e 4 f 3 g 12 h 14
i 15 j 13

2 a 5 b 7 c 3 d 3
e 5 f 6 g 4 h 4
i 6 j 5

3 B

4 a 8 b 3 c 4 d 2
e 3 f 4 g 5 h 2
i 5 j 3

5 B

6 a 3 b 4 c 5 d 2
e 3 f 3 g 4 h 6
i 9 j 4

7 a 3.5 b 2.2 c 2.83 d 3.2
e 2 f 3.44 g 6.83 h 4.17
i 3.33 j 2.5

8 D

9 a 5 b 3 c 6 d 3 e 7
f 11 g 4 h 2 i 2.5 j 1.5

10 a 2 b 2 c 5 d 7 e 2.5
f 5 g 6 h 7.5 i 5.5 j 7.5

11 a 4 b 8 c 2 d 4
e 2 f 3 g 10 h 4
i 9 j 3

12 a 0 b 4 c 6.2 d 14
e 1.5 f 3.5 g 100 and 102
h 45 i 11.2 j 15

13 D

14 a 32 b 24.1

15 C 16 85% 17 104 shoes

18 47 students 19 38

20 a Tom's 75.0 Sarah's 72.8
b Sarah's sells cheaper petrol.
c Tom may have seen the price on a very expensive day. There's a greater range of values at Sarah's petrol station.

21 89.5% 22 72

23 a Mode = 12
b The mode shows which size to order more of. The mean and median would not show the more common sizes and would give only an indication of the middle of the range.

24 a 1.75 m b 221 m
c Based on her past performances and her past range of values, she cannot jump this high.

25 a 101.9
b Both the mode (115) and median (112) give a better impression of how the team has performed, even though they could give a 'misleading' impression of the team's performance.

Summary

1 nominal, ordinal 2 discrete, continuous
3 census, sample, random
4 distribution table, score, tally
5 Frequency, result 6 title, labelled, scaled
7 evenly, columns, equal 8 horizontal bars
9 Line, points, straight 10 change, period
11 Pie, sector, vertical, axes
12 multiply 13 360°
14 range 15 mean
16 median, order 17 common, mode

Chapter review

1 A 2 D 3 A

4 a Coke b tea

5 a 26 b 14 c 18 d 14 e 12

6 a

Cars	Frequency
0	6
1	4
2	8
3	5
4	1
Total	24

b 6 c 14 d 2 e 4 f $\frac{1}{3}$

7 a 10°C b Day 5 c Day 7
d Day 6 e 10°C f Days 6 and 7

8 a Science marks b Percentage
c Names d 1 unit = 10%
e John f Peter
g 59% h 89%

9 a Days of week b Air quality rating
c Geelong region d Wednesday
e Eastern region

10 a Months b Number of flies
c 10 000 d 28 000
e 1 cm = 5 thousand flies (if beginning from 0 flies) or 1 cm = 2 thousand flies (if beginning from 10 000 flies).

f 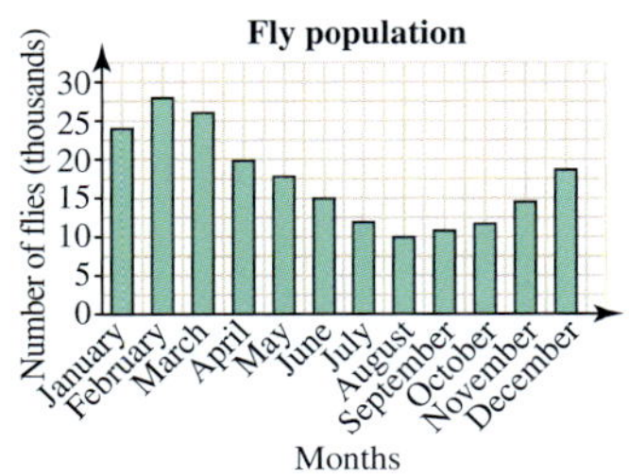

11 a 12 **b** 18
c First aid and fitness
d China in focus, Consumer studies
e Beginners' photography, First aid and fitness, Outdoor education
f 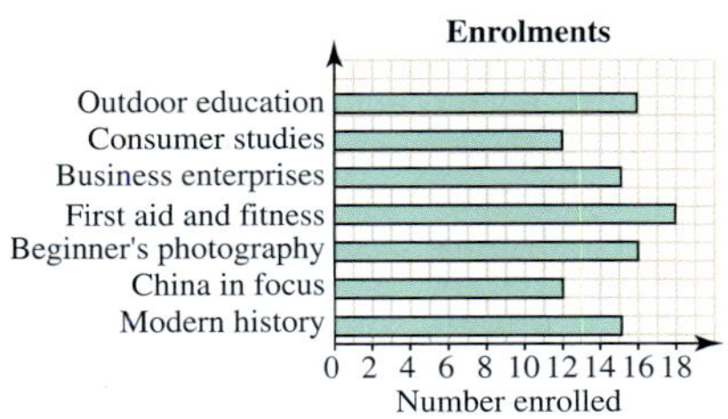

12 a mm^3 **b** Minutes **c** 1000 mm^3
d 11 minutes **e** 3 minutes **f** 700 mm^3

13 a 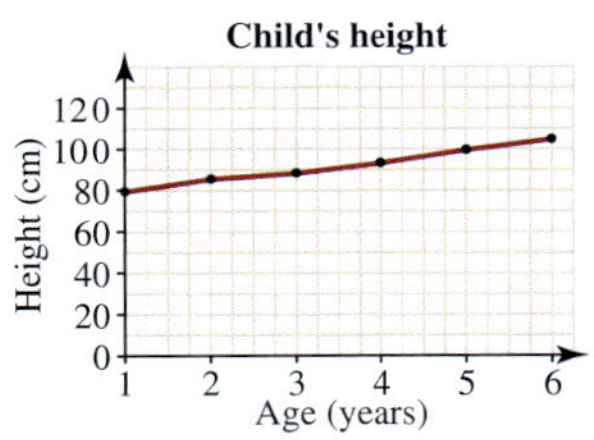

b Horizontal — as it represents time
c Vertical

14 a Pie graph **b** Spouse lives here
c They were probably too small to show on the graph as separate sectors.
d 14.76°
e A value of zero cannot be shown on a pie graph because it would have to take up zero space which is not possible — even a single line must take up space and therefore represent a percentage of the total.

15 a

Type of animal	Number on farm	Fraction	Angle size (°)
Horse	5	$\frac{1}{24}$	15
Cow	25	$\frac{5}{24}$	75
Duck	15	$\frac{1}{8}$	45
Dog	5	$\frac{1}{24}$	15
Cat	10	$\frac{1}{12}$	30
Pig	40	$\frac{1}{3}$	120
Sheep	20	$\frac{1}{6}$	60
Total	120	1	360

b 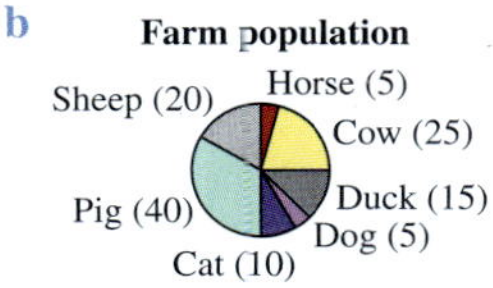

16

Occupation	Number of people	Fraction	Angle size (°)
Bank clerk	10	$\frac{1}{10}$	36
Dentist	5	$\frac{1}{20}$	18
Doctor	15	$\frac{3}{20}$	54
Engineer	5	$\frac{1}{20}$	18
Manager	20	$\frac{1}{5}$	72
Real estate agent	10	$\frac{1}{10}$	36
Secretary	20	$\frac{1}{5}$	72
Solicitor	5	$\frac{1}{20}$	18
Teacher	10	$\frac{1}{10}$	36
Total	100	1	360

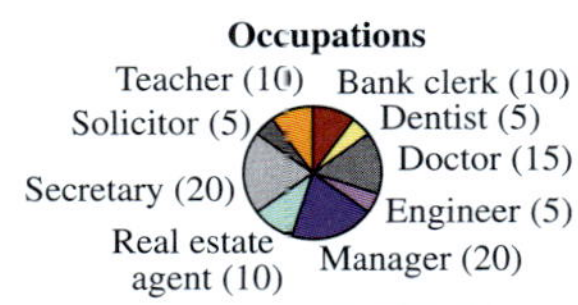

17 a

x	f
0	4
1	5
2	7
3	3
4	0
5	1
Total	20

b 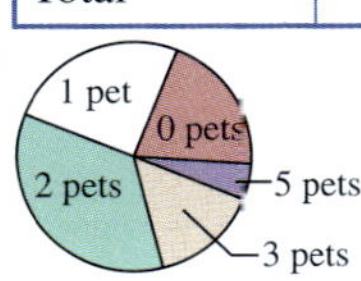

c 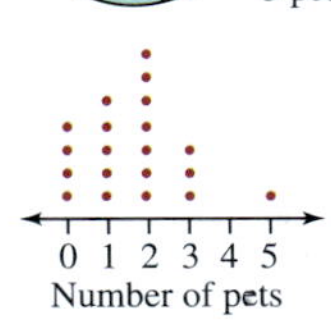

d 2, 2, 1.65

18 a i 3.14 **ii** 4 **iii** 3
b i 3.71 **ii** 4 **iii** 4
c i 4.71 **ii** 6 **iii** 5
d i 4.86 **ii** 5 **iii** 5
e i 2.14 **ii** 1 **iii** 1
f i 5.2 **ii** 5 **iii** 5
g i 4.4 **ii** 5 **iii** 5
h i 4.14 **ii** 2 **iii** 4
i i 5.43 **ii** 5 **iii** 5
j i 5.78 **ii** 8 **iii** 6

19 a 81.56 b 74.22 c Test 1 d 6
e 4 f 85 g 63 h 85
i 68
j John 89.5, Peter 74, Mark 76.5, James 74.5, David 85.5, Rachel 73.5, Mary 81, Eve 78.5, Esther 68.
k 21.5

20 a 41 cm b 154.8 cm c 156 cm
d 156 cm e Yes
f There are quite a lot of values which are very similar, so the mean, median and mode are fairly close.

Chapter 14 Chance

Are you ready?

1 a
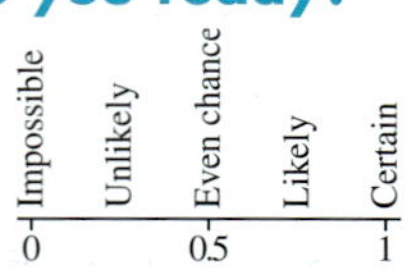

b Check with your teacher.

2 b, a, c

3 a More than a half b Less than a half
c More than a half

4 a {Head, Tail} b {1, 2, 3, 4, 5, 6}
c {Sunday, Monday, Tuesday, Wednesday, Thursday, Friday, Saturday}

5 a 52 b Hearts, diamonds, clubs, spades
c 4

6 a $\frac{3}{5}$ b $\frac{1}{6}$

Exercise 14A — The language of chance

1 a Certain b Likely c Impossible
d Impossible e Certain f Likely
g Unlikely h Unlikely i Impossible
j Even chance k Certain l Certain
m Impossible n Even chance o Likely

2 D, C, A, B

3 Class discussion or check with your teacher.

4 A 5 A

6 a $\frac{1}{2}$ (or 0.5) b 0 c $\frac{1}{5}$ (or 0.2)
d $\frac{1}{2}$ (or 0.5) e $\frac{1}{2}$ (or 0.5) f $\frac{1}{4}$ (or 0.25)
g $\frac{4}{5}$ (or 0.8) h 1 i 1
j 0.3–0.7 Any value within this range is reasonable. Discuss.

7 a 1 b 0.75 c 0.25 d 0.75
e 0.25 f 1 g 0 h 0.25
i 1 j 0.25 k 0 l 0.75
m 0.5 n 0.5

Exercise 14B — The sample space

1 S = {1, 2, 3, 4, 5, 6, 7, 8, 9, 10}

2 a S = {heads, tails}
b S = {A, B, C, D, E}
c S = {win, loss, draw}
d S = {ace of clubs, ace of spades, ace of hearts, ace of diamonds}
e S = {A, B, C, D, E, F}

3 a S = {clubs, spades, diamonds, hearts}
b S = (ace, 2, 3, 4, 5, 6, 7, 8, 9, 10, jack, queen, king}
c S = {red, black}

4 S = {red, green, orange}

5 2

6 a 52 b 45 c 16
d 365 (or 366 in a leap year) e 26 f 180 000
g 4 h 3

7 E

Exercise 14C — Simple probability

1 a 12: 1, 2, 3, 4, 5, 6, 7, 8, 9, 10, 11, 12
b 5: 1, 2, 3, 4, 5
c 2: C, L
d 3: red, blue, black
e 5: black, blue, green, red, yellow
f 3: 2, 4, 6
g 2: 4, 6
h 10: 1, 3, 5, 7, 9, 11, 13, 15, 17, 19

2 a 1, 2, 3, 4, 5, 6. All equally likely.
b Heads, Tails. Each is equally likely.
c White, black. Each is equally likely.
d White, blue, red. It is more likely you will get white rather than red or blue, because it takes up a bigger area.
e 1, 2, 3, 4, 5. It is more likely that a five will be rolled because it appears twice on the die.
f Blue, green, red. Each is equally likely.
g 2, 4, 6, 8, 10. It is more likely that an odd number will be obtained as there are 6 odd numbers and 5 even numbers.
h a, e, i. It is more likely that a consonant would be obtained as there are 7 consonants and 4 vowels.

3 a 1, 2, 3, 4, 5, 6, 7, 8, 9, 10
b i $\frac{1}{10}$ ii $\frac{3}{5}$ iii $\frac{2}{5}$ iv $\frac{7}{10}$
v $\frac{3}{10}$ vi 0 vii $\frac{3}{10}$ viii $\frac{1}{5}$

4 a 20
b i $\frac{3}{10}$ ii $\frac{1}{5}$ iii $\frac{7}{20}$ iv $\frac{3}{20}$
c 10 d $\frac{1}{2}$ e $\frac{1}{2}$ f $\frac{13}{20}$
g 1
h When the probability of a particular result is certain, (in this case, that the marble must be either blue, red, green or black), it equals 1.

5 $\frac{2}{5}$

6 a $\frac{1}{100\,000}$ b $\frac{1}{5000}$ c $\frac{1}{2000}$ d 1
e It would not be a very wise purchase because the total cost would be $500 000 unless of course the car was worth more than $500 000.

7 E 8 D

9 a $\frac{1}{2}$ b $\frac{1}{13}$ c $\frac{1}{4}$ d $\frac{3}{13}$
e $\frac{1}{52}$ f $\frac{3}{4}$ g $\frac{12}{13}$ h 1
i $\frac{3}{4}$ j $\frac{1}{26}$

10 Quick Questions 1

1 Unlikely 0.13
Highly likely 0.89
Impossible 0
Certain 1
Fair chance 0.5

2 Red, white; both equally likely
3 Boy, girl; girl is more likely
4 2, 3, 4; 4 is more likely
5 Green, red, black, pink, orange
6 $\frac{1}{14}$ 7 $\frac{1}{7}$ 8 $\frac{4}{7}$ 9 $\frac{1}{2}$ 10 0

Exercise 14D — Using tables to show sample spaces

1 a {1, 2, 3, 4, 5, 6}
b {1, 2, 3, 4, 5, 6, 7, 8, 9, 10}
c {fruit salad, cheesecake, mudcake, cheese platter}
d {5, 10, 15, 20, 25, 30, 35, 40, 45, 50}
e {Victoria, New South Wales, Queensland, Northern Territory, South Australia, Tasmania, Western Australia, Australian Capital Territory}
f {true, false}
g {king of hearts, king of diamonds, king of spades, king of clubs}
h {guitar, drum, saxophone, piano, trumpet}

2 a

Die \ Spinner	A	B	C
1	A1	B1	C1
2	A2	B2	C2
3	A3	B3	C3
4	A4	B4	C4
5	A5	B5	C5
6	A6	B6	C6

{(A, 1), (A, 2), (A, 3), (A, 4), (A, 5), (A, 6), (B, 1), (B, 2), (B, 3), (B, 4), (B, 5), (B, 6), (C, 1), (C, 2), (C, 3), (C, 4), (C, 5), (C, 6)}
b 18
c i $\frac{1}{3}$ ii $\frac{1}{6}$ iii $\frac{2}{3}$ iv $\frac{1}{3}$
v $\frac{1}{2}$ vi $\frac{1}{6}$ vii 1 viii $\frac{5}{6}$
ix $\frac{1}{9}$ x 0

3 a

Coin 2 \ Coin 1	Head	Tail
Head	H H	T H
Tail	H T	T T

b 4 c 1 d 1 e 2
f $\frac{1}{4}$ g $\frac{1}{4}$ h $\frac{1}{2}$

4 a

10c Coin \ 5c Coin	Head	Tail
Head	H H	T H
Tail	H T	T T

b 4 c $\frac{1}{4}$ d $\frac{1}{4}$ e $\frac{1}{4}$ f $\frac{1}{4}$

5 a

Die \ Coin	Head	Tail
1	H 1	T 1
2	H 2	T 2
3	H 3	T 3
4	H 4	T 4
5	H 5	T 5
6	H 6	T 6
7	H 7	T 7
8	H 8	T 8
9	H 9	T 9
10	H 10	T 10

b 20 c $\frac{1}{20}$ d $\frac{1}{4}$ e $\frac{1}{2}$
f $\frac{3}{20}$ g $\frac{1}{5}$ h $\frac{1}{2}$ i $\frac{1}{4}$
j 1

6 a

Blue die \ Red die	1	2	3	4	5	6
1	1 1	2 1	3 1	4 1	5 1	6 1
2	1 2	2 2	3 2	4 2	5 2	6 2
3	1 3	2 3	3 3	4 3	5 3	6 3
4	1 4	2 4	3 4	4 4	5 4	6 4
5	1 5	2 5	3 5	4 5	5 5	6 5
6	1 6	2 6	3 6	4 6	5 6	6 6

Note: Your table may list the blue die along the top and therefore it would be listed first.

b 36 c $\frac{1}{36}$ d $\frac{1}{36}$ e $\frac{11}{36}$
f $\frac{1}{4}$ g $\frac{1}{4}$ h $\frac{1}{4}$ i $\frac{5}{36}$
j 0 k 0

7 a

Coin \ Marble colour	Red	Blue	Green
Head	R H	B H	G H
Tail	R T	B T	G T

b 6 c $\frac{1}{6}$ d $\frac{1}{6}$ e $\frac{1}{6}$
f $\frac{1}{3}$ g $\frac{1}{3}$ h $\frac{1}{2}$
i The probability of picking out a red, blue or green marble and getting a Head = $\frac{1}{2}$, which is the same as the result you get if you were just to toss a coin. This is because you must get one of these colour marbles, so the only choice in the experiment here is whether you get a Tail or a Head, so Pr(Head) = $\frac{1}{2}$.

8 C 9 C

10 *Note:* You may have chosen the opposite number for rows and columns. This is also correct.
  a i 31 rows, 2 columns ii 62
  b i 2 rows, 3 columns ii 6
  c i 72 rows, 6 columns ii 432
  d i 10 rows, 6 columns ii 60
  e i 30 rows, 6 columns ii 180
  f i 100 rows, 2 columns ii 200

Exercise 14E — Experimenting with chance

1 a

Exp. No.	Heads		Tails	
	Tally	Count	Tally	Count
1	IIII	4	𝍸 I	6
2	I	1	𝍸 IIII	9
3	𝍸 I	6	IIII	4
4	𝍸 III	8	II	2
5	IIII	4	𝍸 I	6
6	𝍸 II	7	III	3
	Total	30	Total	30

Note: This is only one possible solution. Answers will differ each time.

b $\frac{2}{5}$ (40%) c $\frac{3}{5}$ (60%)

d The theoretical value for both results is $\frac{1}{2}$ (50%). They differ by 10%.

e Refer to table.

f The combined result equals the theoretical value.

2 Values will differ for each group. As you complete more trials, you will probably notice the values you get experimentally are closer to those you would expect from theoretical probability ($\frac{1}{2}$ or 50%). If this is not occurring, though, you will probably need more trials.

3 a $\frac{1}{5}$

b

Colour	1	2	3	4	5
Number of times it occurs		II	II	IIII	II

c Pr(1) = 0 Pr(2) = $\frac{1}{5}$ Pr(3) = $\frac{1}{5}$

Pr(4) = $\frac{2}{5}$ Pr(5) = $\frac{1}{5}$

Note: This is the only possible solution. Answers will differ each time.

d Your spinner may not be evenly balanced and this may lead to an increased likelihood of getting one result or another.

e–f Values will differ for each group. As you complete more trials, you will probably notice the values you get experimentally are closer to those you would expect from theoretical probability ($\frac{1}{5}$ or 20%). If this is not occurring, though, you will probably need more trials.

4 4

5 Divide the spinner into 10 sections; 3 of these sections will be shaded in 1 colour and 7 sections will be shaded in another colour.

6 $\frac{3}{8}$

7 See worked example 9.
Note: These values will differ each time the experiment is performed.
Spreadsheets could also be used.

8 a Each die would need to be thrown about 120 times to get some impression of whether it is biased or not. The same person will need to roll the die each time in the same manner and onto the same surface.

b

Number on die	Black	White
1	20	19
2	19	21
3	20	20
4	21	20
5	19	20
6	21	20
Total	120	120

Note: These values will differ each time the experiment is conducted.

c

Number on die	Probability of black	Probability of white
1	$\frac{20}{120} = \frac{1}{6}$	$\frac{19}{120}$
2	$\frac{19}{120}$	$\frac{21}{120} = \frac{7}{40}$
3	$\frac{20}{120} = \frac{1}{6}$	$\frac{20}{120} = \frac{1}{6}$
4	$\frac{21}{120} = \frac{7}{40}$	$\frac{20}{120} = \frac{1}{6}$
5	$\frac{19}{120}$	$\frac{20}{120} = \frac{1}{6}$
6	$\frac{21}{120} = \frac{7}{40}$	$\frac{20}{120} = \frac{1}{6}$
Total	1	1

The long-term trend suggests that the probability of obtaining each value on either dice will be $\frac{1}{6}$.

d The 2 dice appear to be fair as each value occurred approximately 20 times, which is what we would expect in 120 throws. More trials however could be conducted. It is very important when conducting an experiment such as this that the devices used are even in shape and size and that one doesn't have an advantage over the other.

9 a Answers will vary for each packet.

b Check with your teacher.

c Smarties must be replaced each time they are removed, otherwise the chance of taking out 1 of a certain colour would be changed.

d You could use a spinner with sections coloured to give the same probability for each colour as you would get for removing the Smarties from the box. (This is only one possible answer.)

10 Quick Questions 2

1 Very unlikely 2 Highly likely 3 Even chance

4

		Die					
		1	2	3	4	5	6
Marble	blue	1 blue	2 blue	3 blue	4 blue	5 blue	6 blue
	green	1 green	2 green	3 green	4 green	5 green	6 green
	black	1 black	2 black	3 black	4 black	5 black	6 black
	white	1 white	2 white	3 white	4 white	5 white	6 white

5 24 6 6 7 $\frac{1}{24}$

8 $\frac{1}{24}$ 9 $\frac{1}{8}$ 10 $\frac{1}{4}$

Maths Quest challenge (page 643)

1 12 different ways:
20, 20, 10; 20, 20, 5, 5; 20, 10, 10, 10;
20, 10, 10, 5, 5; 20, 10, 5, 5, 5, 5; 20, 5, 5, 5, 5, 5, 5;
10, 10, 10, 10, 10; 10, 10, 10, 10, 5, 5;
10, 10, 10, 5, 5, 5, 5; 10, 10, 5, 5, 5, 5, 5, 5;
10, 5, 5, 5, 5, 5, 5, 5, 5; 5, 5, 5, 5, 5, 5, 5, 5, 5, 5

2 a 6 socks. Worst case scenario is first picking 4 blue socks, then 2 black socks.
b 7 socks. Worst case scenario is first picking 5 black socks then 2 blue socks.
c 7 socks. Minimum needed to guarantee a pair of black socks is 6 (from part a). Picking one more guarantees a pair of blue socks will be included (from part b).

3 SC 378 CS 378
SC 387 CS 387
SC 738 CS 738
SC 783 CS 783
SC 837 CS 837
SC 873 CS 873

Exercise 14F — Fair games

1 a Samira: {2, 4, 6}, Liana: {4, 5, 6}, game is fair.
b Samira: {1, 3, 5}, Liana: {1, 2}, game is not fair.

2 a No b Yes c Yes
d No e No f Yes
g Yes h No i Yes
j No

3 D

4 Results will vary for each student. Check with your teacher.

5 Results will vary for each student. Check with your teacher.

Summary

1 chance, likely, happen
2 Probability, event
3 includes, 0, certain
4 scale, numerical
5 result
6 Favourable outcomes
7 number, possible
8 sample space, experiment, brackets
9 two-way, logical
10 ordered pair
11 long-term trend, experimental, theoretical
12 coins, dice, cards, imitate
13 simulation, predict, complex

Chapter review

1 a Certain b Unlikely c Even chance
d Impossible e Likely f Impossible

2 Discuss in class.

3

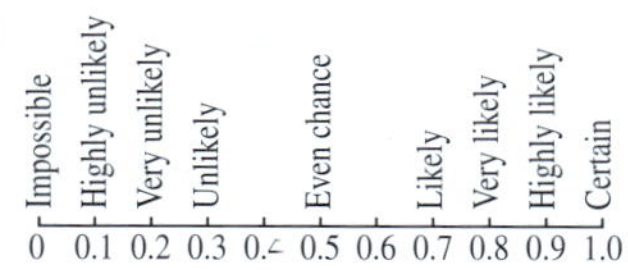

4 a {Head, Tail}
b {red marble, green marble, blue marble}
c {1, 2, 3, 4, 5, 6}
d {1, 2, 3, 4, 5, 6, 7, 8, 9, 10}
e {4 of clubs, 4 of diamonds, 4 of hearts, 4 of spades}
f {true, false} or {correct, incorrect}
g {1, 2, 3, 4, 5, 6, 7, 8, 9}
h {Year 8, Year 9, Year 10, Year 11, Year 12}

5 a 5000 b 26 c 5 d 31

6 a $\frac{1}{6}$ b $\frac{1}{3}$ c $\frac{1}{3}$ d $\frac{2}{3}$
e $\frac{2}{3}$ f $\frac{1}{3}$ g 0 h 0

7 a $\frac{1}{4}$ b $\frac{1}{4}$ c $\frac{3}{13}$ d $\frac{3}{13}$

Note: Do not count aces as a number.

8

		Coin	
		Head	Tail
Die	1	H 1	T 1
	2	H 2	T 2
	3	H 3	T 3
	4	H 4	T 4
	5	H 5	T 5
	6	H 6	T 6

a $\frac{1}{6}$ b $\frac{1}{4}$ c $\frac{1}{4}$ d $\frac{1}{4}$

9

		Spinner									
		1	2	3	4	5	6	7	8	9	10
Die	1	1 1	2 1	3 1	4 1	5 1	6 1	7 1	8 1	9 1	10 1
	2	1 2	2 2	3 2	4 2	5 2	6 2	7 2	8 2	9 2	10 2
	3	1 3	2 3	3 3	4 3	5 3	6 3	7 3	8 3	9 3	10 3
	4	1 4	2 4	3 4	4 4	5 4	6 4	7 4	8 4	9 4	10 4
	5	1 5	2 5	3 5	4 5	5 5	6 5	7 5	8 5	9 5	10 5
	6	1 6	2 6	3 6	4 6	5 6	6 6	7 6	8 6	9 6	10 6

a $\frac{1}{2}$ b $\frac{1}{4}$ c $\frac{4}{15}$ d $\frac{1}{5}$
e $\frac{3}{20}$ f $\frac{23}{50}$

10 Theoretical probability is the expected chance of a certain outcome, knowing the number of favourable outcomes out of the total possible number of outcomes. Experimental probability is the probability of a particular result obtained from a physical trial of the experiment.

11 a $\frac{1}{2}$

b Since both opponents have not played chess before and are given the same instructions on the rules of the game, it is safe to say they will be evenly matched. Neither has an advantage over the other.

c $\frac{1}{4}$

d No, the probabilities are not the same; they have changed because Jane has practised and developed skills while her father has not.

12 Danielle will need to obtain a fair die with 6 sides to represent each card. She will need to roll the die in the same manner and on the same surface each time. Danielle must continue to roll the die until each number 1, 2, 3, 4, 5, 6 has appeared. This will represent one experiment. She will need to repeat this process at least 20 times until she can obtain the average number of cereal packets she must purchase in order to obtain each of the cards. Danielle may also use a circular spinner with 6 equal sectors or 6 tennis balls labelled 1 to 6 in a box and follow the same procedure outlined above. She may also generate random numbers using the calculator or Excel spreadsheet.

13 a Nadia: {1, 2}, Alex: {3, 4, 5, 6}, game is not fair.

b Cassandra: {1, 2, 3}, Jessica: {4, 5, 6}, game is fair.

c John: {3, 6}, Jennifer: {1, 2, 4, 5}, game is not fair.

d Paola: {2, 4, 6}, Serge: {1, 3, 5}, game is fair.

Strategies for problem solving

Strategies for investigation and problem solving

Create a table, then look for a pattern or a result

There are many possible tables. Check if your table has the features of the tables shown.

1

	Time of call to the nearest half minute (minutes)												
Time of call (s)	**0**	$\frac{1}{2}$	**1**	$1\frac{1}{2}$	**2**	$2\frac{1}{2}$	**3**	$3\frac{1}{2}$	**4**	$4\frac{1}{2}$	**5**	$5\frac{1}{2}$	**6**
132					✔								
10	✔												
43		✔											
71			✔										
243									✔				
52			✔										
142						✔							
332												✔	
6	✔												
38		✔											
63			✔										
32		✔											
132					✔								
32		✔											
43		✔											
52			✔										
352													✔
101				✔									
124					✔								
28		✔											
153						✔							
10	✔												
243									✔				
34		✔											
Total	3	7	4	1	3	2	0	0	2	0	0	1	1

The most popular length of phone call (to the nearest half minute) is half a minute.

2

Distance (m)	Tally	Frequency	Number of people who would walk at least this distance
100	III	3	24
200	II	2	21
300	IIII	4	19
400	I	1	15
500	III	3	14
600	I	1	11
700	I	1	10
800	III	3	9
900	I	1	6
1000	I	1	5
1100		0	4
1200	II	2	4
1300		0	2
1400	I	1	2
1500	I	1	1
	Total	24	

Three-quarters of 24 people is 16 people and 16 people will walk at least 300 m.

3

Speed (km/h)	Tally	Frequency	Number cycling at 30km/h or more
10	II	2	0
15	IIII	4	0
20	卌 卌 II	12	0
25	卌 卌 I	11	0
30	卌	5	5
35	卌	5	10
40	卌	5	15
45	II	2	17
50	IIII	4	21
	Total	50	21

The data shows that 21 people out of 50 cycled at 30 km/h or more. This figure is not at least half which is 25, so the coach's statement is not correct.

4

Dress size	Tally	Frequency
8	II	2
10	卌 II	7
12	卌 卌 I	11
14	卌 I	6
16	II	2
18	II	2
	Total	30

Numbers of size 8, 16 and 18 dresses are all equal to 2.
Number of size 10 = 7
Number of size 14 = 6
Number of size 10 and 14 dresses is approximately three times the number of size 8, 16 and 18 dresses.
Number of size 12 dresses = 11
This is approximately five times the number of size 8, 16 and 18 dresses.
So, Rebecca's strategy seems a reasonable one.

5

Daily bottle sales	Tally	Frequency	Number of days selling at least 45 bottles
5	IIII	4	0
10	I	1	0
15	I	1	0
20	III	3	0
25	II	2	0
30	IIII	4	0
35	IIII	4	0
40	III	3	0
45	III	3	3
50	IIII	4	7
55	II	2	9
60	III	4	13
	Total	35	13

Three-quarters of the bottles is $\frac{3}{4}$ of 60 = 45

Two days per week over a period of 5 weeks = 10 days
So, the gymnasium should sell at least 45 drinks for at least 10 days over the 5 weeks. The table shows that the gymnasium sells at least 45 drinks for 13 days of the 5 weeks. This means that the gymnasium is wise to keep the drink machine.

Draw a diagram, then look for a pattern or a result

1 Nadia can communicate in 6 different ways.
2 Michael has 15 choices.
3 There are 12 different seating arrangements.
4 There are only 4 different ways in which the blocks can be stacked.

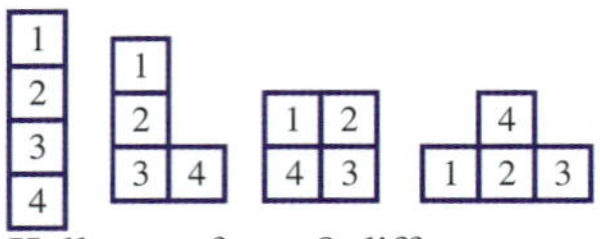

5 Kelly can form 8 different rectangles. Their measurements are:
1 cm × 15 cm, 2 cm × 14 cm, 3 cm × 13 cm,
4 cm × 12 cm, 5 cm × 11 cm, 6 cm × 10 cm,
7 cm × 9 cm, 8 cm × 8 cm.

Use a pattern of numbers, making use of technology such as a computer spreadsheet

1 Three students attend both swimming practice and lifesaving.
2 Twelve students study both French and German.
3 Three people do not work Monday to Sunday.
4 There are 6 chocolate bars which contain both fruit and nut, 6 which contain nuts only and 8 which contain fruit only.
5 Three students voted twice.

Work backwards from the answer

1 $39.45 − 2 × $13.60 = $12.25
$12.25 ÷ $0.25 = 49 calls
2 $138 − 3 × $15.20 = $92.40
$92.40 ÷ $0.22 = 420 calls
3 $14 − 2 × $2.80 = $8.40
$8.40 ÷ 6 = $1.40
A packet of chips cost $1.40.
4 $18.50 − 8 × $1.75 = $4.50
$4.50 ÷ 3 = $1.50
A bottle of juice cost $1.50.
5 The jug originally contained 0.4 L of water.

Use a process of elimination

1 French L2, German K1, Visual Arts L1
2 Two 2-litre containers to house number 5; three 1-litre containers to house number 7; skim milk to house number 18.
3 Phone, Eugenie; SMS, Florence; email, Anthea; mail, Kim.
4 Alan has a $2 coin, Bob a 50c piece and Cam a $1 coin.
5 Penny's brother is Matt, Teresa's brother is Fred and Raelene's brother is Steve.

Look at similar but simpler problems

1 $7 + 7 + 6 + 5 + 4 + 3 + 2 + 1 = 35$
2 $(10 - 2) \times 180° = 1440°$
3 10
4 There could be 260 different numberplates issued under this system.
5 It requires 45 handshakes so that each person can shake hands with everyone exactly once.

Trial and error (guess and check), making use of technology such as a computer spreadsheet

1 22 years old
2 $10\frac{3}{4}$ hours
3 $1215
4 The two numbers are 19 and 13.
5 A hamburger costs $3.75 and a drink costs $2.45.

Communicating, reasoning and reflecting

1 There are two possible solutions: 2 packets containing 3 ice-creams and 5 packets containing 5 ice-creams or 7 packets containing 3 ice-creams and 2 packets containing 5 ice-creams.
With further investigation of the pattern of numbers obtained by summing 3s and 5s, it appears that there are combinations of 3 packs and 5 packs which will amount to any quantity of ice-creams equal to or larger than 8.
2 The two sides of the fencing are 6-m long and the other side is 12 m, providing an area of 72 m^2.
3 Bronwyn's 1st birthday will be on a Thursday unless the next year is a leap year, in which case her 1st birthday will be on a Friday.
There is more chance that the answer will be a Thursday because leap years come around only every 4 years. Since she was born in a non-leap year, then there is a 1-in-3 chance of the following year being a leap year.
4 8 students play volleyball only.
5 Alexander's 1st birthday is on a Tuesday, in January or February, or it is on a Wednesday in any of the other months.
Depending on the month in which he was born, Alexander may have to wait the extra day, 29th February. What would happen if Alexander had been was born on the 29th of February of a leap year? Does he have to wait 4 years for a birthday?
6 The two numbers are 34 and 35. Their sum is 69.
7 In Queensland, Peta needs to change the time on her watch so that it shows one hour earlier, as clocks are set to the actual time rather than to daylight saving time.
It would seem to Peta that there is an extra hour in the day. If the time on her watch showed 9 am when she arrived in Queensland, it would actually be 8 am and she would have the hour from 8 am to 9 am all over again. If Peta walked in the other direction, from Queensland to NSW, she would lose a whole hour, because she would have to change her watch so that it was later by one hour.
8 **a** This system would allow 100 000 000 different telephone numbers.
b This system would allow 20 000 000 different telephone numbers.

Index